ESSENTIALS OF GEOLOGY

Born in 1942, Ed Mell grew up surrounded by the austere and natural beauty of the southwest. After a successful but demanding career as an illustrator in New York's advertising industry during the 1960s and 1970s, Mell returned to the serenity of his home state of Arizona. Bringing with him the simplicity and refinement of form he had admired in the art deco style, Mell began his transition from commercial artist to landscape painter.

Mell paints the landscape of the southwest with an intensity and luminosity of color that captures the living interplay between land, sky, and sunlight. By breaking down the landscape to almost architectural components, he is able to convey the monumental grandeur of each canyon, butte, and mesa. Through the numerous exhibits of his landscapes across America, Mell continues to share his dramatic vision of the land and the story it tells.

ESSENTIALS OF GEOLOGY

Second Edition

Stanley Chernicoff
University of Seattle, Washington

Haydn A. "Chip" Fox
Texas A & M University, Commerce

Houghton Mifflin Company Boston New York

Editor-in-Chief: Kathi Prancan
Senior Associate Sponsor: Susan Warne
Senior Project Editor: Chere Bemelmans
Editorial Assistant: Joy Park
Senior Production/Design Coordinator: Jill Haber
Manufacturing Manager: Florence Cadran
Executive Marketing Manager: Andrew Fisher

Cover Design: Diana Coe / ko Design Studio.

Cover Painting: *Morning Light on Meeks Mesa* by Ed Mell, 1996, oil on canvas, 30″ x 96″.
Suzanne Brown Galleries, Scottsdale, Arizona.

Photo credits begin on page C-1.

Printed in the U.S.A.

Library of Congress Catalog Card Number: 99-71988

ISBN: 0-395-97055-5

2 3 4 5 6 7 8 9-VH-03 02 01 00 99

To my wife, Dr. Julie Stein, and my sons, Matthew and David, who have tolerated my obsession with text writing with extraordinary grace and humor (and endured an uncountable number of interrupted family dinners). Much thanks and appreciation for your remarkable support and encouragement.

Stan Chernicoff

To my wife, Jannie B. Fox, whose patience and encouragement were of limitless value in the writing of this book and in life in general.

Chip Fox

About the Authors

Born in Brooklyn, New York, Stan Chernicoff began his academic career as a political science major at Brooklyn College of the City University of New York. On graduation, he intended to enter law school and pursue a career in constitutional law. He had, however, the good fortune to take geology as his last requirement for graduation in the spring of his senior year, and he was so thoroughly captivated by it that his plans were forever changed.

After an intensive post-baccalaureate program of physics, calculus, chemistry, and geology, Stan entered the University of Minnesota–Twin Cities, where he received his doctorate in Glacial and Quaternary Geology under the guidance of one of North America's preeminent glacial geologists, Dr. H. E. Wright. Stan launched his career as a purveyor of geological knowledge as a senior graduate student teaching physical geology to hundreds of bright Minnesotans.

Stan has been a member of the faculty of the Department of Geological Sciences at the University of Washington in Seattle since 1981, where he has won several teaching awards. At Washington, he has taught Physical Geology, the Great Ice Ages, and the Geology of the Pacific Northwest to more than 20,000 students, and he has trained hundreds of graduate teaching assistants in the art of bringing geology alive for nonscience majors. Stan studies the glacial history of the Puget Sound region and pursues his true passion, coaching his sons and their buddies in soccer, baseball, and basketball. He lives in Seattle with his wife, Dr. Julie Stein, a professor of archaeology, and their two sons, Matthew (the midfielder, second baseman, two-guard) and David (the striker, second baseman, point guard).

Born in Grand Rapids, Michigan, Haydn A. "Chip" Fox received a bachelor's degree in theology and journalism from Ambassador College in Big Sandy, Texas, in 1971. He spent the next 15 years in several nonacademic pursuits, such as managing convenience stores, driving a truck, and working for a newspaper. In 1986, he returned to college with the intent of becoming an earth science teacher. His interests in geology, earth science, environmental science, and science education were immediately sparked, and in 1992 he received a Ph.D. in Geological Sciences from the University of South Carolina.

Chip has served on the faculty of departments of Earth Science at Southeast Missouri State University and Clemson University. He is currently a member of the faculty at Texas A & M University in Commerce, Texas, where he teaches numerous courses in earth science and the environmental sciences. Having spent several years in nonacademic careers, Chip understands the challenges his students will face when they graduate from college. He enjoys an excellent rapport with his students, a group that includes several future geologists, environmental professionals, and public school teachers.

Chip lives with his wife, Jannie, in an old farmhouse in Texas, where they are busy remodeling and trying to make a tenuous paradise of their 54 acres.

Contents in Brief

Contents

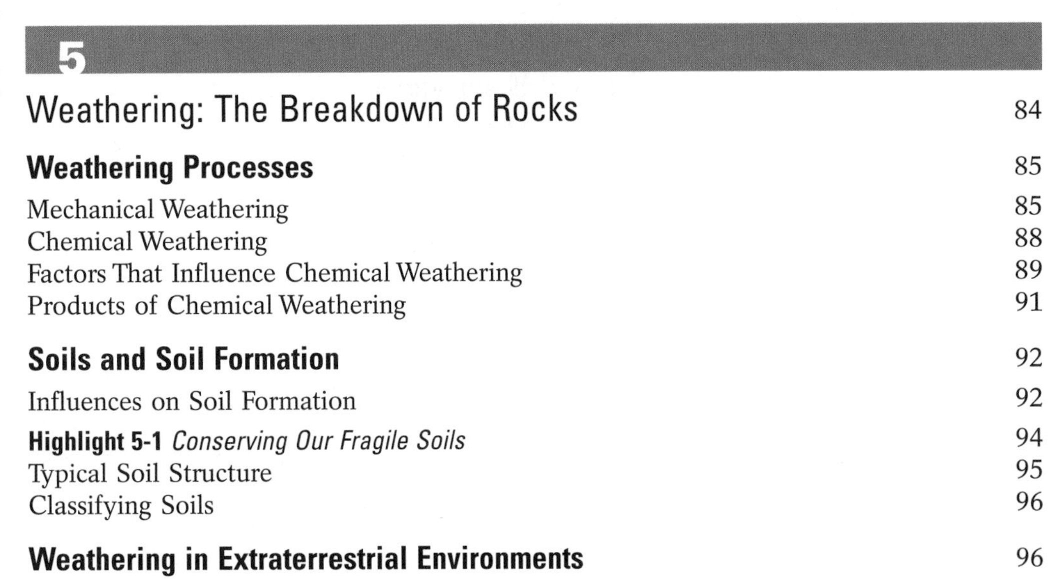

Preface

The introductory course in physical geology, taken predominantly by nonscience majors, may be the only science course some students will take during their college years. What a wonderful opportunity this provides us to introduce students to the field we love and show them how fascinating and useful it is. Indeed, much of what students will learn in their Physical Geology course will be recalled throughout their lives, as they travel across this and other continents, dig in backyards, walk along a beach, or sit by a mountain stream. For this reason, our book team—authors, illustrators, photo researchers, and editors—have expended the best of our abilities to craft an exciting, stimulating, and enduring introduction to the field.

The Book's Goal

The book's goal is basic—to teach what everyone should know about geology in a way that will engage and stimulate. The book embodies the view that this is perhaps the most useful college-level science class a nonscience major can take—one that we believe all students should take. Physical Geology can show students the essence of how science and scientists work, at the same time as it nurtures their interest in understanding, appreciating, and protecting their surroundings. In this course they can learn to prepare for any number of geologic and environmental threats, and see how our Earth can continue providing all of our needs for food, shelter, and material well-being as long as we don't squander these resources.

Content and Organization

Essentials of Geology extracts the most important concepts from Chernicoff's *Geology,* second edition, as they are typically taught in a nonmajors' geology course. All of the concepts are included to provide students with a general understanding of geology, but a more direct route is taken with each of the basics, rather than delving into the more detailed derivations of concepts, alternate theories, and additional details that encompass the vast body of knowledge related to every concept of geology.

The unifying themes of plate tectonics, environmental geology and natural resources, and planetary geology are introduced in Chapter 1 and discussed in their proper context within nearly every chapter. Chapter 1 also presents the three groups of rocks, the rock cycle, and geologic time—building a foundation for the succeeding chapters. Chapters 1 through 8 introduce the basics—minerals, rocks, and time. Part 2, Chapters 9 through 11, discuss structural geology, earthquakes, the Earth's interior, and the details of plate tectonics. Part 3, Chapters 12 through 19, presents the principal geomorphic processes of mass movement, streams, groundwater, and glaciers, as well as processes that occur in desert regions. The final two chapters, Chapters 18 and 19, tie together earlier discussions from throughout the book by discussing human use of Earth's resources and a brief history of Earth and its life forms.

New to the Second Edition of *Essentials of Geology*

A second edition is a wonderful opportunity to build on the first:

- To weave in the latest discoveries in the geosciences.
- To offer up-to-the-minute examples of exciting geological processes, such as the most recent volcanic eruptions and earthquakes.
- To rethink how concepts have been presented in the first edition—to clarify and illustrate them more effectively.

This—the second edition of *Essentials of Geology*—attempts to accomplish all of these goals, all to ensure that our students have the very best introductory experience with our science. Toward these ends, *Essentials* has included coverage of exciting areas like the following:

- An expanded discussion of the proposed origin of the Moon from a collision between the Earth and a Mars-sized impactor.
- Updates on recent and ongoing eruptions in the Caribbean (Montserrat), Mexico City, and New Zealand.
- The moment magnitude scale—an alternative to the Richter Scale.
- Using the global positioning system to track plate motion.
- Global warming, sea level, and coastal destruction.

These topics and many more constitute a substantial effort to ensure that a new edition of *Essentials of Geology* brings new ideas to its readers.

This edition of *Essentials of Geology* also benefits significantly from a change in text design. The new two-column format has enabled the book's designers to offer much-enlarged photos and illustrations—a concern from the first edition. Well over a hundred new photos have been selected (under the outstanding direction of Photo Researcher Townsend P. Dickinson) that illustrate most vividly the processes described in the text.

The Artwork

The drawings in this book are unique. Ramesh Venkatakrishnan is an experienced and respected geology professor and consultant. He is also a highly gifted artist. His drawings evolved along with the earliest drafts of the manuscript, sometimes leading the way for the text discussions.

As you will see when you leaf through this book, the art explains, describes, stimulates, and teaches. It is not schematic; it shows how the Earth and its geological features actually look. It is also not static; it shows geological processes in action, allowing students to see how geological features evolve through time. Every effort has been made to illustrate accurately a wide range of geological and geomorphic settings, including vegetation and wildlife, weathering patterns, even the shadows cast by the Sun at various latitudes. The artistic style is consistent throughout, so that students may become familiar with the appearance of some features even before reading about them in subsequent chapters. For example, the stream drainage patterns appearing on volcanoes in Chapter 4, Volcanoes and Volcanism, set the stage for the discussion of drainage patterns in Chapter 13, Streams and Floods. The colors used and the map symbols keyed to various rock types follow international conventions and are consistent throughout.

The second edition builds on the strengths of the art program of the first. The images in this edition have been enhanced digitally by renowned geology illustrators George Kelvin and John Woolsey to sharpen their focus, deepen their colors, and lend additional clarity and simplicity to their subjects. For this edition of *Essentials*, the maps have been rerendered by Patti Isaacs, Parrot Graphics, to add topographical relief where appropriate to give students a sense of context, and to make them brighter and cleaner and the labels easier to read.

Pedagogy

Nearly every chapter contains one or more Highlights—in-depth discussions of topics of popular interest that provide a broader view of the relevance of geology. In many cases, the Highlights comprise a late-breaking story that also shows the reader that the Earth's geology and its effects on us are changing daily.

To help readers learn and retain the important principles, every chapter ends with a Summary, a narrative discussion that recalls all of the important chapter concepts. Key terms, which are in boldface type in the chapter, are listed at the chapter's end and also appear in boldface in the Summary. Also at the end of every chapter are two question sets: *Questions for Review* helps students retain the facts presented, and *For Further Thought* challenges readers to think more deeply about the implications of the material studied.

The authors and illustrators have tried to introduce readers to world geology. This book emphasizes, however, the geology of North America (including the offshore state, Hawai'i), while acknowledging that geological processes do not stop at national boundaries or at the continent's coasts. Wherever data are available—from the distribution of coal to the survey of seismic hazards—we have tried to show our readers as much of this continent, and beyond, as feasible. Photos and examples have been selected from throughout the United States and Canada and from many other regions of the world.

The metric system is used for all numerical units, with their English equivalents in parentheses, so that U.S. students can become more familiar with the units of measurement used by virtually every other country in the world.

The Supplements Package

Essentials of Geology is accompanied by an array of materials to enhance teaching and learning.

Students who wish additional help mastering the text can use the Study Guide by W. Carl Shellenberger (Montana State University—Northern). For each chapter, the Guided Study section helps students focus on and review in writing the key ideas of each section of the chapter as they read. The Chapter Review, arranged by section and composed of fill-in statements, enables them to see if they have retained the ideas and terminology introduced in the chapter. The Practice Tests and the Challenge Test, which consist of multiple-choice, true/false, and brief essay questions, test their mastery of the material. All answers are accompanied by page references for easy review.

The Instructor's Resource Manual by Chip Fox features an outline lecture guide with teaching suggestions embedded in it and student activities and classroom demonstrations. Answers to the end-of-chapter questions in the textbook are also provided. Also included is a comprehensive Test Bank, compiled by Chip Fox, that contains more than one thousand questions. There are at least 40 multiple-choice questions per chapter, classified as either factual or conceptual/analytical. There are also ten short essay questions, complete with answers, for each chapter. A computerized version of the Test Bank is available in both IBM and Macintosh formats.

Also available with this edition is the *Geology Laboratory Manual* by James D. Myers, James E. McClurg, and Charles L. Angevine of the University of Wyoming. This inexpensive manual is closely tied to the text and offers twenty physical geology labs on topics such as maps, plate tectonics, sedimentary and metamorphic rocks, streams, and groundwater. Each lab contains multiple activities to develop and hone students' geological skills. Worksheets are designed to be torn from the manual and submitted for grading.

More than 130 of the text's diagrams and photographs are available for classroom use as full-color slides or transparencies.

The book is supported further by its award-winning web site, GEOLOGYLINK (found at www.geologylink.com), maintained and updated regularly by its web master, Rob Viens of the University of Washington. This site will tell you what of geological import has happened overnight while you slept. It also contains expanded discussions of "hot topics" in the field of geology and an exhaustive encyclopedia of links to all things geological. For the second edition of *Essentials of Geology,* GEOLOGYLINK contains chapter quizzes and tutorials as well as an on-line version of the Peterson's *Field Guide to Rocks and Minerals* by Frederick Pough. These outstanding teaching and learning aids help the student learn physical geology through multimedia technology, study physical geology in a stimulating, yet thoughtful way, and master the principles of physical geology.

Acknowledgments

Some remarkably talented, dedicated people have helped us accomplish far more than we could have done alone. A "committee" of top-flight geologists has been assembled who have dramatically clarified definitions and explanations, eliminated ambiguities, corrected factual errors and fuzzy logic, and, in general, helped the authors hone the manuscript in countless ways and helped the illustrator select what to show and how best to do it. Special thanks must go to Kurt Hollocher of Union College and L. B. Gillett of SUNY-Plattsburgh for their extremely insightful critiques of the first edition. In addition, for their constructive criticism at various stages along the way, we wish to thank these excellent reviewers:

From the first edition of *Geology:*

Gail M. Ashley, *Rutgers University, Piscataway*

David M. Best, *Northern Arizona University*

David P. Bucke, Jr., *University of Vermont*

Michael E. Campana, *University of New Mexico*

Joseph V. Chernosky, Jr., *University of Maine, Orono*

G. Michael Clark, *University of Tennessee, Knoxville*

W. R. Danner, *University of British Columbia*

Paul Frederick Edinger, *Coker College (South Carolina)*

Robert L. Eves, *Southern Utah University*

Stanley C. Finney, *California State University, Long Beach*

Roberto Garza, *San Antonio College*

Charles W. Hickox, *Emory University*

Kenneth M. Hinkel, *University of Cincinnati*

Darrel Hoff, *Luther College (Iowa)*

David T. King, Jr., *Auburn University*

Peter T. Kolesar, *Utah State University*

Albert M. Kudo, *University of New Mexico*

Martin B. Lagoe, *University of Texas, Austin*

Lauretta A. Miller, *Fairleigh Dickinson University*

Robert E. Nelson, *Colby College (Maine)*

David M. Patrick, *University of Southern Mississippi*

Terry L. Pavlis, *University of New Orleans*

John J. Renton, *West Virginia University*

Vernon P. Scott, *Oklahoma State University*

Dorothy Stout, *Cypress College (California)*

Daniel A. Sundeen, *University of Southern Mississippi*

Allan M. Thompson, *University of Delaware*

Charles P. Thornton, *Pennsylvania State University*

From the second edition of *Geology:*

William W. Atkinson, *University of Colorado, Boulder*

Joseph Chernosky, *University of Maine, Orono*

Cassandra Coombs, *College of Charleston*
Peter Copeland, *University of Houston*
Katherine Giles, *New Mexico State University*
L. B. Gillett, *SUNY-Plattsburgh*
Kurt Hollocher, *Union College*
Kathleen Johnson, *University of New Orleans*
Judith Kusnick, *Cal-State Sacramento*
Bart Martin, *Ohio Wesleyan University*
Ronald Nusbaum, *College of Charleston*
Meg Riesenberg
Roger Stewart, *University of Idaho*
Donna L. Whitney, *University of Minnesota*

From the second edition of *Essentials of Geology:*

Andrew Buddington, *Spokane Community College*
Timothy L. Clarey, *Delta College*
Dale H. Easley, *University of New Orleans*
Terry Engelder, *Pennsylvania State University*
Brice M. Hand, *Syracuse University*
Lindley S. Hanson, *Salem State College*
Richard Robinson, *Santa Monica College*
Robert D. Shuster, *University of Nebraska at Omaha*
David W. Valentino, *State University of New York at Oswego*

We would also like to thank William A. Smith (Charleston Southern State University) for his sharp eye in reviewing the art for accuracy.

At Houghton Mifflin, developmental editors Virginia Joyner and Marjorie Anderson, working under oppressive time constraints, performed the arduous task of reining in the authors' long-windedness with extraordinary grace and intelligence and brought organization wherever they found disorder. Senior Associate Sponsor Sue Warne, Senior Project Editor Chere Bemelmans, Art Editor Charlotte Miller, Copyeditor Jill Hobbs, and Editorial Assistant Joy Park polished each chapter of prose and every rough sketch, working with all the elements of the book until they formed a coherent whole. Photo research was handled masterfully by Photo Researchers Townsend P. Dickinson and Mardi Welch Dickinson. The book's pleasing appearance was created under the supervision of Senior Production/Design Coordinator Jill Haber, Associate Production/Design Coordinator Jodi O'Rourke, and Layout Designer Penny Peters. We very much appreciate Editor-in-Chief Kathi Prancan's support and behind-the-scenes hard work and Executive Marketing Manager Andy Fisher's energetic marketing support. Thanks are due also to Associate Editor Marianne Stepanian, who coordinated and edited the supplements.

Finally, we also wish to acknowledge with deep appreciation the role of Ron Pullins (formerly of Little, Brown and now of Focus Publishing) and Kerry Baruth (formerly of Worth Publishers and now of Houghton Mifflin), who championed the cause of *Essentials of Geology* with their respective companies during its early gestation.

After they leave our classrooms, students may well forget some specific facts and terminology of geology, but they will still retain the general impressions and attitudes they formed during our course. We hope that our words and illustrations will help advance the goals of those teaching this course and contribute to their classes. We have used our teaching experiences to craft a textbook that we think our own students will learn from and enjoy. We hope your students will, too. We invite your comments: please send them to the authors, whose e-mail addresses are sechern@u.washington.edu and haydn_fox@tamu-commerce.edu.

Stan Chernicoff
Haydn A. "Chip" Fox

To the Student

Only a very few years ago, I was an undergraduate "non-traditional" student in college being exposed to geology for the first time. As occurs with a surprisingly large number of students, the spark for geology and the other earth sciences was kindled almost immediately. I have little doubt that you, too, will find your study of geology to be most interesting.

One of the appeals of geology is its application to everyday life. All around you are hills or mountains, valleys, coastlines, and soils; it is geology that tells you why these things are there and how they formed. Every local area has it own geology, which is part of a much larger picture of regional and even world geology.

On my van is a bumper sticker that reads, "If it can't be grown, it has to be mined." Everything we have comes either from growing things or from finding them somewhere within the ground. It is in the realm of geology to determine where the products we mine are located, and to understand how and under what situations they form. As if this weren't a large enough body of knowledge to encompass, it is also in the realm of geology to explore the inner workings of volcanoes and earthquakes, the interior structure of the Earth, and the functions of rivers, glaciers, and a host of other natural phenomena.

Geology is the basis for much of our understanding of the environment. If you become interested in pursuing a career in environmental sciences, you will find geology at the cornerstone. Most environmental consultants, and many people involved in environmental compliance (working for governments, large corporations, or industry) have come from the field of geology.

This book presents a brief survey of the whole field of geology, from rocks and minerals to the formation of entire continents and the processes that act upon and under them. This overview will provide you with a basis for understanding many different aspects of our planet, and it may even entice you to explore them further. If so, you will discover that there is a world of knowledge behind every topic mentioned in this book, each of which could lead to an exciting and challenging career.

Haydn A. "Chip" Fox

Features of

Essentials of GEOLOGY,

Second Edition

The rich detail and technical accuracy of the illustrations help convey complex concepts to introductory students.

Photos are often paired with art to emphasize a point.

106 Part I Forming the Earth

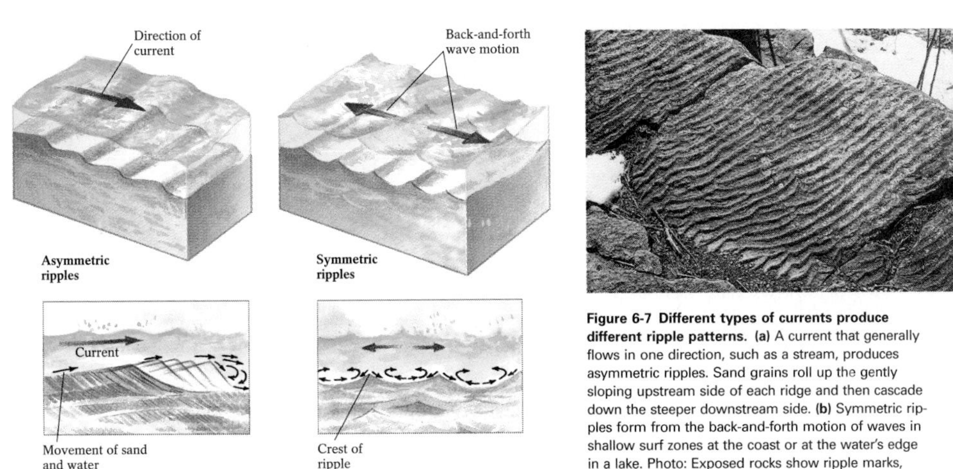

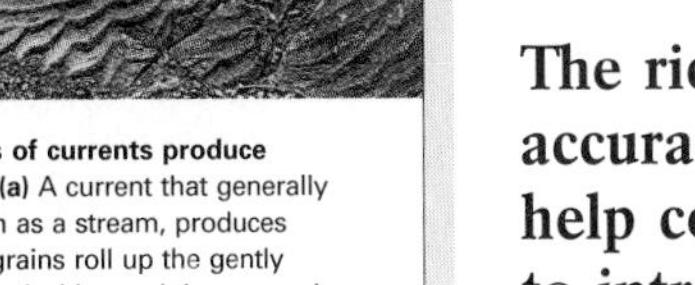

Figure 6-7 Different types of currents produce different ripple patterns. (a) A current that generally flows in one direction, such as a stream, produces asymmetric ripples. Sand grains roll up the gently sloping upstream side of each ridge and then cascade down the steeper downstream side. **(b)** Symmetric ripples form from the back-and-forth motion of waves in shallow surf zones at the coast or at the water's edge in a lake. Photo: Exposed rocks show ripple marks, evidence of past current flow, either water or wind.

series of shallow curving ridges. The configuration of these ridges, which are often visible on sandy surfaces, reflects the nature of the current that produced them (Fig. 6-7).

Mudcracks are fractures that develop when the surface of wet fine-grained sediment (mud) dries and contracts (Fig. 6-8). Because these structures form only at the top of a layer of muddy sediment and narrow progressively downward, geologists can study mudcracks to determine whether a layer of sedimentary rock has been overturned.

Lithification: Turning Sediment into Sedimentary Rock

When a sediment layer is deposited, it buries all previous layers deposited at that location. Eventually, the continuing deposition may enable a sedimentary pile to become several kilometers deep. Such deep burial may convert sediments into solid sedimentary rock by the process of **lithification** (from the Greek *lithos,* meaning "rock," and Latin *facere,* meaning "to make"). During lithification, sediment grains become compacted, often cemented, and sometimes recrystallized.

Compaction is the process by which the volume of buried sediment, either detrital or chemical, becomes diminished by pressure exerted by the weight of overlying sediments. Expulsion of air and water from the sediment and the reduction of the spaces between grains combine to

Artwork shows geologic features in a naturalistic context, to give students a sense of how these features actually look.

Chapter 8 Telling Time Geologically 147

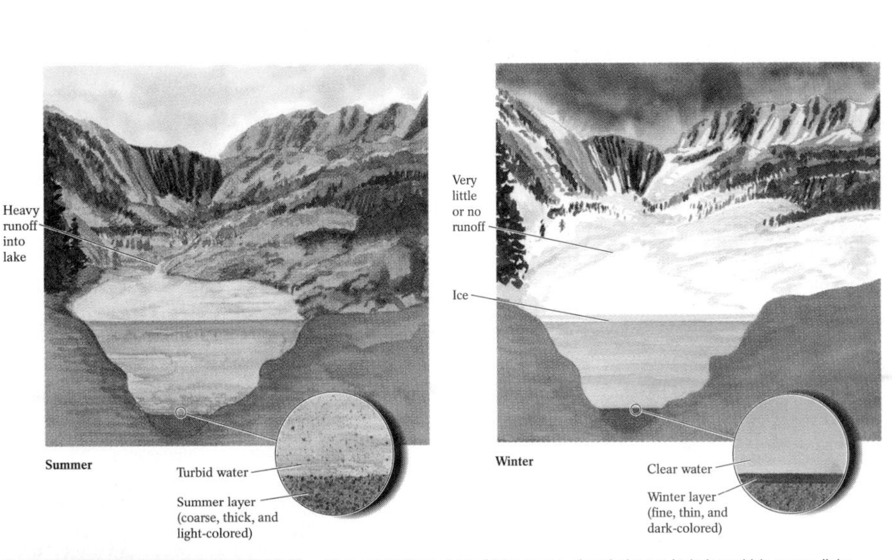

Figure 8-13 The origin of lake varves. A typical varve includes a thick, coarse, light-colored summertime layer produced during high runoff (from snowmelt and spring storms) and high sediment influx, plus a thin, fine dark-colored wintertime layer produced during low runoff and low sediment influx (or no influx, if the lake is frozen). Note the varves in the photo. Why do you think the varves vary so noticeably in thickness?

Figure 8-14 Lichen colonies on a granite boulder. The light-colored areas of rock have been bleached by chemicals in the lichen. The sizes of such colonies can provide clues as to how long the rock surface has been exposed.

deposited in summer and a thin, fine, dark-colored layer deposited in winter. By applying *varve chronology,* in which they study the number and nature of the varves underlying a lake, geologists can determine how long ago the lake formed and identify events, such as landslides, that affected sedimentation in the area (Fig. 8-13).

Lichen (pronounced "LIE-ken"), colonies of simple, plant-like organisms that grow on exposed rock surfaces, are the basis of a dating method known as *lichenometry.* Lichen grow extremely slowly; given similar rocks and climatic conditions, the larger the lichen colony, the longer the period of time since the growth surface was exposed. Study of these organisms can yield accurate dates for young glacial deposits, rockfalls, and mudflows—all events that expose new rock surfaces on which lichen can grow (Fig. 8-14).

Highlight 12-2 How to Choose a Stable Home Site

If you have not already done so, you may eventually wish to purchase your own home. Having studied physical geology, you will want to ensure that your dream house doesn't fall victim to a geological nightmare.

Suppose you're exploring southern California's scenic beachfront locales. You happen upon the mosaic sign for the town of Portuguese Bend and notice that the beautiful ceramic signpost is cracked into two pieces. You check the local real estate listings and find a house priced at $50,000 that should be worth $500,000. Your knowledge of geology, along with your common sense, immediately warns you that something may be wrong here. What else should you look for? How can you tell—in southern California or anywhere else in the world—if you're in mass-movement country?

Examine both the property itself and the entire neighborhood. Do you see any signs of an old mud or debris flow? Is there evidence of a slump scar upslope where a block may have broken away? As you drive through the community and its environs, look for fences that are out of alignment, and for power and telephone lines that seem too slack in some places and too taut in others (Fig. 1).

Next, look carefully at the house itself and, if possible, at neighboring ones. Search for large cracks in the foundation (small cracks may be due to initial drying and settling of the concrete). Doors and windows that stick may indicate that once-linear structural features are now out of line, although poor craftsmanship or high moisture content may also be responsible. A cracked pool lining might explain why a swimming pool doesn't retain water. Finding only one such problem may not indicate danger, but the presence of several problems should send a strong warning signal. If the geology, topography, and hydrology of a home site all raise questions about slope stability, the site may well be prone to progressive slope failure. To confirm your suspicions, try checking newspaper accounts and the records of the local housing authority, contacting the state geological survey, and interviewing the property's neighbors.

Water leakage from swimming pool
Taut power lines
Tension cracks at head of slide mass
Landslide scarp
Displaced fence
Slide/slump plane
Tree trunks bent by soil creep
Cracks in road surface at edge of slide

Figure 1 Various signs of past, current, and potential future mass movement in an urban area. If the power lines in a neighborhood are very loose, it suggests that the poles that hold them have moved closer together, as one might expect at the toe of a slide where the slope is bunching up like a rumpled carpet. At the head of the slide, taut lines may indicate that the poles on the slide mass are moving away from those upslope. The poles in the middle of a slide mass may keep their original spacing, because the mass may not be deforming much internally.

Highlight boxes introduce high-interest topics, making geology relevant to students' lives.

Highlight 13-2 Did We Help Cause the Midwestern Flood of 1993?

In the beginning of this chapter, we discussed one of North America's worst floods, which took place along the upper portions of the Mississippi River in the summer of 1993 (Fig. 1). Did human activity—65 years of "managing" the Mississippi River to prevent floods—actually contribute to the magnitude and tragedy of the flood?

Before the last 100 years or so, the Mississippi and its tributaries determined their own boundaries. During periods of high water, the rivers broke through or overflowed natural levees, flooding tens of thousands of square kilometers of the surrounding, largely uninhabited, lands. In the twentieth century, however, millions of people migrated to the region, building cities, towns, and large farms along the rivers' banks. When a flood in 1927 took 214 lives, Congress enacted the first Mississippi River Flood Control Act, assigning the U.S. Army Corps of Engineers the daunting task of confining the river to its channel.

The Corps of Engineers' efforts consist of about 300 dams and reservoirs and thousands of kilometers of artificial levees and concrete flood walls, all designed to prevent the river from spilling onto its natural floodplain. The system also contains numerous pumping stations, spillways, and diversion channels designed to divert water for storage in temporary holding basins. But the events of 1993 showed that such structures simply cannot contain an extraordinary flood. By confining such great discharges to a channel, the artificial retention structures actually caused the swollen rivers to flow more rapidly and violently, thus damaging the very structures designed to restrain them. By denying the river access to its natural floodplains, the structures caused the streams to rise higher than they would have otherwise, ensuring that once they did breach the levees, the floods would cause greater damage. Furthermore, the existence of artificial levees and flood walls had encouraged the growth of cities, towns, and farms closer to the riverbanks than was really safe.

What does the future hold for the residents of the upper Mississippi valley? Certainly more flooding, but perhaps less human interference with the river's natural behavior. Some communities have proposed that all flood-retention systems be eliminated and that zoning limit future development within the river's floodplain. Others look longingly at St. Louis's 16-meter (52-foot)-high concrete flood wall, which saved that city's downtown business district when the Mississippi reached its record crest at 14.2 meters (47 feet). The debate continues between those who believe we can tame the mighty Mississippi and those who believe we cannot.

Figure 1 Flooding at Kaskaskia, Illinois.

Chapter Summary

Volcanism, the set of processes that results in extrusion of molten rock, begins with the creation of magma by the melting of preexisting rock and culminates with the ascent of this magma to the Earth's surface through fractures, faults, and other cracks in the lithosphere. **Volcanoes** are the landforms created when molten rock escapes from vents in the Earth's surface and then solidifies around these vents. Volcanoes may be active, dormant, or extinct.

Because of its high temperature and relatively low silica content, mafic magma has low viscosity (is highly fluid). It generally erupts (as basaltic lava) relatively quietly, or effusively, because its gases can readily escape and do not build up high pressure. Felsic magma, with its high silica content and relatively low temperature, is highly viscous and generally erupts (as rhyolitic lava) explosively.

The nonexplosive volcanic eruptions characteristic of basaltic lava produce lava flows that, when they solidify, are associated with distinctive features such as pahoehoe- and 'a'a-type surface textures, basaltic columns, lava tubes, and pillow structures. The explosive volcanic eruptions characteristic of rhyolitic lavas typically eject **pyroclastic** material—fragments of solidified lava and shattered preex[...] rock ejected forcefully into the atmosphere. The variou[...] ticles produced when lava cools and solidifies as it falls to the surface are collectively called **tephra.** Explosive [...] tion of pyroclastic material is usually accompanied by a [...] ber of life-threatening effects, such as **pyroclastic flow[...]** **nuée ardentes** (high-speed, ground-hugging avalanches [...] pyroclastic material), and **lahars** (volcanic mudflows).

Nearly all volcanoes have the same two major co[...] nents: (1) a mountain, or **volcanic cone,** built up of the [...] ucts of successive eruptions; and (2) a bowl-shaped de[...] sion, or **volcanic crater,** surrounding the volcano's ve[...] enough lava erupts to empty a volcano's subterranean [...] voir of magma, the cone's summit may collapse, form[...] much larger depression, or **caldera.**

Effusive eruptions, which usually involve basaltic [...] form gently sloping, broad-based cones called **shield v[...] noes.** Basaltic magma reaching the surface through lin[...] ear cracks, or fissures, in the Earth's crust spreads to pr[...] nearly horizontal lava plateaus.

Explosive **pyroclastic eruptions** involve viscous, us[...] gas-rich magmas and so tend to produce great amou[...] solid volcanic fragments rather than fluid lavas. Felsic [...] olitic) lavas are often so viscous that they cannot flow [...] a volcano's crater; they therefore cool and harden within [...] craters to form **volcanic domes.** Ash-flow eruptions oc[...] the absence of a volcanic cone; they are produced whe[...] tremely viscous, gas-rich magma rises to just below th[...] face bedrock, stretching and collapsing it.

The characteristic landform of pyroclastic erupti[...] the **composite cone,** or **stratovolcano,** which is compos[...]

alternating layers of pyroclastic deposits and solidified lava. Pyroclastic eruptions may also produce **pyroclastic cones** or **cinder cones,** created almost entirely from the accumulation of loose pyroclastic material around a vent. All pyroclastic-type volcanoes produce steep-sided cones, because the materials they eject—solid fragments and highly viscous lavas—do not flow far from the vent.

Various types of volcanic eruptions are associated with different plate tectonic settings. Explosive pyroclastic eruptions of felsic (rhyolitic) lava generally occur within continental areas characterized by plate rifting or atop intracontinental hot spots. Most intermediate (andesitic) eruptions take place near subducting oceanic plates. Effusive eruptions of (mafic) basalt generally occur at divergent plate margins and above oceanic intraplate hot spots.

Humans can minimize damage from volcanoes by zoning against development in the most hazardous areas, building lava dams, diverting the path of a flowing lava, and learning to predict eruptions accurately. Techniques used to predict eruptions include measuring changes in a volcano's slopes, recording related earthquake activity, and tracking changes in the volcano's external heat flow.

Volcanism is not restricted to the Earth. It has occurred in the past on the Moon, and relatively recent volcanic ac[...]

End-of-chapter summaries present an overview of the content in narrative form to help students review.

granites are formed within continents by partial melting of the lower portions of the continental crust; these types of igneous rocks are often associated with subduction-produced mountains.

Igneous rocks are also found on the Moon. Lunar igneous rocks differ fundamentally from those on Earth, in that they contain no water and their formation involved neither plate tectonics nor subsurface heat. The Moon's surface consists of highlands composed largely of anorthosite, a coarse-grained plutonic igneous rock, and vast areas of basalt known as maria.

Igneous rocks are valued for the gemstones and precious metals they contain. They are also used for a variety of practical purposes, such as road construction, architectural design, and household abrasives.

Key Terms

igneous rocks (p. 41)
magma (p. 42)
lava (p. 42)
intrusive rocks (p. 43)
plutonic rocks (p. 43)
extrusive rocks (p. 43)
volcanic rocks (p. 43)
peridotite (p. 45)
basalt (p. 45)
gabbro (p. 46)
andesite (p. 46)
diorite (p. 46)
granite (p. 46)
rhyolite (p. 46)
partial melting (p. 46)
Bowen's reaction series (p. 48)
fractional crystallization (p. 49)
plutons (p. 50)
dike (p. 51)
sill (p. 53)
laccolith (p. 53)
lopoliths (p. 53)
batholiths (p. 53)
andesite line (p. 55)

Questions for Review

1. Briefly describe the textural difference between phaneritic and aphanitic rocks. Why do these rocks have different textures?

2. Some igneous rocks contain large visible crystals surrounded by microscopically small crystals. What are these rocks called? How does such a texture form?

3. What elements would you expect to predominate in a mafic igneous rock? In a felsic igneous rock?

4. Name the common *extrusive* igneous rocks in which you would expect to find each of the following mineral types: calcium feldspar; potassium feldspar; muscovite mica; olivine; amphiboles; sodium feldspars. Which *plutonic* igneous rock contains abundant quartz and muscovite mica, but virtually no olivine or pyroxene?

5. What factors, in addition to heat, control the melting of rocks to generate magma?

6. What is the basic difference between the continuous and discontinuous series of Bowen's reaction series?

7. Briefly describe three things that might happen to an early-crystallized mineral surrounded by liquid magma.

8. How do a sill and a dike differ? A laccolith and a lopolith? A lopolith and a batholith?

9. Briefly discuss two specific types of plate tectonic boundaries and the igneous rocks that are associated with them.

10. What is the basic difference between a mid-ocean ridge basalt and an oceanic island basalt?

For Further Thought

1. What type of igneous feature is shown in the photo below?

2. Felsic rocks such as rhyolite often occur together with basaltic rocks near rifting continents. Give one possible explanation for this pairing.

3. Why do we rarely find batholiths made of gabbro?

4. How might the distribution of the Earth's igneous rocks change when the Earth's internal heat is exhausted and plate tectonic movement stops?

5. Why are there virtually no granites or diorites on the Moon? How might small volumes of such felsic rock form under the geological conditions believed to be responsible for the Moon's igneous rocks?

The Key Term list is a tool for quick review and gives the page number for the full discussion, for students who need to reread the material. (The terms also appear in the glossary.)

Questions for Review help students review the factual content of the chapter, and For Further Thought questions encourage them to think critically about the implications of the information they have learned.

Geologylink.com, an award-winning web site that contains a wealth of resources, updated regularly by Stanley Chernicoff and Rob Viens. Includes:

- The Earth Today—Read about geological events as they happen around the globe.
- In the News—Find reports on the latest discoveries and news in the geo-community.
- Hot Topics—Research and discuss the hottest topics in earth science.
- Inside Geology—Link to hundreds of class lectures, web sites, references, news items, organizations, and glossaries on every aspect of geology.

Geologylink.com also includes:

- Virtual Classroom—Find links to physical geology courses from around the world.
- Virtual Field Trips—Take a virtual field trip or link to local geologic surveys and information from around the world. See geologylink.com's new and revised field trip pages for North America.
- Quizzes—Use Houghton Mifflin's A.C.E. testing for each chapter of Chernicoff's *Geology,* 2nd ed.

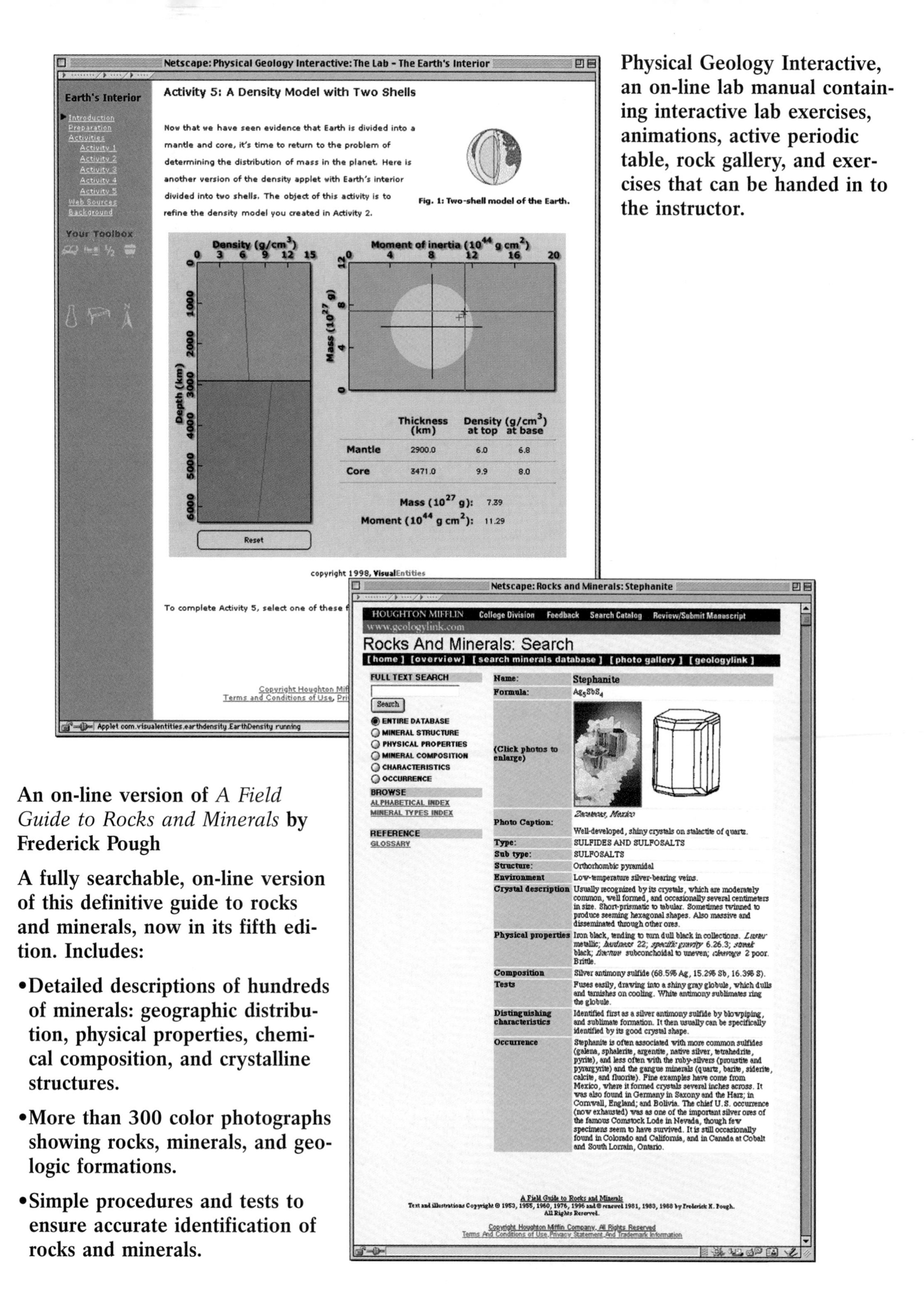

Physical Geology Interactive, an on-line lab manual containing interactive lab exercises, animations, active periodic table, rock gallery, and exercises that can be handed in to the instructor.

An on-line version of *A Field Guide to Rocks and Minerals* by Frederick Pough

A fully searchable, on-line version of this definitive guide to rocks and minerals, now in its fifth edition. Includes:

- **Detailed descriptions of hundreds of minerals: geographic distribution, physical properties, chemical composition, and crystalline structures.**
- **More than 300 color photographs showing rocks, minerals, and geologic formations.**
- **Simple procedures and tests to ensure accurate identification of rocks and minerals.**

ESSENTIALS OF GEOLOGY

Part 1 Forming the Earth

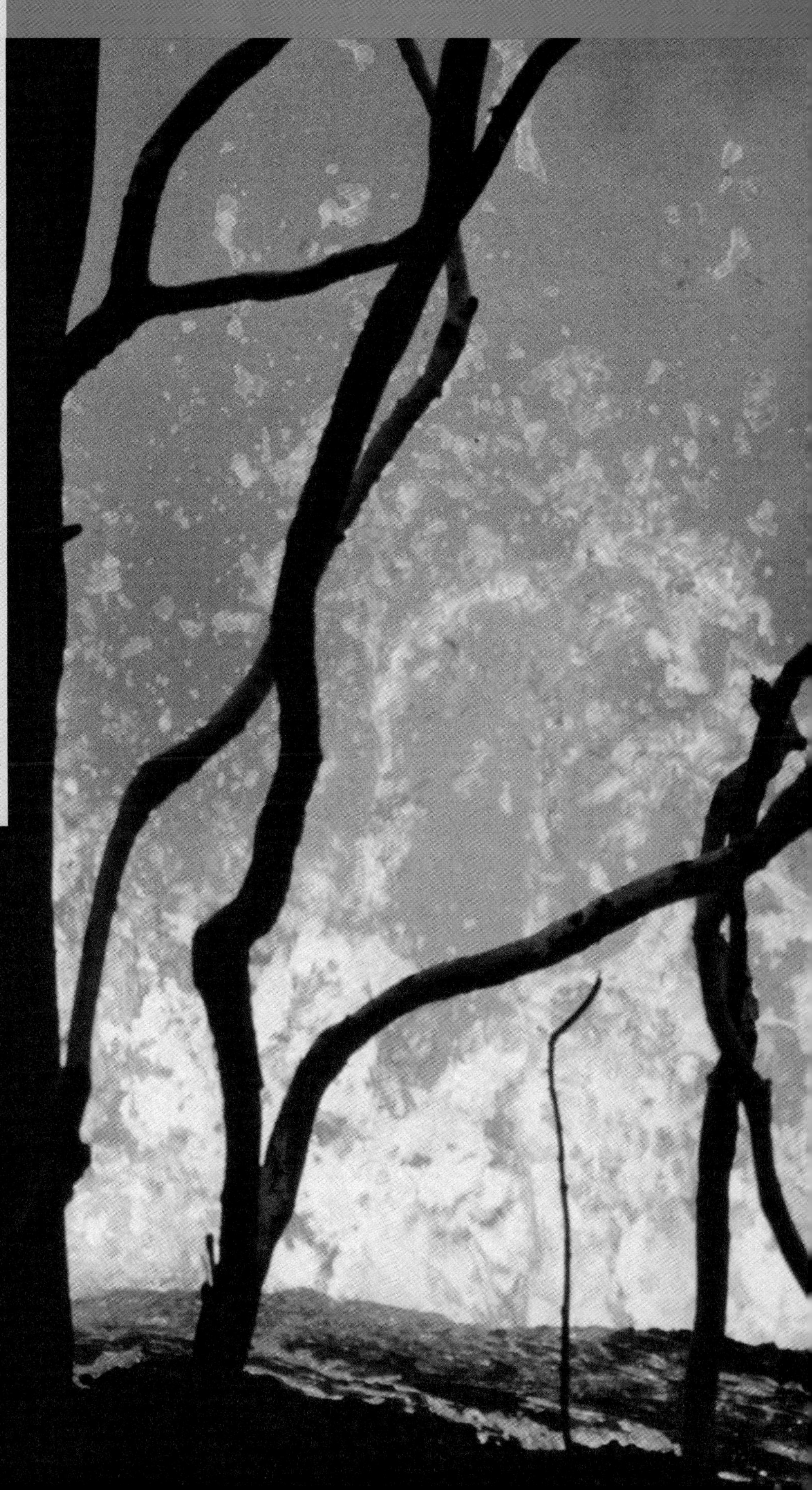

1

A First Look at Planet Earth

At 5:03 P.M. Pacific daylight time on October 17, 1989, baseball fans across North America were settling down in front of their television sets to watch Game Three of the World Series from San Francisco. Minutes later, violent movement along a small segment of California's San Andreas fault (Fig. 1-1) had caused widespread destruction, taking the lives of scores of Bay Area residents and injuring hundreds more. Instead of baseball, millions viewed live broadcasts of grim scenes, including collapsed buildings and freeways (Fig. 1-2). Four years later, at 4:31 A.M. Pacific standard time on January 17, 1994, Southern Californians were jolted awake by a powerful earthquake that took 57 lives, buckled numerous freeways, and proved to be one of the most expensive natural disasters ever in the United States, with estimated cleanup and repair costs of more than $15 billion. Exactly one year later, the effects of the 1989 Loma Prieta and 1994 Northridge earthquakes in California were put into a new perspective by another earthquake—this one more than 9000 kilometers (5700

Figure 1-2 The collapse of the Nimitz Freeway in Oakland, California, during a major earthquake along the San Andreas fault on October 17, 1989. The San Andreas fault, a fracture in the Earth's crust that cuts northwest-southeast across much of California, is responsible for some of North America's most powerful earthquakes.

Figure 1-1 The San Andreas fault, as seen from the air over Carrizo Plain in California.

Highlight 1-1 What Caused the Extinction of the Dinosaurs?

Paleontologists (geologists who study ancient life forms) have long wondered what might have caused more than 75% of all the forms of life then on Earth to vanish about 65 million years ago. The most dramatic loss involved the extinction of the dinosaurs, a group of animals that had roamed the planet for 150 million years, but numerous other life forms vanished as well—large and small, water- and land-dwelling, plant and animal. Many species—whether living in freshwater lakes, in rivers, in saltwater oceans, or on land—became extinct at roughly the same time.

Some early hypotheses focused on only one kind of organism to explain these extinctions. Some proposed that epidemic diseases eliminated dinosaur populations or that the rise of egg-stealing mammals ravaged dinosaur nests. But neither of these hypotheses accounted for the loss of two-thirds of all marine animal species, which led some scientists to propose that the oceans became lethally salty (though this idea did not explain why some marine creatures survived). To explain the extinction of gigantic terrestrial reptiles, tiny marine organisms, and many life forms in between, a number of hypotheses invoked global environmental change. Did the Earth suffer from a period of drastic cooling 65 million years ago? Did a shift in the planet's protective magnetic field allow harmful solar radiation to reach land and sea, eliminating a wide variety of life forms? Did a nearby star explode, bathing the Earth in cosmic radiation? Surely, each of these events would have affected all life on Earth simultaneously. Why, then, did 25% of the planet's species remain unaffected?

Several hypotheses agree that wholesale extinction followed some catastrophic disruption of the global food chain. One group of scientists has proposed that massive volcanic eruptions of India's Deccan plateau may have been such an event. The basalts from these eruptions have been dated to 65 million years ago, coinciding perfectly with the extinction of the dinosaurs. The eruptions consisted of hundreds of lava flows covering an area of 10,000 square kilometers (3861 square miles) that produced more than 10,000 cubic kilometers (2390 cubic miles) of basalt. It is postulated that the eruptions sent a cloud of volcanic ash and gas around the Earth, blocking out sunlight, cooling the planet, and leading to a worldwide decline in vegetation, including microscopic marine plants. Without the plants on which their diets were based, many plant-eating animals would have died, and their extinction would in turn have wiped out the meat-eaters, such as *Tyrannosaurus rex,* that were their predators.

Another group of scientists, led by geologist Walter Alvarez and his father, Nobel prize–winning physicist Luis Alvarez, has proposed another scenario: A meteorite at least 10 kilometers (6 miles) in diameter plowed into the Earth, releasing a shower of pulverized rock into the atmosphere. The resulting dust veil would have blocked out sunlight (in much the same way as volcanic ash would have), cooled the planet, and led to an "impact winter" that may have lasted for decades—long enough to devastate the global food chain. The strongest evidence to support this impact hypothesis is a 2.5-centimeter (1-inch)-thick

miles) away near Kobe, Japan—that killed more than 5000 people, injured 30,000, and left 300,000 homeless.

Why do areas such as California and Japan suffer from periodic and often devastating earthquakes while other areas are spared? The answer to this question, as well as to questions about why volcanoes, landslides, and other catastrophic events occur, can be found in the science of geology. **Geology** is the scientific study of the Earth: the materials that compose it; the processes, such as mountain building and the creation of ocean basins, that shape its surface; and the events, such as floods and glaciation, that sculpt its landscapes. Geologists examine the origin of the Earth and its evolution through its 4.6-billion-year history, and even the geological processes of the other planets in our solar system.

Everything we use comes from the Earth, so geology has an enormous practical impact on our daily lives. Through geological knowledge we are able to locate natural resources, such as the oil, gas, and coal that fuel our cars and heat our homes, the iron and other metals upon which so much of our civilization relies, and even new sources of clean groundwater that are essential to life and agriculture in many areas. Geological study also helps us to predict and avoid some of nature's life-threatening hazards—for example, by identifying slopes that are too unstable to support buildings, warning us away from eroding shorelines, or even predicting where and when earthquakes might occur. In addition, geologists probe the most fundamental mysteries of our planet: How old is the Earth? How did it form? When did life first appear? Why do some areas suffer from devastating earthquakes or volcanoes, while others are spared? Why are some regions endowed with breathtaking mountains and others with fertile plains?

We begin our study of geology by describing how the science of geology operates and by introducing some basic concepts and standards that underpin this discipline. We discuss how our planet may have formed and speculate about how it has changed over the millennia. Finally, we examine some of the Earth's large-scale geological processes and determine how geologists deduced the nature of these processes.

layer of clay found around the world in rocks that date from approximately 65 million years ago (Fig. 1). The clay contains iridium, an element that is extremely rare in rocks of terrestrial origin, but quite common in meteorites. Mineral grains shattered by very high pressures—as would occur if they had been struck by a meteorite—have also been found at the proposed impact sites. The Alvarezes and their associates contend that the iridium-rich layer resulted from the global fallout of pulverized meteorite dust. Fossils of numerous species, including many now-extinct organisms, have been found in the rocks that formed just before the iridium-rich layer was deposited, whereas only about one-fourth as many species are represented in the rocks that formed just after this layer was deposited. This evidence suggests that many extinctions occurred during the time of deposition.

Just as hypotheses may be discarded, modified, or elevated to theory status, they are also sometimes combined. One group of scientists has recently proposed that the Earth was indeed struck by a meteorite 65 million years ago somewhere in the Western Hemisphere, and that reverberations from the impact initiated massive volcanism on the opposite side of the globe. The material spewed into the atmosphere by both events may have combined to devastate the Earth's food chain, thus bringing about the demise of the dinosaurs.

As yet, no extinction hypothesis has achieved theory status. Analysis of the Earth's 65-million-year-old deposits continues today, as scientists seek to document further the proportions of organisms that became extinct at that time and search for additional evidence of a meteorite strike or of a catastrophic volcanic eruption that coincides with the time of the extinctions.

Figure 1 An iridium-containing layer of clay (marked by coin) found by Walter Alvarez in Gubbio, Italy. The Alvarezes believe that this clay, which is found around the world in rock of this age, may have been deposited after a meteor impact about 65 million years ago.

The Methods of Science and Geology

Scientists make one basic assumption: The world works in an orderly fashion in which natural phenomena will recur given the same set of conditions. As scientists understand it, every effect has a cause.

The Scientific Method

The principal objective of science is to discover the fundamental patterns of the natural world. In trying to find the reasons underlying natural phenomena, scientists use a distinctive strategy called the **scientific method.** First, they gather all available information bearing on their subject, such as measurements and descriptions taken in the field and the results of laboratory experiments. Then, they develop a **hypothesis,** a tentative explanation that fits all the data collected and is expected to account for future observations as well. Often, a number of different competing hypotheses are proposed to explain the same set of data. Highlight 1-1, for example, describes the various hypotheses put forth to explain the mysterious extinction of the dinosaurs and many other species 65 million years ago.

Hypotheses are tested over time as scientists conduct further experiments and make further observations. If a hypothesis does not explain subsequent findings, it must be modified or abandoned. The history of science is littered with disproved hypotheses that were once quite popular—such as the suggestion that the Earth is the center of our solar system. A hypothesis that is repeatedly confirmed by extensive observation and experimentation is retained and may become a **theory,** an explanation that has remained consistent with all the data and gained wide acceptance within the scientific community.

Even after a hypothesis survives testing and becomes a theory, sometimes new data become available—perhaps as a result of updated technology—that are not consistent with the theory. Scientists then propose new hypotheses, modifying or completely replacing the established theory. A theory that continues to meet rigorous testing over a long period of time may be declared a **scientific law.** For example, it has been observed repeatedly that when an object is dropped, it falls toward the Earth's surface. The invariability of this observation has led scientists to accept the *law of gravity* as a scientific law.

Geologists, like other scientists, use field observations and laboratory experimentation in much of their work. However, because most geological processes are imperceptibly slow and their scale unimaginably large from a human perspective, they can't always test hypotheses through direct observation or experimentation in the same way that chemists or physicists can. To supplement their field and laboratory work, geologists sometimes use scaled-down models to study large-scale geological phenomena or rely on the power of computers to create mathematical models.

The Development of Geological Concepts

Almost two centuries of observation and hypothesis formation and testing have contributed to our current understanding of how the Earth developed. Prior to the mid-eighteenth century, the common belief was that the Earth's geological evolution had taken place through a series of immense worldwide upheavals such as volcanic eruptions, monumental earthquakes, and worldwide floods. This belief, called **catastrophism,** was called upon to explain the existence of mountains, valleys, fossils, and all other geological features found on the Earth. Inherent in catastrophism was the belief that the Earth was only a few thousand years old, a concept intrinsic to many Christian theologies.

During the latter part of the eighteenth century, the Scottish naturalist James Hutton (1726–1797) proposed that the processes that anyone could see changing the Earth in small ways during his or her lifetime must have operated in a similar manner throughout the planet's history. His hypothesis, called **uniformitarianism,** proposed that current geological processes could be used to explain long-past geological events. Hutton recognized that slow processes, such as rivers cutting through valley floors and loose soil creeping down gentle slopes, acting over a vast amount of time, may have had a greater effect on the Earth than did occasional catastrophic events (Fig. 1-3). Because it assumed that the Earth was much older than the few thousand years most people believed it to be, Hutton's hypothesis met with great resistance. By the 1830s, however, after much debate, uniformitarianism had prevailed over catastrophism. Its acceptance has been hailed as the birth of modern geology, and James Hutton is widely considered to be the "father of modern geology."

James Hutton maintained that "the present is the key to the past." Indeed, geologists today recognize that the Earth's present appearance results from millions of years of the same physical processes, although probably acting at varying rates.

(a) (b)

Figure 1-3 Gradual change of the Earth. Even processes that occur at very slow rates can change the Earth's appearance dramatically over long periods of time. These two photos of the Grand Canyon, taken from the same perspective 100 years apart, show little geological difference between the earlier scene **(a)**, photographed in 1873, and the later scene **(b)**, photographed in 1972. Fossils of ancient marine creatures found in some rocks of the Grand Canyon, however, show that these rocks—found today near the top of a hot, dry plateau 2300 meters (7500 feet) above sea level—once lay at the bottom of an ocean. Over millions of years, the mud at the bottom of this ocean gradually solidified into rock, was uplifted to its present position, and was cut through and exposed by the Colorado River.

They also know, however, that some geological events are indeed catastrophic, and that much geological change does occur during brief spectacular events. A great earthquake may shift a land area more than 6 meters (20 feet) in a single moment. In 1989, Hurricane Hugo eroded more of the Carolina coast in one day than had the preceding century of slow, steady wave action. Both slow, consistent processes and catastrophic events continuously shape our planet.

Modern geology is concerned with more than just the surficial features of the Earth. It is also concerned with the formation, subsequent changes, numerous characteristics, and fantastic variety of rocks and minerals. It is concerned with the deep unseen interior of the Earth. It is also concerned with the Earth's characteristics relative to other planets in our solar system and with the very origin of the universe.

The Earth in Space

The Earth is a slightly flattened sphere with an average radius of 6371 kilometers (3957 miles), orbiting approximately 150 million kilometers (93 million miles) from the medium-sized star we call the Sun. Our Sun is only one of about 100 billion stars in the Milky Way galaxy, a pancake-shaped cluster of stars that itself is only one of about 100 billion such galaxies in the observable universe. Despite Earth's relative insignificance compared to the universe as a whole, it is perfectly positioned to receive just the right amount of the Sun's radiant energy to support life. Because of its composition and geologic past, the Earth has manufactured a watery envelope and protective atmosphere on which countless living species have relied for millions of years. But how did the Earth become what *may* be the only life-sustaining planet in the solar system?

The Probable Origin of the Sun and Its Planets

Cosmologists (scientists who study the origin of the universe) have proposed that the universe began as a very small, very hot volume of space containing an enormous amount of energy. Many scientists believe that the birth of all the matter in the universe occurred when this space expanded rapidly with a "Big Bang" roughly 12 billion years ago. Immediately after the Big Bang, they suggest, the universe began to expand and cool, which it continues to do today. About a million years after the Big Bang, when the universe had cooled sufficiently to allow the first atoms to form, the universe consisted of about 75% hydrogen gas and 25% helium gas, just as it does today. As the universe continued to expand, pockets of relatively high gas concentrations began to form because of gravitational attraction among the gas particles. Where enough gas gathered, the resulting gas clouds collapsed inward from the force of gravity and created galaxies and clusters of galaxies.

Within each galaxy, such as our own Milky Way, some gas clouds collapsed further to form stars. Even today, stars continue to be born in this way in all galaxies, including our own. (Through a telescope, you can see a "star nursery" in the belt of the constellation Orion, for example.) As the gas within each star collapsed under gravity, sufficient heat was generated to fuse together particles within the core, a phenomenon known as *nuclear fusion.* Such nuclear fusion produces the light we see when we look at stars, including the Sun.

Stars are not only born, but also die—some slowly and some rapidly. Stars die when they begin to exhaust their supply of nuclear fuel and collapse under their own gravitational force. A star that is dying very rapidly is called a *nova* (Latin for "new") because it appears as a very bright new star in the heavens. Dying stars are important because they generate so much heat that new nuclear reactions occur, producing the nuclei of the heavier elements of which the Earth and other planetary bodies are composed.

Long after our galaxy was created, the remnants of earlier stars that had died contributed to the gas cloud, or *nebula,* that eventually developed into our solar system (Fig.1-4). Our nebula was probably originally dispersed across a vast area of space, extending well beyond what would become the orbit of our solar system's outermost planet, Pluto. About 5 billion years ago, this nebula began to collapse inward, perhaps due to a shock wave from a nearby exploding star. As its component materials were drawn by gravity toward its center, they collided in nuclear reactions and generated heat in the same way as other stars, forming the infant Sun.

As heat became concentrated in the center of this new star, material in the outer nebula surrounding it began to cool and condense into infinitesimally small grains of matter. Uncondensed substances nearby were swept outward by strong solar winds, consisting of streams of matter and energy that flowed from the infant Sun. In this way, the first solid materials to form in our solar system became separated into a hot inner zone of denser substances, such as iron and nickel, and a cold outer zone of low-density gases, such as hydrogen and helium. Ultimately, this compositional partitioning would evolve into the four rocky inner planets and the five gaseous outer planets.

As the first bits of matter condensed, they collided and coalesced, forming aggregates that grew to a few kilometers or larger in diameter. These planetary seeds, or *planetesimals,* formed the cores of the developing planets. Cosmologists originally believed that this process of planetary growth, or *accretion,* was slow and gradual, much as one might create a large aluminum-foil ball by the steady addition of small lumps. Recently, however, our view of planetary accretion has changed drastically. Cosmologists now believe that as they grew, huge planetesimals—easily the size of Mercury or even Mars—collided violently. Such violent collisions ejected great masses of molten material into space, perhaps forming some of the moons that today orbit the planets of our solar system.

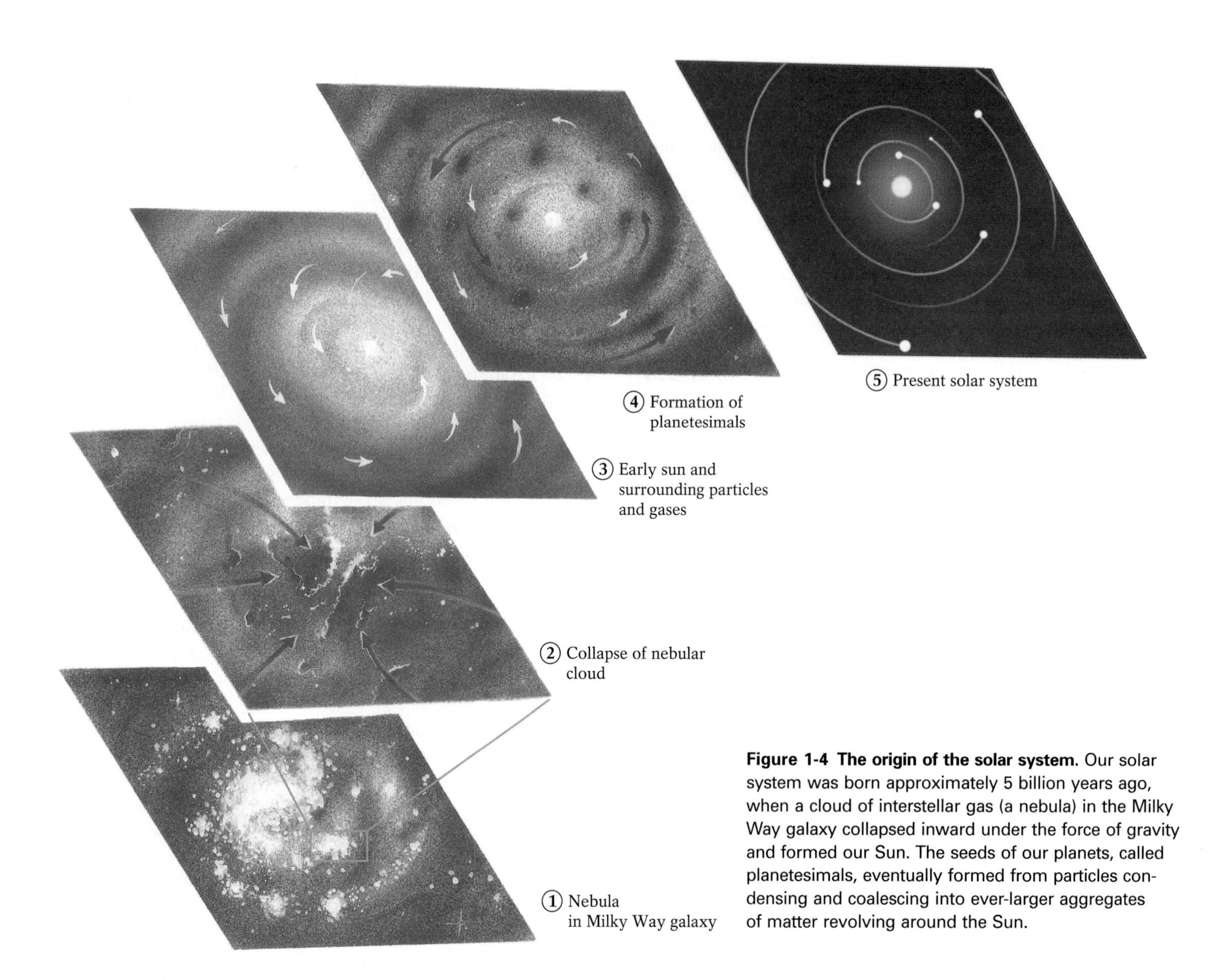

Figure 1-4 The origin of the solar system. Our solar system was born approximately 5 billion years ago, when a cloud of interstellar gas (a nebula) in the Milky Way galaxy collapsed inward under the force of gravity and formed our Sun. The seeds of our planets, called planetesimals, eventually formed from particles condensing and coalescing into ever-larger aggregates of matter revolving around the Sun.

Intense solar radiation warmed the four planets closest to the Sun, causing their surface temperatures to rise. Nearly all of their lighter gases vaporized and were carried away by solar winds. The matter remaining in these four small, dense, inner planets—Mercury, Venus, Earth, and Mars—consisted primarily of iron, nickel, and silicate minerals, which contain a large amount of silicon and oxygen. Much farther from the Sun's heat, the outer planets—Jupiter, Saturn, Uranus, Neptune, and Pluto—formed primarily from the now-frozen lighter gases, hydrogen, helium, ammonia, and methane.

The Earth's Earliest History

The Earth, the largest of the four inner planets, began as a mostly solid, homogeneous body of rock and metal. This temporary state was later changed by the extreme violence and chaos that characterized the Earth's first 20 million years. Proto-Earth's collisions with other planetesimals converted the enormous energy of motion to thermal energy upon impact. Some of this energy was retained and "buried" by succeeding collisions and accretions. "Compressional heating" also resulted from the accumulation of the mass of overlying rocks. In addition, *radiogenic* heating occurred as the atoms of radioactive substances, such as uranium, released heat as their nuclei split apart, a process known as *fission.* The heat from these two sources caused the planet's internal temperature to rise tremendously and set in motion the process that created a layered Earth.

During the Earth's first 10 to 20 million years, the planet's internal temperature rose to the melting point of iron. As a result, much of the iron liquefied. Because it was more dense than the surrounding materials, the iron sank to the proto-Earth's center by the pull of gravity. As it sank, less dense materials rose and became concentrated closer to the planet's surface. Thus the matter that had originally made up a homogeneous Earth became separated into three major concentric zones of differing densities. (This separation process is called **differentiation.**) The densest materials, probably iron and nickel, formed a core at the planet's center. Lighter materials, composed largely of silicon and oxygen as well as other relatively light elements, formed the Earth's outer layers (the mantle and crust). The arrangement of these three layers is somewhat like that of a hard-boiled egg with its thin shell, extensive white, and central yolk. The egg model, however, does not show a number of important sublayers that are fundamental to our understanding of our dynamic Earth. Even lighter materials—gases that had been trapped in the interior—escaped, combining to form the Earth's first atmosphere and oceans.

A Glimpse of the Earth's Interior As the Earth differentiated into its major concentric layers (Fig. 1-5), some upwelling material reached the surface, where it cooled and solidified, forming the Earth's earliest **crust.** Among these low-density substances were oxygen and silicon, which combined to form the silicate minerals that abound in the Earth's crust and upper mantle. Some heat-producing radioactive substances, such as uranium and thorium, also moved toward the surface; because of the heat radiating from these elements, crustal rocks are repeatedly remelted and re-formed into a wide variety of rock types.

Underlying the crust is the **mantle,** a thick layer of denser rocks. The outer 100 kilometers (60 miles) of the Earth, encompassing both the crust and the uppermost portion of the mantle, is a solid, brittle layer known as the **lithosphere** ("rock layer," from the Greek *lithos,* meaning "rock"). Underlying the lithosphere is the **asthenosphere** ("weak layer," from the Greek *aesthenos,* meaning "weak"), a zone of heat-softened rock located in the upper mantle roughly 100 to 350 kilometers (60–220 miles) beneath the Earth's surface. Although it remains solid, the heat-softened rock of the asthenosphere actually flows slowly—a phenomenon that drives much of the planet's geological activity. The lithosphere and asthenosphere are where such large-scale geological processes as mountain building, volcanism, earthquake activity, and the creation of ocean basins originate.

Below the mantle and at the Earth's center is the **core,** the densest layer of all. The core is divided into a liquid outer core and a solid inner core, both consisting primarily of iron and nickel.

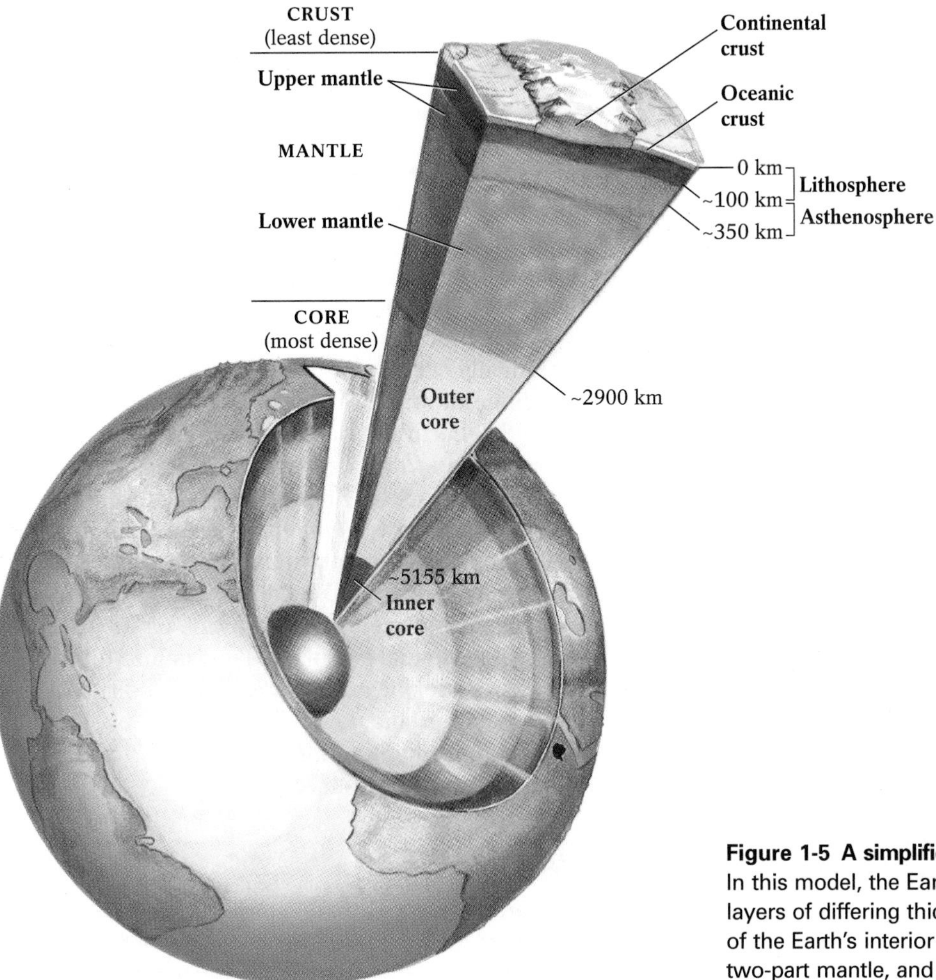

Figure 1-5 A simplified model of the Earth's interior. In this model, the Earth is composed of concentric layers of differing thicknesses and densities. A slice of the Earth's interior reveals a thin crust, a massive two-part mantle, and a two-part core.

The Origin of the Moon The birth of our Moon has sparked lively debate for centuries. Did it form as a companion planet coalescing independently from the solar nebula at the same time as Earth? Did it form elsewhere, only to be drawn into Earth's orbit by our planet's relatively strong gravity? Or was the Moon once part of the Earth?

The answer may lie in the Moon's composition. It is 36% less dense than the Earth and apparently contains much less iron. This difference rules out independent accretion from the solar nebula, for if the Moon did form in the same way as the

Earth, its composition would be similar. The Moon's composition, confirmed in part by the rock-collecting efforts of U.S. Apollo astronauts, is actually quite similar to that of the Earth's mantle, a fact that has led many scientists to suggest that the Moon was formed in a cataclysmic collision between the Earth and another planetesimal.

By roughly 4.55 billion years ago, the Earth had probably attained much of its current size and had become layered, with most of its iron having migrated toward the center to form the core. With the Earth's relatively large gravitational pull, it may have attracted a Mars-sized planetesimal. With the planetesimal traveling toward the Earth perhaps as fast as 14 kilometers per second (31,500 miles per hour), its impact would be quite literally Earth shattering (Fig. 1-6). At the moment of impact, the Earth's young atmosphere would have been blown away, replaced by a rain of molten iron blobs, remnants of the planetesimal's iron core. Such a collision would have vaporized much of the crust and mantle of both the Earth and the planetesimal. Jets of the vaporized crust and mantle would be shot into orbit around the Earth, where the material could eventually coalesce to form the Moon.

Where are the "wounds" of this great collision? Unfortunately, the Earth's dynamic internal processes—related to volcanism, earthquakes, and mountain building—would have eradicated much of the evidence, and erosion would have eliminated the rest. A search for the evidence of the greatest collision in the Earth's history is unlikely to yield a clue, and the story of the formation of the Earth's Moon must remain only a hypothesis.

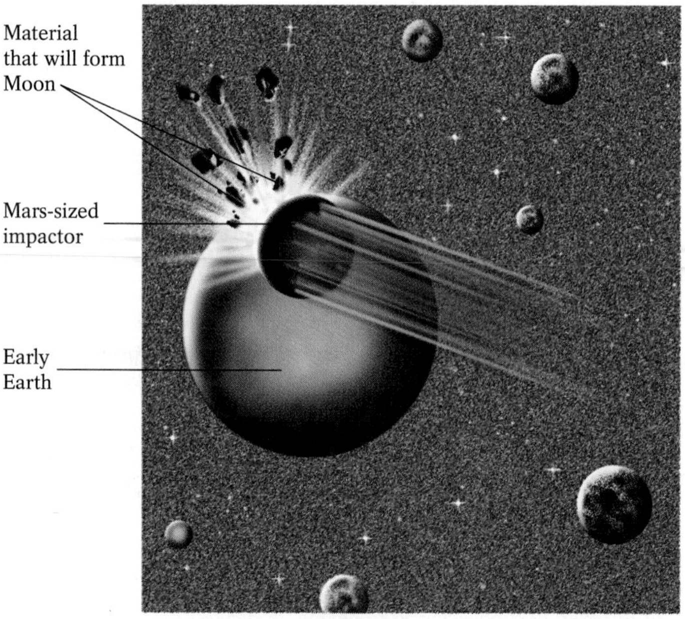

Material that will form Moon

Mars-sized impactor

Early Earth

Figure 1-6 The origin of the Moon. Many scientists now believe that a catastrophic impact between the proto-Earth and a Mars-sized planetesimal spawned the Earth's Moon.

Rocks and Geologic Time

The phrase *geologic time* refers to the time elapsed between the formation of the Earth and the beginning of human history—almost all of the planet's 4.6 billion years of existence. This phrase can also refer to the long spans of time over which geological processes such as mountain building occur. In human terms, the extent of geologic time is almost inconceivably long, and the processes that occur over geologic time are almost inconceivably slow. Fortunately, rocks can bear witness to the effects of even the slowest processes and the oldest events, and almost all of our knowledge about our planet's past reflects the fact that it was preserved, in some way, in rock.

Rock Types and the Rock Cycle

A **rock** is a naturally formed aggregate of one or more minerals. Three types of rocks exist in the Earth's lithosphere and at its surface, with each type reflecting a different process of origin. **Igneous rocks** form from the cooling and crystallization of molten material that has migrated from the Earth's interior to, or just beneath, the Earth's surface. **Sedimentary rocks** form when preexisting rocks become broken down into fragments that accumulate and then become compacted and cemented together. They may also form from the accumulated and compressed remains of certain plants and animals, or from chemical precipitates of materials previously dissolved in water. **Metamorphic rocks** form when heat, pressure, or chemical reactions with circulating fluids change the chemical composition and structure of any type of preexisting rock in the Earth's interior.

Over the great extent of geologic time and through the dynamism of Earth's processes, rocks of any one of these basic types may gradually be transformed into either of the other types, or into a different form of the same type. Rocks of any type exposed at the Earth's surface can be worn away (or *weathered*) by rain, wind, crashing waves, flowing glaciers, or other means, and the resulting fragments transported elsewhere to be deposited as new *sediment;* this sediment might eventually become new sedimentary rock. Sedimentary rocks may become buried so deeply in the Earth's hot interior that they may be changed into metamorphic rocks, or they may melt into magma and eventually cool and recrystallize to form new igneous rocks. Under heat and pressure, igneous rocks can also become metamorphic rocks. The processes by which rocks can form and re-form into different types over time are illustrated in the **rock cycle** (Fig. 1-7).

Time and Geology

An important part of reconstructing our geologic past involves dating layers of rock. *Relative dating* determines which rocks are older than others by referring to spatial relationships between the rocks. For example, the geologic

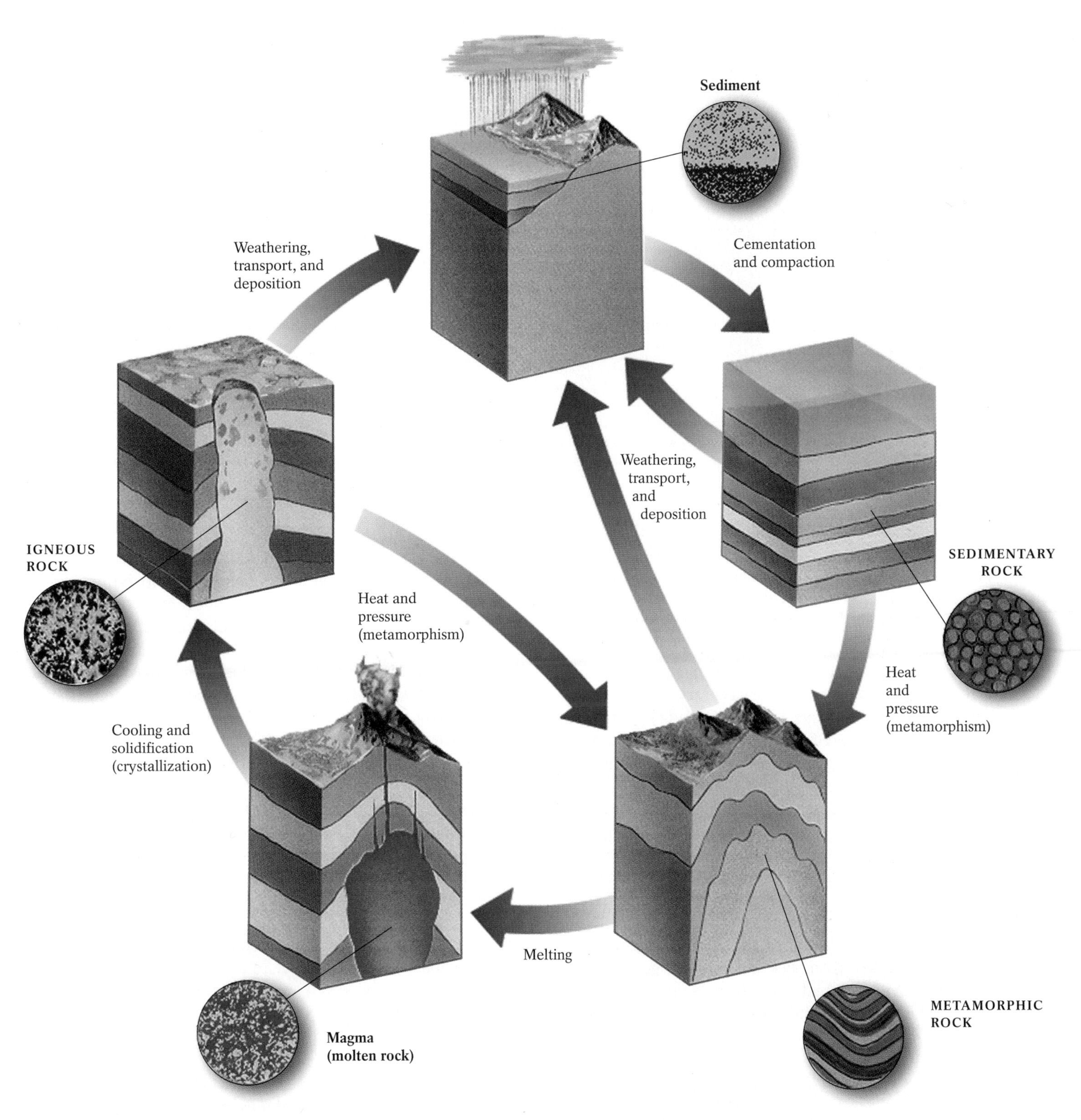

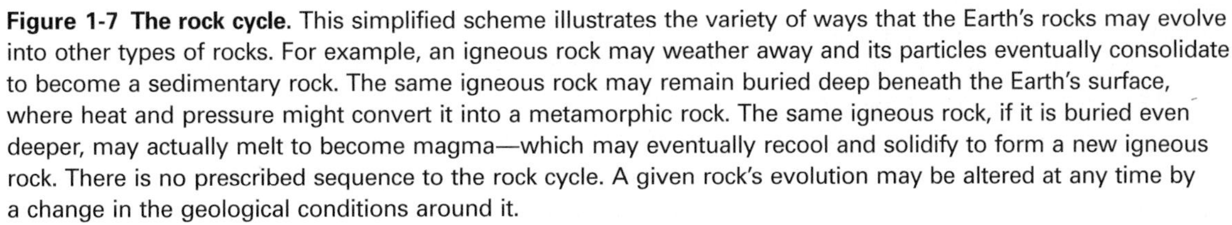
Figure 1-7 The rock cycle. This simplified scheme illustrates the variety of ways that the Earth's rocks may evolve into other types of rocks. For example, an igneous rock may weather away and its particles eventually consolidate to become a sedimentary rock. The same igneous rock may remain buried deep beneath the Earth's surface, where heat and pressure might convert it into a metamorphic rock. The same igneous rock, if it is buried even deeper, may actually melt to become magma—which may eventually recool and solidify to form a new igneous rock. There is no prescribed sequence to the rock cycle. A given rock's evolution may be altered at any time by a change in the geological conditions around it.

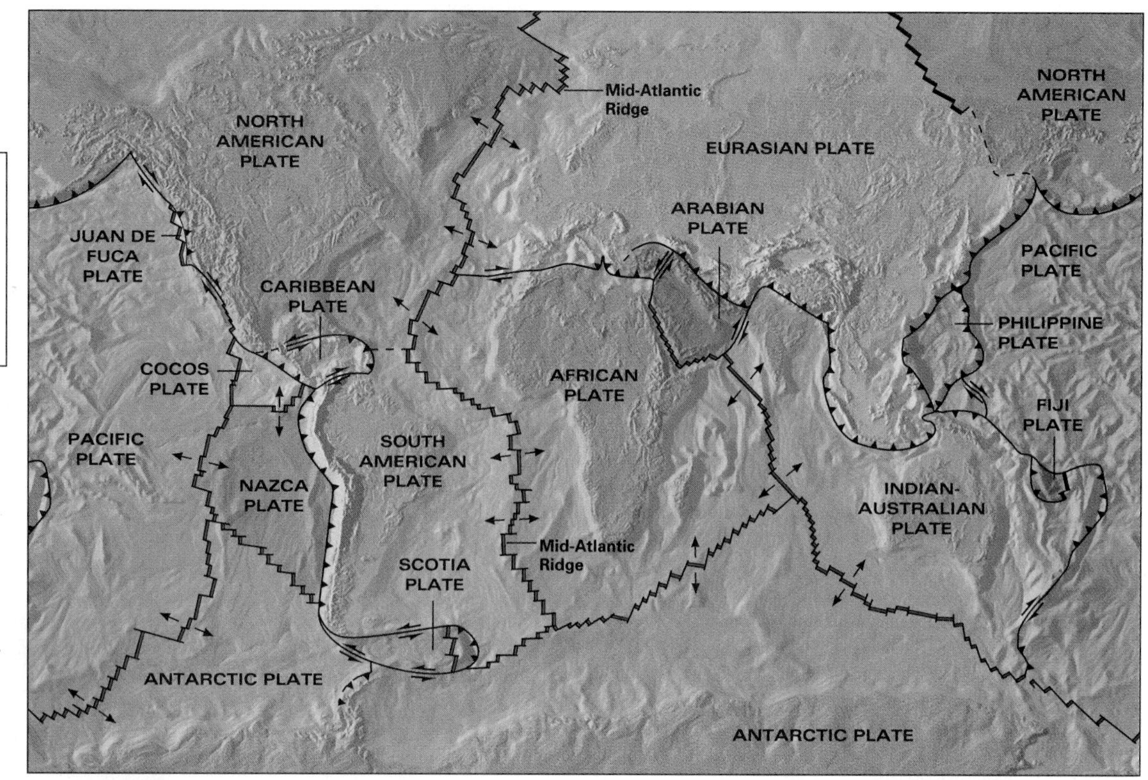

Figure 1-8 A world map showing the Earth's tectonic plates. Note that some plates, such as the North American plate, are composed of both continental and oceanic lithosphere. The Pacific plate is made up almost exclusively of oceanic lithosphere.

principle of superposition states that, where layers of sedimentary rocks have not been disturbed since their deposition, younger rocks overlie older rocks. Equally important is *absolute dating* of rocks, which determines a rock's age in years. Using rock-dating technology developed during the twentieth century, which is based on the constant decay of radioactive elements, geologists can determine the *absolute* ages of rocks with a high degree of accuracy. Earth's oldest rocks, found near Yellow Knife Lake in Canada's Northwest Territories, have been dated using the known decay rate of uranium. They are 3.96 billion years old! We will discuss the various dating methods in detail in Chapter 8.

Our knowledge of past events also relies a great deal on the presence of *fossils*—traces or remains of long-dead plants and animals—in some rocks. These can tell us not only about the structures and activities of extinct organisms, but also—depending on their relative placement in the rock—approximately when the organisms lived. In conjunction with the various rock-dating techniques, the information provided by the sequence of fossils in rock has allowed geologists to develop the geologic time scale, which serves to organize the history of the Earth and its life forms (see Chapter 19).

Plate Tectonics

Today's geologists understand that the Earth behaves in many ways as a single, dynamic system. In earlier centuries, however, geologists were unable to combine the explanations for various geological phenomena, such as the existence of mountain ranges, ocean basins, earthquakes, and volcanoes, into a general explanation of how the Earth functions. Their hypotheses were tailored to specific locations and could not be applied elsewhere. In short, geologists did not have an underlying theory by which *all* geological phenomena could be explained.

In the 1960s, an exciting new hypothesis emerged that provided an elegant unifying explanation for all geological processes, past and present. This hypothesis, called **plate tectonics,** revolutionized the way that geologists viewed the world as dramatically as, a century earlier, the theory of evolution changed how biologists thought about living things. After only a few decades of observation and testing, the hypothesis of plate tectonics has become a widely accepted theory because it provides answers to questions that earlier hypotheses could not resolve. For example, it has enabled us to understand processes such as mountain building, predict such potential catastrophes as earthquakes and volcanic eruptions, and find underground reservoirs of oil, natural gas, and precious metals. Finally, the plate tectonic theory enables us to fit our observations about the ancient past into the same conceptual framework as our understanding of the geological phenomena occurring today.

Basic Plate Tectonic Concepts

The theory of plate tectonics relies on four basic concepts:

1. The outer portion of the Earth—its crust and uppermost segment of mantle (that is, its lithosphere)—is composed of large rigid units called plates.
2. The plates move slowly in response to the flow of the heat-softened asthenosphere beneath them.
3. Most of the world's large-scale geological activity, such as earthquakes and volcanic eruptions, occurs at or near plate boundaries.

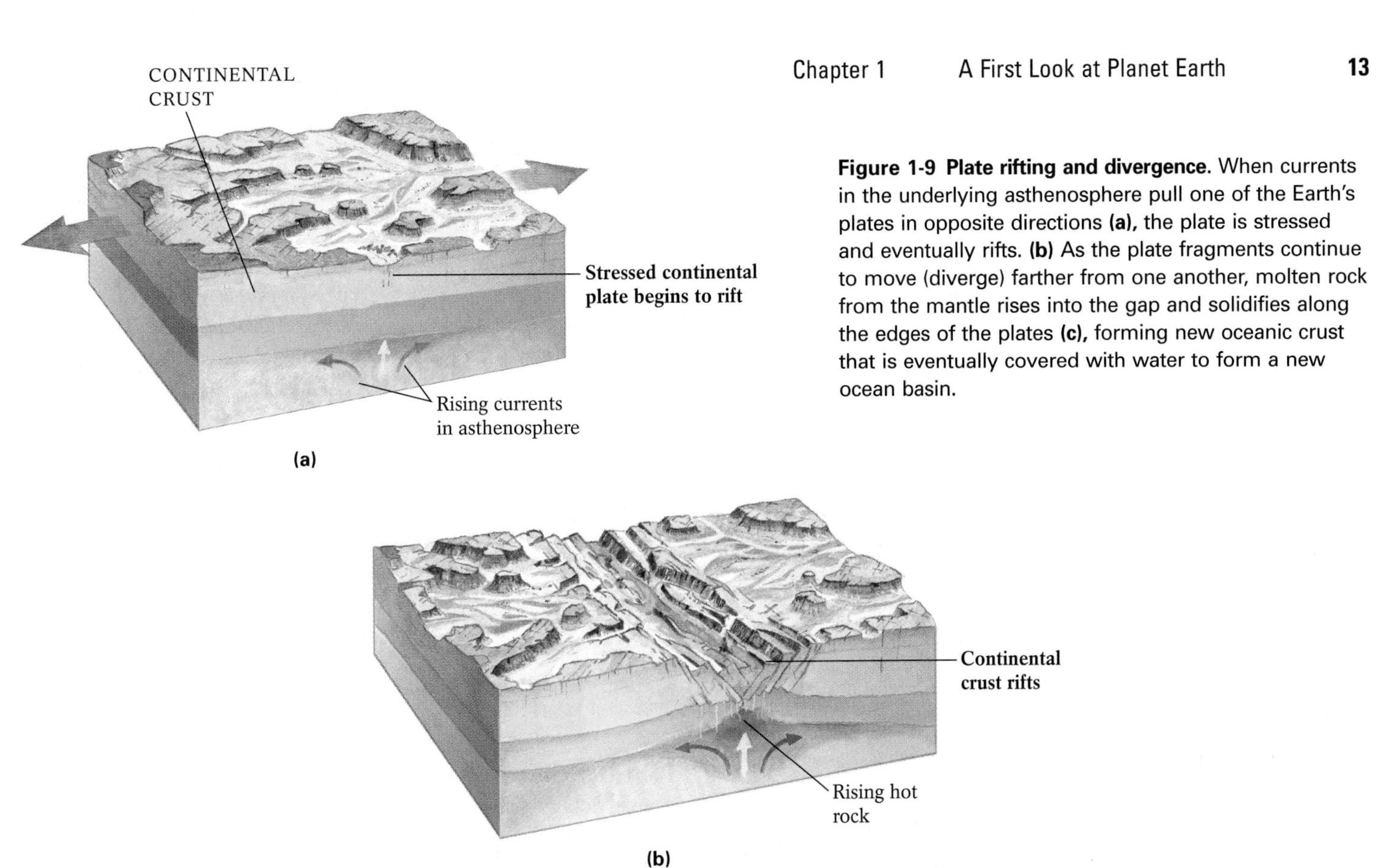

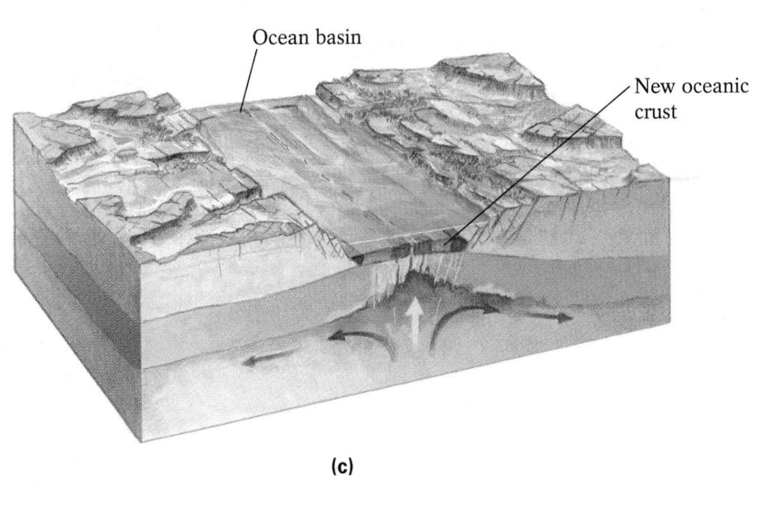

Figure 1-9 Plate rifting and divergence. When currents in the underlying asthenosphere pull one of the Earth's plates in opposite directions **(a)**, the plate is stressed and eventually rifts. **(b)** As the plate fragments continue to move (diverge) farther from one another, molten rock from the mantle rises into the gap and solidifies along the edges of the plates **(c)**, forming new oceanic crust that is eventually covered with water to form a new ocean basin.

4. The interiors of plates are relatively quiet geologically, with little volcanic activity and far fewer and usually milder earthquakes than occur at plate boundaries.

Figure 1-8 shows the Earth's seven large, or major, plates and a number of its smaller ones. Note that the continents themselves are not plates, but instead are usually parts of composite plates that contain both continental and oceanic portions. As these plates move, everything on them—including continents and oceans—moves with them. For example, as the North American plate moves westward away from the Eurasian plate, both the continent of North America and the western half of the Atlantic Ocean are moving farther away from Europe.

Although the fact that entire continents and ocean basins are in motion is rather astounding, the rate at which they move is not; it is comparable to the rate of growth of your fingernails—only a few centimeters per year. If Columbus were crossing the Atlantic Ocean today, 500 years after his famous voyage, he would have to sail only an extra 30 to 50 meters (100–160 feet) to reach shore. Over the vast course of geologic time, however, the rate of plate movement has been sufficient to open and close the Atlantic Ocean several times.

Plate Movements and Boundaries

The Earth's plates move relative to each other in several ways, and plate boundaries are categorized according to which type of movement they demonstrate. Three major types of boundaries exist: divergent plate boundaries, where plates move apart; convergent plate boundaries, where plates move together; and transform plate boundaries, where plates move past one another in opposite directions.

Rifting and Divergent Plate Boundaries Within plate interiors, comparatively little geological activity takes place. Nevertheless, plate interiors may become geologically active if slow-flowing currents in the Earth's asthenosphere generate a pulling-apart motion that tears a preexisting plate into two or more smaller plates. This process, discussed in detail in Chapter 11, is known as **rifting.** The Great Rift Valley of East Africa, where the African plate has been coming apart, is a prime example of early rifting.

Once a plate has been rifted, the resulting smaller plates may continue to separate from one another, a type of plate motion known as **divergence** (Fig. 1-9). Divergence typically

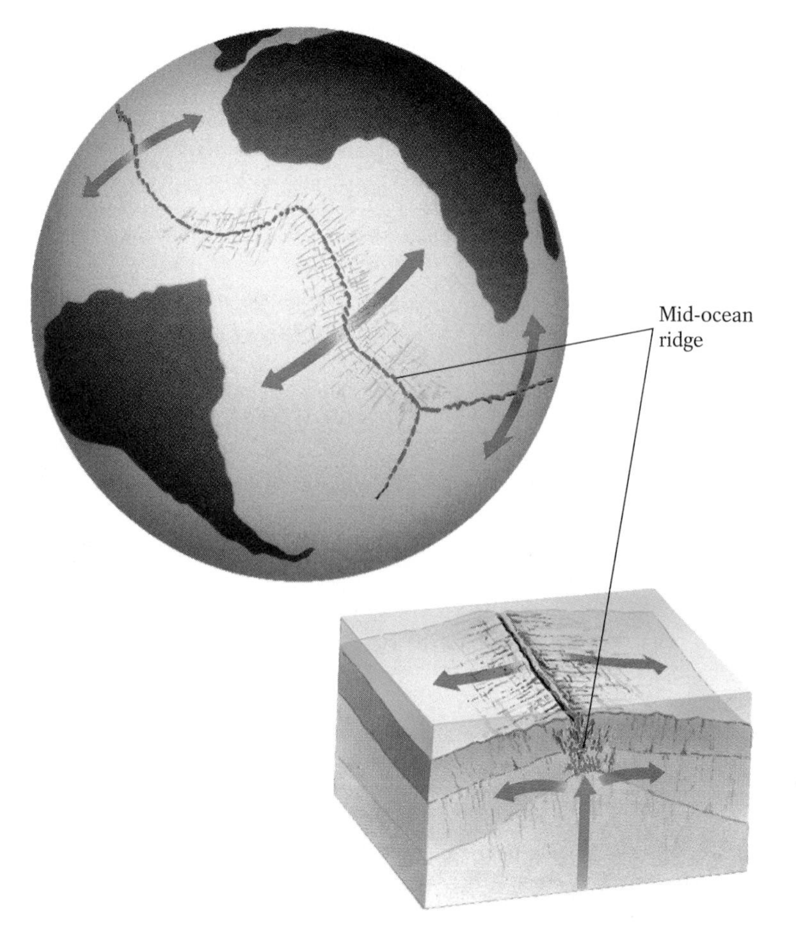

Figure 1-10 **The mid-ocean ridge.** This region, where molten rock from the Earth's interior erupts and cools to become new lithosphere, forms the Earth's longest mountain range, extending for more than 64,000 kilometers (40,000 miles).

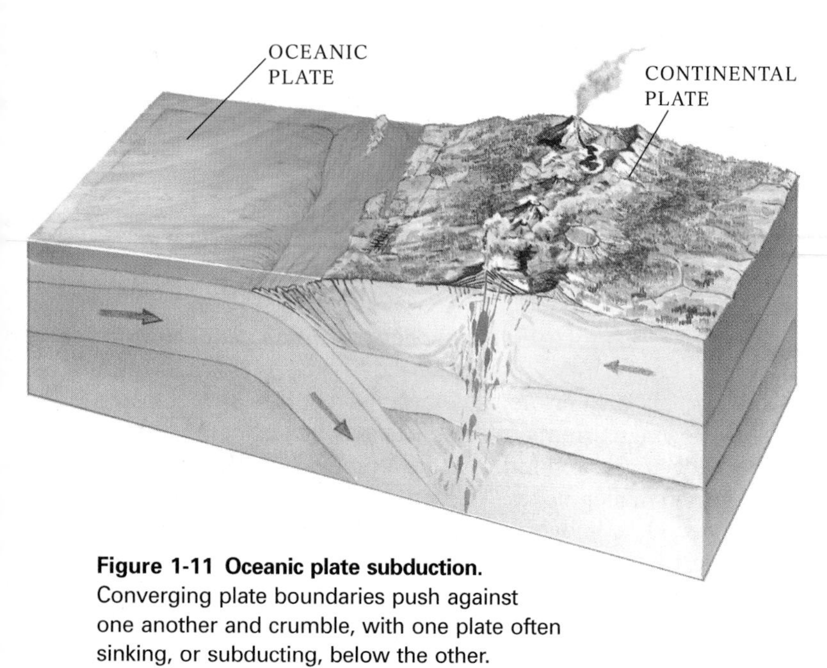

Figure 1-11 **Oceanic plate subduction.** Converging plate boundaries push against one another and crumble, with one plate often sinking, or subducting, below the other.

proceeds at a rate of about 1 to 10 centimeters (0.5–4 inches) per year, as molten rock rises into the thousands of fractures between the rifted plates, cools and solidifies, and becomes attached to the edges of the rifted plates. Meanwhile divergence continues, further separating the older rifted segments and eventually forming an ocean basin. The new ocean basin fills with seawater as further rifting opens new connections to other oceans.

Throughout the period of divergence, erupting molten rock expands the ocean basins by creating new oceanic crust. The center of this volcanic activity is the *mid-ocean ridge,* a continuous chain of submarine mountains that meanders around the globe like the stitches on a baseball (Fig. 1-10). The process of plate growth at mid-ocean ridges is known as **sea-floor spreading.** If the young Red Sea between the African and Arabian plates (see Fig. 1-8) continues to grow at its present rate, it may eventually become a full-blown ocean like the Atlantic or Pacific.

Plate Convergence and Subduction Boundaries Plate **convergence** occurs when plates move toward each other. Convergence may involve two continental plates, two oceanic plates, or one of each type. When two oceanic plates converge, the denser of the two plates sinks beneath the other and is reabsorbed into the Earth's interior, a process known as **subduction.** Because plates of oceanic lithosphere are always denser than those of continental lithosphere, when these two types of plates converge the oceanic plate always subducts (Fig. 1-11).

Subduction itself produces a number of other geological phenomena. For example, friction between the two plates often produces earthquakes. Partial melting of the subducting plate as it reaches the asthenosphere creates a magma that may rise to form volcanoes. This phenomenon is observed in the northern Pacific, where the Pacific plate descends beneath the oceanic edges of the North American plate to form the earthquake-wracked volcanic Aleutian Islands of Alaska. The 1996 eruption of Mount Pavlov in Alaska provided evidence of ongoing subduction in this region.

Continental plates are generally too buoyant to subduct into the denser underlying mantle. Thus, when two continental plates collide, neither plate can subduct completely, although both plates' edges may be temporarily dragged down to depths of perhaps 200 kilometers (120 miles) before being thrown back toward the Earth's surface. Instead of subducting, colliding continental plates become welded together, pushing the colliding edges upward and forming a much larger single plate. Such convergence of continental plates, called **continental collision** (Fig. 1-12), has created many of the Earth's largest mountain ranges. Northern Africa's Atlas Mountains and southern Europe's Alps were formed by past collisions of the African and Eurasian plates, the Himalayas from the ongoing collision of the Indian and Eurasian plates, and North America's Appalachians from a three-way collision of the African, Eurasian,

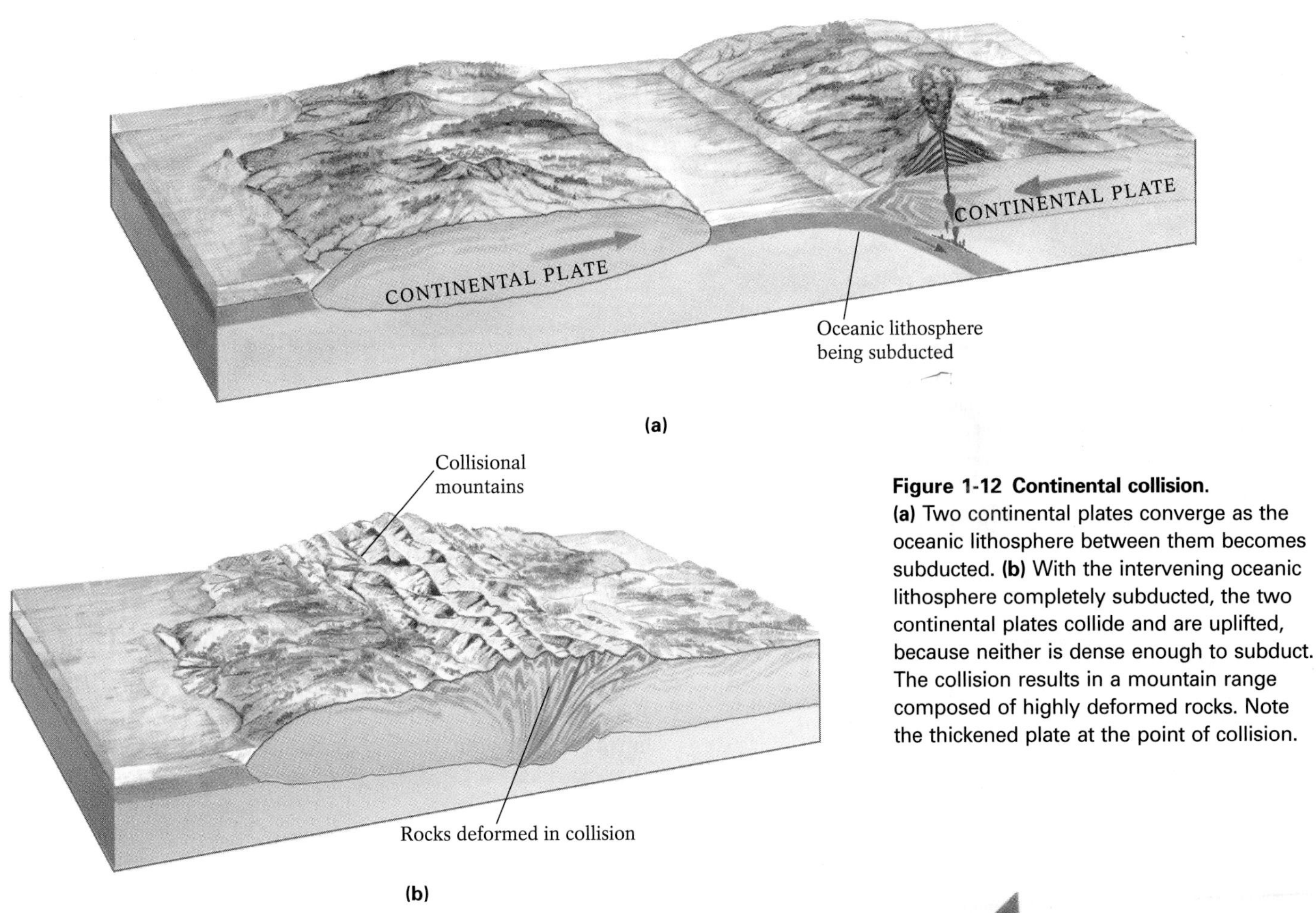

Figure 1-12 Continental collision. **(a)** Two continental plates converge as the oceanic lithosphere between them becomes subducted. **(b)** With the intervening oceanic lithosphere completely subducted, the two continental plates collide and are uplifted, because neither is dense enough to subduct. The collision results in a mountain range composed of highly deformed rocks. Note the thickened plate at the point of collision.

and North American plates that took place between about 400 and 250 million years ago.

Transform Motion and Transform Plate Boundaries The third major type of plate boundary occurs where two plates, either oceanic or continental, move past one another in opposite directions, a process known as **transform motion** (Fig. 1-13). Because great friction results as the moving plates grind past each other, these boundaries produce earthquakes, and any community situated near such a boundary may be periodically devastated. San Francisco and Los Angeles, for example, are both located within the San Andreas transform zone between the North American and Pacific plates.

The Driving Force Behind Plate Motion

Geologists believe that heat-driven currents within the Earth's mantle are principally responsible for plate movements. These currents, known as **convection cells,** develop when portions of the asthenosphere are heated, become less dense than the surrounding material, and rise toward the surface. When the moving material encounters the solid lithosphere, it spreads

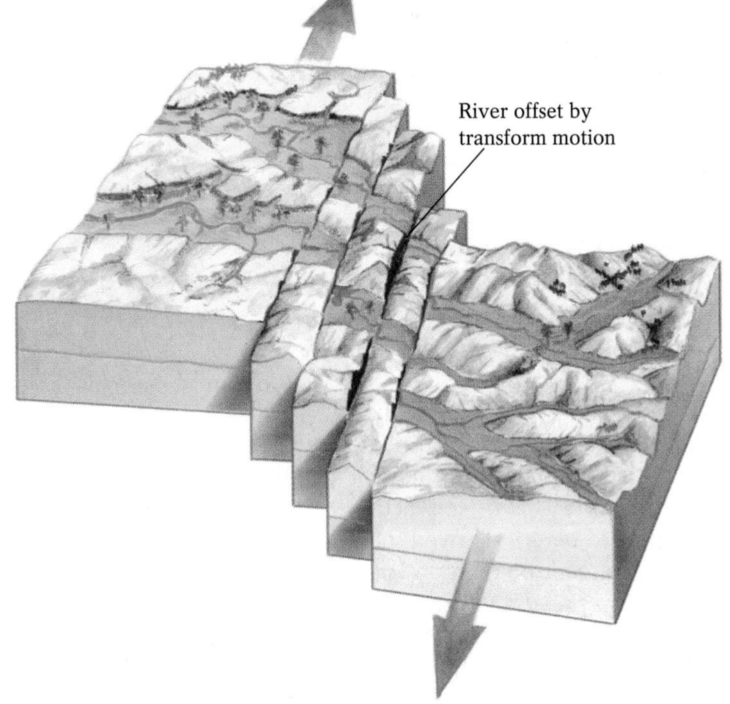

Figure 1-13 Transform motion. When the Earth's plates move past one another in opposite directions, friction builds up at their edges but the plates are neither uplifted nor subducted.

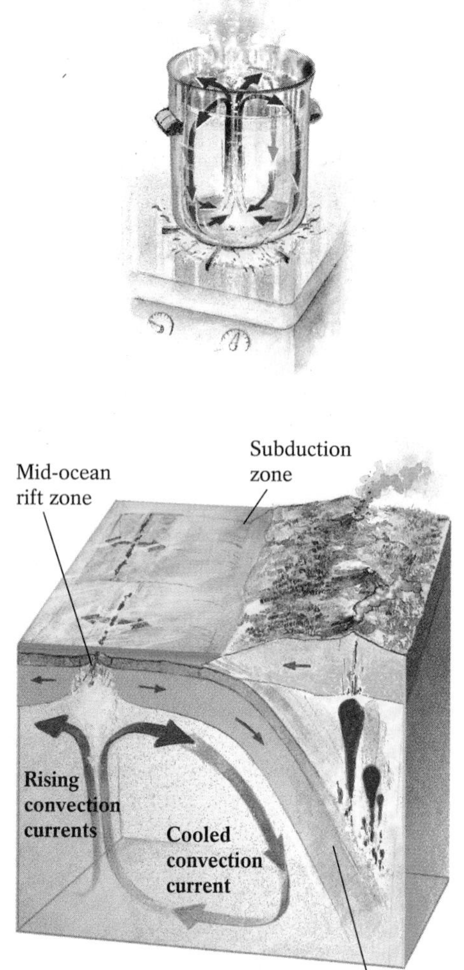

Figure 1-14 Convection cells and plate motion. Heat within the Earth's mantle results in rising ("convecting") currents of warm mantle material, which drag the lighter lithospheric plates along with them as they flow beneath the Earth's surface. As rising mantle material spreads beneath the lithospheric plates, it cools, becomes denser, and sinks back to the deeper interior, where it is reheated to rise again. Such a cycle, known as a convection cell, may be the principal driving mechanism of plate tectonics.

laterally beneath it and drags the lithospheric plates along in a conveyor-belt–like fashion (Fig. 1-14).

In some places, neighboring convection cells pull the lithosphere in opposite directions with enough force to rift it; the rifted plates are then carried along in opposite directions by the slowly convecting currents. A relatively small amount of mantle-derived material escapes to the surface at such divergent plate boundaries, forming a mid-ocean ridge; sea-floor spreading occurs as the plates continue to diverge and mantle material rises to the surface and cools to form new lithosphere. Meanwhile, the leading edges of the diverging plates eventually encounter the edges of other plates at convergent zones. Here, the plate edges are either welded together in continental collisions or subducted back into the asthenosphere to reenter the cycle.

Although convection cells alone may drive plate tectonic processes, some geologists believe that gravity assists the process by literally dragging the plates into the interior of the Earth at subduction zones. Thus both heat-driven divergence and gravity-driven subduction may work together to cycle the Earth's plates.

A Preview of Things to Come

The brief introduction presented in this chapter should prepare you for the more detailed study of geology throughout the rest of this text. In the remainder of Part 1, we examine the details of the rock cycle, looking at the minerals that make up rocks and how the three types of rocks form. We also take a closer look at geologic time. In Part 2, we investigate plate tectonics in depth and discuss how this concept relates to earthquakes and to mountain building. In Part 3, we explore the processes that transform the landscapes of the Earth, looking at the power of gravity, streams, groundwater, glaciers, wind, and waves as they sculpt the surface of the Earth. In Chapter 18, we examine Earth's resources—the many products of geologic processes that have found use in human society. Finally, in Chapter 19, we take a brief look at the geological and biological history of the Earth.

Chapter Summary

Geology is the scientific study of the Earth. In applying the **scientific method,** geologists systematically collect data derived from experiments and observations. They analyze and interpret their findings and develop **hypotheses** to explain how the forces of nature work. A hypothesis that is consistently supported by further study and investigation may be elevated to the status of a widely accepted explanation, or **theory.** A theory that withstands rigorous testing over a long period of time may be declared a **scientific law.**

The hypothesis of **catastrophism,** which was popular until the mid-eighteenth century, held that the Earth had been formed through a series of immense worldwide upheavals. **Uniformitarianism,** which by 1830 had replaced catastrophism, suggests that the Earth has been formed predominantly by slow, gradual, small-scale processes that still operate today.

The universe is believed to have begun with the Big Bang about 12 billion years ago. The Sun, which is a star, formed nearly 5 billion years ago from the collapse of a gas cloud, the center of which heated up as particles drawn inward by gravity collided and produced nuclear reactions. As the outer region of the gas cloud cooled, the Earth and other planets developed (with the Earth being formed about 4.6 billion years ago) by accretion of colliding bits of matter. During the Earth's first several million years, the impact of these accreted masses warmed its interior until the accumulated heat was sufficient to melt most of the planet's constituents. At this time, **differentiation** took place, in which the Earth's densest elements (primarily iron and nickel) sank toward its interior while its lightest elements moved upward to its surface. Today, the Earth has three concentric layers of different densities: a thin, least dense outer layer, called the **crust;** a thick, more dense underlying layer, called the **mantle;** and a much smaller **core,** which is the densest of Earth's layers. The Earth's **lithosphere,** a composite layer made up of the crust and the outermost segment of the mantle, is solid and brittle; below it lies the flowing, heat-softened rock of the **asthenosphere.** Cosmologists hypothesize that the Earth's Moon formed roughly 4.55 billion years ago as a result of a cataclysmic collision between the Earth and a Mars-sized planetesimal.

Rocks, which are defined as naturally occurring aggregates of one or more minerals, are categorized according to the way in which they form. The three basic rock groups are as follows: **igneous rocks,** which solidify from molten material; **sedimentary rocks,** which are compacted and cemented aggregates of fragments of preexisting rocks of any type; and **metamorphic rocks,** which form from any type of rock when its chemical composition is altered by heat, pressure, or chemical reactions in the Earth's interior. The continual transformation of the Earth's rocks from one type into another over time is called the **rock cycle.**

The modern theory of **plate tectonics** states that the Earth's lithosphere is composed of seven major and a dozen or more minor plates, which move in response to the flow of the asthenosphere below them. The plates move relative to each other in three ways: away from one another, by **divergence;** toward one another, by **convergence;** or past one another in opposite directions, by **transform motion.** Most large-scale geological activity, such as earthquakes and volcanoes, occurs at plate boundaries.

Divergence is preceded by **rifting,** a process in which a large preexisting plate is torn into two or more smaller plates. New oceanic lithosphere forms by the process of **sea-floor spreading,** in which molten rock from the Earth's mantle wells upward, cools, and solidifies between rifting plates. Convergence involving either two oceanic plates or one oceanic plate and one continental plate results in **subduction,** in which the denser of the two plates sinks below the other and is reabsorbed into the Earth's mantle. A **continental collision** occurs when both converging plates are continental, in which case neither is dense enough to subduct and the plates' edges are welded together and uplifted into mountains by the pressure of the collision.

The primary force that drives tectonic plates appears to be heat-driven **convection cells** within the Earth's mantle. Near the surface, these convection cells drag the lithospheric plates along, causing plate movement.

Key Terms

geology (p. 4)
scientific method (p. 5)
hypothesis (p. 5)
theory (p. 5)
scientific law (p. 6)
catastrophism (p. 6)
uniformitarianism (p. 6)
differentiation (p. 9)
crust (p. 9)
mantle (p. 9)
lithosphere (p. 9)
asthenosphere (p. 9)
core (p. 9)
rock (p. 10)
igneous rocks (p. 10)
sedimentary rocks (p. 10)
metamorphic rocks (p. 10)
rock cycle (p. 10)
plate tectonics (p. 12)
rifting (p. 13)
divergence (p. 13)
sea-floor spreading (p. 14)
convergence (p. 14)
subduction (p. 14)
continental collision (p. 14)
transform motion (p. 15)
convection cells (p. 15)

Questions for Review

1. Briefly explain the difference between a scientific hypothesis, a scientific theory, and a scientific law.

2. Contrast the principles of catastrophism and uniformitarianism.

3. How did the originally homogeneous Earth become differentiated into concentric layers?

4. Draw a simple sketch of the major layers that make up the Earth's interior. Which of these layers form the Earth's plates?

5. Describe the three major types of rocks in the Earth's rock cycle.

6. List the four basic concepts of the theory of plate tectonics.

7. Draw simple sketches of divergent plate boundaries, two kinds of convergent plate boundaries, and transform plate boundaries.

8. At which type of plate boundary does oceanic lithosphere form? At which type of plate boundary is oceanic lithosphere consumed?

9. Why are there so few earthquakes in Minneapolis and Indianapolis?

10. Draw a simple sketch of a convection cell and explain how it might drive plate movement.

For Further Thought

1. When geologists find ancient glacial deposits in equatorial Africa, they usually interpret them as polar deposits that have drifted from a cold place to a warm place. Formulate another hypothesis to explain this phenomenon.

2. Why is there no current volcanic activity along the eastern coast of North America?

3. Describe how plate tectonic activity might affect the rock cycle.

4. Find the Ural Mountains on a map of Eastern Europe. Briefly explain how they might have formed.

5. As recently as 5 million years ago, South America and North America were completely separated, unattached by Central America. Using Figure 1-8, speculate about how the Central American connection that links the Western Hemisphere might have formed.

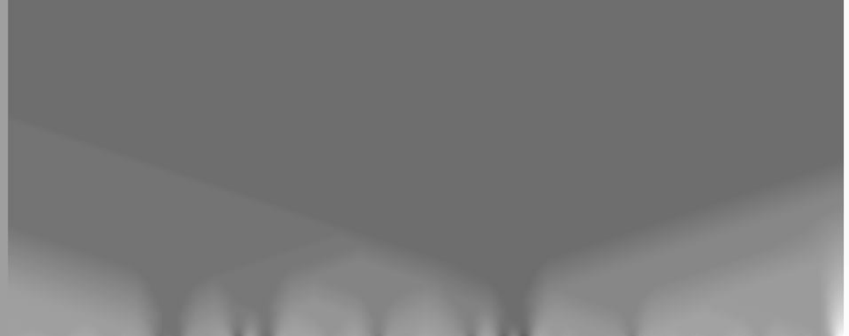

Minerals

Many cultures have long valued minerals for their sheer visual appeal—for their stunning colors or lusters, or their often-perfect symmetry (Fig. 2-1). But the importance of minerals is hardly just aesthetic and is not restricted to the "pretty" specimens found in museums or jewelers' showcases. From the simple flint hand scrapers made by our ancestors hundreds of thousands of years ago to the quartz crystals in modern clocks and watches, minerals have played a fundamental role in human life (Fig. 2-2). Every day we use a vast array of minerals, all derived from the Earth, in a remarkable number of ways. For example, a common absorbent mineral called talc (found in talcum powder) is used to dry and soothe our skin. Other minerals provide the sulfur used to manufacture fertilizers, paints, dyes, detergents, explosives, synthetic fibers, and books of matches, and the fluorine that helps refrigerate food and cool homes and offices. Aluminum,

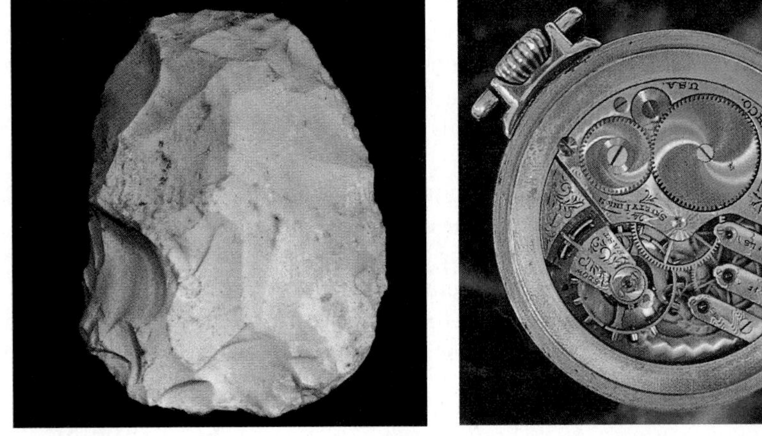

(a) (b)

Figure 2-2 Uses of minerals. Minerals have been providing us with essential tools for hundreds of thousands of years. **(a)** This flint hand scraper was used by Native Americans more than a thousand years ago, probably to scrape flesh and hair from animal skins. **(b)** This elegant personal timepiece runs with great accuracy thanks to the constant vibration of a quartz crystal.

Figure 2-1 Crystals of quartz. These samples illustrate the beauty and symmetry for which many minerals are valued.

commonly derived from the mineral bauxite, is an ideal component for space shuttles, garden furniture, window frames, and beer and soft-drink cans because of its lightness, strength, and resistance to corrosion. When you're laid low by a common intestinal problem, you may run for a spoonful of Kaopectate, a remedy whose active ingredient is the mineral kaolinite.

Many of the nutrients our bodies need come from minerals in the soil that become incorporated into the fruits and vegetables we eat. These minerals include calcium, phosphorus, and fluorine, which give our bones and teeth their hardness, sodium and potassium, which regulate our blood pressure, and iron, which helps our blood carry life-sustaining oxygen.

In this chapter, we examine what minerals are and how they are formed, and discuss methods of identifying the different kinds of minerals. We also look at the distinctive characteristics of several important minerals and mineral groups.

Figure 2-3 Granite, an igneous rock. One can clearly see the individual mineral components in this piece of granite.

The Chemistry of Minerals

Minerals are naturally occurring inorganic solids consisting of one or more chemical elements in specific proportions, whose atoms are arranged in a systematic internal pattern. Diamond, calcite, and quartz are examples of minerals. Because minerals are *naturally* occurring solids, the thousands of synthetic compounds produced in laboratories do not qualify as minerals. Because minerals are *inorganic,* coal, which is composed of heated and compressed remnants of plants, is not considered a mineral. Because minerals have a *systematic internal organization,* substances whose atoms do not follow such a pattern, such as the gemstone opal, are not considered true minerals. **Rocks** are naturally occurring aggregates, or combinations, of one or more minerals, with each mineral retaining its own discrete characteristics. For example, the rock granite contains minerals such as quartz, plagioclase, and hornblende (Fig. 2-3).

Geology students sometimes wonder why so much discussion about minerals relates to their chemistry. The principal reason is that all minerals are composed of combinations of chemical elements, and these chemical structures determine the minerals' distinctive characteristics. These characteristics, in turn, determine their uses and value to society. To know why diamonds are hard, why gold can be pounded into wafer-thin leaves, and why we build skyscrapers with skeletons of titanium steel, we must understand the chemical makeup of minerals.

Minerals are composed of one or more elements in specific proportions. An **element** is a form of matter that cannot be broken down into a simpler form by heating, cooling, or reacting with other chemical elements. Aluminum and oxygen are two common elements. **Atoms** (from the Greek *atomos,* meaning "indivisible") are the smallest particles of an element that retain all its chemical characteristics. All atoms of a given element are identical in certain fundamental ways and differ in these ways from the atoms of every other element.

More than 112 elements are known, of which 92 occur naturally and 20 are laboratory creations. Every element can be represented by its chemical symbol, a one- or two-letter abbreviation. These symbols usually consist of the first letter or letters of the English or Latin name of the element, such as O for oxygen, Al for aluminum, and Na (from the Latin *natrium*) for sodium. Chemists have arranged all the known elements into the Periodic Table of Elements (Fig. 2-4).

Atoms of two or more elements may combine in specific proportions to form chemical **compounds.** The fixed proportions of atoms that make up a compound are expressed by combinations of symbols in a *chemical formula;* for example, the silicon and oxygen compound that makes up the mineral quartz has the chemical formula SiO_2. The subscript numerals denote the ratio of atoms of each element in the chemical compound; thus its formula reveals that quartz contains two atoms of oxygen for every atom of silicon.

The Structure of Atoms

An atom is incredibly small, approximately 0.00000001 centimeter (one hundred-millionth of a centimeter) in diameter. This line of type would contain about 800,000,000 atoms laid side by side. As small as an atom is, it consists of even smaller particles: **protons** and **neutrons,** located in the central region, or **nucleus,** of an atom, plus **electrons,** moving

Strong tendency to lose outermost electrons to complete outer energy level

Tendency to fill outer energy level both by electron sharing and gain and loss of electrons

Strong tendency to gain electrons to complete outer energy level

Inert gases; no tendency to gain, lose, or share electrons

Chemical symbol — **Atomic number** (key: 12 Magnesium Mg)

Transition elements (heavy metals)

1 Hydrogen H																	2 Helium He
3 Lithium Li	4 Beryllium Be											5 Boron B	6 Carbon C	7 Nitrogen N	8 Oxygen O	9 Fluorine F	10 Neon Ne
11 Sodium Na	12 Magnesium Mg											13 Aluminum Al	14 Silicon Si	15 Phosphorus P	16 Sulfur S	17 Chlorine Cl	18 Argon Ar
19 Potassium K	20 Calcium Ca	21 Scandium Sc	22 Titanium Ti	23 Vanadium V	24 Chromium Cr	25 Manganese Mn	26 Iron Fe	27 Cobalt Co	28 Nickel Ni	29 Copper Cu	30 Zinc Zn	31 Gallium Ga	32 Germanium Ge	33 Arsenic As	34 Selenium Se	35 Bromine Br	36 Krypton Kr
37 Rubidium Rb	38 Strontium Sr	39 Yttrium Y	40 Zirconium Zr	41 Niobium Nb	42 Molybdenum Mo	43 Technetium Tc	44 Ruthenium Ru	45 Rhodium Rh	46 Palladium Pd	47 Silver Ag	48 Cadmium Cd	49 Indium In	50 Tin Sn	51 Antimony Sb	52 Tellurium Te	53 Iodine I	54 Xenon Xe
55 Cesium Cs	56 Barium Ba	57 Lanthanum La	72 Hafnium Hf	73 Tantalum Ta	74 Tungsten W	75 Rhenium Re	76 Osmium Os	77 Iridium Ir	78 Platinum Pt	79 Gold Au	80 Mercury Hg	81 Thallium Tl	82 Lead Pb	83 Bismuth Bi	84 Polonium Po	85 Astatine At	86 Radon Rn
87 Francium Fr	88 Radium Ra	89 Actinium Ac	104 Rutherfordium Rf	105 Hahnium Ha	106 Seaborgium Sg	107 Bohrium Bh	108 Hassium Hs	109 Meitnerium Mt	110 Ununnilium Uun	111 Unununium Uuu	112 Ununbium Uub						

Lanthanides	58 Cerium Ce	59 Praseodymium Pr	60 Neodymium Nd	61 Promethium Pm	62 Samarium Sm	63 Europium Eu	64 Gadolinium Gd	65 Terbium Tb	66 Dysprosium Dy	67 Holmium Ho	68 Erbium Er	69 Thulium Tm	70 Ytterbium Yb	71 Lutetium Lu
Actinides	90 Thorium Th	91 Protactinium Pa	92 Uranium U	93 Neptunium Np	94 Plutonium Pu	95 Americium Am	96 Curium Cm	97 Berkelium Bk	98 Californium Cf	99 Einsteinium Es	100 Fermium Fm	101 Mendelevium Md	102 Nobelium No	103 Lawrencium Lr

Figure 2-4 The Periodic Table of Elements. The periodic table groups all elements by similarities in their atomic structures, which result, in turn, in similarities in their chemical properties.

about outside the nucleus. Figure 2-5 depicts a simplified model of an atom.

Each proton carries a single positive charge, expressed as +1, and has a mass of 1.67×10^{-24} gram, which for convenience is referred to as an *atomic mass unit* (AMU) of 1. Neutrons are nearly identical to protons in size and mass, but, as their name suggests, they have no charge—they are neutral. Neutrons do, however, contribute to the **atomic mass** of the atom, which is the total mass of all protons and neutrons within an atom's nucleus. Thus an atom containing one proton and one neutron has an atomic mass of 2 AMU. The third type of atomic particle, the electron, stays in motion around the nucleus at speeds so great that if one were orbiting the Earth, it would do so in less than a single second.

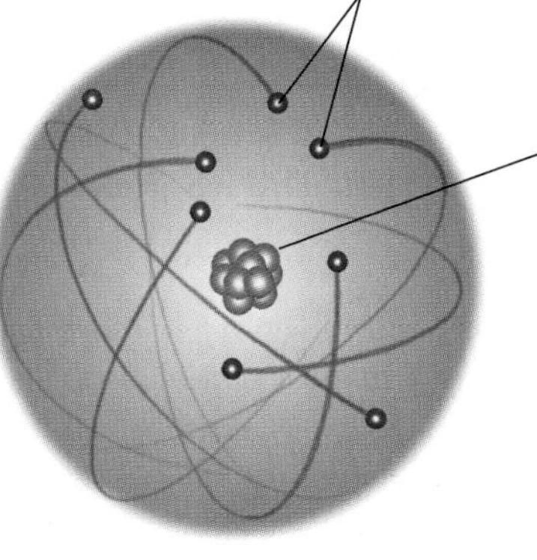

Figure 2-5 A simplified model of an atom. Protons and neutrons compose the nucleus; electrons move around the nucleus at high speed.

Each electron carries a single negative charge, expressed as −1, and has a mass of about 1/1836 that of a proton or neutron; thus an electron makes a negligible contribution to an atom's mass.

The number of protons in an atom's nucleus, its **atomic number,** is constant for each element and determines an atom's identity. The number of neutrons in an atom's nucleus, however, can vary. Atoms of the same element that have different numbers of neutrons in their nucleus, and therefore different atomic masses, are called **isotopes.** For example, the element oxygen (atomic number = 8) always contains eight protons in its nucleus, but it has three isotopes: $^{16}_{8}O$, with eight neutrons in its nucleus; $^{17}_{8}O$, with nine neutrons; and $^{18}_{8}O$, with ten neutrons. The subscript numeral in these notations is the atomic number, and the superscript numeral is the atomic mass. Some isotopes of certain elements contain nuclei that break down spontaneously and emit some of their particles. Such isotopes are described as *radioactive.* Two common radioactive isotopes are $^{235}_{92}U$ (uranium-235) and $^{14}_{6}C$ (carbon-14).

The number of electrons in an atom is usually the same as the number of protons. For example, hydrogen (atomic number = 1) has one proton and one electron, whereas iron (atomic number = 26) has 26 protons and 26 electrons. Because the number of an atom's protons equals the number of its electrons, its positive charge exactly balances its negative charge and the atom therefore has no net charge. All of an atom's positive charge is concentrated in the protons in its nucleus, whereas its negative charge is distributed among the electrons in its periphery.

Most of the time, each electron moves within a specific region of space around the nucleus, called an **energy level.** A maximum number of electrons must generally fill an atom's lowest, or first, energy level, before any can enter the higher energy levels, which are more distant from the nucleus. The lowest energy level in any atom always has a maximum capacity of two electrons. The second energy level can hold a maximum of eight electrons, and succeeding energy levels can each hold eight or more electrons. Figure 2-6 provides a schematic of the number and energy-level positions of some atoms' electrons.

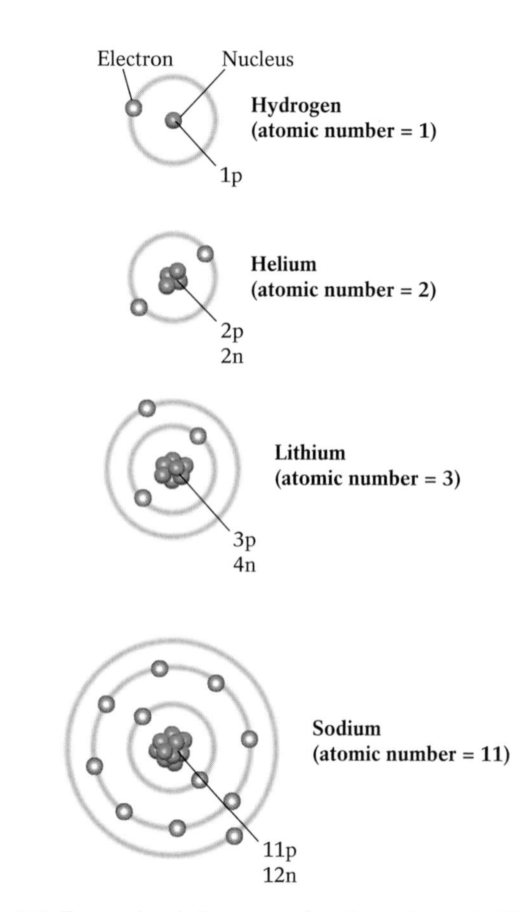

Figure 2-6 Energy-level diagrams of various elements. The nucleus contains the protons (p) and neutrons (n); electrons are shown as balls orbiting the nucleus along concentric circular tracks representing energy levels. (Electron size is exaggerated for clarity; electrons are actually much smaller than protons and neutrons and do not orbit the nucleus in neat little tracts. Hydrogen and helium, because they have two or fewer electrons, have only one energy level; lithium and sodium require multiple energy levels. Note how the number of electrons in the atoms equals the number of protons.

Bonding of Atoms

Atoms frequently combine, or **bond,** to form chemical compounds. Two key factors determine which atoms will unite with which other atoms to form compounds: Each atom tends toward chemical stability, and an electrically neutral compound is more stable.

An atom achieves chemical stability when its outermost energy level is filled with electrons. Thus atoms bond by transferring or sharing electrons so as to attain full outermost energy levels. In the case of hydrogen and helium atoms, which have only the lowest energy level, two electrons must be present in the outermost energy level to achieve stability; for all other atoms, this state requires eight electrons in the outermost energy level.

Atoms of some elements tend to lose their outer electrons, whereas others tend to gain electrons, depending largely on the number of electrons in their outermost energy levels. Elements with one or two electrons in their outermost energy levels, for example, have a strong tendency to give up those electrons. In contrast, elements with six or seven electrons in their outermost energy levels tend to acquire electrons. Most other elements tend to share electrons with other atoms instead of transferring or receiving them. Elements whose outer energy levels are already full are very chemically stable, or *inert*—they do not lose, gain, or share electrons, and they are unlikely to bond with other atoms.

Because of the diversity of electron configurations among elements, various types of bonding are possible. The atoms that make up the vast majority of the Earth's minerals are most often linked by ionic bonding, covalent bonding, or metallic bonding. Another type of bonding—intermolecular bonding—affects minerals principally because of the way it bonds compounds that react with minerals.

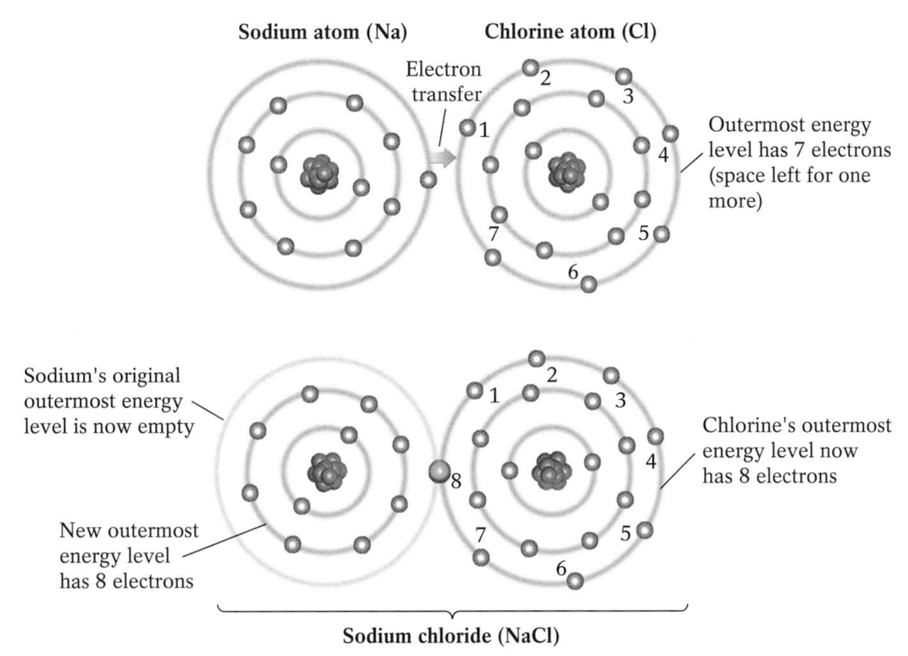

Figure 2-7 Ionic bonding of sodium (Na) and chlorine (Cl). When a sodium atom (with one electron in its outer energy level) donates its outermost electron to a chlorine atom (with seven electrons in its outer energy level), the outer energy level in the new configuration of each atom has eight electrons. The two resulting ions (Na^+ and Cl^-) unite to form sodium chloride (NaCl), a neutral ionic compound.

Ionic Bonding An atom does not change its identity when it loses or gains an electron, but it does lose its electrical neutrality, thereby becoming a positively or negatively charged **ion.** For example, when a chlorine atom gains a single electron, it becomes chemically stable, but it also becomes a negatively charged ion, symbolized by Cl^-. When a potassium atom loses one electron and a calcium atom loses two electrons, they become positively charged ions, symbolized by K^+ and Ca^{2+}, respectively.

When an atom with a strong tendency to lose electrons, such as sodium, comes in contact with an atom with a strong tendency to gain electrons, such as chlorine, electrons are generally transferred so that each atom achieves the chemical stability of a full outer energy level (Fig. 2-7). Atoms that lose one or more electrons become positively charged ions; atoms that gain one or more electrons become negatively charged ions. These oppositely charged ions then attract each other to form **ionic bonds.** The result is an electrically neutral, chemically stable compound.

The physical and chemical properties of compounds differ from those of their component elements. Pure sodium, for instance, is a soft, silvery metal that reacts vigorously when mixed with water and may even burst into flame; chlorine usually occurs as Cl_2, a green poisonous gas that was used as a weapon during World War I. Combining the two results in sodium chloride, which neither ignites nor poisons—it is the white crystalline mineral halite (table salt), a substance that regulates some of the biochemical processes essential to all life.

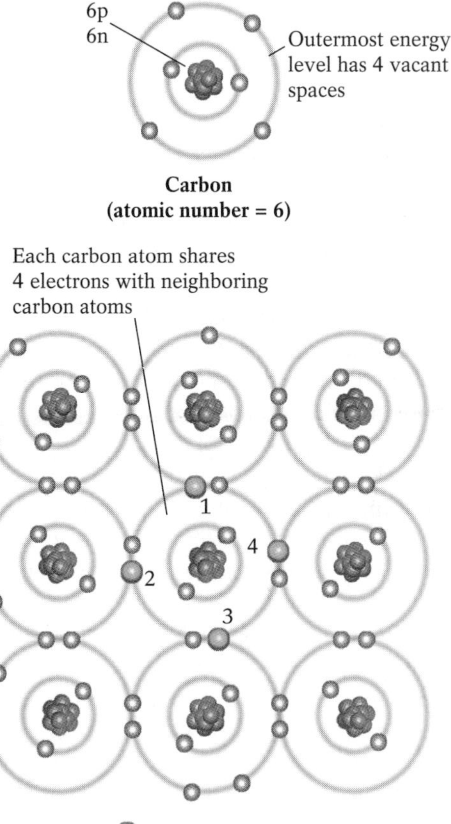

Figure 2-8 Covalent bonding in diamond, a mineral that consists entirely of the element carbon. Each carbon atom in diamond is bonded covalently to four neighboring carbon atoms; the great strength of these bonds accounts for the fact that diamond is the hardest known substance on Earth.

Covalent Bonding Atoms whose outer energy levels are approximately half full (containing three, four, or five electrons) tend to achieve chemical stability by sharing electrons with other atoms. In such a case, both atoms fill their outer energy levels with the shared electrons rather than transferring electrons from one to the other. Sharing electrons produces a **covalent bond,** in which the outer energy levels of the atoms overlap. Covalent bonds are generally the strongest type of bond. In some cases, two or more atoms of a single element may bond covalently with each other. For example, overlapping carbon atoms bond covalently in diamond, an all-carbon mineral (Fig. 2-8).

Metallic Bonding The atoms of some electron-donating elements tend to pack closely together, with each typically surrounded by either eight or twelve others. This arrangement produces a cloud of electrons that roam independently among the positively charged nuclei, unattached to any specific nucleus. In this phenomenon known as **metallic bonding,** the roaming electrons in the metallically bonded substances allow the metals to function as efficient conductors of electricity.

Intermolecular Bonding Certain compounds (but not minerals) exist as *molecules,* stable groups of bonded atoms that are the smallest particles identifiable as compounds. For example, one molecule of water, H_2O, consists of two atoms of hydrogen bonded to one atom of oxygen. Molecules are often attached weakly to other molecules by *intermolecular bonds.* Intermolecular bonding results from the relatively weak positive or negative charges that develop at different locations within a molecule due to the uneven distribution of its moving electrons.

For geologists, the most important type of intermolecular bonding involves water. The uneven distribution of electrons in a water molecule causes it to have a slightly negative side (near its oxygen atom) and a slightly positive side (near its hydrogen atoms). These charged regions in water molecules can attract oppositely charged ions from the surface of some minerals, forming weak **hydrogen bonds** with them. The combined effect of many such bonds can break the internal bonds of some minerals; as a consequence, table salt and many other minerals dissolve in water (see Chapter 5).

Because of the circumstances under which they form, minerals generally do not occur as discrete molecules. Nevertheless, because the groups of atoms that compose minerals have uneven charge distributions, they are subject to intermolecular bonding. One type of intermolecular bond found in minerals, called a **van der Waals bond,** forms when a number of electrons are momentarily grouped on the same side of an atom's nucleus, giving a slight negative charge to that side of the atom and a slight positive charge to the electron-poor side. The positive side may briefly attract electrons of neighboring atoms, and the negatively charged side may fleetingly attract the nuclei of neighboring atoms. Although weak, van der Waals bonds can bond atoms or layers of atoms together in certain minerals. In graphite, for example, sturdy layers of covalently bonded carbon atoms are weakly bonded to one another by van der Waals forces (Fig. 2-9); when you write with a graphite pencil, pressure on the point breaks these bonds, leaving a trail of carbon layers on the paper.

Mineral Structure

When a mineral grows unrestricted in an open space, it develops into a regular geometric shape known as a **crystal** (from the Greek *kyros,* meaning "ice"). This shape is the external expression of the mineral's internal **crystal structure,** the orderly arrangement of its ions or atoms into a latticework of repeated three-dimensional units. As noted earlier, this systematic internal organization is a defining characteristic of all minerals.

Because the internal structure of a given mineral is always the same, the shape of its crystals will be the same in every well-formed, unbroken sample of the mineral. A perfect quartz crystal from Herkimer, New York, for instance, is identical to a perfect quartz crystal from Hot Springs, Arkansas. Most minerals form in restricted growing spaces, however, which prevents them from developing into perfect crystals. As a result, although all samples of a given mineral possess the same internal crystal structure, they may not take the same shape externally.

Sometimes molten rock may cool too rapidly for its atoms to form even a semblance of an orderly arrangement. The resulting solids, called **mineraloids,** lack a specific crystal

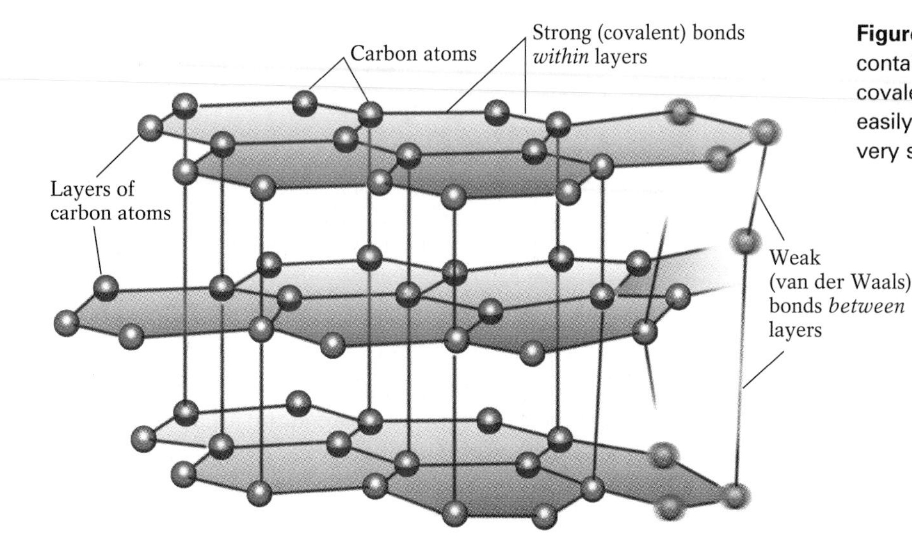

Figure 2-9 Graphite, another form of carbon. Graphite contains weak van der Waals bonds between layers of covalently bonded carbon atoms. Because graphite is easily broken at the sites of these weak bonds, it is a very soft mineral.

structure and thus do not qualify as true minerals. The volcanic rock obsidian, a type of natural glass, is an example of a mineraloid.

Determinants of Mineral Formation

The kinds of minerals that form in a particular time and place depend on the relative abundances of the available elements, the relative sizes and other characteristics of those elements' atoms and ions, and the temperature and pressure at the time of formation.

Only eight of the 92 naturally occurring elements in the Earth's continental crust are relatively abundant, with oxygen and silicon dominating (Table 2-1). Thus most minerals in the crust are oxygen- and silicon-based compounds. Most minerals in the upper mantle are oxygen-silicon-iron-magnesium–based compounds.

In addition to the relative abundance of available elements, mineral formation depends on how readily these elements interact with one another to form a crystalline structure: Given two elements of equal abundance, the element that will contribute more readily to mineral formation will be the one that "fits" better with the other elements present. Atoms and ions in minerals tend to become packed together as closely as their sizes permit. In an ionically bonded mineral, each ion attracts as many oppositely charged ions as can fit around it; their relative sizes therefore determine how many negative ions will surround a positive ion, and vice versa. In the mineral halite (NaCl), for example, one relatively small sodium ion (Na^+) is always surrounded by six larger chlorine ions (Cl^-) (Fig. 2-10).

Table 2-1 The Most Abundant Elements in the Earth's Continental Crust

Element	Proportion of Crust's Weight (%)
Oxygen (O)	45.20
Silicon (Si)	27.20
Aluminum (Al)	8.00
Iron (Fe)	5.80
Calcium (Ca)	5.06
Magnesium (Mg)	2.77
Sodium (Na)	2.32
Potassium (K)	1.68
	98.03
Other elements	1.97
Total	100.00

Ionic Substitution Certain ions of similar size and charge can replace one another within a crystal structure, depending on which is most readily available during the mineral's formation. As a result of such *ionic substitution,* some minerals that have the same internal arrangement of ions may

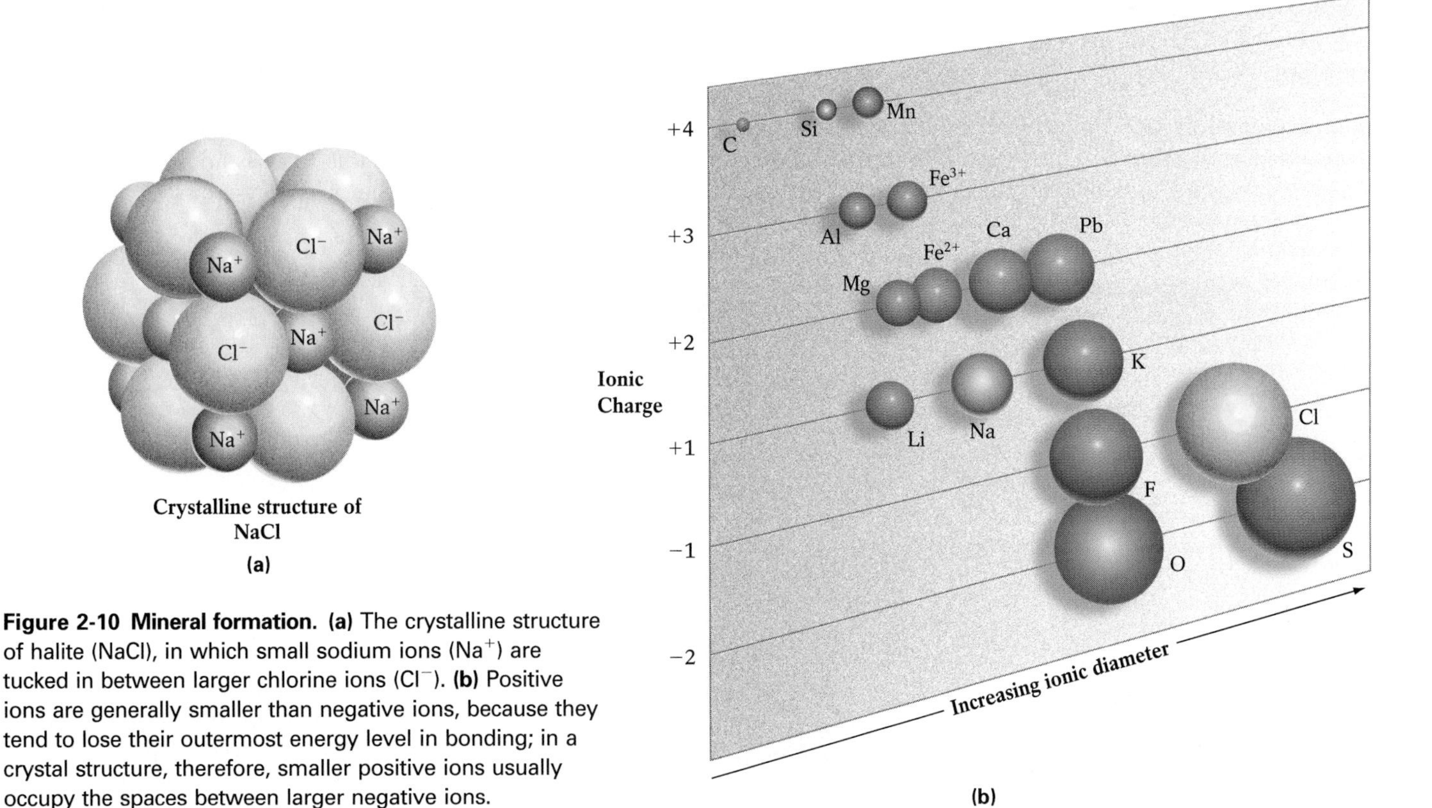

Figure 2-10 Mineral formation. (a) The crystalline structure of halite (NaCl), in which small sodium ions (Na^+) are tucked in between larger chlorine ions (Cl^-). **(b)** Positive ions are generally smaller than negative ions, because they tend to lose their outermost energy level in bonding; in a crystal structure, therefore, smaller positive ions usually occupy the spaces between larger negative ions.

exhibit minor variations in composition. For example, iron (Fe^{2+}) and magnesium (Mg^{2+}), which are nearly identical in size and charge, substitute freely for one another in the mineral olivine ($(Fe,Mg)_2SiO_4$). The color, melting point, and other physical characteristics of olivine differ depending on whether Fe^{2+} or Mg^{2+} is predominant, though olivine's chemical stability and crystal structure remain unaffected.

Polymorphism Two minerals may have the same chemical composition but different crystal structures because they formed under different temperature or pressure conditions. Such minerals are known as **polymorphs** ("many forms"). Graphite (the "lead" in your pencil) and diamond, for example, are polymorphs that consist entirely of carbon. Graphite forms under the low pressure prevalent at shallow depths (only a few kilometers below the Earth's surface), whereas diamond's much more compressed structure results from its formation under intense pressure at depths greater than 150 kilometers (90 miles).

Identification of Minerals

A mineral's chemical composition and crystal structure give it a unique combination of chemical and physical properties that geologists can use to distinguish it from all other minerals. A mineral can seldom be identified accurately on the basis of only one of these properties; usually several must be established before a conclusive identification is made.

In the Field

Many of a mineral's properties are instantly apparent or can be ascertained with minimal effort or rudimentary technology. Most geologists and dedicated rockhounds can identify a great many minerals in the field by merely examining them with the naked eye and performing some very simple tests.

Color Although color may be the first thing you notice about a mineral, it is perhaps the least reliable identifying characteristic. Many minerals occur in a variety of colors due to impurities in their crystal structures (Fig. 2-11). Other minerals with similar colors have completely different compositions. For example, specimens of quartz, calcite, fluorite, halite, and gypsum may appear nearly identical. Because a mineral's color is rarely unique to that mineral, and because it can vary greatly, color is not by itself a reliable criterion for mineral identification.

Luster Luster describes how a mineral's surface reflects light. Minerals can exhibit metallic or nonmetallic luster. Minerals with a metallic luster are either shiny (like car bumpers or aluminum foil) or have the appearance of an oxidized metal (for example, limonite and hematite often look like rusted lumps of metal) (Fig. 2-12). When any light shines on a nonoxidized metal, the light energy stimulates the metal's loosely held electrons and causes them to vibrate. The vibrating electrons emit a diffuse light, giving the metallic surface its characteristic shiny luster. Minerals with nonmetallic lusters are more varied; they can be vitreous (glassy), pearly, silky, adamantine ("like a diamond"), dull, or earthy (Fig. 2-13).

Streak Streak is the color of a mineral in its powdered form, obtained when the mineral's surface is pulverized by rubbing it across an unglazed porcelain slab known as a streak plate. The color of a mineral's streak may differ from the color of the intact mineral sample. Streak is often a more accurate indicator of identity, because this color is not affected by trace impurities in the sample. The steel-gray—or often dark red—

Figure 2-11 Two mineral samples of quartz, which is composed of silicon dioxide (SiO_2). Their colors differ because they each contain minute traces of different impurities.

Figure 2-12 Two types of metallic luster. **(a)** Shiny, as in gold. **(b)** Nonshiny, as in limonite.

mineral hematite (Fe_2O_3), for instance, always has a distinctive reddish-brown streak (Fig. 2-14).

Hardness Geologists define hardness as a mineral's resistance to scratching or abrasion. To test hardness, the unknown mineral is scratched with a series of minerals or other substances of known hardness. (Geological hardness is *not* a function of how easily a mineral breaks—a solid rap with a hammer will easily shatter a diamond, the world's hardest natural substance.) Because every scratch mark represents the removal of atoms from the surface of the mineral, and thus the breakage of the bonds holding these atoms, a mineral's hardness indicates the relative strength of its bonds. Graphite, whose layers of covalently bonded carbon atoms are only weakly attached to each other by van der Waals bonds, is one of the softest minerals.

The Mohs Hardness Scale, named for its developer, German mineralogist Friedrich Mohs (1773–1839), assigns relative hardnesses to several common and a few rare and precious minerals. An unknown mineral that can be scratched by topaz but not quartz has a hardness between 7 and 8 on

Figure 2-13 Two types of nonmetallic luster. **(a)** Vitreous (glassy), as in rose quartz. **(b)** Pearly, as in feldspar.

Figure 2-14 Using streak to identify hematite. Though samples of hematite (Fe_2O_3) are usually steel-gray, hematite's streak is always reddish brown.

Table 2-2 The Mohs Hardness Scale

Mineral	Hardness	Hardness of Some Common Objects
Talc	1	
Gypsum	2	
		Human fingernail (2.5)
Calcite	3	
		Copper penny (3.5)
Fluorite	4	
Apatite	5	
		Glass (5–6), Pocketknife blade (5–6)
Orthoclase (potassium feldspar)	6	
		Steel file (6.5)
Quartz	7	
Topaz	8	
Corundum	9	
Diamond	10	

the Mohs scale. Table 2-2, which lists the minerals and common testing standards used in the Mohs scale, explains why geologists are often found with a few copper pennies, a pocketknife, and well-worn fingernails.

Cleavage Cleavage is the tendency of some minerals, when hammered or struck, to break consistently along distinct planes in their crystal structures where their bonds are weakest. The resulting mineral fragments possess smooth, flat surfaces, with consistent angles between adjacent surfaces. When a mineral that tends to cleave is struck along a plane of cleavage, every fragment that breaks off will have the same general shape. For example, shattering a cube of halite (table salt) produces numerous smaller cubes of halite, all with six sides at right angles to each other. Geologists often use the distinctive number of cleavage surfaces, and particularly the angles by which adjacent surfaces are joined, in identifying minerals (Fig. 2-15).

Fracture Minerals that do not cleave—because all of their bonds are equally strong—will break at random, or *fracture.* Unlike a straight, smooth-faced cleavage surface, a fracture appears as a jagged irregular surface or as a curved, shell-shaped (*conchoidal;* pronounced "kon-KOID-al") surface. In the mineral quartz, which is composed exclusively of silicon and oxygen, all of the atoms are bonded covalently in a three-dimensional framework with equal bond strengths in all directions. Thus, when a crystal of quartz is struck, it fractures (Fig. 2-16). Knowing that quartz fractures instead of cleaves enables geologists to distinguish it from similar-looking minerals, such as halite or calcite, that cleave.

Smell and Taste Experienced geologists occasionally sniff and lick rocks to help identify minerals that have a distinctive smell or taste. Some sulfur-containing minerals, for example, emit the familiar rotten-egg stench associated with hydrogen sulfide gas (H_2S). Halite's salty taste distinguishes it from similar-looking minerals such as quartz and calcite; sylvite (KCl) is distinctively bitter. Kaolinite absorbs liquid rapidly—when licked, it absorbs saliva and sticks to the tongue. Novices should be wary of tasting unknown minerals, however, as some can be dangerous. Realgar and orpiment, for example, which smell like garlic, especially when

Figure 2-15 Distinguishing minerals by their cleavage. (a) Halite has three mutually perpendicular cleavage planes, forming cubes. **(b)** Mica has one perfect cleavage plane, forming sheets.

Figure 2-16 A conchoidal fracture surface on a quartz crystal. Quartz, with equally strong covalent bonds in all directions, has no planes of weakness. It therefore fractures irregularly instead of cleaving.

heated, have the poisonous metal arsenic as their major element.

Effervescence Certain minerals—particularly those that contain carbonate ions (CO_3^{2-})—effervesce, or fizz, when mixed with an acid. A few drops of dilute hydrochloric acid (HCl) on calcite ($CaCO_3$) produce a rapid chemical reaction that releases carbon dioxide gas (in the form of bubbles) and water; this property helps distinguish calcite from similar-looking minerals, such as quartz and halite, that do not effervesce.

Crystal Form Although minerals usually do not grow into perfect crystals, crystal form can nevertheless be an important factor in identifying a mineral. Its role is especially critical in the case of some minerals that grow into particularly distinctive forms, such as rosette-shaped barite and needle-shaped stibnite (Fig. 2-17).

In the Laboratory

Although geologists can positively identify many minerals in the field, some minerals have such similar physical characteristics that even experienced geologists will initially make only an educated guess as to their identity and then bring samples back from the field to test them in the laboratory. There, special equipment is available to analyze a variety of physical and chemical properties with greater precision than is ever possible in the field.

Specific Gravity Specific gravity is the ratio of a substance's weight to the weight of an equal volume of pure water. For example, a mineral that weighs four times as much as an equal volume of water has a specific gravity of 4. The precise specific gravity of an unknown mineral is generally determined in a laboratory, although relative specific gravity can also be helpful in the field in distinguishing between two apparently similar minerals. Minerals with a markedly higher specific gravity will feel much heavier relative to their size than minerals with a lower specific gravity will. For example, gold, with a specific gravity of 19.3, feels much heavier than "fool's gold" (the mineral pyrite), which has a specific gravity of only 5.

Other Laboratory Tests When exposed to ultraviolet light, certain minerals glow in distinctive colors. This property, called *fluorescence,* characterizes fluorite, calcite, scheelite ($CaWO_4$), and willemite (Zn_2SiO_4), among other minerals

(a)

(b)

Figure 2-17 Unusual crystal aggregates. (a) Stibnite needles. **(b)** Barite rosettes.

(Fig. 2-18). A mineral that continues to glow after removal of the ultraviolet light exhibits *phosphorescence.*

An *electron probe* is used to rapidly analyze extremely small mineral samples. This instrument beams electrons at the sample and then analyzes the distinctive X-rays emitted by the sample.

In *X-ray diffraction,* X-rays passed through a mineral sample become scattered, or *diffracted,* producing distinctive patterns (diffractograms) on an X-ray film. These patterns are determined by the arrangement of the atoms and ions in a mineral's crystal structure, and so are unique to each mineral.

Some Common Rock-Forming Minerals

Of the 92 naturally occurring elements in the Earth's continental crust, the eight most abundant (see Table 2-1) form the vast majority of minerals that compose the Earth's rocks. Most such **rock-forming minerals** are classified into one of six categories based on their chemical composition: the silicates, carbonates, oxides, sulfides, sulfates, and native elements.

(a)

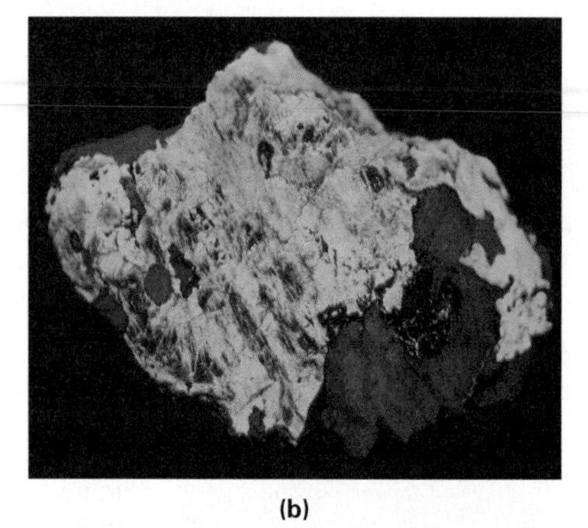

(b)

Figure 2-18 Fluorescence. The fluorescent minerals willemite and calcite in this rock specimen **(a)** glow bright green and red, respectively, while exposed to ultraviolet light **(b)**.

Silicates

Most minerals in the Earth's crust and mantle are oxygen- and silicon-based compounds called **silicates.** Because silicon and oxygen are so abundant and unite so readily, the silicates—encompassing more than 1000 different minerals—make up more than 90% of the mass of the crust. They are the dominant component of most rocks, whether igneous, sedimentary, or metamorphic.

The crystal structure of all silicates contains repeated groupings of four negatively charged oxygen ions congregated around a single positively charged silicon ion to form a four-faced structure called a **silicon-oxygen tetrahedron** (SiO_4). As Figure 2-19 shows, the four oxygen ions in a silicon-oxygen tetrahedron each have a -2 charge, and the one silicon ion has a $+4$ charge. As a result, the tetrahedron carries an overall charge of -4. To form an electrically neutral compound, a silicon-oxygen tetrahedron must either acquire four positive charges by bonding with positive ions or disperse its negative charge by sharing its oxygen ions with neighboring tetrahedra.

The crystal structures of all silicate minerals are derived from one of five principal arrangements of the silicon-oxygen tetrahedra: independent tetrahedra, single chains, double chains, and three-dimensional framework structures. Each of these arrangements represents a different means of sharing oxygen ions.

Independent tetrahedra bond with positive ions of other elements to neutralize their charge. They share no oxygen ions, and so have a silicon-to-oxygen ratio of 1:4 (Fig. 2-20a). The most prominent mineral of this type is olivine,

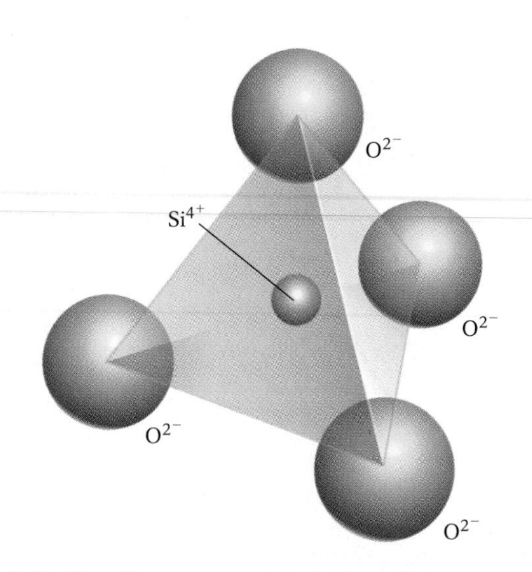

Figure 2-19 The silicon-oxygen tetrahedron. Four oxygen ions occupy the corners of this structure, with a lone silicon ion embedded in the open space at the center.

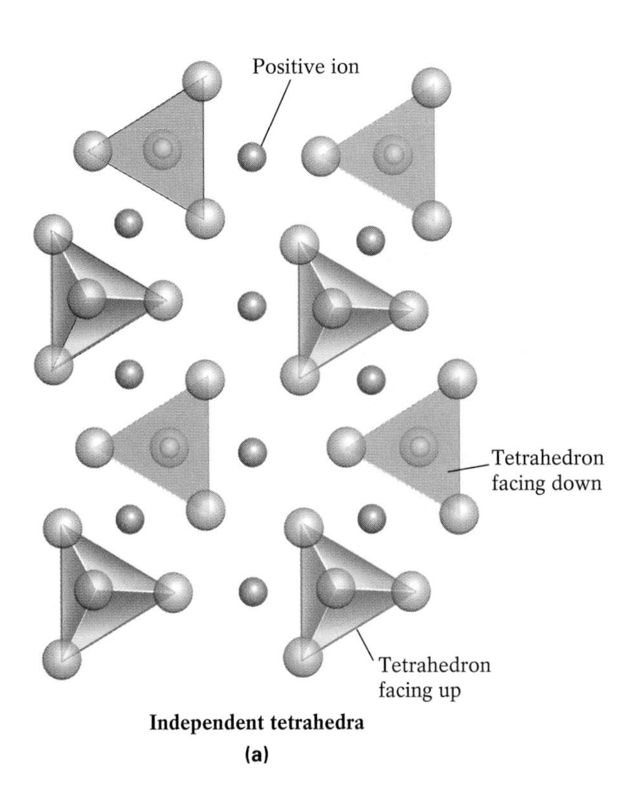

Figure 2-20 (a) Independent tetrahedra. Positive ions are positioned between tetrahedra such that each tetrahedron, with a −4 charge, bonds to two positive ions, each with a +2 charge, thereby neutralizing their combined charge. No oxygen ions are shared between the tetrahedra; therefore the silicon-to-oxygen ratio is 1:4. Photo: Olivine. As is typical of silicates with independent tetrahedral structures, this mineral fractures rather than cleaving.

in which iron (Fe^{2+}) and/or magnesium (Mg^{2+}) ions balance the negative charge of the tetrahedra.

Single chains of tetrahedra form when each tetrahedron shares two corner oxygen ions, producing a silicon-to-oxygen ratio of 1:3 (Fig. 2-20b). The most prominent group of minerals of this type encompasses the pyroxenes, which typically contain iron and/or magnesium ions that bind the chains together.

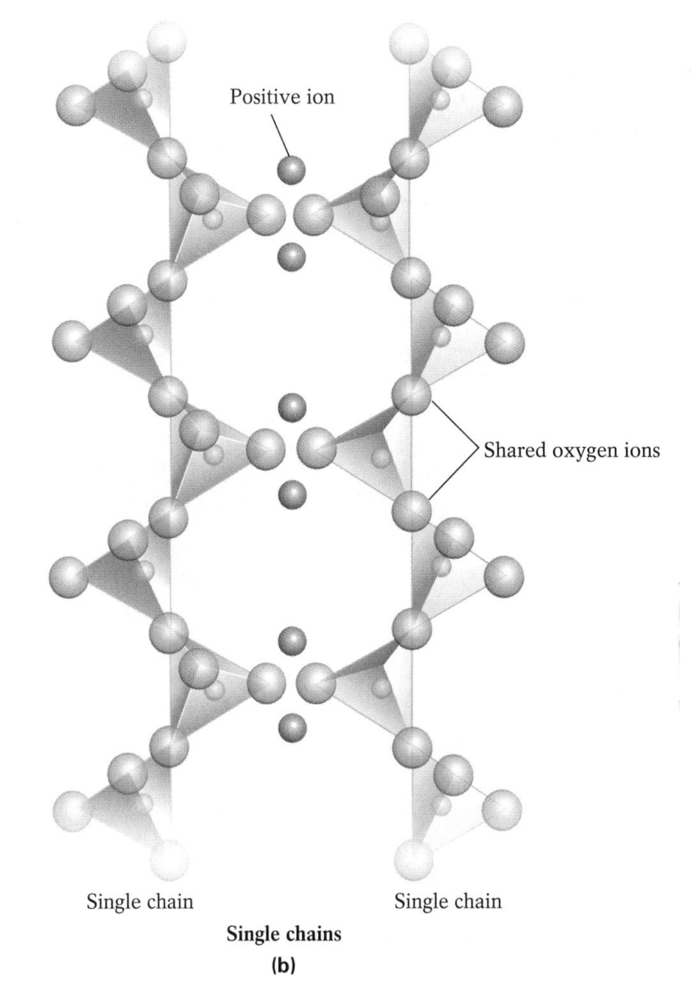

Figure 2-20 (b) Single chains. Each tetrahedron shares two of its corner oxygen ions with adjacent tetrahedra, forming a linear chain of tetrahedra with a silicon-to-oxygen ratio of 1:3. Because each tetrahedron still has a −2 charge after sharing two of its oxygen ions, the cumulative negative charge on such chains attracts a variety of positive ions that bind between them, neutralizing the negative charge of the chains and joining them loosely together. Photo: Pyroxene, showing the 90° cleavage angles characteristic of single-chain silicates.

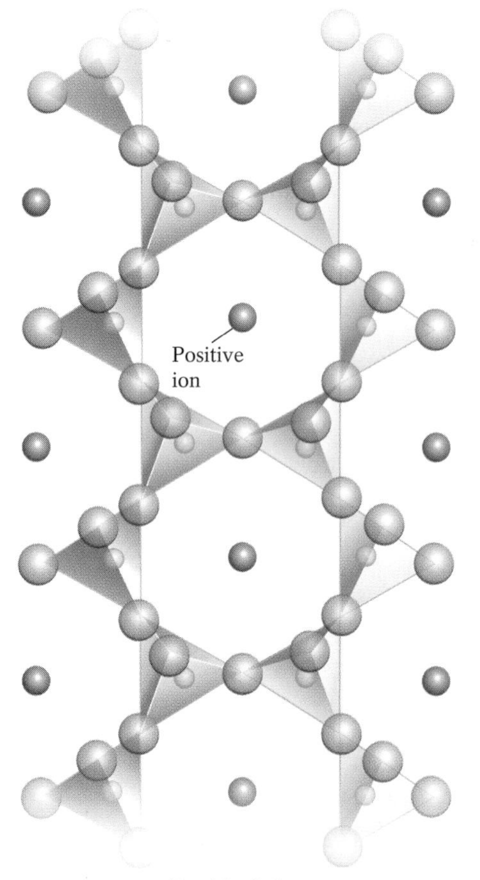

Figure 2-20 (c) Double chains. Each tetrahedron shares two of its corner oxygen ions with adjacent tetrahedra, forming a linear chain; in addition, some tetrahedra share a third oxygen ion with tetrahedra in an adjacent chain, thus joining the chains together. Some positive ions are interspersed between the single chains, as well as between adjoining double chains. The silicon-to-oxygen ratio of double chains is 1:2.75. Photo: Hornblende (an amphibole), showing the 56° and 124° cleavage angles characteristic of double-chain silicates.

Figure 2-20 (d) Sheet silicates. Each tetrahedron shares all three of its corner oxygen ions, forming a sheet of adjoined tetrahedra. The fourth oxygen ion in each tetrahedron extends upward to bond with positive ions and, subsequently, another sheet. (Other positive ions bind two-sheet pairs to adjacent two-sheet pairs.) The silicon-to-oxygen ratio of sheet silicates is 1:2.5. Photo: Muscovite mica, showing the planar cleavage of the sheet silicates.

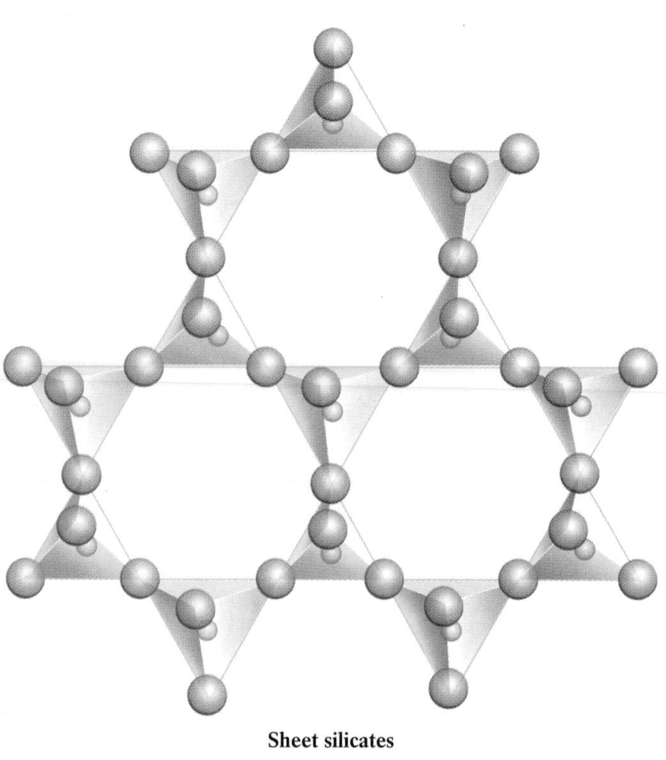

Figure 2-20 (e) Framework silicates. Each tetrahedron shares all four of its oxygen ions, forming a three-dimensional framework structure. Because the charge on each tetrahedron is neutralized by the sharing of all its negative charges, no positive ions bond with the structure. The silicon-to-oxygen ratio of framework silicates is 1:2. Photo: Quartz crystals. Due to its sturdy framework structure, quartz does not cleave.

Double chains of tetrahedra form when adjacent tetrahedra share two corner oxygen ions and some share a third oxygen ion with a tetrahedron in a neighboring chain, forming a silicon-oxygen ratio of 1:2.75 (Fig. 2-20c). The resulting minerals are known as amphiboles, the most common of which is hornblende.

Sheet silicates, which can extend indefinitely in the two dimensions of a plane, form when all three oxygen ions at the base of a tetrahedron are shared with other tetrahedra. The fourth oxygen ion, at the peak of each tetrahedron, projects outward from the sheet and bonds sheets together (Fig. 2-20d). Sheet silicates, such as the common minerals muscovite and biotite mica, have silicon-oxygen ratios of 1:2.5.

Framework silicates form when all four oxygen ions of the silicon-oxygen tetrahedra are shared with adjacent tetrahedra. The resulting silicon-to-oxygen ratio is 1:2 (Fig. 2-20e). This structure appears in the two most abundant minerals of the Earth's crust, quartz and feldspar.

Quartz, the second most abundant mineral in the continental crust, is the only silicate composed entirely of silicon and oxygen. Thanks to its strong covalent bonds, quartz ranks as the hardest of the common rock-forming minerals (7 on the Mohs scale). Because all of its bonds are equally strong, this mineral has no weak planes and does not cleave, but rather breaks by fracturing. When quartz has unrestricted room in which to grow, it assumes a characteristic crystal shape—perfect six-sided prisms with pyramids on top and bottom. Although pure quartz is transparent and colorless, the mineral often traps a few stray ions as it crystallizes. Such impurities can give quartz a wonderful range of colors, as evidenced in the gemstones amethyst, smoky quartz, rose quartz, milky quartz, and tigereye.

In the feldspars, which account for approximately 60% of the volume of continental crust, aluminum atoms often replace silicon atoms. In addition, potassium, sodium, and calcium ions may occupy open spaces between tetrahedra. Two types of feldspars exist: plagioclase feldspars, in which the positive ions between the tetrahedra are sodium or calcium, and alkali or potassium feldspars, in which the positive ions are potassium. Table 2-3 shows the compositions, structures, and properties of the common rock-forming silicates.

Nonsilicates

Nonsilicate minerals constitute only 5% of the Earth's continental crust, but they include some of our most important minerals. The useful metals iron, aluminum, and copper, for instance, as well as the precious metals gold, silver, and platinum and the gemstones diamonds, rubies, and sapphires, are all nonsilicate minerals. The most common nonsilicate groups are the carbonates, oxides, sulfides, sulfates, and native elements (Table 2-4).

Carbonates **Carbonates** are composed of one central carbon atom, three oxygen atoms, and typically one or more positive

Table 2-3 Common Rock-Forming Silicates

Silicate	Formula	Silicon-to-Oxygen Ratio (silicate structure)	Properties
Quartz	SiO_2	1:2 (framework)	Hardness of 7; breaks by fracture; six-sided prismatic crystals; specific gravity 2.65
Alkali feldspars	$KAlSi_3O_8$	1:2 (framework)	Hardness of 6.0–6.5; strong cleavage in two directions at right angles; pink or white in color; specific gravity 2.5–2.6
Plagioclase feldspars	$(Ca,Na)AlSi_3O_8$	1:2 (framework)	Hardness of 6.0–6.5; strong cleavage in two directions at right angles; white to bluish-gray in color; specific gravity 2.6–2.7
Muscovite mica	$K_2Al_4(Si_6Al_2O_{20})(OH,F)_2$	1:2.5 (sheet)	Hardness of 2–3; perfect cleavage in one direction; colorless and transparent to light green-gray; specific gravity 2.8–3.0
Biotite mica	$K_2(Mg,Fe)_6Si_3O_{10}(OH)_2$	1:2.5 (sheet)	Hardness of 2.5–3.0; perfect cleavage in one direction; black to dark brown in color; specific gravity 2.7–3.2
Amphiboles	$(Na,Ca)_2(Mg,Al,Fe)_5(Si,Al)_8O_{22}(OH)_2$	1:2.75 (double chain)	Hardness of 5–6; cleaves in two directions at 56° and 124°; black to dark green in color; specific gravity 3.0–3.3
Pyroxenes	$(Mg,Fe,Ca,Na)(Mg,Fe,Al)Si_2O_6$	1:3 (single chain)	Hardness of 5–6; cleaves in two directions at about 90°; black to dark green in color; specific gravity 3.1–3.5
Olivine	$(Mg,Fe)_2SiO_4$	1:4 (independent tetrahedron)	Hardness of 6.5–7.0; green in color; breaks by fracture; specific gravity 3.2–3.6

ions such as calcium (to form calcite) or calcium and magnesium (to form dolomite). They dissolve readily in acidic water, a fact that makes identification of carbonates relatively easy and has important geological consequences. For example, most of the world's caves were formed as acidic groundwater slowly dissolved channels and caverns in limestone, the Earth's most abundant carbonate rock. We discuss other consequences in subsequent chapters.

Oxides **Oxides** form when negative oxygen ions combine with one or more positive metallic ions. The resulting minerals include our major sources of iron—hematite and magnetite—as well as tin, titanium, and uranium.

Sulfides and Sulfates With six electrons in its outer energy level, the element sulfur can either accept or donate electrons. When sulfur accepts electrons, it becomes a negative ion that bonds with various positive ions to form the **sulfides.** Sulfides include copper sulfides (such as chalcocite), lead sulfides (such as galena), and other economically important metals. When sulfur donates electrons, it becomes a positive ion that bonds with oxygen to form the negative **sulfate** ion complex. Sulfates include such minerals as gypsum, which is used to manufacture sheetrock and plaster of Paris.

Native Elements **Native elements** are elements that do not combine with others in nature. Minerals composed of native elements consist of only one element; they are usually called by that element's name. For example, the mineral gold is composed exclusively of the element gold (Au). Other examples of native element minerals include silver, platinum, and diamond.

Gemstones

Several minerals lead a glamorous life as gemstones—minerals that display particularly appealing color, luster, or crystal form and can be cut or polished for ornamental purposes. Some gems, such as diamonds and emeralds, are quite rare. Others form from common minerals but develop as unusu-

Table 2-4 Common Nonsilicate Minerals

Mineral Type	Composition	Examples	Uses
Carbonates	Metallic ion(s) plus carbonate ion complex (CO_3^{2-})	Calcite ($CaCO_3$)	Cement
		Dolomite ($CaMg(CO_3)_2$)	Cement
Oxides	Metallic ion(s) plus oxygen ion (O^{2-})	Hematite (Fe_2O_3)	Iron ore
		Magnetite (Fe_3O_4)	Iron ore
		Corundum (Al_2O_3)	Gems, abrasives
		Cassiterite (SnO_2)	Tin ore
		Rutile (TiO_2)	Titanium ore
		Ilmenite ($FeTiO_3$)	Titanium ore
		Uraninite (UO_2)	Uranium ore
Sulfides	Metallic ion(s) plus sulfur (S^{2-})	Galena (PbS)	Lead ore
		Pyrite (FeS_2)	Sulfur ore
		Cinnabar (HgS)	Mercury ore
		Sphalerite (ZnS)	Zinc ore
		Molybdenite (MoS_2)	Molybdenum ore
		Chalcopyrite ($CuFeS_2$)	Copper ore
Sulfates	Metallic ion(s) plus sulfate ion (SO_4^{2-})	Gypsum ($CaSO_4 \cdot 2H_2O$)	Plaster
		Anhydrite ($CaSO_4$)	Plaster
		Barite ($BaSO_4$)	Drilling mud
Native elements	Minerals consisting of a single element	Gold (Au)	Jewelry, coins, electronics
		Silver (Ag)	Jewelry, coins, photography
		Platinum (Pt)	Jewelry, catalyst for gasoline production
		Diamond (C)	Jewelry, drill bits, cutting tools

ally well-formed crystals containing a few stray ions that impart color. The semiprecious gem amethyst, for example, is a common gemstone whose appealing purple tones are produced by small numbers of iron atoms scattered throughout the crystal structure of quartz. The precious gems rubies and sapphires both represent forms of the mineral corundum, an aluminum oxide (Fig. 2-21). Although naturally formed gemstones are considered the most valuable, gemstones can be synthesized in the laboratory as well (as discussed in Highlight 2-1).

Minerals of gemstone quality form under conditions that promote the development of perfect, large crystals. This development happens most often in two ways: when molten rock cools and crystallizes deep underground, or when preexisting rock becomes subjected to extraordinary pressure and heat.

(a) (b) (c)

Figure 2-21 Ruby and sapphire. Corundum minerals all have the same chemical formula Al_2O_3. Their different colors are produced by trace impurities in their crystal structure. **(a)** Ruby embedded in corundum. **(b)** Sapphire. **(c)** Ruby.

Highlight 2-1 Synthetic Gems: Can We Imitate Nature?

Because the most valuable gemstones are rare, people have tried for centuries to duplicate nature's feat and produce synthetic gems. In the twentieth century, they have had some success, even managing to surpass nature in some cases. Artificial emerald crystals, first created in the 1930s, are more transparent, richer in color, and more perfect in shape than natural ones, which are often marred by gas bubbles and other impurities. The high quality of synthetic emeralds contributes to their great market value, which, at several hundred dollars per carat, is still far less than the price of the extremely rare natural ones.

Even diamonds can now be made in the laboratory, by subjecting carbon to extreme heat and pressure. (Almost any carbon-rich substance will do as a starting point—even sugar or peanuts.) On December 12, 1954, scientists in the General Electric research lab in Schenectady, New York, created the first tiny synthetic diamonds by subjecting carbon to great pressure and temperatures exceeding 3000°C (5400°F). Today, more than 20 tons of industrial-grade synthetic diamonds are produced each year, destined for such practical uses as drill bits in oil-well drilling and modern dentistry (Fig. 1).

In 1970, the first gem-quality diamonds were synthesized. Now, some synthetics even outshine the original. Strontium titanate, a synthetic mineral sold under the trade names of Fabulite and Wellington Diamond, glitters four times more vividly than a real diamond. Being only moderately hard, between 5 and 6 on the Mohs scale, it is not very durable, however. And because its creators can manufacture tons of it, this synthetic gem is only as rare as they choose. Another synthetic, cubic zirconia (Fig. 2), can be manufactured in batches of 50 kilograms (110 pounds) and sold wholesale for a few cents per carat. Its optical qualities are virtually indistinguishable from those of nature's diamonds, and cubic zirconia is quite durable. Only the fact that its specific gravity is higher than that of natural diamond reveals its identity.

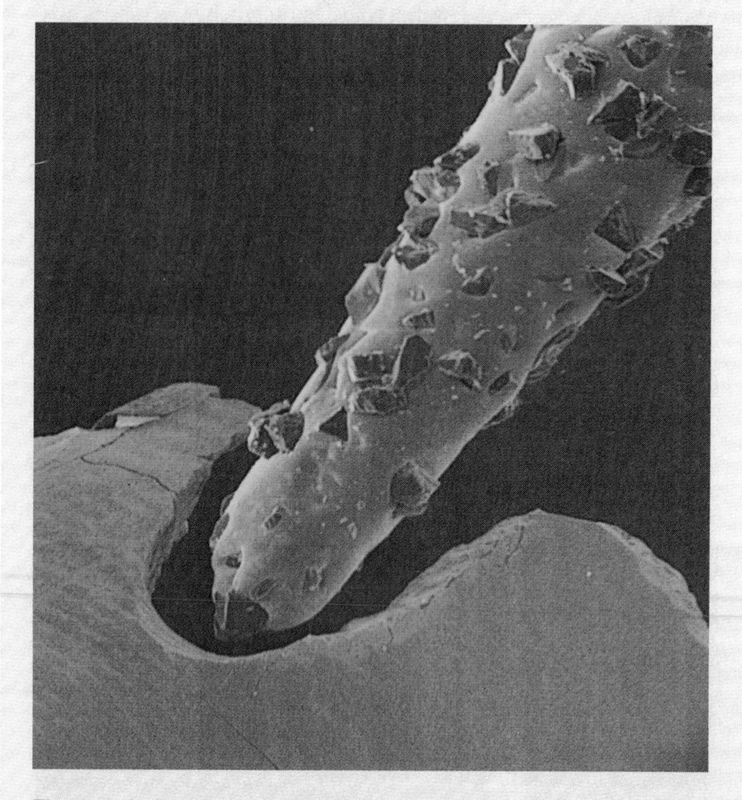

Figure 1 Industrial-grade synthetic gems. A drill bit studded with synthetic diamonds cuts swiftly through the relatively soft enamel of a human molar composed largely of the calcium-phosphorus mineral, apatite. (Magnified 25×)

Figure 2 Cubic zirconia, a synthetic diamond. Cubic zirconia may look exactly like a real diamond (compare with Fig. 2-22), but it is manufactured in bulk and quite cheaply.

(a) (b)

Figure 2-22 Diamond before and after cutting. (a) A raw, uncut diamond. **(b)** A cut and faceted diamond. Actually, diamonds are generally cleaved (not "cut") along the planes of their weakest bonds. Knowing the sites of these weaker bonds, an experienced diamond cutter can cleave a dull, irregular-shaped diamond into a glittering, perfectly symmetrical jewel.

Molten rock often migrates into fractures in surrounding cooler rocks, where its ions and atoms crystallize in reasonably large spaces, producing perfect crystals and, if the space is large enough, enormous crystals. For example, a single pyroxene crystal excavated in South Dakota was more than 12 meters (40 feet) long, 2 meters (6.5 feet) wide, and weighed more than 8000 kilograms (8 tons). The same crystallization process also produces such complex silicate gemstones as topaz, tourmaline, and beryl.

Alternatively, gemstones may form when the heat and pressure applied along the edges of colliding tectonic plates cause the ions and atoms in their rocks to migrate and recombine, creating new minerals that are more stable under the new conditions. For example, heating and compression of carbonate rocks that contain aluminum ions can cause the aluminum to combine with oxygen liberated from calcite, forming the aluminum oxides we know as rubies and sapphires.

Diamonds are transformed from carbon under extreme pressure. These gems are most often found where superheated gas has propelled molten rock rapidly from depths greater than 150 kilometers (90 miles) to the surface, carrying the diamonds along with it. The resulting diamond-rich structures, known as kimberlite pipes (named for Kimberley, South Africa), are typically a few hundred meters to a kilometer across and many kilometers deep. Kimberlite pipes are found in Siberia, India, Australia, Brazil, the Northwest Territories of Canada, southern and central Africa, and the Rockies of Colorado and Wyoming. Most, however, do not yield gem-quality diamonds.

When gem-quality diamonds are found, they exist as raw diamonds. The ability to cut raw diamonds into the valuable multifaceted gemstones with which we are more familiar is a technically difficult and exacting skill. It requires steady hands, finely honed tools, and a precise knowledge of diamonds' crystalline structure (Fig. 2-22). Diamonds are also sawed (a slower but safer way to cut a stone) using ultrafine diamond-edged blades rotating at high speeds.

Now that we have introduced the basic structure of minerals, the methods by which they form, and techniques for identifying them, we can begin to examine more closely the common types of rocks that make up the Earth's crust. In the next five chapters, we discuss the formation of igneous, sedimentary, and metamorphic rocks (introduced in Chapter 1 with the rock cycle) and describe how the minerals in these rocks can be used to interpret past geologic events.

Chapter Summary

Minerals are naturally occurring inorganic solids with specific chemical compositions and specific internal structures. **Rocks** are naturally occurring aggregates of one or more minerals. Minerals are composed of one or more chemical **elements,** the form of matter that cannot be broken down to a simpler form by heating, cooling, or reacting with other elements. Each element, in turn, consists of **atoms,** infinitesimally small particles that retain all of an element's distinguishing chemical characteristics. When atoms of two or more elements combine in specific proportions, they form chemical **compounds,** which have properties that differ from the properties of any of their constituent elements individually. All minerals are chemical compounds.

In the center of an atom is its **nucleus,** which contains both positively charged particles called **protons** and uncharged particles of equal mass called **neutrons.** An element's

atomic mass is the sum of the masses of its protons and neutrons. An element's **atomic number** is determined by the number of protons in its nucleus; every atom of a given element has the same number of protons and thus the same atomic number. Atoms of a given element may differ in atomic mass, however, because the number of neutrons in their nuclei may vary. Atoms of a given element that contain different numbers of neutrons are called **isotopes** of that element.

An atom's nucleus is surrounded by a cloud of negatively charged particles called **electrons** that move about the nucleus at high speed. Each electron occupies a specific region of space called an **energy level;** an atom may have one or more energy levels.

Atoms combine to form chemical compounds in a variety of ways known as **bonding.** Two key factors determine which atoms will unite with which other atoms to form compounds: Each atom tends to achieve chemical stability by having its outermost energy level filled with electrons, and an electrically neutral compound is more stable. To attain a full outermost energy level, an atom may donate or acquire electrons, thereby becoming an electrically charged particle called an **ion.** An atom that donates electrons becomes positively charged; an atom that acquires electrons becomes negatively charged. Oppositely charged ions attract one another, forming an **ionic bond.**

Sometimes atoms share the electrons in their respective outer energy levels, forming a **covalent bond.** In **metallic bonding,** electrons move continually among numerous closely packed nuclei. Intermolecular bonds such as **hydrogen bonds** and **van der Waals bonds** form from weak attractions between molecules or groups of atoms, caused by uneven distribution of their moving electrons.

As a mineral forms through chemical bonding, all of its ions or atoms occupy specific positions to create a **crystal structure,** a three-dimensional pattern repeated throughout the mineral. When mineral growth is not limited by space, a **crystal** may form with a regular geometric shape that reflects the mineral's internal crystal structure. When molten rock cools too quickly for its atoms to form an orderly arrangement, the resulting solids, called **mineraloids,** lack a crystal structure and thus are not true minerals.

The types of minerals that will form at a given time and place are determined by which elements are available to bond, the charges and sizes of their ions, and the temperature and pressure under which the minerals form. **Polymorphs** are minerals that have the same chemical composition but different crystal structures because they form under different temperature or pressure conditions.

Geologists identify minerals primarily by noting external characteristics, such as color, luster, streak, hardness, cleavage, and fracture. They also measure physical properties such as specific gravity.

The Earth's crust is primarily composed of only eight elements, which combine to produce the **rock-forming minerals.** The two most prominent elements, oxygen and silicon, combine readily to form the **silicon-oxygen tetrahedron.** It serves as the basic building block of the Earth's most abundant group of minerals, the **silicates,** which make up more than 90% of the mass of the Earth's crust. Silicon-oxygen tetrahedra may be linked in a variety of crystal structures: independent tetrahedra (such as in olivine), single chains of tetrahedra (such as in pyroxene), double chains (such as in amphibole), sheet structures (such as in mica), and framework structures (such as in feldspar and quartz).

A number of nonsilicates are also common rock-forming minerals. They include the **carbonates,** the **oxides,** the **sulfides** and **sulfates,** and the **native elements.**

Gemstones are minerals that are valued for their particularly appealing color, luster, or crystal form. Some gemstones, such as diamonds and emeralds, are quite rare; others are unusually well-formed specimens of relatively commonplace minerals, usually containing trace impurities in their structures that impart distinctive color to the crystals.

Key Terms

minerals (p. 20)
rocks (p. 20)
element (p. 20)
atoms (p. 20)
compounds (p. 20)
protons (p. 20)
neutrons (p. 20)
nucleus (p. 20)
electrons (p. 20)
atomic mass (p. 21)
atomic number (p. 22)
isotopes (p. 22)
energy level (p. 22)
bond (p. 22)
ion (p. 23)
ionic bonds (p. 23)
covalent bond (p. 23)
metallic bonding (p. 24)
hydrogen bonds (p. 24)
van der Waals bond (p. 24)
crystal (p. 24)
crystal structure (p. 24)
mineraloids (p. 24)
polymorphs (p. 26)
rock-forming minerals (p. 30)
silicates (p. 30)
silicon-oxygen tetrahedron (p. 30)
carbonates (p. 33)
oxides (p. 34)
sulfides (p. 34)
sulfate (p. 34)
native elements (p. 34)

Questions for Review

1. What is a mineral? How does a mineral differ from a rock?

2. Briefly describe the structure of an atom. What is an isotope? What is an ion?

3. How does an atom achieve chemical stability? How does a chemical compound achieve electrical neutrality?

4. Describe three types of chemical bonding.

5. What is a mineral crystal? Describe two circumstances under which a mineral probably will not grow into a well-formed crystal.

6. Define and give an example of mineral polymorphs.

7. Describe four properties of minerals that could help you to identify an unknown mineral.

8. Briefly discuss why the silicate minerals are the most abundant in nature.

9. List four different silicate structures and give a specific mineral example of each.

10. List two types of common nonsilicates and give a specific mineral example of each.

11. Describe the connections between at least two different geological environments and the formation of gemstones.

For Further Thought

1. If some sodium (Na^+) substitutes for calcium (Ca^{2+}) in the plagioclase feldspars, why must some aluminum (Al^{3+}) replace some silicon (Si^{4+}) in the mineral's structure?

2. Sulfur forms a small ion with a high positive charge. Why doesn't sulfur unite universally with oxygen to form the basic building blocks of most crustal minerals?

3. Why doesn't the mineral quartz exhibit the diagnostic property of cleavage? Considering its physical beauty, why isn't a quartz crystal a more valuable gemstone?

4. The photos at the right show crystals of real gold and "fool's gold" (pyrite). How would you distinguish between these two similar-looking minerals?

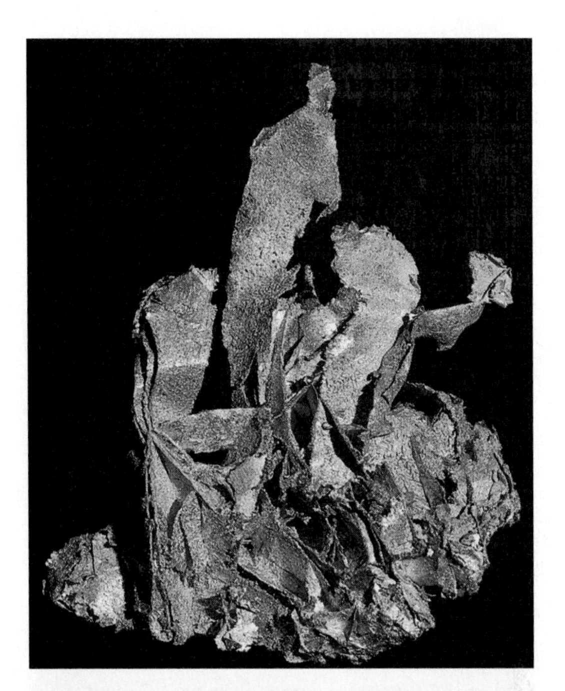

3

Igneous Processes and Igneous Rocks

On the island of Hawai'i, you can watch a volcano erupt and later touch the warm rock that passed through the volcano in a molten state just hours or days earlier. In the Sierra Nevada mountains of California, you can walk on rocks that cooled 80 million years ago from molten rock 20 kilometers (12 miles) below the Earth's surface. As we saw in Chapter 1, such rocks, which cooled and crystallized directly from molten rock, either at the surface or deep underground, are called **igneous rocks** (from the Latin *ignis,* meaning "fire"). More than 95% of the Earth's outer 50 kilometers (30 miles) consists of igneous rocks. In fact, the remains of ancient volcanic eruptions and vast uplifted regions of formerly subsurface igneous rocks can be found in almost every state and province of North America.

Geologists can observe molten material spewing from a volcano, but they cannot see it moving underground—and almost all igneous rock solidifies far beneath the surface. One way that geologists investigate underground igneous processes is to look for regions where erosion has removed surface rock layers (that is, by physically wearing away rocks by natural agents such as wind, water, and ice), thus opening up windows through which we can see the subsurface. Igneous rocks that solidified underground appear in some of North America's most scenic places, from Mount Katahdin in northern Maine to the Yosemite Valley of eastern California (Fig. 3-1). Some of our continent's most ancient rocks are the igneous roots of one-time mountains that have long since eroded to expose the underlying rocks. Such rocks can be seen in northern Minnesota, Ontario, and Quebec.

Some geologists investigate igneous processes by simulating them in laboratories. This practice allows geologists to observe the effects of pressure, temperature, composition, and other factors on the melting and crystallization points of sample rocks and minerals.

Our discussion of the Earth's igneous processes and rocks first describes some of the characteristics of molten rock, before turning to the types of rocks that form when molten material cools and solidifies. We also examine how and why rocks melt and crystallize, paying special attention to how plate tectonics affects the origin and distribution of igneous rocks on Earth (and how Moon rocks differ in this

Figure 3-1 Igneous rocks. These rocks exposed along the western end of Lake Superior are part of the Duluth lopolith, a huge expanse of igneous rock that solidified underground millions of years ago.

respect). Finally, we investigate the economically valuable materials—such as gold, silver, and copper—that originate from igneous processes.

What Is Magma?

Magma is molten rock that flows within the Earth. It may be completely liquid or, more commonly, a fluid mixture of liquid, solid crystals, and dissolved gases. When magma reaches the Earth's surface, we call it **lava,** or molten rock that flows above ground.

Magma forms when underground temperatures become high enough to break the bonds in some minerals, causing the minerals to melt. The rock then changes from a crystalline solid to a fluid mix containing freely moving ions and atoms as well as some still-solid crystalline fragments. Different minerals melt out of the rock at different temperatures as the heat gradually increases, with the minerals having the highest melting points remaining the longest as still-solid fragments in the magma. At the same time, the composition of the magma changes as each newly molten mineral enters and enriches it.

When heat dissipates from a magma, its bonds no longer break and new bonds start to form. First, some of the free atoms and ions in the liquid bond to form tiny crystals. Additional ions and atoms bond at prescribed sites in the crystal structures. The crystals grow until they touch the edges of adjacent crystals. As cooling progresses, different minerals crystallize from the magma, again changing the magma's composition. If cooling continues long enough, the entire body of magma will become solidified as igneous rock.

Classification of Igneous Rocks

Igneous rocks are classified based on their two most obvious properties: their texture, which is determined by the size and shape of their mineral crystals and the manner in which these grew together during cooling, and their composition, which is determined by the minerals that they contain.

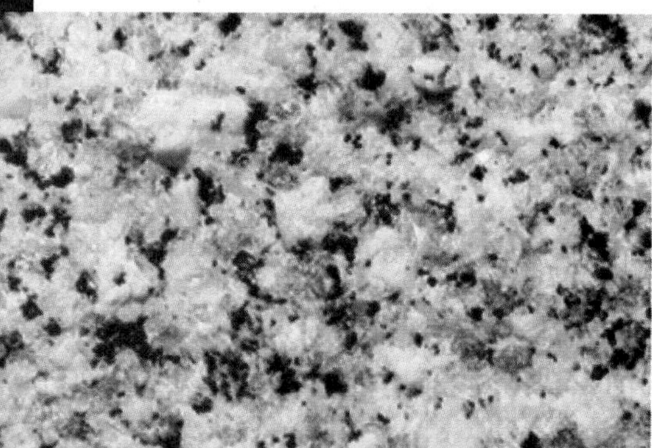

Figure 3-2 Phaneritic rocks. Rocks that solidify slowly underground, as did this granite in Yosemite National Park, California, have phaneritic (coarse-grained) textures.

Igneous Textures

A rock's *texture* refers to the appearance of its surface—specifically the size, shape, and arrangement of the rock's mineral components. The most important factor controlling these features in igneous rocks is the rate at which a magma or lava cools. When a magma's minerals crystallize slowly underground over thousands of years, crystals have ample time to grow large enough to be seen clearly with the unaided eye. The resulting texture is called *phaneritic* (pronounced "FAN-er-i-tic"; from the Greek *phaneros,* meaning "visible") (Fig. 3-2). Slow cooling occurs when magmas enter, or *intrude,* preexisting solid rocks; thus rocks with phaneritic textures are known as **intrusive rocks.** They are also called **plutonic rocks** (pronounced "ploo-TON-ic"; for Pluto, the Greek god of the underworld).

Some igneous rocks develop at relatively low temperatures from ion-rich magmas containing a high proportion of water. Under these conditions, ions move quite readily to bond with growing crystals, enabling the crystals to become unusually large (sometimes several meters long). Rocks with such exceptionally large crystals are called *pegmatites* (from the Greek *pegma,* meaning "fastened together") (Fig. 3-3). In western Maine, near the towns of Bethel and Rumford, some pegmatitic rocks contain 5-meter (17-foot)-long crystals of the mineral beryl.

Some igneous rocks solidify from lava so quickly that their crystals have little time to grow. These *aphanitic* (pronounced "af-a-NIT-ic"; from the Greek *a phaneros,* meaning "not visible") rocks have crystals so small that they can barely be seen with the naked eye (Fig. 3-4). Rocks with aphanitic textures are called **extrusive rocks,** because they form from lava that has flowed out, or been *extruded,* onto the Earth's surface. They are also known as **volcanic rocks,** because lava is a product of volcanoes (named for Vulcan, the Roman god of fire).

In some igneous rocks, large, often perfect, crystals are surrounded by regions with much smaller or even invisible

Figure 3-3 Pegmatites. Extremely coarse-grained pegmatites, such as the one shown here, form from ion-rich magmas having a high water content.

Figure 3-4 Volcanic rocks. Volcanic rocks, such as this basalt, typically have aphanitic (very small-grained) textures, because they solidify rapidly above ground.

Figure 3-5 Porphyritic rocks. Some rocks have a porphyritic texture, marked by large crystals surrounded by an aphanitic matrix.

Figure 3-6 Glassy volcanic rocks. Obsidian **(a)** and pumice **(b)** contain no crystals because they solidify instantaneously. Pumice, which forms from lava foam, commonly has so many tiny air-filled cavities that it can float in water.

grains (Fig. 3-5). These *porphyritic* (pronounced "por-fa-RIT-ic") textures are believed to form as a result of slow cooling followed abruptly by rapid cooling. First, gradual underground cooling produces large crystals that grow slowly within a magma. Next, the mixture of remaining liquid magma and the early-formed crystals is forced close to the surface or actually escapes into the air. There the liquid cools rapidly to produce the body of smaller grains that envelops the larger crystals.

When lava from a volcano erupts into the air or flows into a body of water, much of it cools so quickly that its ions don't have time to become organized into any crystals. The texture of the resulting rock is described as *glassy*. Two common types of volcanic glass exist, both produced by instantaneous cooling of lava: dark-colored *obsidian* and light-colored, cavity-filled *pumice* (Fig. 3-6). The latter forms from bubbling, highly gaseous lava foam.

Igneous Compositions

The Earth's magmas consist largely of the most common elements: oxygen, silicon, aluminum, iron, calcium, magnesium, sodium, potassium, and sulfur. The relative proportions of these components found at any given time within a body of magma give the magma its distinctive characteristics and ultimately determine the mineral content of the rocks it will form. Igneous rocks and magmas are classified into four main compositional groups—ultramafic, mafic, intermediate, and felsic—based on the proportion of silica (oxygen and silicon) they contain (Table 3-1). Figure 3-7 shows how the mineral content of igneous rocks varies in these categories.

Ultramafic Igneous Rocks The term "mafic" is derived from *ma*gnesium and *f*errum (Latin for "iron"). Ultramafic igneous rocks are dark in color and very dense, because they are dominated by the iron- and magnesium-containing silicate

Table 3-1 Common Igneous Compositions

Composition Type	Percentage of Silica	Other Major Elements	Relative Viscosity of Magma	Temperature at Which First Crystals Solidify	Igneous Rocks Produced
Felsic	>65%	Al, K, Na	High	~600 – 800°C (1100–1475°F)	Granite (plutonic) Rhyolite (volcanic)
Intermediate	55–65%	Al, Ca, Na, Fe, Mg	Medium	~800–1000°C (1475–1830°F)	Diorite (plutonic) Andesite (volcanic)
Mafic	40–55%	Al, Ca, Fe, Mg	Low	~1000–1200°C (1830–2200°F)	Gabbro (plutonic) Basalt (volcanic)
Ultramafic	<40%	Mg, Fe, Al, Ca	Very low	>1200°C (2200°F)	Peridotite (plutonic) Komatiite (volcanic)

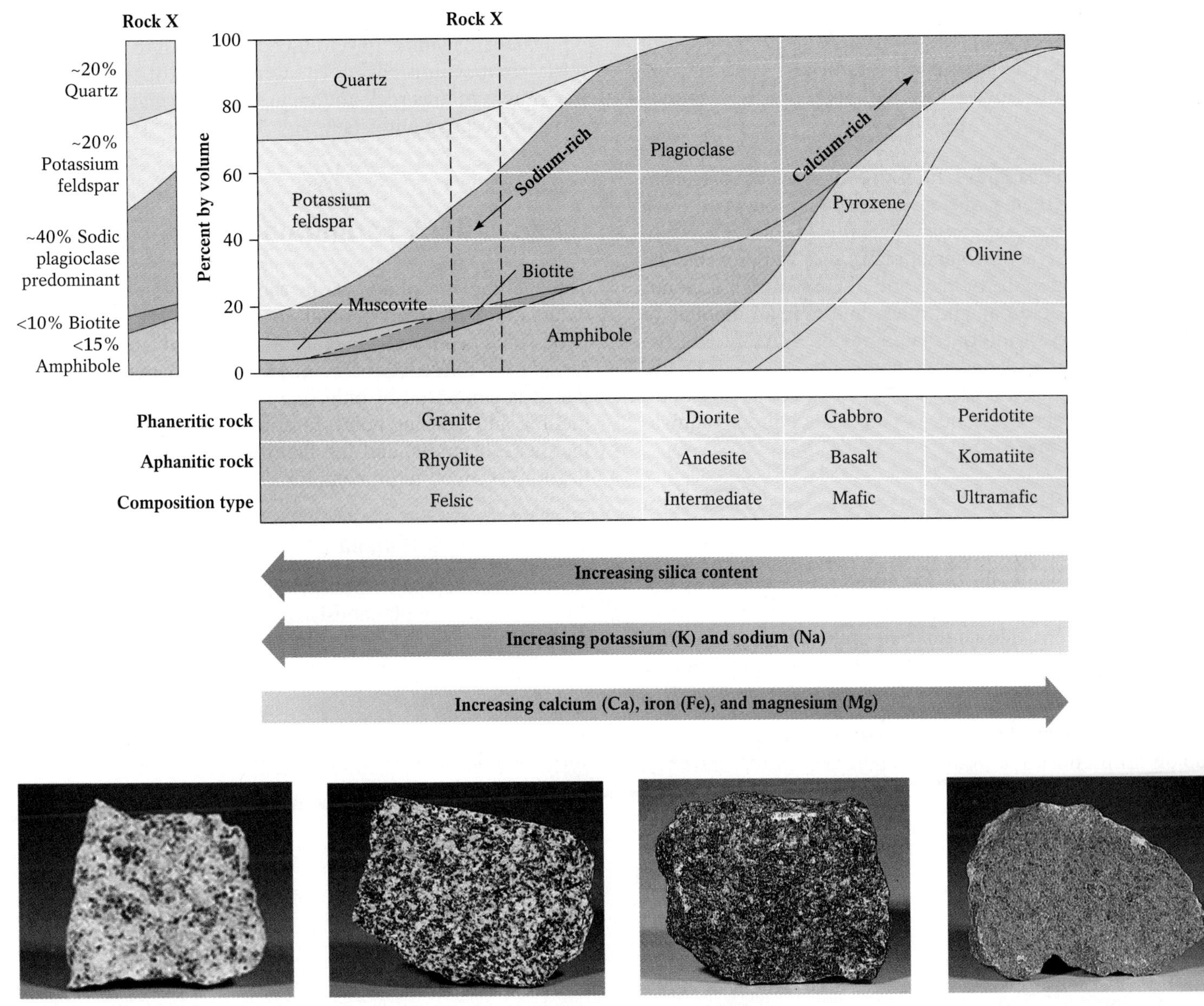

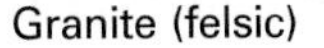
Granite (felsic)

Diorite (intermediate)

Gabbro (mafic)

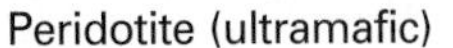
Peridotite (ultramafic)

Figure 3-7 An igneous rock classification chart. Compositional types among the igneous rocks range from felsic to ultramafic. The mineral components of the rocks are indicated by colored areas in the body of the chart. (The sample segment shows how to interpret the chart, using as an example a rock falling between granite and diorite in composition.)

minerals (called *ferromagnesian* minerals) olivine and pyroxene and contain relatively little silica (less than 40%). The most common ultramafic rock, **peridotite** (pronounced "pe-RID-o-tite"), contains 40% to 100% olivine. Ultramafic rocks generally crystallize slowly deep within the Earth's interior and appear at the Earth's surface only where extensive erosion has removed overlying crustal rocks. They are most likely to be found near continental collision plate boundaries, where deep rocks have been uplifted.

Mafic Igneous Rocks Mafic igneous rocks have a silica content ranging between 40% and 55%, with the principal minerals being pyroxene, calcium feldspar, and a minor amount of olivine. They are the most abundant rocks of the Earth's crust; the aphanitic volcanic rock **basalt** (pronounced "ba-SALT") is the single most abundant of them. Most of the ocean floor and many islands, such as the Hawai'ian chain, are composed of basalt. Basalt also constitutes vast areas of our continents and is found in Brazil,

India, South Africa, Siberia, and the Pacific Northwest of North America (Fig. 3-8). The plutonic equivalent of basalt is **gabbro;** it has the same composition as basalt but, because it cools more slowly deep within the Earth, is coarse-grained.

Intermediate Igneous Rocks Intermediate igneous rocks contain more silica than mafic rocks, including between 55% and 65% silica, and are generally lighter in color. They typically consist of some ferromagnesians, such as pyroxene and amphibole, along with sodium- and aluminum-rich minerals such as sodium feldspar and mica, and a small amount of quartz. Examples of intermediate igneous rocks include the aphanitic volcanic rock **andesite** (named for the Andes Mountains of South America, where this type of igneous rock dominates the local geology) and its phaneritic plutonic equivalent **diorite** (see Fig. 3-8).

Felsic Igneous Rocks The term "felsic" is derived from *fel*dspar and *si*lica. Felsic igneous rocks contain more silica—65% or more—than either mafic or intermediate igneous rocks. They are generally lighter in color because they are poor in iron, magnesium, and calcium silicates, but rich in potassium feldspar, aluminum-rich (muscovite) mica, and quartz. Examples of felsic igneous rocks include the common plutonic phaneritic rock **granite** and its volcanic aphanitic equivalent, **rhyolite** (pronounced "RYE-uh-lite"). Rocks of felsic composition have a greater variety of textures than any other igneous rock and include several glassy rocks and ultracoarse pegmatites (see Fig. 3-8).

Igneous Rock Formation

The melting of solid rock in the Earth's interior to form magma and the recrystallization of the magma to form igneous rock is not as simple as the melting of ice to form water and the refreezing of water to form ice cubes. Ice is simple; it is composed of only one substance and has a single melting and freezing point. Interior rock is composed of numerous substances, and the melting and refreezing (crystallization) is much more complex.

The Creation of Magma

As rocks are heated within the Earth, not all minerals within the rocks melt simultaneously. A body of rock undergoes **partial melting** when it is heated to the melting points of some but not all of its component minerals. The magma produced

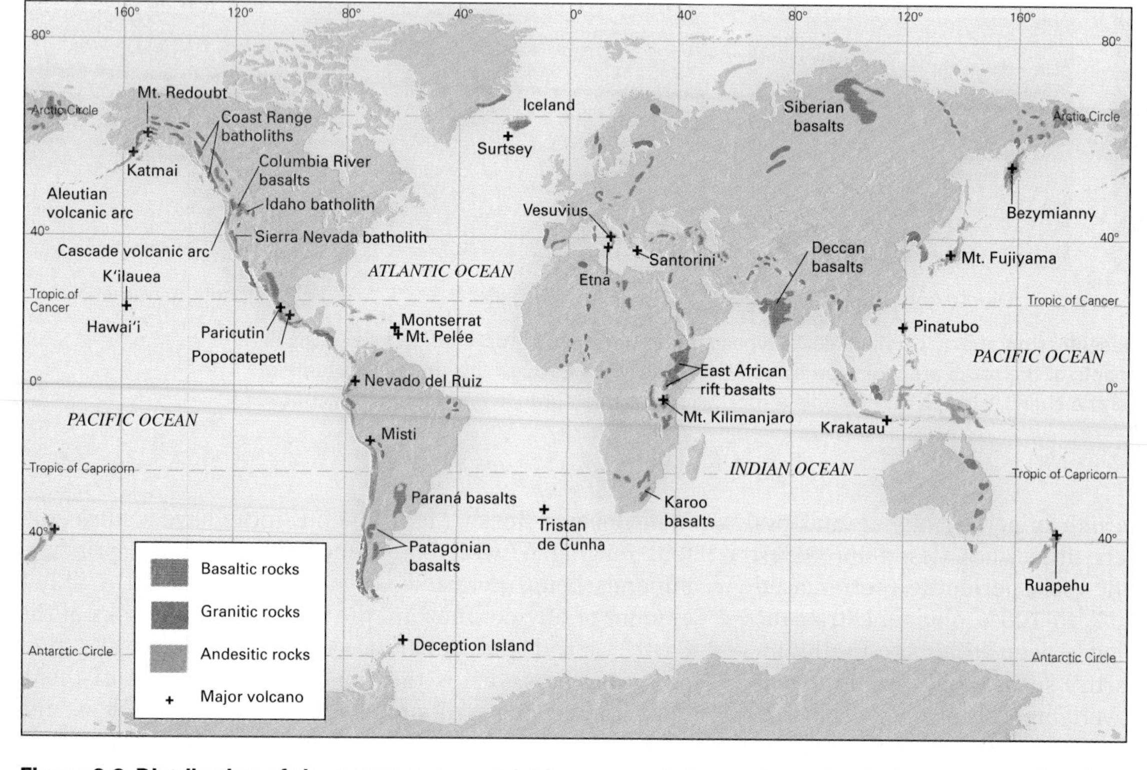

Figure 3-8 Distribution of the common terrestrial igneous rock types. As well as being the most abundant igneous rock of the ocean floors, the mafic igneous rock basalt is widespread on the continents.

by partial melting contains both molten minerals and still-solid chunks of other minerals with higher melting points. In particular, felsic minerals melt at a lower temperature than mafic minerals, so partial melting of mafic rocks may produce felsic or intermediate magmas. If temperatures are subsequently raised, a progressive melting of minerals with higher melting points will occur; these minerals will be added to the magma, thereby changing its composition.

If partial melting determines the composition of a magma, then several other factors determine where and when the magma will form. Heat, pressure, and the amount of water in rocks all combine to influence the point at which the rocks melt. As we will see, these factors affect the type of rocks that eventually form from magmas and their locations on Earth.

Heat The heat in the Earth's interior comes from three primary sources: the heat produced during the formation of the planet, which is still rising from the Earth's core; the heat liberated continuously by decay of radioactive isotopes; and the frictional heat produced as the Earth's plates move against one another and over the underlying asthenosphere. These heat sources ensure that temperatures in the Earth increase with depth, at a rate referred to as the *geothermal gradient*. The geothermal gradient is steepest from about 50 to 250 kilometers (30–150 miles) of depth. Here, temperatures exceed the 700°C (1300°F) required to melt most felsic minerals and the 1300°C (2400°F) required to melt most mafic minerals. Thus most magmas tend to form at these depths, particularly when the right combination of pressure and fluids accompanies the high temperatures.

Pressure Heat melts minerals by causing the ions and atoms in their crystal structures to vibrate, stretching and eventually breaking the bonds between them. When a crystal structure is under pressure, however, its ions remain in place longer. Higher temperatures are then required to break their bonds. Thus, the higher the pressure on a mineral, the higher its melting point. Because rocks located far beneath the Earth's surface experience great pressure from the weight of overlying rocks, higher temperatures are needed to melt them (Fig. 3-9). If the pressure on a rock somehow becomes reduced or removed—as happens when tectonic plates rift and diverge—its melting point drops below its current temperature and it begins to melt.

Water Water, even a small amount, weakens the bonds within minerals (see "Intermolecular Bonding," in Chapter 2) and thus lowers the melting point of rocks. Under high pressure, water has an even greater effect on the melting point of a mineral. Whereas dry rocks become more resistant to melting with greater depth, wet rocks become less resistant, because high pressure drives more water into the rocks. The combination of high pressure and high water content represents an important factor in the production of magmas at subducting plate boundaries.

The Crystallization of Magma

As magma cools and solidifies, it undergoes a number of significant changes in composition. Minerals that melt last (at the highest temperatures) during heating—the more mafic minerals—are the first to crystallize during cooling. A partially

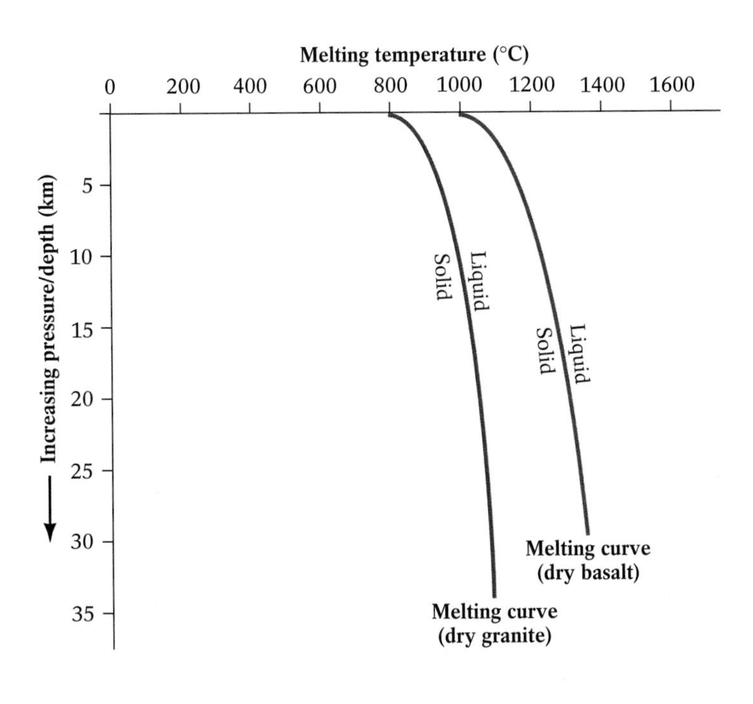

Figure 3-9 Melting-temperature curves for dry basalt and dry granite. (Adding water to rock changes its melting curve.) For both, melting temperatures increase with increasing depth, because the pressure at greater depths stabilizes rock's crystal structure, thereby raising its melting point.

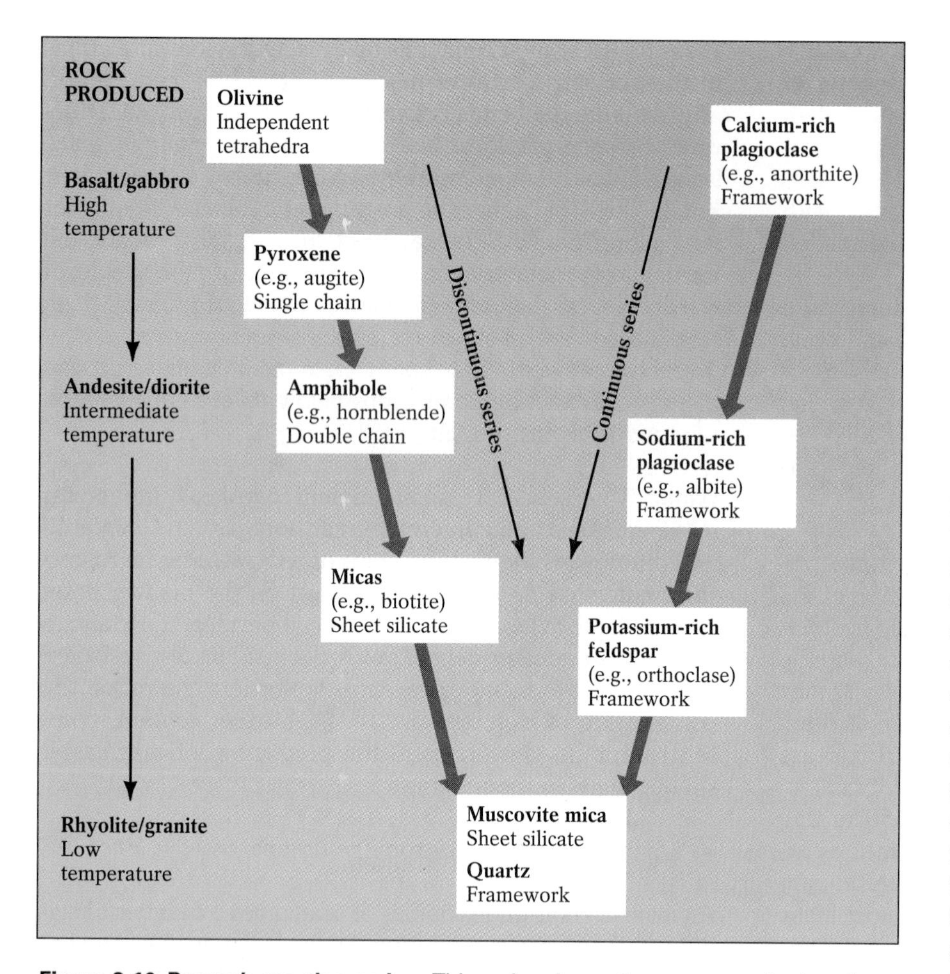

Figure 3-10 Bowen's reaction series. This series shows the sequence of minerals that crystallize as an initially mafic magma cools under conditions in which the early-forming crystals remain in contact with the still-liquid magma.

cooled magma may therefore contain solid crystals of mafic minerals along with liquid of a more felsic composition. As the magma continues to cool, additional ions and atoms crystallize, leaving behind progressively less liquid.

Bowen's Reaction Series In 1922, Canadian geochemist Norman Levi Bowen and his colleagues at the Geophysical Laboratory of the Carnegie Institution in Washington, D.C., determined the sequence in which silicate minerals crystallize as magma cools. Their work made it possible to summarize a complex set of geochemical relationships, called **Bowen's reaction series,** in a single diagram (Fig. 3-10), and demonstrated that a full range of igneous rocks, from mafic to felsic, could be produced from the same, originally mafic magma. The early-forming crystals remain in contact with the still-liquid parent magma, continue to react with it, and so evolve into new minerals.

Bowen's reaction series shows that the silicate minerals can crystallize from mafic magmas in two ways—in a discontinuous series or in a continuous series. Ferromagnesian minerals (the iron- and magnesium-rich silicates) crystallize one after another in a specific sequence. Because each successive type of ferromagnesian mineral crystallized differs in both composition and internal structure from the one before, Bowen called this progression the *discontinuous series.* As mafic magma cools, the first ferromagnesian mineral to crystallize is olivine, which has a low silica content and a relatively simple structure of independent tetrahedra. The crystallization of olivine removes iron and magnesium atoms and ions from the liquid portion of the magma, increasing the proportion of the other major ions in it. Meanwhile, the scattered olivine crystals growing in the magma continue to incorporate silica, and their tetrahedra become linked in the single-chain structure of the pyroxenes. The evolution of the ferromagnesian minerals continues as pyroxene crystals acquire more silica and become transformed into the double-chained amphiboles. Eventually, the series culminates in the formation of the complex sheet silicate biotite mica, the last ferromagnesian mineral to form. By then, the iron and magnesium ions and atoms in the liquid have been depleted. As a result, any minerals that crystallize after biotite will contain no iron or magnesium.

At the same high temperatures at which olivine and the pyroxenes crystallize, calcium feldspar crystallizes as well. As these early-forming crystals continue to interact with the remaining liquid, sodium ions from the liquid magma gradually replace the calcium ions in the calcium feldspar. Eventually, the growing crystals are completely converted to sodium feldspar. Because one type of ion is being replaced by a very similar ion, the internal structure of these minerals does not change; therefore Bowen called this progression the *continuous series.* The resulting sequence of plagioclase feldspars ranges from calcium-rich anorthite through a variety of intermediate calcium–sodium mixtures, to sodium-rich albite.

After both the ferromagnesian minerals (discontinuous series) and the plagioclase feldspars (continuous series) have crystallized completely from an initially mafic magma, less than 10% of the original liquid remains. Depending on its initial composition, this liquid may now contain high concentrations of silica, potassium, and aluminum. In such a case, potassium (alkali) feldspar, potassium-aluminum mica (typically muscovite), and quartz are the last minerals to crystallize.

Cooling-Related Changes in Magma Bowen's reaction series, which was developed in the laboratory, assumes an ideal

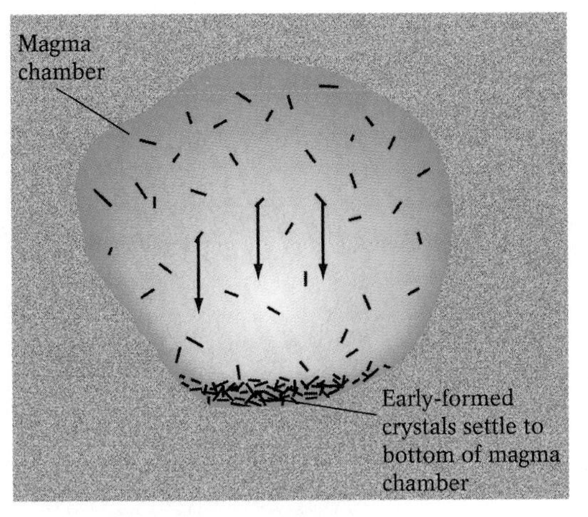

(a)

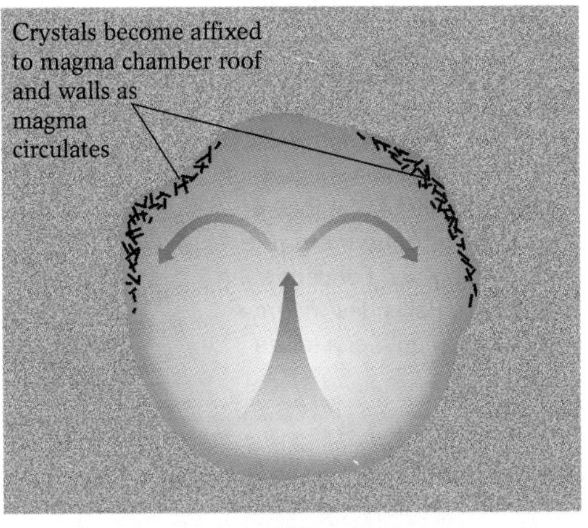

(b)

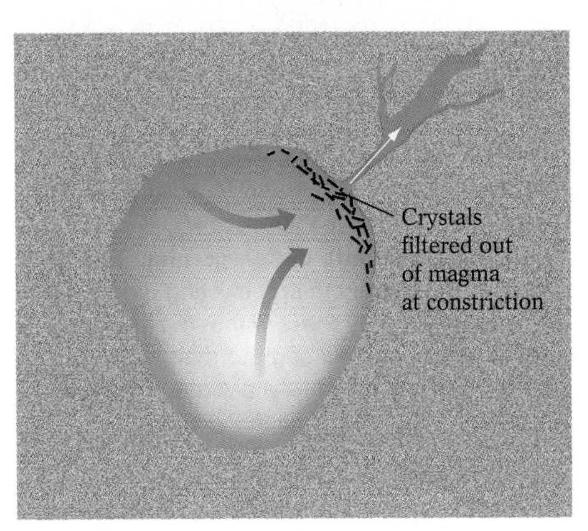

(c)

Figure 3-11 Early-forming crystals. Such crystals do not always remain in contact with the liquid magma, as Bowen's reaction series assumes. Instead, the crystals may **(a)** settle to the bottom of the magma chamber, **(b)** become affixed to the walls and roof of the magma chamber as magma circulates within the chamber, or **(c)** be filtered out of the magma as it is pressed through small fractures in the surrounding rock.

condition in which early-forming crystals remain in contact with the liquid magma, enabling them to interact and evolve continually until crystallization is complete. In nature, however, this condition rarely applies. As magma cools, some crystals may remain suspended, continuing to exchange ions and atoms with the remaining liquid. On the other hand, early-forming crystals might also be physically removed from the magma and have no further chemical interactions with the remaining liquid (Fig. 3-11). Crystals that are denser than the surrounding liquid may sink to the bottom of the magma chamber and become buried under later-settling crystals. In addition, crystals may be plastered against the walls or ceiling of the magma chamber by the hot rising liquid. The largest crystals may even be filtered out of the melt entirely, as the remaining liquid component of the magma flows into fractures too narrow to allow them to pass. Because their ions are no longer available to interact with the magma, removal of crystals in any of these ways limits the types of minerals that may later crystallize from the magma, thereby affecting the composition of any rocks that may form from it.

A magma from which crystals have been removed at various stages of its cooling has, in effect, become separated into a number of independently crystallizing bodies; the rocks that form from such a magma differ in composition both from each other and from the original magma. Because the silica-rich minerals are the last to crystallize, each successive body crystallized is more silica-rich (more felsic) than the last. By this process, called **fractional crystallization,** a single parent magma can produce a variety of igneous rocks of different compositions. The Palisades cliffs of northern New Jersey, on the west bank of the Hudson River, are a classic example of this phenomenon (Fig. 3-12).

Processes other than fractional crystallization can also significantly alter the composition of a moving magma. Blocks of rock from the walls of the magma chamber may break free, melt or partially melt, and become assimilated into the magma. In addition, two or more different bodies of magma may flow together and form a magma of hybrid composition. The 1912 volcanic eruption in Alaska's Aleutian Islands produced rocks containing both felsic and mafic minerals, suggesting that two distinct bodies of magma had combined to fuel the eruption.

Intrusive Rock Structures

Magmas tend to rise, for several reasons. When two materials of different densities occupy a space together, gravity pulls the denser down and the lighter is forced upward. Just as cream rises within nonhomogenized milk, magma rises because it is less dense than the solid rock surrounding it.

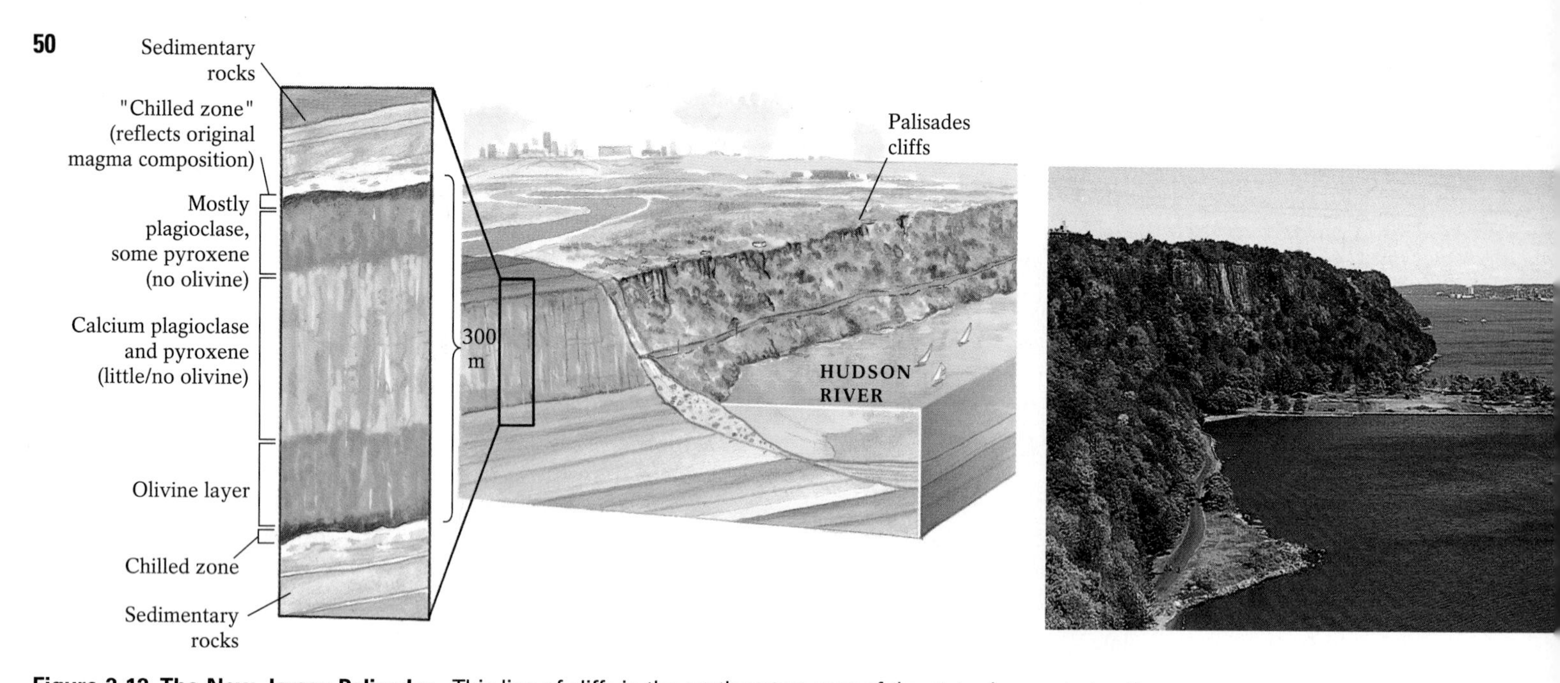

Figure 3-12 The New Jersey Palisades. This line of cliffs in the northeastern part of the state demonstrates the result of fractional crystallization. The rocks of the Palisades crystallized from a 300-meter (1000-foot)-thick body of magma that intruded preexisting rocks at temperatures of at least 1200°C (2200°F). The top and bottom of the Palisades solidified very rapidly without undergoing fractional crystallization, probably because the magma came into contact with cold surrounding rocks; they therefore provide us with a glimpse of the magma's original composition. The bottom third of the Palisades has a high concentration of olivine crystals, the central third is a mixture of calcium plagioclase and pyroxene with no appreciable olivine, and the upper third consists largely of plagioclase with no olivine and little pyroxene. Early-forming olivine apparently crystallized and then settled to the bottom of the magma body; pyroxene and plagioclase crystallized next, with the denser pyroxenes settling and concentrating in the center, and the lighter plagioclases occupying the uppermost section. Because the entire Palisades magma cooled and solidified fairly quickly, it left no residual magma from which later-forming minerals could crystallize.

In addition, because the pressure on a magma decreases with decreasing depth, the gases in the magma expand as it rises, helping to drive the material upward. Finally, magma rises when surrounding rocks press on it and squeeze it upward, much as toothpaste oozes out when you squeeze the tube.

As magma rises, it moves forcefully into cracks in preexisting rocks, pushing the rock aside to create its own space. Sometimes it may force overlying rocks to bulge upward, creating a domed intrusion within other rocks, known as a *diapir* (pronounced "DIE-uh-peer"). When moving magma incorporates some preexisting rock as it rises, some of the incorporated rocks may melt and become assimilated into the magma; Other rocks may remain unmelted and be carried within the magma. When this magma eventually solidifies, such "foreign" rocks appear as distinctly different rock masses called *xenoliths* (Fig. 3-13).

Figure 3-13 A dioritic xenolith within granite. The granitic magma encompassed the preexisting dioritic rock seen here as a dark gray mass within the lighter gray granite) but was not hot enough to melt it; as a result, the diorite was preserved as a discrete mass when the magma eventually solidified.

As magma cools and slowly crystallizes underground, it produces distinctive bodies of igneous rock that are referred to as **plutons.** These sometimes dramatic structures are often exposed at the surface after erosion removes the overlying rocks—in some areas, erosion has exposed thousands of kilometers of solidified intrusive magma. Plutons may be classified by their position relative to the preexisting rock, called *country rock,* surrounding it: *Concordant* plutons lie parallel to layers of country rock; *discordant* plutons cut across layers of country rock. Plutons of both varieties come in a range of shapes and sizes, as shown in Figure 3-14.

Tabular Plutons

Tabular plutons are slablike intrusions of igneous rock that are broader than they are thick, much like a table top. If magma flows into a relatively thin fracture in country rock or pushes between sedimentary rock layers, it will form a tabular pluton

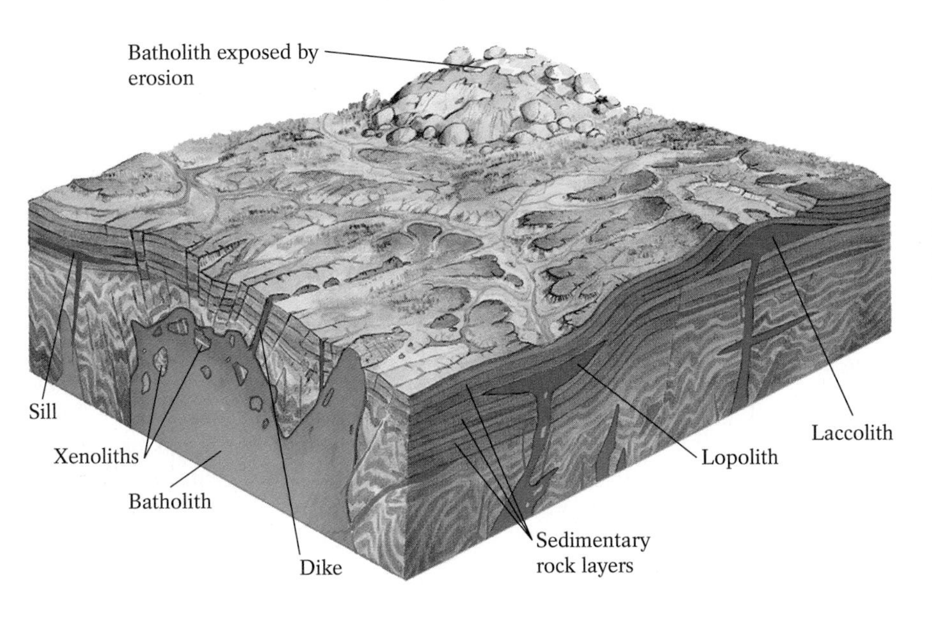

Figure 3-14 Plutonic igneous features. Sills are concordant tabular plutons; dikes are discordant tabular plutons. Laccoliths and lopoliths are larger concordant plutons, and batholiths are even larger discordant plutons.

when it cools. Tabular plutons may be as small as a few centimeters thick or as large as several hundred meters thick.

A **dike** is a discordant tabular pluton cutting across preexisting rocks. Dikes are generally steeply inclined or nearly vertical, suggesting that they formed from rising magma, which tends to follow the most direct route upward. These types of plutons often occur in clusters where magma infiltrated and solidified in a network of fractures.

Many dikes are created when magma rises into volcanoes and then solidifies; we can see these rocks when the less erosion-resistant material surrounding them wears away. Some dikes diverge like the spokes of a bicycle wheel from a *volcanic neck*, a vertical pluton remaining in what was once a volcano's central magma pathway. Along Route 64, running through the Navajo and Hopi lands of the Four Corners (the intersection of Colorado, New Mexico, Arizona, and Utah), we can see more than 100 such volcanic necks, remnants of ancient volcanic plumbing (Fig. 3-15).

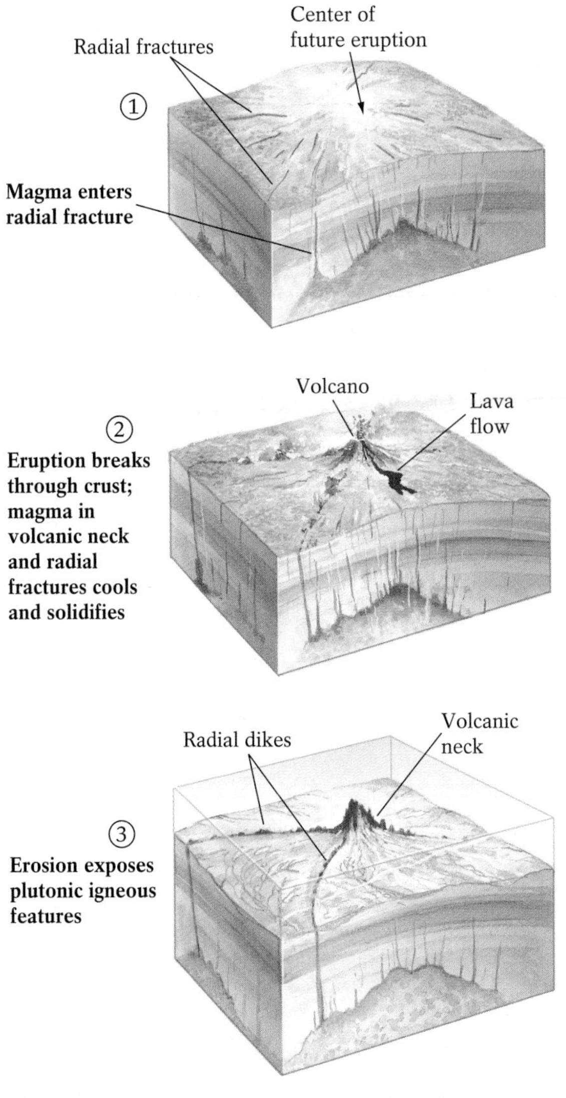

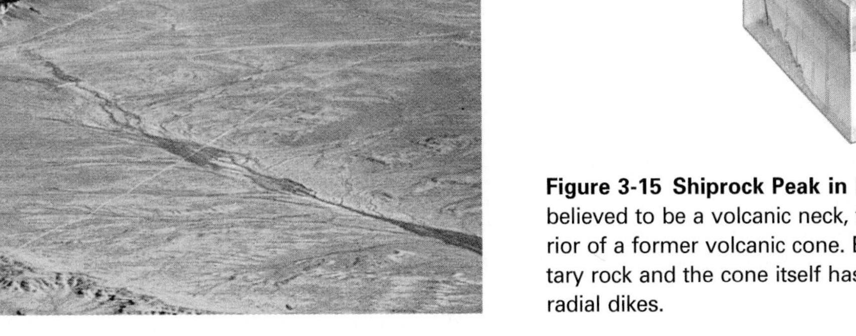

Figure 3-15 Shiprock Peak in New Mexico. This structure is believed to be a volcanic neck, the congealed lava from the interior of a former volcanic cone. Erosion of the surrounding sedimentary rock and the cone itself has exposed this volcanic neck and radial dikes.

Highlight 3-1 Tabular Plutons Save the Union

The Battle of Gettysburg, which lasted three days and took the lives of tens of thousands of Civil War soldiers, was effectively won by the Union on a hot July 3 in 1863. On this day, Confederate troops ventured forth from their outpost on a narrow dike of resistant basalt called Seminary Ridge to charge against the Union stronghold on the equally resistant, but thicker basaltic sill called Cemetery Ridge (Fig. 1). (This offensive would become known as "Pickett's charge.") The forward slope of the Cemetery Ridge sill impeded the Confederate charge, and a protective wall constructed from basaltic boulders by Union troops concealed them and repelled Confederate shots. Thus, with an assist from a well-placed basaltic sill, Union forces defeated the Confederate offensive at Gettysburg, a turning point in the American Civil War.

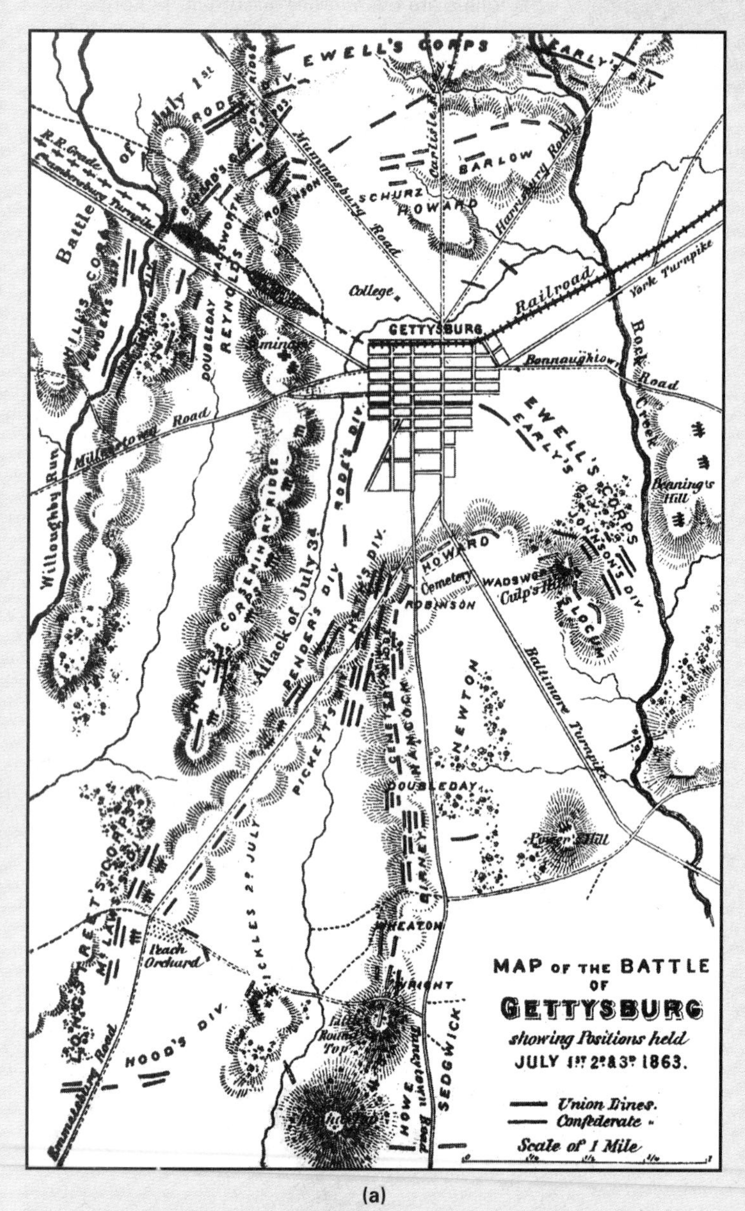

(a)

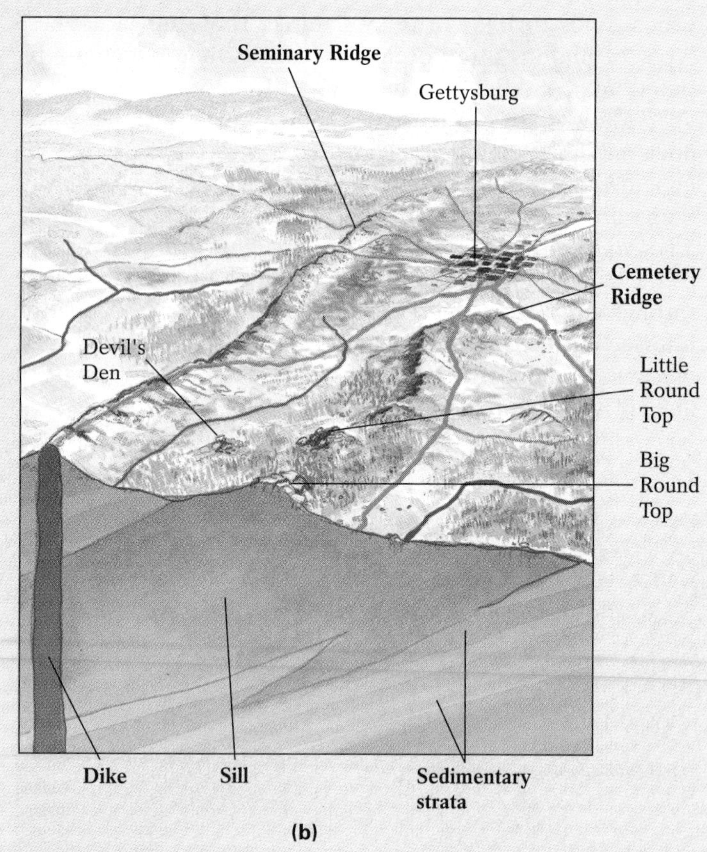

(b)

Figure 1 Geology and the Battle of Gettysburg. (a) A Civil War-era map of the site of the Battle of Gettysburg. Seminary Ridge appears in the upper left; Cemetery Ridge is to its lower right. **(b)** A contemporary artist's rendering of the relevant topographic features.

A **sill** is a concordant tabular pluton lying parallel to layers of preexisting rocks. Sills are produced when intruding magma enters a space between layers of rock, melting and incorporating adjacent sedimentary material. Sills can form only within a few kilometers of the Earth's surface, however—at greater depths, overlying rocks would compress and close off any spaces into which magma might flow. A sill and a dike in the south-central Pennsylvania town of Gettysburg provided the setting for an event that affected the course of American history, as recounted in Highlight 3-1.

Batholiths and Other Large Plutons

Large concordant plutons are commonly several kilometers thick and tens or even hundreds of kilometers across. They may be mushroom-shaped or saucer-shaped, close to the surface or deep beneath it. When thick, viscous magma intrudes between two parallel layers of rock and lifts the overlying one, it eventually cools to form a mushroom-shaped or domed concordant pluton, or **laccolith** (from the Greek *lakkos,* meaning "reservoir") (see Fig. 3-14). Laccoliths tend to form at relatively shallow depths, where little pressure acts to keep the overlying rock in place. They are typically granitic, formed from felsic magma that flows so slowly that it tends to bulge upward instead of spreading outward, raising the overlying rock to form a dome; erosion of this overlying dome may later expose the igneous rock below (Fig. 3-16). Sills form in a similar way but are usually basaltic and relatively flat, because they form from faster-flowing mafic magmas that can enter small spaces readily.

Unlike upward-bulging laccoliths, saucer-shaped concordant plutons called **lopoliths** (from the Greek *lopas,* meaning "saucer") sag downward (see Fig. 3-14). They are probably produced when dense mafic magma sinks as it intrudes, depressing the country rocks below to create a magma-filled basin. One such structure is evident along the western shore of Lake Superior, where the surrounding country rock has been eroded away (see Fig. 3-1).

Some igneous intrusions are even more vast than these large structures. **Batholiths** (from the Greek *bathos,* meaning "deep") are massive discordant plutons with surface areas (when exposed) of 100 square kilometers (40 square miles) or more (see Fig. 3-14). Most batholiths form at a depth of about 30 kilometers (20 miles) deep and are shaped somewhat like a human tooth—reaching their widest point at depth and then tapering to a point. They are generally found in elongated mountain ranges where erosion of overlying rocks has exposed deep cores of plutonic rocks, as in the White Mountains of New Hampshire and the Yosemite Valley of California's Sierra Nevada (Fig. 3-17).

Plate Tectonics and Igneous Rock

The worldwide distribution of igneous rock is not random. Certain structures and compositional types are found consistently in some geological settings, but not in others. Plutonic structures, for example, tend to form at or near the boundaries of diverging or converging tectonic plates, where

Figure 3-16 A laccolith along Route 95, through the Henry Mountains of southeastern Utah. The laccolith is the massive gray structure in the background.

Figure 3-17 Dioritic rocks. Virtually all the rock exposed in Kings Canyon National Park is composed of only one rock type—diorite. The panorama of dioritic rock here is part of the Sierra Nevada batholith.

fracturing rock provides openings in which magma can intrude. Smaller plutonic features, such as dikes and sills, generally appear in divergent or rifting zones, where mafic magmas rise as the Earth's brittle outer layers become stretched and pulled apart. Intermediate and granitic batholiths are found where oceanic plates have subducted, marking the sites of both modern and ancient plate boundaries. Oceanic rocks carried by subduction down into the asthenosphere are partially melted, generating the vast quantities of magma that form coastal batholiths. The chain of batholiths in western North America, which stretches from British Columbia through the California Sierras to Baja California, developed through more than 200 million years of oceanic-plate subduction.

Like igneous structures, the compositions of the common igneous rocks—mafic through felsic—are associated with specific geological settings. Most igneous rocks form where the three factors that create magma—heat, reduced pressure, and water (see page 47)—come into play most dramatically: at active plate boundaries. The type of rocks produced depends largely on the type of plate boundaries involved.

Basalts and Gabbros

The mafic volcanic rock basalt and its plutonic equivalent gabbro are the only igneous rocks found in oceanic crust. Basalts are also relatively common in continental crust. The low viscosity of mafic magma allows much of it to flow to the Earth's surface, where it erupts as basalt. Consequently, gabbros are rarely seen at the Earth's surface, and most of what we know about the origin of mafic rock comes from basalts. Basalts are the most abundant and most variable igneous rocks, occurring in a number of different tectonic settings and in a range of compositions. The compositions of the various basalts apparently depend on whether their parent magmas derived from deep- or shallow-mantle sources (Fig. 3-18).

Mid-ocean ridge basalts are the most abundant volcanic rock, accounting for 65% of the Earth's surface area. They are produced by eruptions at oceanic divergent boundaries and probably originate from partial melting of the upper mantle. *Ocean island basalts*—such as the ones that formed the Hawai'ian Islands—do not occur at divergent plate boundaries but rather appear at "hot spots," volcanic zones (generally intraplate) that overlie deep mantle sources.

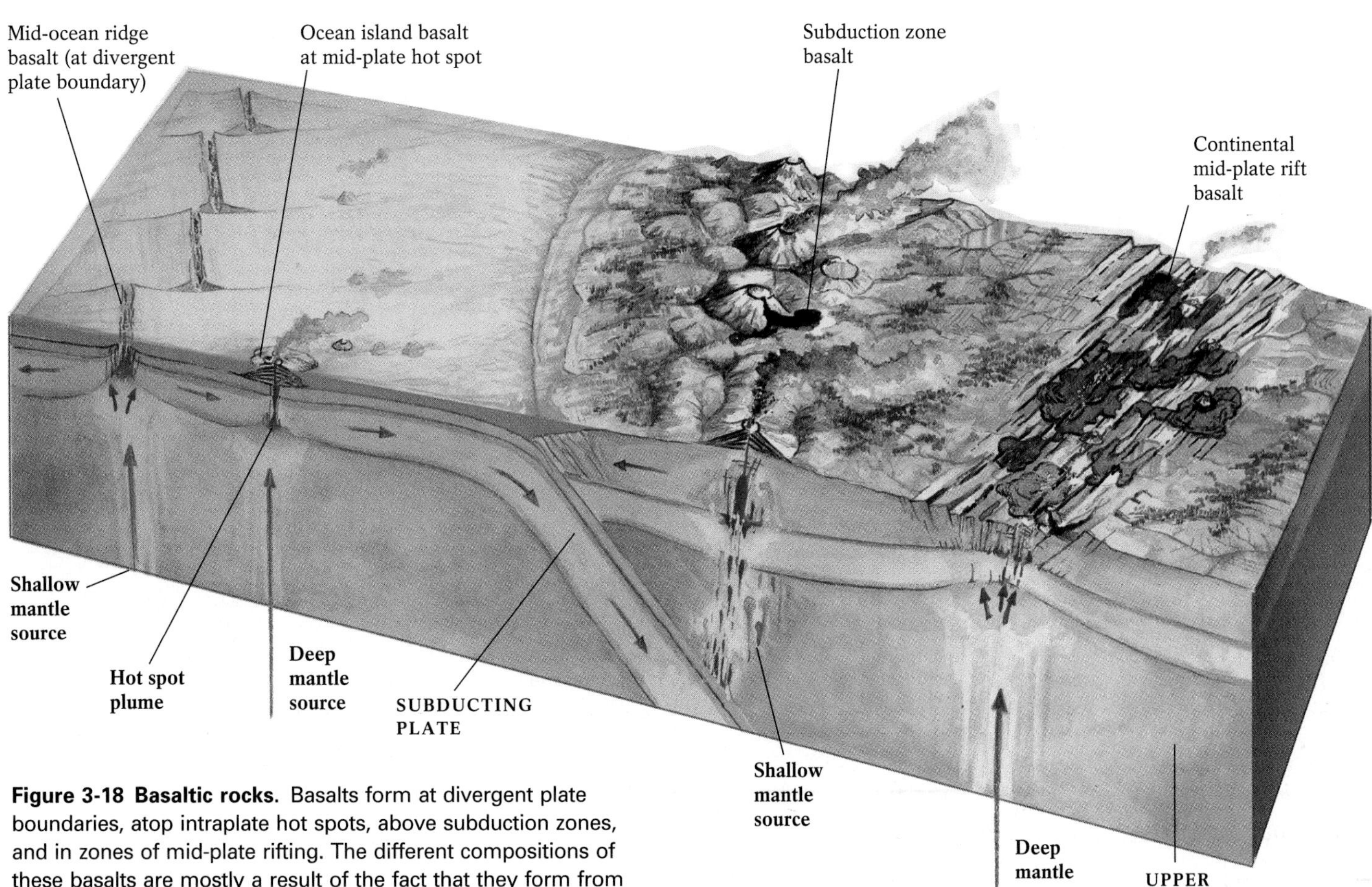

Figure 3-18 Basaltic rocks. Basalts form at divergent plate boundaries, atop intraplate hot spots, above subduction zones, and in zones of mid-plate rifting. The different compositions of these basalts are mostly a result of the fact that they form from magmas originating at varying depths in the Earth's mantle.

Basalts in continental settings vary more in composition than do oceanic basalts. Those associated with continental rifting are most likely derived from deep-mantle sources; others, which arise at subduction zones, tap shallower sources. Both types of basalt form, however, as hot mafic magma rises through tens of kilometers of continental crust, incorporating many of the materials in its path and gradually changing in composition.

Andesites and Diorites

The less mafic (intermediate) rocks andesite and diorite are commonly found along the subductive margins of continents and on volcanic islands formed through subduction of oceanic plates. Regions of andesitic rock are found on virtually all the lands that border the Pacific Ocean (see Fig. 3-8); this **andesite line** follows the nearly continuous ring of subduction zones surrounding the Pacific Ocean basin (Fig. 3-19).

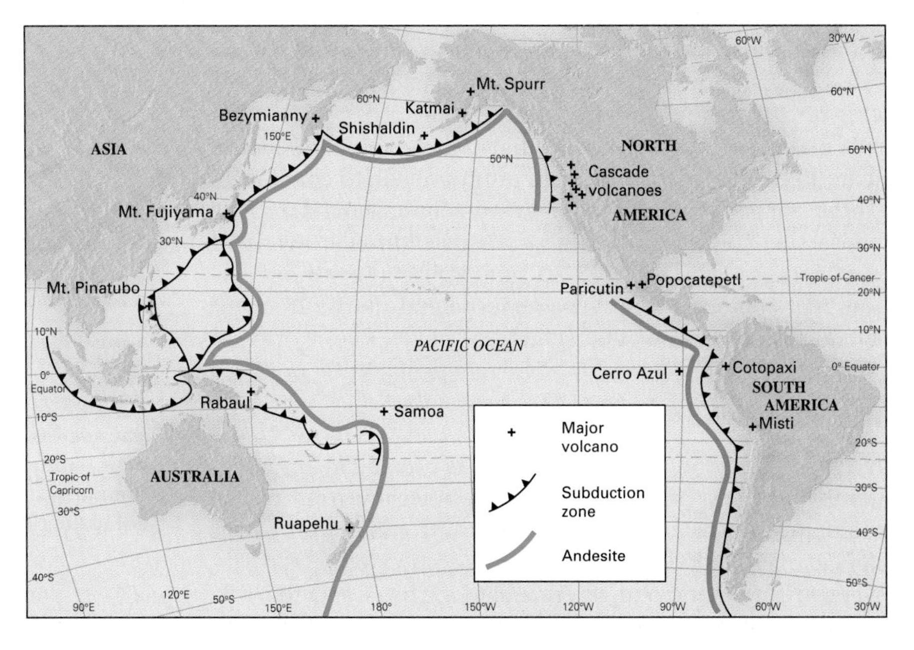

Figure 3-19 The andesite line. Subduction-produced andesitic and dioritic rocks make up most of the surface geology surrounding the Pacific Ocean basin.

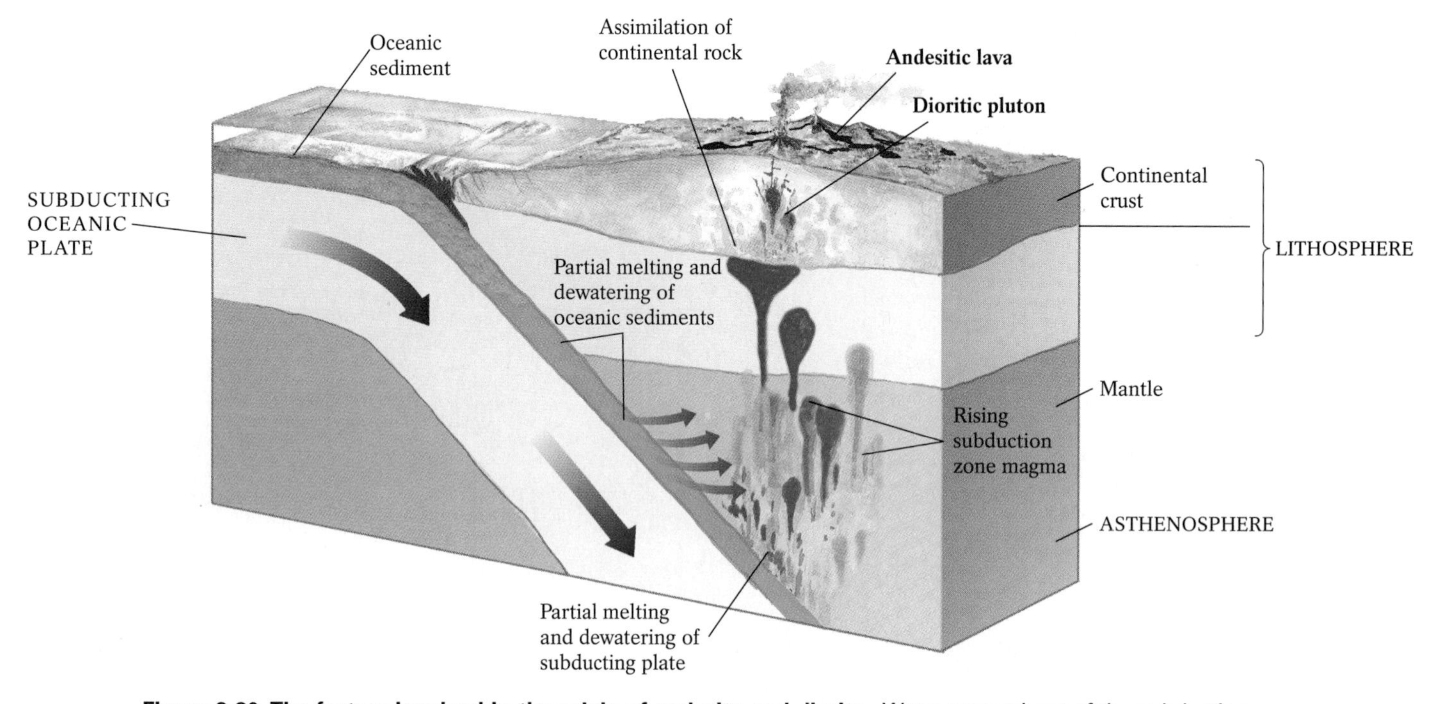

Figure 3-20 The factors involved in the origin of andesite and diorite. Water pressed out of the subducting plate and its associated oceanic sediment enters the mantle rock and lowers its melting point, causing it to melt and rise as basaltic magma. The composition of this initially mafic magma is made more intermediate by its mixing with partially melted felsic oceanic sediment and oceanic crust from the subducting plate, as well as with felsic country rock assimilated by the magma as it rises up through the continental crust of the non-subducting plate.

A number of processes are believed to combine to produce rocks of intermediate composition from subducting oceanic plates, beginning with the production of an initially mafic magma. As an oceanic plate subducts, increasing pressure drives water from the plate and forces its overlying sediments into the ultramafic mantle. The water lowers the melting point of rocks within the mantle (see page 47), producing a basaltic magma. As this magma rises, it melts and assimilates a portion of the mostly felsic oceanic sediments—averaging 200 meters (650 feet) thick on the ocean floor—carried on the subducting plate. Partial melting of the subducting plate itself may also contribute an intermediate component to the magma. Finally, a subduction-zone magma may melt and assimilate felsic materials as it rises through overlying continental rocks. Figure 3-20 summarizes the various factors that combine to produce andesite and diorite from the subduction of oceanic lithosphere.

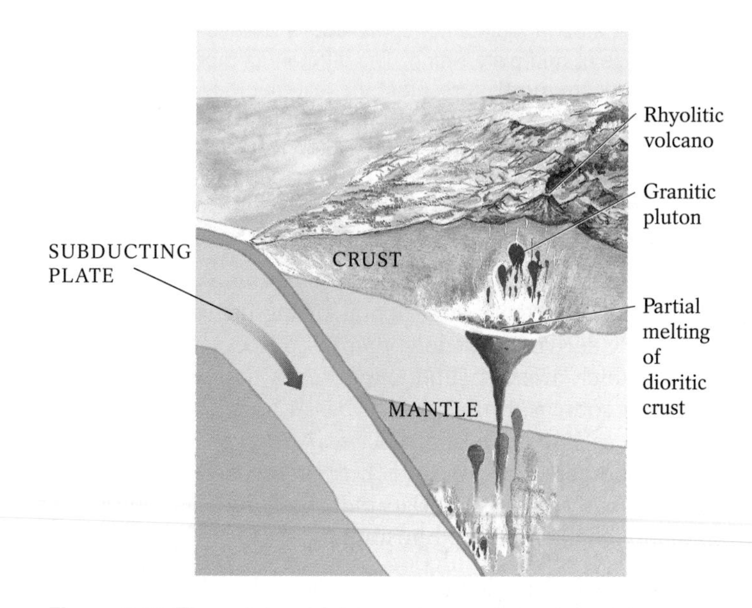

Figure 3-21 The origin of felsic rocks at subducting-plate boundaries. Hot rising mafic and intermediate magmas partially melt dioritic rocks in the lower continental crust, producing granitic plutons (and, occasionally, granite's rare volcanic equivalent, rhyolite).

Rhyolites and Granites

Nearly all rhyolitic and granitic rocks are found on continents, and they probably originate principally from partial melting of lower continental crust. Most granitic intrusions appear near modern or ancient subduction plate margins, where rising mafic and intermediate magmas and the frictional heat that accompanies subduction caused partial melting of dioritic rocks at the base of plate-edge mountain belts (Fig. 3-21). Partial melting of intermediate rocks produces predominantly felsic magma. Because these magmas are typically very viscous, they tend to rise slowly and cool at depth, producing the felsic plutonic rock granite. When such magmas reach the surface (as they sometimes do when their water content is high), they erupt as rhyolite.

Igneous Rocks on the Moon

Since 1969, geologists who study igneous rocks have extended the reach of their rock hammers some 400,000 kilometers (240,000 miles) to the Moon. The Moon appears to be fundamentally different from the Earth in terms of its geology. First, although recent studies suggest that water may actually exist in isolated, protected spots near the Moon's poles, for all practical purposes, the Moon is completely waterless. If any substantial amount of water ever existed on the Moon, it became heated, vaporized, and escaped the Moon's gravitational field very early in the Moon's evolution. Second, the origin of the Moon's igneous rocks seems unrelated to plate tectonics—most lunar geologists believe that the Moon has never had moving plates. Likewise, the Moon's igneous activity seems unrelated to its internal heat, much of which dissipated long ago.

Moon rocks collected by Apollo astronauts in the 1970s indicate that the Moon's surface contains at least two distinct types of geological/geographical provinces—the highlands and the *maria* (pronounced "MAR-ee-a"; plural of Latin *mare*, meaning "sea"). The lunar highlands probably crystallized as the Moon's earliest crust some 4.5 to 4.0 billion years ago, when the early Moon's interior was hot enough to develop a multilayered structure in a process similar to the one that created the Earth's interior. Its rocks consist principally of anorthosite, a type of coarse-grained plutonic igneous rock composed almost exclusively of the calcium plagioclase mineral anorthite.

The lunar maria, or "seas," are actually vast solidified basaltic flows (Fig. 3-22). (Galileo and other early astronomers, who used the crude telescopes of the time, believed that they were true seas—hence their name.) The maria

(a)

(b)

Figure 3-22 Geological provinces of the Moon. **(a)** A telescopic view of the near side of the Moon, showing its highlands and maria. **(b)** The formation of lunar maria from meteor impacts.

probably formed 4 to 3.85 billion years ago, when intense meteorite activity gouged numerous craters in the Moon's surface. The impacts fractured the lunar crust, providing subsurface magmas with easy pathways to the surface; they may also have raised the temperature of the stricken rocks to their melting points, thus generating new magma. Basaltic lava flowed into and filled the craters, forming the lunar maria.

The Economic Value of Igneous Rocks

The practical uses of igneous materials range from the glittering (gemstones and precious metals) to the utilitarian (crushed basalt for road construction). Any urban center displays one of the principal applications for plutonic igneous rock—the decorative building stone that adorns the exteriors and lobbies of many banks and office buildings. The same appealing polished granites and diorites can be found in cemeteries, where they serve as durable tombstones. On a smaller scale, glassy pumice serves as the abrasive in grease-removing cleansers and is used to remove calluses from hands and feet. Until recently, pumice was used as an ingredient in toothpaste, thanks to its ability to remove dental stains and plaque; because it also claimed its share of tooth enamel, however, it has since been replaced by milder abrasives. Some other familiar and useful minerals, such as the diamonds found in ultramafic rocks and the emeralds and topazes in felsic pegmatitic rocks, are also of igneous origin. Gold and silver are often found in or around granitic rocks, as are the less shiny but still valuable ores of copper, lead, and zinc.

We can now build on our general knowledge of the Earth's igneous processes and rocks and expand our discussion to cover igneous activity occurring above ground. The next chapter focuses on readily observable igneous phenomena—the volcanic eruptions and rocks produced when magma reaches the Earth's surface and escapes into the air.

Chapter Summary

Igneous rocks, the most abundant type of rock in the Earth's crust and mantle, form when molten rock cools and crystallizes. **Magma** is molten rock that flows within the Earth. When magma reaches the Earth's surface, it becomes **lava,** molten rock that flows above ground. **Intrusive,** or **plutonic,** igneous rocks form from magma that cools slowly underground. These rocks are generally coarse-grained, because their ample cooling time allows crystals to grow to relatively large sizes. **Extrusive,** or **volcanic,** igneous rocks form when lava cools quickly at the Earth's surface. These rocks are generally fine-grained, because their rapid cooling limits crystal growth.

The most common igneous rocks include ultramafic **peridotite** (containing less than 40% silica), mafic **basalt** and **gabbro** (40%–55% silica), intermediate **andesite** and **diorite** (55%–65% silica), and felsic **granite** and **rhyolite** (65% or more silica).

Magmas are produced in the Earth's interior when preexisting rocks undergo **partial melting**—that is, the minerals with lower melting points liquefy first and start to flow as a molten mass; this mass includes still-solid crystals of minerals that melt at higher temperatures. Other factors that influence the creation of magma include heat, pressure, and water content.

As magma cools, different minerals crystallize from it at different temperatures. The silicate minerals, as a group, crystallize in two specific sequences, known collectively as **Bowen's reaction series.** In the discontinuous series, the ferromagnesian (iron- and aluminum-rich) minerals evolve in distinct steps, with both their composition and their internal crystal structures changing at each step. In the continuous series, the plagioclase feldspars evolve as sodium ions gradually replace calcium ions in the developing crystals, without any accompanying change in the minerals' internal crystal structures.

The removal of early-forming crystals from a magma means that their ions are no longer available to interact with the magma, limiting the types of minerals that may later crystallize from it. As a result of this process, called **fractional crystallization,** the rocks ultimately produced by the magma will differ in composition from those that would have been produced by the original, unseparated magma. The composition of a magma may also be modified by assimilation of preexisting rocks or by mixing with another body of magma having a different composition.

Bodies of magma that cool underground form **plutons,** igneous structures that are distinct from the surrounding rocks. Plutons are classified by their shapes, sizes, and orientation relative to the rocks they intrude. Concordant plutons occur parallel to the preexisting rock layers; discordant plutons cut across the preexisting rock layers. Tabular plutons are relatively thin, igneous structures, much like a tabletop. Discordant tabular plutons are called **dikes;** concordant tabular plutons are called **sills.** Large concordant plutons include mushroom-shaped **laccoliths** and saucer-shaped **lopoliths;** large discordant plutons are called **batholiths.**

The principal igneous rock types are typically associated with specific plate tectonic settings. Basalts and gabbros are found at oceanic divergent boundaries (the mid-ocean ridge basalts), atop intraplate hot spots (the ocean island basalts), at continental margins where oceanic plates have subducted, and near rifting continental plates. Andesites and diorites appear where oceanic plates have subducted, both at continental margins and on volcanic islands. The nearly continuous ring of subduction-produced andesites that surrounds the Pacific Ocean is called the **andesite line.** Rhyolites and granites are formed within continents by partial melting of the lower portions of the continental crust; these types of igneous rocks are often associated with subduction-produced mountains.

Igneous rocks are also found on the Moon. Lunar igneous rocks differ fundamentally from those on Earth, in that they contain no water and their formation involved neither plate tectonics nor subsurface heat. The Moon's surface consists of highlands composed largely of anorthosite, a coarse-grained plutonic igneous rock, and vast areas of basalt known as maria.

Igneous rocks are valued for the gemstones and precious metals they contain. They are also used for a variety of practical purposes, such as road construction, architectural design, and household abrasives.

Key Terms

igneous rocks (p. 41)
magma (p. 42)
lava (p. 42)
intrusive rocks (p. 43)
plutonic rocks (p. 43)
extrusive rocks (p. 43)
volcanic rocks (p. 43)
peridotite (p. 45)
basalt (p. 45)
gabbro (p. 46)
andesite (p. 46)
diorite (p. 46)
granite (p. 46)
rhyolite (p. 46)
partial melting (p. 46)
Bowen's reaction series (p. 48)
fractional crystallization (p. 49)
plutons (p. 50)
dike (p. 51)
sill (p. 53)
laccolith (p. 53)
lopoliths (p. 53)
batholiths (p. 53)
andesite line (p. 55)

Questions for Review

1. Briefly describe the textural difference between phaneritic and aphanitic rocks. Why do these rocks have different textures?

2. Some igneous rocks contain large visible crystals surrounded by microscopically small crystals. What are these rocks called? How does such a texture form?

3. What elements would you expect to predominate in a mafic igneous rock? In a felsic igneous rock?

4. Name the common *extrusive* igneous rocks in which you would expect to find each of the following mineral types: calcium feldspar; potassium feldspar; muscovite mica; olivine; amphiboles; sodium feldspars. Which *plutonic* igneous rock contains abundant quartz and muscovite mica, but virtually no olivine or pyroxene?

5. What factors, in addition to heat, control the melting of rocks to generate magma?

6. What is the basic difference between the continuous and discontinuous series of Bowen's reaction series?

7. Briefly describe three things that might happen to an early-crystallized mineral surrounded by liquid magma.

8. How do a sill and a dike differ? A laccolith and a lopolith? A lopolith and a batholith?

9. Briefly discuss two specific types of plate tectonic boundaries and the igneous rocks that are associated with them.

10. What is the basic difference between a mid-ocean ridge basalt and an oceanic island basalt?

For Further Thought

1. What type of igneous feature is shown in the photo below?

2. Felsic rocks such as rhyolite often occur together with basaltic rocks near rifting continents. Give one possible explanation for this pairing.

3. Why do we rarely find batholiths made of gabbro?

4. How might the distribution of the Earth's igneous rocks change when the Earth's internal heat is exhausted and plate tectonic movement stops?

5. Why are there virtually no granites or diorites on the Moon? How might small volumes of such felsic rock form under the geological conditions believed to be responsible for the Moon's igneous rocks?

4

Volcanoes and Volcanism

One summer day in 1883, the Indonesian island of Krakatoa all but disappeared in a spectacular, massive volcanic eruption. Krakatoa, an uninhabited volcanic island in the Sunda Straits of the southwest Pacific Ocean, had for many years served as a landmark for clipper ships carrying tea from China to England. The volcano, which had been inactive for more than two hundred years, stood 792.5 meters (2601 feet) high. On the morning of August 27, it suddenly erupted in one of modern history's most violent volcanic eruptions, with a force equivalent to the explosion of 100 million tons of TNT. No one, as far as we know, perished directly from the destruction of Krakatoa; however, between 36,000 and 100,000 lives were lost as the resulting ocean waves, up to 37 meters (121 feet) high, pounded coastal villages on the nearby islands of Java and Sumatra. The sound of the explosion was heard in places as distant as central Australia, 4802 kilometers (2983 miles) away, which is akin to the residents of San Diego, California, hearing an explosion in Boston, Massachusetts. The eruption produced a black cloud of volcanic debris that rose to an altitude of 80 kilometers (50 miles), blocked out all sunlight, and plunged the region into darkness for three days. The cloud's finest particles, swept aloft by wind currents, reduced incoming solar radiation by as much as 10% worldwide, causing a drop of more than 1°C (1.8°F) in global temperatures and leading to years of spectacular crimson sunsets.

Volcanoes are the landforms created when molten rock escapes from the Earth's interior through openings, or *vents,* in the Earth's surface and then cools and solidifies around the vents. An estimated 600 volcanoes have erupted in the past 2000 years, some of them many times over. In a single year, approximately 50 volcanoes erupt around the world (Fig. 4-1). As the Krakatoa example illustrates, powerful volcanic eruptions and their aftereffects can be among the Earth's most destructive natural events. On the other hand, volcanoes also provide some of the world's most breathtaking scenery. Each year, millions are drawn to the slopes of Mount Rainier in Washington state, Mount Fuji in Japan, and Mount Vesuvius in Italy. Volcanic activity also adds to the

Figure 4-1 Popocatepetl ("El Popo"). Mexico City's neighboring volcano is one of several volcanoes currently erupting around the world and threatening surrounding communities.

Figure 4-2 Anak Krakatau ("Child of Krakatoa"), the small island that emerged from the remains of the volcanic island Krakatoa during eruptions in the 1920s. The original Krakatoa volcano was demolished in a monumental eruption in 1883.

Earth's inventory of habitable real estate: Iceland, Japan, Hawai'i, Tahiti, many islands of the Pacific and Caribbean, and nearly all of Central America are products of volcanic activity. A new volcanic island is even growing where Krakatoa once stood (Fig. 4-2).

The geological processes that result in the expulsion of molten rock as lava at the Earth's surface are collectively known as **volcanism.** Volcanoes and volcanism are simultaneously a great hazard and a great boon to humankind. In this chapter, we explore the causes and characteristics of the different types of volcanoes. We describe the threats they pose and our strategies for coping with them. Finally, we examine volcanism on some of our neighboring planets.

The Nature and Origin of Volcanoes

Volcanoes and other, less dramatic manifestations of volcanism offer windows into the Earth, providing us with information and materials that would otherwise remain inaccessible. Ascending magma carries subterranean rock fragments to the surface, giving us a glimpse of actual rocks from the Earth's interior. We owe the air we breathe and much of the Earth's water to volcanic eruptions, which have released useful gases from the Earth's interior throughout the planet's existence. Volcanic terrains often become prime agricultural lands as fresh volcanic ash replenishes the nutrients in nearby soils. Volcanism can even be a source of inexpensive, clean energy. Iceland, for example, has an abundant underground hot water supply, thanks to the molten rock that fuels its volcanic activity; by tapping the scalding water just meters beneath their feet, Icelanders can heat more than 80% of their homes and businesses.

Volcano Status

Whether a volcano poses an imminent threat to human life and property depends on its current status as either an active, dormant, or extinct volcano. An *active* volcano is one that is currently erupting or has erupted recently (in geological terms). Certain active volcanoes, such as K'ilauea on Hawai'i or Stromboli in the eastern Mediterranean, erupt almost continuously. Others erupt periodically, such as Lassen Peak in northern California, whose last activity occurred in 1917. Active volcanoes can be found on all continents except Australia and on the floors of all the major ocean basins. Indonesia, with 76 active volcanoes, Japan, with 60, and the United States, with 53, are the world's most volcanically active nations.

A *dormant* volcano is one that has not erupted recently (within the past few thousand years) but is considered likely to do so in the future. The presence of relatively fresh (less than 1000 years old) volcanic rocks in a volcano's vicinity suggests that it is still capable of erupting (Fig. 4-3). Signs of rising magma, such as the presence of hot water springs or small earthquakes occurring near a volcano, may indicate that the volcano is stirring to wakefulness.

A volcano is considered *extinct* if it has not erupted for a very long time (perhaps for tens of thousands of years) and is considered unlikely to do so in the future. One indication that a volcano is probably extinct is that extensive erosion has taken place on its slopes since its last eruption. A truly extinct volcano is no longer fueled by a magma source. Volcanoes can, however, surprise us. Residents of the Icelandic island of Heimaey believed their Mount Helgafjell to be extinct until it came to life in a spectacular eruption in 1973, its first in 5000 years.

The Causes of Volcanism

Volcanism begins when magma created by the melting of preexisting rock (discussed in Chapter 3) travels through fractures in the lithosphere to reach the Earth's surface. The distribution of the Earth's lithospheric cracks, which are usually associated with tectonic plate boundaries and with intraplate hot spots, determines where most volcanoes will form. A magma will erupt if it flows upward rapidly enough to reach the surface before it can cool and solidify. Two characteristics of a magma determine its potential to achieve eruption: its gas content and its viscosity.

(a) (b)

Figure 4-3 Which of these two volcanoes is more likely to erupt? (a) The slopes of Washington state's Mount Rainier are deeply scored by erosion and appear not to have received a fresh covering of lava in thousands of years. **(b)** The slopes of Mount St. Helens (shown here before its 1980 eruption) are relatively uneroded. Although intermittently dormant for hundreds or thousands of years, Mount St. Helens' volcanic cone has continued to grow.

Gas in Volcanic Magma Magmatic gases make up 1% to 9% of most magmas. The principal gases are water vapor and carbon dioxide, though smaller quantities of nitrogen, sulfur dioxide, chlorine, and a few others may also be present. Tens of kilometers underground, these gases remain dissolved in magma, trapped by the pressure of the surrounding rocks. As the magma rises toward the surface, the decreased pressure causes the gases to begin to leave the solution. The released gases, which are less dense than the surrounding magma, migrate upward, pushing any overlying magma before them. The higher the gas content of a magma, the faster it will rise and the greater its chances of reaching the surface before solidifying.

Gases become concentrated near the top of a rising magma body and press against the overlying rock. When these volcanic gases are completely prevented from escaping, perhaps by a plug of congealed lava blocking the passage to the surface, they accumulate and exert even greater pressure against the overlying rock. Ultimately, the overlying rock shatters. The pent-up gases then expand instantaneously, much as the gases in an agitated soft-drink bottle fizz and bubble out when you open the bottle. The initial blast removes any overlying obstructions, hurling masses of older rock skyward. Shreds of the liberated, gas-driven lava are sprayed violently into the air as the gases expand. The eruption may continue violently for hours or even days as the gases escape. The eruption may later settle down to a relatively placid outpouring of degassed magma, or it may cease altogether.

Magma Viscosity A magma's viscosity (resistance to flow) generally decreases with heat and increases with its silica content. Felsic magma tends to be relatively cool (because it crystallizes at low temperatures) and has a high silica content; thus it is very viscous. Conversely, because mafic magma is hot and has a low silica content, it is much less viscous and flows easily. For this reason, mafic magmas are more likely to rise to the surface and erupt than are felsic magmas, which tend to cool underground and form plutonic rocks.

The viscosity of magma has a direct effect on the explosiveness of a volcanic eruption. In more fluid, mafic magmas, rising gases meet with little resistance and therefore escape readily and relatively quickly when the magma reaches the surface. In highly viscous felsic magmas, the slower movement of gases (analogous to air bubbles that tend to rise more slowly in a thick milkshake than they do in water) allows gas pressures to build within the molten material. Thus felsic magmas tend to erupt explosively.

The Products of Volcanism

Volcanic eruptions range from the quiet oozing of basaltic lava, such as flows from K'ilauea volcano in Hawai'i, to Krakatoa-type cataclysmic explosions. Depending largely on the composition of the magma that feeds it, a volcanic eruption can produce a flowing stream of red-hot lava, a shower of ash particles as fine as talcum powder, a hail of volcanic blocks the size of automobiles, or any number of intermediate-sized products. In this section, we first examine the different types of lava flows and then describe the various forms in which volcanic material is deposited on the surface.

Types of Lava Flows

As noted earlier, because mafic magmas are extremely hot and relatively fluid, they are more likely to rise to the surface and erupt than are felsic magmas. Thus basaltic lava is the most common type of lava. The tendency of basaltic magma to become volcanic explains why we find much more basalt than gabbro (basalt's plutonic equivalent) in the Earth's crust. Magmas of felsic composition tend to be cooler and much more viscous, only rarely reaching the surface before solidifying. For this reason, crustal rocks contain much more granite than rhyolite. Andesitic lavas are intermediate between basaltic and rhyolitic lavas in both composition and fluidity; they erupt much more frequently than rhyolitic lava but are less common than basalt.

Basaltic Lava Much of what we know of basaltic lava comes from observing the nearly continuous eruptions on the Big Island of Hawai'i. The temperature of the Hawai'ian flows can reach as high as 1175°C (2150°F). Such hot, low-viscosity lava cools to produce two principal types of basalt, *pahoehoe* (pronounced "pa-HOY-hoy") and *'a'a* (pronounced "AH-ah"). Pahoehoe, which means "ropy" in a Polynesian dialect, is aptly named. Highly fluid basaltic lava moves swiftly down a steep slope at speeds that may exceed 30 kilometers (20 miles) per hour, spreading out rapidly into sheets about 1 meter (3.3 feet) thick. The surface of such a flow cools to form an elastic skin that is dragged into ropelike folds by the continuing movement of the still-fluid lava beneath. The ropy surface of pahoehoe is generally quite smooth (Fig. 4-4).

Figure 4-4 Pahoehoe lavas. These ropy lavas from Hawai'i's K'ilauea volcano cooled only days or even hours before these tourists began trekking around on them.

Figure 4-5 'A'a lavas. This relatively slow-moving basaltic lava cools to form a blocky, jagged, 'a'a-type surface texture.

Native islanders refer to it as "ground you can walk on barefoot," and most old Hawai'ian foot trails follow ancient pahoehoe flows.

As basaltic lava flows farther from its vent, it cools and becomes increasingly viscous. A thick brittle crust develops at its surface and slowly continues to move forward, carried along by the warmer, more fluid lava below it. The molten interior of the flow advances more rapidly than the cooler outer region, breaking it up to produce a rough surface having numerous jagged projections sharp enough to cut animals' hooves. Flows having these features are called 'a'a flows (Fig. 4-5). (*'a'a* is a local term of unknown origin that may recall the cries of a barefoot islander who strayed onto its surface; ancient foot trails meticulously avoid 'a'a fields.) 'A'a is often found downstream from pahoehoe, the product of the same flow.

Basaltic flows may produce several other distinctive features as they cool. For example, gas still present in the lava often migrates to its surface leaving small pea-sized voids, or *vesicles*, which may be preserved at the top of the basalt when it cools. Such vesicle-rich basalt is known as *scoria.* As basaltic lava cools and solidifies into rock, it often contracts in size, producing a polygonal pattern of cracks. The cracks extend from the top and bottom surfaces of the flow into its interior as cooling progresses, creating six-sided columns of rock (Fig. 4-6). In North America, the Devil's Postpile in California's Sierra Nevada and Devil's Tower in northeastern Wyoming (site of the climax of the film *Close Encounters of the Third Kind*) are spectacular examples of basaltic columns. Basaltic flows may also contain large *lava tubes*, which form when lava solidifies into a crust at its surface but continues to flow inside, forming a tunnel that is eventually drained

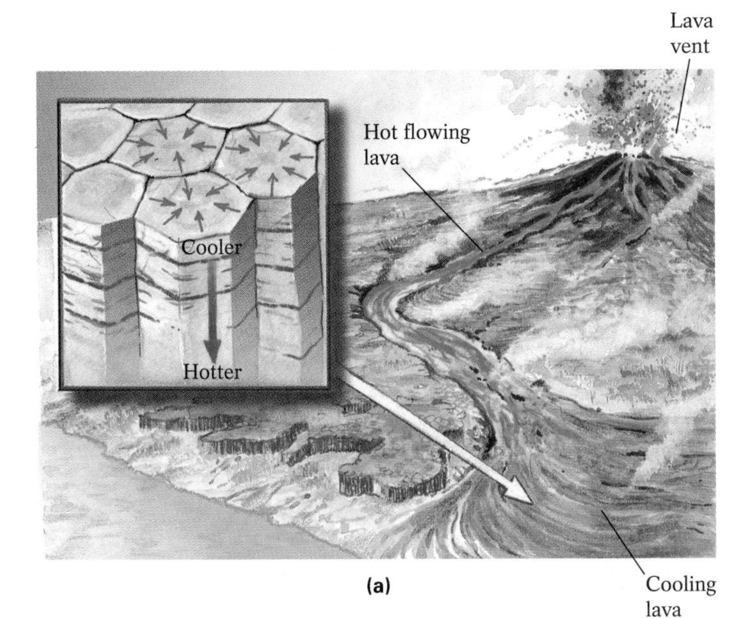

Figure 4-6 Basaltic lava columns. Contraction of basaltic lava flows as they cool **(a)** sometimes produces geometrically patterned columns. **(b)** Such structures can be found in North America in eastern Washington, eastern Oregon, southern Idaho, eastern California, and eastern Wyoming, such as those shown here at Devil's Tower National Monument.

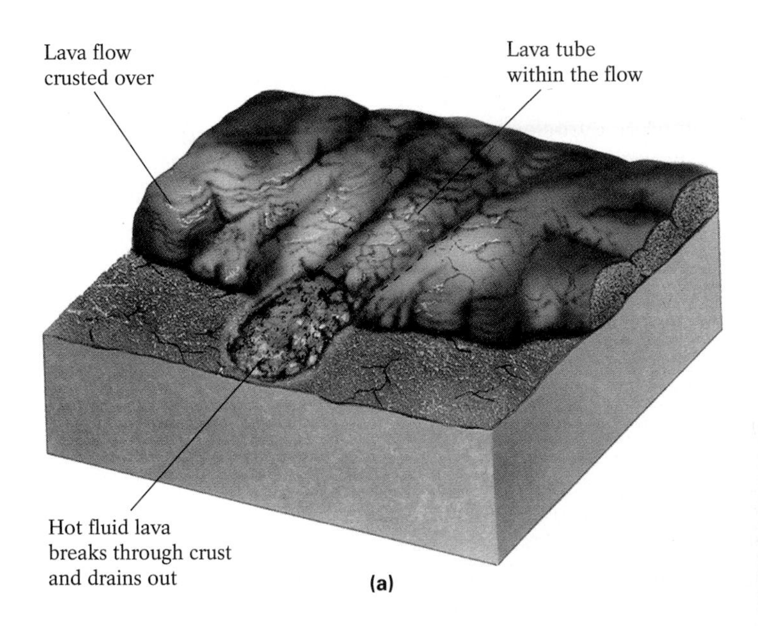

Figure 4-7 Lava tubes. **(a)** Lava tubes form when a lava flow's surface cools and solidifies, but the lava continues to flow under the surface in tunnels. When the lava drains from the tunnel, it leaves behind the empty tube. **(b)** A lava tube forming at K'ilauea.

and left hollow (Fig. 4-7). Lava Beds National Monument in northeastern California contains 300 or more lava tubes in its pahoehoe flows.

When basaltic lava erupts beneath the sea, the contact with cold water instantly chills its surface, which immediately solidifies into a thin, deformable skin. As hot lava enters

Figure 4-8 Basaltic pillow lava in the Galapagos. This form was photographed from a deep-sea submersible.

under it, this skin stretches to form a distinctive *pillow structure* (Fig. 4-8).

Andesitic and Rhyolitic Lavas Andesitic lava, being more felsic and thus more viscous than mafic basaltic lava, flows more slowly than basaltic lava and solidifies before traveling as far from its vent. Such lavas develop many of the same structures as basaltic lava. We rarely see pahoehoe-type andesitic flows, however, because these lavas are too viscous to stretch into a ropy structure. The more felsic andesitic lavas can even be viscous enough to impede the passage of rising gases and erupt in major volcanic explosions.

Felsic magma, being the coolest and most highly viscous form of molten rock, moves so slowly that it tends to cool and solidify underground as plutonic granite. It therefore rarely erupts as rhyolitic lava at the Earth's surface. When rhyolite does erupt, it usually explodes violently, producing an enormous volume of solid airborne fragments. Because it is so viscous, this type of lava never flows far from the vent and does not produce the structures that are typical of less viscous lavas. Felsic magmas with high water and gas content may bubble out of a vent as a froth of lava that quickly solidifies into the glassy volcanic rock known as *pumice.*

Pyroclastics

An explosive eruption expels lava forcefully into the atmosphere, where it cools rapidly and solidifies into countless fragments of various sizes and shapes. Such an eruption might also produce *volcanic blocks,* chunks of preexisting rock ripped from the throat of the volcano during an eruption. Volcanic blocks tend to be angular and range from the size of a baseball to that of a house. Blocks weighing as much as 100 tons have been found as far as 10 kilometers (6 miles) from the volcano that spewed them out. All such fragmental volcanic products, whether they are composed of cooled lava or of preexisting rocks, are termed **pyroclastics** (from the Greek *pyro,* meaning "fire," and *klastos,* meaning "fragments"). Pyroclastic materials may travel through the air as dispersed particles or they may hug the ground as dense flows.

Tephra Pyroclastic particles that cool and solidify from lava that is propelled through the air are called **tephra.** Tephra particles are classified by size, ranging from a fine dust to massive chunks (Fig. 4-9).

(a) (b)

Figure 4-9 Tephra. Tephra particles range in size from fine dust to large boulders. **(a)** This range can sometimes be found within a single deposit, as here at Mono Craters, California. **(b)** A volcanic bomb, the largest type of tephra.

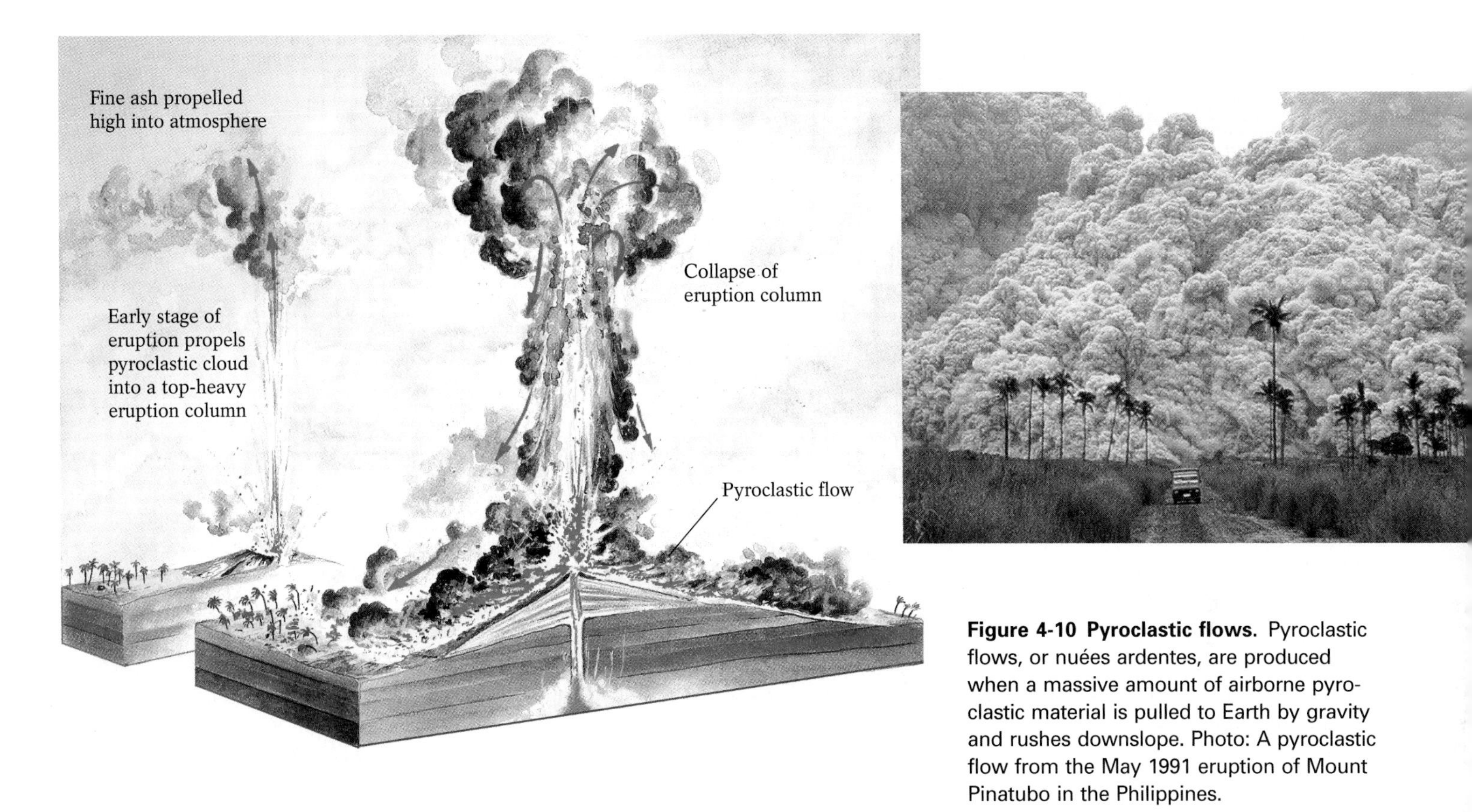

Figure 4-10 Pyroclastic flows. Pyroclastic flows, or nuées ardentes, are produced when a massive amount of airborne pyroclastic material is pulled to Earth by gravity and rushes downslope. Photo: A pyroclastic flow from the May 1991 eruption of Mount Pinatubo in the Philippines.

Volcanic dust particles are only about one-thousandth of a millimeter and have the consistency of cake flour. Because it is so fine, volcanic dust can travel great distances downwind from an erupting volcano and remain in the upper atmosphere for as long as two years. In sufficient quantity, it can diminish the amount of solar radiation reaching the Earth and lower the Earth's temperature by as much as 2° to 3° Celsius, an effect that may last for more than a decade. The spectacular eruption of Indonesia's Mount Tambora in 1815, which took an estimated 50,000 lives and decapitated the volcano's peak, was followed by what became known as the "year without a summer."

Somewhat grittier than volcanic dust is *volcanic ash,* particles of which are less than 2 millimeters in diameter, ranging from the size of a grain of fine sand to that of rice. Ash generally stays in the air for a few hours or days. During the 1980 eruption of Mount St. Helens in Washington state, ash covered the land downwind of the eruption, clogging automobile carburetors and fouling the bearings of farm machinery.

Cinders, or *lapilli* (Italian for "little stones"), range from about the size of peas to that of walnuts (2 to 64 millimeters in diameter). *Volcanic bombs* are large (64 millimeters or more in diameter), streamlined chunks of rock formed when sizable blobs of lava solidify in mid-air while being propelled by the force of an eruption. These coarser types of tephra are pulled to Earth by gravity; hence they fall sooner and closer to the volcanic vent than do dust and ash.

Pyroclastic Flows When the amount of pyroclastic material expelled by a volcano is so great that gravity almost immediately pulls it down onto the volcano slope, this material rushes downslope as a **pyroclastic flow,** or **nuée ardente** (pronounced "noo-AY AR-dent"; French for "glowing cloud"). Because the flow also contains trapped air and magmatic gases, it flows with little frictional resistance along the ground and may reach speeds in excess of 150 kilometers (100 miles) per hour, even on gentle slopes (Fig. 4-10). With temperatures of 800°C (1475°F) or more, a nuée ardente is capable of destroying any life that lies within its path. As the flow travels downslope, its gases escape and the warm particles finally come to rest. At this point, they may still be soft enough to fuse with one another, forming a volcanic rock called *welded tuff.*

Volcanic Mudflows Pyroclastic material that accumulates on the slope of a volcano may become mixed with water to form a volcanic mudflow, or **lahar.** A lahar is often produced when an explosive eruption occurs on a snow-capped volcano, and hot pyroclastic material melts a large volume of snow or glacial ice. One such devastating lahar buried the

Figure 4-11 **Lahars.** Some lahars occur when pyroclastic eruptions melt snow and ice on volcanic slopes, producing torrents of mud. This lahar resulted when Colombia's Nevado del Ruiz volcano erupted in 1985, melting about 10% of the snow on the slopes of the volcano and producing a 40-meter (137-foot)-high wall of mud that buried the town of Armero and killed approximately 23,000 of its residents. Armero was located approximately 50 kilometers (30 miles) from the summit of Nevado del Ruiz.

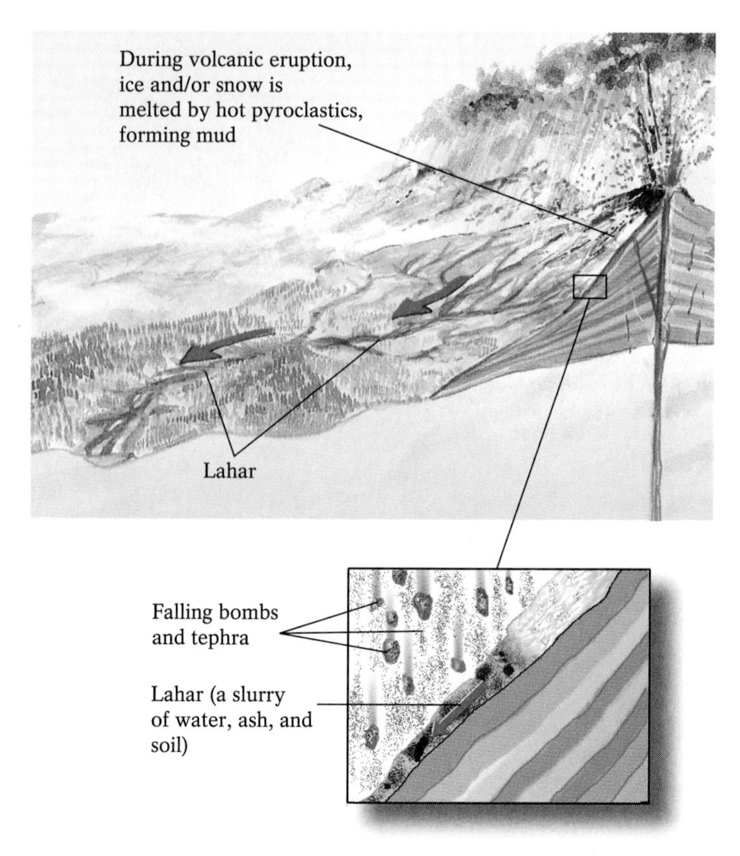

highland town of Armero, on the slopes of Colombia's Andes Mountains, when that country's Nevado del Ruiz erupted on November 13, 1985 (Fig. 4-11).

Eruptive Styles and Associated Landforms

Nearly all volcanoes have the same two major components: (1) a mountain, or **volcanic cone,** constructed by the products of numerous eruptions over time; and (2) a steep-walled, bowl-shaped depression, or **volcanic crater,** surrounding the vent from which those volcanic products emanate. A volcano's crater forms following an eruption, when lava and pyroclastics that have accumulated in the area around the vent are left somewhat unsupported and subside to form a depression. If enough lava erupts to completely or partially empty the volcano's subterranean reservoir of magma, the unsupported summit of the volcanic cone may collapse inward, forming a much larger summit depression called a **caldera.** Volcanic cones and craters take a variety of shapes and dimensions, depending on the types of eruptions and the composition of the volcanic products that formed them (Fig. 4-12).

Effusive Eruptions

Effusive eruptions are relatively quiet, nonexplosive events that generally involve basaltic lava. Basaltic lava, which is highly fluid, flows freely from central volcanic vents, as well as from elongated cracks on land and at submarine plate boundaries.

Central-Vent Eruptions In central-vent eruptions, basaltic lava flows out in all directions from one main vent, solidifying in more or less the same thickness all around. Because of its fluidity, the lava does not develop into a steep mountain. Instead, through successive flows over time, it builds a low, broad, contact lens-shaped structure known as a **shield volcano** (Fig. 4-13). Mauna Loa, one of five such shield volcanoes that form the island of Hawai'i, has a circumference of 600 kilometers (400 miles) and is composed of thousands of layers of lava flows that have erupted over the past 750,000 years.

At the start of an effusive eruption, lava begins to accumulate in the volcanic crater, forming a lava lake that may eventually overflow the rim of the crater. Sometimes the summit collapses to form a caldera, and the weight of the collapsed summit closes off the central vent; in such an event, any remaining magma is diverted laterally, producing a *flank*

Figure 4-12 Volcanoes come in various shapes and sizes. (a) Lofty, symmetrical Mount Augustine, in Alaska's Cook Inlet. **(b)** Small, stubby volcanoes from the Flagstaff area of northern Arizona.

eruption from the side of the volcano. Flank eruptions also occur when the central vent becomes plugged by congealed lava or when the volcanic cone grows so high that rising magma seeks a lower, more direct route to the surface. Hawai'i's K'ilauea is a volcanic cone built up by flank eruptions on Mauna Loa's southeastern slope.

Fissure Eruptions Rising, highly fluid basaltic lava may erupt through linear fractures, or *fissures,* in the Earth's crust that often develop at diverging plates. As lava flows away from a fissure, it may spread out over thousands of square kilometers; successive flows may build up immensely thick *lava plateaus,* or *flood basalts,* such as the 15-million-year-old

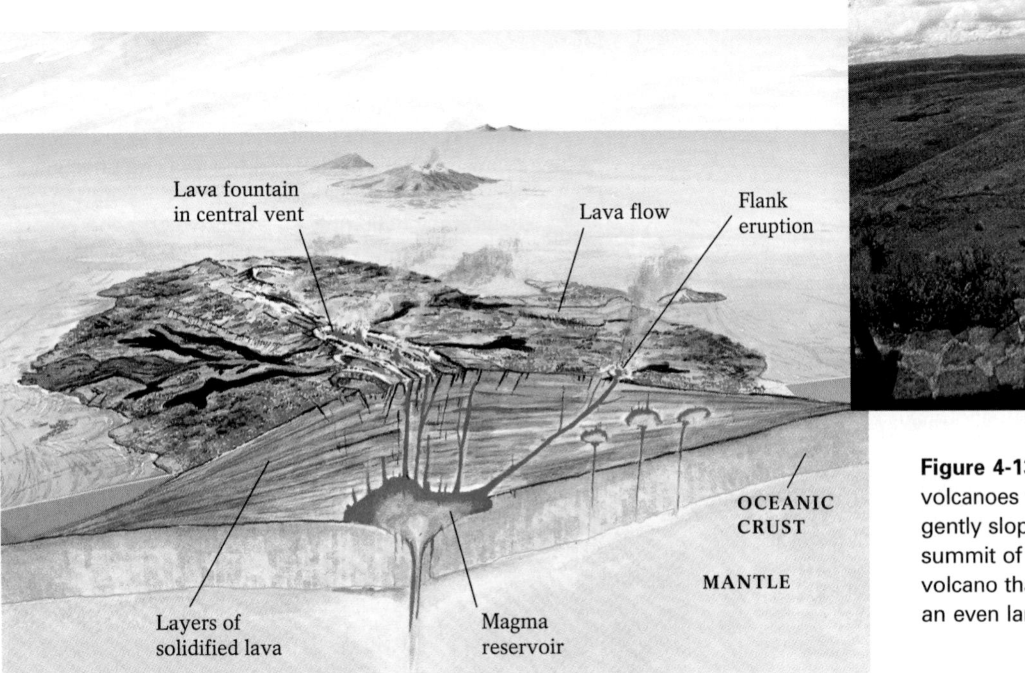

Figure 4-13 Shield volcanoes. Low, broad shield volcanoes form by the gradual accumulation of gently sloping basaltic lava flows. Photo: The summit of Hawai'i's K'ilauea volcano, a shield volcano that actually developed on the flank of an even larger shield volcano, Mauna Loa.

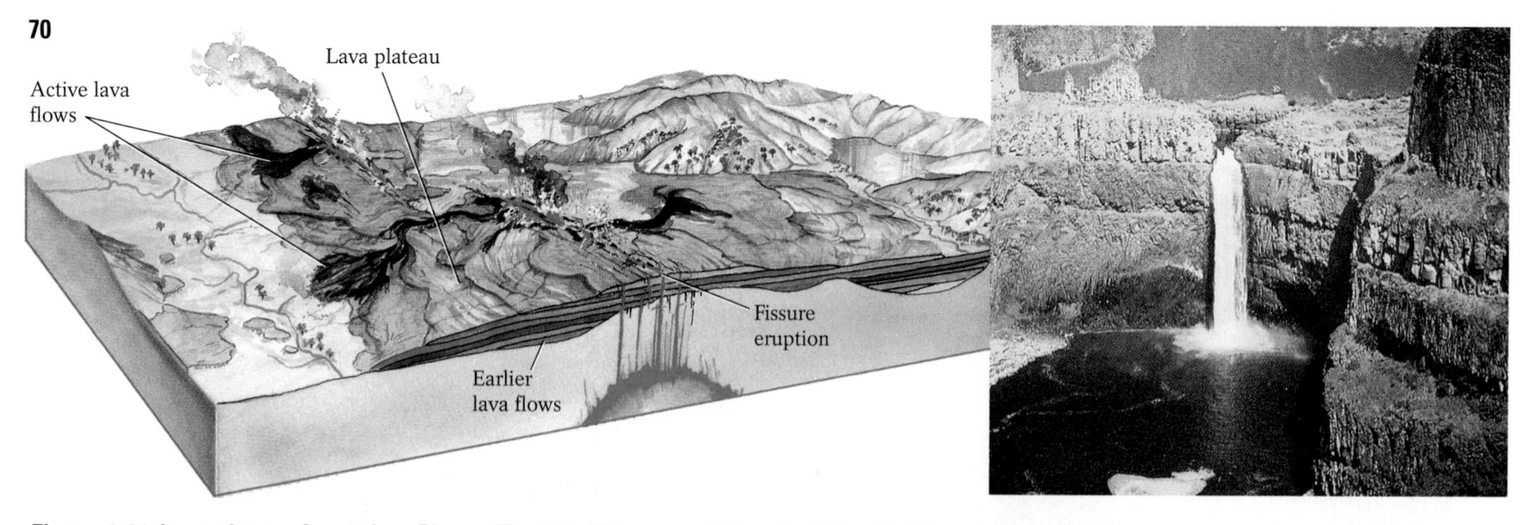

Figure 4-14 Lava-plateau formation. Photo: The 15-million-year-old basaltic Columbia River plateau of eastern Washington state. Geologists have identified more than 60 individual lava flows here, in some places totaling more than 2 kilometers (1.2 mile) in thickness. The lava from just one of these flows could pave Interstate 90 from Boston to Seattle to a depth of 175 meters (575 feet), about the height of the Washington Monument in Washington, D.C.

basaltic Columbia River plateau of eastern Washington state (Fig. 4-14).

Submarine Eruptions Most submarine eruptions are quiet and effusive. At depths greater than 300 meters (1000 feet), water pressure prevents the gases and water vapor in lava from expanding or escaping. Effusive submarine eruptions of basalt typically produce pillow structures such as those shown in Figure 4-8. A more explosive type of submarine eruption may occur when seawater enters a magma chamber through the ruptured walls of an island volcano. As the cold water comes in contact with the red-hot magma, a cloud of superheated steam is produced that may expand violently, shatter the volcanic cone, and propel magma and cone fragments skyward (Fig. 4-15).

Pyroclastic Eruptions

Pyroclastic eruptions, which usually involve viscous, gas-rich magmas, range from moderately to spectacularly explosive and tend to produce a large volume of solid volcanic fragments. On the Italian island of Stromboli in the eastern Mediterranean, a moderately explosive volcano erupts almost continuously, covering the region with a cloud of ash and hurling volcanic bombs as far as 3 kilometers (2 miles) from its crater. An even more spectacular example of a pyroclastic eruption occurred in the eastern Mediterranean: the historic eruption of Mount Vesuvius in A.D. 79, which devastated the prosperous Roman resort town of Pompeii. The entire city of Pompeii was entombed beneath a layer of more than 6 meters (20 feet) of volcanic ash, which preserved the most minute

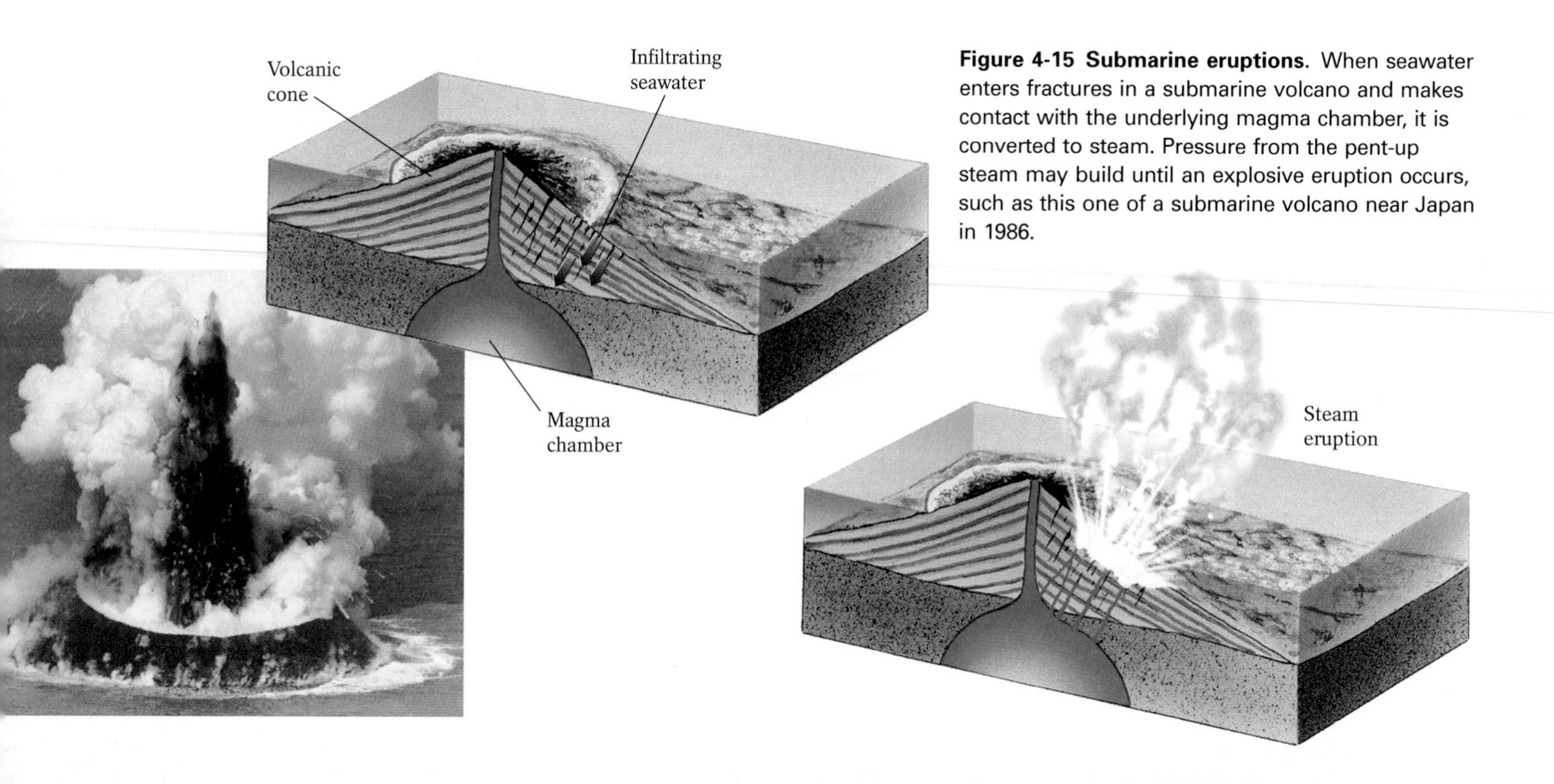

Figure 4-15 Submarine eruptions. When seawater enters fractures in a submarine volcano and makes contact with the underlying magma chamber, it is converted to steam. Pressure from the pent-up steam may build until an explosive eruption occurs, such as this one of a submarine volcano near Japan in 1986.

Trapped gases explode out, destroying volcanic dome

(b)

Figure 4-16 Volcanic domes. Such domes are created when extremely viscous lavas solidify before they leave their crater, plugging the volcanic vent. Buildup of pressure beneath a dome may result in gases escaping from the flank of the volcano **(a)** or eventual explosive destruction of the dome **(b)**.

details of Roman life, and was not seen again until archaeologists uncovered the site in the mid-eighteenth century.

When volcanoes such as Mount Vesuvius are fed by large quantities of extremely gas-rich magma, they may produce towering columns of tephra rising tens of kilometers into the atmosphere. When fed by magma containing less gas, such volcanoes generally produce dense clouds of superheated ash and pumice that, instead of billowing high into the air, fall back to the surface almost immediately and race downslope as nuée ardentes.

If it contains only a very little gas or water, a felsic magma may become a lava so viscous that it does not even flow out of the volcano's crater. Instead, it may cool and harden to form a **volcanic dome** that caps the vent, trapping the volcano's gases and building pressure toward another eruption; this pressure might be relieved by the escape of gases laterally through the flank of the volcano, or it might build to a point that it finally shatters the volcanic dome in a particularly explosive eruption (Fig. 4-16).

The life of an explosive volcano may encompass thousands of years of repeated eruptions marked by towering tephra columns and fiery pyroclastic flows and end in an extremely energetic culminating eruption involving a full range of eruptive styles. The final explosion often begins with a titanic blast that produces a massive vertical tephra column and a ground-hugging blanket of pyroclastic debris. As the escaping gases lose force, the tephra column may collapse. Additional pyroclastic flows may follow. Finally, the summit of the cone may subside into the emptied magma chamber, forming a large caldera. Highlight 4-1 recounts one such violent event, which formed beautiful Crater Lake in Oregon.

Ash-Flow Eruptions Some spectacular and extremely dangerous pyroclastic eruptions have occurred despite the absence of a recognizable volcanic cone. An *ash-flow eruption* results when a large reservoir of highly viscous, gas-rich magma rises to just below the Earth's surface, stretching the crust and forcing it up into a bulge marked by a series of ringlike fractures (Fig. 4-17). The thinned bedrock over such a magma reservoir may collapse, forcing magma into the fractures and thereby

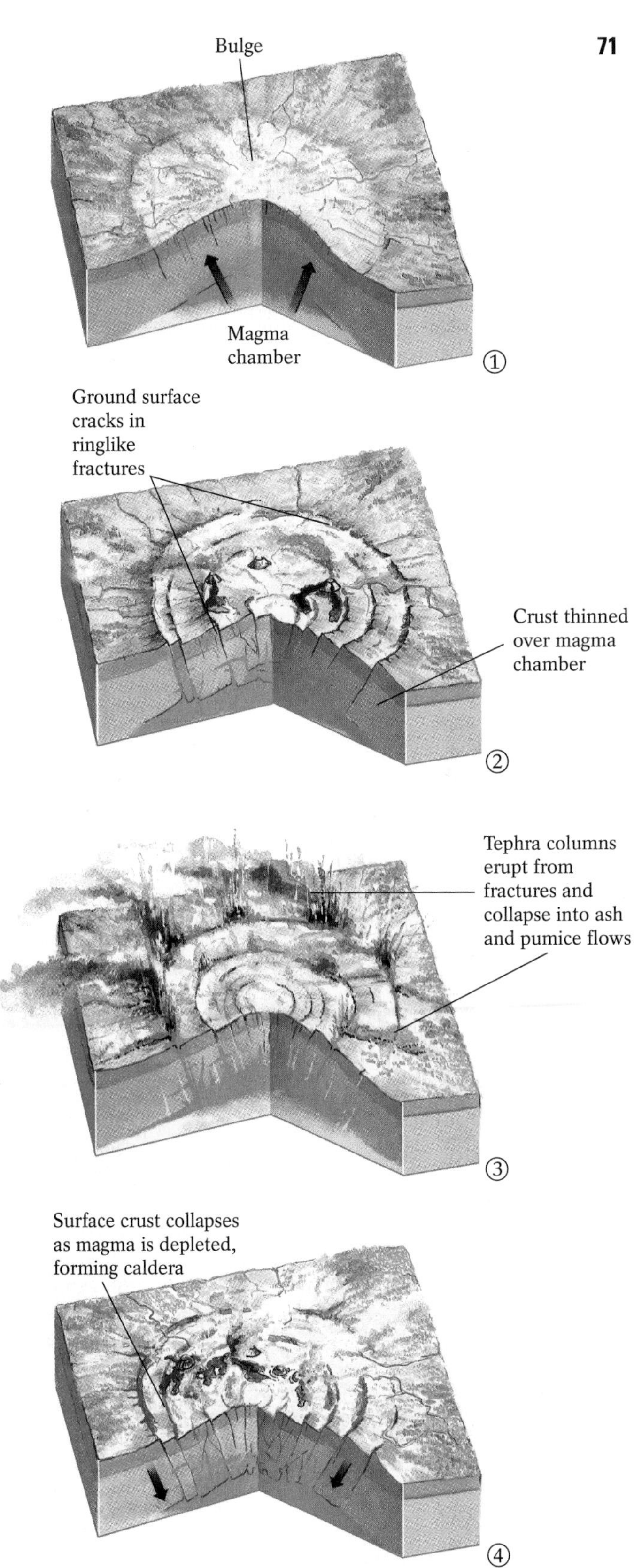

Figure 4-17 Ash-flow eruptions. Such eruptions begin when rising viscous magma causes the surface crust to bulge. Cracks develop in the crust, allowing gas-rich tephra columns to erupt and then collapse into hot swirling ash and pumice flows. The partial emptying of the magma chamber weakens its roof and causes the surface to collapse, expelling more pumice in spectacular ash flows and forming a caldera. Later, new magma may refill the chamber and force the caldera upward again until another eruption takes place.

Highlight 4-1 Mount Mazama and the Formation of Crater Lake

The last eruption of Mount Mazama, a volcano in the southern Oregon Cascades, occurred approximately 6700 years ago. The event buried thousands of square kilometers with tephra and pumice and lowered the mountain's height by more than a mile. Before this culminating eruption, the mountain is believed to have been 3700 meters (12,136 feet) high and glacier-covered, comparable in majesty to Mount Rainier or Mount Shasta. Today, it is a sawed-off, goblet-shaped mountain reaching only 1836 meters (6058 feet) above sea level.

Because very little rock from the shattered cone of the volcano has been found in the areas surrounding Mount Mazama—including Oregon, Washington, and British Columbia—geologists have concluded that the summit of the volcano must have collapsed inward during its last eruption, rather than exploding outward, and remains to this day buried beneath the caldera produced by its collapse (Fig. 1). This caldera now contains Crater Lake, North America's deepest lake (more than 600 meters [1900 feet] deep).

The eruption 6700 years ago may not have ended Mount Mazama's excitement, however. Eruptions during the past 1000 years have produced three smaller volcanic cones within the lake, two of which remain below water level. The third, Wizard Island, rises above the surface near the western shore of the lake. Recent dives to the lake bottom reveal that hot water and steam are being vented continuously, suggesting that Mount Mazama may be entering a new eruptive sequence.

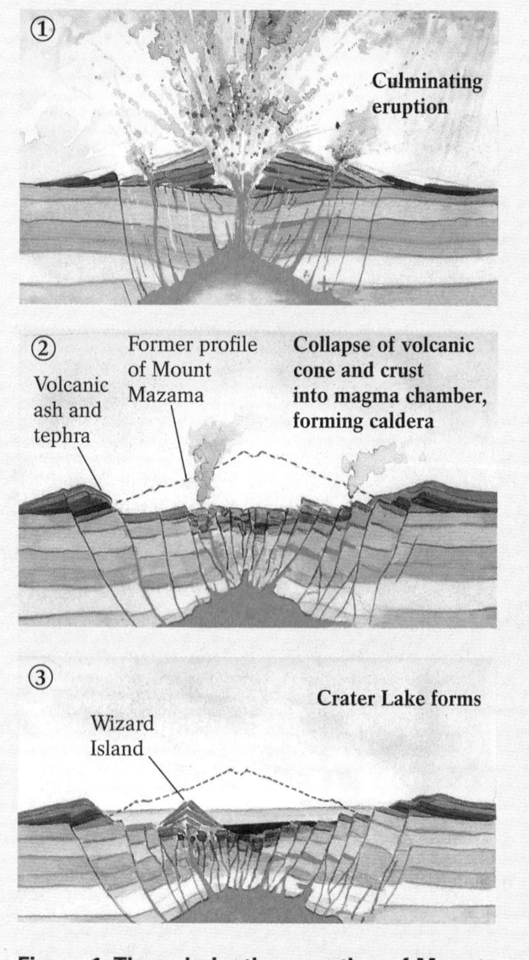

Figure 1 The culminating eruption of Mount Mazama and the creation of Crater Lake.

producing a circular pattern of tremendous tephra columns. These columns soon collapse to form numerous flows of hot, swirling ash and pumice. As the magma reservoir continues to empty, the surface crust becomes further undermined until it collapses to form a huge caldera, sometimes as large as 5000 square kilometers (1800 square miles).

At least three times in the last 2 million years, the Yellowstone plateau in Wyoming has erupted in devastating ash flows after being pushed up like a blister by an enormous mass of felsic magma. One of these events showered thousands of square kilometers with hot ash and pumice and created the caldera that now contains beautiful Yellowstone Lake. Today, only a few kilometers below Yellowstone's caldera, a new huge mass of felsic magma may be accumulating. Yellowstone National Park's thermal features—its geysers, hot springs, and gurgling mudpots (pools of boiling mud)—are all heated by this shallow subterranean magma reservoir.

The most immediate threat of an ash-flow eruption in North America is near Mammoth Lakes, the popular ski resort in eastern California. During the past 20 years, the

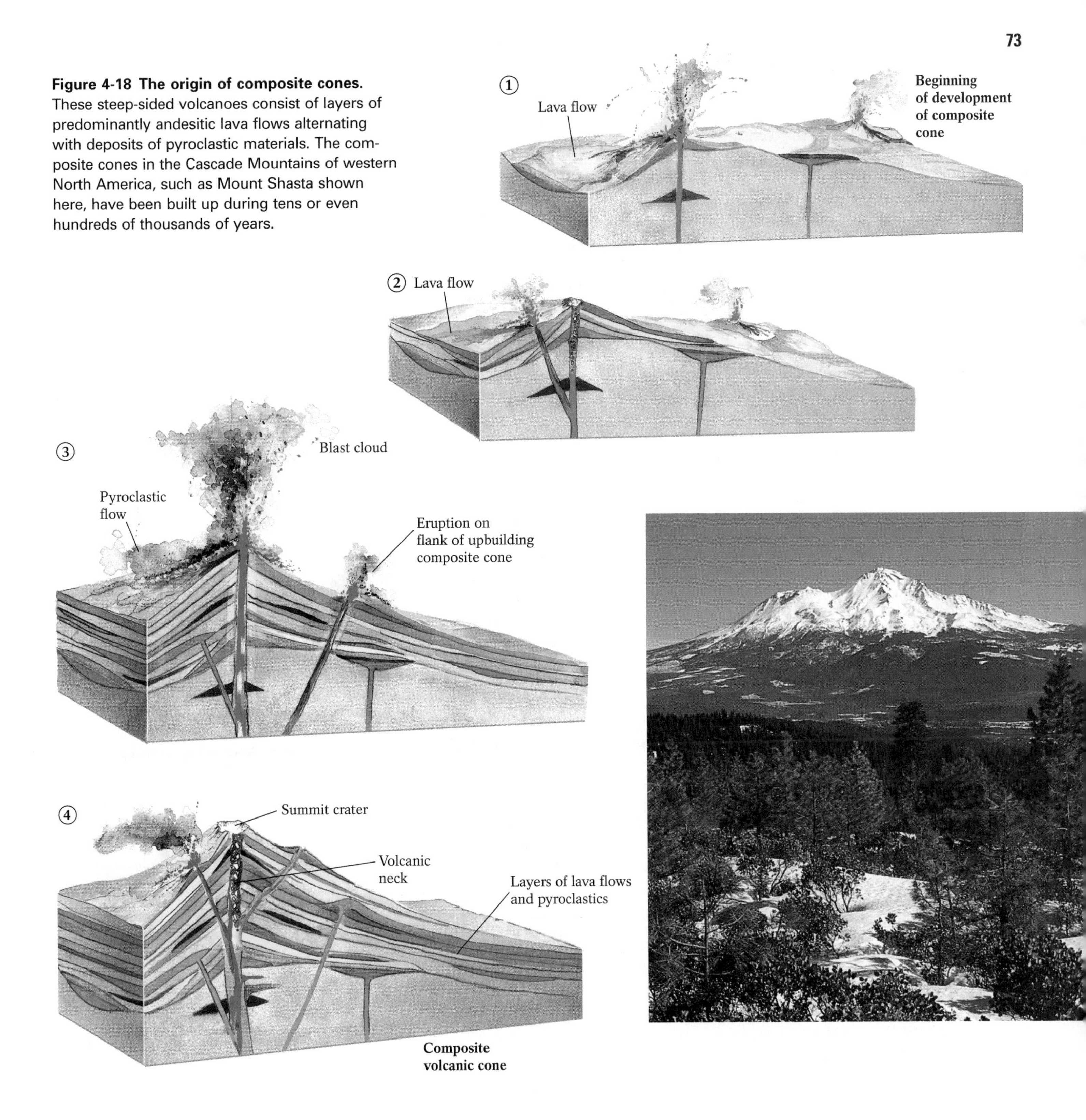

Figure 4-18 The origin of composite cones. These steep-sided volcanoes consist of layers of predominantly andesitic lava flows alternating with deposits of pyroclastic materials. The composite cones in the Cascade Mountains of western North America, such as Mount Shasta shown here, have been built up during tens or even hundreds of thousands of years.

United States Geological Survey has watched the floor of the caldera rise more than 25 centimeters (9 inches). In 1982, it designated the area as a potential volcanic hazard, requiring heightened scientific vigilance and a regional plan for coping with an eruption.

Types of Pyroclastic Volcanic Cones Because they are fairly viscous, felsic and intermediate lavas solidify relatively close to the vent. The composition of the parent magma and thus the volcano's eruptive style may change over time, however, so that the cone intermittently ejects a large quantity of tephra instead of lava. The larger tephra particles from these eruptions fall near the summit to form steep cinder piles, which then become covered by the next lava flow. The characteristic landform of pyroclastic eruptions—the **composite cone,** or **stratovolcano** ("strato" means layered)—is built up from such alternating layers of lava and pyroclastics (Fig. 4-18). Because each pyroclastic deposit produces a steep slope that is then protected from erosion by a successive layer of lava, composite cones grow to be very large and have 10° to 25° slopes. They are among the Earth's most picturesque volcanoes, as evidenced by Mount Fuji in Japan and Mounts Rainier and Shasta in the western United States.

Unlike composite cones, **pyroclastic cones** are built up almost entirely from the accumulation of loose pyroclastic material. When the dominant pyroclastics are cinders, **cinder cones** form. Pyroclastic cones are typically the smallest volcanoes (less than 450 meters [1500 feet] high) and are generally steep-sided, because pyroclastics can be piled up stably to form slopes between 30° and 40°. Because no intervening lava flows bind their loose pyroclastics, however, pyroclastic cones are readily eroded.

Plate Tectonics and Volcanism

Approximately 80% of the Earth's volcanoes surround the Pacific Ocean, where several of the Pacific basin's oceanic plates are subducting beneath adjacent plates. Another 15% are similarly situated above subduction zones in the Mediterranean and Caribbean seas. The remainder are scattered

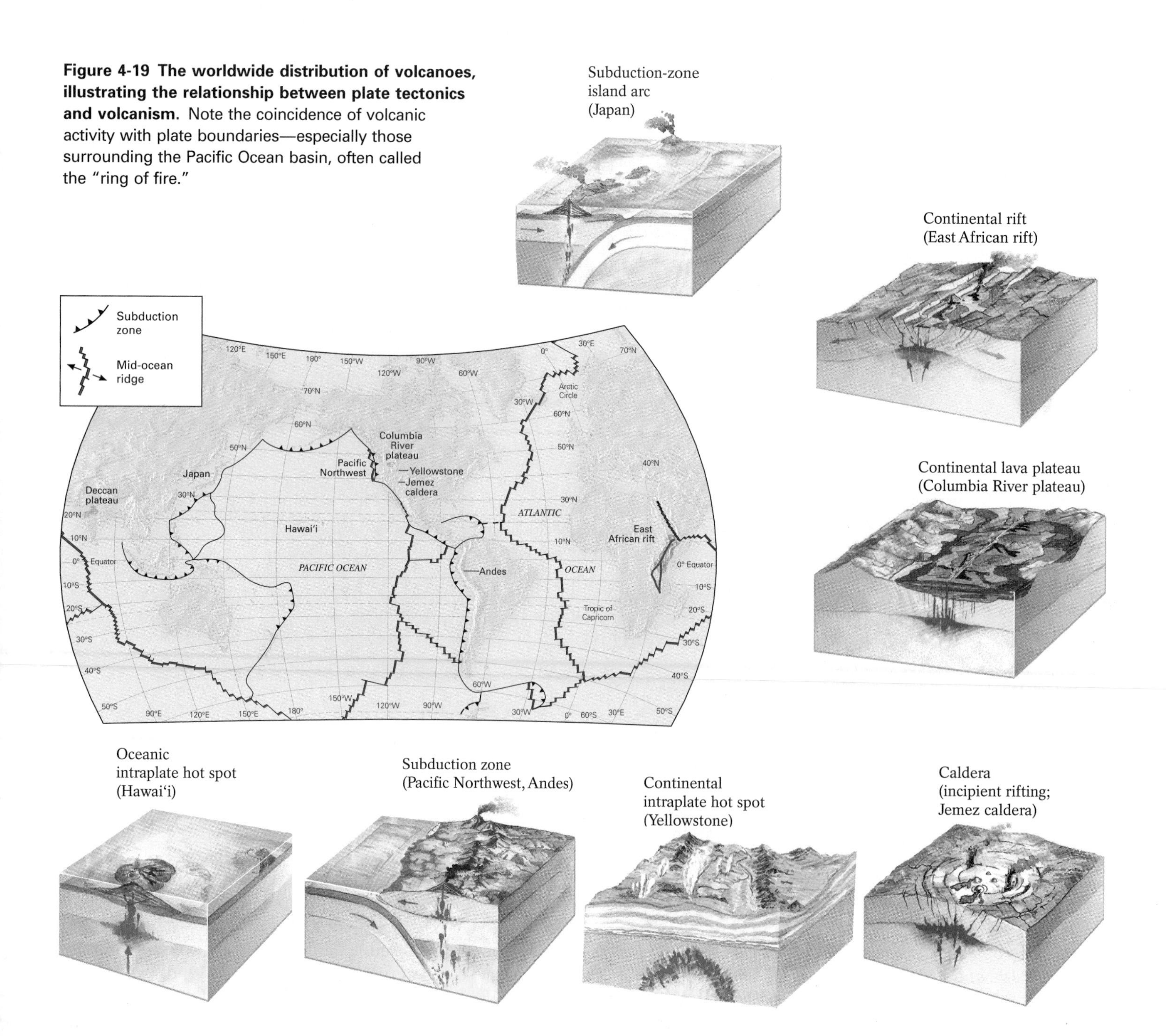

Figure 4-19 The worldwide distribution of volcanoes, illustrating the relationship between plate tectonics and volcanism. Note the coincidence of volcanic activity with plate boundaries—especially those surrounding the Pacific Ocean basin, often called the "ring of fire."

along the ridges of divergent plate boundaries and atop continental and oceanic intraplate hot spots. Each type of plate tectonic setting determines the eruptive style and physical appearance of its volcanoes (Fig. 4-19).

Subduction zones foster explosive pyroclastic volcanism, because partial melting of both the subducting plates and their overlying silica-rich sediments produces intermediate and felsic magmas. Subduction zones tend to produce steep-sided composite cones composed primarily of andesite.

Explosive volcanism also occurs where felsic continental plates are stretched thin and begin to rift, as well as at intracontinental hot spots. In both cases, hot mafic magma rising from the mantle comes in contact with and melts large volumes of felsic rock at the base of thick continental crust. Eruptions of viscous felsic magmas create the volcanic domes, calderas, and ash-flow deposits characteristic of intracontinental rift zones and hot spots. Rift zones and hot spots may also produce mafic lava, although not usually at the same time as felsic lava. An example of this type of volcanism can be found in the Basin and Range region of southwestern North America, where powerful blasts of felsic ash and pumice have alternated with quiet effusions of basaltic lava throughout the last 20 million years.

At divergent zones and above oceanic intraplate hot spots, effusive eruptions of low-viscosity basaltic lava generally produce near-horizontal lava plateaus and gently sloping shield volcanoes. Table 4-1 shows the relationships between lava types, eruptive styles, types of volcanic cones and products, and plate tectonic settings.

Coping with Volcanic Hazards

Because we can't slow subduction, stop divergence, or chill intraplate hot spots, we will probably never be able to prevent volcanic eruptions. Nevertheless, geologists continue to work to identify areas where volcanic eruptions are possible and to develop ways to predict them more accurately so as to prevent casualties and minimize danger.

In North America, the most immediately threatening volcanoes lie in the Cascade Mountains. Several volcanoes within these mountains have erupted in relatively recent times (Fig. 4-20), and remain active where the North American plate overrides the subducting Juan de Fuca and Gorda plates and where it sits atop the continent's scattered intraplate hot spots. The most recent example of the Cascades' potential for explosive volcanism was the 1980 eruption of Mount St. Helens in Washington—although relatively unspectacular compared with some volcanic events elsewhere, it was the greatest volcanic disaster in North America's recent history (see Highlight 4-2 on pages 78–79).

Other Cascade volcanoes threaten major metropolitan areas in southern British Columbia, Washington, Oregon, and northern California. For example, volcanologists keep a close watch on Mount McLoughlin, 50 kilometers (30 miles) from Medford and Klamath Falls in Oregon, and Mount Hood near Portland, which last erupted in 1865, for signs that they are awakening. Mount Baker, near Washington's Canadian border, resumed its intermittent rumbling and steam emissions

Table 4-1 Lava Types, Associated Volcanic Features, and Plate Tectonic Settings

Lava Type	Eruptive Style	Typical Volcanic Landforms	Common Volcanic Products and Effects	Common Plate Tectonic Setting	North American Example(s)
Basaltic (mafic composition)	Quiet, effusive	Lava plateaus, shield volcanoes, occasional cinder cones	ʻA'a lava, pahoehoe lava, vesicular basalts, pillow lavas, columnar basalts	Divergent plate boundaries (such as the mid-Atlantic ridge), oceanic intraplate hot spots (such as underlies Hawai'i), intraplate rifts (such as the East African rift)	Columbia River lava plateau (Washington and Oregon), Belknap Crater (eastern Oregon), Craters of the Moon (Idaho)
Andesitic (intermediate composition)	Fairly explosive, pyroclastic	Composite cones, cinder cones	Relatively viscous lava, lahars, tuffs (from airborne ash), welded tuffs (from pyroclastic flows)	Subduction zones	Cascades (British Columbia, Washington, Oregon, northern California), Aleutians (Alaska)
Rhyolitic (felsic composition)	Very explosive, pyroclastic	Volcanic domes, calderas	Extremely viscous lava, ash-flow deposits, welded tuffs (from pyroclastic flows)	Subduction zones, especially at continental margins, intracontinental rifts, intracontinental hot spots	Yellowstone plateau (Wyoming, Montana), Jemez Mountains (Rio Grande rift, New Mexico), Long Valley caldera (eastern Sierra Nevada, California)

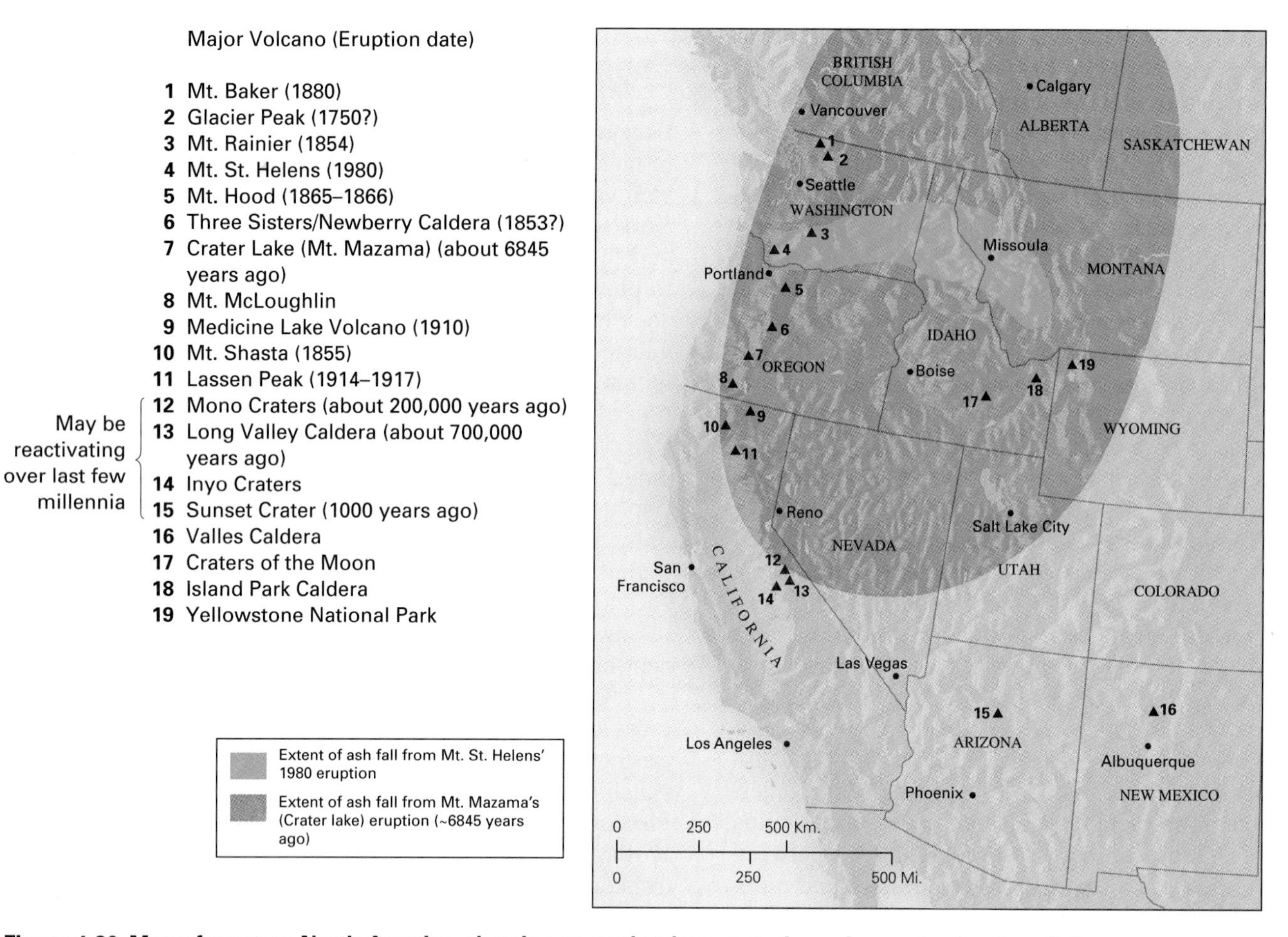

Figure 4-20 Map of western North America, showing areas that have experienced recent volcanic activity. These sites may be considered some of the most likely for future volcanic events.

so actively that in 1975 the U.S. Geological Survey predicted—albeit incorrectly—that it would be the next Cascade volcano to erupt. Mount Rainier, the Cascade's grandest peak, last erupted in 1882 with a small whiff of brown ash; in a future eruption, the greatest threat would come from large lahars spawned by steam and tephra emissions onto its glacier-clad slopes. Mount Garibaldi, just north of Vancouver, British Columbia, at the northern end of the Cascade volcanic chain, has eruptive potential as well.

Another region with eruptive potential is the Valles caldera near Sante Fe, New Mexico, one of North America's largest volcanoes. This massive crater—more than 25 kilometers (15 miles) in diameter—has experienced numerous explosive ash-flow eruptions in the past few hundred thousand years, probably due to the presence of a large reservoir of magma found some 20 kilometers (12 miles) below the Rio Grande region. Some geologists believe the area may someday become the site of a continental rift zone.

Defense Plans

An effective plan to avert volcanic disasters must start with keeping people out of harm's way by zoning against development in the most hazardous regions, building lava dams, diverting the path of a flowing lava, and providing ample warning to allow for timely evacuations. Hawai'i, for example, practices *volcanic zoning,* as do other western states. Under such a policy, areas with great potential for danger are set aside as national parks, monuments, and recreation areas, which remain closed to residential and commercial development.

Where preventive zoning is impractical, other means can avert or reduce potential damage, especially from lava flows and lahars. For example, a carefully positioned explosive device can disperse a lava flow over a wider area, ensuring that it thins out and cools and solidifies more rapidly. The hardening lava will block the path of still-flowing lava behind it, forcing it to accumulate upstream and flow along a less damaging route. In 1935, a strategically placed bomb coaxed a flow to detour around the Hawai'ian town of Hilo. In 1973, Icelanders on the coastal island of Heimaey, after contemplating bombing, decided instead to cool a flow with seawater pumped by 47 large barge-mounted pumps anchored in a nearby harbor; their effort was successful in diverting the dangerous flow. Japanese engineers have designed steel-and-concrete dams that can trap large boulders and slow mudflows so as to minimize damage and gain time for evacuation. Even a simple lava wall—a hand-built barrier of boulders or rocks—can sometimes effectively protect an individual homestead by guiding lava away.

Predicting Eruptions

In any given year, roughly 50 of the Earth's active volcanoes erupt—usually with some warning. Before they blow, they typically shake, swell, warm up, and belch a variety of gases. Because developing countries rarely have the necessary equipment to monitor an awakening volcano, they often seek help from the Volcano Disaster Assistance Program (VDAP)—a scientific SWAT team that rushes to volcanoes to assess their potential for violence and to predict when they might ignite. The VDAP, created a decade ago after the Armero, Colombia, disaster, is based at the U.S. Geological Survey's Cascade Volcanic Observatory in Vancouver, Washington. Team members wait for a call that sends them jetting around the world armed with lasers, seismometers, and other devices used to monitor volcanoes. The role played by the VDAP was dramatically depicted in the Hollywood film, *Dante's Peak*.

Geologists have enjoyed fair success in predicting individual eruptive episodes when they concentrate on a specific volcano *after* an eruptive phase has begun. These monitoring efforts involve measuring changes in a volcano's surface temperature, watching for the slightest expansion in its slope, and keeping track of regional earthquake activity (Fig. 4-21). A laboratory at the University of Washington in Seattle is staffed 24 hours a day to monitor the rumblings of Mount St. Helens. Even with the advances brought by today's technology, however, the art of volcano prediction has not been fully mastered. The U.S. Geological Survey missed the call on Mount St. Helens' 1980 blast despite the fact that the mountain was being watched closely by a large team of scientists armed with the latest in prediction technology. It did successfully predict the eruption of Mount Pinatubo in the Philippines, evacuating virtually everyone within 25 kilometers (15 miles) before the volcano's powerful blast on May 17, 1991.

Before a volcano erupts, hot magma rises toward the surface, so any local manifestation of increasing heat may signal an impending event. Ongoing surveys can identify new surface hot springs and take the temperature of the water and steam in existing ones. If the escaping steam isn't much hotter than the boiling point of water, then surface water is probably seeping into the mountain and being heated by contact with hot subsurface rocks, and all is well for the time being. If the steam is superheated, with temperatures as high as 500°C (900°F), then it probably derives from shallow water-rich magma, a sign that an eruption may be brewing. As magma rises, the volcanic cone itself begins to heat up.

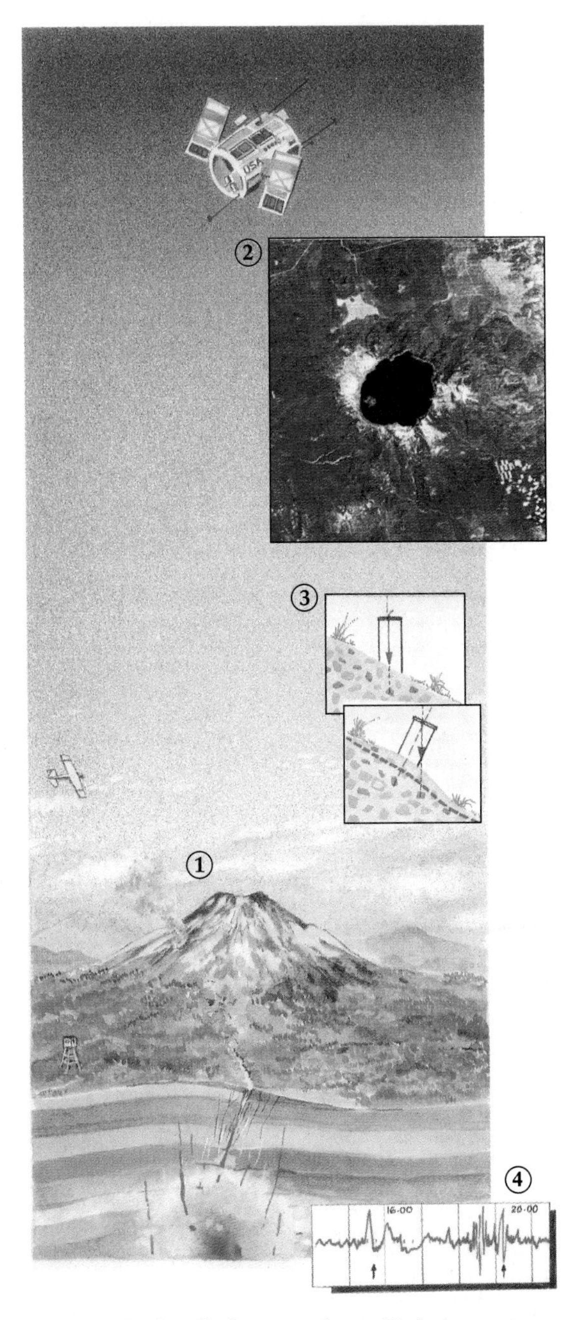

Figure 4-21 Predicting eruptions. Techniques commonly used to predict volcanic eruptions include watching for escape of superheated steam (1); satellite monitoring of volcanic cone temperature (here, Crater Lake in southern Oregon) (2); detecting volcanic cone bulges, using tiltmeters (3); and locating increased tremor activity, using seismographs (4).

Highlight 4-2 Mount St. Helens

Mount St. Helens' cone, the youngest in the Cascades at less than 10,000 years of age, came to life with an audible boom on March 30, 1980, after 123 years of silence. Weeks of public anxiety and scientific watchfulness followed, culminating on May 18 when an eruption blasted about 400 meters (1300 feet) of rock from the summit, changing what many had considered North America's most beautiful peak into a squat gray crater (Fig. 1).

Pre-eruption underground rumblings had signaled the rise of magma into Mount St. Helens' cone. Probably blocked on its way to the central vent by a plug of congealed lava from an earlier eruption, the ascending magma took a sharp turn northward instead. A bulge appeared on the volcano's northern slope and, growing at the rate of 1.5 meters (5 feet) per day, protruded an ominous 122 meters (400 feet) by May 17.

At 8:31 on the morning of May 18, a powerful earthquake beneath the mountain released internal pressure that had been building for weeks. It dislodged the bulge and sent the material hurtling down the mountain as an avalanche of debris traveling at 400 kilometers (250 miles) per hour. A northward-directed jet of superheated (500°C [900°F]) ash and gas immediately erupted as a pyroclastic flow, racing downslope with hurricane force (at speeds greater than 300 kilometers [200 miles] per hour) and cutting a swath of complete destruction 30 kilome-

(a)

(b)

(c)

Figure 1 Before and after the eruption. Mount St. Helens' pre-eruption symmetry made it one of the world's most beautiful and photographed volcanoes **(a).** Its eruption on May 18, 1980, opened a crater on the volcano's north side **(b)**, changing its appearance dramatically **(c).**

ters (20 miles) wide. The blast and subsequent nuée ardente buried the nearest 12 kilometers (7 miles) of forest land beneath meters of pyroclastics, blew down entire stands of mature trees like matchsticks to a distance of 20 kilometers (12 miles) (Fig. 2), and singed the forest beyond for an additional 6 kilometers (4 miles). More than 26 kilometers (16 miles) away, the heat scalded fishermen, who plunged into lakes and streams. On Mount Adams, 50 kilometers (30 miles) away, climbers felt a gust of intense heat just before being bombarded with hot, ash-blasted pine cones.

Meanwhile, a tephra column rose from the summit vent to an altitude of 25 kilometers (15 miles). Swept eastward by the prevailing winds, the dense cloud of gray ash began to fall on the cities and towns of eastern Washington. Yakima, located 150 kilometers (100 miles) to the east of Mount St. Helens, received 10 to 15 centimeters (4–6 inches) of ash. In Spokane, farther east, visibility fell to less than 3 meters (10 feet) and automatic street lights switched on at noon. Proceeding across the continent, the ash cloud dusted every state in its path. Hundreds of downwind communities would later spend millions of dollars cleaning up.

Several hours after the eruption began, snow and large chunks of glacial ice trapped within the dislodged debris from the mountain's northern slope began to melt. This enormous volume of meltwater mixed with loose material and the eruption's fresh pyroclastics to produce a lahar that rushed 28 kilometers (17 miles) westward down the Toutle River valley at 80 kilometers (50 miles) per hour, picking up logging trucks and hundreds of thousands of logs along the way. It eventually buried 123 homes beneath 60 meters (200 feet) of mud.

In all, 60 human lives were lost, along with 500 blacktail deer, 200 brown bear, 1500 elk, and countless birds and small mammals. The only survivors were burrowers such as frogs and salamanders, which fled into the soft sands of lake shores and stream banks. The human toll would have been much higher if state officials had not heeded the warnings issued in March by the U.S. Geological Survey, after the volcano's initial reawakening, and evacuated the area of most of its year-round residents and closed it to seasonal residents and spring hikers. However, if the eruption had occurred one day later, on Monday, hundreds of loggers at work would have been buried beneath the debris avalanche.

Figure 2 The awesome power of volcanoes. The forest north of Mount St. Helens was blown down for miles by the force of the volcano's pyroclastic flow and covered with volcanic ash. The area shown in this photo is 12 kilometers (7 miles) from the crater.

Figure 4-22 The remains of San Juan Parangaricutiro, Mexico. This town was engulfed during June and July of 1944 with lava from the eruption of Mexico's Paricutin volcano (visible in the background). The eruption, which lasted nine years, began in February 1943 with the sudden appearance of a small cinder cone in a cornfield about 300 kilometers (200 miles) west of Mexico City. Because the flow that buried the town moved only a few feet per hour, the Mexican Army was able to evacuate the 4000 or so residents before any lives were lost.

The overall temperature of a volcanic cone can be monitored from an orbiting satellite equipped with infrared heat sensors to detect the slightest change in surface temperature. This high-altitude technology serves as a simultaneous early-warning system for most of the Earth's 600 or so active volcanoes. Impending eruptions may also be preceded by increased gas emissions from rising magmas. For this reason, volcanologists continuously monitor sulfur dioxide and carbon dioxide emissions from potentially active volcanoes.

Active volcanoes expand in volume as they acquire new supplies of magma from below. As a result, an increase in the steepness or bulging of a volcano's slope may signal an impending eruption. To detect the inflation of a volcanic cone, a *tiltmeter*, a device like a carpenter's level, is used.

As magma rises, it pushes aside fractured rock, enlarging the fractures as it moves. Because this type of fracturing causes earthquakes, eruptions are often preceded by a distinctive pattern of earthquake activity called *harmonic tremors*, a continuous rhythmic rumbling. Sensitive equipment that monitors the location of these tremors can measure the increased height of rising magma. The rate at which the magma rises provides an estimate of when an eruption may occur. Indeed, it served as the principal means by which scientists accurately predicted recent eruptions of Mount St. Helens.

Efforts to predict eruptions are thwarted, however, when we are unaware of a site's volcanic potential. Occasionally a new volcano appears suddenly and rather unexpectedly, as was the case in 1943 when the volcano Paricutin developed literally overnight in the Mexican state of Michoacan, 320 kilometers (200 miles) west of Mexico City (Fig. 4-22). The area *was* known to be volcanic, lying just northeast of a subducting segment of the Pacific Ocean.

Extraterrestrial Volcanism

Volcanism in our solar system occurs (or has occurred in the past) not only on Earth but also on the Moon, Mars, Venus, and the moons of Jupiter and Neptune. Although no volcanoes remain active on the Moon, at least one-fourth of the Moon's surface is covered by ancient flood basalts known as

Figure 4-23 An artist's rendition, based on images from NASA's Viking mission to Mars, of the Olympus Mons volcano. This enormous shield volcano is large enough to cover most of the northeastern section of the United States.

lunar *maria* ("seas"). As we saw in Chapter 3, early in its existence, the Moon was struck repeatedly by large meteorites that left deep craters and fractures in its crust. Basaltic lava flowed through those cracks, filling the craters and forming the maria. The Moon's surface also includes a few distinct shield volcanoes, such as those in the Marius Hills.

As much as 60% of Mars' surface is covered by volcanic rock derived from approximately 20 centers where volcanic activity was concentrated. Virtually all Martian volcanoes are shields of incredible size. Olympus Mons is approximately 23 kilometers (14 miles) high, more than twice the height of Mount Everest; its diameter is roughly equal to the width of the state of California (Fig. 4-23). The size of this and other Martian shields suggests that they have remained stationary over underlying hot spots, allowing enormous mountains of volcanic rock to accumulate. Either Mars' lithosphere does not move or it moves far more slowly than the Earth's plates do.

Venus also contains large shield volcanoes, some stretching in long chains along great faults in the surface. Radar measurements taken by NASA's Magellan satellite indicate that the volcanoes in the Maxwell Montes region are more than 11 kilometers (6.5 miles) tall, higher than Mount Everest. The area of the shield volcano Rhea Mons is large enough that it would cover all of New Mexico and much of adjacent parts of Colorado, Texas, and Arizona. Radar also indicates that molten lava lakes may still exist on the Venusian surface. A series of lava flows, in what appear to be rift valleys similar to those observed in East Africa, suggest a long sequence of multiple volcanic events on Venus. We have not yet determined the ages of these flows and whether they record plate activity on Venus.

One of Jupiter's moons, Io, and Neptune's largest moon, Triton, are believed to be the only other bodies in our solar system showing direct evidence of volcanic activity. NASA's Voyager and Galileo probes have detected eight volcanoes so far on Io, seven of which erupted in a recent four-month period (Fig. 4-24). Unlike the basaltic lavas produced by the volcanoes of the inner rocky planets, molten sulfur and enormous clouds of sulfurous gas erupt from Io's volcanoes; this material is propelled at speeds approaching 3200 kilometers (2000 miles) per hour to heights as great as 500 kilometers (300 miles) above the surface. Consequently, Io is noted for its yellow-red snowfalls of sulfur, lakes of molten sulfur, and huge multicolored lava flows of black, yellow, orange, red, and brown. Geologists believe that volcanism on Io may result from the frictional heat generated as its surface rises and falls in response to the enormous gravitational pull of Jupiter.

We have now examined igneous activity at relatively shallow depths and its impact on the Earth's surface. In later chapters, we explore the more deep-seated processes that generate such near-surface activity. In the next few chapters, however, we continue to look at the Earth's outermost layer and the variety of rocks that exist there. Chapter 5 reveals how these rocks are affected by conditions at the planet's surface.

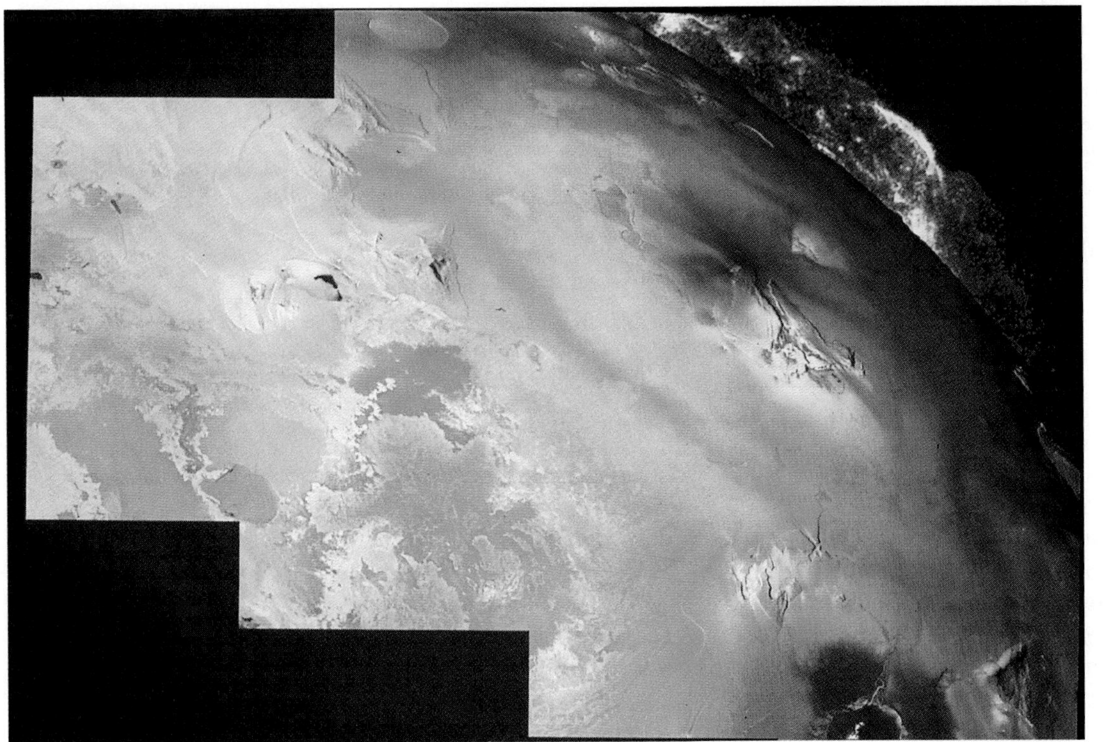

Figure 4-24 A computer-enhanced photo of a volcanic eruption on Jupiter's moon, Io, from data gathered by NASA's Voyager 1 probe.

Chapter Summary

Volcanism, the set of processes that results in extrusion of molten rock, begins with the creation of magma by the melting of preexisting rock and culminates with the ascent of this magma to the Earth's surface through fractures, faults, and other cracks in the lithosphere. **Volcanoes** are the landforms created when molten rock escapes from vents in the Earth's surface and then solidifies around these vents. Volcanoes may be active, dormant, or extinct.

Because of its high temperature and relatively low silica content, mafic magma has low viscosity (is highly fluid). It generally erupts (as basaltic lava) relatively quietly, or effusively, because its gases can readily escape and do not build up high pressure. Felsic magma, with its high silica content and relatively low temperature, is highly viscous and generally erupts (as rhyolitic lava) explosively.

The nonexplosive volcanic eruptions characteristic of basaltic lava produce lava flows that, when they solidify, are associated with distinctive features such as pahoehoe- and ʻaʻa-type surface textures, basaltic columns, lava tubes, and pillow structures. The explosive volcanic eruptions characteristic of rhyolitic lavas typically eject **pyroclastic** material—fragments of solidified lava and shattered preexisting rock ejected forcefully into the atmosphere. The various particles produced when lava cools and solidifies as it falls back to the surface are collectively called **tephra.** Explosive ejection of pyroclastic material is usually accompanied by a number of life-threatening effects, such as **pyroclastic flows,** or **nuée ardentes** (high-speed, ground-hugging avalanches of hot pyroclastic material), and **lahars** (volcanic mudflows).

Nearly all volcanoes have the same two major components: (1) a mountain, or **volcanic cone,** built up of the products of successive eruptions; and (2) a bowl-shaped depression, or **volcanic crater,** surrounding the volcano's vent. If enough lava erupts to empty a volcano's subterranean reservoir of magma, the cone's summit may collapse, forming a much larger depression, or **caldera.**

Effusive eruptions, which usually involve basaltic lava, form gently sloping, broad-based cones called **shield volcanoes.** Basaltic magma reaching the surface through long linear cracks, or fissures, in the Earth's crust spreads to produce nearly horizontal lava plateaus.

Explosive **pyroclastic eruptions** involve viscous, usually gas-rich magmas and so tend to produce great amounts of solid volcanic fragments rather than fluid lavas. Felsic (rhyolitic) lavas are often so viscous that they cannot flow out of a volcano's crater; they therefore cool and harden within their craters to form **volcanic domes.** Ash-flow eruptions occur in the absence of a volcanic cone; they are produced when extremely viscous, gas-rich magma rises to just below the surface bedrock, stretching and collapsing it.

The characteristic landform of pyroclastic eruptions is the **composite cone,** or **stratovolcano,** which is composed of alternating layers of pyroclastic deposits and solidified lava. Pyroclastic eruptions may also produce **pyroclastic cones** or **cinder cones,** created almost entirely from the accumulation of loose pyroclastic material around a vent. All pyroclastic-type volcanoes produce steep-sided cones, because the materials they eject—solid fragments and highly viscous lavas—do not flow far from the vent.

Various types of volcanic eruptions are associated with different plate tectonic settings. Explosive pyroclastic eruptions of felsic (rhyolitic) lava generally occur within continental areas characterized by plate rifting or atop intracontinental hot spots. Most intermediate (andesitic) eruptions take place near subducting oceanic plates. Effusive eruptions of (mafic) basalt generally occur at divergent plate margins and above oceanic intraplate hot spots.

Humans can minimize damage from volcanoes by zoning against development in the most hazardous areas, building lava dams, diverting the path of a flowing lava, and learning to predict eruptions accurately. Techniques used to predict eruptions include measuring changes in a volcano's slopes, recording related earthquake activity, and tracking changes in the volcano's external heat flow.

Volcanism is not restricted to the Earth. It has occurred in the past on the Moon, and relatively recent volcanic activity has been detected on Mars, Venus, and the moons of Jupiter and Neptune.

Key Terms

volcanoes (p. 61)
volcanism (p. 62)
pyroclastics (p. 66)
tephra (p. 66)
pyroclastic flow (p. 67)
nuée ardente (p. 67)
lahar (p. 67)
volcanic cone (p. 68)
volcanic crater (p. 68)
caldera (p. 68)
shield volcano (p. 68)
pyroclastic eruptions (p. 70)
volcanic dome (p. 71)
composite cone (p. 73)
stratovolcano (p. 73)
pyroclastic cones (p. 74)
cinder cones (p. 74)

Questions for Review

1. What criteria do geologists use to designate a volcano as active, dormant, or extinct?

2. Briefly compare basaltic, andesitic, and rhyolitic lava, in terms of their composition, viscosity, temperature, and eruptive behavior.

3. Within a single basaltic lava flow issuing from a Hawaiʻian volcano, why is pahoehoe lava found closer to the vent than ʻaʻa lava?

4. Contrast the nature and origin of nuée ardentes and lahars.

5. Describe the three basic types of effusive eruptions (central-vent, fissure, submarine) and the volcanic landforms associated with each.

6. How does a composite cone form? What type of lava is associated with a composite cone? How could you distinguish a composite cone from a pyroclastic cone?

7. What types of volcanoes and volcanic landforms are associated with subduction zones? With divergent plate boundaries?

8. Identify three sites in North America that pose a volcanic threat to nearby residents.

9. Describe three techniques that geologists use to predict volcanic eruptions.

10. Compare the volcanism on the Moon to volcanism on Io, Jupiter's moon.

For Further Thought

1. Look at the photograph below and speculate about the plate tectonic setting where this volcano is found; the composition of the rocks that make up the volcano; and whether eruptions of this volcano tend to be explosive or effusive.

2. The 1980 eruption of Mount St. Helens made a lot of headlines but had no discernible effect on global climate. Conversely, the eruption of the Philippines' Mount Pinatubo in 1991 caused a 1°C drop in global temperature. What differences between these eruptions might explain why one had a sharp effect on climate, whereas the other did not?

3. Why do we find andesite throughout the islands of Japan, but not throughout the islands of Hawai'i?

4. How would you explain the origin of a volcanic structure composed of 10,000 meters of pillow lava covered by 3000 meters of basalt containing vesicles, basaltic columns, and 'a'a and pahoehoe structures?

5. Under what circumstances might active volcanism resume along the east coast of North America? Within the Great Lakes region of North America?

5

Weathering: The Breakdown of Rocks

Much of the Earth's most spectacular and unique scenery has been created by quite ordinary environmental factors acting upon rock. These factors, which act constantly and everywhere on Earth, remove or alter individual mineral grains in rock, yielding end products that look much different from the original rocks. The processes by which environmental agents at or near the Earth's surface cause rocks and minerals to break down is called **weathering.** Weathering is a slow but potent force to which even the hardest rocks are susceptible (Fig. 5-1).

Rocks that have been weakened by weathering are more vulnerable to **erosion,** the process by which gravity, moving water, wind, or ice transports pieces of rock and deposits them elsewhere. The fragments removed from rock by erosion have usually been loosened by weathering, although a high-energy erosive event, such as a flood, may dislodge even unweathered rock. Like many geological processes, weathering and erosion are interrelated and often work in tandem. Together, they produce *sediment,* the loose, fragmented surface material that serves as the raw material for sedimentary rock (discussed in Chapter 6).

Weathering plays a vital role in our daily lives, with both positive and negative outcomes. It frees life-sustaining minerals and elements from solid rock, allowing them to become incorporated into our soils and finally into our foods. Indeed, we would have very little food without weathering, as this process produces the very soil in which much of our food is grown. But weathering can also wreak havoc on the structures we build. Countless monuments—from the pyramids of Egypt to ordinary tombstones—have suffered drastic deterioration from freezing water, hot sunshine, and other climatic forces.

Figure 5-1 The beauty of weathering. Monument Valley in northeastern Arizona shows the results of millions of years of exposure to the Earth's environment. The rocks that were removed to produce these landforms were less resistant to weathering and erosion than are the hardy rocks that remain.

Weathering Processes

Rocks can be weathered in two ways. **Mechanical weathering** breaks a mineral or rock into smaller pieces (*disintegrates* it) but does not change its chemical makeup.

Changes induced by mechanical weathering affect physical characteristics, such as the size and shape of the weathered structure. **Chemical weathering** changes the chemical composition of minerals and rocks that are unstable at the Earth's surface (*decomposes* them), converting them into more stable substances; minerals and rocks that are chemically stable at the Earth's surface are resistant to chemical weathering.

Mechanical and chemical weathering take place constantly and simultaneously in most environments. Mechanical weathering renders rocks more susceptible to chemical weathering by making more surface area available for chemical attack, much as crushing a sugar cube with a spoon causes it to dissolve more rapidly in hot water. By mechanically disintegrating the sugar cube, you vastly increase its surface area; crystal surfaces previously hidden inside the larger cube therefore become exposed to the hot water. Similarly, the area of a boulder that is vulnerable to weathering agents consists only of its outer surface until mechanical weathering increases its surface area (Fig. 5-2).

Mechanical Weathering

A number of natural processes reduce rocks to smaller sizes without changing their chemical makeup. In any given location, several of these processes may be working at the same time.

Frost Wedging Water expands in volume by about 9% when it turns to ice. When water enters pores or cracks in a rock and then the temperature falls below 0°C (32°F, the freezing point of water), the force of the expanding ice greatly exceeds that needed to fracture even solid granite. As a result, the cracks become enlarged, often loosening or dislodging fragments of rock (Fig. 5-3). This enlargement allows even more water to enter the crack, and the process repeats. This process, called **frost wedging,** is the fastest, most effective type of mechanical weathering. It is most active in environments where surface water is abundant and temperatures often fluctuate around the freezing point of water.

Crystal Growth Along rocky shores, where salty sea spray soaks into rocks and evaporates, it leaves behind growing salt crystals within cavities in the rocks. The growth of the crystals applies great pressure to the walls of the cavities, prying them farther apart. This process has significantly damaged not only rocks, but also some well-known stone structures. Among them is Cleopatra's Needle, a granite obelisk that stood for more than 3000 years in Egypt without much loss of the finely engraved hieroglyphics on its sides; when the obelisk was relocated to New York City, salt crystals that had

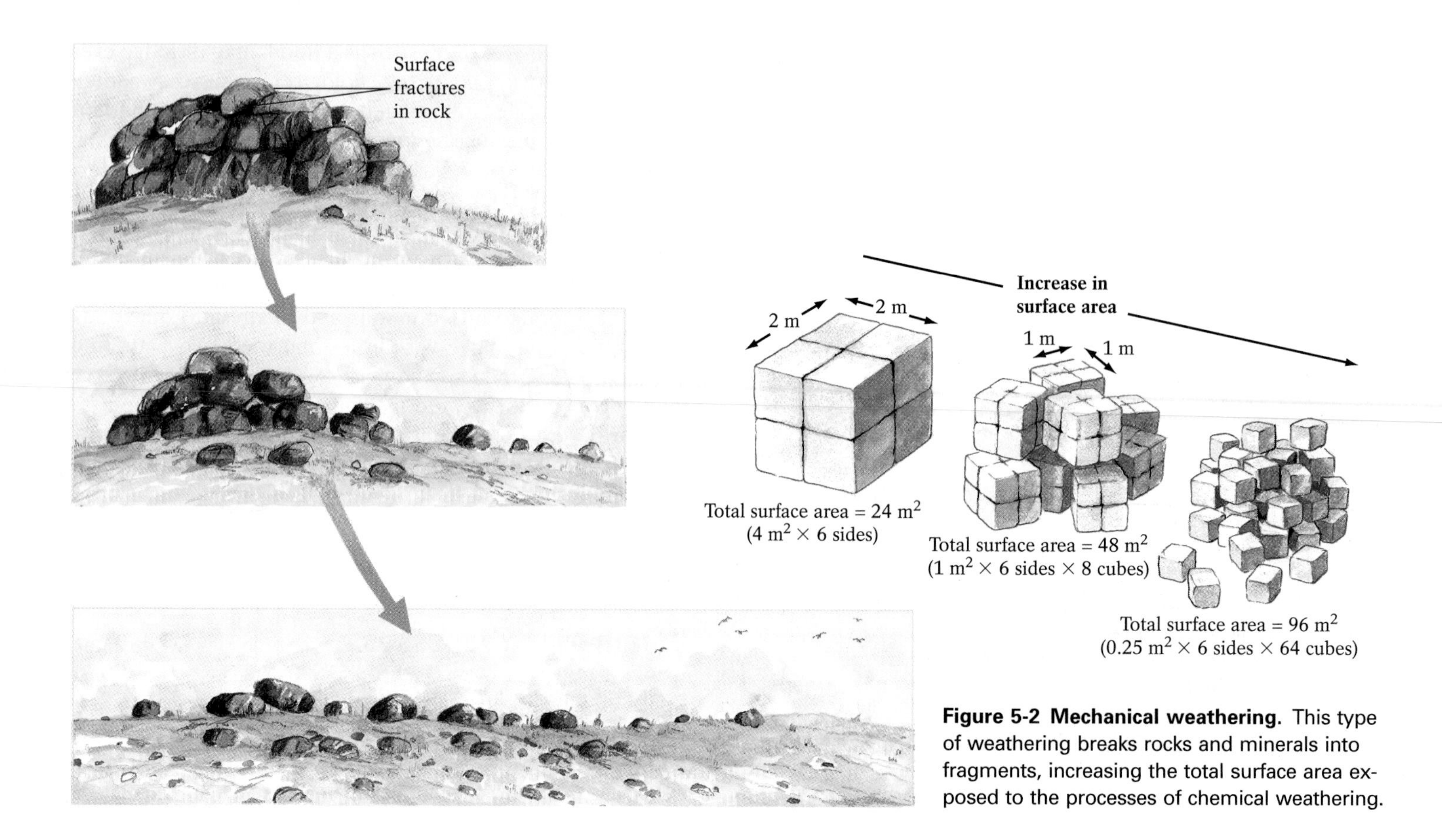

Figure 5-2 Mechanical weathering. This type of weathering breaks rocks and minerals into fragments, increasing the total surface area exposed to the processes of chemical weathering.

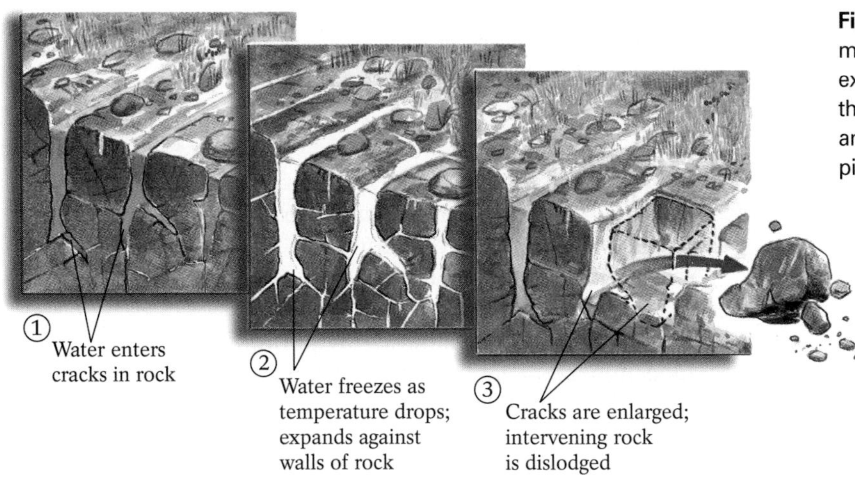

Figure 5-3 Frost wedging. In this type of mechanical weathering, water freezes and expands within cracks in rock, enlarging them. The cracks get bigger with each frost and may eventually dislodge intervening pieces of rock.

already infiltrated small cracks in the needle absorbed water from the humid air and expanded, accelerating its weathering (Fig. 5-4).

Thermal Expansion and Contraction If you have ever sat around a campfire, you may have noticed that thin layers of rock material tend to fall off the cracked surfaces of rocks close to the fire. This deterioration occurs because of **thermal expansion,** the enlargement of a mineral's crystal structure in response to heat. Each of the different minerals within a rock expands to a different degree when heated. For example, quartz grains expand about three times as much as grains of plagioclase feldspar that are exposed to the same heat. If a rock that contains both minerals becomes heated, the expanding quartz grains will push against neighboring feldspar grains, loosening and eventually dislodging them. Rocks, however, are poor conductors of heat. Heat does not penetrate very deeply or very quickly into the rock; instead, the heated outer portion of a rock tends to break away from the cool inner portion. Although

Figure 5-4 Cleopatra's Needle. The structure as it looked for about 3500 years in Egypt **(a),** and how it looks today, after a century of weathering in New York City's Central Park **(b).**

thermal expansion and contraction generally works extremely slowly, in regions such as deserts, where rocks are exposed to high daytime and low nighttime temperatures, it may cause a significant amount of weathering over long periods of time.

Mechanical Exfoliation When erosion of overlying rock or soil exposes a wide area of a large plutonic mass, pressure on the mass is reduced and it expands; because most of the structure remains underground, where surrounding rock continues to exert pressure on it, it expands upward. As the rock expands, it fractures into sheets parallel to its exposed surface. These sheets may then break loose and fall from the sloping surface of the structure, a weathering process known as **mechanical exfoliation.** Many mountains exhibit a dramatic "stepped" appearance because large slabs of rock, several meters in thickness, have exfoliated from underlying rocks (Fig. 5-5).

Other Mechanical Weathering Processes Almost all rocks contain some cracks and crevices. Plants and trees may take root in such cracks in surface rocks. Although rock would seem strong enough to withstand it, the force applied to a crack by a growing tree root is surprisingly powerful and quite capable of enlarging the crack. The buckled and broken sidewalks of some tree-lined boulevards convincingly demonstrate the weathering ability of such root growth.

Animals also contribute to mechanical weathering. For example, larger animals may step on stones or pebbles and crush them into smaller particles. Some birds ingest stones, which are used as grinding tools in their digestive processes and eventually released when worn down in size. The cumulative activities of even very small creatures, such as earthworms and insects, can also help break rocks into smaller fragments.

Mechanical weathering by **abrasion** occurs when rocks and minerals collide during transport or when loose transported material scrapes across the exposed surfaces of stationary rock. For example, sediment fragments carried in swirling streams or by gusty winds collide and grind against both each other and other rock surfaces, breaking into smaller particles. Similarly, rocks carried along at the base of a glacier scrape against underlying rocks and are ground to even smaller sizes.

Chemical Weathering

Chemical weathering alters the composition of minerals and rocks, principally through reactions involving water. Water is the single most important factor controlling the rate of

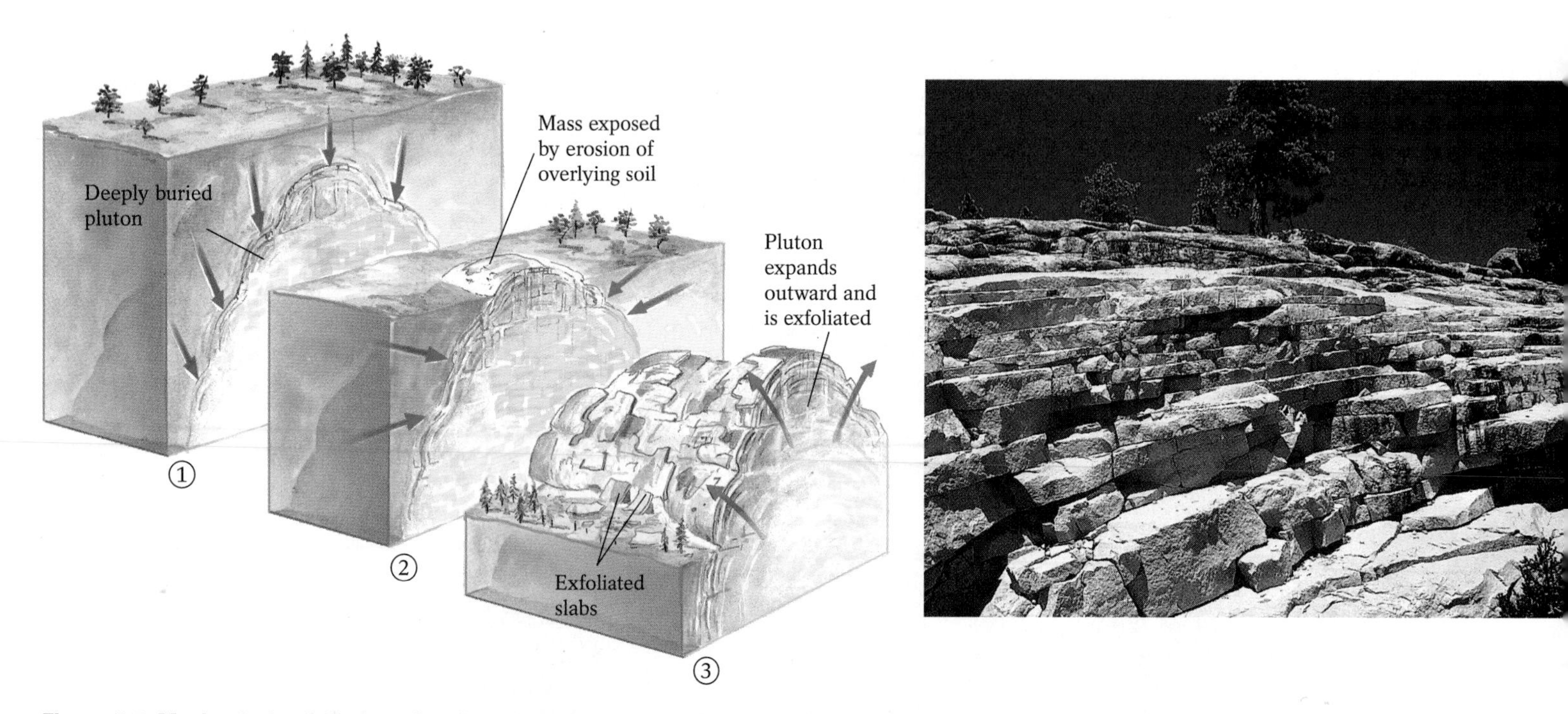

Figure 5-5 Mechanical exfoliation of a plutonic igneous mass. Removal of overlying rocks allows the mass to expand upward, fracturing into thin slabs of rock parallel to its exposed surface. Photo: The "steps" of this mountain in California's Yosemite National Park were produced by mechanical exfoliation of the mountain's granitic rock.

chemical weathering, because it carries ions to the reaction site, participates in the reaction, and then carries away the products of the reaction. Rocks weather chemically by three major processes: dissolution, oxidation, and hydrolysis.

Dissolution In **dissolution,** ions or ion groups from a mineral or rock are removed and carried away by water. As we learned in Chapter 2, water molecules can dissolve minerals by attracting and removing charged ions from the surfaces of the minerals. For example, dissolution of halite (NaCl) occurs when the positive sides of water molecules attract and dislodge chloride ions (Cl^-) and the negative sides of water molecules attract and dislodge sodium ions (Na^+).

Some mineral deposits, such as those containing gypsum or halite, are readily dissolved by water. In many cases, however, water itself does not remove ions from a mineral or rock, but rather reacts with another compound in the environment to form a substance—usually an acid—that does. This reaction most commonly involves the reaction of water with carbon dioxide (CO_2) to form carbonic acid (H_2CO_3). Carbonic acid, in turn, reacts with and dissolves limestone. Most of the world's caves were created by this process (Fig. 5-6).

Ions that have been removed from a rock or mineral by water and then carried away in the water are said to be *in solution*. When the water eventually evaporates, these substances remain behind and are said to have *precipitated* from the solution. As we will see in Chapter 6, such precipitated materials represent an important kind of sediment and produce many sedimentary rocks.

Oxidation In **oxidation,** a mineral's ions combine with oxygen ions. For example, when iron ions (Fe^{2+}) in mafic rocks bond with oxygen in the atmosphere, they form the iron oxide Fe_2O_3 (hematite). Other iron oxides, such as yellow-brown goethite and yellowish limonite, form whenever iron-rich minerals and rocks come into contact with water. Iron oxides, commonly called *rust,* may stain the surfaces of minerals that have iron contents, such as olivine, pyroxene, and amphibole, as well as the rocks containing these minerals, such as basalt, gabbro, and peridotite. Rust is very stable at the Earth's surface. Indeed, anyone who owns an older car is well aware that rust does not dissolve and wash off in the local car wash.

Hydrolysis In **hydrolysis,** H^+ or OH^- ions from water molecules displace other ions from a mineral's structure. Aluminum-rich silicates such as the feldspars, the most abundant minerals in the Earth's crust, are weathered primarily by hydrolysis. In a typical hydrolysis reaction, the potassium ions in potassium feldspar ($KAlSi_3O_8$) are replaced with water's hydrogen ions, forming the stable clay mineral kaolinite ($Al_2Si_2O_5(OH)_4$). The clays formed during hydrolysis represent an important component of many soils and of oceanic mud.

Figure 5-6 Weathered and unweathered limestone boulders. In humid environments, such as those in most southeastern states, limestone dissolves extensively along cracks and crevices, leaving behind thousands of cavities that range from minor surface depressions to large underground cave systems. In this photo, the limestone boulder above the hammer is weathered; the one below is not.

Factors That Influence Chemical Weathering

The effectiveness of water as a chemical weathering agent may be enhanced by local climate, the activity of living organisms, and the amount of time that the mineral or rock remains exposed to weathering. The weathering rate for any particular mineral or rock also depends upon its own chemical stability or that of its components.

Climate We have already seen the importance of water in all three types of chemical weathering. Another key factor that accelerates virtually all chemical reactions is heat. In general, the rate of chemical reaction doubles with every 10°C (18°F) increase in temperature.

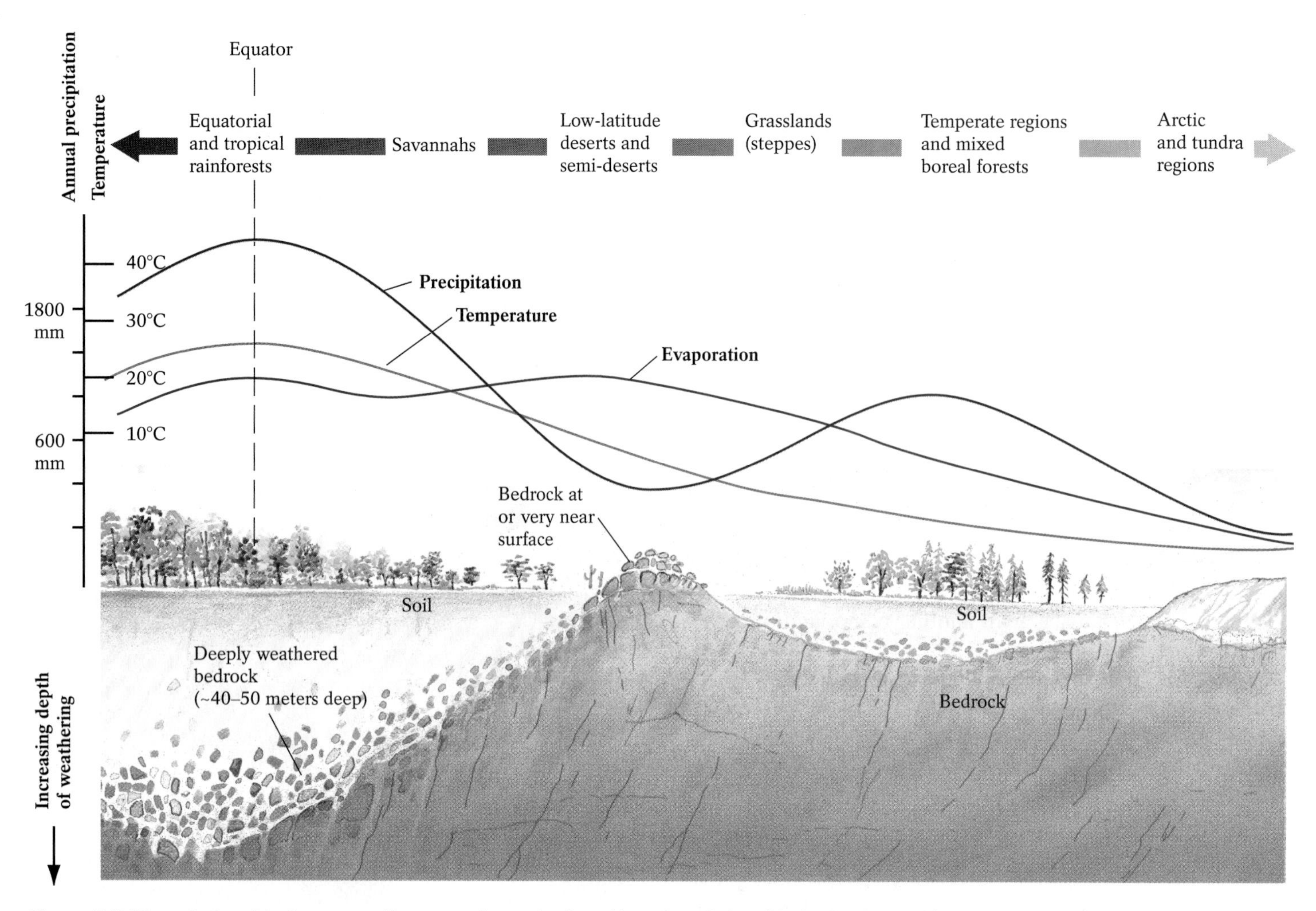

Figure 5-7 The relationship between climate and weathering. Here the relationship is shown as a function of the depth to which a region's bedrock (the solid rock immediately underlying the surface soil) is typically weathered. The deepest weathering occurs in the warm, moist tropical zone near the equator, where temperature and precipitation are greatest. Weathering is minimal in the arctic and in the deserts, where water is in short supply as an agent of chemical weathering.

Because water and heat combine to accelerate the deterioration of rocks, chemical weathering occurs more readily in warm, moist climates, such as Puerto Rico, than in arid, cold climates, such as Antarctica (Fig. 5-7). Warm, moist climates also promote the growth of vegetation and other living organisms, which themselves act as weathering agents.

Living Organisms Organisms—plants, animals, even bacteria—can significantly affect the rate of weathering of rocks and minerals. Burrowing animals such as groundhogs, prairie dogs, ants, and especially earthworms commonly transport unweathered materials from below ground to the surface, where they can be weathered (Fig. 5-8). The burrows of such animals, as well as channels left by the roots of plants, are also instrumental in weathering, as they allow the circulation of air and water in the soil. Finally, when plants and animals die, their decomposition by bacteria and fungi releases organic acids that aid chemical weathering.

Time The longer a rock or mineral remains exposed to a weathering environment, the more it will decompose. In North America's Great Lakes region, for example, some hills of sediment

Figure 5-8 Animal activities as a weathering tool. Earthworms churn up soil as they move through it, bringing underground minerals to the surface to be exposed to weathering.

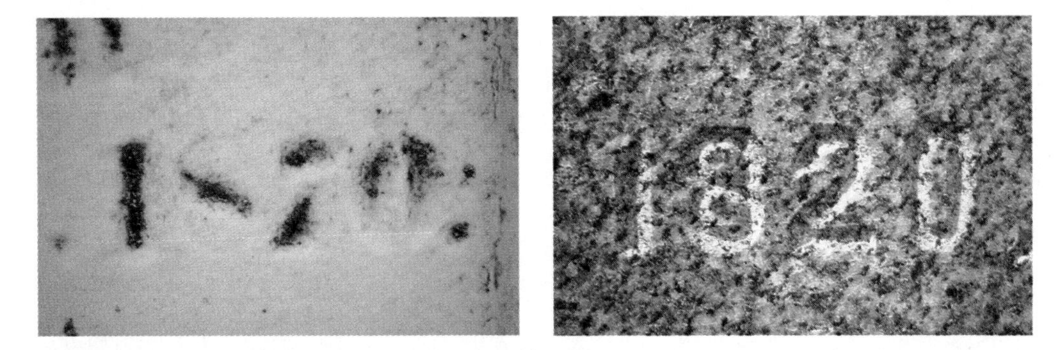

Figure 5-9 Two gravestone inscriptions from the same cemetery in Williamstown, Massachusetts, showing differential weathering of rock types. The marble gravestone (left), though exposed to the same climate as the granite gravestone (right) and erected 50 years *later,* has suffered noticeably more weathering damage. Marble, composed predominantly of the chemically reactive mineral calcite, is much more susceptible to chemical weathering than is granite, a rock composed of very stable minerals such as quartz and orthoclase.

left behind by ice-age glaciers contain pebbles so thoroughly weathered that they can be rubbed between one's fingers into balls of soft clay; other pebbles of the same composition within this same region are still completely fresh and solid, having been exposed to weathering for a much shorter time.

Mineral Composition The effect of chemical weathering on rock is determined largely by the *stability* (resistance to chemical change) of the rock's component minerals (Fig. 5-9). As a general rule, the greater the difference between the conditions under which a mineral crystallized and conditions at the Earth's surface, the greater will be the mineral's susceptibility to chemical weathering. Thus minerals such as olivine and pyroxene, which crystallize at high temperatures and pressures, offer far less resistance to chemical weathering than low-temperature minerals such as quartz and mica (Fig. 5-10).

Products of Chemical Weathering

After dissolution, oxidation, and hydrolysis remove the soluble ("dissolvable") components in rocks, a combination of relatively insoluble, unreactive, and stable new products of chemical weathering remain. These materials include the clay minerals and several economically valuable metal ores. In addition, chemically weathered surface rocks are typically found in the form of rounded boulders.

Clay Minerals Several varieties of clay minerals exist, most resulting from the hydrolysis of feldspars. Each clay mineral has a distinctive chemical composition and set of physical properties that determine its practical uses; the type produced from a given rock depends on the climatic conditions under which the hydrolysis occurs. For example, kaolinite, the active ingredient in Kaopectate and the clay used to manufacture fine porcelain, is commonly produced by hydrolysis of feldspars in a warm, humid climate. To produce kaolinite, H^+ ions from water must completely displace the large positive ions (K^+, Na^+, Ca^{2+}) in the parent feldspars. Smectite, a highly absorbent clay that is a key ingredient in cat litter and that has utility in filtering impurities from beer and wine, forms from partial hydrolysis of micas and amphiboles in drier, cooler climates. Other types of clays are used as agricultural fertilizers, lubricants on oil-drilling rigs, in the manufacture of bricks and cement, and in paper production. For instance, the slick finish of the paper on which this book is printed is a clay product.

Metal Ores *Ores* are rock or soil units that contain one or more minerals that have economic value and can be profitably extracted. Most economically valuable minerals—usually metals—rarely occur in pure form. Aluminum, for example, is quite abundant in the Earth's crust, though it is

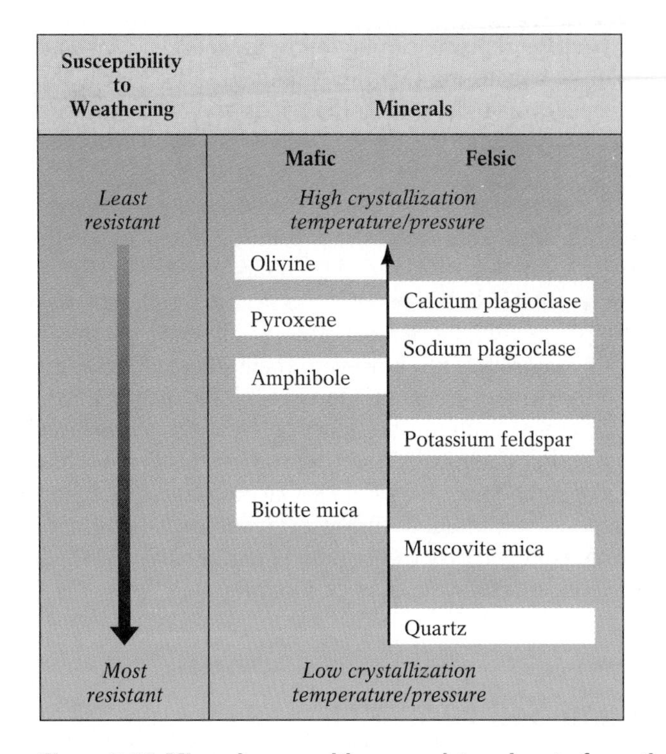

Figure 5-10 Mineral composition as a determinant of weathering resistance. Mafic igneous rocks, which crystallize at high temperatures, tend to weather fairly rapidly in the Earth's relatively cooler surface environment. Felsic rocks, which crystallize at lower temperatures—closer to those at the Earth's surface—tend to weather more slowly.

Figure 5-11 Spheroidal weathering. Rounded boulders, such as these in Australia, evolve from angular ones as the rocks' protruding corners and edges are gradually eroded by chemical and mechanical weathering.

generally widely dispersed in clays, feldspars, and micas. Intense chemical weathering of feldspar-rich rocks in hot, moist climates produces the valuable aluminum ore, or bauxite, which forms when soluble minerals leach away, leaving behind a relatively higher concentration of insoluble aluminum. Ancient bauxite deposits can be found in Georgia, Alabama, and Arkansas, suggesting that the climate in those places was once warmer and more moist than it is today. Weathering processes also produce or enrich much of the world's minable iron, copper, manganese, nickel, and even silver deposits.

Rounded Boulders One interesting result of chemical weathering is the formation of rounded boulders through a process called **spheroidal weathering.** Mechanically weathered and broken rocks are initially quite angular; in an active chemical-weathering environment, however, their jagged edges are quickly transformed. Because the corners and edges of angular rock masses offer more surface area, they are more likely to weather chemically and to do so at a faster rate than the rocks' planar faces (Fig. 5-11). Once attained, a spheroidal shape is maintained through the uniform action of weathering agents across the boulder's entire surface, which weathers concentric layers of rock in turn.

Soils and Soil Formation

Mechanical and chemical weathering of both sediment and the solid rock beneath the sediment, known as *bedrock,* commonly produce **regolith** (pronounced "REG-uh-lith"; meaning "rock blanket"), the fragmented material covering much of the Earth's land surface (Fig. 5-12). Geologists refer to the upper few meters of regolith, which contain both mineral and organic material, as **soil.** Soils are the most common product of weathering on land (oceanic mud is more plentiful) and may well be the most valuable natural resources (besides clean water). People everywhere depend on fertile soils to support plant growth and provide life-sustaining nutrients in foods (Highlight 5-1).

Soils differ depending on the weathering processes, local climatic conditions, and preexisting materials that produced them. As a result, they are complex and often difficult to classify. In this section, we discuss the factors that interact to form soils and then examine the structure of a typical soil. Finally, we briefly consider the various types of soils.

Influences on Soil Formation

Many of the factors that affect the rate of weathering consequently help determine the nature of the soil produced through the weathering process. Five factors that are particularly important are parent material, climate, topography, vegetation, and time.

Parent Material A soil's **parent material** consists of the bedrock or sediment from which the soil develops. The parent material's mineral content determines both the nutrient richness of the resulting soil and the amount of soil produced (Fig. 5-13). The relationship between parent material and soil development is readily evident on Java and Borneo, two neighboring Indonesian islands with the same climate. The parent materials for Java's soils are largely fresh, nutrient-rich volcanic ash deposits; in contrast, Borneo's soils are derived from numerous granitic batholiths, gabbroic intrusions, and andesite flows. Ongoing volcanism on Java replenishes

Figure 5-12 A roadcut in Sarawak, Borneo, showing layers of regolith (weathered surface sediment and bedrock). Extreme chemical weathering can produce regolith that is more than 60 meters (200 feet) deep.

nutrients such as potassium, magnesium, and calcium in its soils, whereas the soils of Borneo, which lacks fresh ash deposits, are depleted of nutrients. Not surprisingly the islands' population densities reflect their dramatically different soil fertility and agricultural productivity: Java supports approximately 460 inhabitants per square kilometer compared with Borneo's sparse 2 inhabitants per square kilometer.

Climate An area's climate—the amount of precipitation it receives and its prevailing temperature—controls the rate of chemical weathering and consequently the rate of soil formation. Climate also regulates the growth of vegetation and the abundance of microorganisms that contribute CO_2 and O_2 to the processes of dissolution, hydrolysis, and oxidation. Chemical weathering and soil formation take place most rapidly in warm, moist climates and most slowly in cold or dry climates.

Topography **Topography** refers to the physical features of a landscape, such as mountains and valleys, the steepness of slopes, and the shapes of landforms. It influences the availability of water and other weathering factors, and thereby the rate of soil accumulation. For example, steep slopes allow rainfall and snowmelt to flow away swiftly, leaving

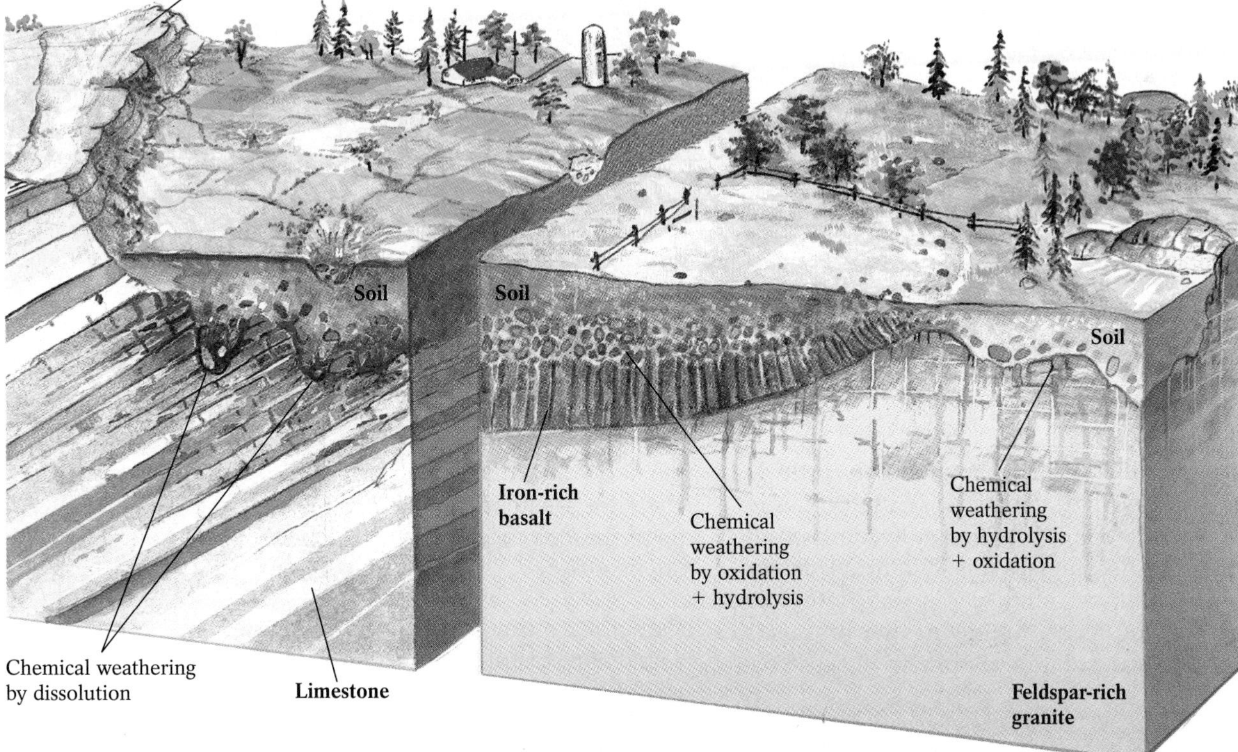

Figure 5-13 The effect of bedrock composition on soil development in a moist, temperate environment. Advanced chemical weathering of iron-rich basalts (oxidation), fractured limestone (dissolution), and feldspar-rich granite (hydrolysis) produces thick soils on these types of rock. Sandstone, however, tends to remain relatively unweathered, because it consists primarily of weathering-resistant quartz grains; as a result, any soil produced on sandstone will be sparse.

Highlight 5-1 Conserving Our Fragile Soils

The United States is not a post-agricultural society. All of us are completely dependent on agriculture to provide the food we need to sustain our lives. Agriculture, in turn, depends on soil, and especially the uppermost, most fertile layer, known as *topsoil*. Yet few of us view the tremendous loss of agricultural topsoil through erosion—about 6 billion tons annually in the United States and many times that amount worldwide—as a major environmental problem (Fig. 1a).

Although soil erosion is a natural process that occurs to all soils, the rate of erosion actually determines the future of agricultural productivity. If topsoil is lost to erosion faster than its nutrients can be replaced by natural weathering processes, a soil's fertility—its capacity to sustain crops—drops dramatically. When land is covered by vegetation, erosion is a relatively slow process. Many of the activities of civilization—urbanization, logging, ranching—remove vegetative cover, however, exposing the land to much more rapid rates of erosion. Likewise, most methods of agriculture operate by clearing the land of all vegetation and plowing the stubble into the ground, which leaves the bare land susceptible to increased rates of erosion by water and wind.

Fortunately, many methods are available to protect valuable topsoil—some of them traditional farming methods that have been in use for hundreds of years. No-tillage farming or minimum-tillage farming uses methods that disturb the soil as little as possible and leave previous crop residue on top of the ground to retain soil (Fig. 1b). For example, specially designed tillers may prepare the subsoil without turning over the topsoil. Other machines simultaneously inject fertilizers, herbicides, and seeds into the soil, disturbing neither the groundcover nor the soil. Although economics, both short-term and long-term, is generally the major factor governing their acceptance, these and other soil conservation methods—such as terracing (converting a slope into a series of steplike platforms), contour farming (planting crops in rows that run perpendicular to the slope of the land), and strip cropping (alternating multi-row strips of two or more crops within an agricultural field)—are becoming more widespread throughout the United States and other parts of the world.

(a)

(b)

Figure 1 The fight against soil erosion. (a) Agricultural soil being eroded by surface water near Moscow, Idaho. **(b)** One weapon against soil erosion, no-tillage farming, minimizes topsoil disruption when preparing the soil for planting.

little water to penetrate the surface. Consequently, little or no soil develops as a result of chemical weathering. Any soil that does form on steep slopes is usually transported downslope before it can accumulate to a significant depth. Conversely, water accumulates and readily infiltrates the ground in level, low-lying areas, improving the prospects for chemical weathering and soil development in those regions (Fig. 5-14).

Vegetation Vegetation produces much of the O_2, CO_2, and H^+ ions involved in chemical weathering reactions and contributes organic matter to soils when it dies. Soils developing on prairie grasslands, for example, receive large quantities of organic matter from surface plant remains and from the decay of extensive subsurface root systems. The increased chemical weathering and organic material associated with plant life enhance both the thickness and fertility of vegetated soils.

Time If all other factors affecting soil formation were equal, a landscape exposed to weathering over a long period of time would have a thicker and more well-developed soil than a younger landscape. Because other factors almost always vary considerably, however, a thicker soil is not necessarily an older one. A thick, fertile soil may develop within a few hundred years in a warm, wet environment, whereas thousands or hundreds of thousands of years may pass before any soil forms in an arid or polar region.

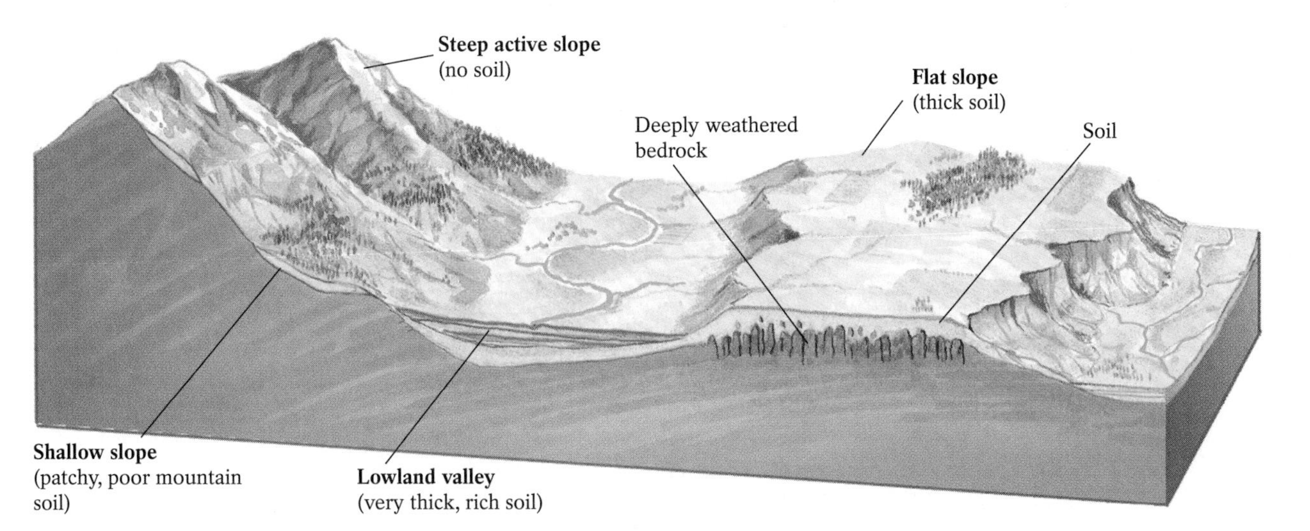

Figure 5-14 The effect of landscape on soil development. Soils are generally thin or nonexistent on steep slopes, because the water required for chemical weathering runs off such slopes and because any soil that does accumulate would wash away downhill. Soils tend to be thickest in lowland valleys, where water and loose material transported from upland come to rest.

Typical Soil Structure

The greatest effects of weathering are seen at a soil's surface, which is directly exposed to the weathering environment. Below the surface, distinct weathering zones, called **soil horizons,** develop as infiltrating water dissolves substances from the upper layers and redeposits them in the lower soil layers. Among the most common layers are the O, A, E, B, and C horizons, which are found virtually everywhere within temperate zones; other horizons are also found in more specialized environments. The vertical succession of soil horizons in a given location, known as the **soil profile** (Fig. 5-15), reflects

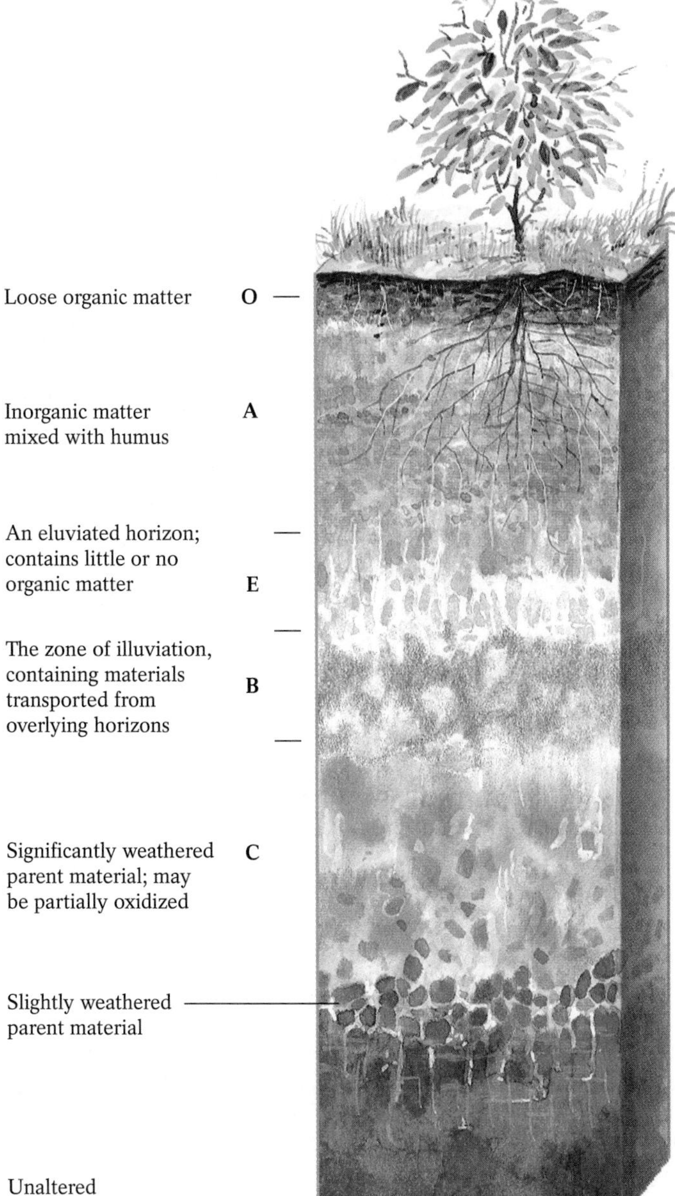

Figure 5-15 Profile of a typical mature soil. The features of a typical temperate-zone soil profile, including the O, A, B, and C horizons. Photo: A vertical succession of soil horizons is clearly visible even in this small section of soil in Australia.

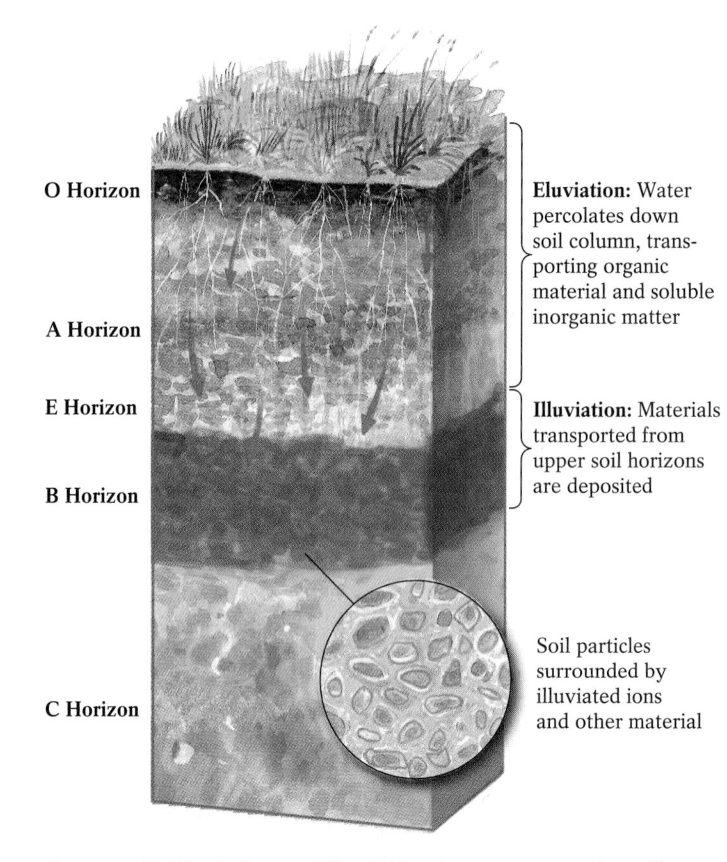

Figure 5-16 Eluviation and illuviation in a mature soil profile. Material is removed, or eluviated, as water passes through the O, A, and E horizons; it is deposited, or illuviated, as water infiltrates the B horizon.

the local soil-forming conditions. Different localities vary in the types of horizons and their depth and degree of development; they therefore exhibit different soil profiles.

The upper portion of a soil profile consists of the O, A, and E horizons. The O ("organic") horizon, which is most distinct in temperate regions such as North America, consists mainly of organic matter, both dead (such as plant fibers) and living (bacteria, fungi, algae, and insects). The A horizon consists mainly of inorganic mineral matter mixed with *humus,* a dark-colored, carbon-rich substance derived from decomposed organic material from the O horizon. The E horizon is a light-colored zone below the A horizon with little or no organic matter. Its light color results from the action of eluviation, the dissolution and removal of minerals—in this case, iron and aluminum—by water that percolates downward through the soil.

The lower part of the soil consists of the B and C horizons—beneath them lies unaltered bedrock. The B horizon contains the materials added by *illuviation,* the process by which materials dissolved or transported mechanically from the upper horizons are deposited in the lower horizons (Fig. 5-16). In arid or semi-arid areas, where surface water evaporates quickly, a distinct carbonate-rich horizon appears within or below the B horizon at the depth to which annual rainfall penetrates. This *caliche* (pronounced "ka-LEE-chee") layer forms when brief heavy rains dissolve calcium carbonate in the upper layers of soil and transport it downward; as the water rapidly evaporates, the carbonate precipitates, forming a markedly white layer in the soil. The C horizon, the lowest zone of significant weathering, consists of parent material that has been partially weathered but still retains most of its original appearance.

Classifying Soils

The complex interaction of parent material, climate, topography, vegetation, and time creates a variety of distinctly different physical and chemical characteristics in soils. Accurate classification of the different types of soils is necessary to ensure that a particular soil is used for the appropriate purpose. Not every soil is suitable as the location of a landfill, building, or septic system, and not every soil can support agricultural cultivation without extensive use of additives or other modifications.

Our current soil classification system, called **soil taxonomy,** names each soil according to its obvious physical characteristics, describes its texture, and indicates its degree of nutrient depletion. Also taken into account in the classification are moisture content, mean annual air temperature, degree of horizon development, soil chemistry, organic matter content, and even the origin and relative age of the soil. The first step in classifying a soil is to assign it to a specific soil order (Table 5-1). Next, each order is divided and subdivided several times, until it includes literally thousands of different soil names. As a result of these fine demarcations, in a detailed soil map of an area, a single farm field may include two or more different soil types.

Weathering in Extraterrestrial Environments

Recent discoveries about surface conditions on the Moon, Venus, and Mars have confirmed that Earth-style weathering does not occur on our celestial neighbors. The Moon, however, does experience a form of mechanical weathering resulting from the impact of meteorites and micrometeorites; because it lacks an atmosphere, water, or biological activity, it has no chemical weathering. As a result, footprints left in the lunar dust by the Apollo 11 astronauts will likely retain their freshness for millions of years (Fig. 5-17).

Venus has a surface temperature of about 457°C (900°F) and an atmosphere composed almost exclusively of CO_2, which traps heat radiating from the planet's surface in an exaggerated "greenhouse effect." Although high temperatures normally promote an increased rate of chemical

Table 5-1 Classification of Soils

Soil Order	General Properties	Typical Geologic or Geographic Setting
Andisols	Soils develop principally on fresh volcanic ash; high fertility resulting from high weathering rates of glass shards	Young surfaces, in volcanic terranes
Entisols	Minimal development of soil horizons; first appearance of O and A horizons; some dissolved salt in subsurface	Young, newly exposed surfaces, such as new flood or landslide deposits, fresh volcanic ash, or recent lava flow; also found in very cold and very dry climates, or wherever bedrock strongly resists weathering
Inceptisols	Well-developed A horizon; weak development of B horizon, which still lacks clay enrichment; some evidence of oxidation in B horizon; little evidence of eluviation or illuviation	Relatively young surfaces; cold climates where chemical weathering is minimal, or on very young volcanic ash in tropics, in resistant bedrock, and on very steep slopes
Mollisols	Thick, dark, highly organic A horizon; B horizon may be enriched with clays; first appearance of E horizon	Semi-arid regions; generally grass-covered areas having adequate moisture to support grasses but not to cause significant dissolution of soluble materials in upper horizons
Alfisols	Relatively thin A horizon overlying clay-rich B horizon; strongly developed E horizon	Many climates, although most common in forested, moist environments
Spodosols	Much eluviation of A and E horizons, leaving a light-colored, grayish topsoil; aluminum/iron-enriched B horizon stained by dissolved organic material	Moist climates, usually on sandy parent materials (which allow water to infiltrate readily); grasses or trees may provide the organic matter
Aridisols	Thin A horizon with little organic matter overlying thin B horizon with some clay enrichment; caliche layer present in B or C horizons	Arid lands with sparse plant growth
Histosols	Wet, organic-rich soil dominated by thick O and A horizons	Found where production of organic matter exceeds addition of mineral matter, generally where surface is continuously water-saturated; often found in coastal environments
Vertisols	Very high clay content; soil shrinks (upon drying) and swells (upon wetting) with moisture variations	Equatorial and tropical areas with pronounced wet and dry seasons
Oxisols	Shows extensive weathering; highly oxidized B horizon is deep red from layer of oxidized iron	Generally older landscapes in moist climates having tropical rainforests
Ultisols	Shows extensive weathering; highly weathered clay-rich B horizons with high concentrations of aluminum	Very moist, lushly forested climates, often subtropical and tropical

Figure 5-17 A footprint left on the lunar surface by one of the Apollo 11 astronauts in July 1969. Because no chemical weathering takes place on the Moon, this footprint will remain for millennia.

reaction, the temperature of Venus is so high that it instantly evaporates every trace of water at the planet's surface, virtually precluding chemical weathering. Mechanical weathering on Venus occurs in the form of thermal expansion and contraction, exfoliation, and abrasion by wind-blown particles.

The thin Martian atmosphere consists largely of CO_2, with small amounts of nitrogen and water vapor being present as well. Because of cool temperatures at the planet's surface, ranging from −108°C to 18°C (−225°F to 63°F), most surface water exists as ice and is generally unavailable for

Figure 5-18 An unenhanced color photo of the surface of Mars, taken by the Sojourner lander of the Pathfinder probe in July 1997. The redness of the Martian regolith most likely derives from oxidation of iron-rich rocks and sediments.

chemical reactions. There is, however, evidence of chemical weathering in Mars' past. The well-known red color of Mars is believed to derive from reddish iron oxides produced by past oxidation of iron-rich bedrock (Fig. 5-18).

Chapter Summary

Weathering is the slow but constant process whereby rocks are gradually broken down by environmental factors at the Earth's surface. **Erosion** occurs when rock fragments are transported and deposited elsewhere by gravity, moving water, wind, or ice. The products of weathering and erosion contribute to the Earth's sediment—the loose, fragmented geologic material that serves as the raw material for sedimentary rock.

Two types of weathering exist: **mechanical weathering,** which results in the physical disintegration of rock into smaller pieces without changing its chemical composition, and **chemical weathering,** which changes the chemical composition of the weathered rock.

Mechanical weathering may be accomplished in several ways: by **frost wedging,** the enlarging of cracks in rock as water in the cracks freezes and expands; by crystal growth within rock cavities, which forces the cavity walls farther apart; by **thermal expansion** and contraction, the alternate enlargement and shrinking of rock as it is repeatedly heated and cooled; by **mechanical exfoliation,** the fracturing and removal of successive rock layers as deep rocks expand upward after overlying rocks have eroded away; by penetration by growing plant roots, which expands existing cracks in rock, and through the activities of animals; and by **abrasion** of transported particles as they collide with each other and with stationary rock surfaces.

The chemical-weathering process of **dissolution,** which works most effectively on soluble rocks such as limestone, occurs when minerals or rocks are decomposed by contact with water or by reaction with naturally occurring acid, and the products are carried off by water. **Oxidation,** the reaction of certain chemical compounds with oxygen, occurs most commonly in iron-rich rocks such as basalt, gabbro, and peridotite. **Hydrolysis,** the replacement of major ions in minerals (particularly the feldspars) with H^+ or OH^- ions from water, produces the most common products of chemical weathering, the clay minerals. The rate at which a given rock or mineral weathers chemically depends on climatic factors (that is, regional temperature and water availability), the activity of living organisms, the length of time during which the material is exposed to weathering, and the chemical stability of its components at the Earth's surface. The products of chemical weathering include clay minerals, metal ores, and **spheroidal weathering.**

Mechanical and chemical weathering of bedrock or sediment produce **regolith,** the fragmented material that overlies much of the Earth's bedrock. **Soil** is the uppermost, organic-rich portion of the regolith. Soil development is governed by five factors: **parent material** (the bedrock or sediment from which a soil develops), climate, **topography** (the physical features of a landscape), vegetation, and time. A developing soil consists of distinct layers having different compositions, called **soil horizons**; the vertical succession of soil horizons in a given location constitutes its **soil profile.** Most temperate-zone soil profiles consist of the typical layers designated (from the surface down) the O, A, E, B, and C horizons. In recent years, soil classification has become increasingly important to land-use decisions. Soil scientists in North America use a system called **soil taxonomy** to categorize soils into eleven different orders and many more suborders.

Weathering takes place on other planets in our solar system and on the Moon, albeit differently than on Earth. No chemical weathering currently occurs on the Moon, Venus, or Mars, although both the Moon and Venus experience a form of mechanical weathering. Mars' red color probably results from past chemical oxidation of iron-rich bedrock.

Key Terms

weathering (p. 85)
erosion (p. 85)
mechanical weathering (p. 85)
chemical weathering (p. 86)
frost wedging (p. 86)
thermal expansion (p. 87)
mechanical exfoliation (p. 88)
abrasion (p. 88)
dissolution (p. 89)
oxidation (p. 89)
hydrolysis (p. 89)
spheroidal weathering (p. 92)
regolith (p. 92)
soil (p. 92)
parent material (p. 92)
topography (p. 93)
soil horizons (p. 95)
soil profile (p. 95)
soil taxonomy (p. 96)

Questions for Review

1. Describe the fundamental difference between mechanical and chemical weathering.

2. Discuss three ways that rocks can be weathered mechanically.

3. Explain how the process of mechanical exfoliation works.

4. Would granite, limestone, or basalt be most susceptible to the chemical-weathering process of oxidation? To the process of dissolution? Which would weather to produce the most clay?

5. What role does climate play in chemical weathering?

6. Why is quartz a more common mineral in sandstones than plagioclase feldspars? (*Hint:* See Figure 5-10.)

7. Discuss how soils vary in a region of irregular topography.

8. Describe the principal characteristics of the major soil horizons, and explain how those characteristics develop.

9. In what ways does the weathering environment of the Moon differ from that of Earth?

For Further Thought

1. Since the Industrial Revolution, we have been burning coal, heating oil, and gasoline at an ever-increasing rate. Combustion of these fuels releases carbon dioxide into the atmosphere. How might burning these fuels affect weathering rates? Explain.

2. Describe what would happen to the physical condition of Cleopatra's Needle (Fig. 5-4) if the obelisk were returned to its original home in Egypt.

3. Imagine that the Earth some day becomes devoid of water. How would the nature of chemical weathering change in polar regions? In the arid subtropical deserts? In the equatorial tropics?

4. Below is a photo of a soil developed on a lava flow in Washington. Judging from the appearance of the soil, what rock type makes up the lava flow? What weathering processes and products are responsible for the color of the soil? Describe the climate that was most likely responsible for this type of weathering.

5. Look around your community, identify the major building stones, and compare their relative states of weathering. Did the local builders make wise choices when selecting their building materials, considering your local climate? Which rock(s) would work best for construction in your area? Which would be poor choices?

6

Sedimentation and Sedimentary Rocks

Sediment, the unconsolidated material that accumulates continuously at the Earth's surface, consists mostly of the physical and chemical products of weathering and erosion discussed in Chapter 5: solid fragments of preexisting rocks, and minerals precipitated out of solution in water. The organic remains of plants and animals also contribute to sediment. After being transported by water, wind, or ice or precipitating out of solution, sediment accumulates virtually everywhere on Earth—from the glaciated summits of the Himalayas, 10 kilometers (6 miles) above sea level, to the deep trenches on the floor of the Pacific Ocean, 10 kilometers below sea level. It is continually deposited throughout the world—in lakes, streams, oceans, swamps, beaches, lagoons, deserts, caves, and at the bases of glaciers.

Much sediment is ultimately converted to solid **sedimentary rock** (Fig. 6-1). Sedimentary rocks make up only a thin layer of the Earth's uppermost crust, accounting for barely 5% of the Earth's outer 15 kilometers (10 miles), but they constitute 75% of all rocks exposed at the Earth's land surface. Sedimentary rocks serve as our principal source of coal, oil, and natural gas, as well as much of our iron and aluminum ores; they also store nearly all of our fresh underground water and represent the source of cement and other natural building materials.

In addition, sedimentary rocks offer clues about the condition of the Earth's surface as it existed in the geologic past. They record the former presence of great mountains in areas now monotonously flat, and they tell tales of vast seas that once covered the now dry interior of North America. Some sedimentary rocks contain the fossil remains of past life, revealing much about how Earth has evolved through its history.

This chapter examines the origins of sedimentary rocks, describes their classification, and explains the ways in which geologists use them to reconstruct past surface environments. It concludes by showing how various sedimentary rocks relate to common plate tectonic settings.

Figure 6-1 Layers of sedimentary rock (Navajo Sandstone) in Zion National Park, Utah. These rocks are composed of the cemented sand grains of ancient sand dunes. The fascinating "cross-bedded" patterns they exhibit are typical of windblown sands.

The Origins of Sedimentary Rocks

Sediments are classified according to the source of their constituent materials as either detrital or chemical (Fig. 6-2). **Detrital sediment** is composed of solid fragments, or *detritus*, of preexisting igneous, sedimentary, or metamorphic rocks. **Chemical sediment** forms from previously dissolved minerals that have either precipitated from solution in water or been extracted from water by living organisms and converted to shells, skeletons, or other organic substances (which are deposited as sediment when the organisms die or discard their shells). Through various processes, each type of sediment may be transformed into sedimentary rock.

Sediment Transport and Texture

The vast majority of sediments are detrital in origin. During transport, detrital sediments are generally carried from high places to low places; they are moved largely by gravity but

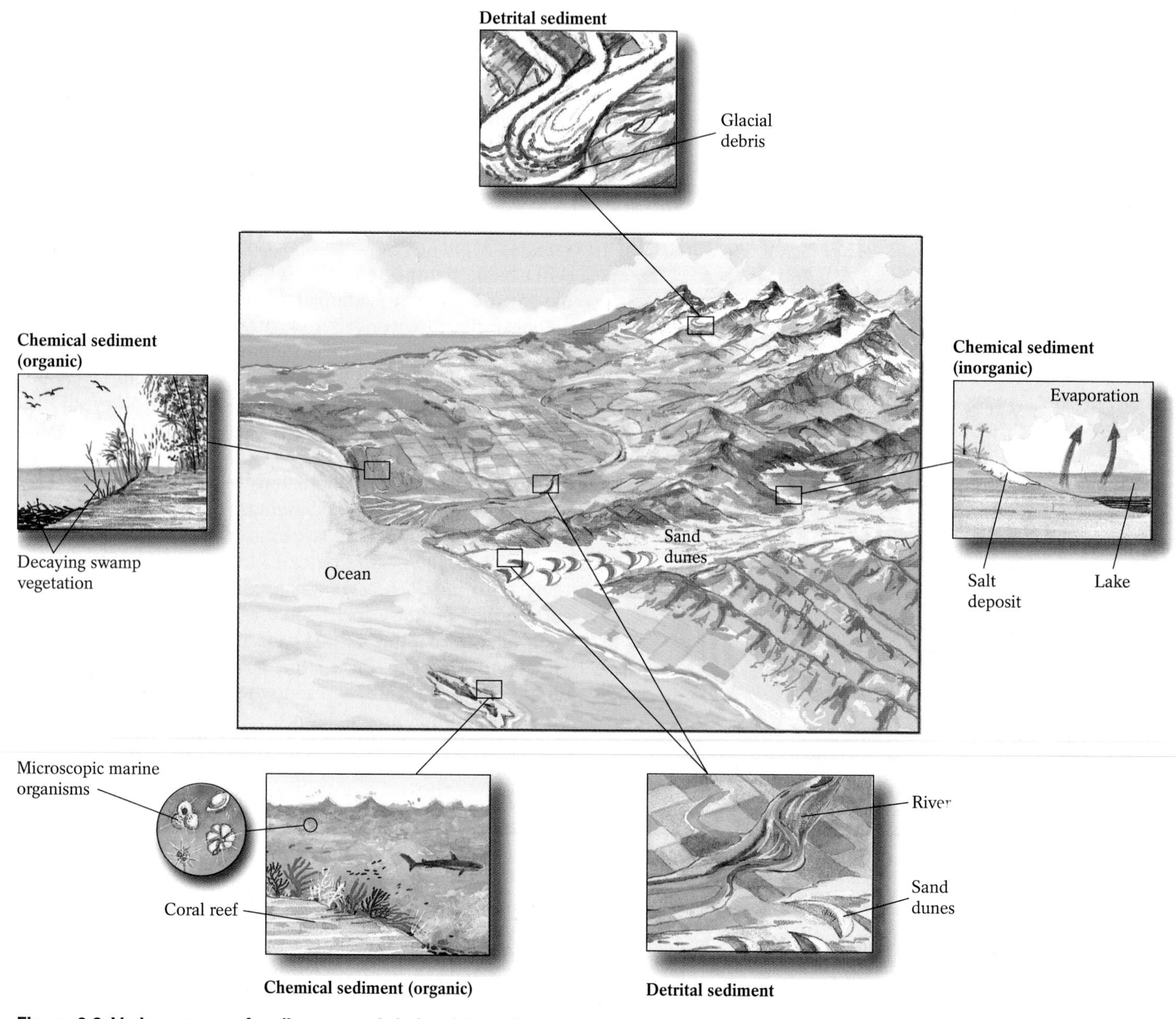

Figure 6-2 Various types of sediments and their origins. Detrital sediments, such as glacial debris or river-channel sand, consist of preexisting rock fragments. Chemical sediments often consist of minerals precipitated directly from water, such as salt deposits produced by evaporation of small temporary lakes; they may also be composed of organic debris, such as partially decayed swamp vegetation or the shells of small marine organisms.

often gain help from running water, wind, or glacial ice. The loose particles continue to move until the transporting medium loses its capacity to carry them, such as when a river ceases to flow upon entering relatively still marine water at a coast. At this point, the particles drop to the bottom and are deposited. Each year, an estimated 10 billion tons of detrital sediment, most of it carried by rivers, is delivered to the world's oceans.

If you've ever held a handful of beach sand or lake-bottom mud, you can appreciate the variety of textures that characterize detrital sediments. Detrital sediment texture—the size, shape, and arrangement of the sediment particles—depends on the source of the particles and the medium that transported them. (Chemical sediments, which we discuss later, are distinguished by their composition more than by their texture.)

Grain Size Because rock fragments continue to be weathered during transport, they are generally reduced in size when they are finally deposited as sediment. The extent to which they become worn down depends partly on their constituent minerals' resistance to weathering and partly on the nature of their parent rocks. Coarse-grained granite, for example, generally weathers to produce grains larger than those created by the weathering of fine volcanic ash. The nature and energy level of the transport medium also influence sediment grain size. For example, a pebble that would be pulverized by a creeping glacier might remain unchanged by an oozing mudflow; it might be worn down by a white-water river but be unaffected by a trickling stream.

Geologists can often determine the medium that transported a sediment by making a quick visual estimate of its sorting. **Sorting** is the process by which a transport medium separates or "selects" particles of different sizes, shapes, or densities. Wind, which generally carries only relatively small particles, is the most selective of the transport media; a deposit of windblown silt, which comprises particles mostly within a narrow size range, is considered *well sorted* (Fig. 6-3a). At the other extreme, glacial ice is unselective, transporting particles with a wide range of sizes. Glacial deposits, which may contain extremely fine particles as well as boulders the size of small office buildings, are said to be *poorly sorted* (Fig. 6-3b). Sorting of sediment can also reflect changing energy levels in the transport medium. When a moving current (wind or water) loses energy, its capacity to carry sediments is diminished. Because it takes more energy to carry them, the larger, heavier particles are deposited first. As energy diminishes further, the finer particles settle out.

Grain Shape Sediment particles may be either angular or rounded depending on the transport medium, the distance the particles are carried, and the hardness of the particle minerals. Some transport media are particularly efficient in rounding particles. Swiftly flowing rivers, for instance, bounce pebbles and sand grains around vigorously, so that they collide with other particles and with the river bottom; as a result, the sediment becomes ever smoother and smaller. A stream that carries particles for long distances will round the grains much more than a stream of comparable energy that carries them for only a short distance. Glaciers, on the other hand, embed some of their sediment particles in hundreds of meters of ice, so that, regardless of the distance they are carried, they remain angular when the glacier finally deposits them.

Grains of softer minerals, such as gypsum and calcite, become rounded more readily than grains of harder ones, such as quartz. In a recent field study, fragments of soft

(a) (b)

Figure 6-3 Differential sediment sorting by transport media. **(a)** Because wind is highly selective regarding the particles it transports, this well-sorted windblown silt near Vicksburg, Mississippi, is limited to very fine sediment particles. **(b)** Glaciers are capable of transporting sediment of all sizes. Their deposits, such as this one in Rocky Mountain National Park, Colorado, are typically very poorly sorted.

sedimentary rock became well rounded after only 11 kilometers (6.6 miles) of stream transport. More durable fragments of granite, transported in the same stream, required 85 to 335 kilometers (53–208 miles) of transport to achieve comparable roundness.

Sedimentary Structures

Sediments often contain **sedimentary structures,** physical features that develop during or soon after deposition and reflect the conditions under which the sediments settled. These structural clues, along with any fossils present (discussed in detail in Chapter 8), make sedimentary rock a crucial source of information about the nature and sequence of past environments on Earth.

Bedding **Bedding,** or **stratification,** is the arrangement of sediment particles into distinct layers (*beds* or *strata*) having different sediment compositions and/or grain sizes. A clear break, or *bedding plane,* is generally visible between adjacent beds. Such bedding planes mark the end of one depositional event and the beginning of the next. For example, a flooding river often deposits a particle load of heavy, coarse-grained sediment on top of finer preexisting sediments in the surrounding area; such a difference in grain size would be visible as a bedding plane in a cross section of the resulting sediment layers (Fig. 6-4).

When a sediment load containing a variety of sediment sizes is suddenly dumped into relatively still water, such as when a sediment-laden stream empties into a lake or ocean, its particles will settle at different rates, depending on their sizes, densities, and shapes. This gradual settling produces a **graded bed,** a single sediment layer (formed by a single depositional event) in which particle size varies gradually, with the coarsest particles found on the bottom and the finest at the top (Fig. 6-5). To demonstrate the principle of grading, drop a handful of unsorted backyard dirt into a tall glass of water. The largest particles will quickly settle to the bottom, while the finest will settle last and land on top of the larger particles. Where graded beds exist, geologists can use them to determine whether tectonic forces have overturned the strata.

Cross-beds are sedimentary layers that, because they are deposited by a moving current instead of settling from still water, form at an angle to underlying beds. Cross-beds are often found in wind-deposited sand dunes (see Fig. 6-1 and 6-6) and in water-deposited sediment at the bottom of a stream. Because cross-beds always slope toward the down-current direction, they record the flow direction of the current that deposited them.

Surface Sedimentary Features The surface appearance of a layer of detrital sediment often provides clues about the environment in which it was deposited or the conditions to which it was exposed after deposition. For instance, a pattern of wavy lines, or **ripple marks,** preserved on top of a sediment bed indicates that wind or water currents shaped its particles into a

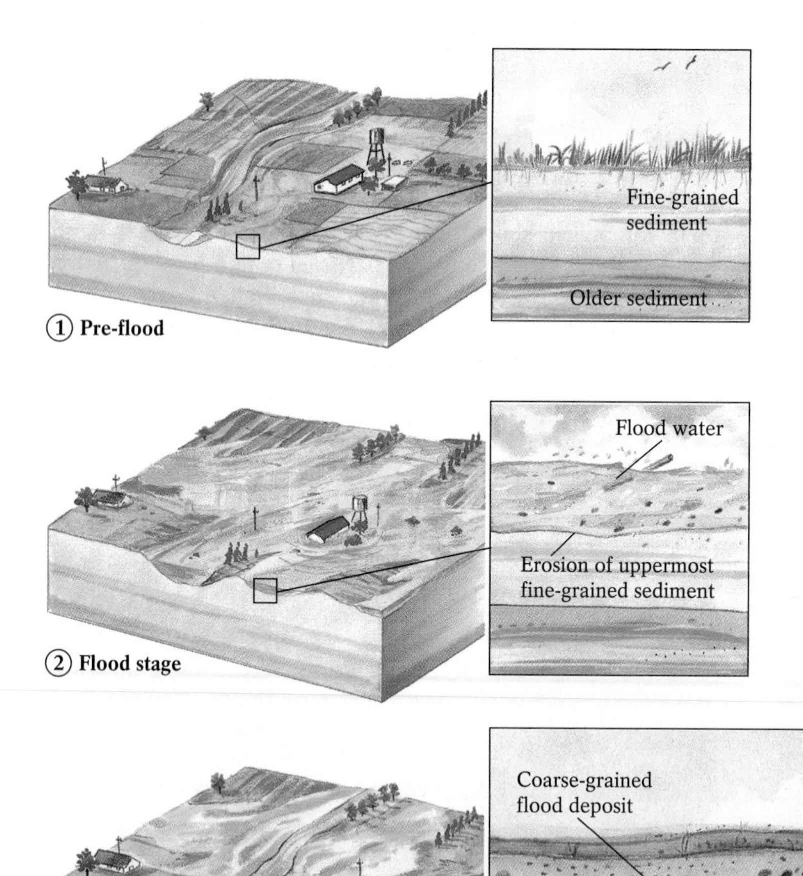

Figure 6-4 Development of a bedding plane due to river flooding. Any depositional event that leaves sediment that differs from the preexisting sediment (in terms of either grain size or composition) leaves a demarcation, a bedding plane, between the resulting sediment layers.

(a)

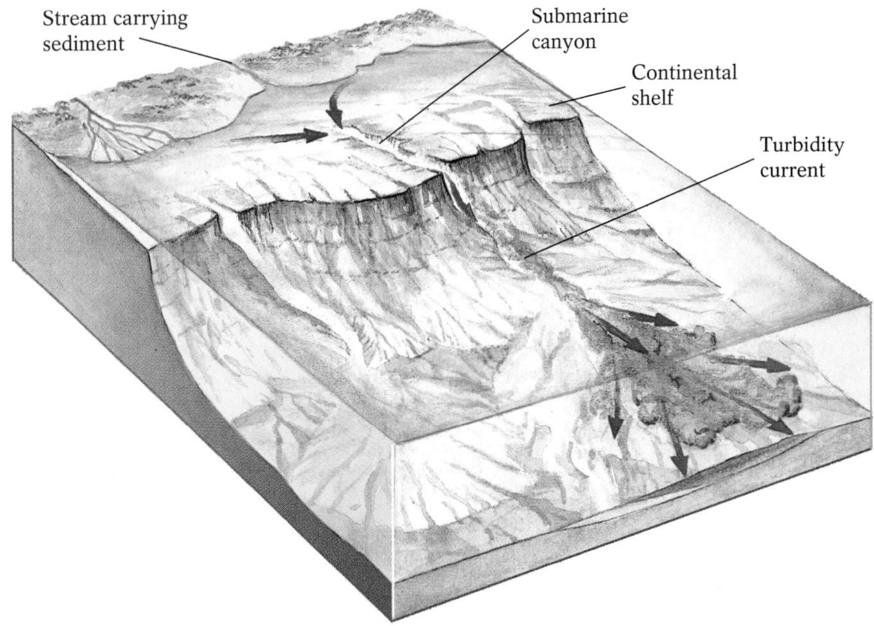

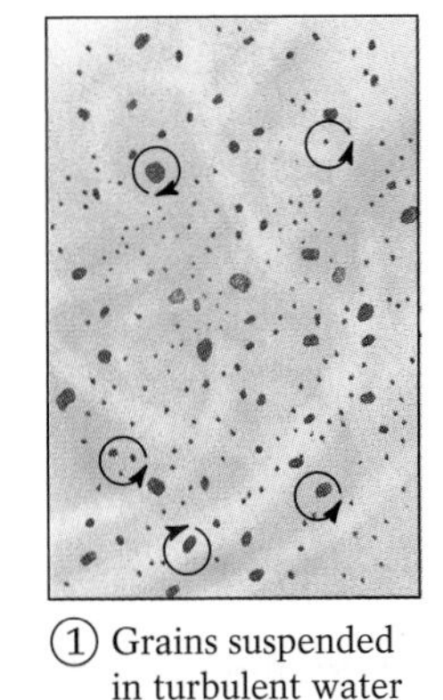

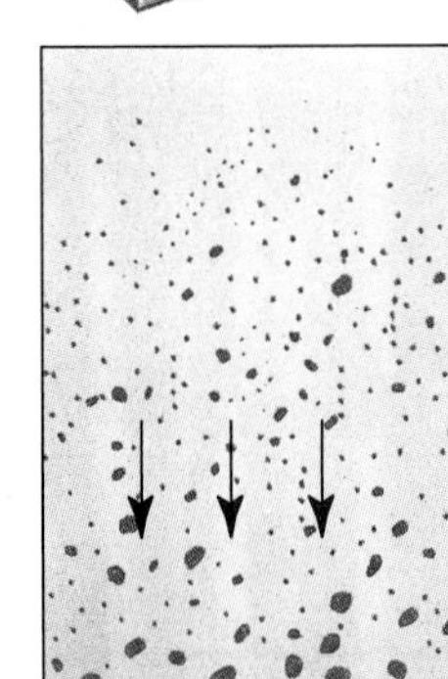

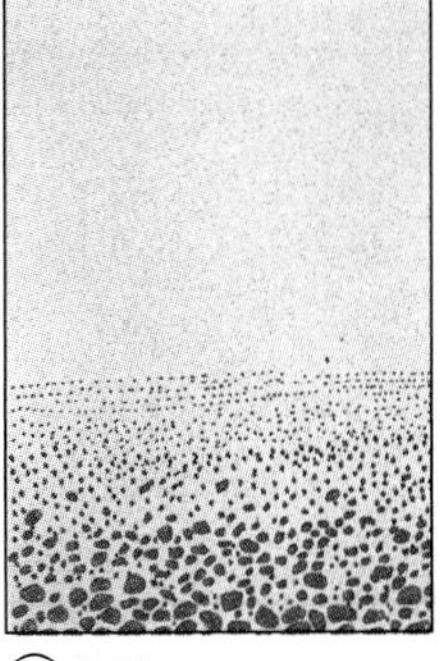

(b)

Figure 6-5 Graded bedding of sediment at Furnace Creek, Death Valley, California. **(a)** Graded beds form as particles of different density, size, and shape settle out of a standing body of water into distinct layers. The larger, heavier particles settle to the bottom first and the smaller, lighter particles settle above them. **(b)** Graded sediment is frequently produced by turbidity currents, offshore sediment flows that abruptly lose their energy and drop their particle loads onto the ocean floor.

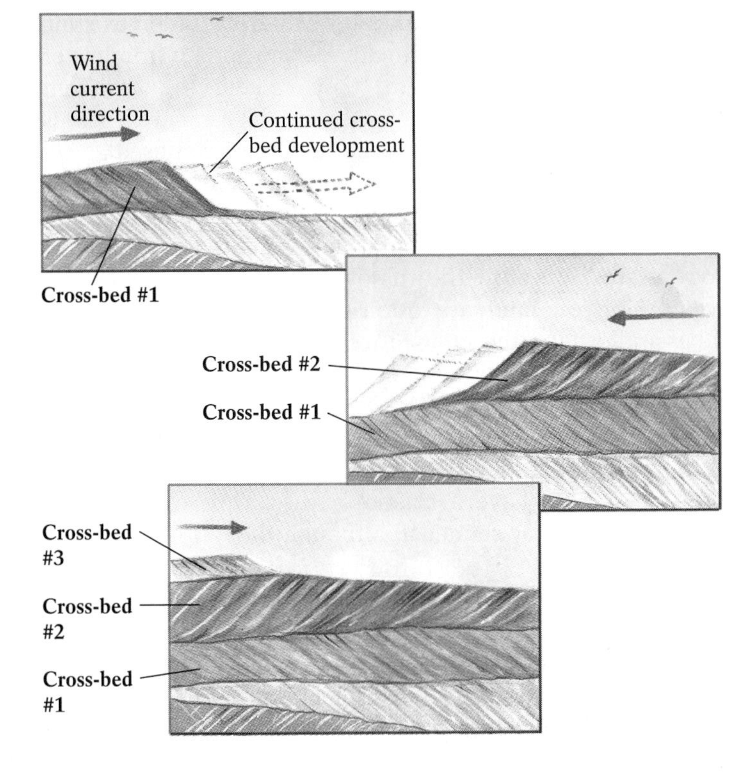

Figure 6-6 The development of cross-bedding in sand dunes. The sets of cross-beds with different orientations seen here form as wind directions shift through time.

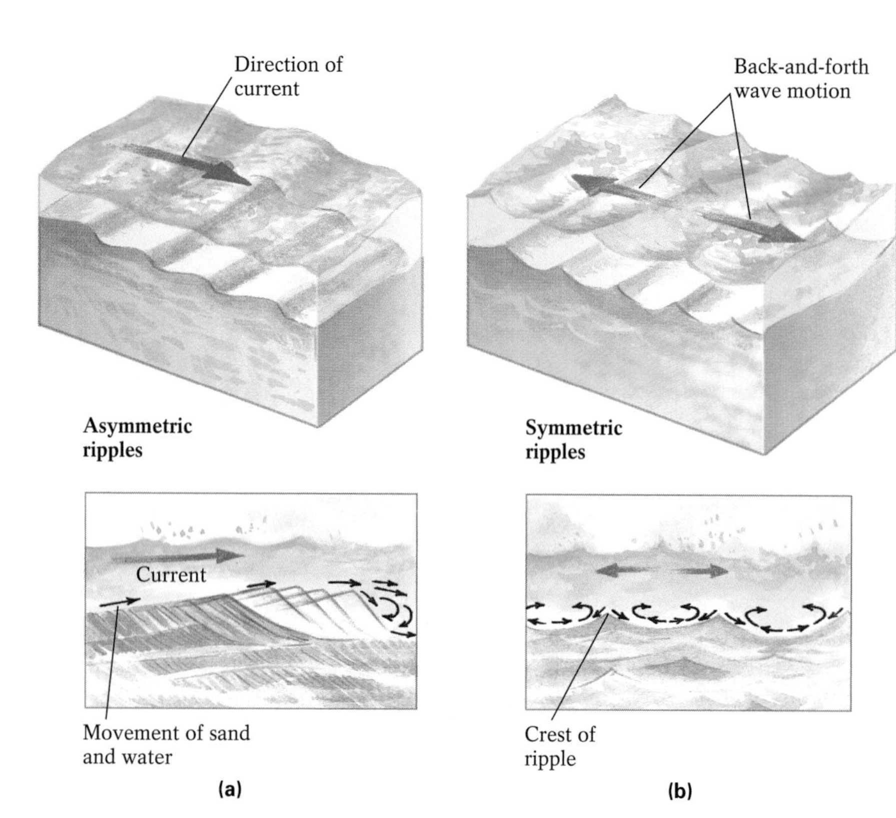

Figure 6-7 Different types of currents produce different ripple patterns. **(a)** A current that generally flows in one direction, such as a stream, produces asymmetric ripples. Sand grains roll up the gently sloping upstream side of each ridge and then cascade down the steeper downstream side. **(b)** Symmetric ripples form from the back-and-forth motion of waves in shallow surf zones at the coast or at the water's edge in a lake. Photo: Exposed rocks show ripple marks, evidence of past current flow, either water or wind.

series of shallow curving ridges. The configuration of these ridges, which are often visible on sandy surfaces, reflects the nature of the current that produced them (Fig. 6-7).

Mudcracks are fractures that develop when the surface of wet fine-grained sediment (mud) dries and contracts (Fig. 6-8). Because these structures form only at the top of a layer of muddy sediment and narrow progressively downward, geologists can study mudcracks to determine whether a layer of sedimentary rock has been overturned.

Lithification: Turning Sediment into Sedimentary Rock

When a sediment layer is deposited, it buries all previous layers deposited at that location. Eventually, the continuing deposition may enable a sedimentary pile to become several kilometers deep. Such deep burial may convert sediments into solid sedimentary rock by the process of **lithification** (from the Greek *lithos,* meaning "rock," and Latin *facere,* meaning "to make"). During lithification, sediment grains become compacted, often cemented, and sometimes recrystallized.

Compaction is the process by which the volume of buried sediment, either detrital or chemical, becomes diminished by pressure exerted by the weight of overlying sediments. Expulsion of air and water from the sediment and the reduction of the spaces between grains combine to convert loose sediment into more cohesive sedimentary rock.

Cementation is the process by which sediment grains become bound together by precipitation of minerals originally dissolved during chemical weathering of preexisting rocks. When these dissolved materials precipitate from water circulating through sediment, they cement the sediment grains together (Fig. 6-9). The most common cementing agents are calcium carbonate (lime—the same material used as cement in concrete), silica (derived principally from the chemical weathering of feldspar), and several iron compounds.

Compaction and cementation can bind more than just rock grains. Because the dissolved products of chemical weathering on land are often ultimately transported to lakes and oceans, the same processes can lithify shells, shell fragments, and other remains of dead organisms that accumulate in these bodies of water. Any rock that consists of preexisting solid particles compacted and cemented together—whether preexisting rock fragments or organic debris—is said to have a **clastic** texture. Thus clastic rocks may be detrital or chemical, although they are most often detrital.

The increased heat and pressure associated with sediment burial also promote **recrystallization,** the development of stable minerals from unstable ones. One common mineral that recrystallizes is aragonite, an unstable form of calcite that many marine organisms secrete to form their shells,

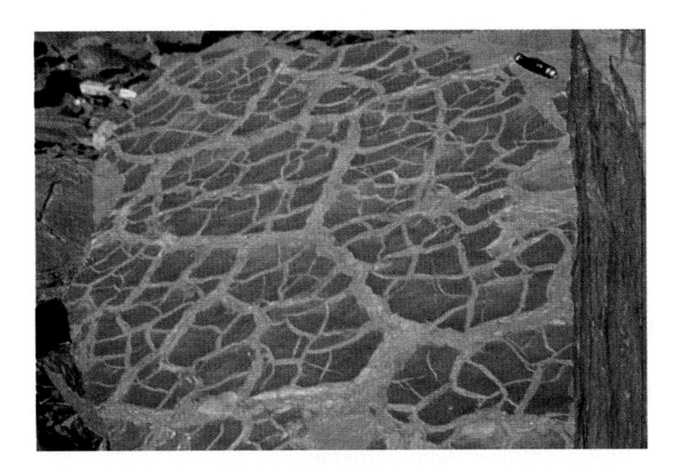

Figure 6-8 The origin of mudcracks. These mudcracks in oxidized red shale in Glacier National Park, Montana, suggest that a body of water once evaporated to dryness within the terrain that is today part of the Montana Rockies.

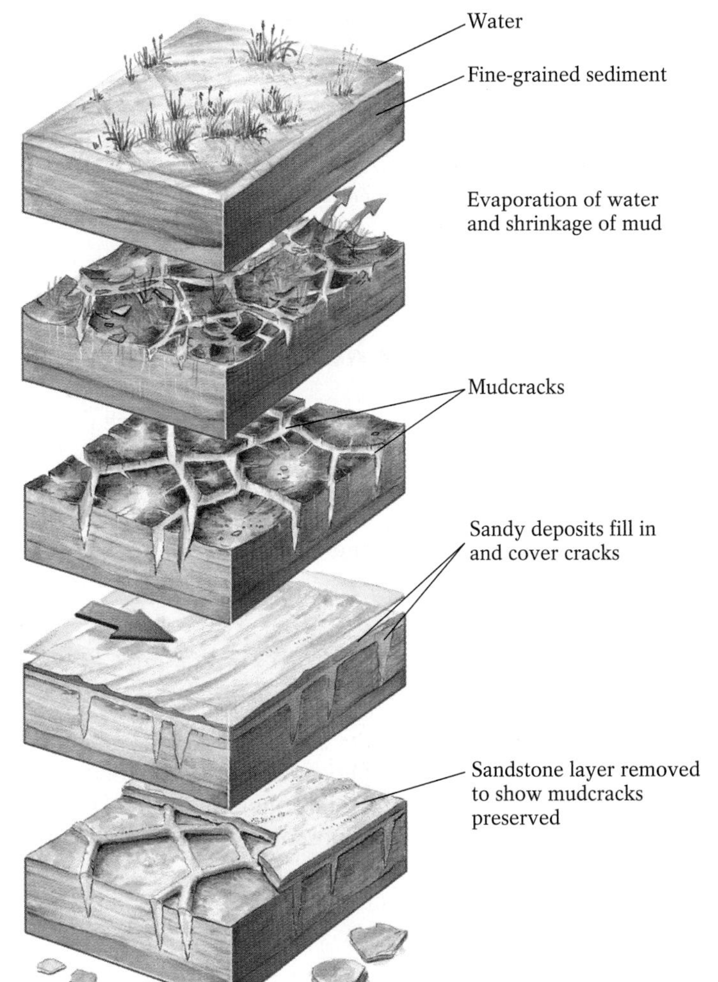

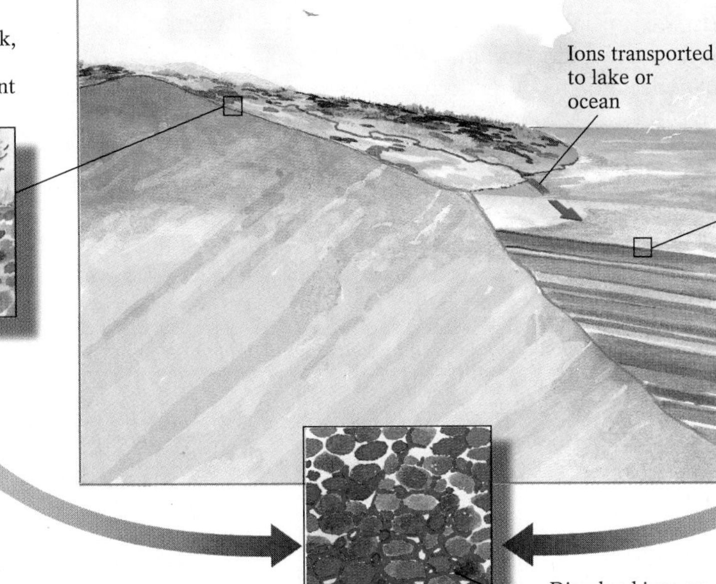

Figure 6-9 Lithification of sediment by cementation. Weathering of source rocks releases ions in solution, which then circulate via groundwater, lake water, or ocean water through coarse-grained sediments. When these dissolved materials precipitate as solid compounds within the spaces around sediment grains (called pore spaces), they form a cement that binds the grains together.

but which eventually recrystallizes as stable calcite. This transformation explains the absence of aragonite in ancient carbonate rocks.

Classifying Sedimentary Rocks

As noted earlier in this chapter, sedimentary rocks are generally classified as either detrital or chemical, depending on their source material. Each of these broad categories, however, encompasses a wide variety of rock types, reflecting the diverse transport, deposition, and lithification processes at work.

Detrital Sedimentary Rocks

All detrital sedimentary rocks are clastic, being composed of solid particles from preexisting rocks or organic debris. Detrital sedimentary rocks are classified on the basis of their particle sizes (Table 6-1). Shales and mudstones are the finest-grained, as they are essentially composed of lithified mud. Sandstones have grains of intermediate size (sand). Conglomerates and breccias (pronounced "BRECH-uhs") contain a variety of larger grains, collectively called gravel.

Mudstones More than half of all sedimentary rocks are **mudstones.** The extremely fine particles in mudstones (less than 0.004 millimeter in diameter) consist largely of clay minerals and mica. Such fine particles settle out only in relatively still waters. Thus mudstones originate in lakes, lagoons, swamps, and deep ocean basins and on river floodplains—the same places where you would expect to find their unlithified form, mud.

When silt and clay particles become buried beneath hundreds of meters of sediment, compaction flattens them into parallel layers. The lithified layers become shale, a common variety of mudstone that can easily be split along these thin layers (Fig 6-10). Shales can be quite colorful (reds, greens, and blacks) as a result of differences in their mineral composition. These rocks are a common source of the clay used in making bricks, pottery, tile, and Portland cement. Some shales also contain abundant oil and may become a key source of energy in the future.

Sandstones **Sandstones** are detrital sedimentary rocks whose grains range from 1/16 millimeter to 2 millimeters in diameter. They account for roughly 25% of all sedimentary rocks. Three major types of sandstones have been identified (Fig. 6-11). *Quartz arenites,* which are generally white, red, or tan in color, are composed of more than 90% quartz grains. Their grains are rounded and well sorted, suggesting that they were transported over a long distance. *Arkoses* are pinkish sandstones containing more than 25% feldspar, with poorly sorted and angular grains. These qualities suggest that the

Table 6-1 Detrital Sediments and Rocks

Particle Size (mm)	Particle Name	Name of Rock Formed
<0.004	Clay* } Mud	Shale } Mudstone
0.004–0.063	Silt } Mud	Siltstone } Mudstone
0.064–2	Sand	Sandstone
2–4	Granule } Gravel	Breccia (if particles are angular)
4–64	Pebble } Gravel	
64–256	Cobble } Gravel	Conglomerate (if particles are rounded)
>256	Boulder } Gravel	

1 mm = 0.039 inch

*Note that the term "clay," when used in the context of sediment size, denotes very fine particles of any rock or mineral (as opposed to the term "clay mineral," which refers to a compositionally specific group of minerals); all clay minerals have clay-sized particles, but not all clay-sized particles are composed of clay minerals.

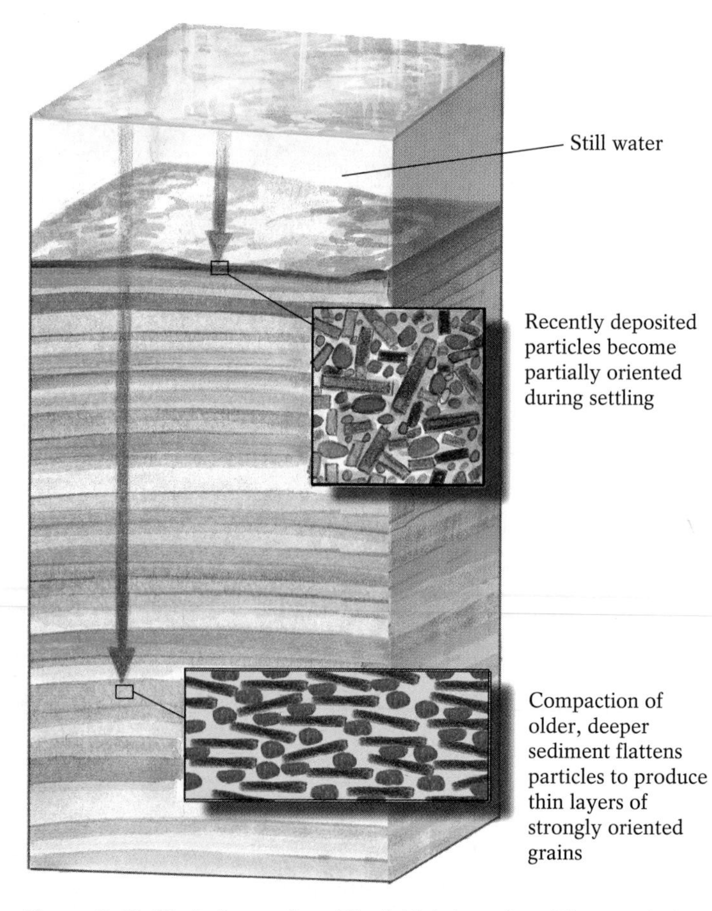

Figure 6-10 Shale formation. The initial deposits of flat or tabular clay and mica grains may be oriented randomly. The weight of subsequent deposits, however, causes these grains to "collapse" into a parallel orientation, producing the typical layered appearance of shales.

Figure 6-11 The three major types of sandstone. (a) Quartz arenite, composed predominantly of quartz. Inset: Quartz micrograph. Scale bar = 1 mm. **(b)** Arkose, containing abundant feldspar. Inset: Arkose micrograph. Scale bar = 1 mm. **(c)** Graywacke, a poorly sorted mixture of quartz and feldspar grains, angular volcanic fragments, and fine clay and mica particles. Inset: Graywacke micrograph. Scale bar = 1 mm.

parent sediments were transported for only a short distance, underwent minimal chemical weathering, and were deposited and buried rapidly. *Graywackes* are dark, gray-to-green sandstones that contain a mixture of quartz and feldspar grains, abundant dark rock fragments (often of volcanic origin), and fine-grained clay and mica particles. Their poor sorting, angular grains, and inclusion of such easily weathered minerals as feldspar suggest that graywackes are deposited rapidly after short-distance transport.

Because sandstones generally contain abundant pore space, fluids can easily flow into and be extracted from them. As a result, sandstones serve as the source of much of our groundwater, crude oil, and natural gas. This type of rock is also a popular building stone used in the construction of Victorian brownstone houses as well as the Gothic-style edifices found on many college campuses.

Conglomerates and Breccias **Conglomerates** and **breccias**, the coarsest of detrital sedimentary rocks, consist of a mixture of both larger (greater than 2 millimeters in diameter) and smaller grains that have been cemented together. In conglomerates, the grains are rounded; in breccias, they are angular. The relatively large size of conglomerate and breccia grains enables geologists to readily identify their parent rocks. Likewise, the shape of the grains provides clues to their transport path: The rounded particles in conglomerates suggest

Figure 6-12 Conglomerates and breccias. The grains in these coarse sedimentary rocks reveal much about their history. **(a)** The roundness of the grains in conglomerates suggests long-distance transport by vigorously moving water. **(b)** The angularity of the grains in breccias suggests short-distance transport.

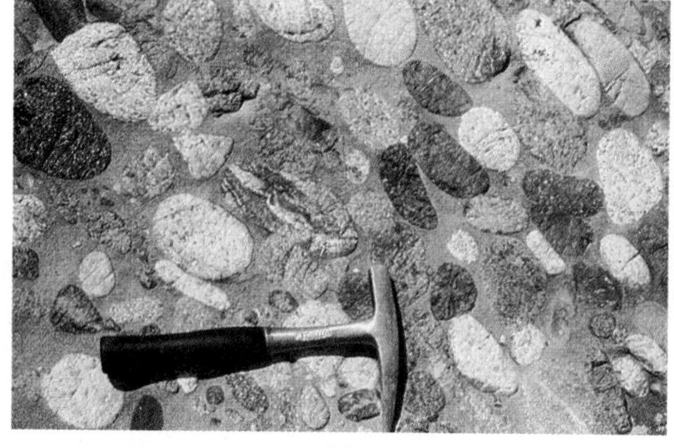

(a)

(b)

lengthy transport by vigorous water currents; the angular grains in breccias suggest brief transport, as when shattered rock debris accumulates at the base of a cliff (Fig. 6-12).

Chemical Sedimentary Rocks

Unlike detrital sedimentary rocks, which always consist of distinct fragments of preexisting rocks or minerals, chemical sedimentary rocks typically comprise an interlocking mosaic of crystals derived from dissolved compounds. Two kinds of chemical sediments exist: *inorganic* sediments, which are precipitated directly from solution in water, and *biogenic* (or organic) sediments, which are produced through the biological activity of plants and animals. Table 6-2 summarizes the various types of chemical sedimentary rocks.

Table 6-2 Chemical Sedimentary Rocks

	Rock Name	Typical Composition
Inorganic	Inorganic limestone	Calcite ($CaCO_3$)
	Evaporites	Halite ($NaCl$), gypsum ($CaSO_4 \cdot H_2O$)
	Dolostone	Dolomite ($CaMg(CO_3)_2$)
	Inorganic chert	Chemically precipitated silica (SiO_2)
Organic	Biogenic limestone	Calcium carbonate remains of marine organisms (e.g., algae, foraminifera)
	Biogenic chert	Silica-based remains of marine organisms (e.g., radiolaria, diatoms, sponges)
	Coal	Compressed remains of terrestrial plants

Inorganic Chemical Sedimentary Rocks Inorganic chemical sedimentary rocks form when the dissolved products of chemical weathering precipitate from solution; this precipitation typically occurs when the water in which these materials are dissolved evaporates or undergoes a significant temperature change. The four common types of inorganic chemical sedimentary rocks are inorganic limestones, dolostones, evaporites, and inorganic cherts.

Limestones, which are composed largely of calcite (calcium carbonate), account for 10% to 15% of all sedimentary rocks. Most limestones forming today are organic. Nevertheless, under certain conditions—such as a rise in water temperature, a decrease in water pressure, or agitation by waves—calcite also precipitates directly from water, producing **inorganic limestone.** Inorganic limestone is currently forming along the Grand Bahama Banks, a shallow submarine platform separated from Florida by the Straits of Florida. It also forms on land in several geologic settings. For example, the stalactites and stalagmites found in caves and the soft, spongy *tufa* formations that develop around natural springs are inorganic limestone structures.

Dolostone is a rock composed of the mineral dolomite, a calcium and magnesium carbonate. Similar to limestone in appearance and chemical structure, dolostone is believed to form when magnesium ions replace calcium ions in a preexisting body of limestone. For instance, ocean water may evaporate to produce a magnesium-rich solution that then circulates through a limestone bed. Consequently, dolostone production is particularly likely whenever a tropical or semi-

Highlight 6-1 When the Mediterranean Sea Was a Desert

Every year, heat evaporates more than 4000 cubic kilometers (960 cubic miles) of water from the Mediterranean Sea. Most of this water is replaced by a massive inflow from the Atlantic Ocean through the Strait of Gibraltar. This pattern has not always been the case, however. Deep-sea exploration in the early 1970s revealed a massive evaporite layer more than 2000 meters (6600 feet) thick beneath the floor of the Mediterranean Sea. What past conditions could have caused such a great salt accumulation?

According to one hypothesis, between 10 and 8 million years ago ongoing convergence between the African and Eurasian plates, which meet beneath the Mediterranean, gradually raised the sea floor in the area of what is now the Strait of Gibraltar, creating a natural limestone dam that blocked the inflow of Atlantic waters (Fig. 1). The Mediterranean rapidly evaporated. Occasional pulses of Atlantic water over or through the Gibraltar barrier, like a leaky faucet, may have provided the now-dry basin with a periodic supply of salty water that, upon evaporating, added to the thick Mediterranean salt deposits.

Analysis of sediment samples from the Mediterranean's floor indicates that deposition of evaporites there ceased about 5.5 million years ago, when a significant amount of water began to flow through a break in the dam. The Rock of Gibraltar, a remnant of that ancient natural dam, serves as a reminder of the time when an intermittently dry Mediterranean basin separated Europe and Africa. If the African and Eurasian plates continue to converge, however, a new rocky barrier may rise to block the inflow of water from the Atlantic. In that event, the Mediterranean basin could again become dry and covered with salt.

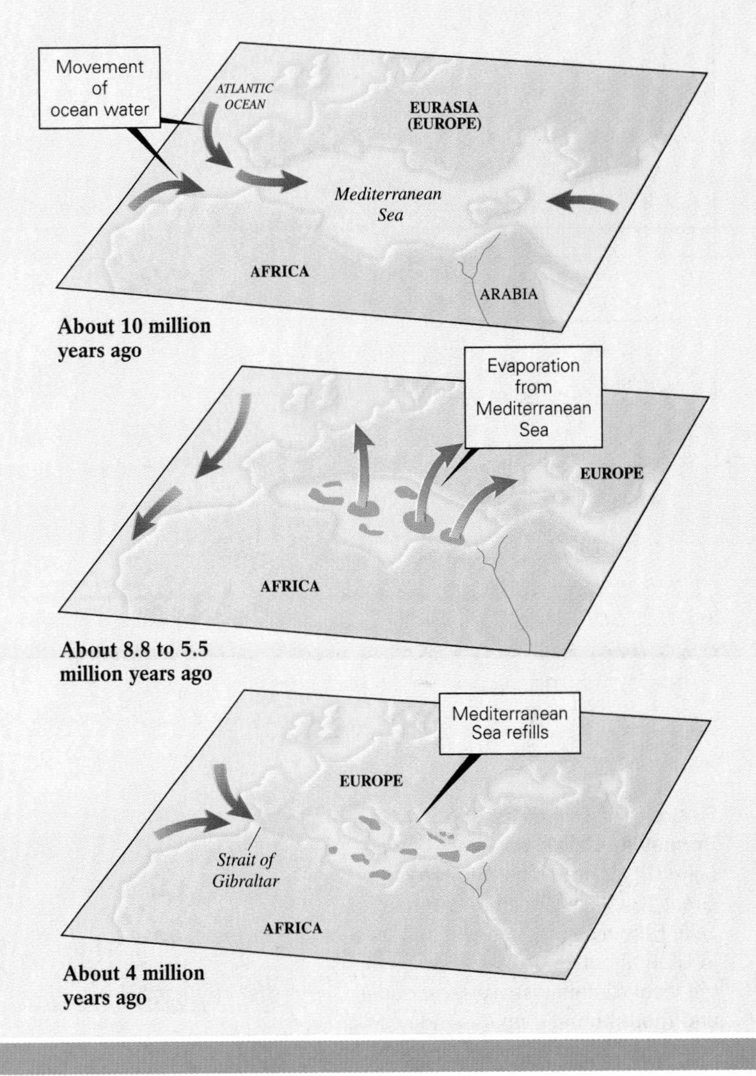

Figure 1 The Mediterranean: sea and desert. The thick evaporite deposits that underlie much of the Mediterranean basin today probably accumulated during a time (8–5.5 million years ago) when a topographic barrier stretched across what is now the Strait of Gibraltar, cutting off the water influx from the Atlantic Ocean. With no water supply to replenish it, the Mediterranean Sea largely evaporated, precipitating vast amounts of salt.

tropical climate promotes evaporation of salty marine water near a porous body of limestone. A thin crust of dolomite crystals is developing today above the low-tide level on some limestone bodies in the Persian Gulf states, the Florida Keys, and the Bahamas. Massive quantities of dolomite are not being produced today as they were in the past, however. Recent discoveries suggest that the presence of a family of sulfate-consuming bacteria, which are today nearly absent from the Earth's oceans, may be a necessary condition for substantial dolostone accumulation.

Evaporites are inorganic chemical sedimentary deposits that accumulate when salty water evaporates. On average, the world's seawater contains, by volume, almost 3.5% dissolved salts. Where marine water is shallow and the climate warm, evaporation increases the concentration of these salts. When evaporation of seawater exceeds the inflow of water into a marine basin, solid crystals precipitate and begin to accumulate at the sea bottom (see Highlight 6-1). The first mineral to precipitate is gypsum, which appears when roughly two-thirds of a volume of seawater has evaporated. Halite (common table

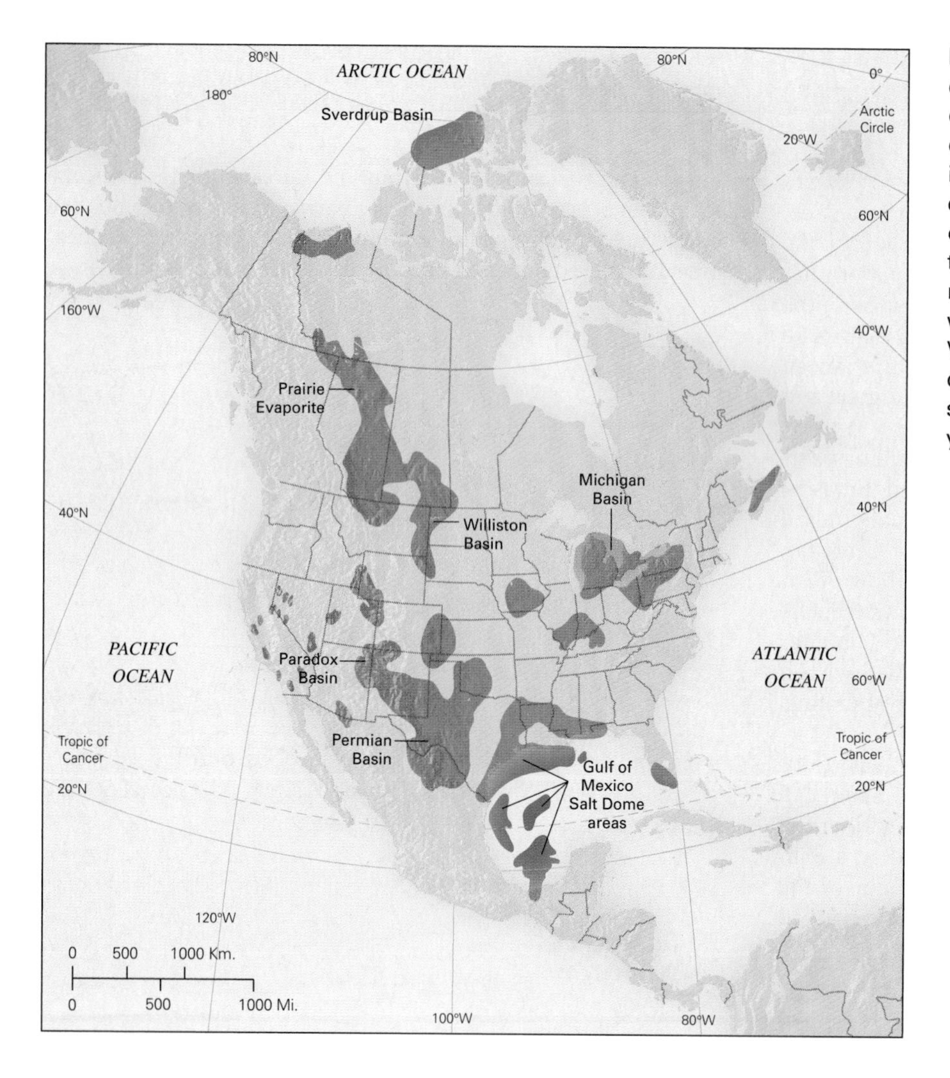

Figure 6-13 The location of major subsurface evaporite deposits in North America. These deposits were produced in the ancient past when, during times of high sea level, salty marine water invaded topographic low spots on the North American continent. These shallow seas later evaporated during periods of climatic warming, leaving behind thick evaporite deposits. For example, about 400 million years ago, evaporating seas occupying what is now part of Michigan, Indiana, Ohio, West Virginia, Pennsylvania, New York, and Ontario produced great deposits of gypsum and halite. Most such deposits were subsequently covered by younger sedimentary rocks.

Figure 6-14 Evaporite deposits at the Bonneville Salt Flats, west of Salt Lake City, Utah. The modern-day Great Salt Lake is a small remnant of Lake Bonneville, a vast lake that existed in Utah about 15,000 years ago, when the local climate was cooler, cloudier, and more humid than it is today. Such salt deposits formed when most of Lake Bonneville evaporated under the aegis of the modern climate, which is warm, clear, and dry.

Figure 6-15 Stone axes made from chert some 150,000 years ago. Taking advantage of the sharp edge that forms when chert is chipped, our ancestors constructed many of their tools from this hard, silica-rich rock.

salt) precipitates when 90% of the water has evaporated. Further evaporation yields several other types of salt.

Thick deposits of salt underlie much of the continental United States, wherever inland seas once existed in the geologic past (Fig. 6-13). In some areas, such as in Colorado, Wyoming, and the arid Southwest, more recent deposits of gypsum and other evaporites appear at the surface, indicating where large salt lakes have evaporated. One well-known example of this phenomenon is the Bonneville Salt Flats, located west of Salt Lake City, Utah (Fig. 6-14).

Chert is the general name for a group of sedimentary rocks that consist of microscopic crystals of silica. Although many cherts have organic origins, most are **inorganic chert;** these rocks form as chemicals precipitate from silica-rich water. Geologists believe that *chert nodules,* fist-sized masses of silica commonly found in bodies of limestone and dolostone, form when portions of these rocks are dissolved by circulating groundwater and replaced by precipitated silica. Because chert is easy to chip and forms sharp edges, early humans often shaped it into weapons, cutting blades, and other tools (Fig. 6-15).

Biogenic Chemical Sedimentary Rocks Organic chemical sedimentary rocks are derived from living organisms and lithified by the same processes (compaction, cementation, and recrystallization) that produce detrital and inorganic chemical sedimentary rocks. The principal rocks of this type are biogenic limestones, biogenic cherts, and coal.

Nearly all **biogenic limestones** consist of calcite derived from accumulations of calcium carbonate shells and external skeletons of ocean-dwelling organisms (Fig. 6-16). Most such limestones form in shallow water along the continental shelves of equatorial landmasses, where warm water, sunlight, and abundant nutrients enable marine life to flourish. Such conditions exist today around most of the Caribbean islands, the Florida Keys, and the east coast of Australia.

Biogenic cherts (such as *flint,* a gray-to-black variety of chert) are found in layered beds, rather than as the nodules

(a)

(b)

Figure 6-16 Biogenic limestones. (a) Chalk is a shallow-marine limestone formed from the carbonate secretions of countless microscopic organisms. **(b)** Accumulation of chalk can eventually produce deposits of impressive size, such as the famed White Cliffs of Dover in England. The cliffs are composed mainly of the skeletons of microscopic marine plants and animals that accumulated about 100 million years ago, when the global sea level was apparently higher and coastal England was under water.

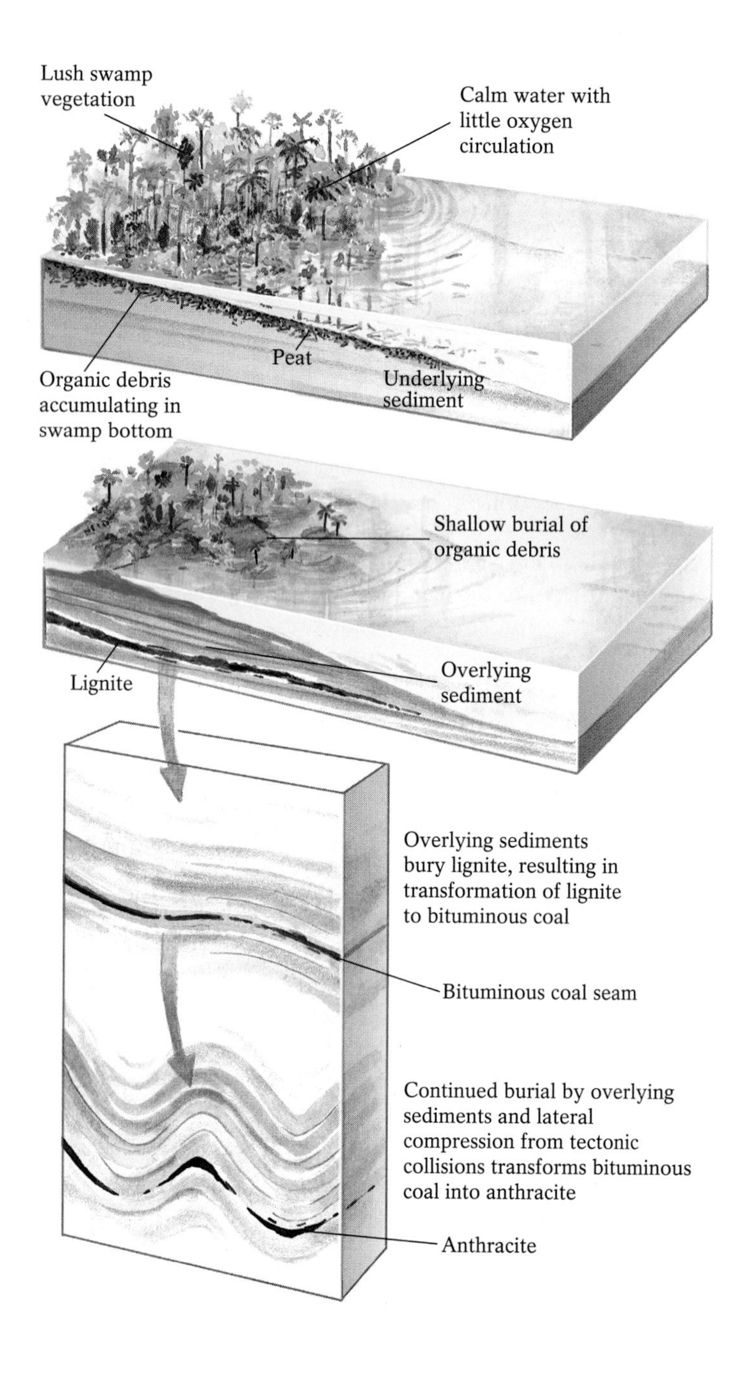

Figure 6-17 The formation of coal from swamp deposits. Abundant organic debris accumulates on the swamp floor and is buried before it can decay in the oxygen-poor swamp water; the weight of subsequent deposits and the increased temperatures at greater depths change the debris to progressively harder forms of coal. Photo: A body of coal (a "coal seam") in central Utah.

characteristic of inorganic cherts. Microscopic examination of organic cherts reveals the remains of silica-based organic debris, such as the shells of single-celled animals called radiolaria, the skeletal structures of single-celled protista called diatoms, and the skeletons of larger, more complex animals such as marine sponges.

Coal is a biogenic sedimentary rock composed mostly of plant remains. Original plant structures, such as fragments of leaves, bark, wood, and pollen, can often be seen in a lump of coal under magnification—and sometimes even with the naked eye. Vegetation generally decomposes quickly at the Earth's surface; in an environment with little oxygen, such as in the stagnant waters in the bottom of a swamp, much of the plant material accumulates relatively undecayed, however. Over the course of millennia, increasing pressure from the weight of overlying sediments and the increased temperatures found at greater depths combine to transform the plant material into a series of coal forms (Fig. 6-17). When much of the original plant structure remains intact, the forces produce a soft brown material called *peat;* increased heat and pressure create progressively harder and more compact forms of coal, ranging from soft brown *lignite* to moderately hard *bituminous* coal. Eventually, the intense temperature and pressure at great depth may produce the dense, lustrous metamorphic form of coal, *anthracite.*

"Reading" Sedimentary Rocks

Sedimentologists—geologists who specialize in the study of sedimentary rocks—can analyze sedimentary rock formations in an area, study their fossils and sedimentary structures, and determine the depositional environment that produced each formation. In this way, they can ultimately deduce the region's unique geologic history—the sequence of events, such as the rise and fall of sea levels, the actions of streams and other agents of erosion and deposition, and even past plate tectonic events, that formed the region's landscape.

Sediment Deposition Environments

Sediment can be deposited at almost any spot on the Earth's surface—from atop the highest peak to the farthest depths of the ocean. Figure 6-18 shows just a sampling of the diversity of depositional environments. These **sedimentary**

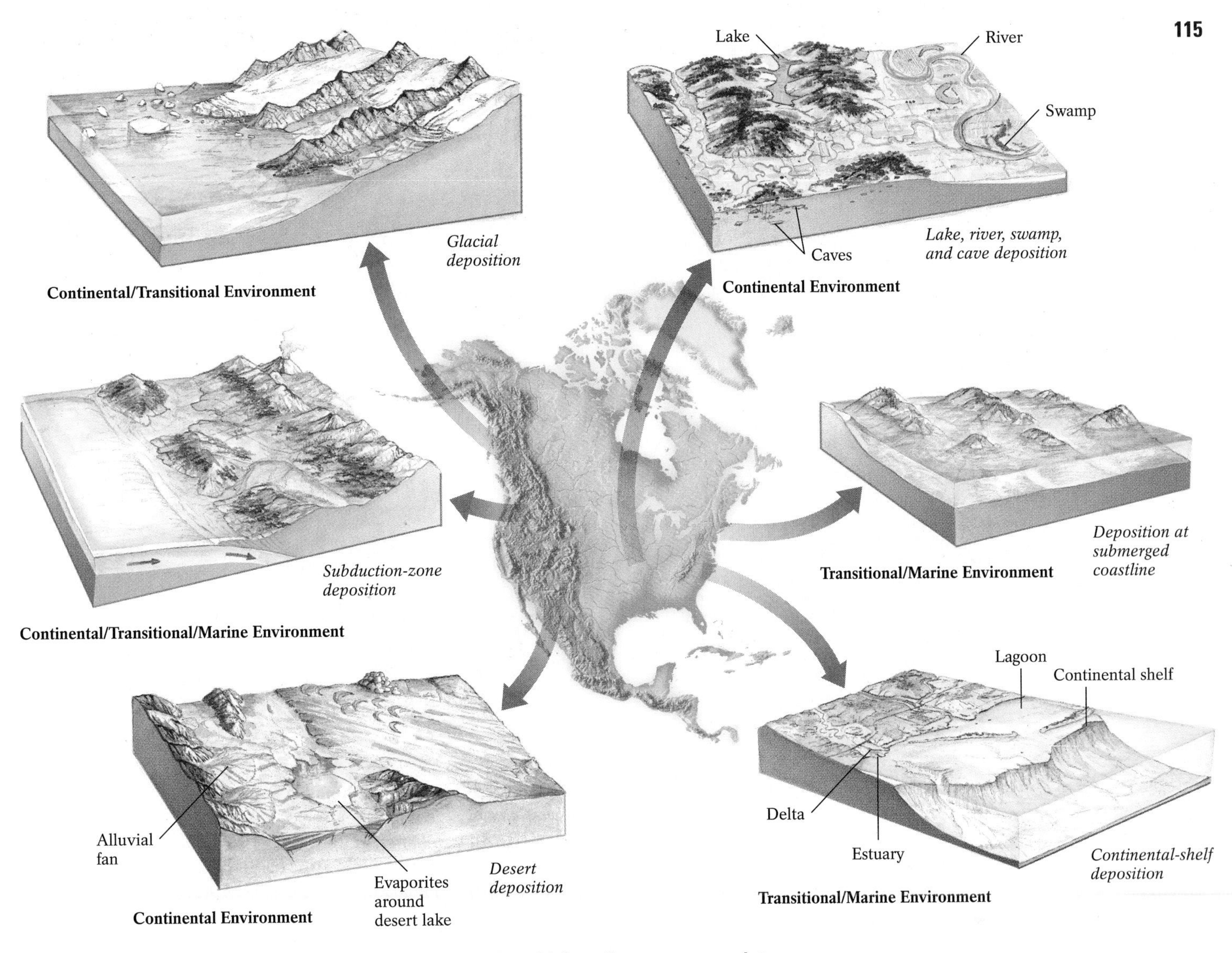

Figure 6-18 Some common geological environments in which sediments accumulate.

environments may be *continental* (on a landmass), *marine* (at sea), or *transitional* (in coastal zones). At any given time, the geological, biological, and climatic conditions in a particular sedimentary environment determine the properties of its sediments and leave telltale features such as sedimentary structures and fossils in the rocks that form there. Such features, found only in sedimentary rocks and a few metamorphic rocks, provide geologists with much information about past environments.

Continental Environments Continental sedimentary environments include river, lake, desert, glacier, and cave environments. Most of the sediment in these environments is detrital. Swift river currents carry and deposit coarse-grained sediments, forming rippled and cross-bedded structures. Sediments deposited by lower-energy currents, such as where rivers have flooded their banks or where they empty into lakes or swamps, tend to be well sorted, fine-grained, and graded. Likewise, sediments deposited in lakes and swamps tend to be well sorted, fine-grained, and graded; they often contain an abundance of organic matter.

Sedimentary deposits in desert environments include poorly sorted *alluvial fans* where mountain streams deposit their loads as they reach the flat desert floor, evaporite deposits left where temporary bodies of water have evaporated, and, most commonly, well-sorted, fine-grained sand dunes deposited and shaped by wind.

Sediments deposited in glacial environments vary greatly in terms of their composition, texture, and structure. Because glaciers transport particles of all sizes, direct deposition by these rivers of ice produces poorly sorted and unstratified sediments. In contrast, sediments carried by meltwater streams and deposited beyond the glacier's margin are generally coarse and well rounded; the streams' swift currents can readily carry away these large particles, and the turbulent flow promotes the forceful collisions that round sediment grains. The remaining finer particles typically become glacial lake deposits. The finest of these grains settle to the lake floor more slowly than the larger ones, resulting in graded and well-sorted sediments.

In caves, calcite precipitated from underground water is deposited as protrusions from the cave's ceilings, walls, and

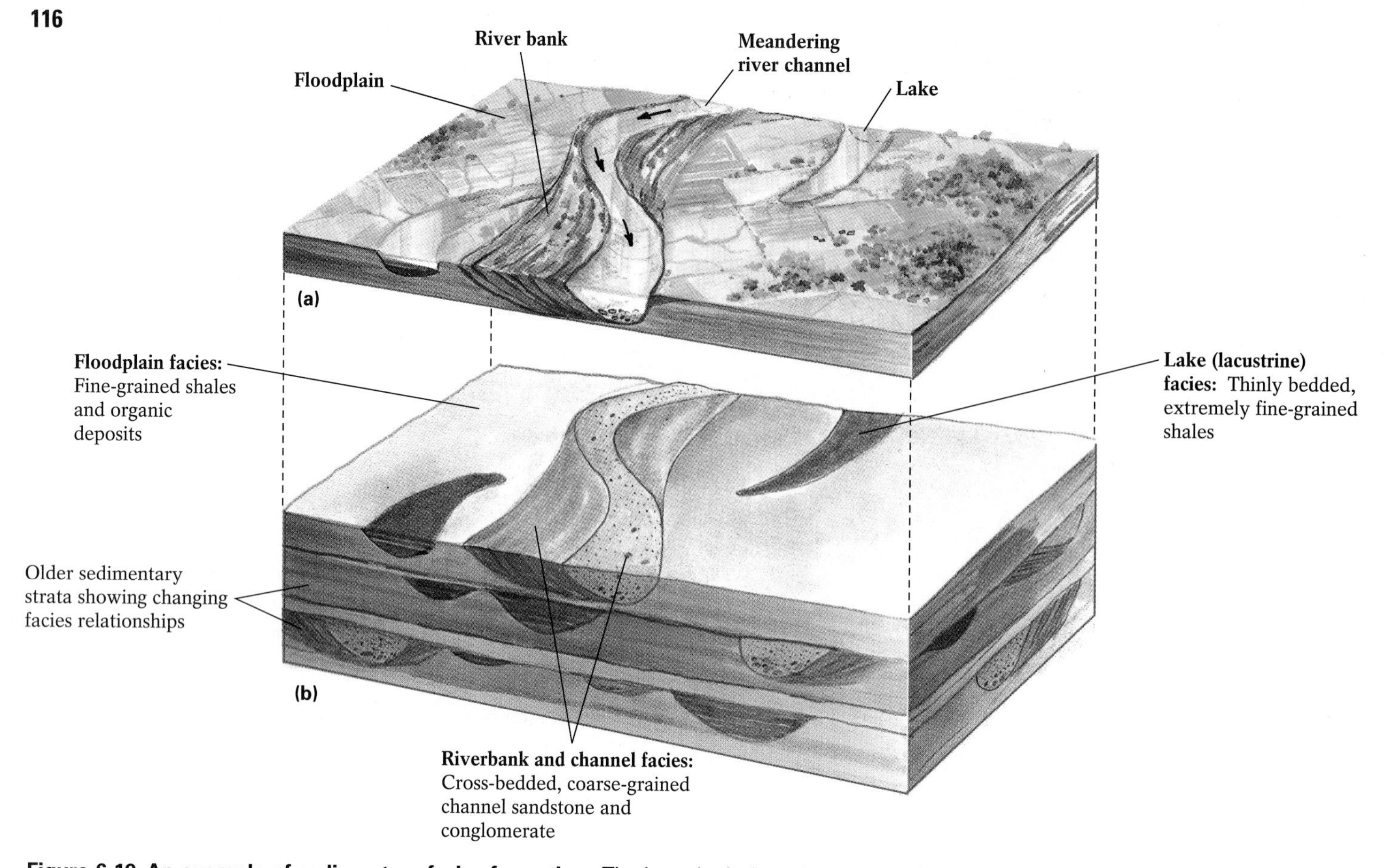

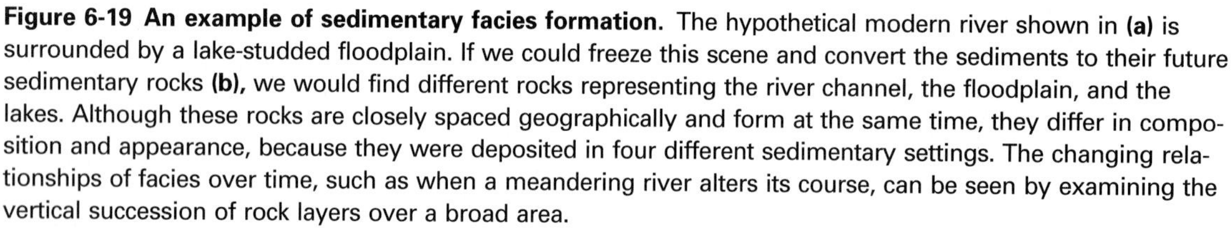

Figure 6-19 An example of sedimentary facies formation. The hypothetical modern river shown in **(a)** is surrounded by a lake-studded floodplain. If we could freeze this scene and convert the sediments to their future sedimentary rocks **(b)**, we would find different rocks representing the river channel, the floodplain, and the lakes. Although these rocks are closely spaced geographically and form at the same time, they differ in composition and appearance, because they were deposited in four different sedimentary settings. The changing relationships of facies over time, such as when a meandering river alters its course, can be seen by examining the vertical succession of rock layers over a broad area.

floors. Cave sediments may also include the bones and droppings of cave-dwelling bats, birds, and other animals.

Marine Environments Shallow-marine environments—those that lie on the continental shelves at the edges of landmasses—receive land-derived sediment carried seaward by waves and tides. Because sunlight penetrates approximately 50 meters (165 feet) below the water's surface, the upper part of the shallow-marine environment abounds in plant and animal life. Thus its sediments often consist of carbonate-rich sands and mud containing the remains of diverse marine life forms.

In contrast, sediments in deep-marine environments—those that lie beyond the continental shelf—consist largely of the remains of calcium carbonate- and silica-secreting microorganisms that have died and settled to the sea floor from the upper 50 meters (165 feet) of the ocean. Also present are land-derived deposits carried from the continental shelves by submarine landslides, clays derived from both continental and submarine volcanoes, and a small amount of meteoritic fragments from outer space.

Transitional (Coastal) Environments Along ocean shores, both continental and marine sedimentary processes are at work. In these regions, breaking waves, tides, and ocean currents pulverize soft mineral grains and shells and sweep fine particles out to sea. Left behind are well-sorted, rounded, sand-sized deposits consisting principally of durable mineral grains, such as quartz.

Where the sea intrudes upon the mouth of a coastal river, an *estuary* forms. Here, the mixture of fresh river water and salty ocean water produces a brackish (somewhat salty) environment that may contain a wide variety of organisms. The resulting sediment is rich in organic material. Where a river empties into the sea, the environment may yield a *delta*, fan-shaped accumulations of well-sorted sediments. *Lagoon* environments—shallow bodies of water between narrow offshore islands and the coast—generally contain fine continental sediments, peat, and organic mud. Transitional deposition also occurs when rising sea levels submerge coastlines, bringing marine environments to continental interiors; in such a case, marine sediments are deposited directly on top of continental sediments.

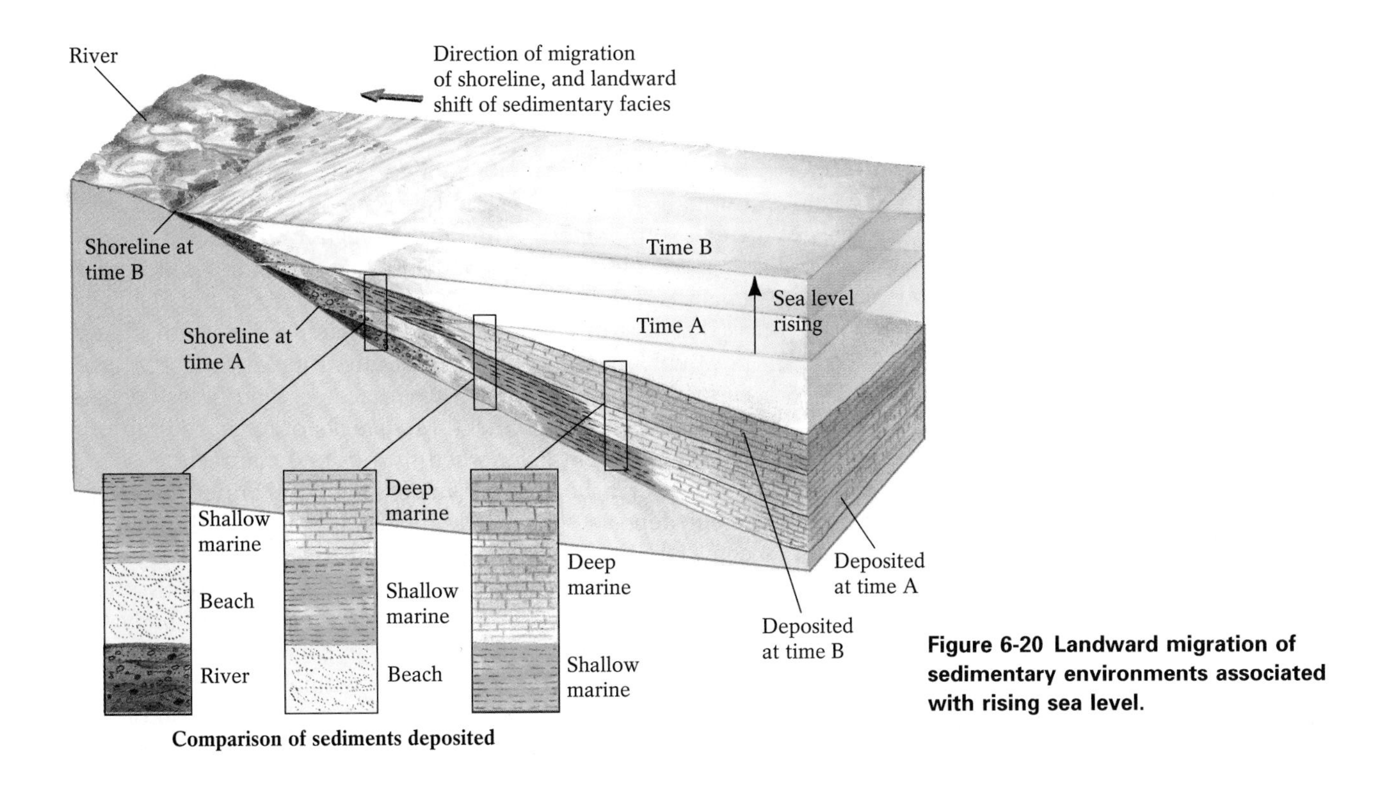

Figure 6-20 Landward migration of sedimentary environments associated with rising sea level.

Sedimentary Facies

The unique set of characteristics (such as mineral content and particle size, shape, and sorting) that distinguishes a sedimentary rock deposit from others deposited under different environmental conditions at the same time is called a **sedimentary facies** (pronounced "FAY-sheez"; from the Latin *facies,* meaning "form" or "aspect"). (The term "facies" may also refer to the rock itself.) For example, if the sediments associated with a modern coastal river became lithified, each of the related depositional settings would appear as a separate facies (Fig. 6-19). In this case, coarse, cross-bedded sandstones and conglomerates would record flowing water in the river's channel; fine shales and organic deposits would mark the river's floodplain; and thinly bedded, graded shales, perhaps containing fossils, would exist as remnants of floodplain lakes.

By examining a *vertical* succession of rock layers encompassing adjacent sedimentary settings, sedimentologists can see how these settings changed in relation to one another over time. Thus the changing positions of the facies in successive rock layers reflect events such as a shift in the course of a river (as depicted in Figure 6-19b). Sedimentologists can use the same method to determine the history of larger geological events, such as the inward migration of shorelines due to rising sea levels (Fig. 6-20).

Sedimentary Rocks and Plate Tectonics

Specific plate tectonic settings are often associated with certain characteristic sedimentary environments. Recently rifted plate margins, such as the East African rift zone, commonly contain large alluvial fan deposits, volcanic graywackes, and extensive lake deposits—all typical of fault areas that have moved downward during rifting. Transform boundaries, such as the San Andreas system in California, are noted for rapid sedimentation of angular, feldspar-rich arkosic sands (formed by the crushing of igneous rock at the plate boundaries). Rapid sedimentation also takes place near sites of continental collisions, such as in the shadow of the still-rising Himalayas, where a 15-kilometer (10-mile)-thick tongue of sediment extends 2500 kilometers (1500 miles) into the Indian Ocean south of Calcutta, India. Volcanoes that rise above a subducting plate supply sediment that tends to be poorly sorted and rich with angular fragments of volcanic rock, testimony to their rapid burial and the contributions of an active sediment source.

Just as recent or current tectonic events have allowed us to learn about the different sedimentary processes, ancient sedimentary rocks can teach us about long-past plate tectonic events. Some of the Earth's most majestic mountains—the Northern Rockies, the Appalachians, the European Alps, the Urals of Russia, and the Himalayas of China, Nepal, and Tibet—contain sedimentary rock strata that were clearly deposited in marine environments. They exhibit ripple marks that record rising and falling tides and coastal waves, evaporites and mudcracks from past sea-level fluctuations, and thousands of meters of limestone. These rocks, all of which were derived from sediments that accumulated in shallow-marine environments, have been uplifted by the mountain-building processes associated with convergent plate boundaries.

Sedimentary rocks also provide evidence that lofty mountains once stood where none stands today. The fairly squat Taconic Hills of western Connecticut and Massachusetts and eastern New York, for example, are thought to be the remnants of mountains that rose when the plates containing what are now Europe, Africa, and North America collided about 375 million years ago. Rivers carried a great volume of coarse detrital sediment from those now-departed mountains into shallow seas to the west; the cross-bedded sandstones derived from these sediments, which were themselves subsequently uplifted and then cut through by streams, now form the Catskill Mountains of east-central New York.

We will encounter sedimentary processes and rocks throughout the remainder of this book. Sedimentary rocks record local, regional, and even global geological events, as well as the evolution of the Earth's animals and plants. The motion of the Earth's plates, for example, can cause horizontal beds of sedimentary rock to be uplifted into continent-long mountain ranges such as the Rockies and Appalachians. In later chapters, we explore how sediment is moved from place to place by gravity, rivers and streams, underground water, glaciers, desert winds, and ocean waves. Finally, we discuss some sedimentary processes that produce many of the Earth's valuable energy and mineral resources.

Chapter Summary

Sediment consists of unconsolidated fragments of preexisting rock, minerals precipitated directly out of solution in water, and the remains of organisms. Much of the sediment that accumulates continuously at the Earth's surface is eventually converted to **sedimentary rocks.** Sedimentary rocks make up a thin layer of the Earth's crust that accounts for approximately 75% of all rocks exposed on land.

Sediments are classified according to the source of their constituent minerals. **Detrital sediment** is composed principally of fragments of preexisting igneous, sedimentary, or metamorphic rocks. **Chemical sediment** consists of minerals —originally derived from dissolved chemical weathering products—that have either precipitated directly out of solution by inorganic chemical processes or been extracted from solution by organisms and ultimately deposited in the form of shells, skeletons, and other organically derived materials.

The texture of detrital sediment—the size, shape, and arrangement of the particles in a deposit—is determined largely by its transport (in addition to its parent material's weathering resistance). During transport, rock fragments generally become reduced in size and rounded. Some sediments undergo **sorting,** a process by which sediment grains are carried or deposited selectively based on their size, density, and shape and on the energy of the transport medium.

Detrital sediments often display **sedimentary structures,** features that develop during or soon after deposition and that reflect the conditions under which they were deposited. **Bedding,** or **stratification,** is the arrangement of sediment particles into distinct layers marking separate depositional events. **Graded bedding** forms as sediment settles through standing water. In this type of bedding, the coarsest grains settle to the bottom first, and grain size decreases gradually toward the top of a layer. **Cross-bedding** refers to sediment layers that are oriented at an angle to underlying beds, as is typical of wind-deposited sediments and sediments deposited by moving currents of water. **Ripple marks** are small surface ridges produced when water or wind flows over recently deposited sediment. **Mudcracks** occur in the top of a muddy sediment layer when it dries and contracts.

After a body of sediment has been buried by subsequent deposits, it may eventually undergo **lithification,** the conversion of its loose particles into solid sedimentary rock. **Compaction** is the process by which the weight of overlying materials compresses the volume of a sedimentary body. **Cementation** of sediment grains occurs when dissolved minerals precipitate in the pore spaces within the sediment. Any rock that is formed by compaction and cementation of sediment particles is said to have a **clastic** texture. **Recrystallization** converts unstable minerals in sediment into new, stable minerals.

Detrital sedimentary rocks are classified by grain size. They include the fine-grained **mudstones** and shales, intermediate-grained **sandstones,** and coarse-grained **conglomerates** (with rounded grains) and **breccias** (with angular grains).

Classification of chemical sedimentary rocks is based on the composition of the sediment rather than grain size. These rocks are further classified as being inorganic or biogenic, depending on how their mineral components were converted from their original dissolved state into solid form. Inorganic chemical sedimentary rocks precipitate directly from water; they include **inorganic limestone, dolostone, evaporites,** and **inorganic chert.** Biogenic chemical sedimentary rocks form when organisms extract dissolved compounds from water and convert them into biological hard parts (such as shells and skeletons) that are ultimately deposited as sediment. They include **biogenic limestone** and **biogenic chert. Coal,** formed from the remains of swamp plants, is also considered a biogenic chemical sedimentary rock.

Sediment accumulates in numerous **sedimentary environments,** which may be continental, marine, or transitional (coastal). Because the properties of any sedimentary rock stem from the specific conditions under which it develops, geologists can distinguish rocks formed in one depositional setting from rocks formed in a different environment. Using this information, they can then deduce how past environments have changed over geologic time. The unique set of characteristics (such as mineral content and particle size, shape, and sorting) that distinguishes a sedimentary rock deposit from others deposited under different environmental conditions at the same time is called a **sedimentary facies.**

Key Terms

sediment (p. 101)
sedimentary rock (p. 101)
detrital sediment (p. 102)
chemical sediment (p. 102)
sorting (p. 103)
sedimentary structures (p. 104)
bedding (p. 104)
stratification (p. 104)
graded bed (p. 104)
cross-beds (p. 104)
ripple marks (p. 104)
mudcracks (p. 106)
lithification (p. 106)
compaction (p. 106)
cementation (p. 106)
clastic (p. 106)
recrystallization (p. 106)
mudstones (p. 108)
sandstones (p. 108)
conglomerates (p. 109)
breccias (p. 109)
inorganic limestone (p. 110)
dolostone (p. 110)
evaporites (p. 111)
inorganic chert (p. 113)
biogenic limestone (p. 113)
biogenic chert (p. 113)
coal (p. 114)
sedimentary environments (p. 114)
sedimentary facies (p. 117)

Questions for Review

1. What are the two major classes of sedimentary rocks? On what basis are they distinguished?

2. Briefly describe how sorting and rounding of detrital sediment vary with different transport media, such as wind, rivers, and glaciers.

3. Describe the differences between graded beds and cross-beds. Which indicates the flow direction of ancient currents? Which can be used to determine if a sedimentary bed is upside down or right side up? Give an example of a setting in which each forms.

4. Explain the processes of lithification, and describe how each affects the physical properties of sediment.

5. Describe the composition and texture of mudstones, sandstones, conglomerates, and breccias.

6. What are evaporites? Describe two sedimentary environments where evaporites form.

7. Name and describe the origins of three different types of organic chemical sedimentary rocks.

8. What is the difference between peat, lignite, bituminous, and anthracite coals?

9. Describe how deposition occurs in each of three different *continental* sedimentary environments. How do sediments in deep-marine and shallow-marine environments differ?

10. Name some types of sedimentary facies associated with each of two different plate tectonic boundaries.

For Further Thought

1. Why do we find much more of the evaporites halite and gypsum than other types of evaporites? Why do we often find dolostone associated with evaporite deposits?

2. Under what circumstances might we find poorly sorted, angular, arkosic sediments in a coastal environment?

3. Study the photo of a modern salt flat above and speculate about the environmental conditions that existed when these sediments were first deposited. How would the appearance of the present sediments change if the climate in this area became very moist?

4. In the figure below, what can we tell about the route taken by the current that deposited the conglomerate?

5. What environmental conditions might create four different types of coal in beds of comparable age?

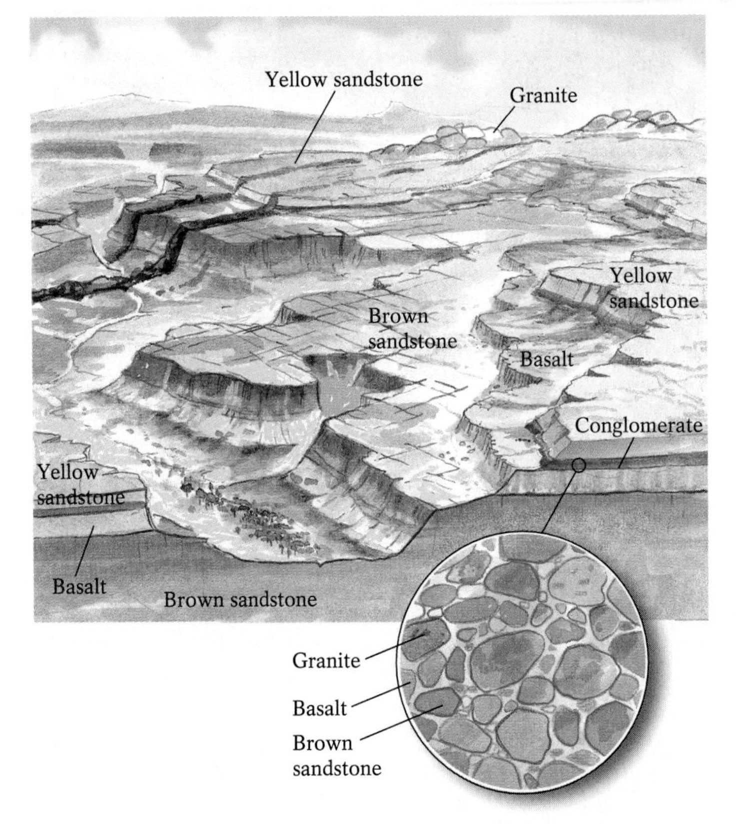

Section of conglomerate

7

Metamorphism and Metamorphic Rocks

We saw in Chapter 3 that high temperatures deep in the Earth's interior create the magmas that eventually cool to form igneous rocks. We saw in Chapter 6 that sediments lithify to become sedimentary rocks in the relatively low-temperature environment near the Earth's surface. The third major type of rock, **metamorphic rocks,** generally forms at conditions between those that produce igneous and sedimentary rocks (Figs. 7-1 and 7-2).

Metamorphism is the process by which temperature, pressure, and chemical reactions deep within the Earth—but above the 50- to 250-kilometer–deep "melting zone" that creates most magmas—alter the mineral content and/or

Figure 7-1 The beauty of metamorphic rocks. These picturesque rocks at Pemaquid Point in coastal Maine have been converted to metamorphic rocks by extreme heat and pressure.

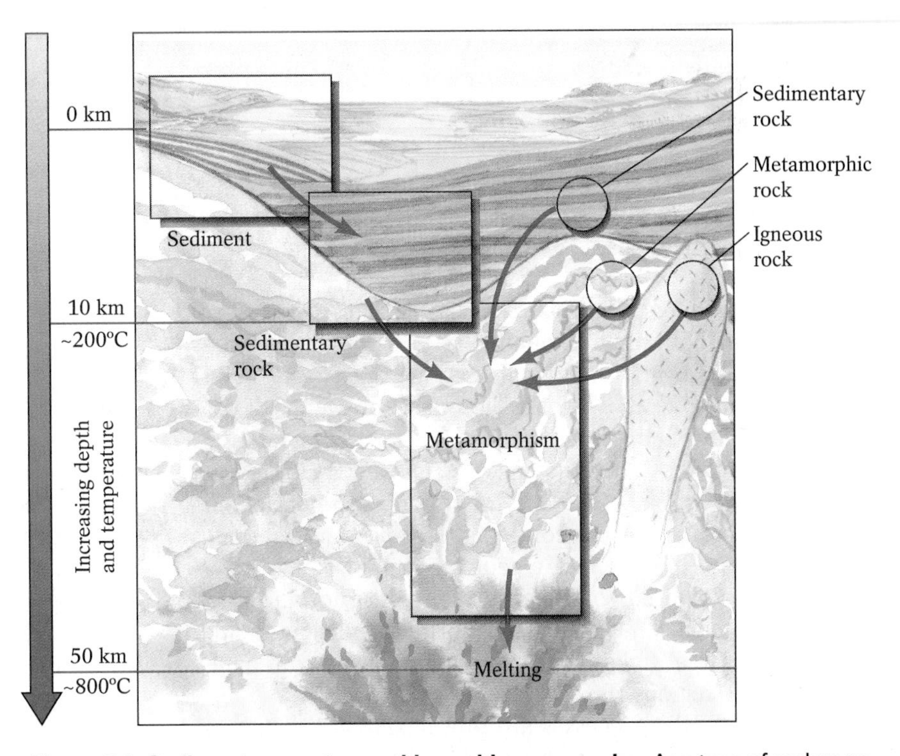

Figure 7-2 Sedimentary, metamorphic, and igneous rocks. Any type of rock may be metamorphosed at temperatures and pressures greater than those that lithify sediment into sedimentary rock, but less than those that melt rock into magma.

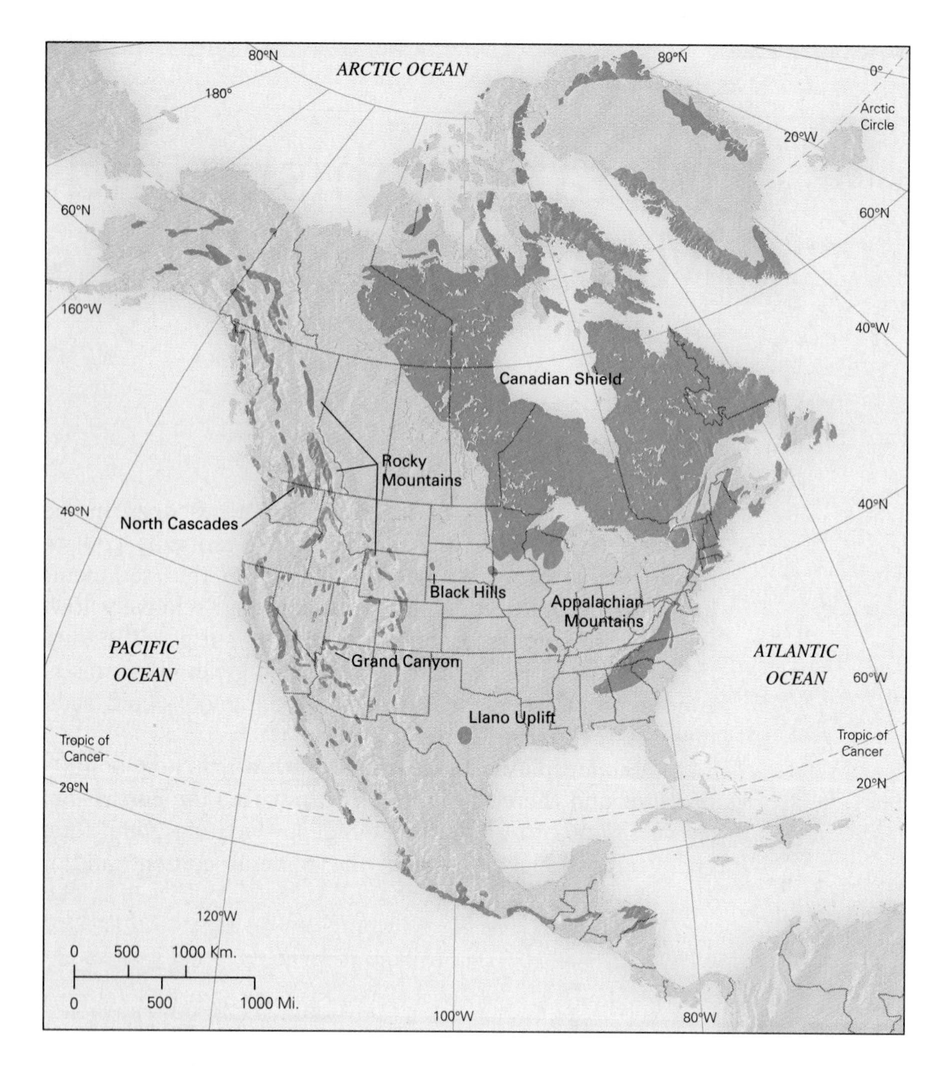

Figure 7-3 Exposed and near-surface metamorphic rocks of North America. Originally buried beneath kilometers of sedimentary rocks, the continent's metamorphic rocks can be seen only where erosion has removed the overlying rock—principally in very deep river valleys, such as the Grand Canyon; in the cores of mountain ranges, such as the Colorado Rockies, the North Cascades of Washington state, and the Appalachians of eastern North America; and in the glacially scoured Canadian Shield region of Ontario, Quebec, and the adjacent Great Lakes states of Minnesota, Wisconsin, and Michigan.

structure of preexisting solid rock *without melting it*. Any rock—whether igneous, sedimentary, or even metamorphic—may be a candidate for metamorphism.

Most of the metamorphic rocks in the United States are buried beneath thousands of meters of sedimentary rock. Near Topeka, Kansas, for example, sedimentary rocks lie just below the surface and extend thousands of meters downward; they formed from shallow-marine sediments left behind by inland seas that invaded North America more than 100 million years ago. The ancient rocks that underlie surface rocks throughout the continent, however, are predominantly metamorphic. In fact, a geologic map of North America reveals exposed or near-surface metamorphic rock throughout much of Canada, the northern portions of the Midwestern Great Lakes states, and many of the continent's mountainous regions (Fig. 7-3).

Metamorphic processes are presumably going on continuously beneath the Earth's surface, but we see the resulting rocks only after uplift and erosion have stripped away the overlying rocks. Most of what we know about the causes of metamorphism has come from laboratory experiments and theoretical models that replicate conditions in the Earth's interior.

Conditions Promoting Metamorphism

A basic geologic principle states that rocks and their constituent minerals are most stable in the environment in which they form and least stable in markedly different environments. For example, as we learned in Chapter 5, feldspars formed by relatively high-temperature igneous processes rapidly break down (weather) chemically at the Earth's surface, an environment characterized by relatively low temperatures and low pressure. Their replacements, clay minerals, are more stable than the feldspars under surface conditions. But if the clays were placed in a new environment of higher temperature and pressure—for example, if they became buried beneath tens of kilometers of overlying sediments and heated to 600°C (1110°F)—they would react to form new minerals, and perhaps even some feldspars, that would be stable under *these* new conditions.

Metamorphic rock is created when heat and pressure combine to break the bonds between some of the ions in an unstable mineral, allowing them to migrate to other sites in the rock and rebond. This migration of ions, which usually occurs via fluids circulating through the rock, results in recrystallization of the mineral: If the ions rebond at different sites within the same mineral, the process produces a new, more stable form of the same mineral; if they rebond with other unstable minerals in the rock, it yields an entirely new set of more stable minerals. During recrystallization, minute grains in a fine-grained rock are usually replaced by an interlocking mosaic of large, visible grains.

Metamorphic processes never break all of the bonds in a rock's minerals—if all bonds were broken, the rock would melt to form a magma (an igneous process). Instead, metamorphism occurs when heat and pressure exceed certain threshold levels, destabilizing the minerals in rocks, but do not reach the heights needed to cause melting.

Heat

Heat, which speeds up the pace of nearly all chemical reactions, is perhaps the most important factor contributing to metamorphism. Metamorphism can sometimes occur in the absence of great pressure, but heat is usually necessary to initiate any significant chemical changes associated with metamorphism. Beneath the Earth's surface, temperature generally increases with depth due to the presence of radioactive isotopes, intruding magma, and friction between moving bodies of rocks. Temperatures sufficient to metamorphose rocks—greater than about 200°C (400°F)—are generally reached at a depth of about 10 kilometers (6 miles), a position still within the crust and upper mantle. Temperatures in this range may also occur nearer to the Earth's surface in rocks close to magmatic intrusions.

Pressure

Like heat, pressure increases with depth. The pressure required for metamorphism is more than 1 kilobar, the prevailing pressure about 3 kilometers (2 miles) beneath the Earth's surface. (A kilobar is a unit of pressure equal to 1020 kilograms/square centimeter or 14,700 pounds/square inch.) Metamorphism will not occur at this shallow a depth, however, unless heat is carried up from greater depths by magma or is induced by friction between tectonic plates.

When pressure is applied to a rock equally on all sides, as occurs at great depths, it is called **lithostatic** or **confining pressure.** When exposed to lithostatic pressure, a deeply buried rock becomes compressed into a smaller, denser form, but its shape remains the same (Fig. 7-4a). In some geologic settings, such as where tectonic plates collide to form mountains, pressure is applied principally in one plane, creating a force known as **directed pressure.** Directed pressure flattens and elongates a rock perpendicular to the direction of pressure (Fig. 7-4b). It may also deform individual components

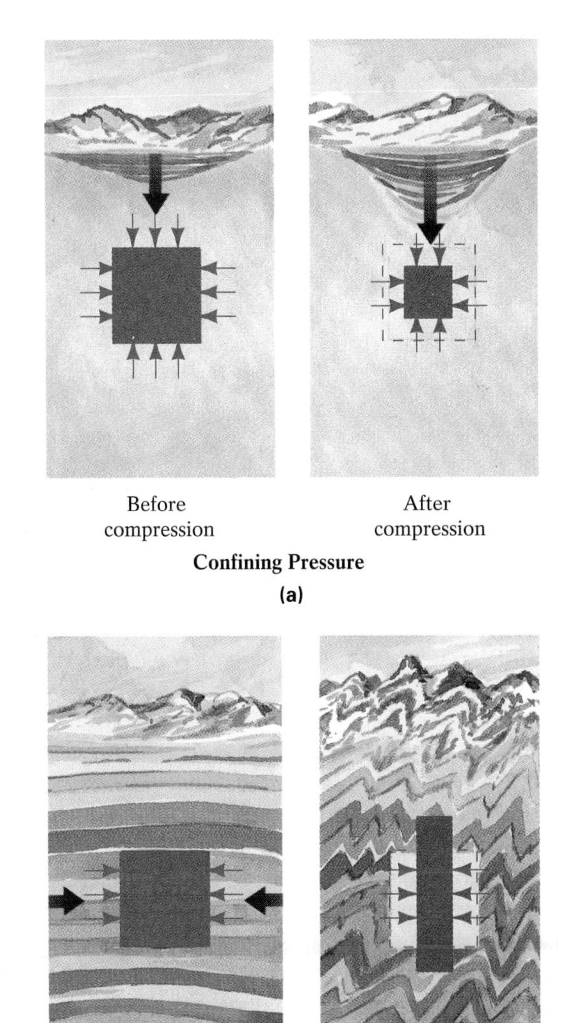

Figure 7-4 Lithostatic versus directed pressure. (a) Deep burial of rocks generally subjects them to lithostatic, or confining, pressure, an inward-pressing force that acts equally from all directions. A rock subjected to confining pressure becomes compressed without changing shape. **(b)** A rock subjected to directed pressure has its shape distorted, becoming thinner in the direction of the greatest pushing, or stress, and elongated in the direction perpendicular to the stress.

Figure 7-5 Folded rock in the Blue Ridge Mountains of North Carolina. This example shows the deformation typical of rocks subjected to directed pressure.

within the rock, such as fossils and mineral grains, causing them to become stretched and folded (Fig. 7-5).

Rocks subjected to either confining or directed pressure change in a number of ways. Both types of pressure close pore spaces between mineral grains, producing a more compact, denser rock. In addition, pressure at the contact points between compressed grains breaks some of their bonds, especially where water is present. The unbonded ions then migrate to lower-pressure sites, where they rebond.

Directed pressure creates one special change in rocks: Recrystallized minerals develop a parallel alignment perpendicular to the directed pressure. This preferred alignment is known as **foliation** (from the Latin *foliatus,* meaning "leaf-like") (Fig. 7-6). If the mineral grains are thin and planar, such as sheet-like mica flakes, their parallel orientation gives a distinctive layered look to the rock.

Circulating Fluids

Underground fluids, such as water and magmatic gases, enhance metamorphism by providing a medium through which unbonded ions can migrate among mineral grains. In addition, most such fluids contain a variety of ions derived from other rocks, which may be transferred to the rocks undergoing metamorphism. Thus fluids not only serve as the medium through which a rock's own unbonded ions migrate, but also contribute foreign ions to metamorphic reactions. As a result, the type of metamorphic rock that forms from a given parent rock depends on both the rock's original mineral content and, to some extent, on the ion content of the fluids that flowed through it.

Figure 7-6 Pressure and foliation. Directed pressure creates foliation in rock.

Types of Metamorphism

Heat, pressure, and chemically active fluids interact in different ways in different geological settings to produce metamorphic rocks. These settings therefore yield several distinctive types of metamorphism. Under some conditions, the metamorphic process affects only a narrow area of rock; in other settings, rocks may be changed over a vast region.

Contact Metamorphism

Preexisting rock touched by the intense heat of migrating magma may undergo **contact metamorphism.** In this process, metamorphic change within the rock develops entirely from the heat from the magma and from hot circulating fluids; pressure is not a significant factor. Because most rocks are poor conductors of heat, the effect of contact metamorphism decreases with increasing distance from the magma. Rocks in direct contact with the intruding magma will be highly metamorphosed; those located farther away, which receive little magmatic heat, will be only slightly altered. The entire zone of contact metamorphism, referred to as a metamorphic **aureole** (Fig. 7-7), may extend only a few centimeters or meters around a small igneous dike or sill, or it may extend up to several kilometers around a major batholith.

Regional Metamorphism

Unlike contact metamorphism, which has relatively local effects, **regional metamorphism** alters rocks for thousands of square kilometers. For example, it created the vast regions of exposed metamorphic rock in central Canada and the tracts of metamorphic rock found in the Appalachians of New England, the Rockies, and the Cascades. Two types of regional metamorphism exist—burial metamorphism and dynamothermal metamorphism.

Burial metamorphism occurs when rocks are overlain by more than about 10 kilometers (6 miles) of rock or

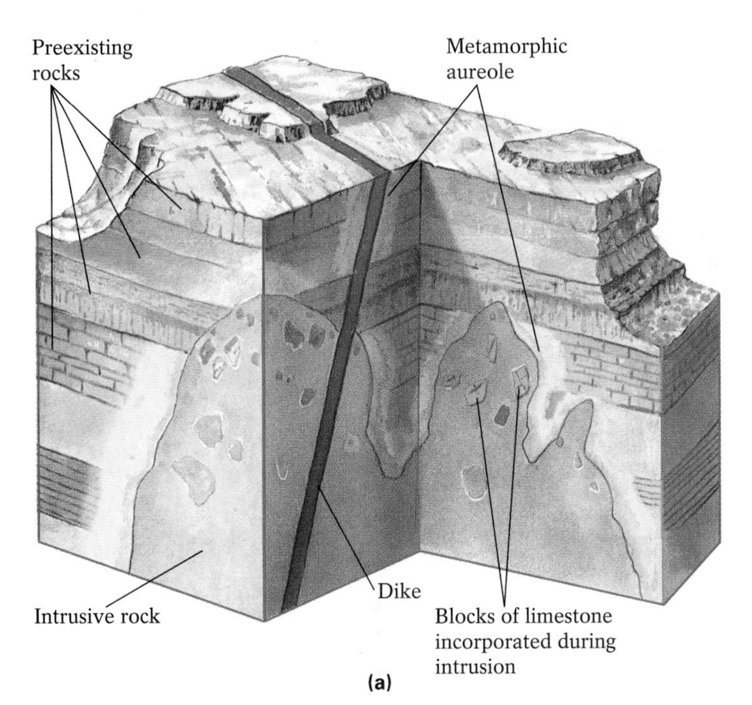

(b)

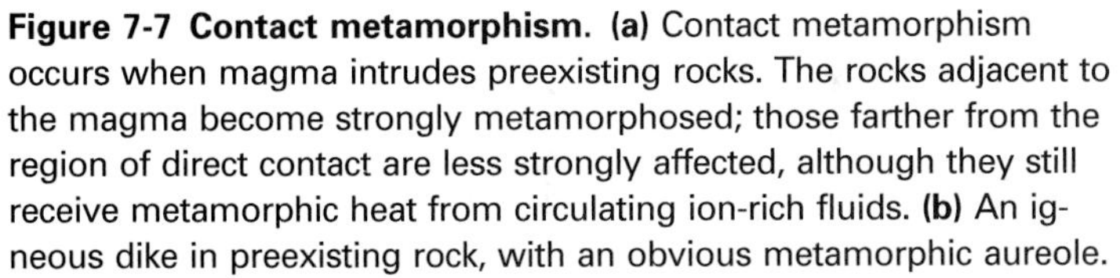

Figure 7-7 Contact metamorphism. **(a)** Contact metamorphism occurs when magma intrudes preexisting rocks. The rocks adjacent to the magma become strongly metamorphosed; those farther from the region of direct contact are less strongly affected, although they still receive metamorphic heat from circulating ion-rich fluids. **(b)** An igneous dike in preexisting rock, with an obvious metamorphic aureole.

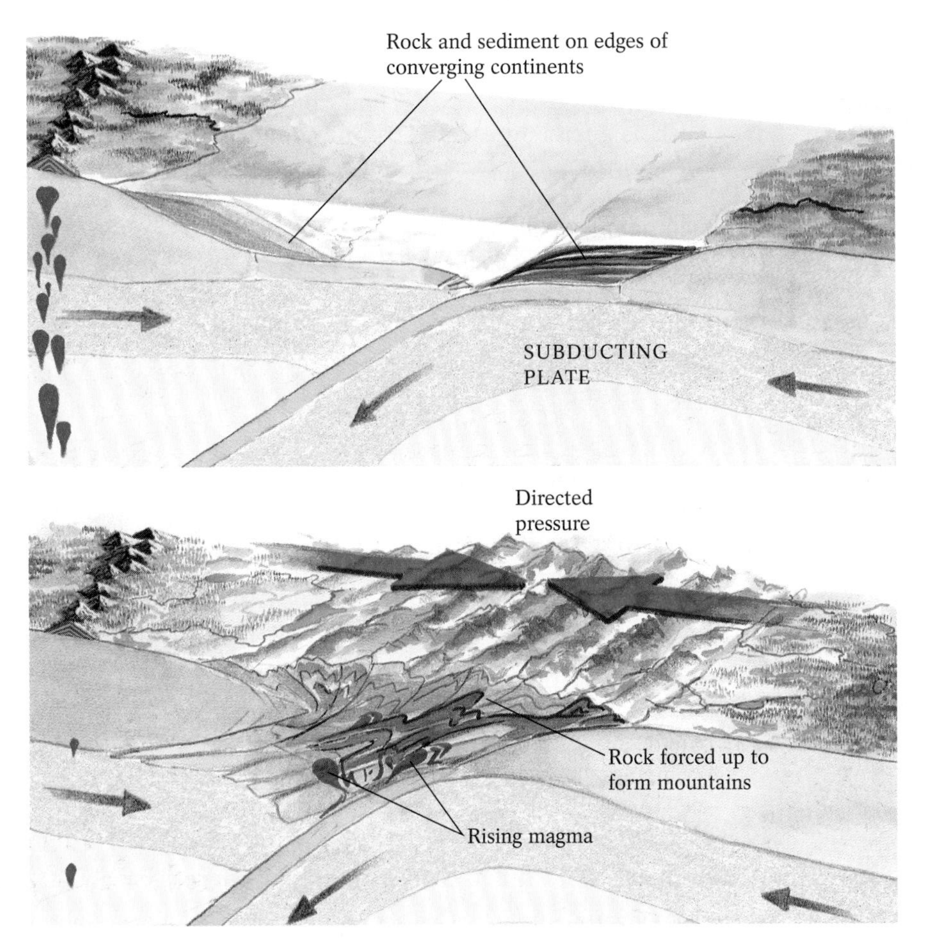

sediment, and the confining pressure and geothermal heat at these depths combine to recrystallize their component minerals. Because the process does not involve directed pressure, burial metamorphic rocks are generally nonfoliated. Samples collected from deep drill holes show that burial metamorphism is occurring today in such locations as the Gulf Coast of Louisiana, where clay minerals lie beneath 12 kilometers (8 miles) of sediment near the bottom of the Mississippi River's deltaic deposits.

Dynamothermal metamorphism occurs when rocks become caught between two converging plates during mountain building (Fig. 7-8). In such a convergent plate setting, the lateral compression—a form of directed pressure—forces some rocks upward, forming mountains, and forces other rocks downward to depths of several to tens of kilometers. The

Figure 7-8 Dynamothermal metamorphism. This type of metamorphism is generally associated with the directed pressures and magmatic heat of convergent plate boundaries.

latter rocks are then subjected to great heat and confining pressure. The resulting metamorphic rocks are typically foliated. Dynamothermal metamorphism created the vast regions of metamorphic rock that lie at the cores of many mountain ranges, including the Alps, Himalayas, and Appalachians. It also formed the roots of more ancient, now-vanished mountains, including those exposed by billions of years of erosion in the northern Great Lakes states of Minnesota, Wisconsin, and Michigan and adjacent parts of southern Ontario and Quebec.

Other Types of Metamorphism

Rocks may be metamorphosed in several ways other than by regional increases in heat and pressure or by contact with intruding hot magmas. *Hydrothermal metamorphism* is the chemical alteration of preexisting rocks by hot water. Most hydrothermal metamorphism occurs within ocean floors, when seawater penetrates cracks near a divergent plate boundary and becomes heated by hot magma. Rising back up through the rock as steam, it promotes the recrystallization of olivine and pyroxene in the oceanic basalt, converting them into the magnesium silicate serpentine. *Fault metamorphism* occurs as rocks grind past one another along a fault, thereby generating a large amount of directed pressure and considerable frictional heat. *Shock metamorphism*, a rare but dramatic type of metamorphism, results when a meteorite strikes the Earth's surface, generating tremendous pressures and extremely high temperatures. Some minerals created in a meteoric impact, such as stishovite, occur in no other geologic setting, and can thus be used to identify such sites long after erosion has removed any impression of a crater. *Pyrometamorphism* sometimes occurs where lightning bolts or burning subterranean coal seams subject rocks to intense heat.

Common Metamorphic Rocks

The first criterion used to distinguish any metamorphic rock is whether it is foliated or nonfoliated. When the parent rock is a simple, single-mineral sedimentary rock such as quartz, the resulting metamorphic rock is usually nonfoliated (or so slightly foliated that the foliation is invisible to the naked eye) and composed predominantly of that same mineral; the key change involves its recrystallization into a more stable configuration. For example, the quartz grains in a pure sandstone recrystallize to form the nonfoliated metamorphic rock quartzite. Dramatic, readily visible foliation tends to develop when a multimineral, mica-rich rock is subjected to progressively greater heat and directed pressure. In this case, the rock's several minerals may recombine in different ways to form a number of new minerals.

Table 7-1 summarizes the classification scheme for metamorphic rocks. Note that some metamorphic rocks, such as gneiss and schist, can be produced from a number

Table 7-1 Classification and Derivation of Some Common Metamorphic Rocks

	Rock	Parent Rock(s)	Key Minerals	Metamorphic Conditions
Foliated	Slate	Shale, mudstone	Clay minerals, micas, chlorite	Relatively low temperature and directed pressure
	Phyllite	Shale, mudstone	Mica, chlorite	Low–intermediate temperature and directed pressure
	Schist	Shale, mudstone, basalt, graywacke sandstone, impure limestone	Mica, chlorite, epidote, garnet, talc, hornblende, graphite	Intermediate–high temperature and directed pressure
	Gneiss	Shale, felsic igneous rocks, graywacke sandstone	Quartz, feldspars, garnet, mica, augite, hornblende, staurolite, kyanite	High temperature and directed pressure
Nonfoliated	Marble	Pure limestone or dolostone	Calcite, dolomite	Contact with hot magma, or confining pressure from deep burial
	Quartzite	Pure sandstone	Quartz	Contact with hot magma, or confining pressure from deep burial
	Hornfels	Shale, mudstone, basalt	Andalusite, mica, quartz	Contact with hot magma; little pressure

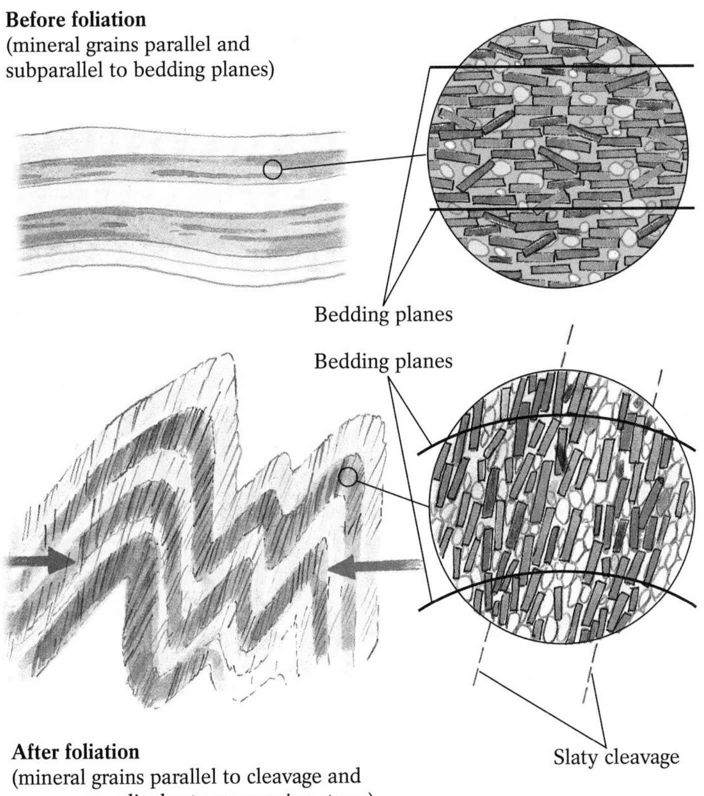

Figure 7-9 Slaty cleavage. The parallel cleavage characteristic of slate is due to foliation, which always occurs perpendicular to the directed pressure applied during metamorphism.

of different parent rocks; geologists distinguish between these rocks based on their appearance and the conditions that formed them, rather than their composition. Other metamorphic rocks, such as quartzite and marble, are produced by a specific type of parent rock and so are classified by their compositions.

Foliated Metamorphic Rocks Derived from Shales or Mudstones

Foliated metamorphic rocks are most often derived from shale (or mudstone). Shale, which consists largely of clay minerals and micas, is derived from the lithification of mud buried within the upper few kilometers of the Earth's surface. With increasing directed pressure and temperatures exceeding 200°C (400°F), the clay minerals in shale gradually recrystallize to form minute but relatively long, flat mica flakes that become aligned perpendicular to the direction of the applied pressure. The resulting metamorphic rock is known as **slate** (Fig. 7-9). Slate tends to break parallel to the mica-rich planes to form relatively uniform thin, flat fragments. This property, known as *slaty cleavage* or *rock cleavage,* renders slate useful for blackboards, floor tiles, roofing tiles, and pool-table tops. (Note that rock cleavage is not the same as mineral cleavage, discussed in Chapter 2. *Mineral cleavage* refers to minerals breaking between planes of atoms or ions in a crystal, whereas *rock cleavage* refers to rocks breaking between planes of minerals.)

At temperatures of about 300°C (575°F), the microscopic mica flakes in slate (as well as chlorite and graphite flakes) grow larger, producing the foliated metamorphic rock **phyllite** (Fig. 7-10). The light reflected off the surfaces of these larger mica grains gives phyllite its characteristic sheen. With increased heat and pressure, the fine mineral flakes grow even more. When the flakes become readily visible to the unaided

Figure 7-10 Phyllite. The shiny appearance of phyllite is due to the light reflected off its mica grains.

Figure 7-11 Schist. This strongly foliated metamorphic rock has large, easily visible flakes of mica.

eye, the metamorphic rock is classified as a **schist** (Fig. 7-11), a medium- to coarse-grained, strongly foliated, usually "glittery" rock. Schists represent a relatively high grade of metamorphism, with any sedimentary structures preserved in the original rock usually being obliterated.

As temperatures approach 500° to 700°C (950°–1300°F), the minerals in schists become segregated into alternating light- and dark-colored layers. During metamorphism, the liberated ions in unstable felsic minerals migrate and recrystallize in distinct light-colored, single-mineral bands containing larger grains of new, stable minerals, such as quartz and feldspar. Meanwhile, ions freed from other minerals, such as biotite and the dark-colored amphiboles and pyroxenes, form intervening bands of dark mafic minerals. This process, known as **metamorphic differentiation,** creates the distinctively banded metamorphic rock **gneiss** (pronounced "NICE"). Gneisses are most often found in the exposed cores of ancient mountain ranges, such as in the Front Range of the Colorado Rockies or the North Cascades of Washington. The 3.7-billion-year-old Morton Gneiss (Fig. 7-12), which is exposed in the Minnesota River valley of southwestern Minnesota, and the 600- to 700-million-year-old Fordham Gneiss, which can be seen in New York's Central Park, represent remnants of such long-departed mountain ranges.

When gneiss is subjected to heat in the 600° to 800°C range (1110°–1470°F), its felsic bands begin to melt, because their melting points have been reached. On the other hand, the mafic bands, whose melting points are higher, remain solid. The mafic bands do begin to deform at these temperatures, creating intricately contorted patterns. If cooling occurs at this stage, the resulting rock—part metamorphic and part igneous—is known as a **migmatite** (Fig. 7-13).

Figure 7-12 Morton Gneiss. The characteristic banding in this block of Morton Gneiss formed under the high directed pressure and high temperatures typical of mountain building at convergent plate boundaries.

Figure 7-13 Migmatite. Migmatites are foliated like metamorphic rocks, but their felsic regions, which cooled from a melted state, contain phaneritic textures like those of plutonic igneous rocks.

Foliated Metamorphic Rocks Derived from Igneous Rocks

Like multimineral sedimentary rocks, such as shales, igneous rocks exposed to high temperatures and high directed pressure may metamorphose to form schists and differentiate to form gneisses. For example, when basalt metamorphoses, its pyroxenes, olivines, and calcium plagioclases become converted to a group of more stable minerals that includes chlorite and epidote. Because both of these minerals are green, the resulting foliated metamorphic rock is called *greenschist*. When granite and diorite metamorphose, they differentiate to form gneisses. Because their parent rocks are felsic, however, these gneisses contain more pronounced quartz and feldspar foliations and fewer mafic foliations than do gneisses derived from shale.

Nonfoliated Metamorphic Rocks

Many nonfoliated metamorphic rocks form from recrystallization of essentially single-mineral sedimentary rocks. They may be produced by the increased heat and high confining pressure ensuing from very deep burial, or they may result from contact with heat from an intruding body of magma. Even under directed pressure, foliation is unlikely to develop, because these rocks lack the flatter (*platy*) mineral grains typical of clay, which can be easily rotated and reoriented. When relatively pure limestones and dolostones metamorphose, they form **marble,** a coarse, nonfoliated mosaic of predominantly calcite and/or dolomite grains. Impurities in the parent rock are excluded during the recrystallization, thereby creating the familiar colored patterning often associated with marble. Pure quartz sandstone metamorphoses to **quartzite,** a particularly durable nonfoliated rock composed of very tightly interlocked quartz grains (Fig. 7-14).

When hot magma intrudes a shale, slate, or basalt, the accompanying heat drives off virtually all mineral-bound water. This development promotes recrystallization and metamorphic reactions that produce anhydrous minerals with more compact structures. The resulting nonfoliated rock, **hornfels,** is dark-colored, dense, and hard.

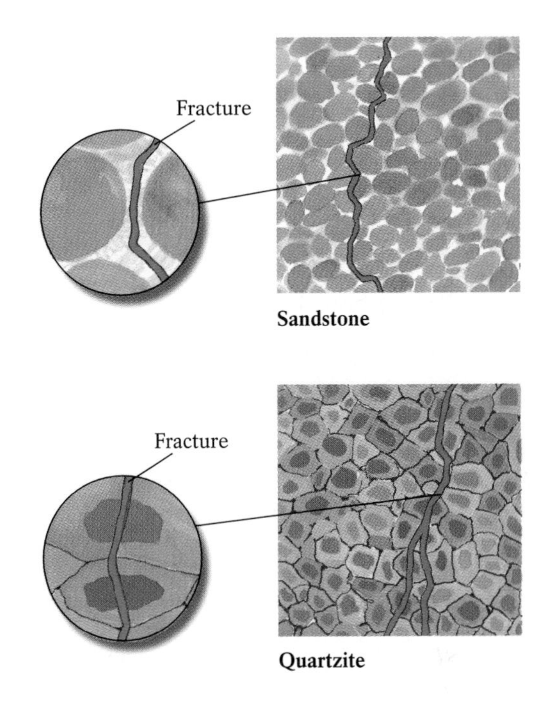

Figure 7-14 Formation of nonfoliated metamorphic rock. When a quartz sandstone fractures, it tends to break within the cement that surrounds the sand grains. After undergoing metamorphism to quartzite, the rock breaks right through the grains and the recrystallized cement.

Metamorphic Grade and Index Minerals

The degree to which a metamorphic rock has been changed from its parent material is designated by its **metamorphic grade.** *Low-grade* metamorphic rocks retain enough of their original character—such as some bedding and other sedimentary structures, fossils, and many of their original minerals—to enable geologists to readily identify their parent rock. Low-grade metamorphism occurs at relatively low temperatures (200°–400°C, or 400°–750°F) and pressures (about 1–6 kilobars). Because they have been subjected to much higher temperatures and pressures, *high-grade* metamorphic rocks lack virtually all of their original structures, fossils, and minerals. High-grade metamorphism occurs at temperatures between 500° and 1000°C (950°–1825°F) and pressures of about 12 to 15 kilobars. A number of intermediate metamorphic grades exist as well.

The specific minerals present in metamorphic rocks offer clues to the temperature and pressure conditions under which they were formed. Some minerals are stable only within a narrow range of temperatures and pressures; they therefore serve as indicators of specific metamorphic environments. Areas of rock containing such **metamorphic index minerals** are designated as distinct **mineral zones.** The presence of a given index mineral throughout a mineral zone indicates that the same set of temperature and pressure conditions were present during the creation of the entire zone, distinguishing it from other zones that formed under different conditions.

Mineral zones are found in rocks produced by both contact metamorphism and regional dynamothermal metamorphism. In areas that have undergone contact metamorphism, the highest-grade metamorphic rocks appear closest to the igneous intrusion and contain index minerals associated with high temperatures. Lower-grade metamorphic rocks are found at progressively greater distances from the intrusion and contain index minerals reflecting lower temperatures. A clear example of mineral zones produced by contact metamorphism can be found at Onawa, Maine. Some 365 million years ago, a granite intrusion invaded and metamorphosed tightly

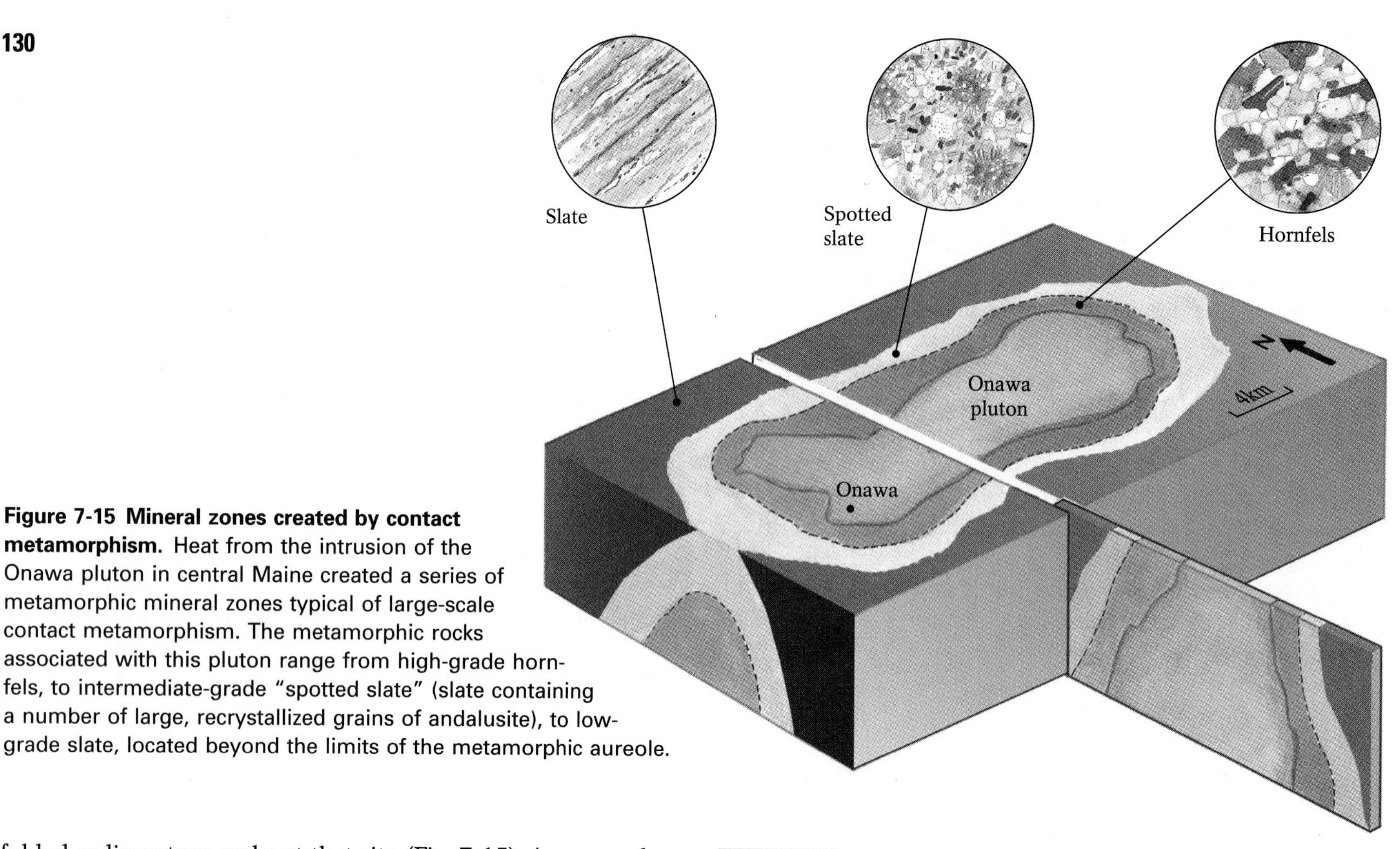

Figure 7-15 Mineral zones created by contact metamorphism. Heat from the intrusion of the Onawa pluton in central Maine created a series of metamorphic mineral zones typical of large-scale contact metamorphism. The metamorphic rocks associated with this pluton range from high-grade hornfels, to intermediate-grade "spotted slate" (slate containing a number of large, recrystallized grains of andalusite), to low-grade slate, located beyond the limits of the metamorphic aureole.

folded sedimentary rocks at that site (Fig. 7-15). An example of mineral zones produced by regional dynamothermal metamorphism is found in the rocks of the northeastern United States. Beginning about 390 million years ago, a series of collisions occurred between the North American, African, and European plates that crumpled the eastern edge of North America (forming the Appalachian Mountains) and compressed the region's sedimentary parent rocks into a sequence of metamorphic rocks that show distinct mineral zones (Fig. 7-16).

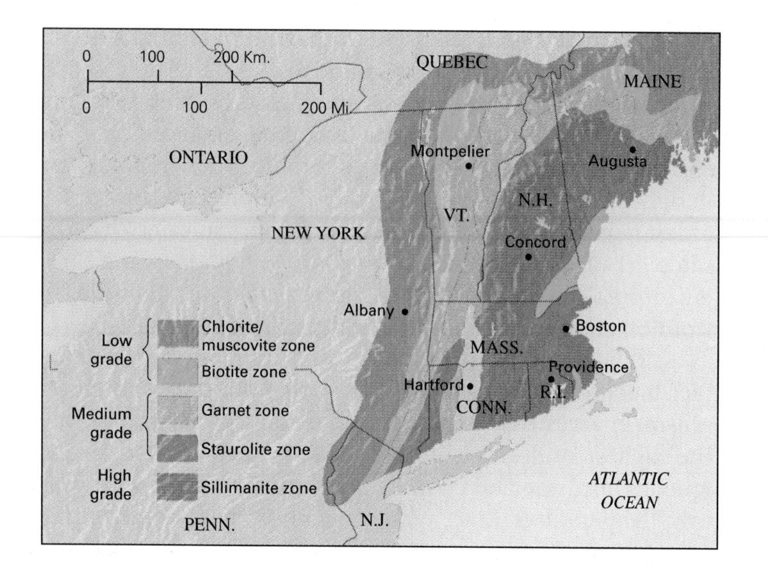

Figure 7-16 The metamorphic mineral zones of the northeastern United States. These zones resulted from the regional dynamothermal metamorphism associated with several episodes of plate convergence between 400 and 250 million years ago.

Metamorphic Facies and Plate Tectonics

Assemblages of different minerals that are commonly found together in metamorphic rocks are called **metamorphic facies.** Like an index mineral, a facies indicates the specific temperature and pressure conditions that characterized the development of these metamorphic rocks. Moreover, the presence of such a group of minerals, all associated with a given set of metamorphic conditions, allows geologists to infer these conditions even when a particular individual index mineral is missing, as occurs when the parent rock did not contain the necessary component elements.

To understand the concept of metamorphic facies, consider a slab of basaltic oceanic crust and an overlying sediment layer subjected to regional dynamothermal metamorphism within a subduction zone, as well as the nearby continental sediment subjected to contact metamorphism from the rising magma (Fig. 7-17). The basalt and the sediments experience the same temperatures and pressures at each location in the subduction zone. They will metamorphose to different minerals, however, because the compositions of their parent rocks differ. Nevertheless, the new minerals emerging at each location constitute a single metamorphic facies because they formed under the same temperature and pressure conditions. The continental sediments, on the other hand, although they may contain the same constituent minerals as the oceanic sediments, will produce a different facies—one associated with low-pressure, high-temperature metamorphism.

Clearly, metamorphic facies serve as useful indicators of ancient subduction zones, where plate movements created the heat, pressure, and circulating hot fluids that produced much of the Earth's metamorphic rock. In fact, directed pressure at convergent plate boundaries causes virtually all regional dynamothermal metamorphism, and new magmas generated at subduction and divergent zones are responsible for most contact metamorphism. We can identify several distinct metamorphic environments in subduction zones. A relatively shallow low-temperature/high-pressure zone appears in the subduction trench, where the subducting rocks are subjected to directed pressure. A high-temperature/low-pressure zone is located above the subducting plate but close to the Earth's surface, where heat from rising magmas—coupled with the relatively low burial pressures associated with shallow depths and the directed pressures at convergent plate boundaries—metamorphoses rocks. In addition, a high-temperature/high-pressure zone occurs above the subducting plate but at greater depth, where high-grade metamorphic facies form. All metamorphic rocks associated with subduction zones are foliated, because the directed pressures of convergence affect the entire region.

Intraplate ocean basins contain virtually no regionally metamorphosed rocks. The little metamorphism that does occur in ocean basins is either relatively low-grade hydrothermal metamorphism or contact metamorphism caused by the heat of rising basaltic magma at mid-ocean ridges. Because they are not subjected to directed pressure, these rocks generally do not become foliated.

Metamorphic Rocks in Daily Life

Metamorphic rocks are generally strong and durable, for several reasons. Heat and pressure eliminate pore spaces in the rocks, increasing their density. In addition, reactions with fluids replace unstable minerals with more stable ones. Finally, recrystallization strengthens the bonds between sediment grains and recrystallized cement.

Figure 7-17 The three main metamorphic environments associated with subduction zones: low temperature/high pressure, high temperature/high pressure, and high temperature/low pressure. Each environment produces a separate metamorphic facies, regardless of the composition of the parent rocks.

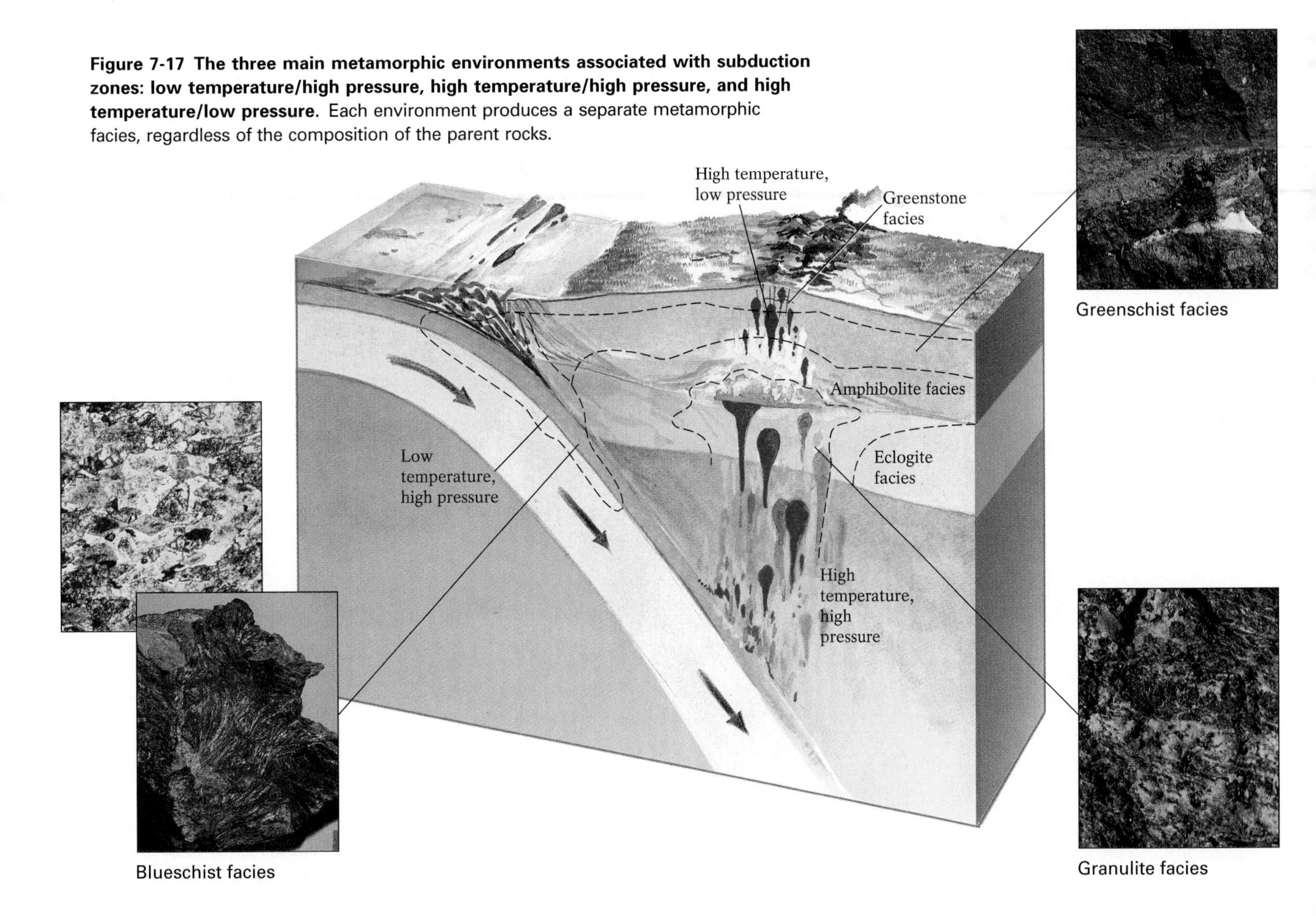

Practical Applications of Metamorphic Rocks

Thanks to its strength and stability, metamorphic rock is a popular material for making weather-resistant office building exteriors and for serving as the foundation stones of bridges and dams. Every year in the United States, building and road construction consumes 1.6 billion metric tons of slate quartzite.

Certain metamorphic rocks are valued highly for their appearance. One of architecture's most prized decorative building stones is serpentinite, the rock produced by hydrothermal metamorphism at divergent plate boundaries (Fig. 7-18). Sculptors prize the metamorphic rock marble for its attractive appearance, softness, and texture (Fig. 7-19). The marble from which Michelangelo fashioned his renowned sculptures was pure white; others, metamorphosed from impure limestones, come in shades of red, brown, black, gray, or green.

Many useful everyday commodities are the products of metamorphism. Soapstone, formed by regional dynamother-

Figure 7-18 Serpentinite. Verd antique, a popular decorative stone, is composed of the hydrothermally metamorphosed rock, serpentinite.

Figure 7-19 The Carrara marble quarry in northern Italy. The purity of Carrara marble, which stems from the purity of its parent limestone, has long made it a favored sculpting material.

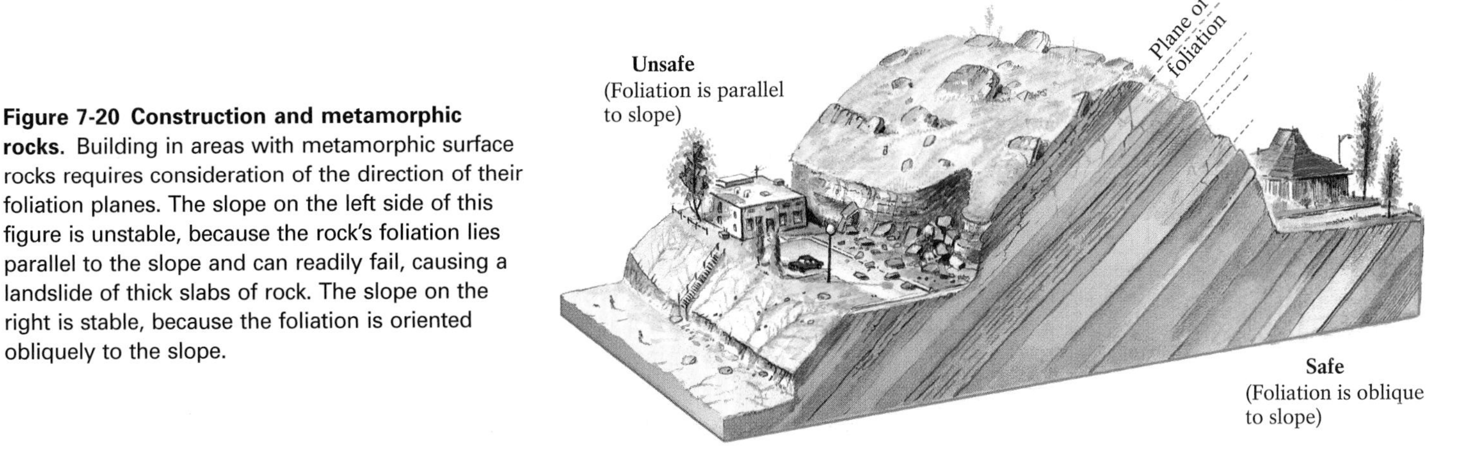

Figure 7-20 Construction and metamorphic rocks. Building in areas with metamorphic surface rocks requires consideration of the direction of their foliation planes. The slope on the left side of this figure is unstable, because the rock's foliation lies parallel to the slope and can readily fail, causing a landslide of thick slabs of rock. The slope on the right is stable, because the foliation is oriented obliquely to the slope.

mal metamorphism, is used for sculpting and for laboratory work counters. Talc, used in talcum powder, and flame-resistant asbestos, used in automobile brake linings and in the safety garb of firefighters, are both products of the low-grade regional metamorphism of ultramafic rocks. Garnet, an industrial abrasive valued for its hardness (and also January's gemlike birthstone), results from intermediate- to high-grade metamorphism. The high-grade metamorphic minerals kyanite and sillimanite are key components of porcelain casings for spark plugs, because the high temperatures at which they crystallize mean that they can withstand intense heat. Important minerals associated with metamorphic rocks include zinc, lead, copper, iron, and gold.

Potential Hazards from Metamorphic Rocks

Although some characteristics of metamorphic rocks have great aesthetic or practical value, others can produce hazardous conditions in natural settings. Foliation, in particular, can prove dangerous. Because it reduces the strength of metamorphic rocks, slaty cleavage can cause slopes to fail and set off landslides, particularly where the cleavage planes occur parallel to steep slopes (Fig. 7-20).

The weak planes of metamorphic foliation are especially susceptible to earthquake damage. Such was the case at midnight on August 17, 1959, near Yellowstone Park in Montana, when a powerful earthquake shook the Madison Canyon area. A deeply weathered body of gneiss and schist, the foliations of which paralleled the canyon's south wall, easily slid away from adjacent rock during the shaking. The landslide accumulated debris as it raced downslope at about 100 kilometers (60 miles) per hour, burying a campground and its 28 visitors with some 45 meters (150 feet) of rock, soil, and vegetation (Fig. 7-21).

Building in foliated metamorphic terrains requires careful planning and thorough geologic investigation to determine the orientation of the foliation. Metamorphic foliation must also be analyzed when designing large-scale engineering projects such as dams. Teams of geologists and engineers routinely study local and regional rock formations for evidence of inherent weakness before beginning any such project.

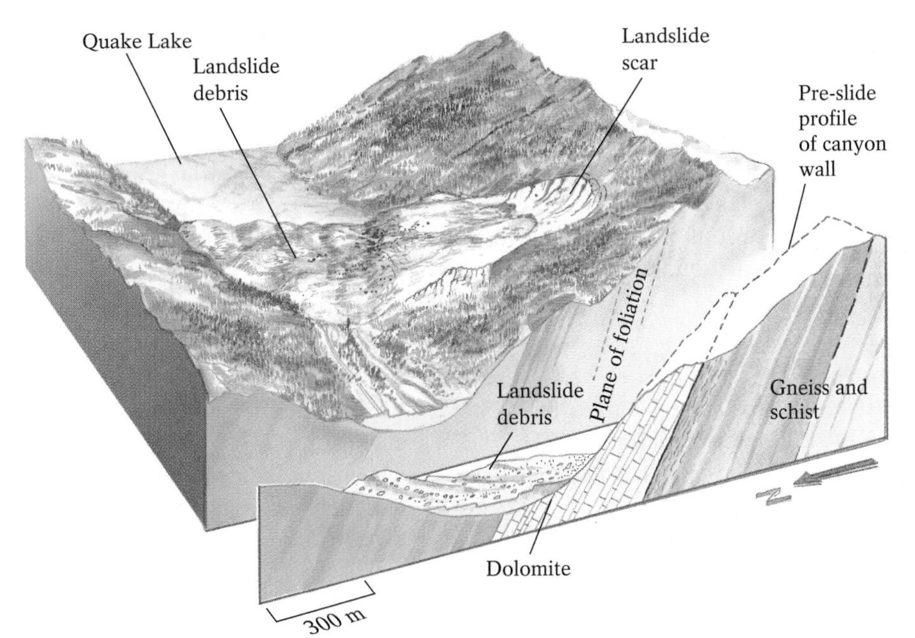

Figure 7-21 The disastrous rockslide of 1959 at Madison Canyon, Wyoming. The rockslide occurred when an earthquake jolted weak, highly fractured gneiss and schist along their foliation planes parallel to the canyon's south wall. Photo: The landslide aftermath, as viewed from the northern canyon wall.

Looking Ahead

We have now examined in detail the three basic types of rocks—igneous, sedimentary, and metamorphic—and the major processes—melting, recrystallization, weathering, lithification, and metamorphism—that form and alter them. We have also seen how plate tectonic processes contribute to the cycling of the Earth's rocks. In the next chapter, we examine how the Earth's rocks tell us about the past. In succeeding chapters, we look more closely at how some of the processes that create rocks also create the Earth's mountains, continents, and ocean floors. We also consider how other geologic processes involving water, wind, and glaciers sculpt many of the Earth's diverse landscapes.

Chapter Summary

Metamorphic rocks, the third major type of rock in the Earth's rock cycle, form at temperatures and pressures between those that form igneous and sedimentary rocks. They are created by **metamorphism** of preexisting rock, during which heat, pressure, and ion-rich fluids alter the mineral content and structure of solid rock without melting it.

Rocks at great depths are subjected to **lithostatic,** or **confining, pressure;** this force pushes in on rock equally from all sides, making it smaller and denser but maintaining its general shape. Some rocks—particularly those at convergent plate boundaries—are subjected to **directed pressure,** which distorts the shapes of rocks and flattens them in a single plane perpendicular to the direction of pressure. The individual mineral components in a rock subjected to directed pressure also tend to align in parallel streaks or bands perpendicular to the plane of the pressure, giving the rock a layered appearance known as **foliation.** All metamorphic rocks are categorized as being either foliated or nonfoliated.

Different types of metamorphism occur under different geological conditions. **Contact metamorphism** takes place where rocks are heated by direct or close contact with magma; the zone of metamorphosed rock surrounding a magma is called a metamorphic **aureole. Regional metamorphism** occurs over a broad area of rock and may be one of two types: **burial metamorphism,** which occurs when geothermal heat and confining pressure are applied to rocks buried beneath more than 10 kilometers of overlying rock, and **dynamothermal metamorphism,** which develops from increased heat and pressure (both directed and confining) in an area of plate convergence. Other types of metamorphism include hydrothermal metamorphism, fault metamorphism, shock metamorphism, and pyrometamorphism.

Increased heat and directed pressure produce progressively greater foliation in rocks. The sequence of foliated rocks that forms from a typical marine shale, for example, begins with the fine-grained, subtly foliated **slate** and **phyllite** and progresses to the coarse-grained, more obviously foliated **schist** and **gneiss.** Foliation in gneisses is marked by the formation of distinct, single-mineral layers of alternating felsic and mafic minerals, the result of **metamorphic differentiation.** When temperature and pressure increase beyond metamorphic conditions to reach the threshold of igneous conditions, they produce the part-igneous, part-metamorphic rock **migmatite**.

Nonfoliated rocks commonly develop in the absence of directed pressure and/or when the parent rock is composed largely of a single mineral. **Marble** forms from the metamorphism of relatively pure calcite-rich limestone or dolostone. **Quartzite** is produced by metamorphism of relatively pure quartz sandstone. **Hornfels** forms where shale or basalt comes in direct contact with hot magma.

The degree to which a metamorphic rock differs from its parent material is described in terms of its **metamorphic grade.** Low-grade metamorphism occurs when rocks are heated to temperatures of approximately 200° to 400°C; high-grade metamorphism occurs at temperatures of about 500° to 1000°C. Some minerals signal the presence of specific metamorphic conditions and are called **metamorphic index minerals;** areas of rock containing them are designated as distinct **mineral zones.** Specific metamorphic environments may also produce diagnostic assemblages of minerals called **metamorphic facies.** Mineral zones and metamorphic facies delineate areas of rock subjected to the same metamorphic conditions.

Different types of metamorphic processes and rocks occur in different plate tectonic settings. Convergent plate interactions cause virtually all dynamothermal metamorphism and thus produce most foliated rocks. In contrast, new magmas generated at subduction and divergent zones produce much of the heat for both contact and regional metamorphism. Because several distinct metamorphic environments prevail at different locations in a subduction zone, these areas are associated with specific metamorphic facies.

Many metamorphic rocks are valued because they are strong, durable, or beautiful. Because foliation produces weak planes in some metamorphic rocks, however, they can be hazardous in natural settings, as they are highly susceptible to earthquake damage and prone to landslides.

Key Terms

metamorphic rocks (p. 121)
metamorphism (p. 121)
lithostatic (confining) pressure (p. 123)
directed pressure (p. 123)
foliation (p. 124)
contact metamorphism (p. 124)
aureole (p. 124)
regional metamorphism (p. 124)
burial metamorphism (p. 124)
dynamothermal metamorphism (p. 125)

slate (p. 127)
phyllite (p. 127)
schist (p. 128)
metamorphic differentiation (p. 128)
gneiss (p. 128)
migmatite (p. 128)
marble (p. 129)
quartzite (p. 129)
hornfels (p. 129)
metamorphic grade (p. 129)
metamorphic index minerals (p. 129)
mineral zones (p. 129)
metamorphic facies (p. 130)

Questions for Review

1. What are three major factors that induce or influence metamorphism?

2. Describe four ways that rocks may change during metamorphism.

3. Define metamorphic foliation, explain how it develops, and list three metamorphic rocks that are foliated.

4. List the sequence of metamorphic rocks that develop during the progression of shale to migmatite. Describe the successive changes in mineralogy and structure that these rocks undergo.

5. Under what conditions do nonfoliated metamorphic rocks generally form? Give examples.

6. List six common types of metamorphism, and discuss the geological conditions under which they occur.

7. What are metamorphic index minerals? How do they indicate specific types of metamorphism?

8. What are metamorphic facies? How do they indicate different environments in the vicinity of a subduction zone?

9. List one way in which we use each of the following metamorphic rocks, and explain what properties of the rock make it suitable for that purpose: slate, marble, quartzite, serpentinite.

10. Under what circumstances might the presence of metamorphic rocks lead to landslides?

For Further Thought

1. Why do quartzites appear to be essentially nonfoliated, even when they have been subjected to the directed pressures of regional metamorphism?

2. Why do we rarely find minerals such as chlorite and amphibole directly next to an igneous intrusion?

3. Referring to Figure 7-7a, suggest why some of the rocks crosscut by the intrusive rocks have not undergone contact metamorphism.

4. The Earth's internal temperature was apparently much higher during the first billion years of its history. Speculate about how metamorphism, metamorphic environments, and the distribution of metamorphic rocks would have been different during that early period.

5. From the photo below, estimate the orientation of the directed pressures that metamorphosed these rocks.

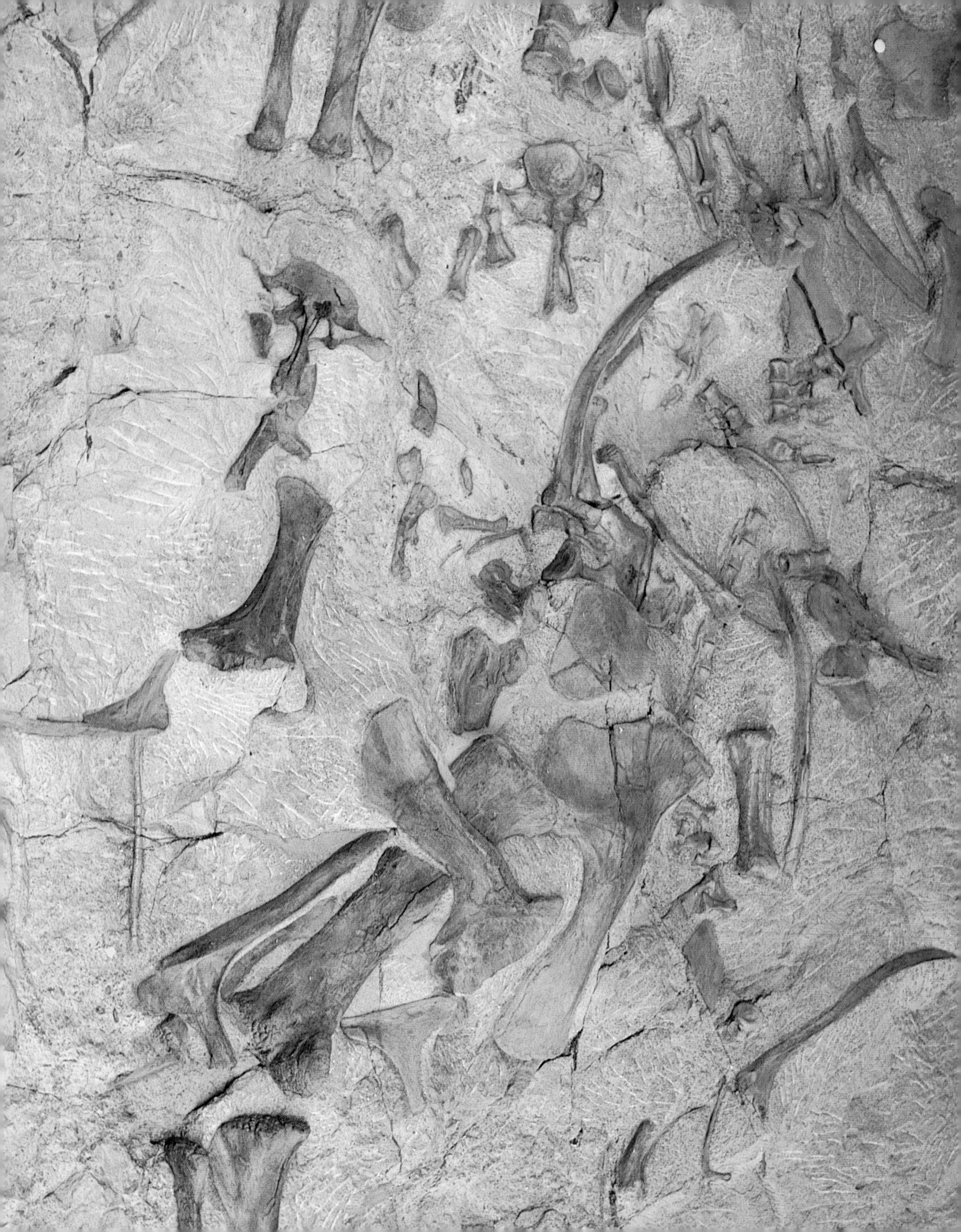

8

Telling Time Geologically

In the first seven chapters of this text, we learned that geological processes affect the Earth today much as they have since the planet's birth 4.6 billion years ago. Many of the major events that affected the Earth—such as the extinction of the dinosaurs 65 million years ago (Fig. 8-1)—began and ended long before the first humans appeared on the Earth. Some were virtually instantaneous in a geological sense, requiring "only" a few million years or so to occur. Others, such as the rise of the Rocky Mountains or the formation of the Atlantic Ocean, required tens or hundreds of millions of years to unfold—still only a small fraction of the Earth's history. The age and time span required for such geological events and processes are the subject of **geochronology,** the study of "Earth time." Geochronology helps us comprehend the exceedingly slow pace of certain geological processes and the relatively rapid pace of others relative to the Earth's total lifetime. This discipline is an important component of the larger field of **historical geology,** which focuses on the origin and evolution of Earth's life forms and geologic structures.

Figure 8-1 Now-extinct life forms preserved in the geologic record. This rich cache of dinosaur bones is exposed at Dinosaur National Monument in northeastern Utah.

Geologic Time in Perspective

How can humans—with our average lifespans of only 70 or 80 years—begin to grasp the magnitude of the age of the Earth, covering a time span of 4,600,000,000 years? Perhaps one way is to picture a timeline across the United States, 5000 kilometers (3000 miles) long, on which each kilometer represents about 1 million years (Fig. 8-2). Although the Earth is believed to be about 4.6 billion years old, no rocks from the Earth's first 600 million years have been identified. The oldest rocks known, which were recently discovered in Canada's Northwest Territories, are dated at 3.96 billion years. The earliest known life forms on Earth were microscopic blue-green algae (cyanobacteria) that lived in muddy ponds 3.77 billion years ago; their fossils have been identified in lithified silt in what is now northwestern Australia.

More than 3 billion years passed after the emergence of blue-green algae before more complex and diverse life forms began to appear. These higher organisms emerged only when

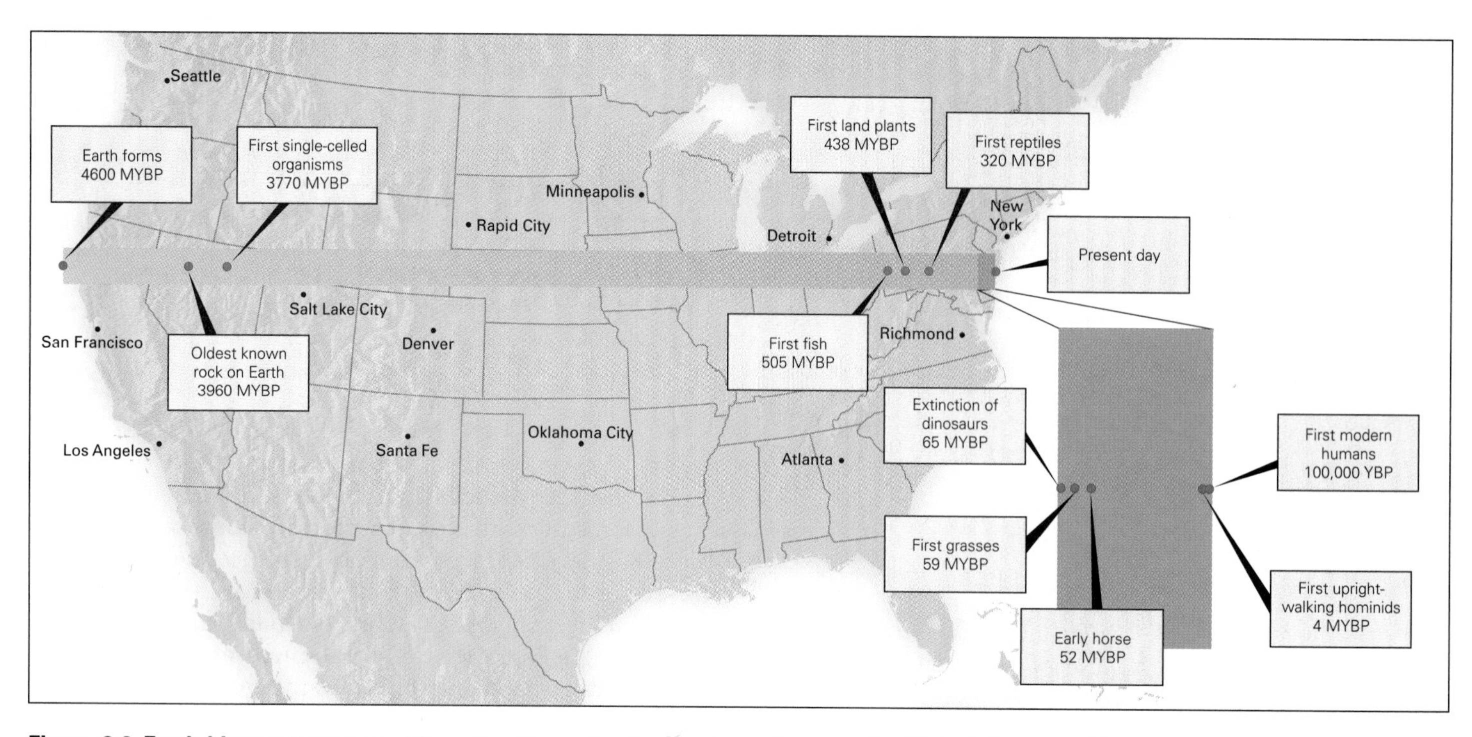

Figure 8-2 Earth history represented by a timeline extending east–west across the United States, with each kilometer corresponding to about 1 million years. Our planet's earliest history remains a mystery, and our knowledge is greatest about events that generally occurred quite recently in geologic time. (MYBP = million years before present.)

the Earth's environment had become more hospitable, in the last few hundred kilometers of our timeline. Fish began to appear about 510 million years ago, land plants some 438 million years ago, and the first dinosaurs only about 245 million years ago. The origin of mammals, including humans, the onset of the Earth's most recent ice ages, and the arrival of humans on the North American continent are all relatively recent events. Indeed, all of the material taught in a typical history course represents only 0.0000011% of Earth's history!

How did geologists determine the age of the Earth and the chronology of its geologic and biological evolution? Geologists examine time in two ways. First, they use **relative dating,** comparing the physical characteristics and positions of two or more rocks, fossils, or other geologic features to determine which is older and which is younger. Second, they employ **absolute dating,** which establishes the actual age (even if only approximate) of a specific geologic feature or event. By combining information from relative and absolute dating, geologists can piece together—and date with some assurance—the sequence of events that led to virtually any geologic feature on Earth.

Determining Relative Age

Geologists determine the sequence of geological events that has produced an area's rocks by applying their knowledge of such fundamental processes as sediment deposition, volcanism, and erosion in conjunction with certain basic principles and assumptions. On this basis, they can usually determine which is older and which is younger, compare the ages of rocks from different geographic regions, and even identify gaps in the geologic record.

Principles Underlying Relative Dating

Most of the principles used to determine the relative ages of rocks involve common sense, basic logic, and understanding of a few spatial relationships, such as where new rock layers form with respect to preexisting layers. Some also require an understanding of the concept of evolution and its effect on the fossil record.

Uniformitarianism The most basic principle used to illuminate Earth history is James Hutton's **principle of uniformitarianism,** which states that the geological processes taking place in the present operated similarly in the past (see Chapter 1). From this concept, we can assume that ancient earthquakes, volcanoes, floods, and other geological events happened in much the same way as they do today, albeit at significantly different rates. Our observations of modern geological phenomena, therefore, can help us interpret ancient events.

Horizontality and Superposition As we saw in Chapter 6, most sediments settle out from bodies of water and are deposited as horizontal or nearly horizontal layers. Similarly, lava flows

Figure 8-3 Horizontal beds of sedimentary rock outside of Salt Lake City, Utah. Using the principle of superposition, we can instantly determine that the rocks at the peak of this formation are the youngest and those at the bottom are the oldest.

generally solidify as horizontal layers. This **principle of original horizontality** is fundamental, because rock layers are often found in nonhorizontal positions, indicating that they have been tilted or deformed by tectonic forces.

The **principle of superposition** states that rock materials are generally deposited on top of earlier, older deposits. Consequently, in any unaltered sequence of rock strata, the youngest stratum will appear at the top and the oldest at the bottom (Fig. 8-3). The principle of superposition, like the principle of original horizontality, commonly applies to sedimentary rocks and lava flows.

When tectonic forces have tilted or even overturned a sequence of rock layers, we must look for readily identifiable sedimentary structures, such as ripple marks, mudcracks, graded beds, or cross-bedding (all discussed in Chapter 6), to identify the upper surface of any one sedimentary layer (Fig. 8-4). Similarly, vesicles—the pea-sized dimples left behind as gas bubbles burst at the tops of some lava flows—may indicate the upper surfaces of some volcanic rocks. Once we have identified the top of a rock layer, we can apply the principle of superposition to date relatively the layers below and above it.

(a)

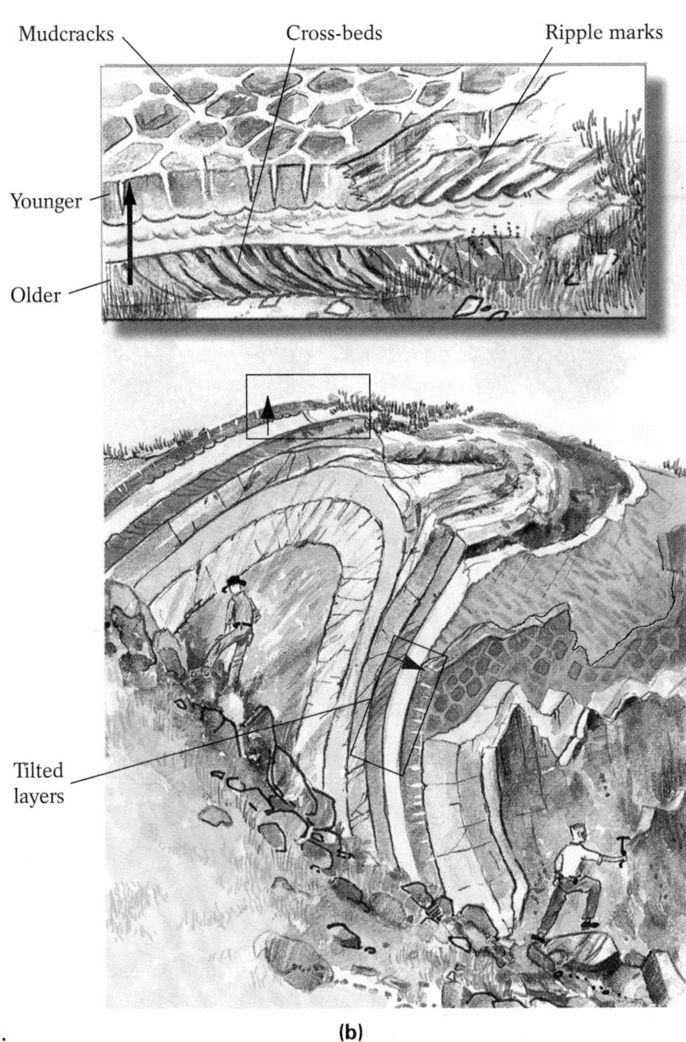

(b)

Figure 8-4 Tectonic tumbling. **(a)** Tilted turbidite beds in Zumaya, Spain. **(b)** When sedimentary rocks have been displaced from their initial horizontal orientation, we must look for sedimentary structures to identify the top of any individual bed; once the top of a single bed has been determined, the initial orientation of all beds can be reconstructed using the principle of superposition.

Cross-cutting Relationships Bodies of igneous rock often appear within other rock types, indicating that the latter were intruded, or cut across, by molten magma. The **principle of cross-cutting relationships** states that any intrusive formation, such as a dike, must be younger than the rock across which it cuts (Fig. 8-5). Cross-cutting relationships also enable geologists to develop relative dates for faults, fractures in rocks along which displacement has occurred; faults must be more recent than the rocks they cut through (Fig. 8-6).

Inclusions The **principle of inclusions** states that fragments of other rocks contained within a body of rock must be older than the host rock. Sedimentary conglomerates, for example, contain pieces of preexisting rock that are necessarily older than the conglomerate itself. Many igneous rocks also contain pieces of preexisting rock, xenoliths, that were broken but not melted by intruding magma (see Chapter 3); a body of granite must therefore be younger than its xenoliths.

Fossils and Faunal Succession **Fossils** are the remains of ancient organisms, or other evidence of their existence, preserved in geologic material. Because fossils form only under special circumstances (see Highlight 8-1), only some 1% of all species that ever existed are believed to have been preserved in this way. When fossils *are* found in rocks (usually sedimentary rocks), they can help date the rock units in which they appear. The **principle of faunal succession** states that over time the organisms of Earth have changed in a definite order, and that this progression is reflected in the fossil record: Rocks containing the fossils of more recent life forms are younger than those containing only older life forms. Furthermore, rocks containing identical groupings (assemblages) of fossils are thought to be identical in age.

Certain varieties of organisms have existed (or did exist) for hundreds of millions of years. Sharks, for example, have inhabited the oceans for more than 400 million years. Other organisms lived only during relatively brief, specific periods of Earth history. Fossils of such short-lived organisms serve as **index fossils,** identifying a fairly specific age for the rocks in which they are found. The species that become the most useful index fossils were geographically widespread during their short time on Earth; they can eventually be used to date many different—even far distant—rock formations.

Unconformities

No single place on Earth contains all of the rock strata composing the entire geologic record. In fact, the sedimentary rock layers of most regions record only 1% to 5% of the Earth's existence. Gaps in the geologic record are marked by **unconformities,** boundaries separating rocks of markedly different ages. Unconformities occur where erosion has removed

Figure 8-5 A cross-cutting relationship. A mafic dike cuts across these layered sedimentary rocks in the Grand Canyon, Arizona.

Figure 8-6 Faulting as a determinant of relative age. The fault that cuts through this sedimentary rock, found in California's Anza Borrego Desert State Park, must be younger than the deposition of any of the rock layers.

rock layers or where no layers were deposited during certain geologic periods.

Geologists have identified three types of unconformities. A *nonconformity* is the boundary (usually an obvious one) between an unlayered body of plutonic igneous or metamorphic rock and an overlying layered sequence of sedimentary rock layers (Fig. 8-7a); rock underlying a nonconformity usually shows evidence of having been eroded before deposition of the overlying rock. Somewhat less obvious is an *angular unconformity,* the boundary between a sedimentary rock that has been tilted and eroded and a later, horizontal rock deposit that overlies it (Fig. 8-7b). The angular unconformity at Siccar Point in Scotland has particular historical significance (Fig. 8-8). The most subtle type of unconformity, a *disconformity,* occurs between parallel layers of sedimentary rock (Fig. 8-7c); such a boundary is revealed to be a disconformity if fossil groupings above and below it are of substantially different ages or if the surface of the lower layer shows evidence of significant erosion.

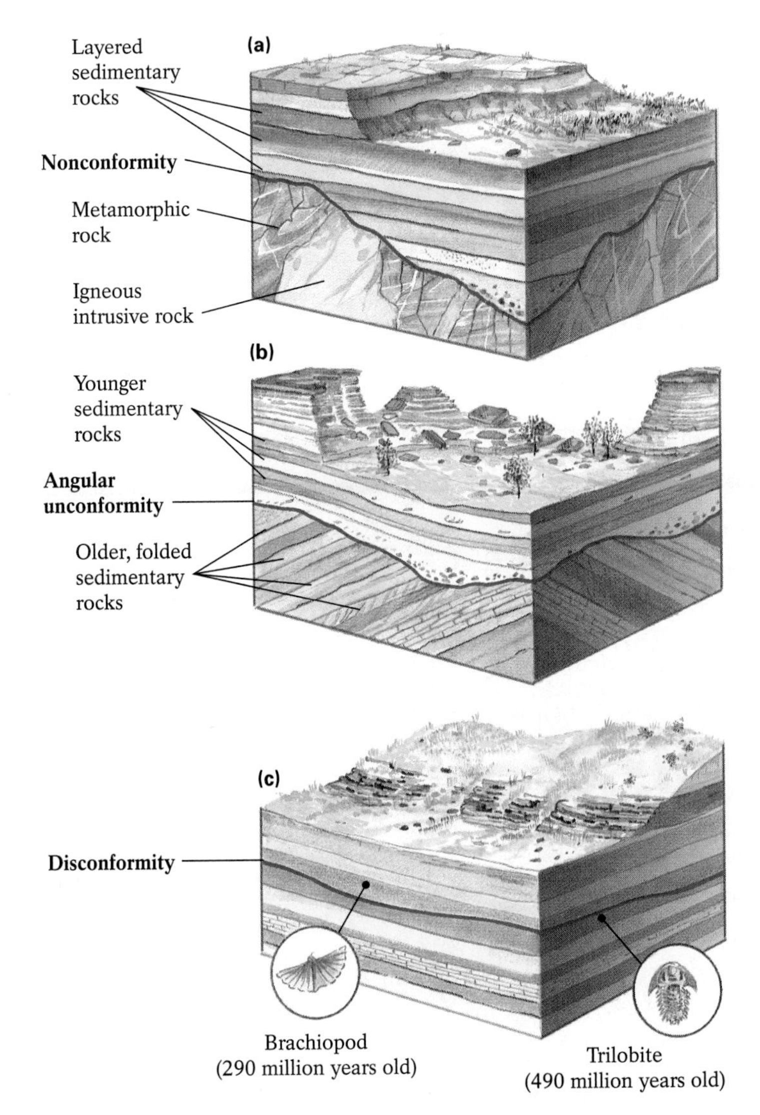

Figure 8-7 The three major types of unconformities between rock layers. (a) A nonconformity, between metamorphic or igneous rock and overlying sedimentary rock; **(b)** an angular unconformity, between older, deformed sedimentary rock and younger, undeformed sedimentary rock; **(c)** a disconformity, between parallel layers of sedimentary rock.

Correlation

The farther apart two rock formations are geographically, the less likely that they will contain identical rock sequences, because different environmental factors would have influenced rock formation and erosion in each area. Nevertheless, even formations that are physically separated—whether by a few meters or by thousands of kilometers—may contain individual layers that are similar, suggesting that these layers originated at the same time. The process of determining equivalence in age between geographically distant rock units is

Figure 8-8 The angular unconformity at Siccar Point, Scotland. James Hutton, the founder of modern geology (Chapter 1), correctly interpreted this formation as showing layers of rock that had been deposited horizontally, lithified, tilted to a nearly vertical position, eroded smooth at the surface, and then overlain with subsequent horizontally deposited layers that were also later tilted—a graphic example of the dynamics that shape the Earth's features, and the amount of time necessary to achieve them.

Highlight 8-1 How Fossils Form

Most fossils form when the hard parts of organisms become buried in layers of sediment, where they decompose so slowly that the rock preserves their shapes as it lithifies around them. *Paleontologists,* scientists who study fossils to learn about the history of life on Earth, have unearthed countless remains of ancient clam shells, many fish skeletons, and a fair amount of dinosaur bones. On the other hand, they have found relatively few remains of worms, slugs, and jellyfish. Organisms with hard parts such as shells and bony skeletons are more likely to remain intact long enough to become preserved in sediment; in contrast, those composed only of soft tissue are more likely to decompose rapidly at the surface in most environments. Even the hard parts of organisms eventually decompose or dissolve after burial if they come into contact with underground water. For that reason, a fossil is seldom an actual remnant of an organism's original biological substance, but rather comprises some type of replacement mineral.

Some fossils consist only of the impressions of the original organism. After circulating groundwater removes the original bones, shells, and other organic materials, a *mold*—in the original shape of the organism's form—remains in the enveloping sediment. At a later time, the mold may become filled with other materials, such as minerals precipitated from groundwater. This process creates a solid *cast* of the mold, which is what we typically find when we go fossil hunting.

Preservation of an organism's actual original parts occurs only under unusual environmental conditions. For instance, the skeletons of thousands of mammoths, mastodons, saber-toothed cats, and other late Ice Age mammals were preserved in the viscous tar of the La Brea tar pits, in what is now downtown Los Angeles. Intact woolly mammoths have been dug up from the tundras of Alaska and Siberia, where they became frozen in Arctic *permafrost* (permanently frozen ground). The original carcasses of entire insects have been recovered from hardened tree sap, or *amber,* in which they became stuck millions of years ago. In each of these cases, the enveloping material did not permit the circulation of water or air, thereby preventing decomposition of the organism's original material.

Even when we don't find fossils showing an organism's physical form, we sometimes find other evidence of its presence. Such *trace fossils* take a variety of forms. If an animal burrowed through sediment, it may have left "tunnels"; if it crawled across soft mud, it may have etched delicate tracks into the sediment. Larger animals may have left behind footprints. We have even discovered rounded, polished rocks, called *gastroliths,* that dinosaurs swallowed to help grind their coarse food, and lithified "droppings," or *coprolites,* from a variety of creatures. Figure 1 depicts some of the many types of fossils.

(a)

(b)

(c)

Figure 1 Various types of fossils. **(a)** A 3.5-million-year-old fly in amber; **(b)** a saber-toothed cat skeleton from the La Brea tar pits; **(c)** ancient worm burrows. Only the fossils in parts **(a)** and **(b)** contain actual remains of the organisms.

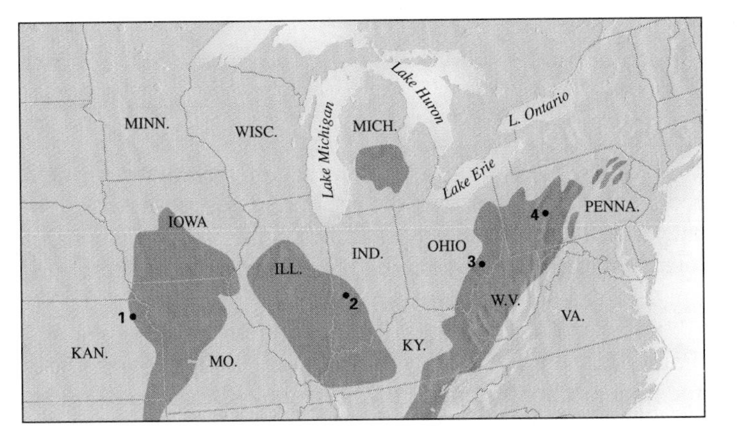

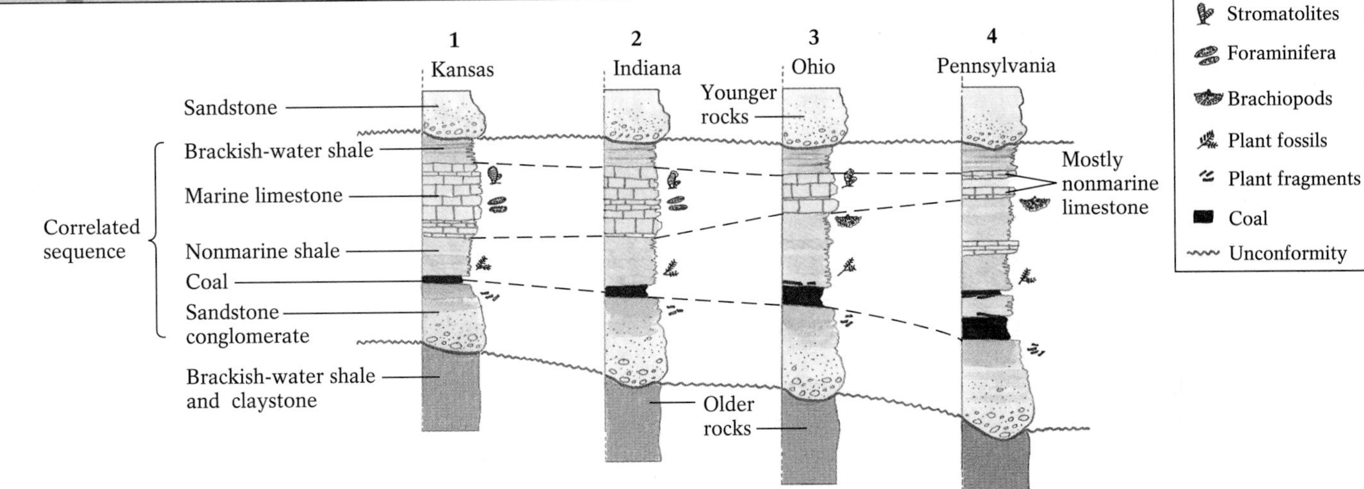

Figure 8-9 Correlation as a relative-dating tool. Correlation of the sedimentary rock sequences and fossil assemblages above and below these four separate coal fields indicates that the coal was deposited at about the same time at all four sites.

known as **correlation** (Fig. 8-9). To establish correlation, geologists often try to identify similarities in fossil assemblages or individual index fossils. Two widely separated rock sequences can also be correlated if both contain the same *key bed,* a distinctive stratum that appears at several locations. A key bed records a geologic event of short duration that affected a wide area. The eruption of Mount Mazama in the Pacific Northwest about 6700 years ago (see Highlight 4-1), for example, produced a whitish tan, highly felsic ash that was scattered so widely that it can be used to correlate the geologic strata of Oregon, Washington, British Columbia, and Alberta.

Determining Absolute Age

Even the combination of several relative-dating principles and techniques can tell us only that rock A is older or younger than rock B. Understandably, geologists also need the more specific "years ago" information yielded by absolute dating methods. All absolute-dating methods depend on some type of "natural clock"—a process, such as radioactive decay, that operates at a constant, quantifiable rate over long periods of time.

Radiometric Dating

The atoms of certain chemical elements exist as isotopes containing different numbers of neutrons in their nuclei (as discussed in Chapter 2). *Radioactive* isotopes are those isotopes whose nuclei spontaneously decay. The decaying radioactive isotope, or **parent isotope,** evolves into a decay product, the **daughter isotope.** This radioactive decay process involves one or more steps in which various subatomic particles (neutrons, protons, and electrons) are either emitted from or captured by the nuclei. The loss or gain of neutrons changes a parent isotope into a daughter isotope of the same element; the loss or gain of protons changes the parent element into an entirely different daughter element with a new set of chemical and physical properties. Eventually, after a number of intermediate steps, an unstable radioactive parent isotope becomes a stable, nonradioactive daughter isotope.

For example, uranium-238, an isotope of the element uranium, has an atomic number of 92 (that is, 92 protons in its nucleus). When an atom of uranium-238 undergoes radioactive decay, it loses 2 protons and 2 neutrons and is converted into an atom of an element with 90 protons—that is, thorium. Additional steps eventually convert the thorium atom into lead-206.

Radioactive isotopes often become incorporated into the crystal structures of the minerals within an igneous rock when its parent magma is cooling. **Radiometric dating** uses the continuous decay of these isotopes to measure the amount of time elapsed since the rock's formation. As time passes, the rock will contain less of its initial radioactive parent isotopes and more of their daughter products.

Radioactive decay rates are constant, remaining unaffected by changes in temperature or pressure or by chemical reactions involving the parent isotope. The time it takes for *half* the atoms of the parent isotope to decay is a constant value, known as the **half-life** of an isotope. For example, if a rock has 12 parent and 12 daughter atoms—a parent-to-daughter ratio of 1:1—the original rock had 24 parent atoms, and the rock has existed for one half-life. After another half-life, the number of remaining parent atoms would halve again, leaving 6 unstable parent atoms and a total of 18 stable daughter atoms—a parent-to-daughter ratio of 1:3. By measuring the ratio of parent-to-daughter isotopes and comparing it with the parent element's known rate of radioactive decay in this way, we can determine the absolute age of the rock (Fig. 8-10). Half-lives of many radioactive isotopes are on the order of billions of years.

Factors Affecting Radiometric-Dating Results Radiometric dating is more accurate and useful for igneous rocks than for other types of rocks. It is difficult to determine the radiometric age of a sedimentary rock, because sedimentary rocks contain mineral fragments of countless preexisting rocks of different ages. Likewise, radiometric dating of metamorphic rocks can prove challenging, because parent-to-daughter ratios in metamorphic rocks may be distorted by the heat, pressure, and circulating fluids associated with metamorphism. Although these factors do not affect the rate at which isotopes decay, they can influence the relative number of parent and daughter isotopes ultimately found in a rock, because they allow atoms to migrate away from their point of origin.

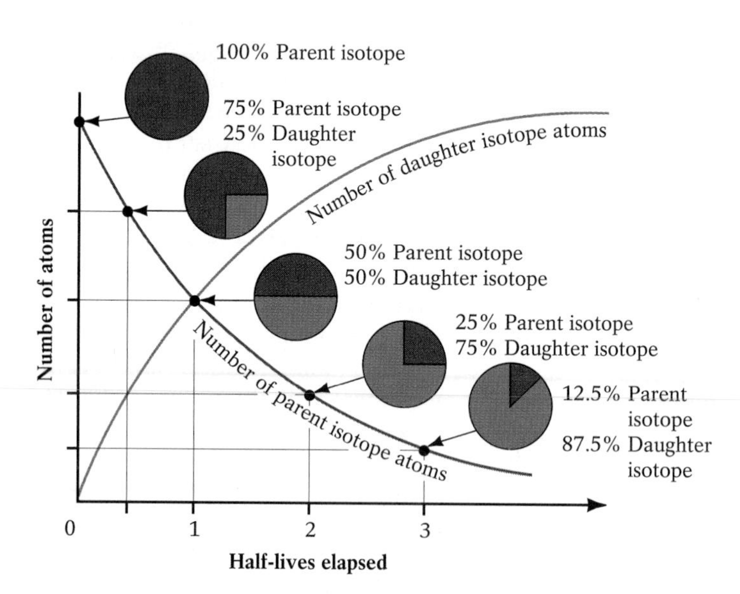

Figure 8-10 Radiometric dating. Radioactive decay converts a radioactive parent isotope to a stable daughter isotope. With the passage of each half-life, the number of atoms of the radioactive isotope is reduced by half, whereas the number of atoms of the daughter isotope increases by the same quantity. If they know the half-life of the isotope, geologists can measure the ratio of parent-to-daughter isotopes to determine the absolute ages of some rocks.

To estimate the age of a sedimentary or metamorphic rock, geologists often determine the age of igneous rocks that cross-cut, underlie, or overlie it and then apply the principles of relative dating. In addition, the presence of fossil assemblages can serve to date sedimentary rocks.

The age or condition of the materials dated can sometimes affect the reliability of radiometric dating. Young rocks and minerals, for instance, may not yet have produced a measurable amount of daughter isotopes. For radioactive isotopes with half-lives on the order of hundreds of millions of years, a rock must be 100,000 or more years old to include a measurable quantity of daughter isotopes. For instance, rocks that contain uranium-238, with a half-life of 4.5 billion years, must be at least 10 million years old to yield a measurable amount of lead-206. In material that is in poor condition, perhaps because of fracturing or intense weathering, migrating groundwater may have dissolved away some parent or daughter isotopes, resulting in inaccurate parent-to-daughter ratios. Fortunately, recently developed technology now makes it possible to analyze minute fractions of some crystals and date their unfractured, unweathered regions.

To ensure the accuracy of radiometric dating, geologists always run tests on multiple samples of the particular rock or mineral. This redundancy minimizes experimental error and diminishes the possibility of sample contamination. Usually, rocks contain more than one radioactive isotope system, allowing geologists to cross-check the dates yielded by each system. Thus, whenever possible, several dating systems are used to date the same sample, and the results of the different systems are checked against one another. Rarely, if ever, is a rock or mineral dated by a single analysis of a single sample.

Isotopes Used in Radiometric Dating Although scores of radioactive isotopes exist, only a few are useful for dating purposes. Some decay too rapidly and some too slowly, others migrate into and out of rocks and minerals too readily under ordinary environmental conditions, and still others are simply too rare. To be effective for dating purposes, isotopes must be present in common Earth materials, not be overly susceptible to gains or losses of parent and daughter isotopes, and have a half-life of at least several thousand years.

Table 8-1 shows the most frequently used radioactive isotope dating systems. Most of these systems—those involving uranium, thorium, rubidium, and potassium parent isotopes—have half-lives on the order of millions or billions of years and are employed to date minerals and rocks hundreds of thousands of years old or older. Unique in the list is carbon-14, with a half-life of only 5730 years. **Carbon-14 dating** can be used to date organic materials from 100 to about 70,000 years old, a span of time that encompasses the most recent glaciation and climate change, the latest rise and fall of worldwide sea levels, and the rise to dominance of our own species, *Homo sapiens.* Carbon-14 is a radioactive iso-

Table 8-1 The Major Isotopes Used for Radiometric Dating

Method	Parent Isotope	Daughter Isotope	Half-Life of Parent (years)	Effective Dating Range (years)	Materials Commonly Dated	Comments
Rubidium–strontium	Rb-87	Sr-87	47 billion	10 million–4.6 billion	Potassium-rich minerals such as biotite, potassium muscovite, feldspar, and hornblende; volcanic and metamorphic rocks (whole-rock analysis)	Useful for dating the Earth's oldest metamorphic and plutonic rocks.
Uranium–lead	U-238	Pb-206	4.5 billion	10 million–4.6 billion	Zircons, uraninite, and uranium ore such as pitchblende; igneous and metamorphic rock (whole-rock analysis)	Uranium isotopes usually coexist in minerals such as zircon. Multiple dating schemes enable geologists to cross-check dating results.
Uranium–lead	U-235	Pb-207	713 million	10 million–4.6 billion		
Thorium–lead	Th-232	Pb-208	14.1 billion	10 million–4.6 billion	Zircons, uraninite	Thorium coexists with uranium isotopes in minerals such as zircon.
Potassium–argon	K-40	Ar-40	1.3 billion	100,000–4.6 billion	Potassium-rich minerals such as amphibole, biotite, muscovite, and potassium feldspar; volcanic rocks (whole-rock analysis)	High-grade metamorphic and plutonic igneous rocks may have been heated sufficiently to allow Ar-40 gas to escape.
Carbon-14	C-14	N-14	5730	100–70,000	Any carbon-bearing material, such as bones, wood, shells, charcoal, cloth, paper, animal droppings; also water, ice, cave deposits	Commonly used to date archaeological sites, recent glacial events, evidence of recent climate change, environmental effects of human activity.

tope of carbon created in the Earth's atmosphere by cosmic-ray bombardment of nitrogen atoms. Along with nonradioactive carbon-12, it finds its way into the cells of most living things through their food or water or directly from the air (Fig. 8-11). When an organism dies, its intake and recycling of carbon stops, and the carbon-14 slowly decays to form nitrogen. By measuring the ratio of carbon-14 to carbon-12 in ancient organic material, scientists can determine the age of the particular material. With this technique, they can accurately date bones, shells, wood, charcoal, plants, peat, paper, cloth, pollen, seeds, or anything else composed of once-living materials.

Other Absolute-Dating Techniques

Several absolute-dating methods other than radiometric dating have been developed. *Fission-track dating* exploits the fact that when a radioactive atom trapped within a mineral crystal undergoes fission (splits apart), its high-speed movement leaves a trace across the crystal lattice. Because the fission proceeds at a constant rate, the number of fission tracks within the crystal reveals its age. Fission tracks can be counted to date minerals that are 50,000 to billions of years old.

In temperate climates, the annual growth of most trees results in concentric sets of dark and light rings that can be

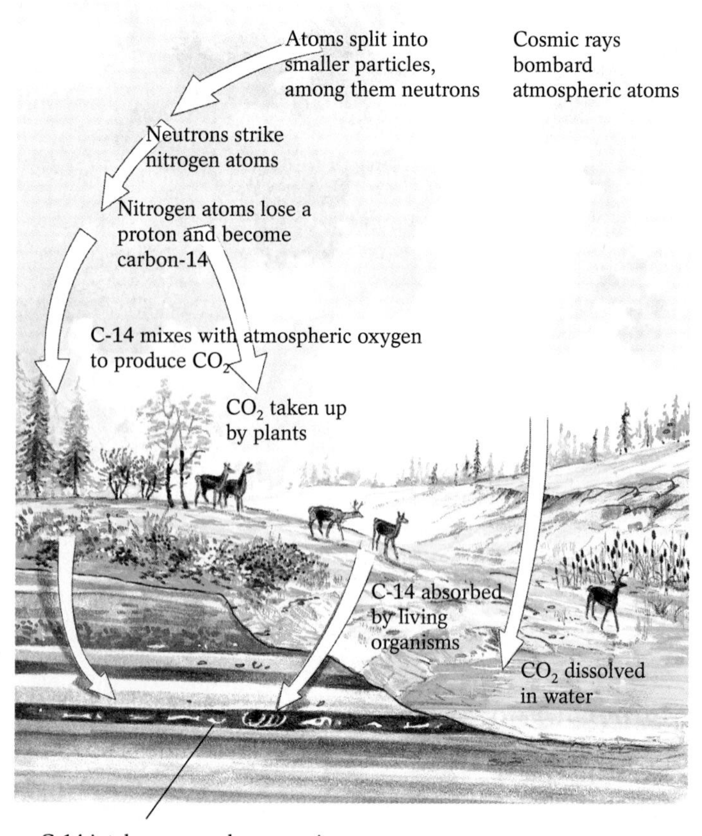

Figure 8-11 Carbon-14 dating. The carbon-14 used to date long-dead organisms and ancient artifacts originates in the Earth's atmosphere. Here, it combines with O_2 to produce CO_2, much of which dissolves in water or is taken up by plants, ultimately being ingested by animals. While alive, an organism constantly replenishes the carbon-14 supply in its body. When it dies, its intake of carbon-14 ceases, and the carbon-14 begins to decay to its daughter isotope, nitrogen-14. The less carbon-14 left in the remains of an organism, the more time has elapsed since it died.

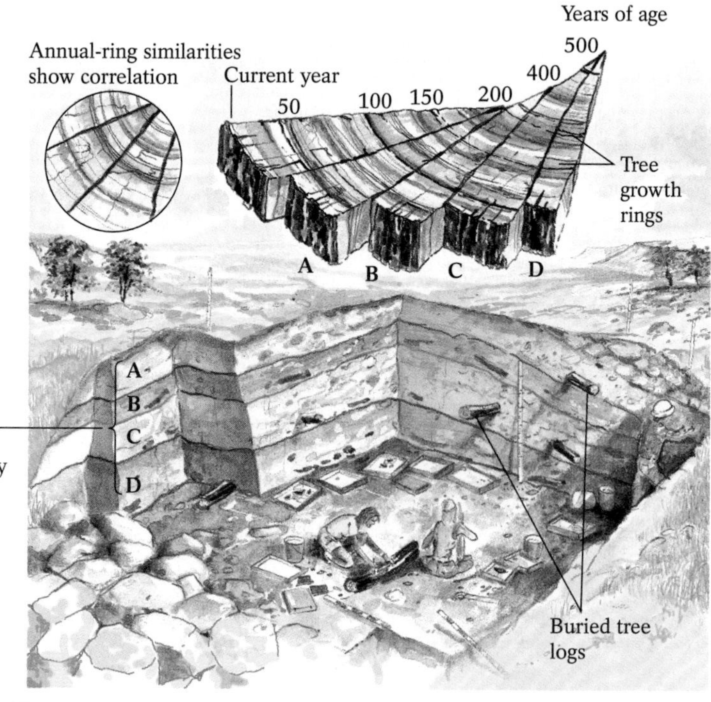

Figure 8-12 Dendrochronology. Correlation of annual-ring sections in trees or wooden artifacts of overlapping life spans can establish ages of thousands of years.

seen in cross sections of their trunks and branches. By counting the number of these rings, we can determine a tree's age—an absolute-dating method known as *dendrochronology*. Dating trees in this way can help us estimate the timing of relatively recent geological events such as landslides or mudflows—wherever trees have become established on top of new surfaces. (The older the trees growing on a landslide deposit, the longer ago the landslide occurred.) In addition, dendrochronology can be used to date much older geologic events and, in certain cases, archaeological artifacts. Changes in climatic conditions, such as prolonged droughts, produce the same type of tree-ring variations in all trees living in the affected area; by correlating sections of the ring patterns from progressively older trees in the same area—including fossilized trees—geologists have successfully established dates as long ago as 9000 years (Fig. 8-12).

Lakes, particularly ones that freeze in the winter and are fed by glacial meltwater in the summer, can also produce countable annual layers. *Varves* are paired layers of sediment, typically consisting of a thick, coarse, light-colored layer

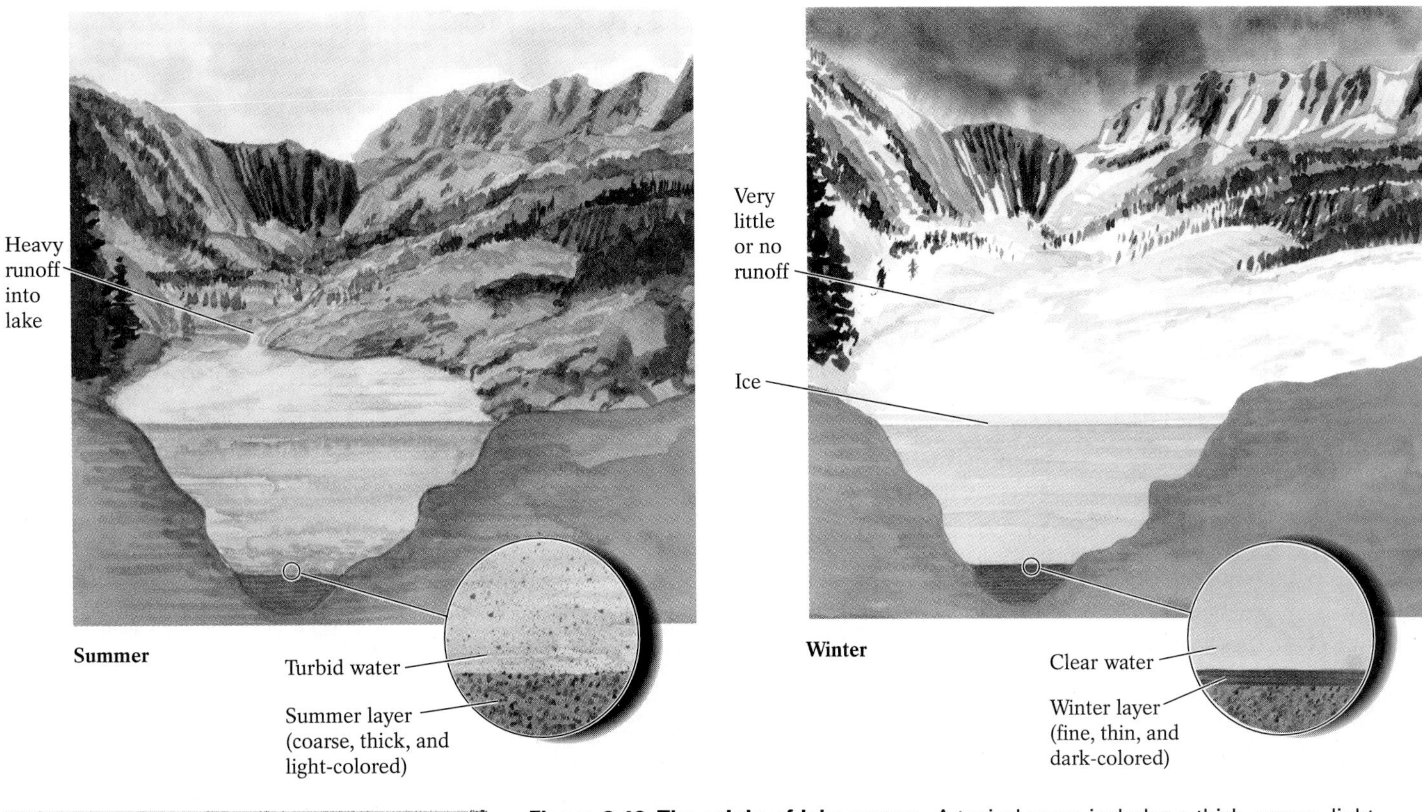

Figure 8-13 The origin of lake varves. A typical varve includes a thick, coarse, light-colored summertime layer produced during high runoff (from snowmelt and spring storms) and high sediment influx, plus a thin, fine dark-colored wintertime layer produced during low runoff and low sediment influx (or no influx, if the lake is frozen). Note the varves in the photo. Why do you think the varves vary so noticeably in thickness?

deposited in summer and a thin, fine, dark-colored layer deposited in winter. By applying *varve chronology,* in which they study the number and nature of the varves underlying a lake, geologists can determine how long ago the lake formed and identify events, such as landslides, that affected sedimentation in the area (Fig. 8-13).

Lichen (pronounced "LIE-ken"), colonies of simple, plant-like organisms that grow on exposed rock surfaces, are the basis of a dating method known as *lichenometry*. Lichen grow extremely slowly; given similar rocks and climatic conditions, the larger the lichen colony, the longer the period of time since the growth surface was exposed. Study of these organisms can yield accurate dates for young glacial deposits, rockfalls, and mudflows—all events that expose new rock surfaces on which lichen can grow (Fig. 8-14).

Figure 8-14 Lichen colonies on a granite boulder. The light-colored areas of rock have been bleached by chemicals in the lichen. The sizes of such colonies can provide clues as to how long the rock surface has been exposed.

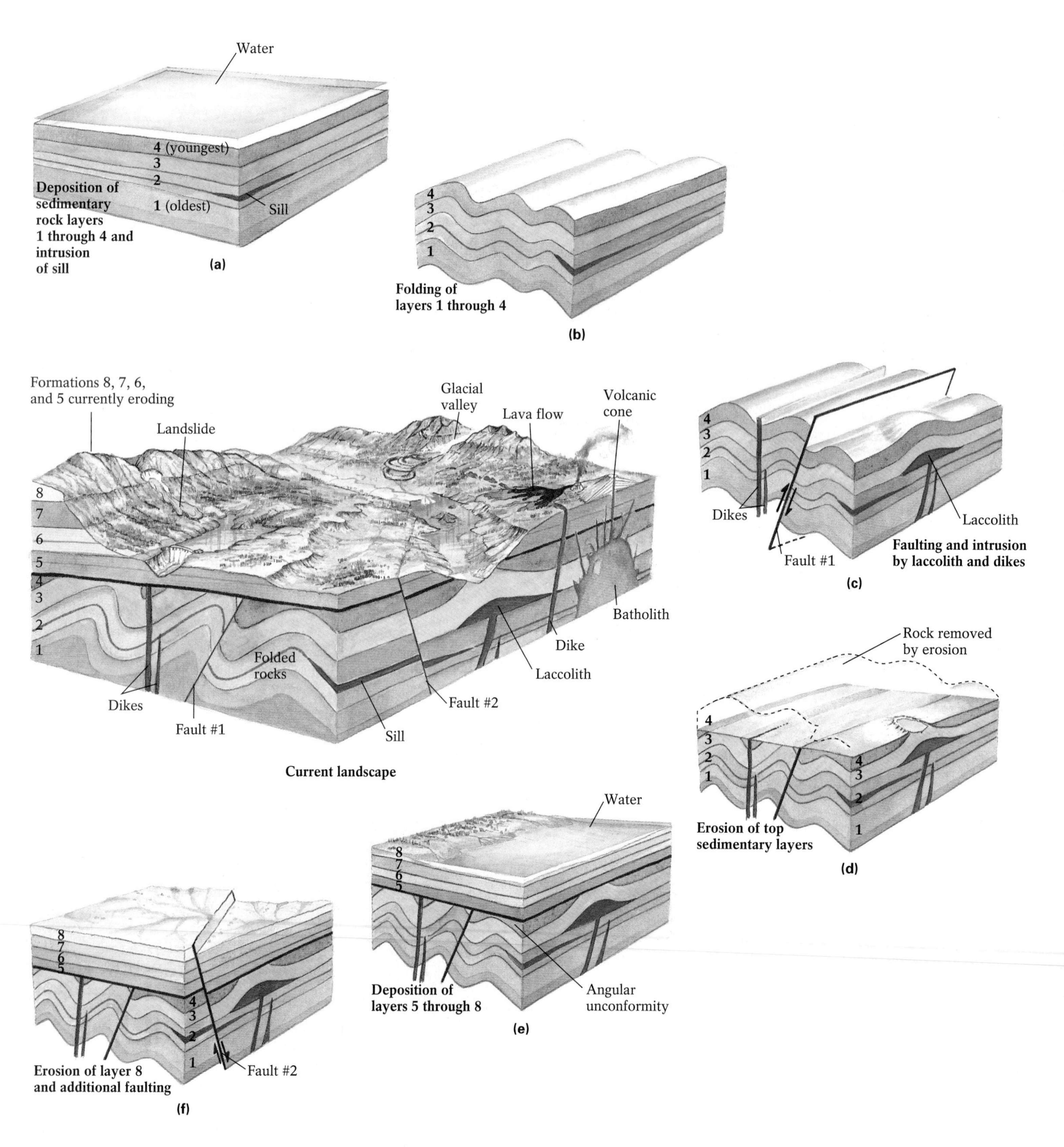

Figure 8-15 A geological biography. The rocks underlying this hypothetical landscape reveal a complex history involving initial horizontal deposition **(a)**, folding **(b)**, faulting and volcanism **(c)**, erosion **(d)**, and subsequent deposition, faulting, and erosion (**e** and **f**). (See Chapter 9 for a discussion of folding and faulting.) The geological activity occurring currently at the surface will be similarly recorded in the area's stratigraphy.

Figure 8-16 The Hoba meteorite, in Namibia. Almost every meteorite that has struck the Earth and been radiometrically dated yields an age of about 4.6 billion years.

Combining Absolute Dating with Relative Dating

When geologists study a region for the first time, they use all dating methods at their disposal—both relative and absolute—to reconstruct its geological history. For example, they can usually write the geological "biography" of any region by radiometrically dating every layer that contains radioactive isotopes and then applying the principles of relative dating to supplement this information. Figure 8-15 shows how the complex history concealed in a region's rocks can be reconstructed, revealing a series of geological events spanning billions of years of Earth's history.

The Age of the Earth

The one critical date that we cannot verify by any direct absolute-dating method is the age of the Earth. The first rocks to form on the planet have since been melted, metamorphosed, or otherwise changed by the Earth's internal heat, recycling of tectonic plates, meteor impacts, and many other geological processes. Although no rocks older than 3.96 billion years have ever been found, most geologists agree that the Earth is about 4.6 billion years old.

Why 4.6 billion? Because all components of our solar system are believed to have formed at about the same time, the ages of extraterrestrial rocks provide indirect evidence of the age of our own planet. Rocks collected on the Moon have been dated by both uranium–lead and potassium–argon systems at about 4.53 billion years. In addition, radiometric dating of numerous meteorites that have struck the Earth, such as the one shown in Figure 8-16, has yielded about the same age—4.6 billion years. Most meteorites are remnants from the creation of the solar system. In all likelihood, they faithfully record the age of the solar system and of the Earth.

We have now examined the vast reaches of geologic time and how we measure it. In the next few chapters, we examine more closely the dynamic tectonic processes that have combined to create the Earth's most massive features, such as its immense interior layers, its continents and ocean basins, and its lofty mountain ranges. We also focus on earthquakes, one of the most dramatic side effects of these dynamic processes.

Chapter Summary

Geochronology is the study of time in relation to the Earth's history. It is part of the larger field of **historical geology,** which focuses on the origin and evolution of the Earth's structures and life forms. Geochronology is based on **relative dating**—determining how old geological features are in relation to their surroundings—and **absolute dating**—determining the actual age of geological features in terms of years.

Relative dating relies on several key principles. The **principle of uniformitarianism** states that modern geological processes are similar to those that operated in the past. The **principle of original horizontality** states that most sedimentary rocks and lava flows are initially deposited in horizontal layers. The **principle of superposition** states that for tectonically undisturbed sedimentary rocks and lava flows, the uppermost layer in a sequence of rocks is the youngest, with those below it being successively older. The **principle of cross-cutting relationships** states that layers cut by other layers and features, such as igneous dikes and faults, must be older than the features that cut across them. Similarly, rocks found within other rocks must be older than the rocks that incorporate them—the **principle of inclusions. Fossils** are the remains or traces of ancient organisms preserved in rock. The **principle of faunal succession** states that layers of rock containing the fossils of more recent organisms are younger than those containing only older organisms. The most helpful fossils, called **index fossils,** come from species that had wide geographic distribution but lived for only a relatively brief period of time.

Unconformities are gaps in the rock record that result either from erosion of entire rock layers or from periods of nondeposition. To determine the relative histories of geographically distant rock sequences, geologists establish age equivalence between similar individual rock layers in the different sequences, a process known as **correlation.**

Whenever possible, geologists determine the absolute date for a rock unit. This type of dating is possible when the rock contains a radioactive **parent isotope** that decays to produce a measurable amount of a nonradioactive **daughter isotope.** The decay occurs at a constant rate, called the **half-life.** The use of radioactive isotopes to date rocks is called **radiometric dating.** Radioactive isotopes used in radiometric dating include uranium, thorium, potassium, and rubidium. **Carbon-14 dating** is used to date organic materials as much as 70,000 years old. Other absolute-dating methods include fission-track dating, dendrochronology, varve chronology, and lichenometry.

Radiometric dating of Moon rocks and meteorites has enabled us to determine that the age of the Earth is approximately 4.6 billion years.

Key Terms

geochronology (p. 137)
historical geology (p. 138)
relative dating (p. 138)
absolute dating (p. 138)
principle of uniformitarianism (p. 138)
principle of original horizontality (p. 139)
principle of superposition (p. 139)
principle of cross-cutting relationships (p. 140)
principle of inclusions (p. 140)
fossils (p. 140)
principle of faunal succession (p. 140)
index fossils (p. 140)
unconformities (p. 140)
correlation (p. 143)
parent isotope (p. 143)
daughter isotope (p. 143)
radiometric dating (p. 143)
half-life (p. 144)
carbon-14 dating (p. 144)

Questions for Review

1. Briefly explain the difference between relative and absolute dating.

2. Discuss three of the basic principles that are the foundation of relative dating.

3. What qualifies a species to become an index fossil? How are index fossils used in the correlation of sedimentary rock strata?

4. Sketch and label two types of unconformities.

5. Name three parent–daughter radiometric dating systems, give the half-lives of each parent isotope, and indicate the rocks or sediments that are most likely to be dated by each.

6. Briefly discuss two potential problems that may diminish the reliability of an isotopically derived date.

7. How does carbon-14 dating differ from other types of radiometric dating?

8. If the oldest rocks ever found on Earth are less than 4.0 billion years old, what evidence suggests that the Earth is actually 4.6 billion years old?

For Further Thought

1. Using a combination of relative- and absolute-dating methods, derive the history of the hypothetical landscape below. (Go slowly, and don't jump to premature conclusions. Consider all the principles discussed in this chapter.)

2. Although geologists claim that "the present is the key to the past" (the principle of uniformitarianism), the Earth has most certainly changed throughout its 4.6-billion-year history. Think of two geological processes that operate differently today than they did in the past, and discuss the differences.

3. Suppose you decided not to accept the 4.6-billion-year age of the Earth that geologists propose (primarily from the ages of Moon rocks and meteorites). Devise an alternative strategy for determining the age of the Earth, assuming that you have unlimited funds.

Part 2 Shaping the Earth's Crust

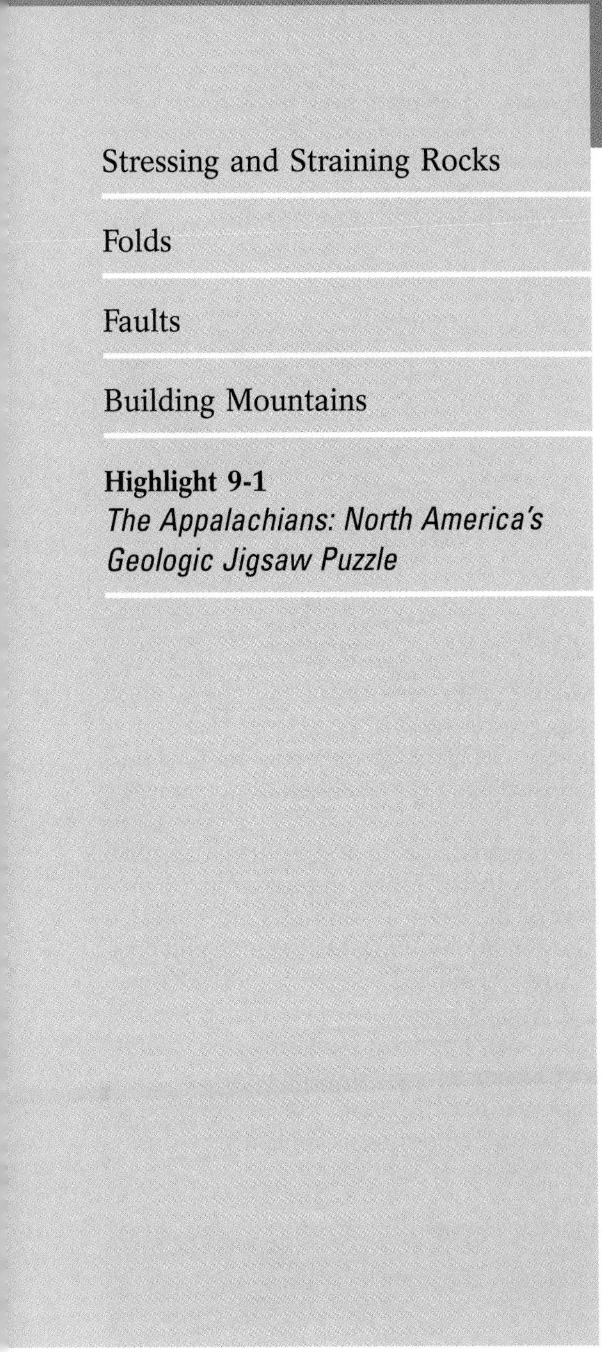

9

Folds, Faults, and Mountains

In the rocks along Sagami Bay near Yokohama, Japan, lives a colony of clams called *Lithophaga,* or "rock eaters." These creatures scoop out small shelters for themselves from the soft rocks at sea level and wait there for high tide to flood their homes, bringing them meals of marine algae. Moments after Japan's great earthquake of 1923, the land at Sagami Bay shifted upward, leaving rows of *Lithophaga* 5 meters (16 feet) above sea level. Several rows of abandoned *Lithophaga* dwellings are found even higher, in the cliffs at Sagami Bay—one that correlates with the area's 1703 tremor and another that correlates with an earthquake occurring in 818. The rocks adjacent to the bay have risen roughly 15 meters (50 feet) during the last 2000 years. At Sagami Bay we are witnessing the building of a mountain.

The Rockies, the Alps, the Andes, and the Himalayas have all been shaped from common rocks during the course of millions of years. In the Pacific Northwest, the Cascade Mountains continue to rise higher even as you read this book. In locales all over the world, we can see rocks that have been twisted and bent or that have broken and shifted position in less dramatic fashion (Fig. 9-1). What enormous forces could distort solid rock in these ways? With few exceptions, all such features owe their existence to plate tectonics. The same forces that carry the Earth's plates can tear the edges of those plates apart, rupturing them into huge displaced blocks, or squeeze plate edges together, crumpling and uplifting them into great folds of rock.

Figure 9-1 Rock distortion. These rocks in the Lower Ugab valley, Namibia, have been tilted and bent by powerful plate-tectonic forces.

Stressing and Straining Rocks

When tectonic plates interact, crustal rocks near the plate boundaries are subjected to a powerful force, or **stress,** that deforms them, changing their shape and often their volume. Rocks may be stressed in three ways, corresponding to three basic types of plate-boundary movements. Rocks at converging plate margins are pushed together, a type of stress

Figure 9-2 The three types of stress applied to rocks, which occur most often at the edges of the Earth's tectonic plates. The edges of converging plates are generally compressed **(a)**, becoming thickened vertically and shortened laterally. The edges of diverging plates are generally subjected to tension **(b)**, becoming thinned vertically and lengthened laterally. The edges of plates at transform boundaries generally undergo shearing stress **(c)**, becoming sliced into parallel blocks of rock.

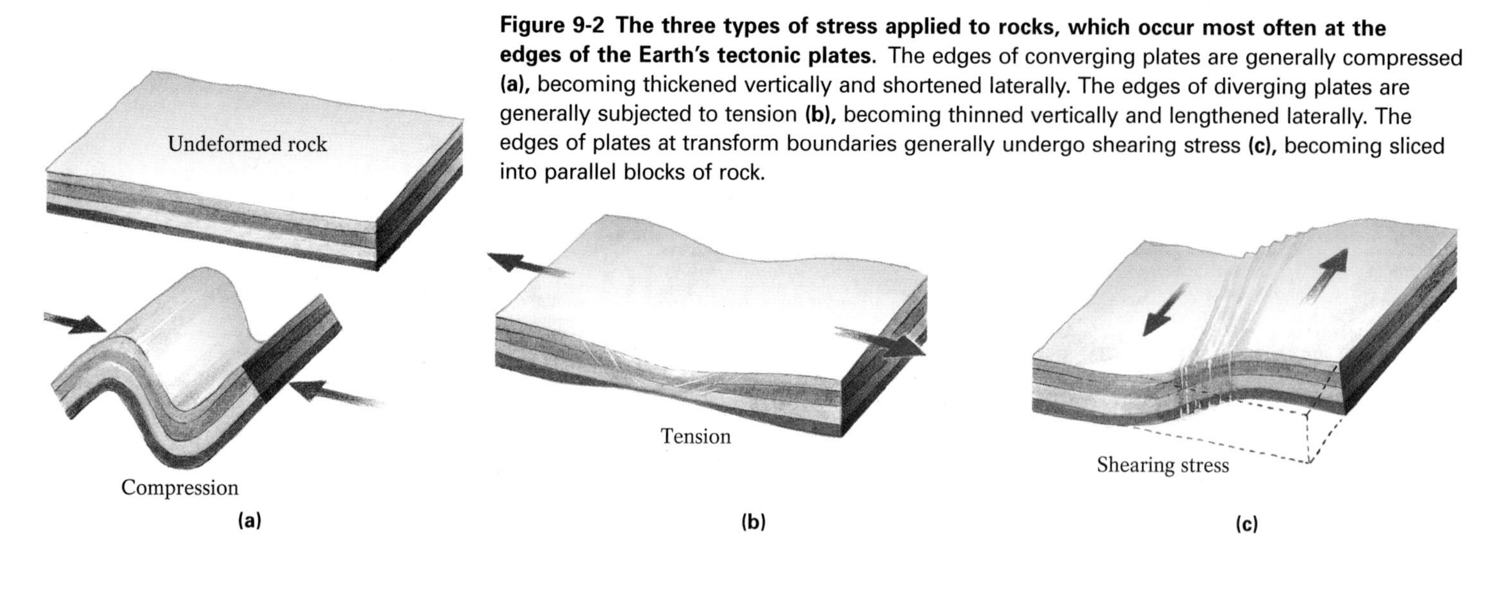

known as **compression** (Fig. 9-2a). Compression generally crumples rocks, causing them to become thickened vertically and shortened laterally. Rocks at diverging or rifting plate margins are pulled apart, a type of stress known as **tension** (Fig. 9-2b). Tensional stress stretches, or extends, rocks so that they become thinner vertically and longer laterally. Rocks at transform plate margins are forced past one another in parallel but opposite directions, known as **shearing stress** (Fig. 9-2c). Shearing stress slices rocks into parallel blocks. The change in shape undergone by rocks under stress is called **strain.**

Types of Deformation

When a rock experiences only minor stress—whether compression, tension, or shearing stress—it may return to its original shape and volume after removal of the stress, much as a stretched rubber band regains its original shape after use. Such a temporary change is described as **elastic deformation.** A rock that is strained elastically is not permanently deformed.

Application of a greater amount of stress to a rock, however, may deform it so that it cannot return to its original shape after removal of the stress. When rocks are subjected to great stress under conditions of low temperature and pressure or when the stress is applied suddenly, the rocks may rupture or crack, a type of permanent deformation known as **brittle failure** (Fig. 9-3a). When rocks are subjected to stress under conditions of higher temperatures or pressures (such as occurs when rocks lie buried several kilometers beneath the Earth's surface) or when the stress is applied very slowly,

Figure 9-3 Types of rock deformation. **(a)** These rocks, from the Sierra Nevada of California, have undergone brittle failure. They have been broken and shifted by powerful rifting-type forces. **(b)** These rocks have undergone plastic deformation. The person is standing next to a fold in layers of strong, well-cemented sandstone rock in California's Borrego Desert.

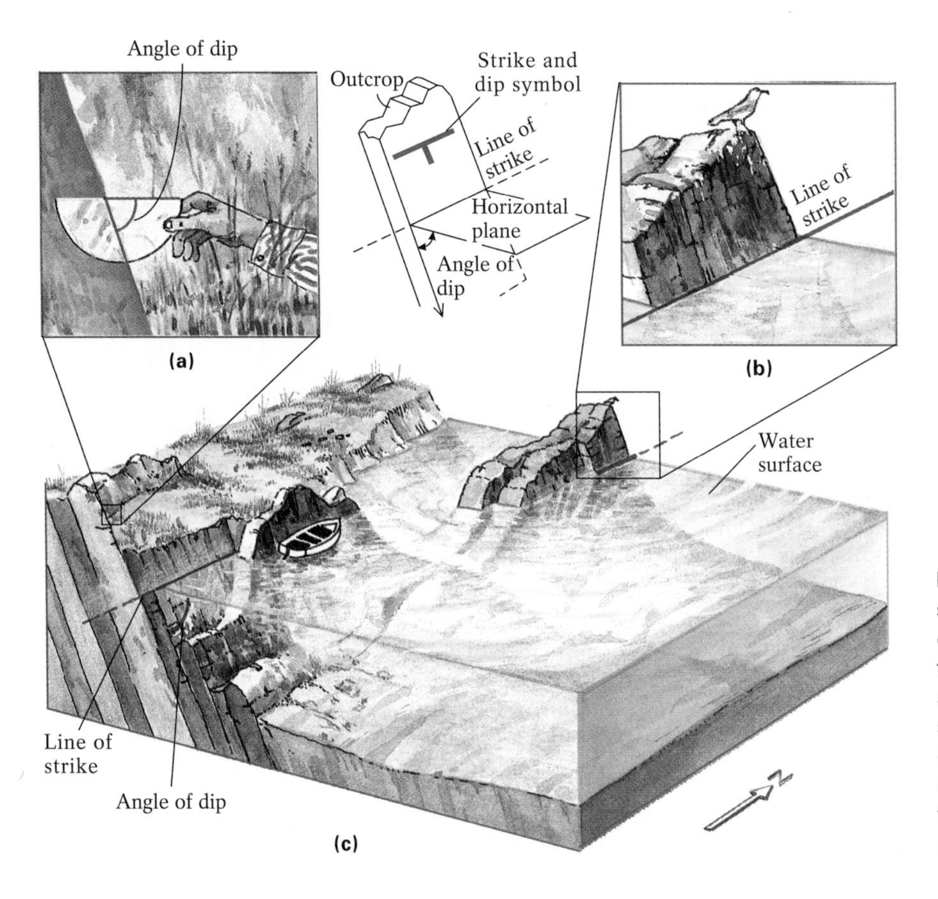

Figure 9-4 Geologic orientation. Because the surface of a body of water is horizontal, it can help determine the orientation of rock outcrops relative to the Earth's surface. The plane of the water traces a strike line across the exposed surface of a partially submerged outcrop and provides a horizontal reference against which to measure a structure's angle of dip (using an *inclinometer*). Note that a structure's strike and dip are always perpendicular to one another.

they may undergo **plastic deformation,** an irreversible change in the shape or volume of the rock (Fig. 9-3b).

In addition to heat, pressure, and time, one other factor governs the way in which rocks respond to stress—rock composition. Thus, under moderate amounts of heat, pressure, or time, a rock may deform elastically, brittlely or plastically, depending on the rock's mineral composition, texture, internal planes of weakness, fluid content (high fluid content tends to weaken a rock), and geologic history (past geologic forces may have weakened or strengthened the rock).

Deformed Rocks in the Field

Deformation is most readily observed in sedimentary rocks, because they were originally horizontal and laterally continuous. When sedimentary layers deform plastically, they display an obvious crumpled appearance. Fractures and displacements from brittle failure are also clearly evident in sedimentary rock, because of the offset of the rock layers.

When faced with deformed rocks, geologists want to find out what type of stress (compression, tension, or shearing) was responsible and how much and in what directions the deformed rocks have moved. This information enables them to determine past plate movements and other geological events. They also want to know how the deformation continues underground, because this information can help in locating subterranean resources such as oil, coal, natural gas, metal ores, and other minerals. Geologists estimate the subsurface direction of rocks by determining the orientation of their surface features.

The orientation of a geologic structure or rock layer is expressed in terms of **strike** and **dip** (Fig. 9-4). A rock's strike is the compass direction (for example, northeast) of an imaginary horizontal line lying on the surface of the rock structure. The dip of the structure is the angle and direction at which it is inclined, or tilted, relative to the horizontal. Measuring the strike and dip of a rock structure establishes its orientation in three-dimensional space. By making hundreds of strike and dip measurements at numerous outcrops and then recording the data on an aerial map (thus producing a *geologic map*), geologists can determine the three-dimensional subsurface form of structures that are barely visible at the surface.

Folds

Folds are bends in rock layers that develop when originally horizontal strata become deformed plastically, usually by compression. When sedimentary rocks are compressed laterally, they fold down and up, forming a series of troughs and arches—much like a rug that bunches up when pushed on a polished hardwood floor. Folds in rock vary in size, ranging from the small crinkles visible in some fist-sized hand speci-

Figure 9-5 Folded rocks of various sizes. **(a)** A hand specimen of folded gneiss. **(b)** Folded rock strata along Route 23 in Newfoundland, New Jersey. **(c)** Folded sedimentary rocks in Dorset, England, which result from strong compressional forces at a former convergent plate margin.

mens, to the obvious folds seen in rockcuts bordering many highways, to mountain-sized folds, such as those seen in the Canadian Rockies and in Dorset, England (Fig. 9-5).

Synclines and Anticlines

The most common types of folded rocks are the tightly folded troughs and arches that occur from compression at active convergent plate margins. The trough-like folds are **synclines** (from the Greek for "inclined together"). The arch-like folds are **anticlines** (from the Greek for "inclined against"). Folding usually produces a series of alternating synclines and anticlines, with the adjacent features sharing a common side, or limb (Fig. 9-6). Each syncline and anticline has an *axial plane,* an imaginary plane that divides the fold into two approximately equal halves. The line where the axial plane intersects the bedding planes of the folded rocks represents the *axis* of the fold. In any group of folded rock layers, the innermost rocks of the synclines are the youngest and the innermost rocks of the anticlines are the oldest, because they represent the layers that were uppermost (that is, youngest) prior to folding.

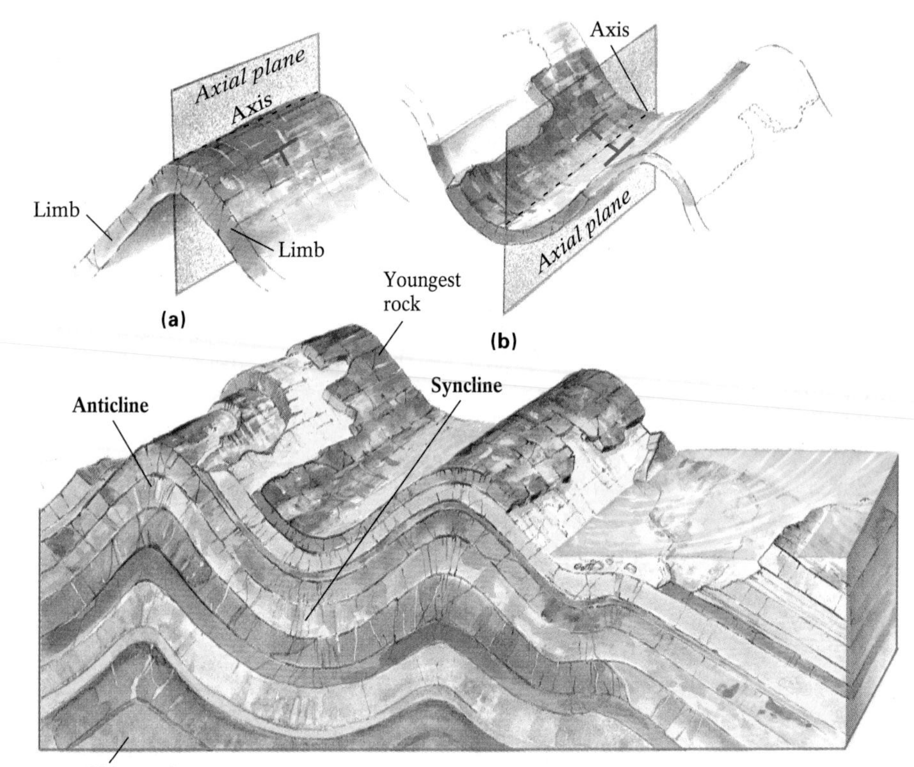

Figure 9-6 The geometry of anticlines and synclines.

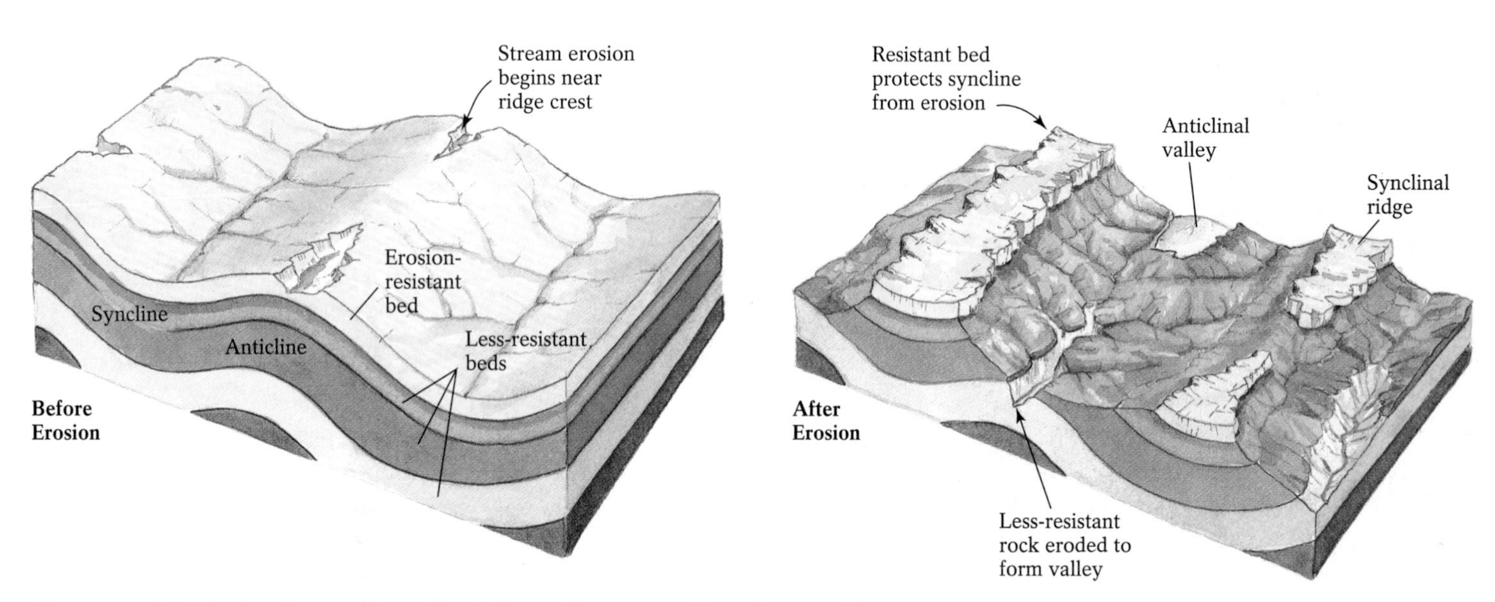

Figure 9-7 Anticlinal valleys and synclinal ridges. The composition of exposed bedrock structures strongly controls their susceptibility to weathering and erosion. If weak, erodible rock crops out along the crest of an anticline, its erosion will produce a valley at that site. If the axis of a syncline consists of strong, resistant rock, its surface expression will be a ridge.

Although diagrams may depict anticlines as hills and synclines as valleys, the two are actually structures in the rocks, not surface landforms. Anticlines do not always form hills or ridges, nor do synclines always underlie valleys; these topographic features are determined largely by differences in the features' resistance to erosion. For example, if a resistant sandstone were exposed in the trough of a syncline and readily eroded rocks formed the crests of adjacent anticlines, structural anticlines could erode to form topographic valleys and structural synclines could form topographic ridges (Fig. 9-7).

Anticlines and synclines may take on different symmetries, depending on the amount of folding they have undergone (Fig. 9-8). Under relatively gentle compression, they appear as broad, *symmetrical* structures with near-vertical axial planes and gently dipping limbs inclined at roughly the same angle. These structures are most often found in tectonically quiet, mid-continent areas. At active convergent plate margins, moderate to intense compression may force one limb to move more than the other, causing folds to become *asymmetrical.* Prolonged directed pressure may rotate an asymmetrical fold until its axial plane becomes sharply angled, producing an *overturned* fold, or even until its axial plane is essentially horizontal, paralleling the Earth's surface. Such *recumbent* folds typify highly deformed mountain belts, such as the Northern Rockies, Appalachians, Himalayas, and the European Alps.

More complex structures develop when a sequence of anticlines and synclines tilts so that the folds' axes intersect the Earth's surface, continuing underground where we cannot see them. We identify such *plunging* folds by the

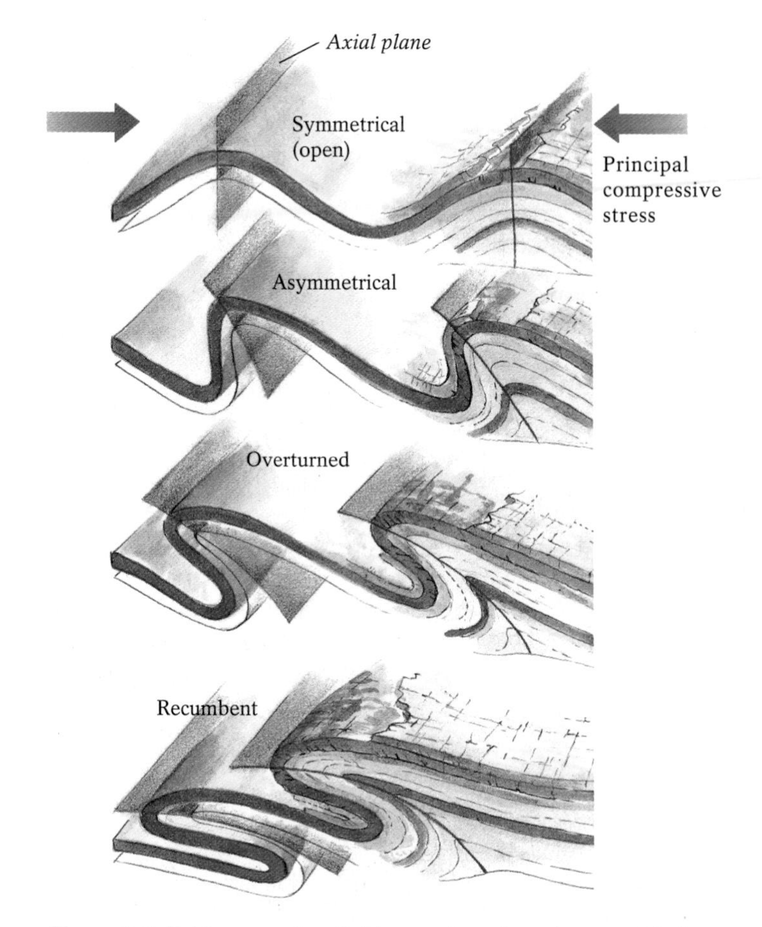

Figure 9-8 Fold symmetry. Folds vary from broad symmetrical structures, with limbs dipping at about the same angles, to overturned and recumbent folds.

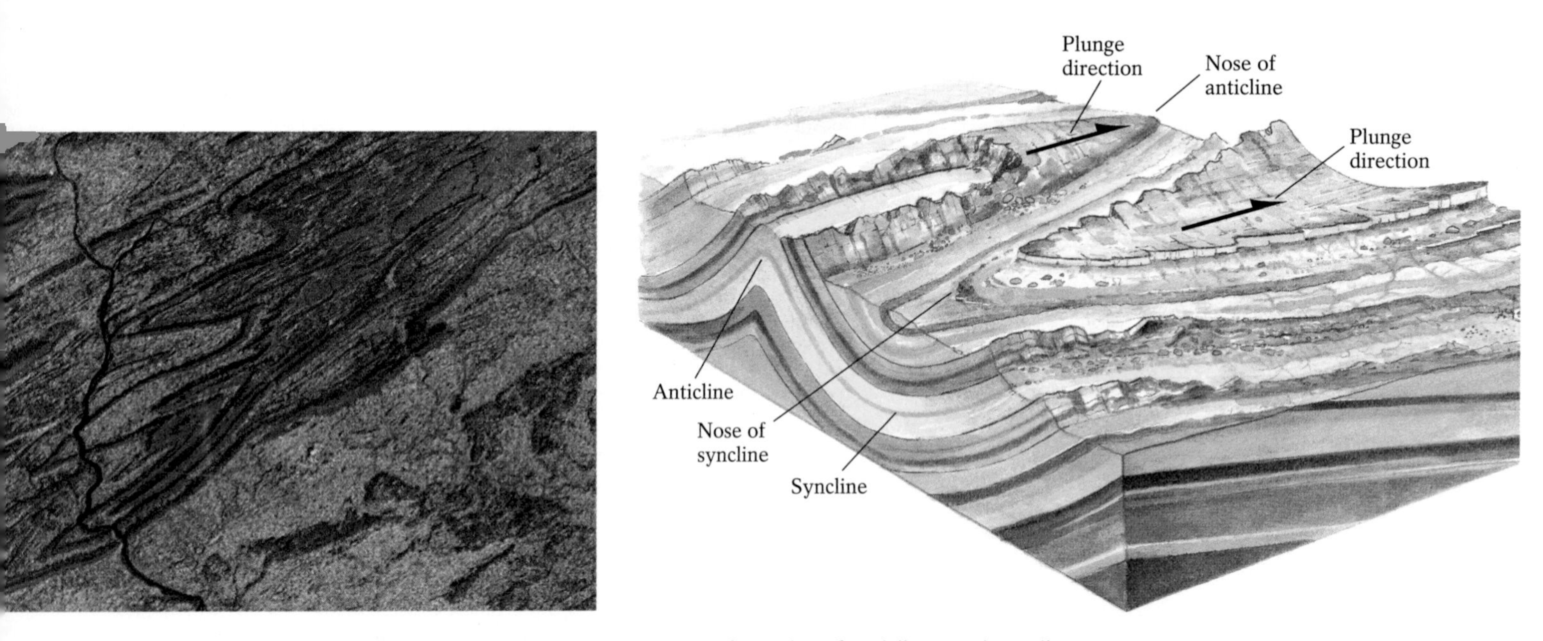

Figure 9-9 Plunging folds. These folds occur when the axes of a series of anticlines and synclines tilt into the Earth. Photo: A satellite image of plunging folds in the Valley and Ridge province of central Pennsylvania near Harrisburg.

characteristic zigzag pattern that emerges as they erode. The Valley and Ridge province of the folded Appalachians of Pennsylvania contains some of North America's best examples of plunging folds (Fig. 9-9).

Domes and Basins

A structural **dome** is an oval-shaped bulge of rock layers that, in cross section, resembles an anticline. As in an anticline, the oldest rocks in a dome appear at its center, and all other layers dip away from them (Fig. 9-10). A structural **basin** is a bedrock depression that, in cross section, resembles a syncline. As in a syncline, the youngest rocks in a basin are located at its center, and all other layers dip toward them. Domes and basins are frequently found in mid-continent, mid-plate locations, leading some geologists to conclude that they do not result primarily from lateral compression associated with plate movement. Instead, they likely result from vertical forces, such as upwelling magma or warm mantle currents that rise toward the surface and push sedimentary strata upward. Some basins appear to form where materials near and just below the surface are dense enough to depress and deform the materials below them.

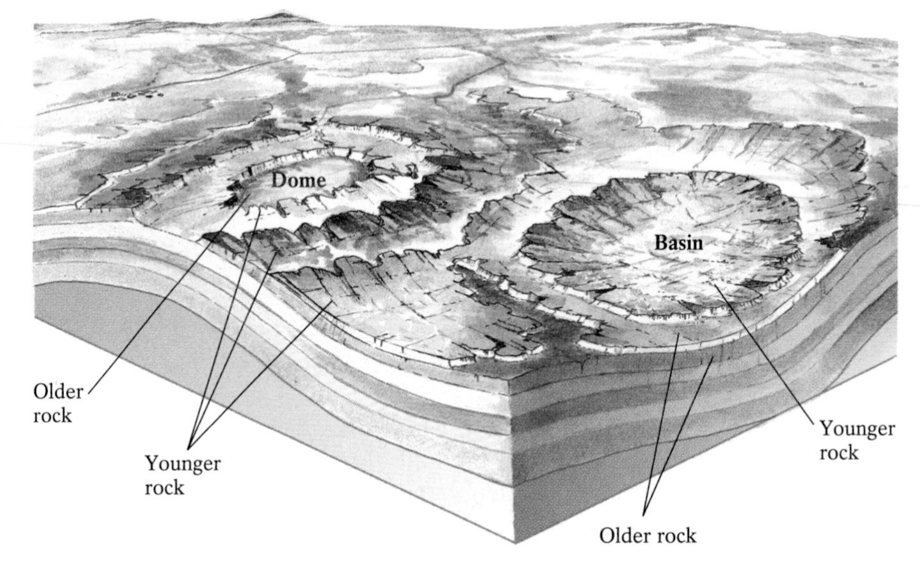

Figure 9-10 Domes and basins. Structural domes and basins resemble anticlines and synclines, respectively, in the relative positions of their variously aged layers. Because these structures often form far from plate boundaries, they may reflect the effects of vertical motion related to crustal density variations, rather than plate-edge activity.

Domes are more readily apparent than basins, standing out prominently at the surface; in contrast, basins usually become filled with sediment and thus obscured. Noteworthy in North America are the Adirondack dome of northern New York, the Ozark dome of the southern Mississippi River valley, the Nashville dome of the Tennessee River valley, the oil-producing domes of eastern Wyoming, and the Black Hills of South Dakota, from which scenic Mount Rushmore was sculpted. Notable structural basins include the oil-and-coal-bearing Williston basin of North Dakota, the Illinois basin of central Illinois, and the Michigan basin, which forms virtually all of lower Michigan (with the campus of Michigan State University at its center).

Figure 9-11 Faults. Faulted rocks are those that have moved relative to each other. The fault in this photo is evident in the displacement of the various colored rock layers. Strike and dip are often used to describe the orientation and relative movement of the fault blocks.

Faults

Because most rocks are brittle at low temperature and pressure, virtually every solid rock and nearly all layers of unconsolidated sediment at or near the Earth's surface contain evidence of brittle failure in the form of cracks, or *fractures.* The orientation of these fractures in surface rocks provides clues to the stresses that have affected the rocks. Some fractures are oriented randomly and occur in rocks that show no evidence of relative movement; they may be produced by a variety of fairly localized stresses, such as those associated with cooling igneous rocks or freezing surface rocks. Other fractures are oriented systematically, appearing at right angles to one another, again with no evidence of relative movement; they form in response to a specific set of stress conditions, such as those at a plate boundary. The most dramatic examples of cracks in rock are **faults,** fractures along which marked relative movement has taken place (Fig. 9-11). *Fault blocks* are the rock masses on either side of the *fault plane,* the approximately planar (flat) surface along which the movement occurs. To determine the orientation of a fault plane and each fault block's relative direction of movement, geologists take strike and dip measurements.

Types of Faults

All faults result from relative movement between adjacent blocks of rock—that is, they occur when one or both blocks move, or *slip,* relative to their original positions. Such slippage is usually caused by the compressional, tensional, or shearing stresses associated with plate tectonic movement. Faults are generally classified according to the direction of the relative movement between fault blocks, which itself relates to the type of stress causing the fault.

Strike-slip faults are caused by shearing stress (usually at transform boundaries); in this case, fault block movement is largely horizontal, occurring parallel to the strike of the fault plane (Fig. 9-12). North America's most famous strike-slip fault,

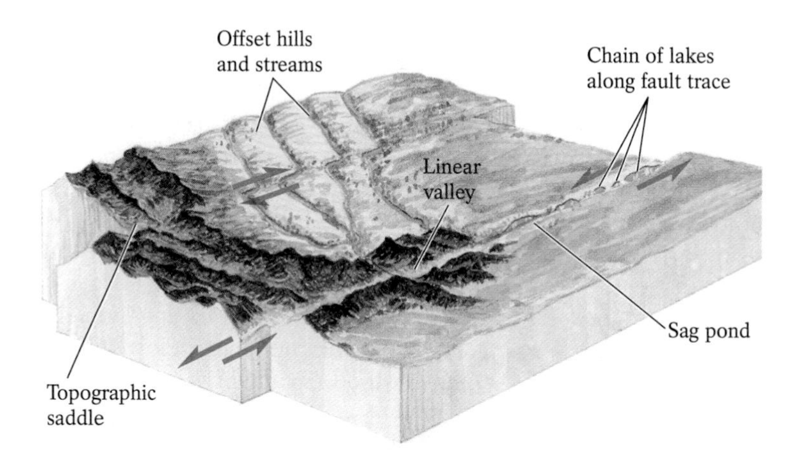

Figure 9-12 Strike-slip faults. Horizontal movement along strike-slip faults creates offset topographic features and distinctive erosional landforms, such as linear valleys, lake chains, and sag ponds.

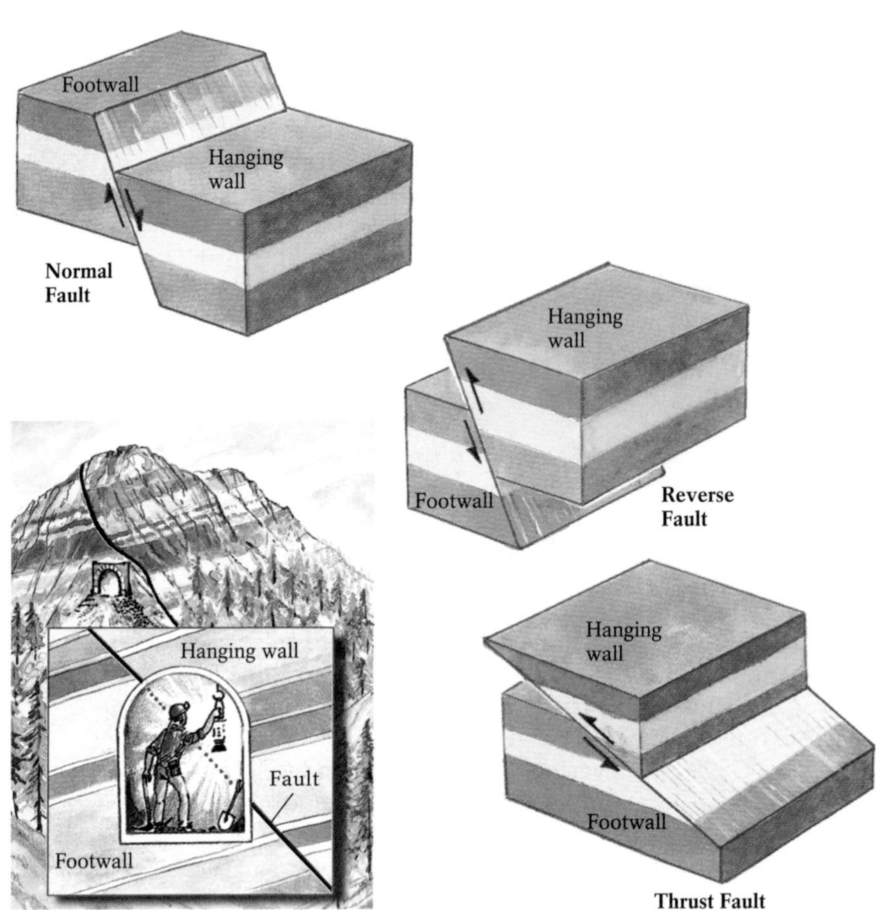

Figure 9-13 Dip-slip faults. This dip-slip fault is located along McGee Creek, south of Mammoth Lakes, California. The rocks in the foreground of the photo (in front of the "step" in the landscape) have slipped downward relative to the rocks in the background. Dip-slip faults are classified by the relative movement of their fault blocks. In normal faults, the hanging wall moves downward relative to the footwall; in reverse faults, the hanging wall moves upward relative to the footwall; thrust faults are low-angle reverse faults.

the San Andreas, is an active fault stretching for 1000 kilometers (600 miles). The San Andreas underwent one of its greatest recorded displacements on April 18, 1906, the day of the great San Francisco earthquake. The extent of displacement during the 1906 quake—7 meters (23 feet)—was evident from offset fences and other linear features along more than 400 kilometers (250 miles) of the fault, from Point Arena to San Juan Bautista. By comparison, California's Loma Prieta earthquake of 1989 produced far less displacement—a maximum of 2.2 meters (7 feet) was measured east of Santa Cruz, and the surface rupture stretched for only a few tens of kilometers.

Relative movement between fault blocks is more often vertical than horizontal, producing **dip-slip faults** (Fig. 9-13). Many old mine shafts can be found along dip-slip faults, because precious metals become concentrated there when hot, mineral-rich solutions migrate along faulted rocks. Geologists identify dip-slip fault blocks by their position relative to the fault plane, using terms derived from mining traditions. The block above the fault plane, from which the miners hung their lanterns, is known as the *hanging wall;* the block below the fault plane, on which the miners stood, is called the *footwall.*

Dip-slip faults are classified as either normal, reverse, or thrust faults. In a **normal fault,** the hanging wall has moved downward relative to the footwall. This type of fault is generally associated with tensional stress, and is most often seen where plates are rifting or diverging. A fault block that drops during normal faulting produces a depression called a **graben** (pronounced "GRAB-en"; from the German for "grave"); the blocks that remain standing above and on either side of a graben are **horsts** (from the German for "height"). We can readily see these features in the normal faults of East Africa's rift valleys, as well as along the Earth's 60,000 kilometers (40,000 miles) of mid-ocean divergence zones (Fig. 9-14).

In a **reverse fault,** the hanging wall has moved upward relative to the foot wall. This type of fault is generally associated with powerful horizontal compression, most often resulting from plate convergence. As the hanging wall moves over the footwall, it carries along deeper, older rocks, transporting them up and over younger rocks to create a rock sequence that violates the principle of superposition (see Chapter 8).

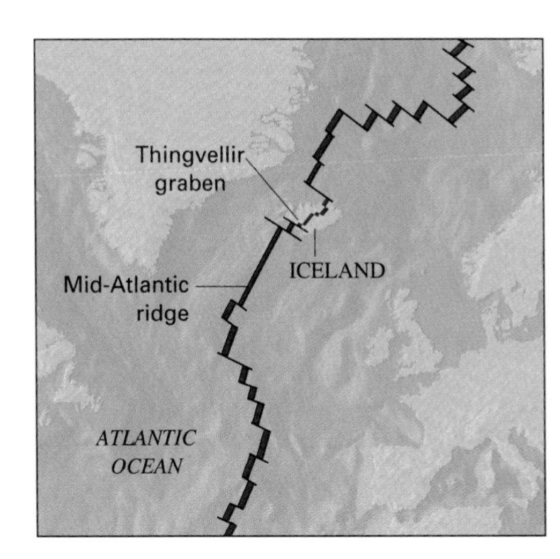

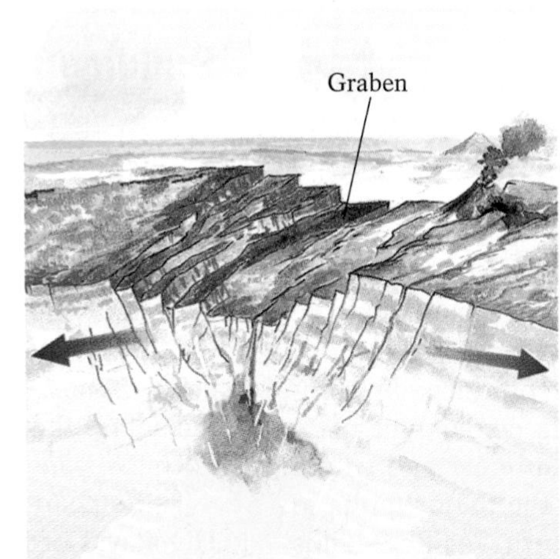

Figure 9-14 Graben formation. Iceland's central valley, the Thingvellir, formed when tensional stress and normal faulting within the diverging mid-Atlantic ridge system created a graben.

A **thrust fault** is a type of reverse fault in which the blocks move at a relatively low angle (less than 45°). An example of an extremely low-angle thrust fault is the Lewis Overthrust in Glacier National Park, Montana, and Waterton-Lakes National Park, Alberta, Canada, where a 3-kilometer (2-mile)-thick slab of Precambrian marine sedimentary rock was transported more than 50 kilometers (30 miles) eastward, coming to rest atop much younger continental river sediments (Fig. 9-15).

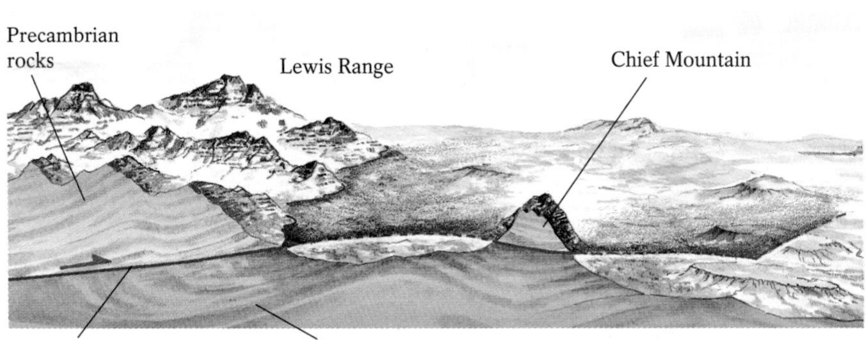

Figure 9-15 Thrust faults. The Lewis Overthrust, in Glacier National Park, Montana, formed when compression moved a slab of 800-million- to 1.1 billion-year-old Precambrian marine sediment over the top of a layer of 150-million-year-old continental sediment. Photo: Chief Mountain, formed by subsequent erosion of both layers, is composed of the Precambrian rock at its top and the younger Mesozoic rock at its base.

Plate Tectonics and Faulting

Each of the three major types of plate margins is associated with a particular type of stress and thus with a particular type of fault (Fig. 9-16). Normal faults are commonly found where tensional stresses pull apart the Earth's lithosphere, such as along the axes of the world's mid-ocean ridges and near rifting continents, such as in the north–south Rio Grande valley in New Mexico, where North America may be splitting apart.

Reverse and thrust faults, which are produced by compressive stresses, are concentrated along convergent plate boundaries, both where oceanic plates subduct and where continental plates collide. For instance, the major faults in Japan, the Philippines, and other western Pacific islands largely derive from the convergence of two ocean plates. In contrast, faults in the Andes of South America and the Cascades of Washington and Oregon have developed where an ocean plate subducts beneath an adjacent continental plate. Faults in the Alps and Himalayas owe their existence to collisions between continental plates. Some of the world's longest continuous faults, such as the San Andreas fault in California and the Anatolian fault in Greece and Turkey, are strike-slip faults that coincide with transform boundaries.

Building Mountains

Every continent has mountains, as does every ocean basin. Some mountains, such as Stone Mountain in Georgia, stand alone, isolated and towering above their surroundings. Others are grouped together in *ranges,* such as the volcanic Cascades of the Pacific Northwest. Some ranges occur in continent-long mountain *systems,* such as the Appalachian Mountain system, which comprises several ranges—including the Great Smokies, the Blue Ridge, and the Poconos—and extends from Alabama to Newfoundland.

Some of the mountains we see today, such as the Himalayas of India, China, and Tibet, are young and still rising. Others, such as the Appalachians and the Urals of central Europe, are so old that the forces that created them ceased to operate hundreds of millions of years ago—today they are subject to only weathering and erosion. Still other mountains are so ancient that they have been completely worn away and are now flat regions in continental interiors.

When we look at mountains, we are truly seeing the handiwork of eons of the planet's past. A mountain and its rocks represent several periods of geologic and tectonic activity. Except in the case of volcanic mountains, the rocks

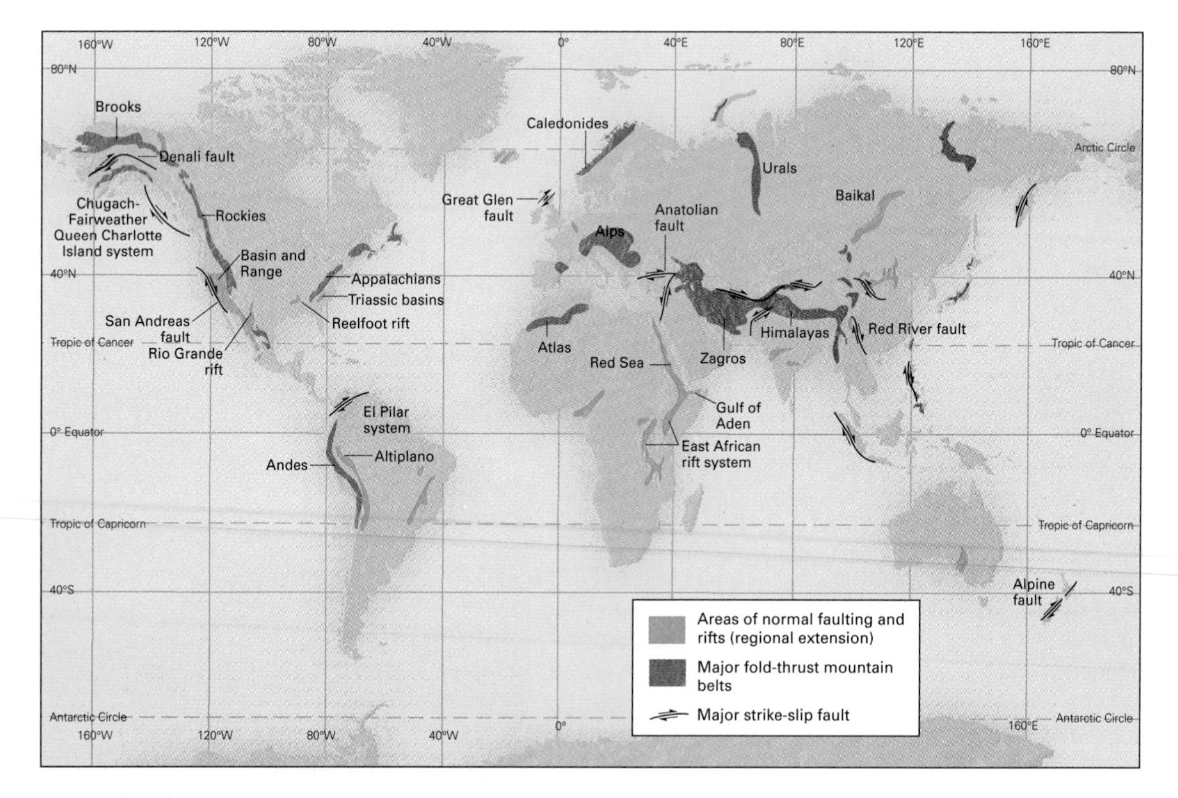

Figure 9-16 The coincidence of different fault types with the Earth's plate boundaries. Normal faults typically occur at rifting and divergent zones. Reverse and thrust faults commonly occur at convergent zones, including subduction zones and zones of continental collisions. Strike-slip faults occur principally at transform boundaries.

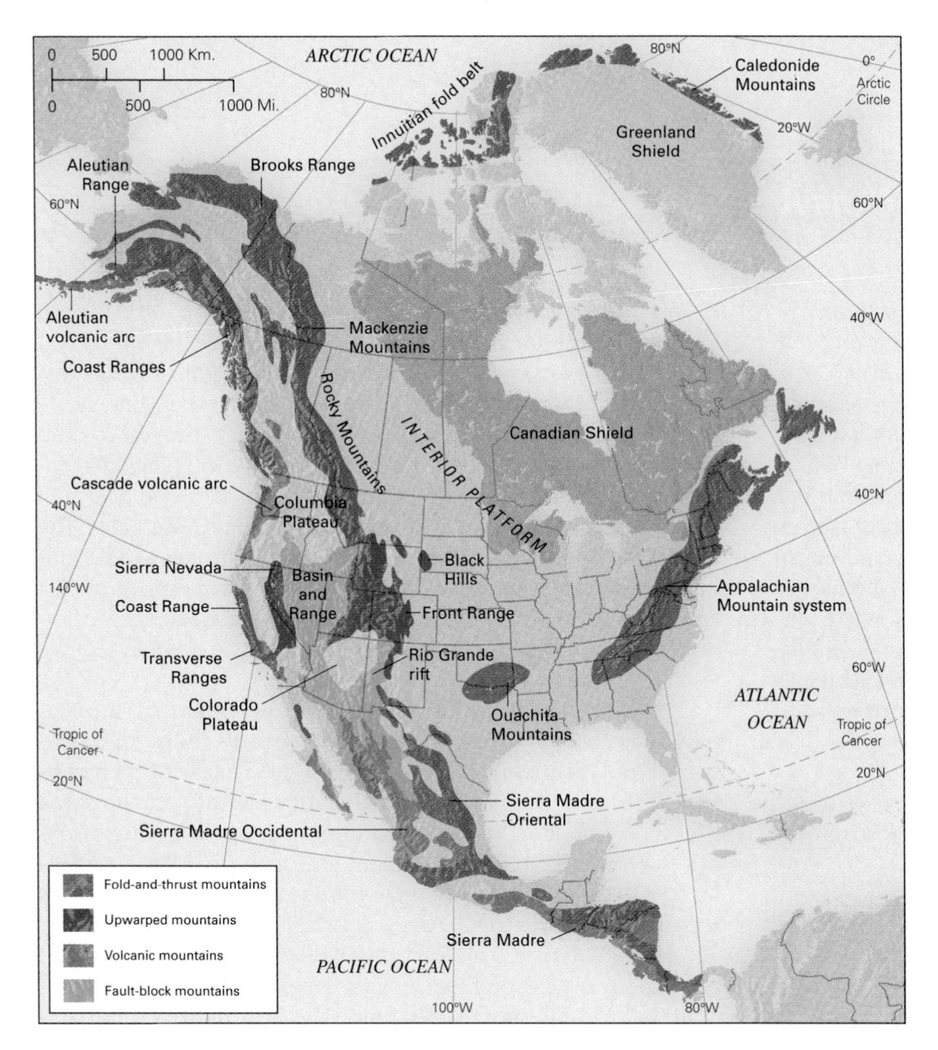

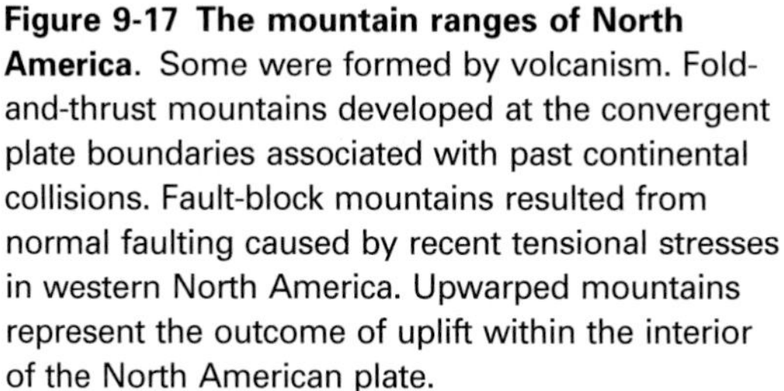
Figure 9-17 The mountain ranges of North America. Some were formed by volcanism. Fold-and-thrust mountains developed at the convergent plate boundaries associated with past continental collisions. Fault-block mountains resulted from normal faulting caused by recent tensional stresses in western North America. Upwarped mountains represent the outcome of uplift within the interior of the North American plate.

within a mountain are usually much older than the mountain itself, having existed for a long time before being uplifted. Today's mountains may have formed by one tectonic process, but their component rocks generally formed much earlier and by entirely different processes.

Types and Processes of Mountain Building

Mountains can form in so many ways that even adjacent ones may have been constructed by completely different processes at different times (Fig. 9-17). Some mountains, such as the Catskill Mountains of upstate New York, are actually just the uplands that remain after streams have cut deep valleys into a plateau. Some, such as Mauna Loa and Mauna Kea of Hawai'i, are undeformed accumulations of basalt that erupted from sea-floor hot spots. Others, especially complex systems such as the Appalachians of eastern North America and the Alps of southern Europe, formed by multiple episodes of sedimentation, intense folding, thrust faulting, plutonism, volcanism, and metamorphism during collisions of broad plate edges. As a consequence of their varied origins, mountains vary in composition and shape, even within the same system.

Geologists refer to the processes of mountain building as **orogenesis** (from the Greek *oro,* meaning "mountain," and *genesis,* meaning "birth"). We examined one of the Earth's principal orogenic processes—volcanism—in Chapter 4. Volcanic mountains, such as Hawai'i's shield volcanoes, the submarine peaks of the mid-Atlantic ridge, South America's Andes, and North America's Cascades, form around volcanic vents by the accumulation of lava flows and pyroclastic materials. In this chapter, we focus on mountains created by processes that deform and uplift materials in the Earth's crust. Three principal nonvolcanic types of mountains exist, each originating in a distinct tectonic setting and identifiable by its geological structure: fold-and-thrust mountains, fault-block mountains, and upwarped mountains.

Fold-and-Thrust Mountains **Fold-and-thrust mountains** develop where continental plates collide. They may produce exceptionally high mountain systems (3 kilometers [2 miles] or taller). Typically, these rocks consist predominantly of marine sediments that have been intensely folded, thrust-faulted, and sometimes intruded and metamorphosed by large plutons. A succession of mountain-building events has deformed most of them more than once. Fold-and-thrust systems include the European Alps, the Himalayas of India, Tibet, and China, the Urals of central Europe, the Northern and Canadian Rockies, and the Appalachians of eastern North America (Highlight 9-1).

Highlight 9-1 The Appalachians: North America's Geologic Jigsaw Puzzle

The Appalachian Mountain system of eastern North America is a mosaic of folded, thrusted, and metamorphosed provinces that evolved over nearly a billion years of Earth history. The North American segment of the system extends 3000 kilometers (2000 miles) from eastern Newfoundland to 75 kilometers (40 miles) east of Birmingham, Alabama. When we reconstruct the Northern Hemisphere's plate boundaries, however, we find that the Appalachians actually extend a good deal farther—to the Caledonides of western Europe and on to Norway. To the west of North America's Appalachian Mountains lies the Appalachian plateau, a region of relatively unfolded, unmetamorphosed, coal-bearing rocks. During Paleozoic time, this plateau consisted of lushly vegetated wetlands located adjacent to the rising Appalachians. To the east lie the more recent deposits of the Atlantic coastal plain.

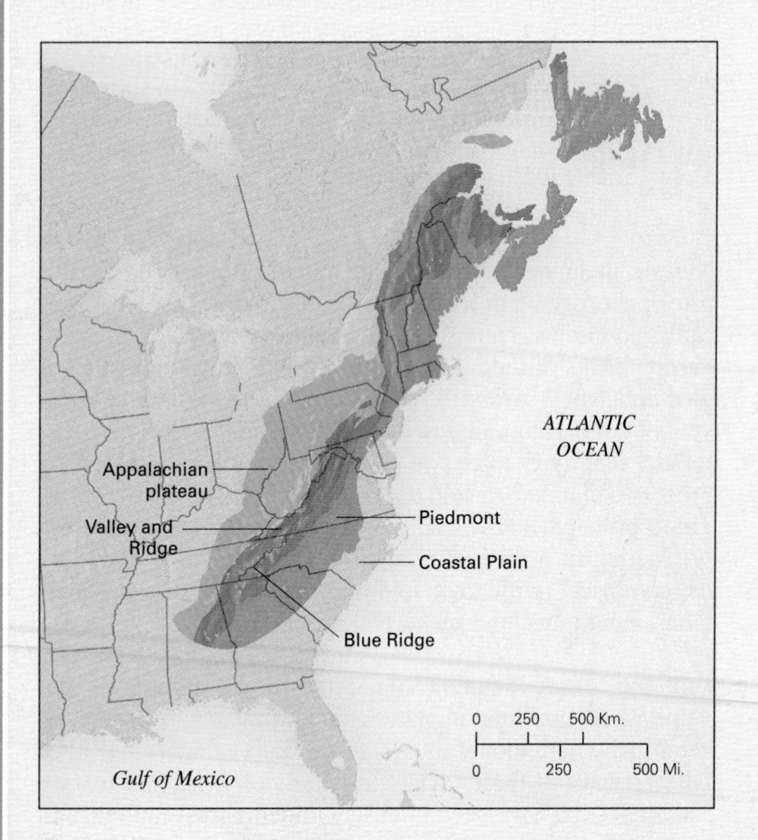

Figure 1 The provinces of the Appalachian mountain system in eastern North America.

The Appalachians' width of 600 kilometers (400 miles) covers three principal provinces, separated by major thrust faults. Each province developed at a different time and has distinct rock formations, but all were affected by the same series of mountain-building episodes. West to east, the Valley and Ridge province consists of a thick sequence of relatively unmetamorphosed Paleozoic sediments that were folded and thrust to the northwest by compression from the southeast. The Blue Ridge province includes highly metamorphosed Precambrian and Cambrian crystalline rocks. The Piedmont province consists of metamorphosed Precambrian and Paleozoic sediments and volcanic rocks that were intruded by granitic plutons (Fig. 1).

The most likely hypothesis for the evolution of the Appalachians begins with Late Precambrian plate collisions (about 1.1 billion years ago) that assembled a pre-Pangaea supercontinent. This Precambrian supercontinent began to break up some 800 to 700 million years ago, with North America rifting from Eurasia and Africa. This movement gave rise to an ancestral Atlantic Ocean and an additional continental fragment that was separated from the North American plate by a marginal sea (Fig. 2a). About 700 to 600 million years ago, subduction began within the ancestral Atlantic Ocean, with a volcanic island arc developing above this subduction zone (Fig. 2b). Between 600 and 500 million years ago, the oceanic crust in the marginal sea also began to subduct, producing a volcanic arc within the continental fragment (Fig. 2c). Meanwhile, subduction continued beneath the island arc, and the ancestral Atlantic Ocean began to shrink.

By about 500 million years ago the marginal sea had entirely subducted, allowing the continental fragment to collide with the eastern edge of the North American plate (Fig. 2d). In this collision, the rocks of the continental fragment were thrust-faulted northwesterly over younger continental and marine sediments along the continental shelf of the North American plate. This first mountain-building episode is called the Taconic orogeny; today's Blue Ridge Mountains eventually developed on the structures formed at this time.

The second mountain-building episode, the Acadian orogeny, occurred 400 to 350 million years ago (Fig. 2e). The ancestral Atlantic Ocean continued to close as Africa began its northwestward convergence. The island arc developed earlier now collided with the eastern margin of the North American plate, pushing the Blue Ridge and western Piedmont (formed from the accumulation of material eroded from the Blue Ridge) on top of the plate and farther to the northwest. Today these rocks form the eastern Piedmont province.

The third and final mountain-building collision, the Allegheny orogeny, occurred between 350 and 270 million years ago (Fig. 2f). The continental edge of Africa approached North America as the ancestral Atlantic Ocean continued to subduct. The orogeny culminated when the continental crust of Africa collided with North America, finally closing the ancestral Atlantic Ocean and forming the supercontinent Pangaea. The collision produced the fold-and-thrust structures of the Valley and Ridge province, and thrust the Blue Ridge and Piedmont provinces farther inland. In Africa, a corresponding orogenic belt is marked by the Mauritanide Mountains.

The last great East Coast tectonic event involved the opening of the modern Atlantic, which occurred some 200 million years ago (Fig. 2g). At this time the African plate was once again separated from the North American plate, but a healthy chunk of the African plate was left attached to eastern North America, stretching from New York City to Florida.

According to this hypothesis, the three plate collisions that produced the Appalachians added to the East Coast an ancient (Precambrian) fragment of North America that had earlier been rifted away, an island arc that originated on the African-European side of the ancestral Atlantic basin, and a substantial slice of the African plate. The Appalachians' complex geology and their long history of multiple orogenic events typify all fold-and-thrust mountain systems.

Figure 2 A model for the evolution of the southern Appalachians. The northern Appalachians developed in part from collision with Europe.

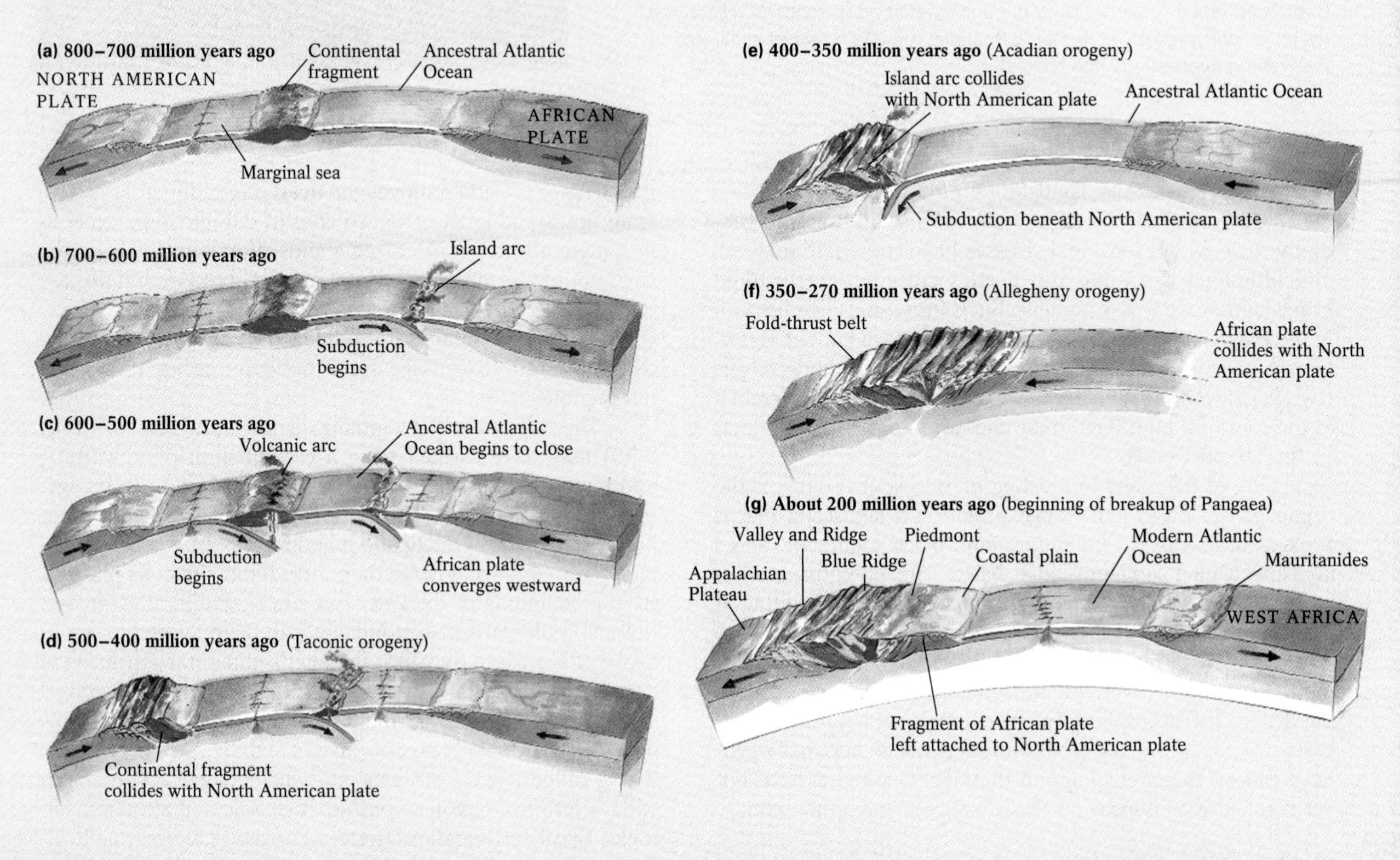

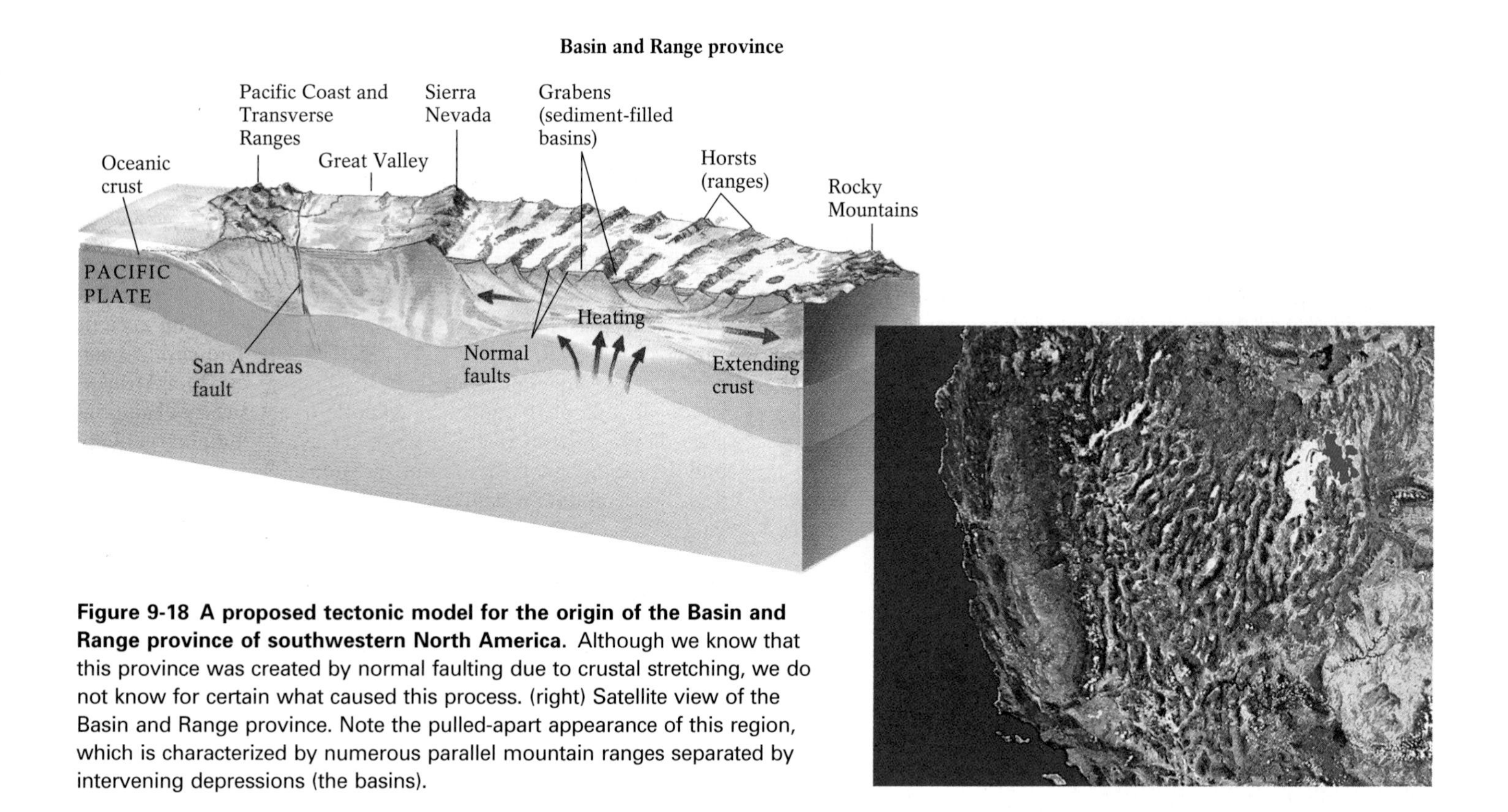

Figure 9-18 A proposed tectonic model for the origin of the Basin and Range province of southwestern North America. Although we know that this province was created by normal faulting due to crustal stretching, we do not know for certain what caused this process. (right) Satellite view of the Basin and Range province. Note the pulled-apart appearance of this region, which is characterized by numerous parallel mountain ranges separated by intervening depressions (the basins).

Fault-Block Mountains **Fault-block mountains** are bounded on at least one side by high-angle normal faults. They generally form where tensional stresses have stretched, thinned, and ultimately fractured the Earth's crust into high, tilted horsts and much lower grabens. Such formations can be seen in the Basin and Range province of the western United States, lying between the Sierra Nevada and Rocky Mountain ranges (Fig. 9-18). Picturesque Jackson Hole, Wyoming, is a graben at the foot of a fault block that rises 2000 meters (7000 feet) to the Grand Tetons.

One of the many interesting mysteries of geology is the origin of the Basin and Range province, a landscape that is broken into scores of individual fault-block mountain ranges and intervening sediment-laden basins. According to one hypothesis, it was created when magma, generated by melting of the Pacific plate as it subducted under the western edge of the North American plate about 30 million years ago, spread laterally beneath the overlying lithosphere, thinning and stretching it until the lithosphere fractured. Another hypothesis suggests that magma rising from an independent hot spot might have caused the stretching and thinning. Geologists have not yet reached a consensus on the province's origin, however.

Upwarped Mountains **Upwarped mountains** form when a large area of the Earth's crust becomes gently bent into broad regional uplifts without much apparent deformation of the rocks. After erosion removes the overlying sedimentary strata from upwarped crust, a rugged core of durable older igneous and metamorphic rocks often stands prominently above the surrounding terrain. Because many upwarped mountains are located far from plate boundaries, geologists believe local vertical forces at plate interiors, such as the ascent of low-density mantle material, create upwarps rather than plate interactions.

The Adirondack Mountains of northern New York (Fig. 9-19) may have formed from a combination of upwarping and plate tectonic forces. Approximately 1 billion years ago, the rocks now at the core of the Adirondacks were compressed, folded, thrusted, and metamorphosed at a collisional plate boundary; they were then intruded by igneous plutons. By the beginning of the Paleozoic Era 600 million years ago, these complex proto-Adirondacks had been eroded down to a fairly flat surface on which younger sedimentary rocks were deposited. Then, for unknown reasons, upwarping occurred during the Cenozoic Era (the last 65 million years) after hundreds of millions of years of relative stability. During the last 2 to 3 million years, streams and glaciers have eroded deep valleys into the region's uplifted Paleozoic and Precambrian rocks. Local earthquake activity occurring as recently as 1989 suggests that uplift continues in the Adirondacks today, even though the area now resides thousands of kilometers from a plate boundary.

Figure 9-19 The Adirondack Mountains of northern New York. The center peak in the background is Mount Marcy, the Adirondack's tallest at 1630 meters (5344 feet).

Mountain Building on Our Planetary Neighbors

Scientists have studied the three extraterrestrial bodies nearest to the Earth—the Moon, Mars, and Venus—using satellites, telescopes, robot landers, and, in the case of the Moon, human visits. The data collected suggest that Earth-like plate tectonics do not operate on the Moon and Mars, and that orogenic processes on these two bodies must therefore differ from those on Earth.

Mountains of the Moon and Mars In the absence of plate tectonics, the Moon's mountains probably developed over billions of years as meteoroid impacts fragmented surface rocks and hurled massive chunks about; this material eventually accumulated as mountains of rock debris. Some of the Moon's major highlands are as lofty as our planet's Mount Everest.

The principal mountain-building process on Mars seems to be volcanism on a grand scale. Olympus Mons, a single mountain in Mars' Tharsis region, is nearly 600 kilometers (400 miles) in diameter (about the size of Ohio) and more than 23 kilometers (14 miles) high.

Mountains of Venus Venus may well be tectonically active. In 1990, the U.S. spacecraft Magellan transmitted radar images of the surface of Venus, revealing a mountainous topography. The pictures showed Venusian highlands that may be analogous to the Earth's continents, and, although Venus has no surface water, its plains may be analogous to the Earth's low-lying oceanic crust. The highlands, which constitute only about 8% of Venus' surface, are mostly plateau-like structures, but a few discrete mountain ranges exist as well (Fig. 9-20). Maxwell Montes on the Ishtar highlands, for example, rises to a height of 11 kilometers (7 miles—about the elevation of Mount Everest) above the surrounding lowlands. Venus also contains structures that have been tentatively identified as 5-kilometer (3-mile)-high shield volcanoes and enlarged tectonic valleys (possibly grabens). Its surface may contain folded and faulted mountain belts as well.

Figure 9-20 Venus' Ishtar mountains and adjacent lowlands, from the high-resolution radar images provided by the Magellan probe in 1990.

Chapter Summary

The rocks that make up the Earth's lithosphere are often subjected to great force, or **stress,** particularly at the edges of tectonic plates. Three principal types of stress act on these rocks: **compression,** which squeezes rocks together; **tension,** which stretches and pulls rocks apart; and **shearing stress,** which forces rocks past one another in opposite directions.

When rocks are subjected to sufficient stress, they may **strain,** changing in shape or volume. When only a minor amount of stress is applied, rocks undergo **elastic** (temporary) **deformation.** When a great amount of stress is applied rapidly or under conditions of relatively low temperature and pressure, the rocks break, undergoing **brittle failure.** When stress is applied gradually to rocks under conditions of higher temperature and pressure, the rocks do not break but undergo **plastic deformation,** an irreversible change in shape or volume. To determine the orientation of a deformed rock structure, geologists measure the rock's **strike,** the compass direction of an imaginary horizontal line lying on the surface of the rock structure, and its **dip,** the angle at which it is inclined relative to the horizontal.

When rock layers deform plastically, they form a series of **folds,** including trough-like **synclines,** arch-like **anticlines,** oval-shaped bulges (called **domes**), and bedrock depressions (known as **basins**). When rocks deform by brittle failure, they develop cracks, or fractures. **Faults** are fractures along which marked relative movement has taken place. Geologists classify faults according to the relative direction in which the affected rocks, or fault blocks, have slipped along the fault plane. **Strike-slip faults,** which are marked by horizontal movement, are generally found at transform plate boundaries. **Dip-slip faults** are characterized by vertical movement along the fault plane. The most common dip-slip faults are **normal faults,** which develop under tensional stress, such as at divergent plate boundaries, and in which the hanging wall (the fault block above the fault plane) moves downward relative to the footwall (the fault block below the fault plane). A fault block that drops during normal faulting produces a depression called a **graben.** The blocks that remain standing above and on either side of a graben are called **horsts. Reverse faults** are dip-slip faults that develop under compressional stress, especially at convergent plate boundaries, in which the hanging wall moves upward relative to the footwall. **Thrust faults** are low-angle reverse faults.

The most dramatic effect of rock deformation is mountain building, or **orogenesis.** Compressional stress produces exceptionally high **fold-and-thrust** mountains, which are both thrust-faulted and folded, usually through a series of mountain-forming events. Tensional stress may cause regional normal faulting, which produces uplifted ranges, or **fault-block mountains. Upwarped mountains** are broad regional uplifts that may form by the rise of low-density material below the Earth's surface.

Mountain building also occurs on some of our planetary neighbors. For instance, the Moon's mountains were probably produced by meteoroid impacts, while Martian mountains have a volcanic origin. The mountains on Venus may have risen in response to Earth-like tectonic forces.

Key Terms

stress (p. 153)
compression (p. 154)
tension (p. 154)
shearing stress (p. 154)
strain (p. 154)
elastic deformation (p. 154)
brittle failure (p. 154)
plastic deformation (p. 155)
strike (p. 155)
dip (p. 155)
folds (p. 155)
synclines (p. 156)
anticlines (p. 156)
dome (p. 158)
basin (p. 158)
faults (p. 159)
strike-slip faults (p. 159)
dip-slip faults (p. 160)
normal fault (p. 160)
graben (p. 160)
horsts (p. 160)
reverse fault (p. 160)
thrust fault (p. 161)
orogenesis (p. 163)
fold-and-thrust mountains (p. 163)
fault-block mountains (p. 163)
upwarped mountains (p. 163)

Questions for Review

1. What are the three principal types of tectonic stress, and at which type of plate boundary does each develop?

2. Compare brittle failure and plastic deformation of rocks. Where does each type of deformation commonly occur within the Earth?

3. Determine which way is north, and then hold this book so that its direction of strike is 45° west of north, and it dips 45° and toward the northeast.

4. Draw an anticline and syncline, and label the limbs, axial plane, and axis of each. Why does the core of an anticline contain the formation's oldest rocks, and the core of a syncline contain its youngest rocks?

5. Explain how hills may be formed from synclines and valleys from anticlines.

6. How does the formation of domes and basins differ from that of anticlines and synclines?

7. What are the directions of relative movement in strike-slip faults and in dip-slip faults? What types of stress are involved in each?

8. Draw two simple sketches illustrating the difference between normal faults and reverse faults. Label the hanging wall and the footwall; use arrows to show their relative direction of movement.

9. Briefly discuss two North American examples of different styles of orogenesis.

10. Describe the evidence indicating that the planet Venus may have undergone plate tectonic processes.

For Further Thought

1. In what types of rocks might you expect to find fractures but no faults?

2. Identify the type of fold and the type of fracture that appear in the photo below.

3. Speculate about how the sedimentary layers in the diagram below became arranged in their present stratigraphic order.

4. Suggest future plate tectonic scenarios that would result in development of normally faulted grabens in Minnesota and the creation of Appalachian-type fold-and-thrust mountains along the Louisiana coast.

5. What would the Moon's surface look like if plate tectonics had been active there for its entire history? How might the Earth's surface look if plate tectonics had never developed here?

Earthquakes and the Earth's Interior

At 5:46 A.M. on January 17, 1995, a powerful earthquake, calculated at 7.2 on the Richter scale, hit western Japan. The earthquake, which was centered near the port city of Kobe on Awajishima Island, left more than 5000 dead and at least 29,000 injured. Approximately 30,000 residences in and around Kobe were destroyed or severely damaged, leaving 310,000 people homeless. Hanshin Expressway, connecting Kobe and Osaka, 31 kilometers (19 miles) away, collapsed in five places. Rail transportation, including the Shinkansen bullet train service, was disrupted for hundreds of kilometers, and at least seven derailments were reported. The cost of cleanup and restoration in the aftermath of the Kobe earthquake was expected to reach as much as $150 billion.

In December 1995, an even more powerful earthquake occurred 5 kilometers (3 miles) off the west coast of Mexico. The quake, which struck some of Mexico's most popular beach resorts, toppled houses and hotels, cracked bridges, opened meter-wide fissures in the main coastal highway, and cut power and phone service throughout the region. Damage was greatest in the resort town of Manzanillo, where the luxury seven-story Costa Real Hotel collapsed, killing at least 20 guests and staff. Even in Mexico City, located 335 kilometers (201 miles) to the east, earthquake vibrations caused skyscrapers to sway violently.

In the United States, two major earthquakes have occurred relatively recently. The powerful Loma Prieta earthquake struck northern California in 1989, toppling buildings, freeways, and bridges, rupturing gas mains and starting fires, and setting off landslides throughout the San Francisco Bay area. In 1994, another earthquake produced severe damage in the Northridge area of southern California. Together, these two quakes resulted in billions of dollars of damage and caused more than 100 deaths.

In this chapter, we will discuss what causes earthquakes (Fig. 10-1), how geologists study, evaluate, and even predict them, and how the study of earthquake waves allows us to investigate the Earth's interior.

Figure 10-1 The power of earthquakes. This enormous rent in the Earth's surface appeared during the February 4, 1998, earthquake in Afghanistan that took thousands of lives.

Causes and Characteristics of Earthquakes

An **earthquake** is a trembling of the ground caused, most often, by the sudden release of energy in underground rocks. Most earthquakes occur where rocks are subjected to the stress associated with tectonic plate movement—that is, near plate boundaries (Fig. 10-2). The application of such stress may cause rocks to deform elastically and to accumulate *strain energy,* which builds until the rocks either shift suddenly along preexisting faults or rupture to create new faults. The result—earthquakes.

The precise subterranean spot at which rocks begin to rupture or shift marks the earthquake's **focus** (Fig. 10-3). Approximately 90% of all earthquakes have a relatively shallow focus, located less than 100 kilometers (60 miles) below the surface; the focus of virtually all catastrophic quakes lies within 60 kilometers (40 miles) of the surface. Large earthquakes seldom occur at greater depth, because heat has softened rocks there and robbed them of some of their ability to store strain energy. A few earthquakes have occurred at depths as great as 700 kilometers (435 miles). Deeper than this level, however, higher temperatures and pressures cause stressed rocks to deform plastically, rather than rupture or shift.

The point on the Earth's surface directly above an earthquake's focus is its **epicenter.** The greatest impact of a quake is generally felt at the epicenter, with the effects decreasing in proportion to the distance from the epicenter. After a major earthquake, the rocks in the vicinity of the quake's focus continue to reverberate as they adjust to their new positions, producing numerous, generally smaller, earthquakes, or *aftershocks.* Aftershocks may continue for as long as one or two years after the main quake, shaking and further damaging already-weakened structures.

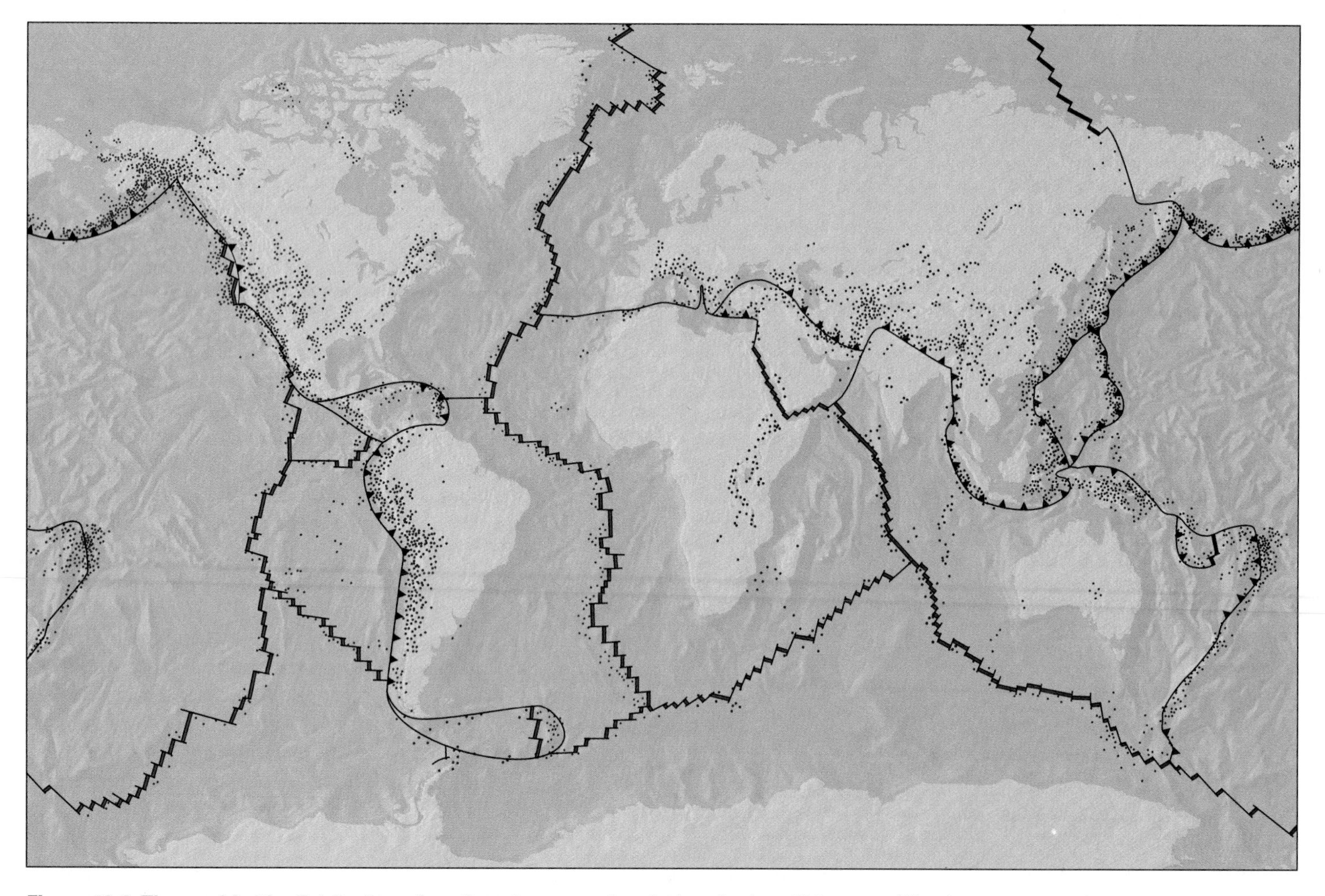

Figure 10-2 The worldwide distribution of earthquakes occurring during the last 100 years. (The dots represent significant earthquakes—greater than magnitude 4.) Notice that earthquakes define both current and ancient plate boundaries.

After an earthquake releases its stored strain energy, the rocks along a fault cease to move and the fault blocks become temporarily locked in place by the friction between them. If the rocks undergo subsequent stress, however, the strain energy again accumulates, sometimes over a period of many decades. Eventually the friction holding the fault blocks is overcome and they lurch, releasing newly accumulated energy in the form of another earthquake.

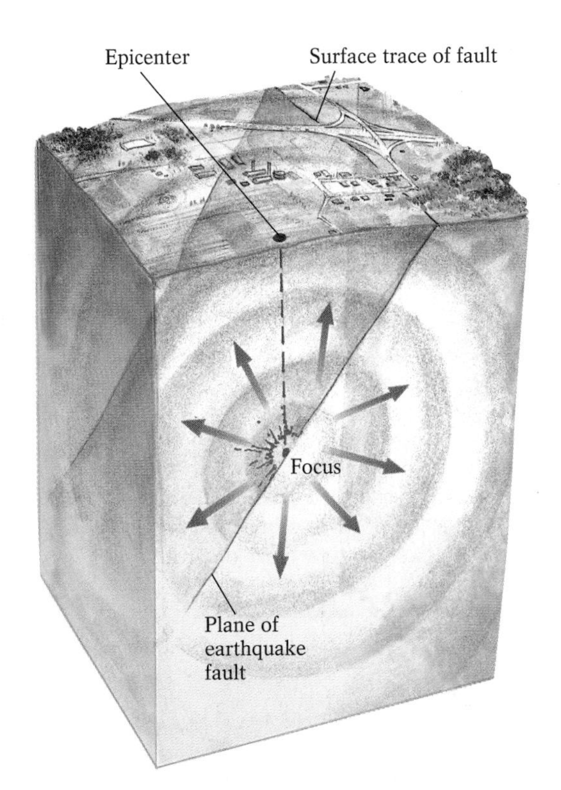

Figure 10-3 The anatomy of an earthquake. At the moment an earthquake occurs, pent-up energy is released at the quake's focus and transmitted through the Earth. The point on the Earth's surface directly above a quake's focus is its epicenter.

Seismic Waves

When you toss a pebble into a placid lake, the energy from its impact creates small, concentric waves that travel through the water in all directions until they eventually die out some distance from the impact point. Similarly, when underground rocks rupture or shift along a fault, the released energy is transmitted as **seismic waves** that travel at great speeds in all directions through the surrounding rocks. The term *seismic* (from the Greek for "shaking") refers to anything involving earthquakes: A *seismograph* is a machine with which geologists determine the magnitude of an earthquake; *seismology* is the study of earthquakes and the Earth's interior.

Seismology relies on information yielded by two main types of seismic waves: **body waves,** which transmit energy through the Earth's interior in all directions from an earthquake's focus, and **surface waves,** which transmit energy along the Earth's surface, moving outward from a quake's epicenter.

Body Waves Two types of body waves are distinguished by the speed and type of motion with which they travel through the Earth. Primary waves, or **P waves,** are the fastest seismic waves—passing through the Earth's crust at a velocity of 6 to 7 kilometers (4 miles) per second—and are the first to arrive at a seismograph following an earthquake. P waves are *compressional:* They are initiated and transmitted when the sudden release of energy compresses rocks near an earthquake's focus. As the wave continues on to compress adjacent rocks, the initially compressed rocks expand elastically past their original volume, and then return to their original shape, only to be compressed again as the next wave of energy passes (Fig. 10-4). During an earthquake, the arrival of P waves at the Earth's surface is marked by a series of sharp jolts, as surface rocks alternately compress and expand with passing waves of energy.

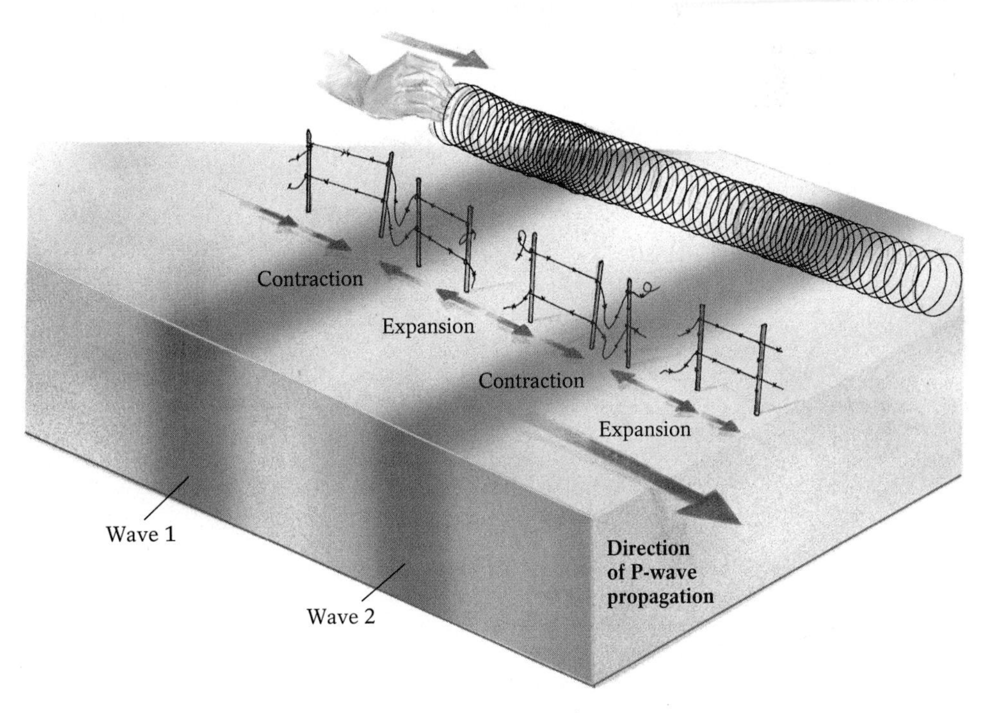

Figure 10-4 P waves. Primary waves are compressional, resulting in the alternate contraction and expansion of rocks in a direction parallel to the direction of wave propagation. The back-and-forth motion that displaces rocks as P waves pass (like the vibrating coils of a Slinky toy) can cause powerlines and fences to snap.

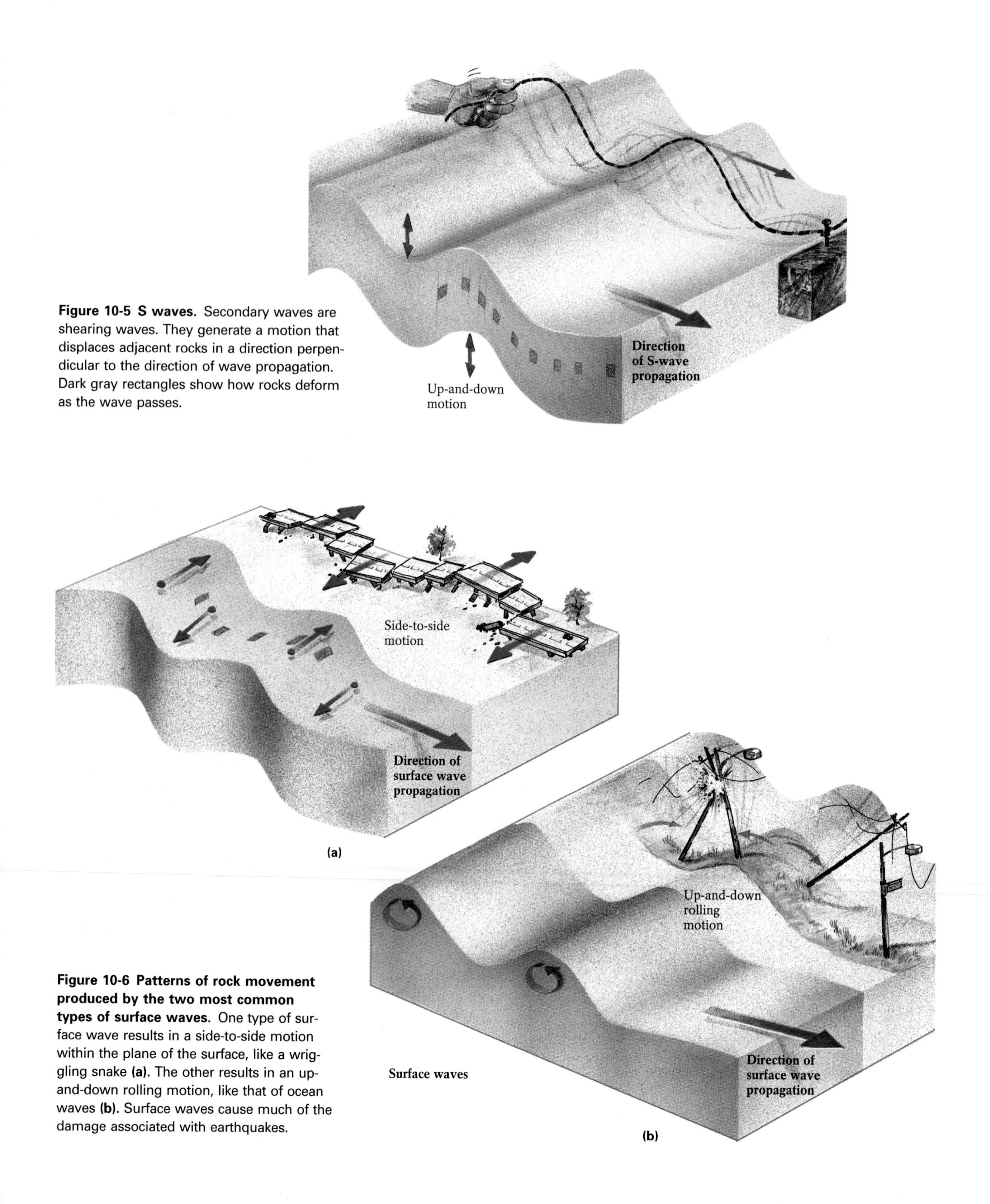

Figure 10-5 S waves. Secondary waves are shearing waves. They generate a motion that displaces adjacent rocks in a direction perpendicular to the direction of wave propagation. Dark gray rectangles show how rocks deform as the wave passes.

Figure 10-6 Patterns of rock movement produced by the two most common types of surface waves. One type of surface wave results in a side-to-side motion within the plane of the surface, like a wriggling snake **(a)**. The other results in an up-and-down rolling motion, like that of ocean waves **(b)**. Surface waves cause much of the damage associated with earthquakes.

Secondary waves, or **S waves,** travel significantly slower than P waves, passing through the Earth's crust at a velocity of about 3.5 kilometers (2 miles) per second. Consequently, they arrive at the seismograph sometime after the arrival of the P waves. Rather than compressing rocks, S waves cause rocks to move up and down, perpendicular to the direction in which the wave travels (Fig. 10-5). The movement of an area of rock creates a drag on neighboring rocks, pulling them along as well. The resultant *shearing* effect on the rocks temporarily changes their shape, but not their volume. During an earthquake, P waves strike with a succession of compressional jolts, but S waves impart a continuous wriggling motion.

Surface Waves When an earthquake occurs, some of the body waves that emanate from the focus move upward toward the epicenter, where they cause the Earth's surface to vibrate. This vibration generates surface waves, which travel within the upper few kilometers of the Earth's crust. The slowest seismic waves, traveling through the crust with a velocity of about 2.5 kilometers (1.5 miles) per second, they are the last to arrive at a seismograph. Two main types of surface waves exist—one having a side-to-side whipping motion (Fig. 10-6a), the other demonstrating a rolling motion that resembles ocean swells (Fig. 10-6b). People who have experienced this second type of surface wave have likened it to a brisk walk across a waterbed, and some have become "seasick." When both types of surface waves occur together, they cause objects to rise and fall while being whipped from side to side; this combination is responsible for most earthquake damage to rigid structures.

Measuring Earthquake Strength

An earthquake's strength depends on how much of the energy stored in the rocks is released. Scientists have developed three scales to measure this strength: the Mercalli intensity scale, the Richter scale, and the moment-magnitude scale.

The Mercalli Intensity Scale The **Mercalli intensity scale,** developed in 1902 by Italian seismologist Giuseppe Mercalli, is subjective; it is based on eyewitness reports of damage to human-made structures (Table 10-1 on page 176). The scale ranges from I (earthquake hardly detected) to XII (maximum destruction). It has several shortcomings. Damage to structures actually depends not only on the strength of the earthquake, but also on the structures' distance from the epicenter, their quality of construction, and the nature and stability of the local soil. Moreover, this scale is not effective in uninhabited areas or under the sea, which lack human-made structures that might sustain damage. Therefore, more objective, qualitative measures of earthquake energy have proved more useful to geologists and, in most cases, to the public as well.

The Modern Seismograph Seismographs measure the magnitude of earthquakes by sensing and recording seismic waves. They produce *seismograms,* visual records of the arrival times of the different waves and the magnitude of the shaking associated with them (Fig. 10-7). A modern seismograph receives signals from a *seismometer,* a device that amplifies wave motion electronically so that the seismograph can detect even weak or distant disturbances. Seismometers are sensitive enough to detect passing traffic, high winds, nearby crashing surf, and underground nuclear explosions. They can even detect earthquakes that occur on the opposite side of the globe.

The Richter Scale In 1935, Charles Richter of the California Institute of Technology developed the **Richter scale,** which is based on the magnitude of earthquakes as determined via

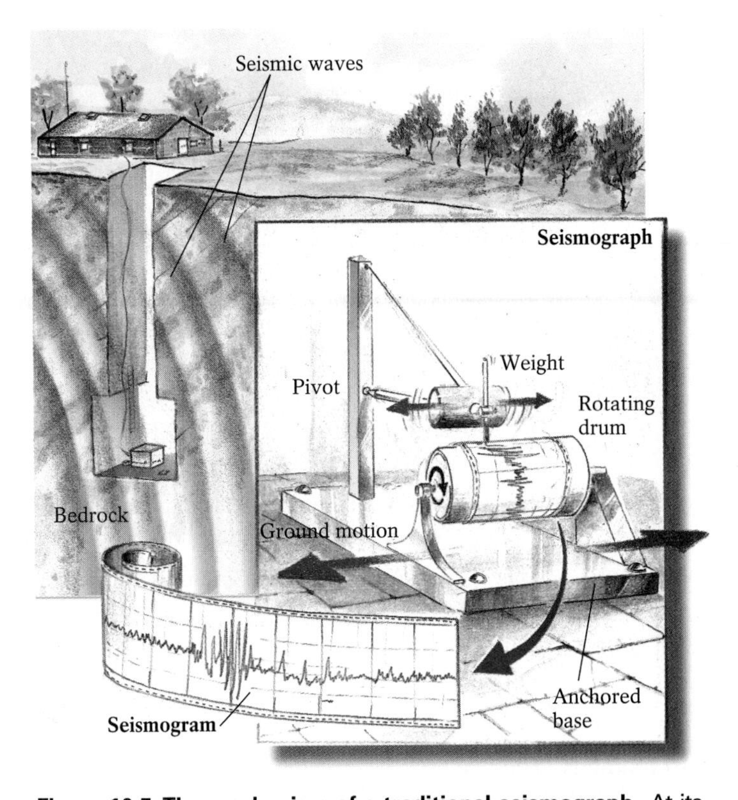

Figure 10-7 The mechanism of a traditional seismograph. At its most basic level, a traditional seismograph consists of a mass suspended by a spring or wire from a base that is firmly anchored. (The entire seismograph is usually encased in a protective box and bolted to the Earth's crust meters below the surface.) The base moves with the Earth during an earthquake, while the suspended mass remains motionless. When a seismic wave jostles the seismograph beneath the suspended stylus, the stylus records the shaking on a roll of paper that turns at a steady rate on a rotating drum anchored to the base, producing a tracing called a seismogram. The amplitude of the seismogram's squiggles is proportional to the amount of energy released by the earthquake. Seismographs can also be oriented vertically, to record vertical ground movement.

Table 10-1 Modified Mercalli Intensity Scale

Intensity Value	Intensity Description
I	Not felt, except by a very few persons under especially favorable circumstances.
II	Felt by only a few persons at rest, especially on upper floors of buildings. Delicately suspended objects may swing.
III	Felt quite noticeably indoors, especially on upper floors of buildings, but many people do not recognize it as an earthquake. Standing automobiles may rock slightly. Vibration resembles a passing truck. Duration estimated.
IV	During the day felt indoors by many people, outdoors by few. At night some awakened. Dishes, windows, doors disturbed; walls make creaking sound. Sensation like heavy truck striking building. Standing automobiles rocked noticeably.
V	Felt by nearly everyone; many awakened. Some dishes, windows, and other objects broken; cracked plaster in a few places; unstable objects overturned. Disturbance of trees, poles, and other tall objects sometimes noticed. Pendulum clocks may stop.
VI	Felt by all; many become frightened and run outdoors. Some heavy furniture moved; a few instances of fallen plaster and damaged chimneys. Damage slight.
VII	Everyone runs outdoors. Negligible damage in buildings of good design and construction; slight to moderate damage in well-built ordinary structures; considerable damage in poorly constructed or badly designed structures; some chimneys broken. Noticed by persons driving cars.
VIII	Slight damage in specially designed structures; considerable damage in ordinary substantial buildings, with partial collapse; great damage in poorly built structures. Panel walls thrown out of frame structures. Chimneys, factory stacks, columns, monuments, walls collapse. Heavy furniture overturned. Sand and mud ejected in small amounts. Changes in well water. Persons driving cars disturbed.
IX	Damage considerable in specially designed structures; well-designed frame structures thrown out of plumb; great damage in substantial buildings, with partial collapse. Buildings shifted off foundations. Ground cracked conspicuously. Underground pipes broken.
X	Some well-built wooden structures destroyed; most masonry and frame structures with foundations destroyed; ground badly cracked. Rails bent. Landslides considerable from river banks and steep slopes. Shifted sand and mud. Water splashed, slopped over banks.
XI	Few (masonry) structures remain standing. Bridges destroyed. Broad fissures in ground. Underground pipelines completely out of service. Earth slumps and land slips in soft ground. Rails bent dramatically.
XII	Damage total. Waves seen on ground surface. Lines of sight and level distorted. Objects thrown into the air.

Source: U.S. Federal Emergency Management Agency (FEMA).

seismographs. The Richter scale correlates the amplitude (height) of the largest peak traced on a seismogram during a quake with the amount of energy released by the quake. On this logarithmic scale, each successive unit corresponds to a 10-fold increase in the amplitude of the seismogram tracings. Thus an earthquake of magnitude 5 is 10 times more powerful than one of magnitude 4, and 100 times greater than a magnitude 3 quake. The actual amount of energy released during tremors of different Richter magnitudes, however, varies even more. For each unit increase in the Richter scale, we see about a 30-fold increase in energy released. Thus the energy released by an earthquake of magnitude 5 is 30 times greater than that associated with one of magnitude 4, and 900 times greater than the energy of a magnitude 3 earthquake.

To understand the logarithmic nature of the Richter scale, consider metropolitan Los Angeles, which experiences an earthquake with a magnitude greater than 8 every 160 years or so. The last quake of this magnitude occurred there in 1857. One might think that an occasional magnitude 6 quake would relieve a significant amount of accumulated strain energy and forestall or prevent the next "Big One." The energy released by a magnitude 8 earthquake, however, is 900 times that released by a magnitude 6 quake. Nine hundred earthquakes of

Table 10-2 Earthquakes and Their Energy Equivalents

Moment Magnitude	Approximate Energy Equivalent
−2	100-watt lightbulb left on for a week
0	1-ton car going 40 kilometers (25 miles) per hour
2	Amount of energy in a lightning bolt
4	1 kiloton (1000 tons) of explosives
6	Hiroshima atomic bomb
8	1980 eruption of Mount St. Helens
10	Annual total U.S. energy consumption (largest recorded earthquake—9.5, Chile, 1960)

Table 10-3 Frequency of Earthquake Occurrence Based on Observations Since 1900

Description	Magnitude	Annual Average Number of Events
Great	8.0 or higher	1
Major	7.0–7.9	18
Strong	6.0–6.9	120
Moderate	5.0–5.9	800
Light	4.0–4.9	6200 (estimated)
Minor	3.0–3.9	49,000 (estimated)
Very Minor	2.0–3.0 1.0–2.0	1000/day 8000/day

magnitude 6 would therefore be needed to dissipate the accumulated strain energy of one magnitude 8 quake. As southern California experiences only one quake of magnitude 6 every 5 years on average, the region can still expect a seismic event 900 times as violent as any in recent memory.

Despite its previously universal acceptance, seismologists have deemphasized the Richter scale during the last decade because of several notable drawbacks:

- Although theoretically without an upper limit, the scale cannot effectively measure quakes with a magnitude of 7.0 or greater.
- More modern devices have replaced the seismographs on which the Richter was based.
- The scale works best for local earthquakes (occurring within a few hundred kilometers of a seismograph) in which seismic waves travel through rocks of fairly homogeneous composition, but is less effective when long-traveling seismic waves pass through regions of differing crustal composition.

The Moment-Magnitude Scale During the past few decades, seismologists have discovered that they can more accurately gauge a large earthquake's total energy by measuring a quantity called the *seismic moment,* defined as:

$$\begin{aligned}\text{Moment} = \; & (\text{total length of a fault rupture}) \\ & \times (\text{depth of a fault rupture}) \\ & \times (\text{total amount of slip along rupture}) \\ & \times (\text{strength of rock})\end{aligned}$$

One principal benefit of the moment-magnitude scale is that it enables seismologists to *directly* measure the size of the earthquake from its "cause" (the rupture in the rocks and the distance that the rocks have moved) rather than solely from its "effect" (the tracing of seismic waves on a seismograph). To determine the length and depth of the rupture and the total slip requires either field observations (if the rupture reached the surface) or analysis of the precise location of the main shock and its aftershocks (if it failed to break the surface). Table 10-2 describes various earthquake magnitudes and the energy they release in "human terms." Table 10-3 indicates how many of these events we can expect annually throughout the world.

Locating an Earthquake's Epicenter

Seismologists can use the different velocities of P and S seismic waves to determine the location of an earthquake's epicenter. Because they travel more rapidly, P waves reach a seismograph station earlier than the S waves following a seismic event. Because we know the velocity at which each type of wave travels, we can use the difference between their arrival times to calculate the distance from seismograph stations to the earthquake's epicenter (Fig. 10-8a). Simply calculating the distance from an earthquake to a single seismic station does not identify the earthquake's epicenter, of course, because the quake could have originated at that distance *in any direction* from the station. Instead, seismologists represent this distance on a map as the radius of a circle centered on the station. If they draw similar circles from three different stations, the intersection of the three circles will precisely pinpoint the earthquake's epicenter (Fig. 10-8b). (Note that the three circles will intersect if the focus of the earthquake lies within a few kilometers of the surface. Deeper foci

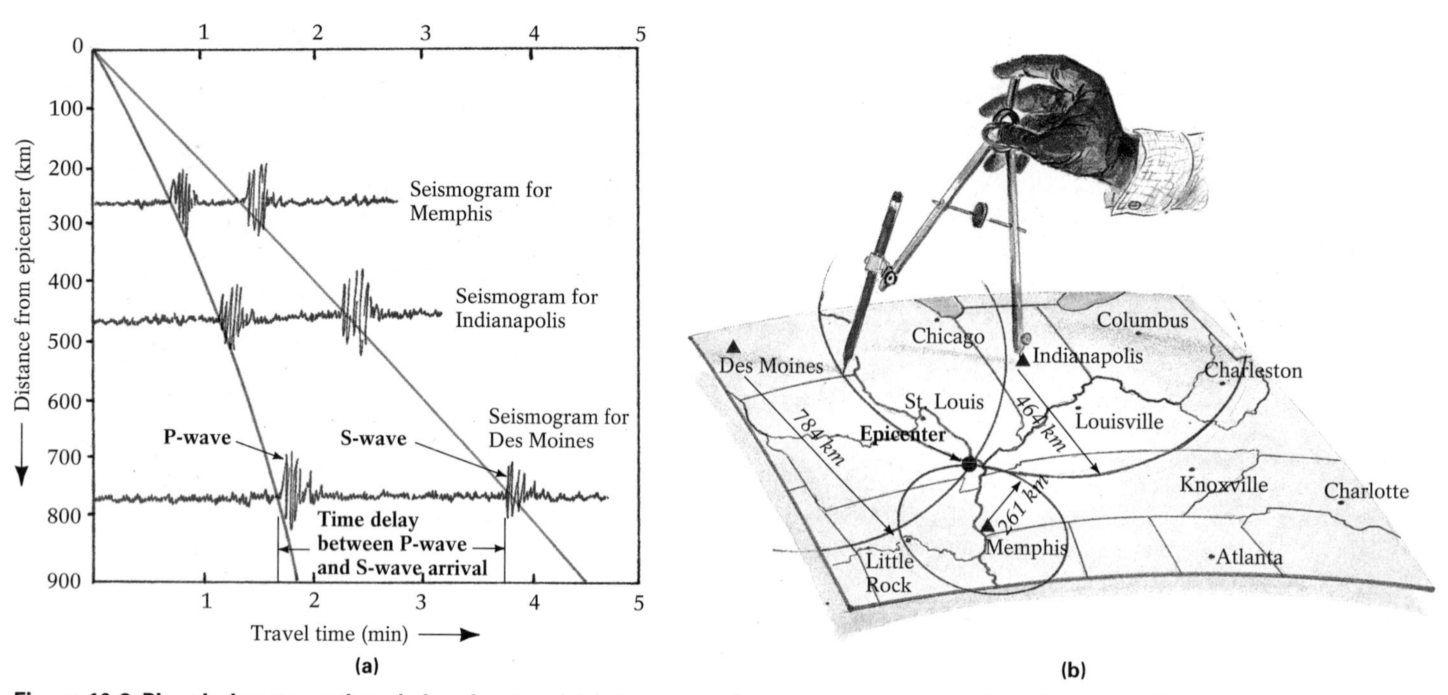

Figure 10-8 Pinpointing an earthquake's epicenter. **(a)** Seismograms from various seismograph stations, recording the arrival times of P and S waves at each station from an earthquake in the New Madrid, Missouri, area. The closeness of the two wave types at the Memphis, Tennessee, recording station indicates that the earthquake occurred nearby. Using our knowledge of the velocities of P and S waves through the Earth's crust, and the 120-second and 186-second delays at Indianapolis, Indiana, and Des Moines, Iowa, respectively, we can determine each station's distance from the earthquake epicenter. **(b)** By drawing a circle around each of these stations on a map, with each circle's radius proportional to the station's distance from the quake's epicenter, we can determine the location of the epicenter—it is marked by the single point at which the three circles intersect on the Earth's surface.

produce a triangle around the quake's epicenter, the size of which is proportional to the earthquake's depth.)

The value of being able to locate an earthquake's epicenter was demonstrated on May 23, 1989, when a magnitude 8.3 earthquake occurred. Moments after the quake took place, a worldwide computerized network of more than 100 state-of-the-art seismological facilities transmitted data to high-speed computers at the U.S. Geological Survey facility at Golden, Colorado. Within minutes, seismologists determined that the earthquake had occurred along the Macquarie ridge, a submarine mountain chain found 800 kilometers (500 miles) southwest of New Zealand. As we will discuss shortly, a powerful quake that strikes the ocean floor often produces enormous sea waves that can devastate coastal regions. A sea-wave alert was issued for coastal communities around the Pacific Ocean basin in enough time to successfully evacuate all people at risk in these areas.

The Effects of Earthquakes

During an earthquake, many geological effects occur simultaneously: Large areas of the ground shift position as displacements occur along faults; the passage of surface waves sets off landslides and mudflows; in places, seemingly solid ground begins to flow as if it were liquid; surfaces of bodies of water rise and fall; groundwater levels fluctuate as the Earth compresses and expands with the passage of P waves; and enormous ocean waves, generated principally by submarine fault displacements, smash against coastlines.

Ground Displacements

Among the most obvious geological effects of an earthquake is large-scale shifting of the landscape (Fig. 10-9). During the 1906 San Francisco quake, a 432-kilometer (270-mile)-long rupture opened along the strike-slip San Andreas fault; along one segment of the rupture, the west side of the fault lurched northward 7 meters (25 feet). Vertical ground shifts also occur, caused by movement along dip-slip faults. During the Good Friday earthquake of 1964, 500,000 square kilometers (200,000 square miles) of Alaska and the adjacent sea floor were thrust upward by reverse faulting. The maximum upward land shift totaled 12 meters (39 feet).

Landslides and Liquefaction

Violent shaking from a large earthquake often jostles and dislodges large masses of unstable rocks and soils from hillsides, spurring landslides. Earthquakes may also cause **liquefaction,** a process in which vibrations from seismic waves increase

Figure 10-9 Landscape shifting. Wallace Creek, on the Carrizo Plain in southern California, has been offset by lateral plate movement along the San Andreas fault.

the water pressure between soil grains, transforming once-cohesive soil into a slurry of mud—essentially quicksand or "quick clay." For example, during Alaska's 1964 quake, the clay sediment beneath the Turnagain Heights neighborhood of Anchorage liquefied (Fig. 10-10). Similarly, liquefaction led to much of the damage to Oakland's Nimitz Freeway in the 1989 Loma Prieta earthquake.

On June 7, 1692, the combined effects of liquefaction and landsliding from a relatively moderate quake sent the entire city of Port Royal, on the Caribbean island of Jamaica, sliding into the sea, where it finally came to rest beneath 15 meters (50 feet) of water. The city, once infamous as a harbor for pirates and other scoundrels, was built on loose, steeply sloping sediment in an area prone to periodic quaking. When marine archaeologists rediscovered the city in 1959, it lay protected and virtually intact beneath 3 meters (10 feet) of marine silt; in one kitchen, a copper kettle still contained the evening's meal of turtle soup.

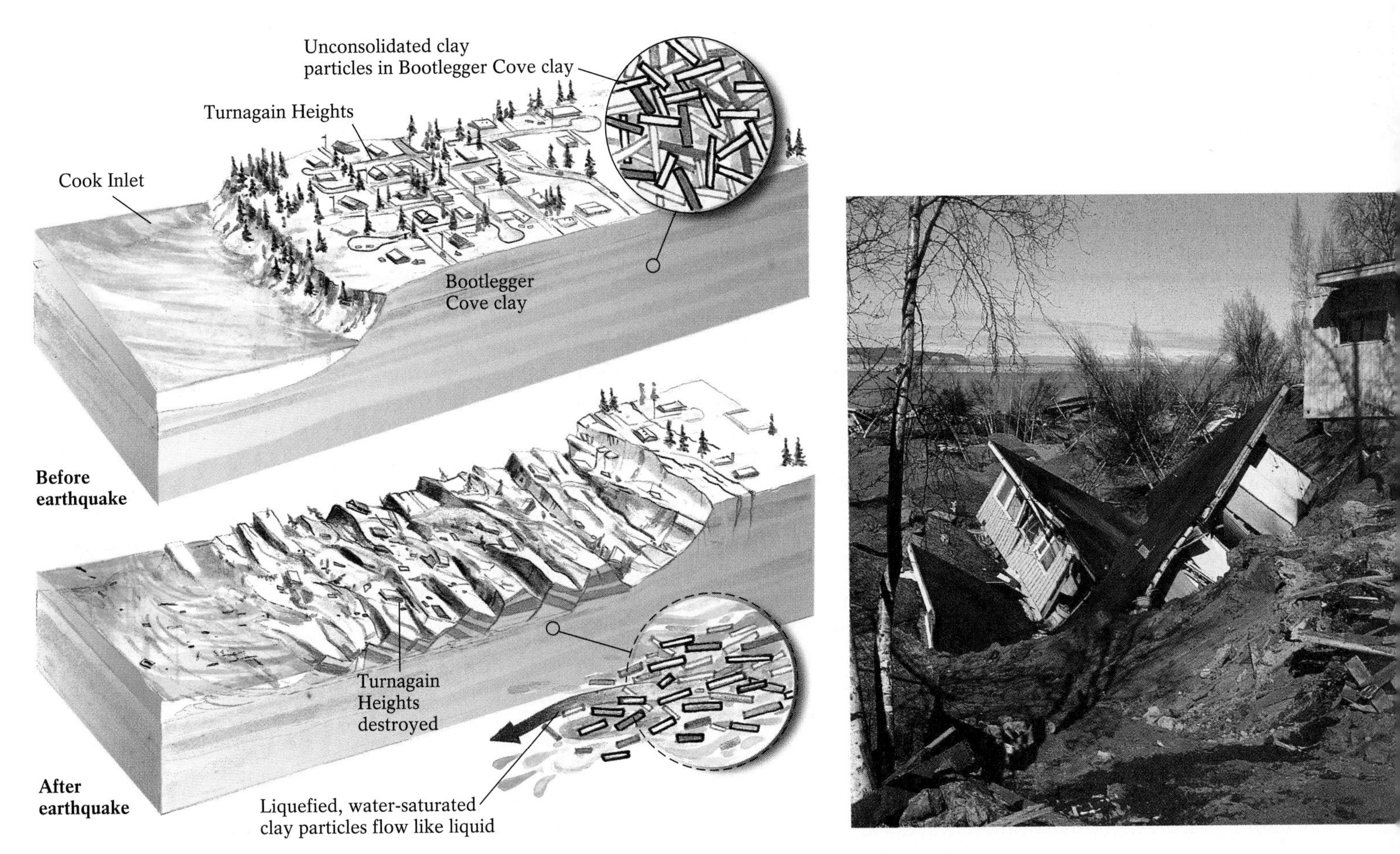

Figure 10-10 Liquefaction. This phenomenon occurs when the grains in a layer of wet, fine-grained sediment are shaken during an earthquake. In 1964, the clay layer beneath the entire Turnagain Heights area of Anchorage, Alaska, liquefied, resulting in a chaotic landscape of jumbled homes and asphalt blocks.

Figure 10-11 Earthquake effects. In the earthquake of 1906, the buildings and homes along San Francisco's Fisherman's Wharf District that were not destroyed by the city's widespread fires were left listing and severely damaged by liquefaction of the underlying wet, loose sediments.

Seiches

Seismic waves cause the water in an enclosed or partially enclosed body of water, such as a lake or bay, to move back and forth across its basin, rising and falling as it sloshes about. This phenomenon is called a **seiche** (pronounced "SAYSH"). In 1964, the water at one end of Kenai Lake, south of Anchorage, Alaska, rose 9 meters (about 30 feet), overflowed its banks, and flooded inland before reversing its direction. The back-and-forth motion of the water stripped the soils around the lake down to bare bedrock. Seiches may develop even at great distances from powerful earthquakes—sometimes so far from the epicenter that no other effects of the quake are felt. The Alaska earthquake, for example, set off seiches in reservoirs as far away as Michigan, Arkansas, Texas, and Louisiana. Even swimming pools in Texas, 6000 kilometers (4000 miles) from the quake's epicenter, developed small seiches.

Tsunami

A powerful submarine earthquake that causes rapid displacement of a portion of the ocean floor often produces enormous sea waves called **tsunami,** commonly (but mistakenly) called "tidal waves." At sea, tsunami are barely perceptible. The crests of successive waves, sometimes traveling more than 800 kilometers (500 miles) per hour are often separated by as much as 160 kilometers (100 miles). The waves create mere bumps on the water's surface, usually less than 1 meter (3 feet) high; unlike most waves (which are caused by wind), however, they extend to the bottom of the ocean. As these waves approach the coastline, they drag along the sea floor, which causes them to bunch up and grow into huge, devastating walls of water. In 1960, a tsunami generated by a catastrophic quake in southern Chile struck the Hawai'ian shore at Hilo seven hours after the seismic event, where it took 61 lives; by 22 hours after the quake, the tsunami had traveled 17,000 kilometers (11,000 miles) to reach Honshu and Hokkaido in Japan, where it claimed 180 lives.

Fires

In inhabited regions, earthquakes may damage gas mains, oil tanks, and electrical power lines, thereby starting fires. Because the events frequently rupture water mains as well, fire fighters may be impeded in battling the blaze, and fires can burn uncontrolled for days. During the 1995 earthquake in Kobe, Japan, more than 200 ruptures in gas lines ignited numerous fires, and interruption of water service severely hampered fire-fighting efforts. During its earthquake of 1906, San Francisco's wooden buildings were tinder for a blaze that caused 80% to 90% of the city's damage, engulfing more than 500 square blocks of the city's central business district (Fig. 10-11). The fire raged uncontrolled for three days, largely because of broken water lines. Ultimately, fire fighters were forced to create a fire break by dynamiting rows of buildings.

The World's Principal Earthquake Zones

As we saw in Figure 10-2, the vast majority of earthquakes are concentrated at plate boundaries. Plate interiors are generally seismically inactive, but major earthquakes do occasionally strike within intraplate regions. Indeed, Charleston, South Carolina, Boston, Massachusetts, and southern Missouri—all located far from plate boundaries—have all experienced a major quake within the past 250 years.

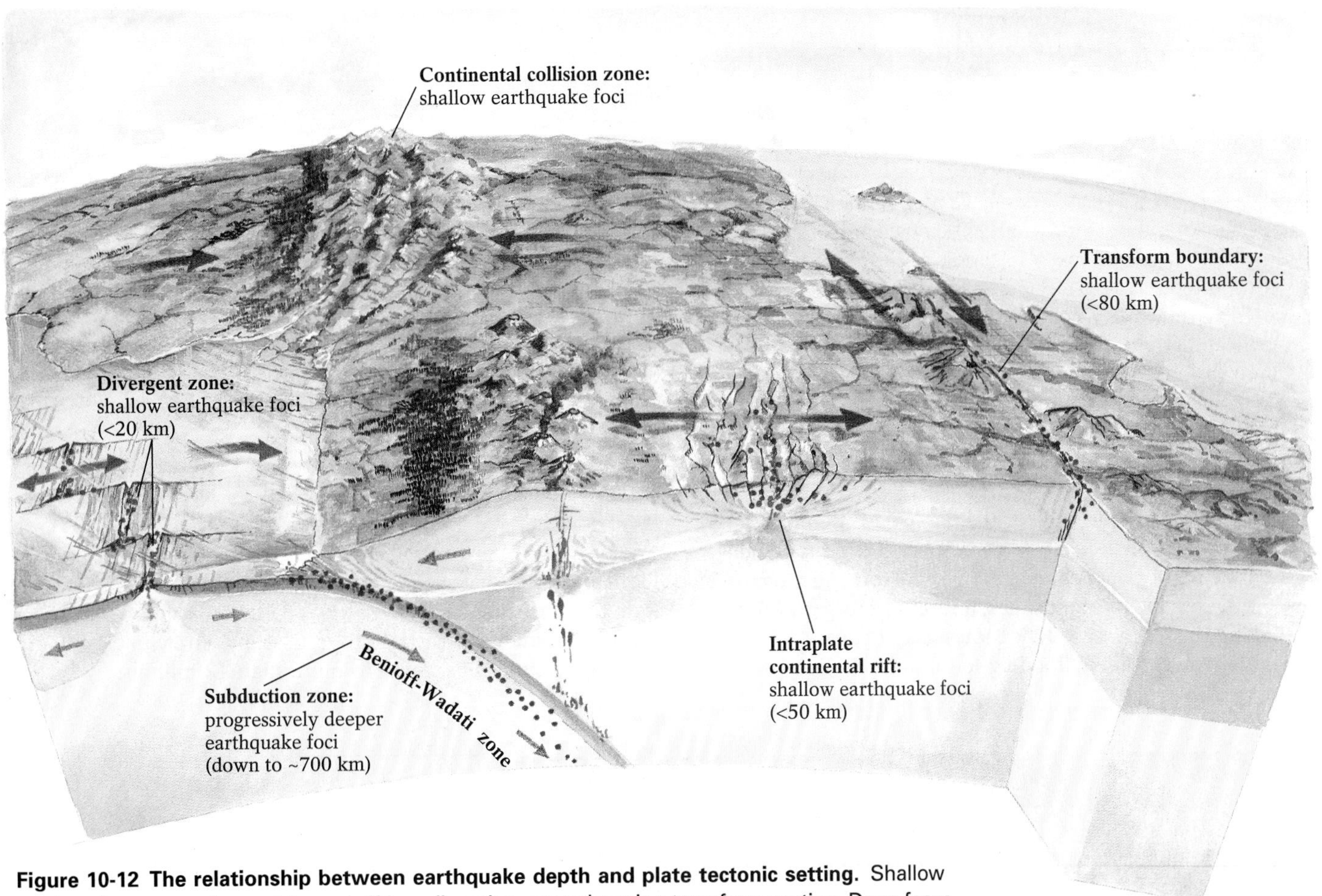

Figure 10-12 The relationship between earthquake depth and plate tectonic setting. Shallow earthquakes occur where plates are rifting, diverging, or undergoing transform motion. Deep-focus earthquakes occur exclusively at subduction zones, within the brittle portion of the descending slab.

Earthquake Zones at Plate Boundaries

About 80% of the world's earthquake energy is released in the Pacific Rim region, where oceanic plates of the Pacific Ocean basin subduct beneath adjacent plates along most of the ocean's perimeter (see Figure 10-2). This region includes the earthquake zones of Japan, the Philippines, the Aleutians, and the west coasts of North, Central, and South America. Most of the remaining 20% of the Earth's earthquake energy is released along the collision zone stretching through Turkey, Greece, Iran, India, and Pakistan to the Himalayas, southern China, and Myanmar.

The depth of earthquakes varies significantly among the different types of plate boundaries (Fig. 10-12). At oceanic divergent zones, most earthquakes are extremely shallow, occurring at depths less than 20 kilometers (12 miles), because strain energy can build up only within the uppermost, brittle portion of an oceanic plate. Deep earthquakes are virtually nonexistent in these zones, because the asthenosphere, which is located about 100 kilometers (60 miles) beneath oceanic plates, consists of partially melted, softer material that deforms plastically instead of rupturing. The same characteristics apply to continental rift zones, continental collision zones, and the areas along continental transform boundaries. Earthquakes in these zones rarely occur at depths greater than about 50 to 80 kilometers (30–65 miles)—the maximum thickness of the brittle portion of continental plates.

At subduction zones, shallow quakes arise where the subducting plate begins its descent into the mantle. Progressively deeper quakes are initiated within the descending slab to a depth of about 700 kilometers (435 miles). From the onset of subduction, intense friction between the descending slab and the overriding plate produces ruptures along the slab's brittle upper surface, resulting in shallow and intermediate-depth earthquakes. These seismic events are generally the most powerful subduction-zone earthquakes. As the slab descends to greater depths, the center of quake activity shifts to the still-brittle interior of the slab, which has not been in the Earth's interior long enough to warm up and soften. By the time the slab has descended to approximately 700 kilometers (435 miles) below the surface, it has been heated sufficiently to soften throughout and thus deforms

Highlight 10-1 The New Madrid Earthquakes of 1811–1812

Beginning on December 16, 1811, the town of New Madrid, Missouri, was hit with a surprising series of earthquakes that lasted for 53 days, ending February 7, 1812. Of the three main shocks and the more than 1500 powerful aftershocks, the most violent was the last—it shook a 2.5-million-square-kilometer (1-million-square-mile) area from Quebec to New Orleans and from the Rocky Mountains to the eastern seaboard. Church bells tolled in Boston 1600 kilometers (1000 miles) away, windows rattled violently in Washington, D.C., and pendulum clocks stopped in Charleston, South Carolina. Judging from the observations of fur trappers and an eyewitness account from naturalist John Audubon, the magnitudes of the three principal tremors may have exceeded 8.5.

Near the estimated epicenter of the New Madrid quake, entire forests were flattened when trees snapped from the violent shaking. Large fissures opened, including some too wide to cross on horseback. Geysers of sand, water, and sulfurous gas erupted as the shaking ground compacted. Thousands of square kilometers of prairie land subsided, flooding to form the St. Francis and Reelfoot lakes in northwestern Tennessee as well as swamps that stretched 300 kilometers (190 miles) from Cape Girardeau, Missouri, to northern Arkansas (the swamps were drained in the 1920s) (Fig. 1). Uplift of other areas exposed lake beds that have since dried out. The Mississippi River became dammed in places by uplifts and landslides but flowed wildly elsewhere, with waves that overwhelmed many riverboats. The flow of the river even reversed direction in portions where uplifts changed lowlands into highlands. Entire river-channel islands disappeared, and vertical offsets along the river produced new cliffs and waterfalls.

Because they occurred at a mid-plate location and produced a great variety of landscape changes, the New Madrid events hold considerable interest for geologists. The study of these events is especially urgent now, because of the possibility that the area could again experience seismicity of the same magnitude. In the nineteenth century, the Mississippi valley was sparsely populated. The same area today, stretching from St. Louis to Memphis, is home to more than 12 million.

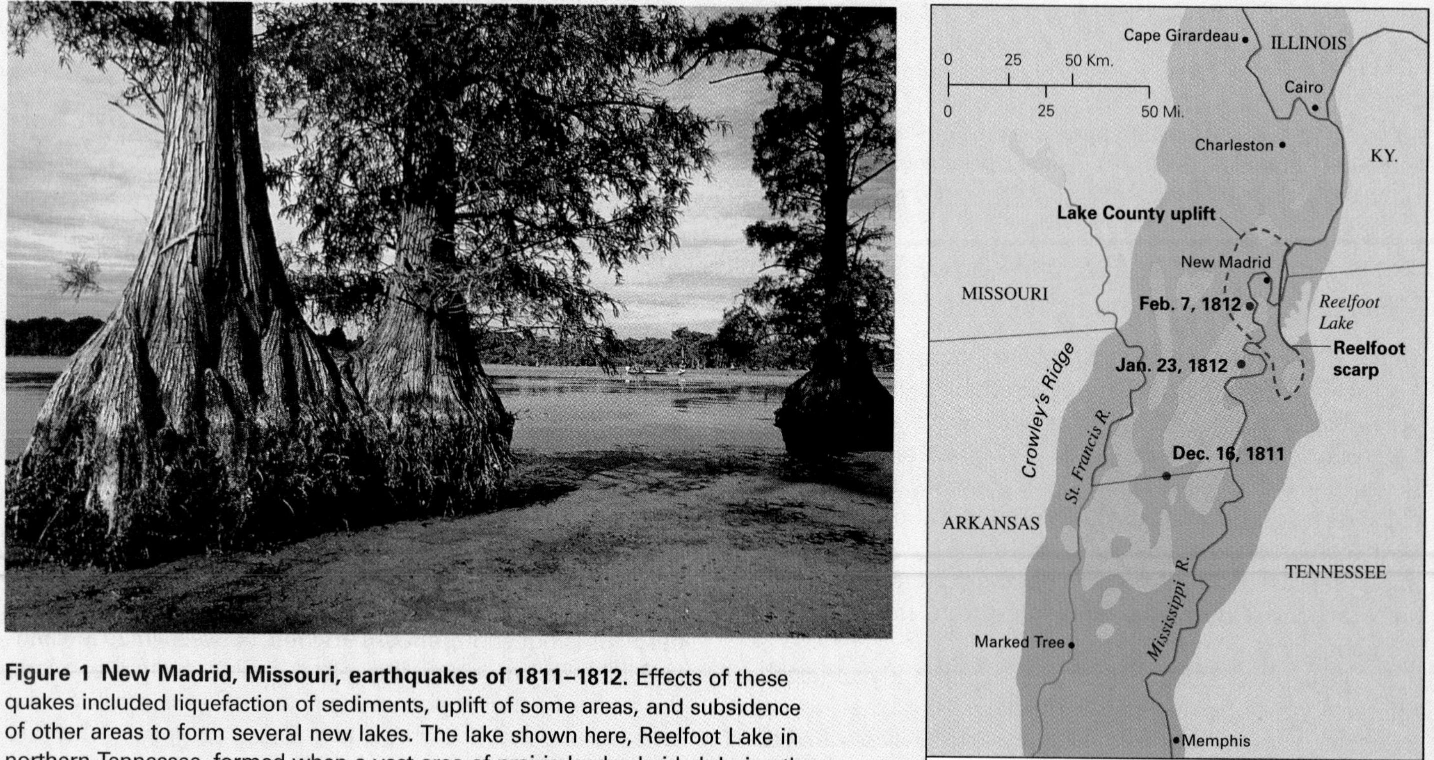

Figure 1 New Madrid, Missouri, earthquakes of 1811–1812. Effects of these quakes included liquefaction of sediments, uplift of some areas, and subsidence of other areas to form several new lakes. The lake shown here, Reelfoot Lake in northern Tennessee, formed when a vast area of prairie land subsided during the New Madrid quakes.

plastically—that is, it no longer produces earthquakes. The region within which earthquakes occur at subducting oceanic boundaries is called the **Benioff-Wadati zone,** named for the American seismologist who first observed that these earthquake foci are progressively deeper the farther inland they occur, and for the Japanese seismologist who pioneered methods of comparing earthquake sizes.

Intraplate Earthquakes

Mid-plate earthquakes are generally shallow, occurring at depths less than about 50 kilometers (30 miles). Their magnitude tends to be lower than the magnitudes of quakes occurring at active plate margins, possibly because less strain energy builds at mid-plate. Nevertheless, the older, colder, and therefore more brittle rocks at mid-plate transmit seismic waves more efficiently than those at active plate margins; hence the former earthquakes are felt over a significantly larger area than are their plate-boundary counterparts. On June 10, 1987, for example, a rare mid-plate earthquake, registering 5.0 on the Richter scale, unnerved the citizens of Lawrenceville, Illinois, and 16 surrounding states and Canadian provinces. Highlight 10-1 describes the most powerful intraplate earthquakes ever recorded in North America, the New Madrid, Missouri, quakes of 1811–1812.

Although hypotheses abound, seismologists have not definitively identified the exact cause of intraplate earthquakes. Seismic-wave studies have shown that the crust of eastern North America is riddled with a deep irregular network of old, near-vertical faults, probably the result of ancient plate-rifting episodes. According to one hypothesis, plate motion creates stress on the already weakened and faulted crust at such sites, causing infrequent but sometimes powerful quakes. Other hypotheses cite processes that are unrelated to plate tectonic motion. One, for example, proposes that after long-term erosion has removed vast quantities of surface materials, the unloaded crust becomes buoyant, rising and generating enough stress to arouse old faults. Another suggests that the load of mid-continent sediment deposition weighs down the crust, creating pressure that reactivates its faults.

Coping with the Threat of Earthquakes

Although we cannot stop earthquakes from occurring, we can take steps to minimize their damage. Most importantly, we can educate the public about the risk from seismic events. Many schools and public agencies in earthquake-prone areas strive to educate people in earthquake preparedness, and the Federal Emergency Management Agency of the U.S. government publishes a checklist of what to do before, during, and after an earthquake hits.

Construction in Earthquake-Prone Regions

Local building codes can be written to ensure that buildings will suffer minimal damage during earthquakes. The most secure structures are solidly constructed buildings made of strong and flexible, but relatively light, materials such as wood, steel, and reinforced concrete (containing steel bars) that tend to move *with* the shaking ground. Perhaps surprisingly, well-designed and engineered skyscrapers may be the safest buildings. In contrast, poorly constructed buildings or those made of unreinforced concrete, heavy masonry, stucco, and adobe (mud brick) are generally inflexible. Their components move independently and in opposition to the shaking, battering one another until the structure collapses. The collapsed buildings in Figure 10-13, for example, were poorly constructed. In addition, projections or decorations not securely attached may loosen or fall during a tremor.

To illustrate the importance of proper construction in earthquake-prone regions, compare the damage suffered by different cities hit by earthquakes of the same magnitude: The earthquake that destroyed several cities in the central Asian country of Armenia in 1988, taking more than 50,000 lives, was of the same magnitude as California's 1989 Loma Prieta quake, which took 65 lives. San Francisco's structures used superior designs, methods, and materials, with the expectation that they would have to survive high-magnitude quakes.

Another important aspect of minimizing earthquake damage involves building on the most stable ground possible. Although solid bedrock shakes in brief sharp jolts, loose

Figure 10-13 Earthquakes' destruction of poorly constructed buildings. The devastation from the January 25, 1999 earthquake in Armenia, Colombia, killed more than 900 persons, injured thousands, and left at least 200,000 people homeless.

Figure 10-14 Differential earthquake damage in Varto, Turkey, in 1953. Variations in the destructiveness of earthquake shaking are often controlled by the underlying geologic materials. The collapsed houses shown here were built on the unconsolidated sands of an old river channel, while the surviving houses rest on a solid bedrock bench.

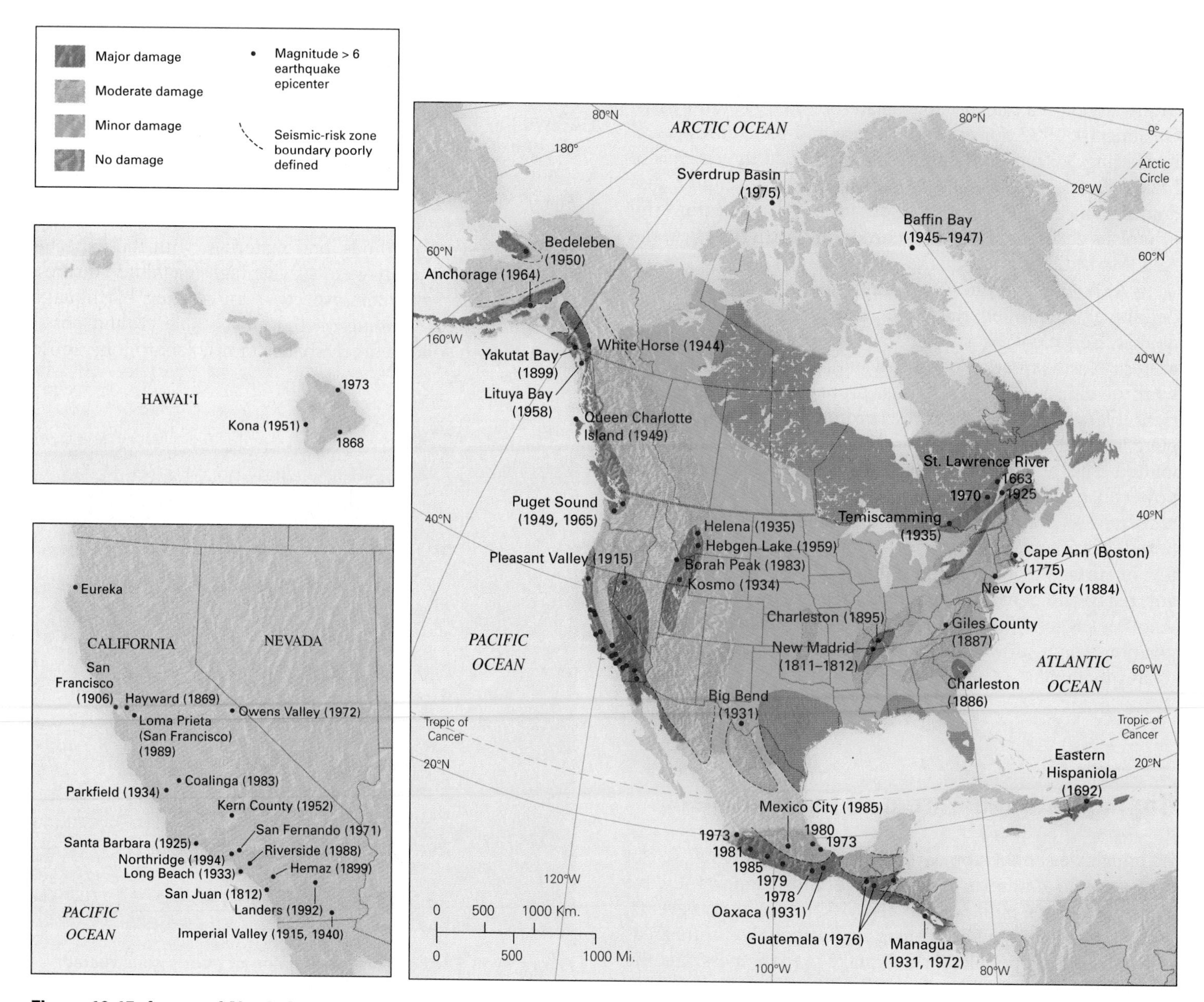

Figure 10-15 A map of North American seismic risk zones. The map includes the locations and dates of some of the continent's major historical earthquakes.

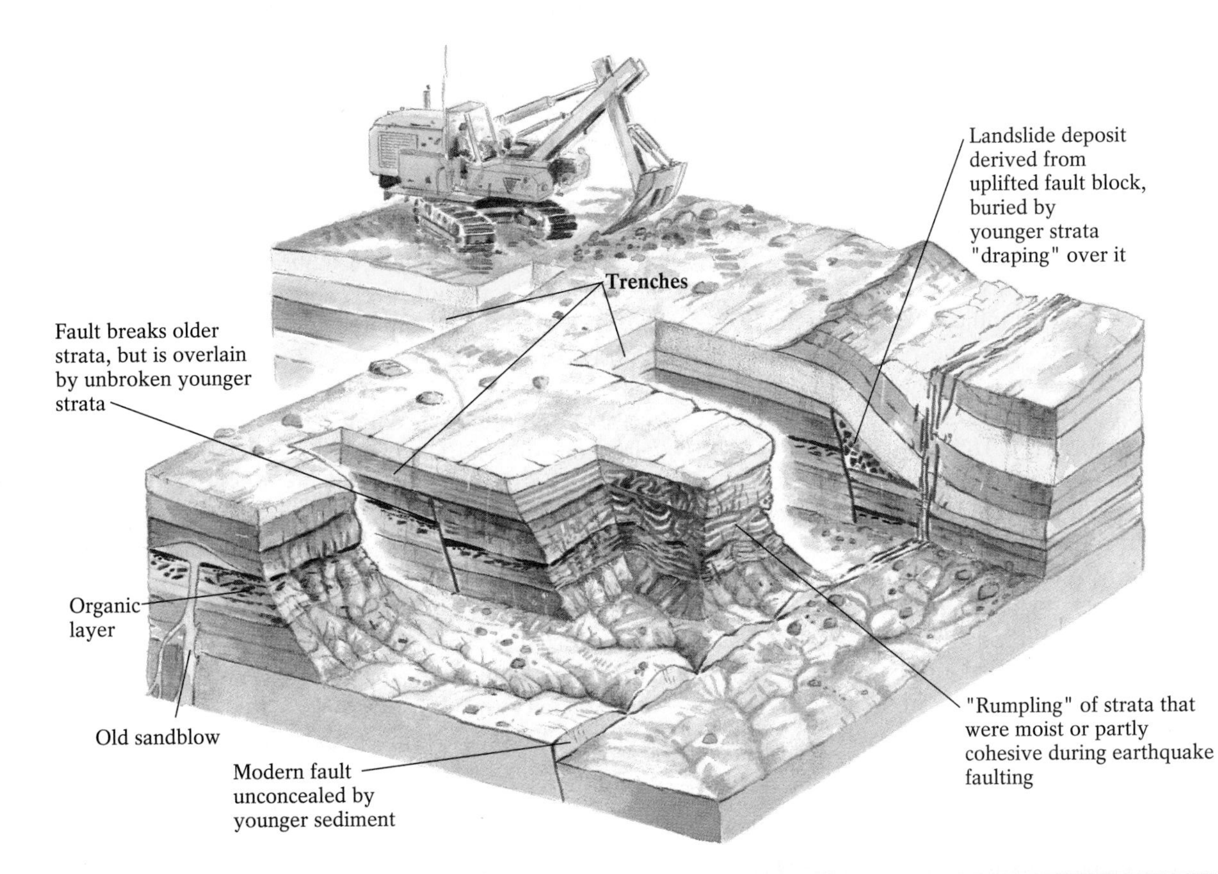

Figure 10-16 Trench excavation across a fault to expose the stratigraphy of the fault zone. According to the principle of cross-cutting relationships, any movement along a fault must be younger than the age of the youngest rock or sediment that it displaced, and older than the oldest rock or sediment that it did not displace. Each of the seismic events shown here can be therefore dated relative to one another, and its absolute age estimated in relation to the organic deposit. Photo: Vertically offset deposits near Palmdale, California, showing evidence of a previous earthquake there.

sediment develops a continuous rolling motion that can magnify a quake's intensity. Structures built on unconsolidated materials therefore typically suffer far more damage than comparable ones located on solid bedrock (Fig. 10-14). During the 1906 San Francisco quake, the buildings constructed on top of wet bay mud and landfill sustained four times as much damage as those sited on nearby bedrock; during the 1989 Loma Prieta earthquake, the waterfront areas again suffered the greatest damage.

Earthquake Prediction: The Best Defense

Seismologists study the earthquake history of an area to determine the statistical probability of future earthquakes there. If you hear that there is a 40% chance of a magnitude 7.5 earthquake striking the Palm Springs section of the San Andreas fault within the next 30 years, for instance, that "prediction" is based largely on the frequency and magnitude of past events in that area.

Maps such as that in Figure 10-15, which are based on historically recorded events, identify areas that have experienced quakes in the past and are thought likely to experience them again. Historically recorded events, however, occurred too recently to tell us much about the pattern of seismic events over thousands of years. To develop a long-term history of North America, we must attempt to date fault and quake activity that preceded the 400 to 500 years of recorded North American history.

To date earlier seismic events in California, for example, we must look for geological evidence of a fault, such as topographic breaks in the landscape or linear features that have been offset by fault movements, like the creek shown in Figure 10-9. When they find a fault, geologists look for evidence of all past seismic events related to the fault and try to date them to determine the fault's *recurrence interval,* the average amount of time between its associated quakes. For example, they examine the geologic strata composing the area. Trenches excavated across the San Andreas fault, for instance, reveal a complex stratigraphy that includes vertically offset organic deposits (Fig. 10-16). Carbon-14 dating and the principle of cross-cutting relationships can be applied to determine when the offsets occurred, enabling geologists to estimate when the related seismic events may have taken place. Evidence northeast of Los Angeles indicates that nine major events have taken place there during the past 1400 years, with an average recurrence interval of 160 years. This fact is unnerving to southern Californians, given that the last great earthquake in that area occurred in 1857.

Long-term prediction (on the order of decades or years) is often used to establish building codes and construction standards and to site critical facilities such as nuclear power plants. To formulate more precise and accurate predictions than are possible based only on an area's seismic history, however, seismologists have identified certain prequake signals that can be monitored in earthquake-prone areas. These signals allow scientists to make short-term predictions on the order of months, weeks, or days.

Long-Term Prediction Although strain energy probably accumulates uniformly in rocks throughout a seismically active zone, it may be released irregularly, with earthquakes occurring in some places but not others. Some fault segments, such as the Hayward fault in Berkeley, California, release their strain energy continuously and do not produce great earthquakes. The fault blocks along such segments move with slow but nearly constant motion, called *tectonic creep,* that prevents a large buildup of strain energy (Fig. 10-17). Fault segments that are locked in place by friction, on the other hand, accumulate substantial amounts of strain energy. According to some geologists, these locked segments, known as **seismic gaps,** represent prime candidates for major earthquakes. Figure 10-18 shows the areas surrounding the Pacific Ocean basin that have not suffered their expected share of earthquakes in recent times and so are considered seismic gaps. After identifying a seismic gap, geologists estimate the rate of the plate movement causing the stress and determine the amount of strain energy built up in the fault segments—the longer the period of seismic "silence," the more strain energy is being accumulated, and the greater the eventual earthquake is likely to be.

Short-Term Prediction When rocks undergo stress, they develop countless minute cracks long before they rupture completely. These cracks begin to appear when accumulated strain energy approaches about one-half the amount needed to produce overall rock failure. As cracking continues, rocks expand in volume, or *dilate.* This **dilatancy** produces a number of side effects—such as series of micro-earthquakes (known as foreshocks) having magnitudes of less than 1, subtle tilts or bulges in surface rocks, and changes in seismic-wave velocity through the rocks—that geologists can monitor to make short-term earthquake forecasts. In addition, the cracks may change the electrical conductivity of the rocks and alter groundwater elevation and chemistry in the area.

Another possible signal of prequake activity is unusual animal activity, such as dogs howling incessantly, cattle, horses, and sheep refusing to enter their corrals, ordinarily timid rats appearing in crowded rooms, fish jumping from water, and snakes leaving their winter hibernation dens. When dilatancy occurs, the cracking rocks evidently emit high-pitched sounds and minute vibrations imperceptible to humans but noticeable by many animals.

With several methods of earthquake prediction available, why can't we generally predict earthquakes more accurately? The main reason is that earthquakes do not always behave as expected. Seismologists figure their statistical probability based on an *average* historical frequency. Rarely do earthquakes occur right on schedule. Likewise, earthquakes do not always exhibit the precursors, such as numerous fore-

Figure 10-17 A low wall bent by tectonic creep along the Calaveras fault in California. The creeping of the fault continuously confounds road-construction engineers, who can't keep the curbs straight.

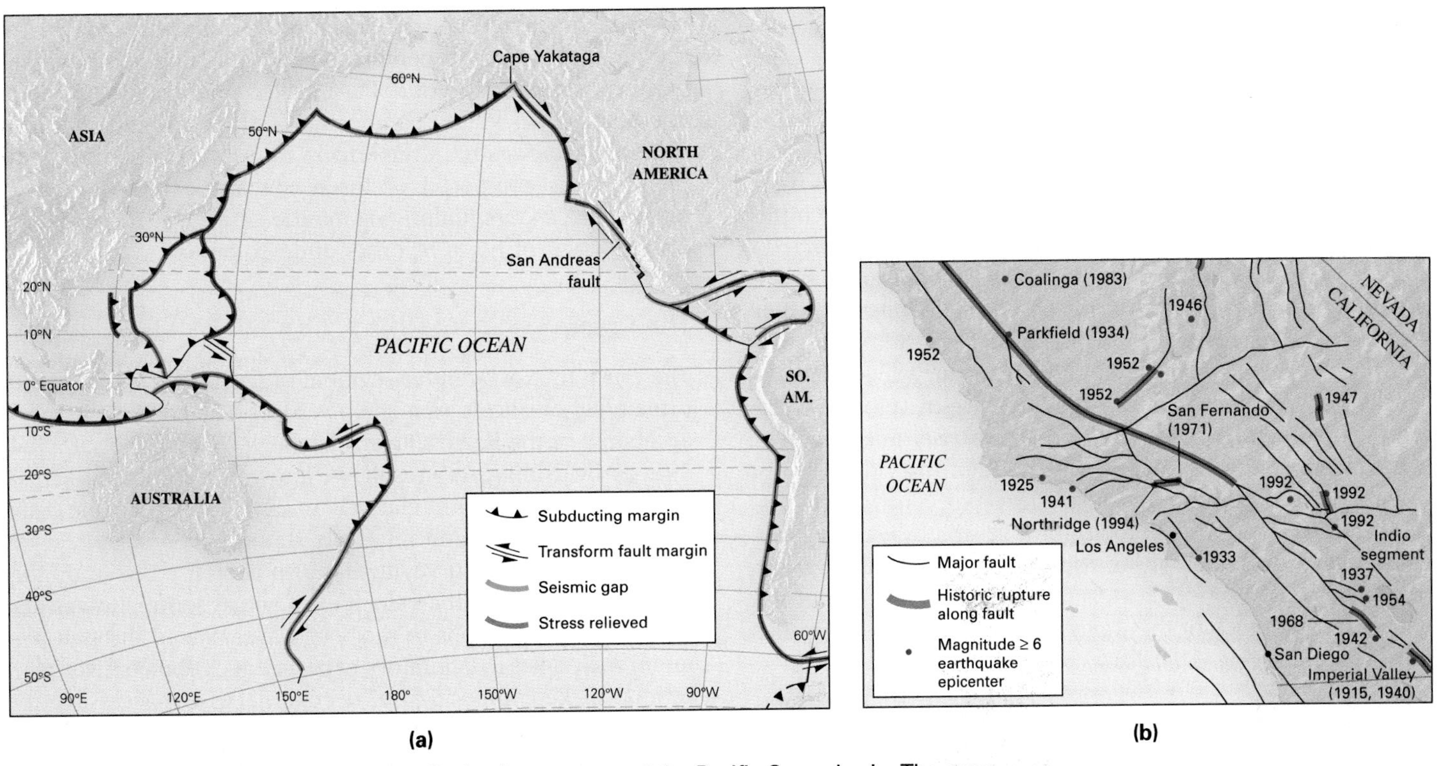

Figure 10-18 Seismic gaps. (a) A map of seismic gaps around the Pacific Ocean basin. The areas in green are overdue for major seismic activity, based on statistical probabilities calculated from past recurrence intervals. **(b)** A closeup map of southern California, showing why the U.S. Geological Survey is closely watching the Indio segment of the San Andreas fault, a seismic gap that stretches from the town of Indio northwest to Palm Springs.

shocks or bulges in rocks, predicted by the dilatancy model. In addition, small earthquakes may sometimes jar loose locked segments, thereby triggering a larger earthquake without warning. Despite all of our knowledge and technology, seismologists have not yet developed the ability to predict earthquakes with accuracy and consistency.

Investigating the Earth's Interior

The study of seismology, or more specifically the study of the behavior of seismic waves as they pass through the body of the Earth, has lead to an entire body of knowledge that would otherwise remain a mystery—our understanding of the interior structure of planet Earth. Most of our knowledge about the Earth's interior comes from analysis of variations in the speed of seismic waves. Like all waves, seismic waves tend to travel in a straight line and at an unchanging velocity as long as they continue to pass through a homogeneous medium of constant temperature and pressure. When comparing data collected at seismograph stations around the world, however, it becomes apparent that seismic waves occasionally slow down or speed up. Their varied speeds indicate that they are passing through materials of varied composition, structure, temperature, and pressure. From this information, we conclude that the Earth's interior is not homogeneous, nor do its temperature and pressure remain the same at all depths.

The Behavior of Seismic Waves

P waves compress the materials through which they pass—whether those materials are solid, liquid, or gas. Different media conduct P waves at different rates, however. After being compressed by a passing P wave, a solid medium tends to react elastically; that is, it returns to its original volume more swiftly than does an inelastic medium such as a liquid or gas. Hence, P waves accelerate when they travel through relatively more elastic solids and slow down when they enter liquids or gases (fluids), which are inelastic.

The inelasticity of fluids also explains why S waves disappear altogether upon entering such media. S waves move through material with a shearing motion, temporarily displacing adjoining areas of material, which return to their original shape after the wave exits. Fluids cannot deform and then return to their original shape—they simply flow away from the shearing stress. Consequently, S waves can pass through solid media only and die out completely when they encounter an inelastic fluid medium within the Earth's interior.

The changing velocities of P and S waves as they pass through media of different elasticities determine the paths they take through the Earth's interior, a fact that serves as the basis of many seismological studies. Seismologists have determined, for example, that if seismic waves approach the boundary between adjoining media—such as the boundaries between the layers of the Earth—at a relatively small angle, they may bounce, or *reflect,* off the surface, much as a flat stone can be made to skip off the surface of a lake. If they approach at a somewhat greater angle, seismic waves may enter the new medium and undergo a change of velocity, causing them to bend, or *refract* (Fig. 10-19a). Seismic-wave velocities also change as waves travel within individual layers of the Earth's interior: Because the rigidity of a layer increases with depth, the velocity of a seismic wave increases as the wave descends within the layer. The deeper segment of a wavefront therefore travels faster than segments that are less deep, and much of the wavefront curves back toward the Earth's surface (Fig. 10-19b).

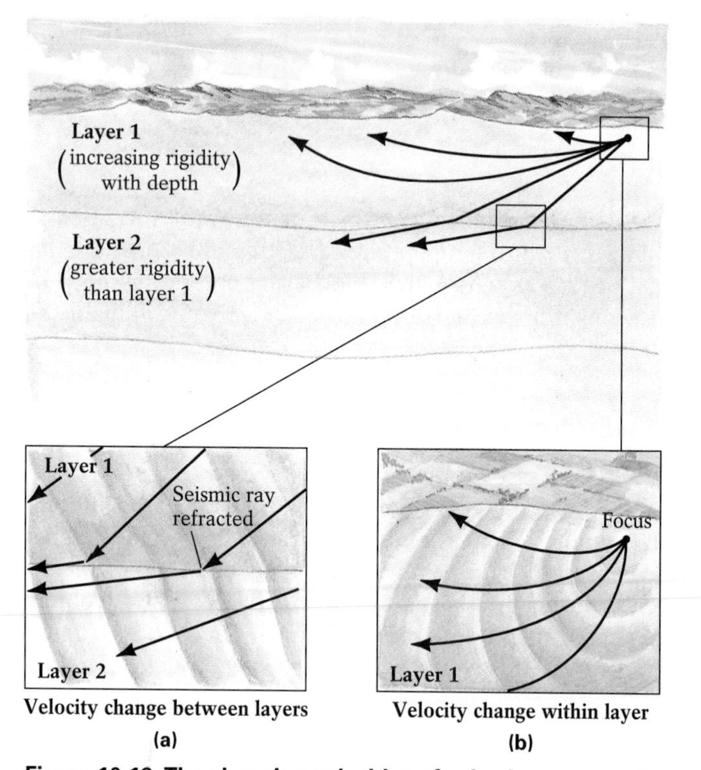

Figure 10-19 The changing velocities of seismic waves as they pass through materials of different elasticities. (a) Seismic waves speed up or slow down as they pass between two layers, causing the waves to bend, or refract. **(b)** Because rigidity increases with increasing depth, seismic waves speed up as they descend within individual layers, causing their paths to curve back toward the surface.

Such knowledge of seismic-wave behavior, applied to the great volume and complexity of data from seismograph readings around the world, has enabled scientists to estimate the density, thickness, composition, structure, and physical state of each portion of the Earth's interior. They believe that the Earth is composed of three concentric layers—a thin outer crust, a large underlying mantle, and a central core (Fig. 10-20). Each of these layers is quite complex.

The Crust

The crust, the rocky outer segment of the Earth, is composed principally of silicate-rich igneous rocks. The crust is the most accessible of the Earth's layers, as much of it can be sampled directly by drilling. As a result, we know that continental crust is typically granitic and relatively thick, whereas oceanic crust is typically basaltic and relatively thin.

Seismological surveying has also been used to study the crust. By comparing the velocities of actual seismic waves detected by seismographs to those of waves passed through different rock types in laboratory experiments, geophysicists have estimated the composition of deep, otherwise inaccessible regions of the crust. In addition, seismological measurement of the shock waves generated by underground nuclear testing during the 1960s provided data that have been used to determine the composition and structure of the Earth's crust.

Continental Crust At its thinnest, where plates are being stretched or rifted, continental crust is generally less than 20 kilometers (12.5 miles) thick. Where continental plates have collided to form mountains, the crust may be as much as 70 kilometers (45 miles) thick. In North America, the thinnest crust—found in the basins between mountain ranges in Nevada, Arizona, and Utah—is only 20 to 30 kilometers (12.5–19 miles) thick. The thickest crust—located in the Rocky Mountains of Montana and Alberta and the Sierra Nevada of California—exceeds 50 kilometers (30 miles) in thickness.

Slight variations in seismic-wave velocities indicate that the composition of the upper portion of continental crust differs from that of the lower portion. The upper portion comprises a complex jumble of granitic intrusions, basalt, andesite, vast subsurface regions of high-grade metamorphic rocks, and a nearly continuous blanket of sediment and sedimentary rock. The deepest parts of continental crust, through which P waves travel slightly faster, are probably composed of metamorphosed gabbro.

Oceanic Crust Whereas a variety of processes form continental crust, only eruptions of basalt and intrusions of gabbro at oceanic divergent plate boundaries generate oceanic crust; deep-sea sedimentation and hydrothermal alteration of plutonic rocks also contribute to the formation of the ocean crust. As a result, ocean crust is remarkably uniform both in composition and in thickness. Beneath an average 200 meters

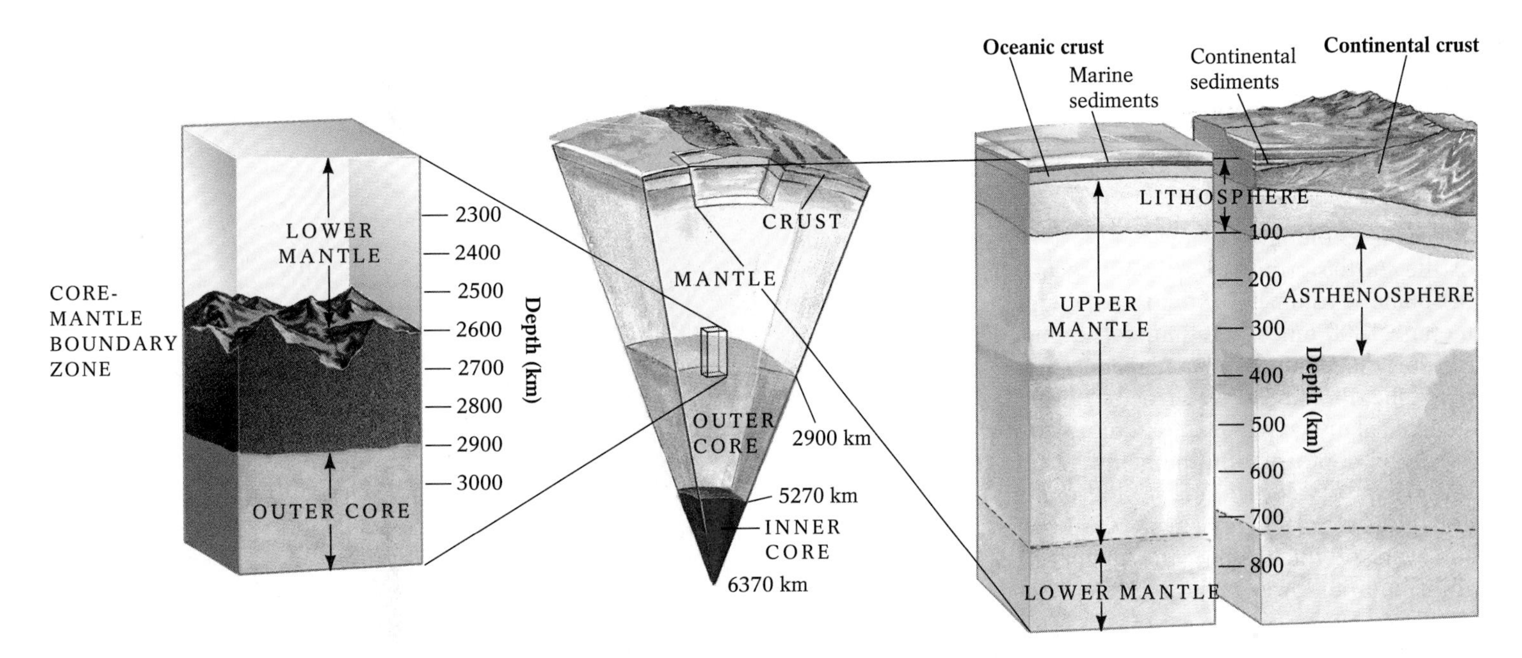

Figure 10-20 The layers of the Earth. As well as delineating the three major components (crust, mantle, and core) of the Earth, seismological studies provide details about the different natures of these layers. For example, continental crust is significantly thicker and less dense than oceanic crust, and the boundary between the core and the mantle is rugged with mountains and valleys.

(700 feet) of marine sediment, it typically consists of a 2-kilometer (1.2-mile)-thick layer of submarine pillow basalts underlain by a 6-kilometer (3.7-mile)-thick layer of gabbro.

The Crust–Mantle Boundary Often, two sets of seismic waves of the same type that emanate from the same source at the same time arrive at a single seismograph station at different times. Such occurrences have led geologists to conclude that such waves must have traveled at different velocities along different paths through different layers of the Earth's interior. One set took a shorter, more direct route through the Earth's crust; the other took a longer route through the mantle, but actually arrived earlier because its velocity increased as it entered and passed through rigid mantle material (Fig. 10-21). (To understand this concept, imagine a motorist who drives a longer distance on an empty freeway rather than a shorter distance on a city street with traffic lights, stop signs, and lower speed limits.) In 1910, Croatian seismologist Andrija Mohorovičić attributed the differing arrival times of seismic waves to a *seismic discontinuity*, an abrupt change in seismic-wave velocity induced by a marked change in the composition of materials in the Earth's interior. The boundary

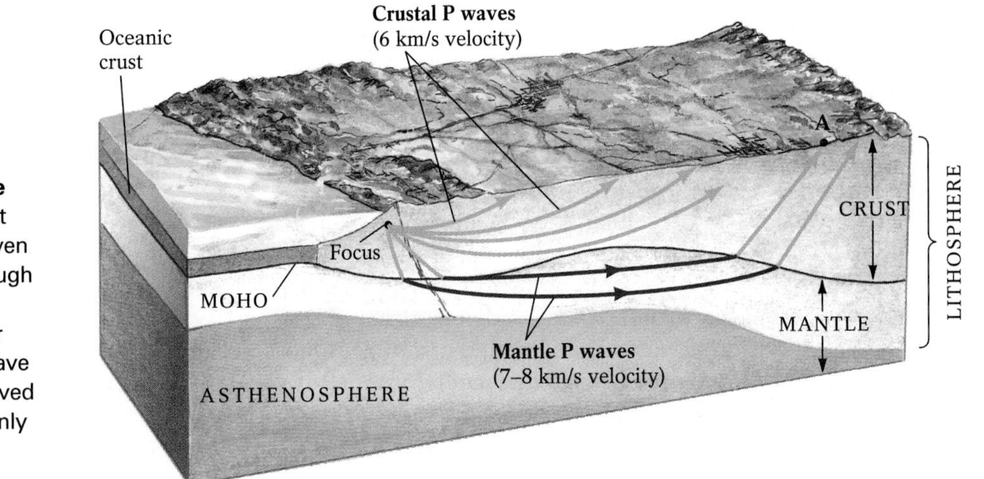

Figure 10-21 Seismic-wave velocity and the crust–mantle boundary. A seismic wave that travels through the mantle may arrive at a given location faster than one that travels only through the crust—even though the latter travels the shorter distance—because waves travel faster through the mantle. Here, the P waves that have traveled through the mantle have already arrived at point A, whereas those that are traveling only through the crust are still en route.

between the base of the crust and the top of the more rigid mantle—known as the Mohorovičić discontinuity, or simply the **Moho**—is named for its discoverer.

The Mantle

The mantle, the layer below the crust, is the largest segment of the Earth's structure. Accounting for more than 80% of the planet's volume, it extends to a depth of about 2900 kilometers (1800 miles). The minerals in mantle rocks are dense silicates that contain iron and magnesium, such as olivine and pyroxene. P- and S-wave velocities, which generally increase steadily as the waves penetrate more deeply, change abruptly at certain depths within the mantle. These abrupt changes in seismic-wave velocity indicate the presence of distinct mineralogical segments and mark the boundaries between the sublayers of the mantle: the upper mantle, the transition zone, and the lower mantle.

The Upper Mantle The upper mantle has relatively uniform composition. It is dominated by peridotite, a coarse-grained metamorphic rock composed largely of olivine, pyroxene, calcium-rich and magnesium-rich garnet, and other metamorphic minerals that form under high-pressure conditions. The increase in temperature and pressure that occurs with depth, however, produces variations in the physical characteristics of mantle material.

Directly beneath the Moho, the mantle consists of brittle, rocky material, extending to about 100 kilometers (62 miles) beneath the Earth's surface. The velocity of descending P waves increases as they pass from the relatively flexible crust into this more rigid region. This upper portion of the upper mantle and the crust together constitute the lithosphere. Beneath the lithosphere, where temperatures increase to the point of melting some rocks, the upper mantle is partially molten to a depth of about 350 kilometers (217 miles). In this asthenosphere, also called the *low-velocity zone,* rocks are capable of flowing plastically; hence P-wave velocity decreases when waves enter this region (Fig. 10-22). Beneath the asthenosphere, extending to a depth of about 700 kilometers (440 miles), intense pressure not only keeps the remainder of the upper mantle solid, but also compresses and structurally alters minerals, increasing their density and therefore their rigidity. P-wave velocity therefore increases again in this region, known as the *transition zone.*

The Lower Mantle The lower mantle extends from 700 to 2900 kilometers (440–1800 miles) below the Earth's surface. Despite the extremely high temperatures found in this region, pressure from the weight of the overlying crust and upper mantle is sufficient to keep lower-mantle materials solid. This pressure further compresses minerals and crystal structures and, with increasing depth, produces a steady increase in the density and rigidity of lower-mantle rocks.

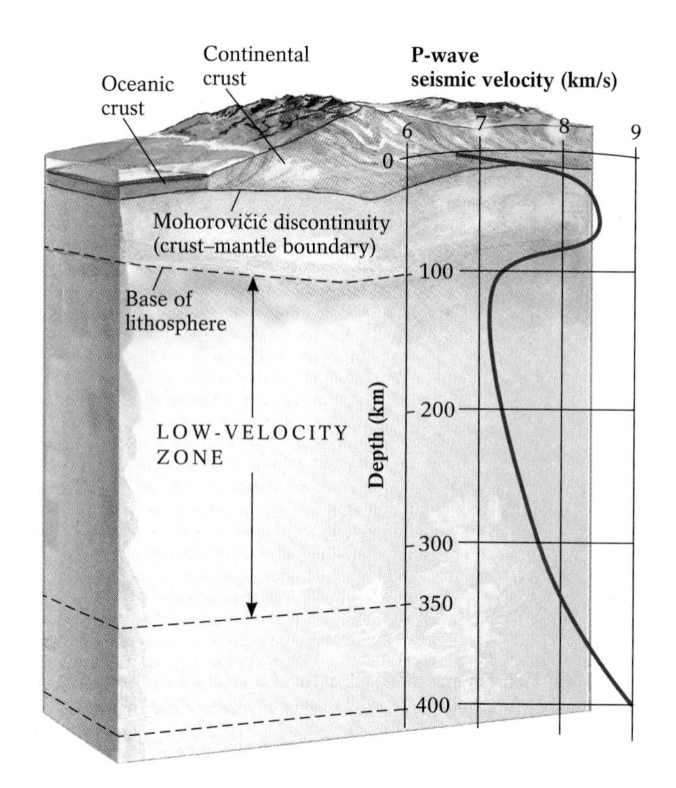

Figure 10-22 Variations in the velocities of seismic waves across the boundaries of the Earth's interior layers. These seismic discontinuities, such as occur at the Mohorovičić boundary (or Moho) and the low-velocity zone within the upper mantle, result from the changing density, rigidity, and elasticity of the Earth at various depths.

The Mantle–Core Boundary The boundary between the mantle and the core appears about 2900 kilometers (1800 miles) below the surface, almost halfway to the center of the Earth. Of all the Earth's internal horizons, this boundary—another prominent seismic discontinuity—has the most dramatic effect on seismic waves. Here, the velocity of P waves suddenly decreases dramatically, and S waves vanish altogether. Because we know that the progress of P waves is slowed substantially by materials of low rigidity and that shearing S waves are not transmitted at all by a liquid medium, we can infer that the outer portion of the core is a liquid.

The mantle–core boundary may be the Earth's most rugged topography, consisting of great *inverted* mountains that hang like cave stalactites from the mantle into the core, and deep *inverted* valleys, reaching upward into the mantle. Expansive slabs of rock, known as "anti-islands" and "anti-continents," reside at the floor of the mantle. The study of the mantle–core boundary and other features within the Earth's interior is one of today's most exciting frontiers in the science of geology.

The Core

The Earth's core, which is slightly larger than the planet Mars, accounts for 3486 kilometers (2167 miles) of our planet's total radius of 6370 kilometers (3959 miles). It most likely consists of a mixture of ultradense metals—at least 90% iron, mixed with nickel—along with a few other elements, such as sulfur, silicon, and oxygen. As a result, although it constitutes only one-sixth of the Earth's total volume, the core is so dense that it accounts for more than one-third of the planet's total mass. Because of the great depth of this layer, pressures within the core are more than 3 million times greater than Earth's atmospheric pressure. Core temperatures are believed to exceed 4700°C (8500°F).

Seismic-wave data suggest that the core has two layers: a liquid outer core that is 2270 kilometers (1411 miles) thick, surrounding a solid inner core with a diameter of 1216 kilometers (756 miles, or slightly larger than the Moon).

The Liquid Outer Core Through monitoring seismic-wave arrivals after earthquakes, geologists have discovered that P waves reach the side of the planet opposite an earthquake's epicenter somewhat later than would be expected judging from their initial speed, and that they do not arrive at all in the zone between 103° and 143° from the epicenter. These phenomena can be explained by known P-wave behavior at solid–liquid boundaries. When a P wave grazes the mantle–core boundary, it reflects and continues on a curved path to the surface, where a seismograph records it as being about 103° from the epicenter. When a P wave enters the liquid outer core, it is refracted so that it always emerges 143° or more from the epicenter (Fig. 10-23). No P waves arrive in the region between 103° and 143° of an epicenter, forming what is called the *P-wave shadow zone*.

S waves cast an even larger "shadow," as they do not appear within an arc of 154° across the globe from an earthquake's epicenter. This region, known as the *S-wave shadow zone*, arises because of S waves' inability to penetrate the Earth's liquid outer core (Fig. 10-24).

The Solid Inner Core Once they had proved the existence of the liquid outer core, geophysicists puzzled over another enigma: Why did the P waves that passed directly through the entire planet arrive on the other side *earlier* than expected if the core was liquid? Because P waves would travel faster in a solid region, they hypothesized that the waves' early arrival indicated that some part of the core was solid.

To test this hypothesis, seismologists analyzed the arrival times of P waves generated by underground nuclear test blasts, for which time and place of origin were known precisely. These studies showed conclusively that the velocity of these waves increased abruptly as they entered the inner core, supporting the hypothesis that the inner core is indeed solid.

Our knowledge of the Earth's interior has grown considerably during recent years, thanks in part to rapid advances in seismologic and other technology. In the next chapter, we see how technological advances are enabling Earth scientists to track the movements of the continents, study the details of the ocean floor, and reconstruct past positions of the Earth's plates.

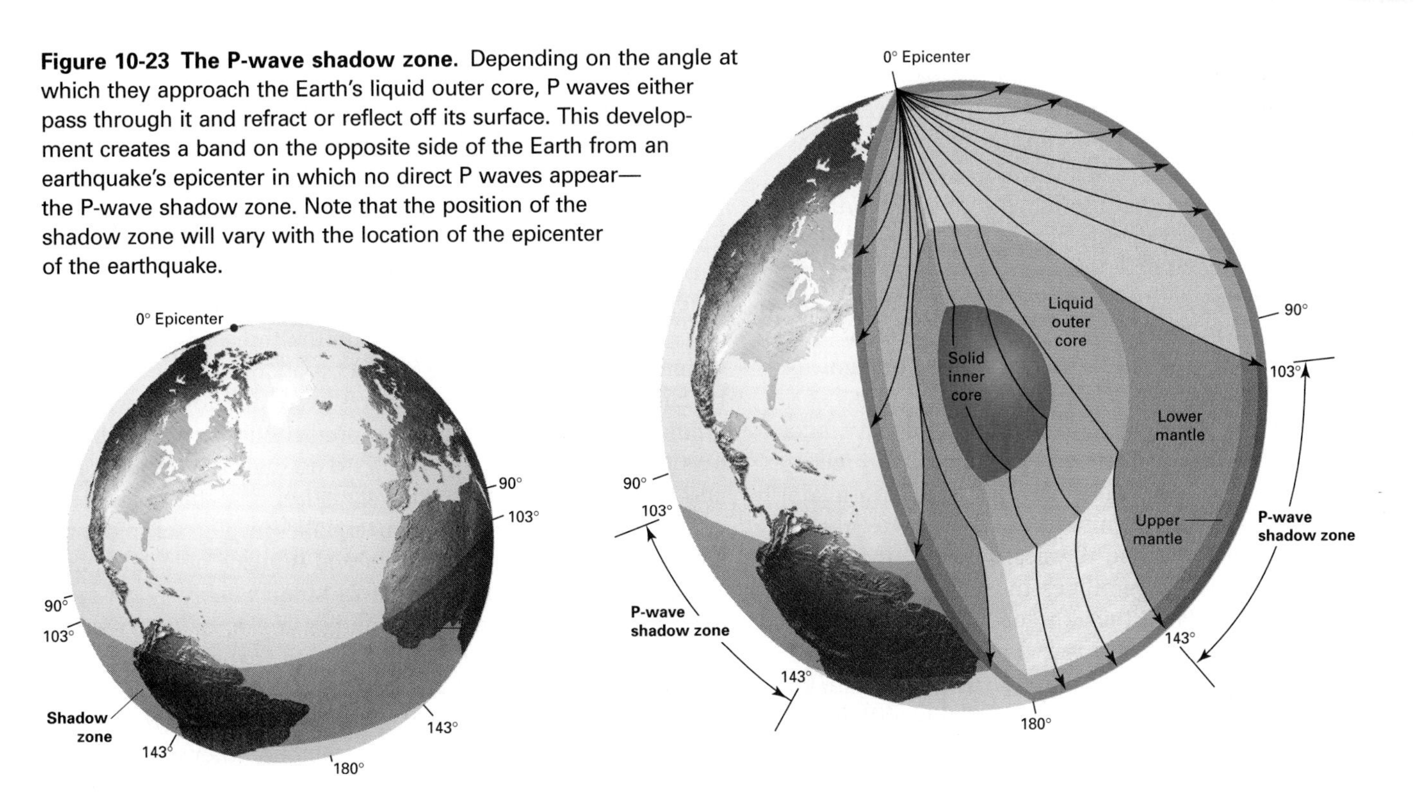

Figure 10-23 The P-wave shadow zone. Depending on the angle at which they approach the Earth's liquid outer core, P waves either pass through it and refract or reflect off its surface. This development creates a band on the opposite side of the Earth from an earthquake's epicenter in which no direct P waves appear—the P-wave shadow zone. Note that the position of the shadow zone will vary with the location of the epicenter of the earthquake.

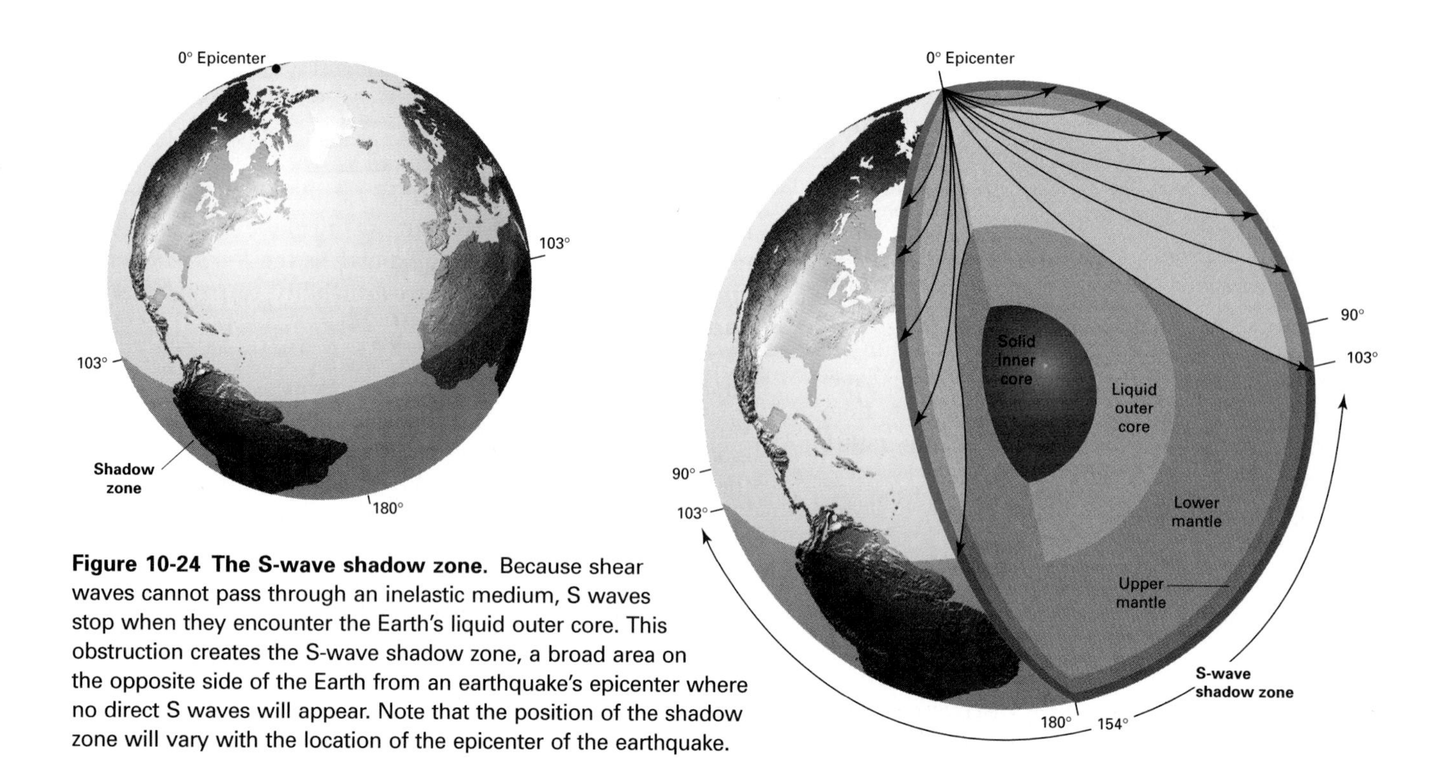

Figure 10-24 The S-wave shadow zone. Because shear waves cannot pass through an inelastic medium, S waves stop when they encounter the Earth's liquid outer core. This obstruction creates the S-wave shadow zone, a broad area on the opposite side of the Earth from an earthquake's epicenter where no direct S waves will appear. Note that the position of the shadow zone will vary with the location of the epicenter of the earthquake.

Chapter Summary

Earthquakes are vibrations of the ground that occur when strain energy in rocks is suddenly released, causing rocks to shift along preexisting faults or to fracture, creating new faults. The precise subterranean spot where rocks begin to shift or fracture is a quake's **focus,** and the point on the surface directly above the quake's focus is its **epicenter.**

Earthquake energy is transmitted through the Earth as **seismic waves.** Two types of seismic waves exist: **body waves,** which transmit energy through the Earth's interior, and **surface waves,** which transmit energy along the Earth's surface. Furthermore, two types of body waves exist: **P waves,** which cause rocks in their path to be alternately compressed and expanded in a direction parallel to the wave's movement, and **S waves,** which cause rocks to shear and move perpendicularly to the wave's direction.

Earthquake measurement utilizes both subjective and objective scales. The **Mercalli intensity scale** subjectively measures earthquakes in terms of damage caused to human-made structures. To measure the magnitude of earthquakes objectively, we use seismographs and physical features of a fault. The **Richter scale** assigns magnitudes to earthquakes based on the amplitude of the largest peak traced on a seismogram. The moment-magnitude scale uses fault length, depth, amount of slip, and rock strength to calculate earthquake magnitude. Knowledge of the different velocities of P and S waves can be employed to pinpoint the epicenter of an earthquake, which can be determined from the seismic waves' differing arrival times at three or more seismograph stations.

The effects of major earthquakes—those of magnitude 6 or greater—include significant ground shifts, both lateral and vertical; landslides; **liquefaction,** the conversion of cohesive soil into flowing mud; **seiches,** the back-and-forth sloshing of water in enclosed bodies of water; **tsunami,** large, fast-moving sea waves; and fire.

The world's principal earthquake zones are located at or near plate boundaries, although some earthquakes occur within the relatively inactive plate interiors. The subduction zones surrounding the Pacific Ocean basin are the world's most seismically active regions. The region within which earthquakes occur at subduction zones is called the **Benioff-Wadati zone;** the deepest earthquakes generally occur here.

Human attempts to cope with the threat of earthquakes include undertaking public education about earthquake risks, building quake-resistant structures, and developing ways to predict earthquakes. Long-term predictions, on the order of tens to hundreds of years, take into account the frequency of past earthquake events and the statistical probability of future quakes. Likely sites of future earthquakes include **seismic gaps,** fault segments that have been locked in place by friction and consequently accumulated a lot of strain energy. Short-term predictions, on the order of months, weeks, or days, center around the phenomenon of **dilatancy,** the expansion of highly stressed rocks as they begin to develop minute cracks. Dilatancy may produce foreshocks, bulging of surface rocks, and changes in seismic-wave velocities, among other effects.

Analysis of a huge volume of seismic wave data has enabled seismologists to estimate the thickness, density, composition, structure, and physical state of numerous layers within the Earth's interior. The thin outer layer of the Earth is the crust. The boundary between the crust and the underlying mantle is called the **Moho.** Beneath the Moho, the mantle is divided into two regions: the upper mantle and the lower mantle. The uppermost region in the upper mantle plus the crust comprise the Earth's lithosphere, which is composed of solid rock. Beneath the lithosphere lies the asthenosphere, where a unique combination of pressure and temperature causes the rock to flow plastically. The remainder of the upper mantle, as well as the lower mantle, consists of more rigid rock. The core is composed of dense metals, mostly iron and nickel. The outer core is molten; the inner core is solid.

Key Terms

earthquake (p. 172)
focus (p. 172)
epicenter (p. 172)
seismic waves (p. 173)
body waves (p. 173)
surface waves (p. 173)
P waves (p. 173)
S waves (p. 175)
Mercalli intensity scale (p. 175)
Richter scale (p. 175)
liquefaction (p. 178)
seiche (p. 180)
tsunami (p. 180)
Benioff-Wadati zone (p. 183)
seismic gaps (p. 186)
dilatancy (p. 186)
Moho (p. 190)

Questions for Review

1. Draw a simple diagram illustrating the difference between the focus and the epicenter of an earthquake.

2. Describe how P and S waves differ.

3. What is the fundamental difference between the Mercalli intensity scale, the Richter scale, and the moment-magnitude scale? How much more powerful than a magnitude 5 earthquake is a magnitude 7 earthquake?

4. Draw a simple sketch to illustrate how geologists use P- and S-wave arrival times to locate the epicenter of an earthquake.

5. Describe how earthquakes cause tsunami and liquefaction.

6. Why are the foci of divergent-zone earthquakes limited to depths of less than 100 kilometers, whereas those of subduction zones extend to 700 kilometers? Why do no earthquakes occur at depths greater than 700 kilometers? If you heard that an earthquake occurred at a depth of 300 kilometers, what could you conclude about its geological setting?

7. Describe the behavior of a seismic wave as it enters a more rigid medium.

8. Describe how oceanic and continental crust differ in composition and thickness.

9. What is the Moho, and how was it discovered?

10. What is the S-wave shadow zone, and what causes it?

For Further Thought

1. Speculate about the nature of the geologic materials (solid bedrock or loose sediment) within the map area in the figure below. Where do you think the major faults lie?

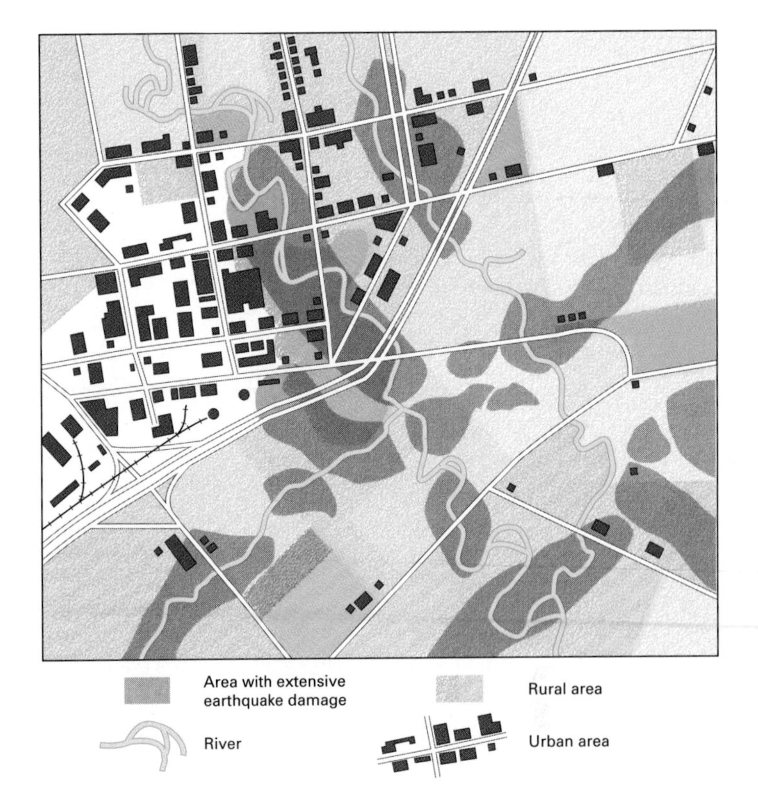

2. What evidence would you need to estimate the magnitude of an earthquake that occurred before the advent of the seismograph?

3. If you were a city planner in an earthquake-prone region, responsible for coordinating plans for an urban center that would include schools, hospitals, fire and police stations, residential areas, parks, commercial areas, a mass-transit system, power plants, dams, and so on, how would you proceed in determining the optimal location of these components of the community?

4. Suppose all earthquake activity on Earth ceased. What set of geological circumstances could explain such a change?

5. The outer core of the Earth's interior is in a liquid state. Why doesn't the liquid rise to the surface and erupt as lava?

11

Plate Tectonics: Creating Oceans and Continents

Virtually every aspect of geology—from the origin of earthquakes, volcanoes, and mountains (Fig. 11-1) to the rise and fall of sea level—is affected by plate motion. In Chapter 1, we made our initial examination of the theory of plate tectonics. In later chapters, we detailed several of the basic assumptions of this theory:

- The Earth's lithosphere consists of rigid plates averaging 100 kilometers (60 miles) in thickness, ranging from about 70 kilometers (43 miles) thick for the oceans to 150 kilometers (90 miles) thick for the continents.
- The plates move relative to one another by divergence, convergence, or transform motion.
- Oceanic lithosphere forms at divergent plate boundaries and is consumed at subduction zones, one type of convergent plate boundary.
- Three basic types of convergent plate boundaries exist: those that occur where ocean plates subduct beneath other ocean plates; those that occur where ocean plates subduct beneath adjacent continents; and those that occur where two continental plates collide.
- Most earthquake activity, volcanism, faulting, and mountain building takes place at plate boundaries.
- Plates generally do not deform internally; that is, the centers of plates tend to be geologically stable.

In this chapter, we take a large-scale view of plate motion and its effects. We examine the rates of plate motion, the origin of both the oceanic and continental portions of the plates, and the driving mechanisms for plate motion.

Figure 11-1 The Himalayas in Tibet. Located at the boundary between India and China, this range is a spectacular example of the mountain-building capacity of colliding lithospheric plates.

Development of the Theory of Plate Tectonics

In 1782, the American scientist-philosopher-statesman Benjamin Franklin hypothesized: "The crust of the Earth must be a shell floating on a fluid interior. Thus the surface of the globe would be capable of being broken and disordered by the violent movements of the fluids on which it rested." Almost 200 years later, geologists would accept that remarkable insight by one of history's finest scientific thinkers as the key concept of the theory of plate tectonics. How was the comprehensive plate tectonic theory developed? What evidence supports a theory that evokes images of opening and closing oceans and movable continents? And what causes these events to happen?

Figure 11-2 A reconstruction of the proposed supercontinent Pangaea, with present-day coastlines and continent names shown for reference. This configuration of plates dates from about 225 million years ago.

Alfred Wegener and Continental Drift

The first conception of the revolutionary theory of plate tectonics accompanied by a full body of supporting evidence was proposed by the German geophysicist and meteorologist Alfred Wegener (1880–1930), in what was called the **continental drift** hypothesis. Despite blistering ridicule from the leading geologists of his day, Wegener proposed that the continents float on the denser underlying interior of the Earth and periodically break up and drift apart. He asserted that all of the Earth's continents had been joined together about 200 million years ago as a supercontinent he called *Pangaea* ("all lands") (Fig. 11-2). According to Wegener, Pangaea had covered roughly 40% of the Earth's surface, most of it in the Southern Hemisphere. During Pangaea's existence, what is now the area of New York City sweltered in the lush tropics near the equator, and most of Eastern Africa and India shivered under a dome of glacial ice near the South Pole. Pangaea was surrounded by a single ocean, which Wegener dubbed *Panthalassa* (named for the Greek goddess of the sea). Over a vast period of time, Pangaea broke up, forming a number of continents that migrated to all regions of the globe. Wegener supported his hypothesis by noting his observations on the shapes of continental margins, patterns of present-day animal life, similarities among far-distant fossils and rocks, and evidence of past climates at odds with the climates at present locations.

Continental Fit In 1620, after seeing the new maps drafted following the global explorations of the sixteenth century, the English philosopher Sir Francis Bacon was the first to note that the outlines of the continents of the world could be pieced together in a jigsaw-puzzle style. This concept reappeared periodically in the scientific community for the next three centuries. Almost 300 years later, Wegener tinkered with the puzzle of continental shapes until he fit them together, forming a model of the Pangaea landmass. Today, precise computer fitting has confirmed Wegener's hypothesis, and we can see that the reunited continents would indeed fit together remarkably well (Fig. 11-3).

Animal Habitats Studies of the distribution patterns of ancient animals, based on fossil evidence and on the habitats of modern animals, helped convince Wegener that now-separated landmasses were once united into a supercontinent. Identical types of animals—clearly not built for long-distance, open-ocean swimming—had somehow appeared on landmasses widely separated by ocean basins. For example, fossils of *Mesosaurus,* a small reptile that lived in shallow lakes and estuaries 240 million years ago, have been found only in Brazil and South Africa, which are separated today by 5000 kilometers (3000 miles) of the southern Atlantic Ocean. Similarly, paleontologists have discovered fossils of the land-dwelling reptile *Lystrosaurus* in regions now separated from each other by large bodies of water (Fig. 11-4). A modern example is the hippopotamus, which is found in only two places—in Africa and on the island of Madagascar 500 kilometers (300 miles) to the east.

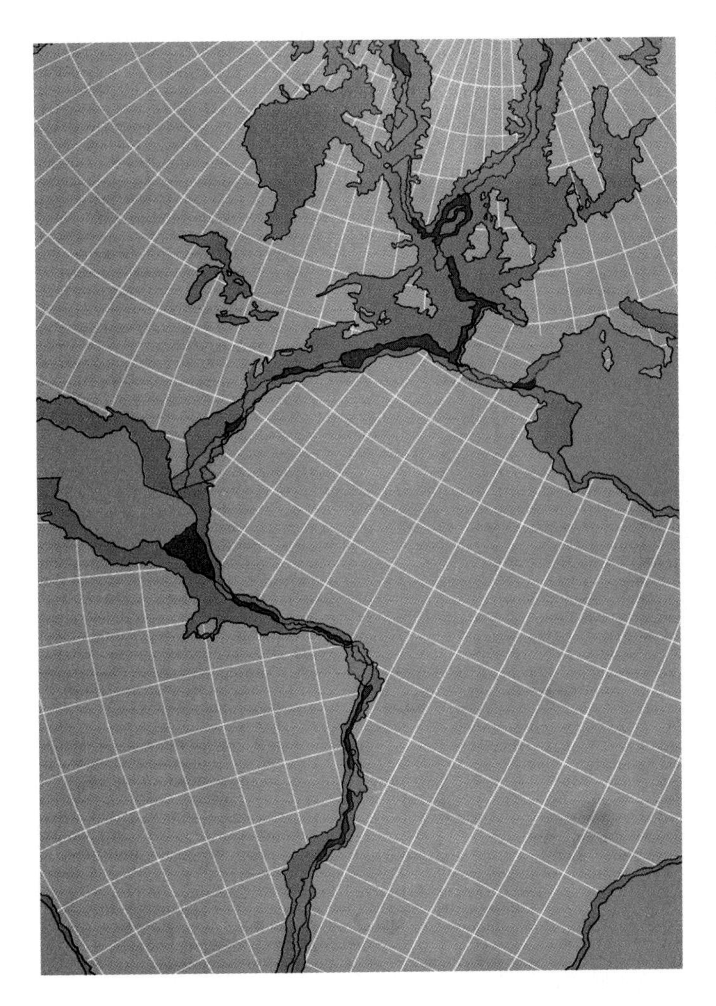

Figure 11-3 Precise matching of the continental shelves of circum-Atlantic continents by computer analysis. (From "The Origin of the Oceans," by Sir Edward Bullard, published 1969. Copyright © 1969 by Scientific American, Inc. All rights reserved.) Because the seaward (below sea level) edges of the continental shelves, shaded dark brown, represent the true edges of the continents, their fitting is more precise than simple shoreline fitting.

Related Rocks According to Wegener, if North America, Africa, and Europe had been joined in the past, a continuous chain of mountains would have stretched from Alabama all the way to Scandinavia. The fact that the 390-million-year-old rocks of the mountains of eastern North America are remarkably similar in mineral composition, structure, and fossil content to same-aged rocks in eastern Greenland, western Europe, and western Africa supports this possibility (Fig. 11-5).

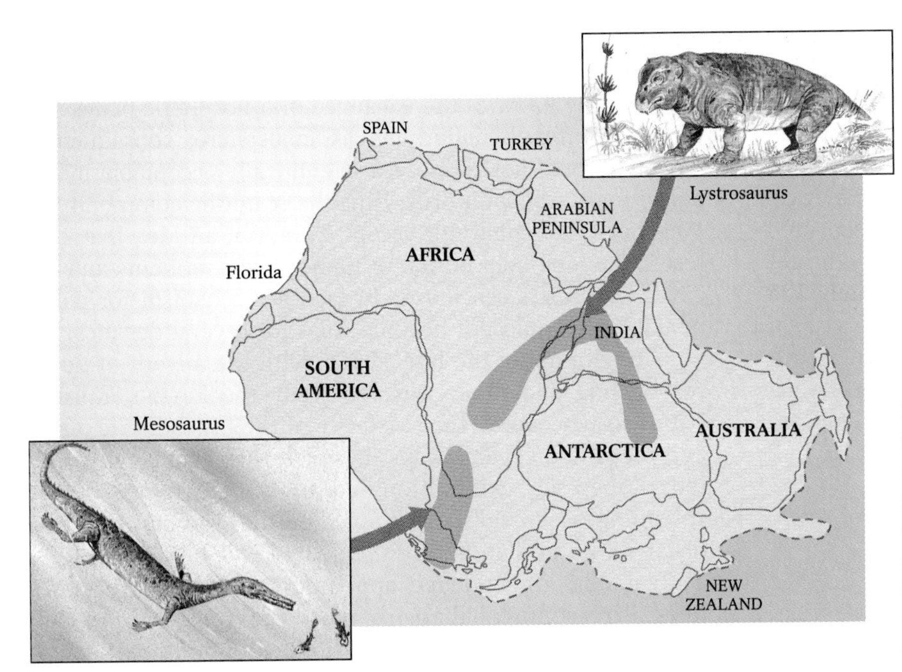

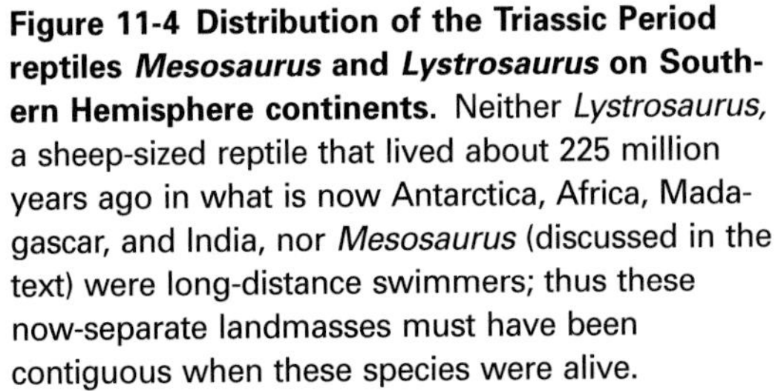

Figure 11-4 Distribution of the Triassic Period reptiles *Mesosaurus* and *Lystrosaurus* on Southern Hemisphere continents. Neither *Lystrosaurus,* a sheep-sized reptile that lived about 225 million years ago in what is now Antarctica, Africa, Madagascar, and India, nor *Mesosaurus* (discussed in the text) were long-distance swimmers; thus these now-separate landmasses must have been contiguous when these species were alive.

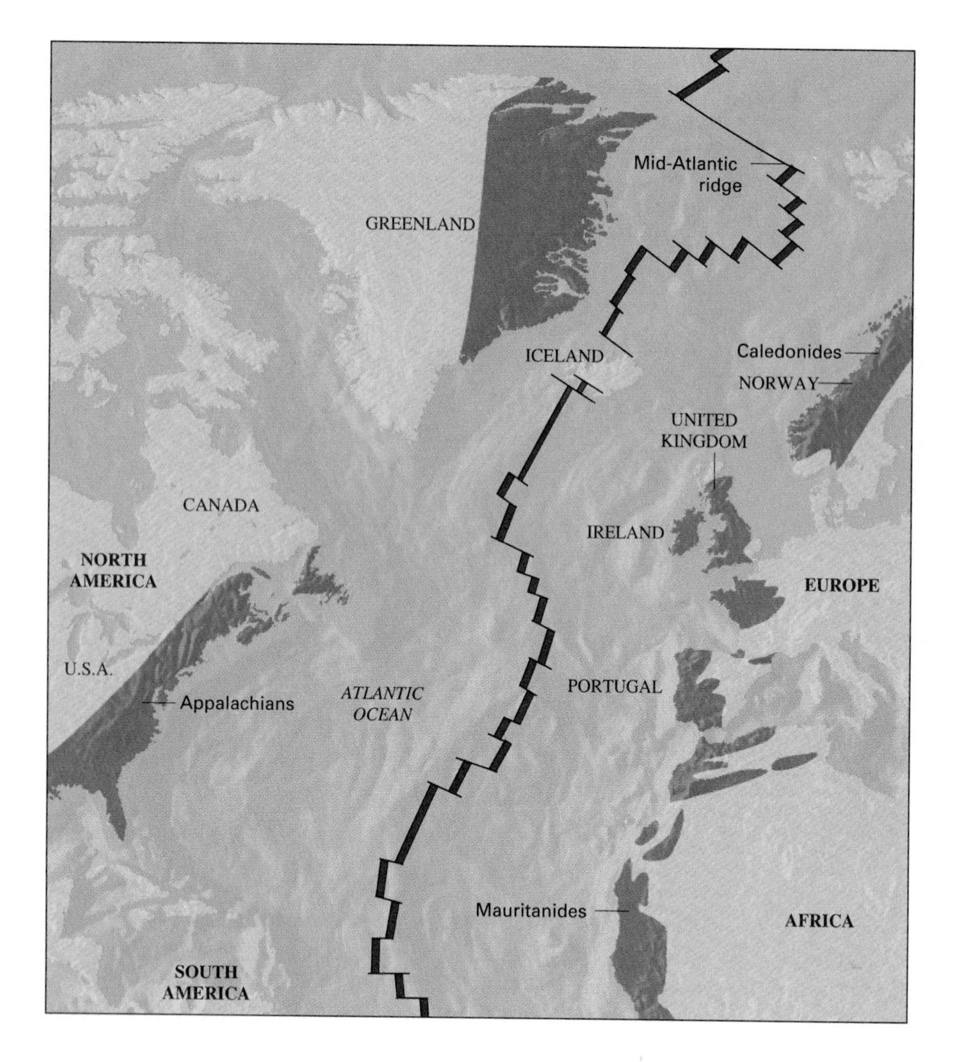

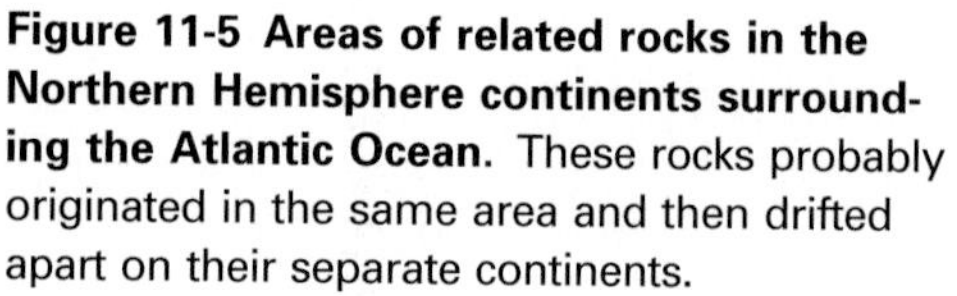
Figure 11-5 Areas of related rocks in the Northern Hemisphere continents surrounding the Atlantic Ocean. These rocks probably originated in the same area and then drifted apart on their separate continents.

Ancient Climates Wegener also pointed out that, in certain locations, geological evidence suggests a past climate very different from the current climate. For example, a distinctive pattern of scratches, typical of the striations etched by glaciers as they drag debris along, appears in the ancient bedrock of such warm places as India, Australia, and South America. In addition, glacially deposited debris has been discovered on the western coast of South Africa. The evidence indicates that these areas were covered by a glacier approximately 225 million years ago, which would have been possible only if they were once much nearer the South Pole. Continental drift would likewise explain why coal, which forms from the accumulation of swamp vegetation in warm, moist environments, has been found surrounded by kilometers of ice in a cliff in Arctic Spitsbergen, Norway.

The Birth of the Plate Tectonic Theory

With such strong evidence to support it, why did the scientific community initially reject Wegener's hypothesis of continental drift? To many of Wegener's peers, his proposal that the continents, driven by the Earth's rotation, plowed through denser oceanic rocks did not seem physically plausible. At the time, no one knew of the existence of lithospheric plates and the heat-softened underlying asthenosphere; consequently, Wegener could not propose a scientifically acceptable mechanism for continental drift. Thus the scientific community of Wegener's day could not accept so radical an idea that contested the long-held notion of immovable continents and ancient, featureless sea floors. In fact, a mechanism for continental drift would not become clear for several decades.

Beginning in the late 1950s, with the advent of deep-sea drilling and other ways to study the Earth's interior, scientists discovered the existence of the lithospheric plates and identified several ways in which they might move. Intense deep-sea exploration led to the discovery of the spreading ocean ridges and revealed the true nature of the ocean's role in plate motion. During the 1960s and 1970s, a great deal of evidence appeared to support Wegener's lines of reasoning, enhancing the acceptability of the Pangaea hypothesis and replacing the idea of continental drift with the theory of plate tectonics. In the 1980s and beyond, new research technologies have increased our knowledge of the physics of the Earth's interior, enabling geologists to propose convincing hypotheses to explain what drives plate tectonics.

Determining Plate Velocity

A plate's velocity is a measure of both its speed and its direction of movement. Geologists can estimate a plate's velocity in absolute terms by using hot spots as fixed reference points, by tracking magnetic field reversals, or by using orbiting satellites.

Plate Movement Over Oceanic Hot Spots

Hot spots are localized regions where plumes of hot mantle material rise from great depths to penetrate the base of the lithosphere. Although hot spots migrate, they move so slowly compared with the rate of plate movements that they are considered stationary. Approximately three dozen hot spots, from which magma rises to form volcanoes, are active today (Fig. 11-6).

Hot spots beneath oceanic plates produce volcanoes on the sea floor directly above them. As a plate moves across a hot spot, it forms a chain of submarine volcanoes, each of which may grow above sea level and become an island. Eventually, active volcanoes become extinct as the plate passes beyond the hot spot, and new volcanoes form at the end of the trail of volcanic islands. We can deduce the direction of the plate's movement from the locations of the extinct volcanoes and determine its speed from the ages of the volcanoes and the distances separating them and the hot spot.

Volcanoes that have not grown above sea level form conical submarine mountains called *seamounts.* Some island volcanoes that originally formed over oceanic hot spots have since been worn flat by weathering, wave action, and stream erosion. Called *guyots* (pronounced "GHEE-owes"), these plateaus are now also found below sea level, where they sank as the lithosphere that carried them cooled, became more dense, and subsided deeper into the mantle.

All oceans contain numerous hot spot islands, seamounts, and guyots, generally arranged in long chains that formed as drifting plates trailed away from hot spots. Most prominent is the Hawai'ian Island–Emperor Seamount chain, which extends first to the northwest, then north, through more than 6000 kilometers (4000 miles) of the central

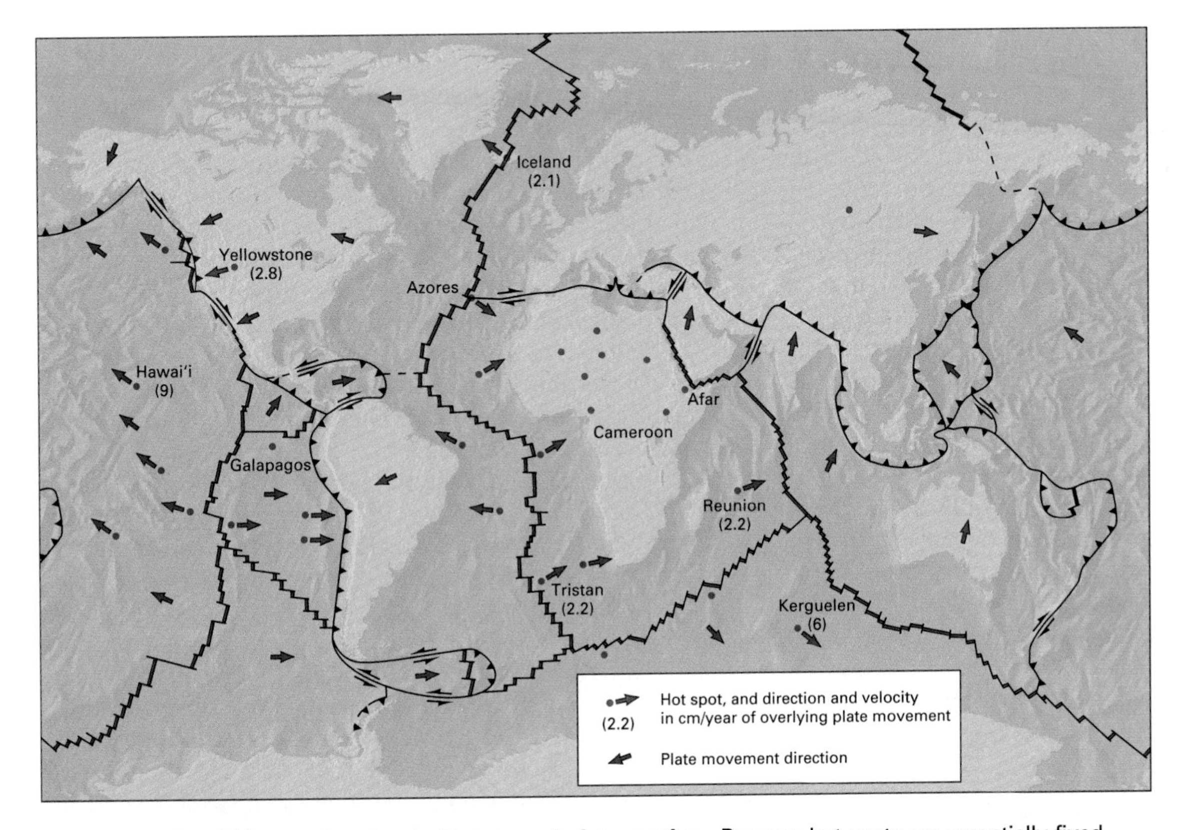

Figure 11-6 A world map showing hot spots and plate motion. Because hot spots are essentially fixed relative to the faster-moving plates above them, they can be used as references to determine the velocities of the world's tectonic plates.

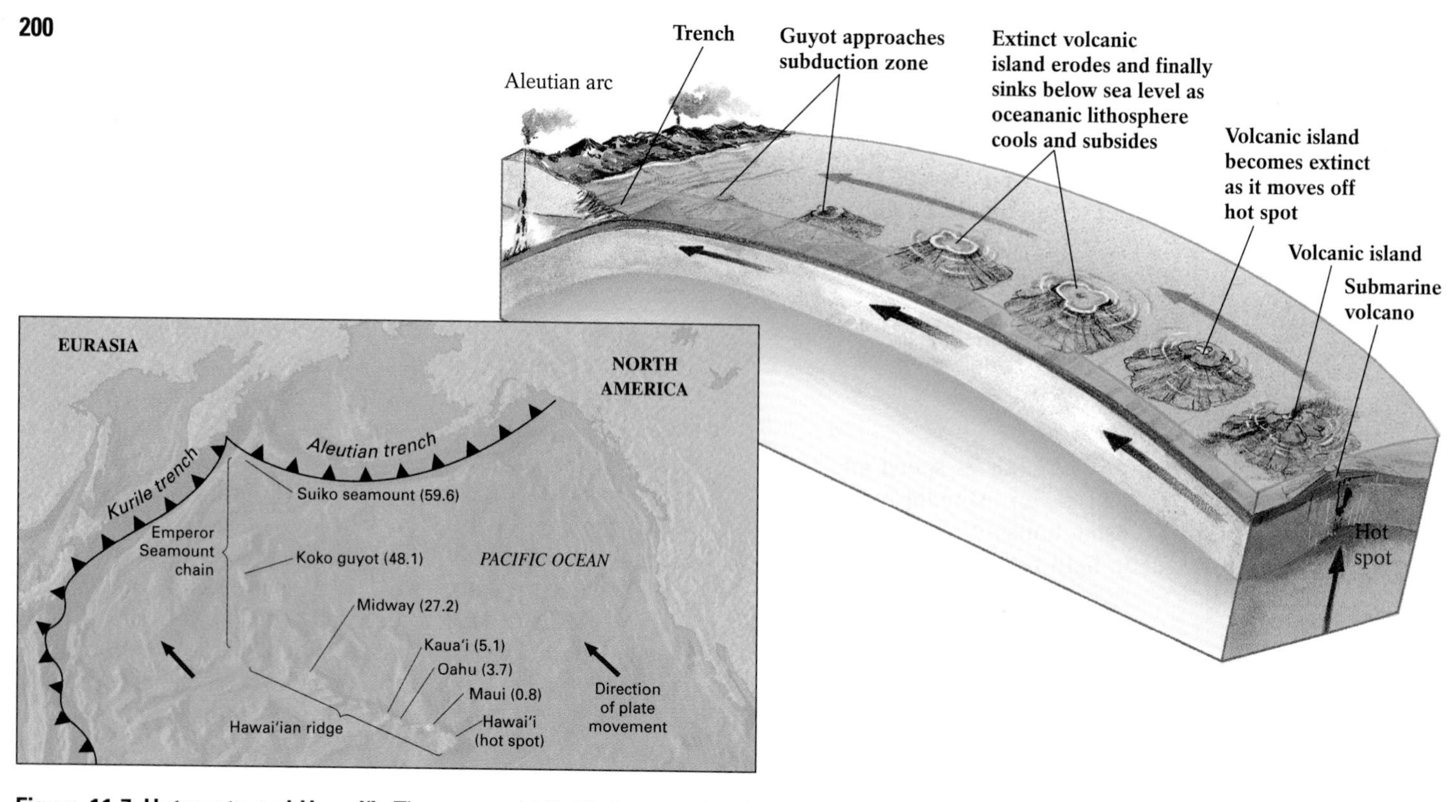

Figure 11-7 Hot spots and Hawai'i. The same mid-Pacific hot spot has fueled the eruptions that produced every volcanic island and submarine mountain in the 6000-kilometer (4000-mile)-long Hawai'ian Island–Emperor Seamount chain. As the moving plate carries each island away from the hot spot, the volcanoes lose their heat source and become extinct. Weathering and erosion gradually reduce their height. Ultimately, they become submerged as the lithosphere cools and subsides. (Volcano ages, in millions of years, are shown in parentheses.)

Pacific, ending at the Aleutian trench (Fig. 11-7). The Hawai'ian Islands are relatively recent features within this chain of volcanoes. Hawai'i, the "Big Island," is the youngest, southeastern-most, and only actively volcanic island in the chain; it experiences nearly continuous eruptions today. The islands and seamounts become sequentially older as they trend toward the Aleutian Trench. To the southeast of Hawai'i is a new rising seamount, the future island of Loihi; its summit remains several thousand meters below sea level.

Several other central-Pacific seamount chains exist as well. All are marked by a noticeable bend, presumably formed after the Pacific plate changed direction from due north to its present northwesterly trend. Basalts from volcanoes at these bends date to roughly 40 million years ago; thus we know the Pacific plate changed direction at that time. To estimate the rate of Pacific plate movement, we can use the Hawai'ian hot spot as a fixed reference point. For example, Midway Island is located about 2700 kilometers (1700 miles) northwest of the hot spot, and its basaltic rocks are 27.2 million years old. Dividing the distance from the hot spot by the age of the rocks, we conclude that the Pacific plate has moved at an average speed of 10 centimeters (4 inches) per year.

Tracking Magnetic Field Reversals

Rates of oceanic plate motion can also be determined from records of the periodic reversals of the Earth's magnetic field. Magnetic north and magnetic south have exchanged places hundreds—and perhaps even thousands—of times during Earth's history. Evidence of such changes in the Earth's magnetic field can be detected in the geologic record, most often in mafic igneous rocks such as basalt. As mafic lava cools, small crystals of the magnetic mineral magnetite form and, like little compasses, orient themselves with the Earth's magnetic field. By the time the entire lava body has solidified, the magnetite crystals have become frozen in place, recording the alignment of the Earth's magnetic field at that time.

Through recordings made by ocean-going *magnetometers* (devices that measure the intensity of a magnetic field), scientists have discovered alternating stripes of slightly stronger and weaker magnetism along the ocean floor parallel to the mid-ocean ridges. In the early 1960s, Fred Vine and Drummond Matthews of Cambridge University proposed that these **marine magnetic anomalies** constituted evidence of sea-floor spreading at divergent plate boundaries. As basaltic lava cools at spreading mid-ocean ridges, it becomes magnetized in the direction of the Earth's magnetic field at the time of eruption; basalts that erupted at times of reversed polarity would consequently display the magnetic reversal (Fig. 11-8). Magnetometers towed above a normally magnetized stripe record a slight strengthening (about 1%) of the magnetic field; towed above a stripe formed under reversed magnetism, they record a slight weakening of this field. (The Earth's magnetic field is enhanced in the vicinity of rocks of normal polarity, but weakened near rocks of reversed polarity.)

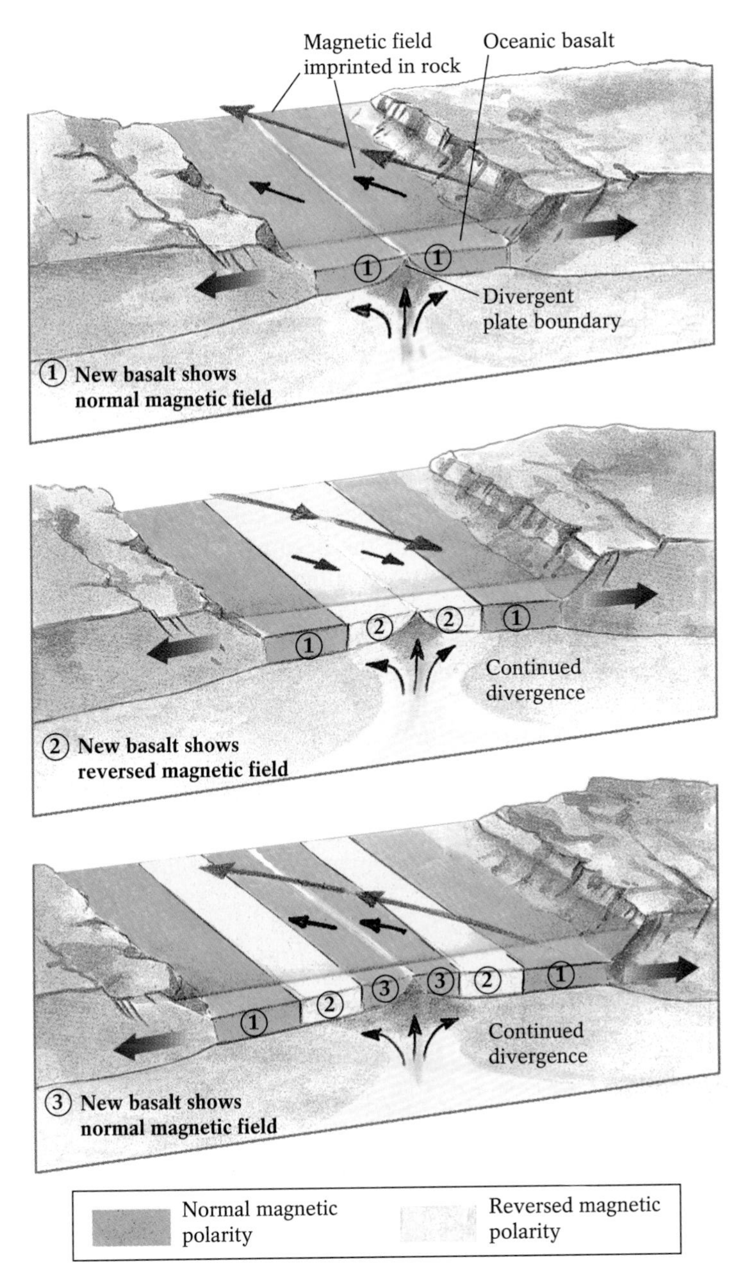

Figure 11-8 Marine magnetic anomalies reflect reversals in the Earth's magnetic field. As basaltic lavas cool and solidify at mid-ocean ridges, their magnetite crystals become aligned like small compasses with the prevailing direction of the field. Each resulting stripe of basalt has either normal (like today's field) or reversed magnetism (opposite to today's field).

By dating the basalt within a marine magnetic anomaly and measuring the distance of the anomaly from the spreading ridge, we can determine the rate of spreading and therefore the rate of plate motion. For example, if a 4.5-million-year-old anomaly is located 90 kilometers (55 miles) from the ridge crest, the spreading rate is 2 centimeters (0.8 inch) per year.

Satellite Tracking

Every year, ground-based lasers bounce concentrated beams of light off the Laser Geodynamics Satellite (LAGEOS), and

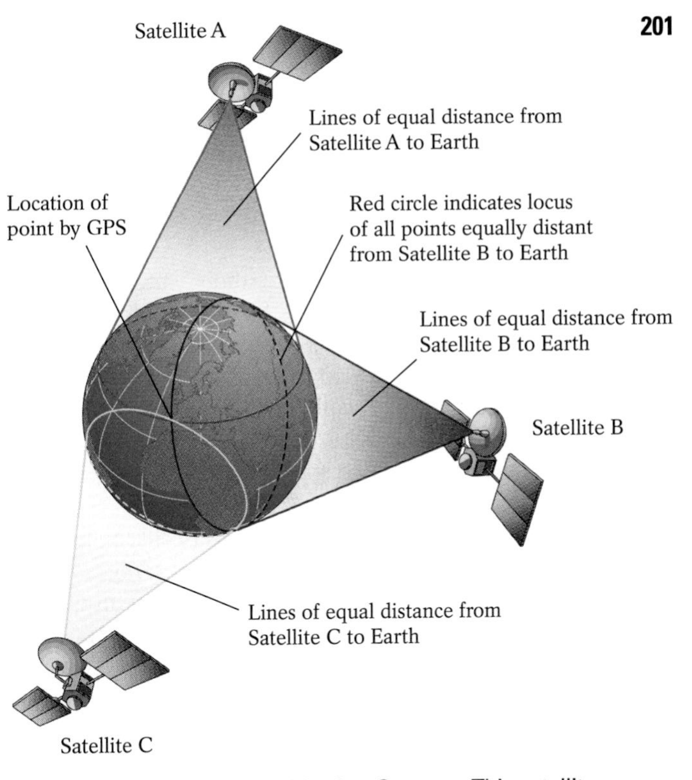

Figure 11-9 The Global Positioning System. This satellite-based system enables geologists to locate changes in precise points on the Earth's surface, confirming that the Earth's plates do indeed move.

scientists record the amount of time required for the beams to make a round-trip journey. Armed with that information, they can calculate the precise distance between the satellite and the ground station. Because LAGEOS speeds around the Earth in the same direction and at exactly the same velocity as the Earth rotates on its axis, the satellite's position over the planet remains fixed. Any change in the laser beam's travel time therefore represents a change in the geographic position of the landmass containing the ground station. This space-age technology has confirmed that the Earth's plates do indeed move and shown that they do not all move at the same rate.

Similarly, a constellation of 24 orbiting satellites that constitutes the Global Positioning System (GPS) can locate precise points on the Earth's moving plates. Much like the triangulation used to locate earthquakes, the overlapping beams from a trio of satellites can locate a single geographic point on the Earth's surface with great accuracy (Fig. 11-9). Geoscientists can then measure the movement of the plate that contains that point to within a few millimeters.

How rapidly, then, do plates move? Combining the information gleaned from relatively stationary oceanic hot spots, magnetic anomalies, and orbiting satellites, we have been able to determine that the Pacific, Nazca, Cocos, and Australian-Indian plates, among the Earth's fastest, move more than 10 centimeters (4 inches) per year. The North and South American, Eurasian, and Antarctic plates, among the slowest movers, travel 1 to 3 centimeters (0.4–1.2 inches) per year.

The Nature and Origin of the Ocean Floor

If the ocean basins were drained of their water, we would see chasms as deep as 11 kilometers (7 miles), volcanic mountain ranges thousands of kilometers long, chains of flat-topped mountains, long linear fractures oozing lava, and faulted cliffs stretching like walls for great distances. Studies of this topography and of the bedrock composition of the ocean floor have enabled us to speculate about the origin and evolution of the ocean basins.

Most of the ocean basin features were discovered and mapped in the early 1940s using *echo-sounding sonar.* A ship using this method emits a sharp pinging noise. The sound waves travel at a rate of about 1500 meters (5000 feet) per second, the speed of sound in seawater. By noting the time that elapses while the sound waves bounce off the sea floor and return to a listening device, geologists can calculate the depth of the ocean bottom and map its topographic features.

Since the 1940s, several other technologically sophisticated methods have been developed by which we can study the ocean floor. For example, we can map subsurface ocean-floor rocks by means of *seismic profiling,* which uses powerful energy waves that can penetrate the surface and reflect off layers of underlying sediment and rock. In addition, between 1965 and 1980, the Deep Sea Drilling Project (DSDP) and a specially equipped research vessel, the *Glomar Challenger,* obtained thousands of deep-sea core samples used for studying the composition of the ocean floor and the origin of the ocean basins. Deep-sea submersibles such as ALVIN have been used to perform field work on the sea floor and have proved instrumental in discovering new life forms and geological processes. Today, scientists are using SEASAT, a satellite dedicated to studying ocean-floor topography, to produce maps of the entire ocean floor (Fig. 11-10).

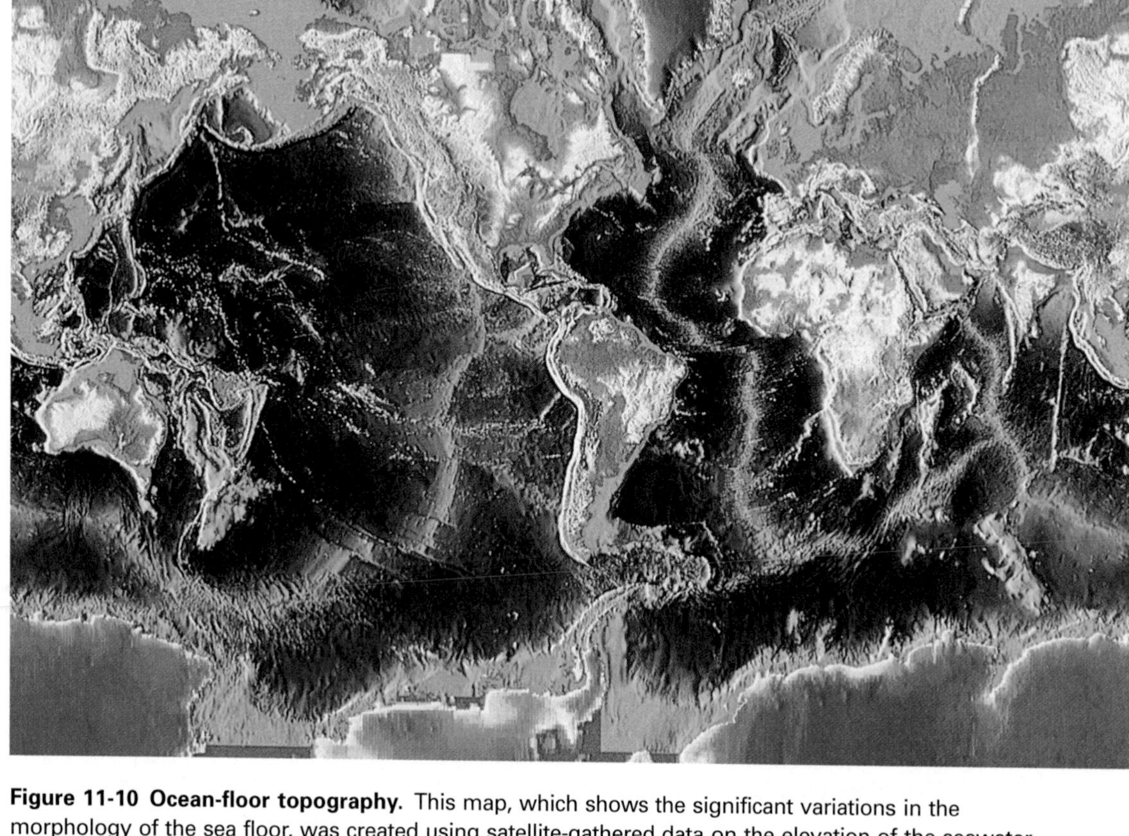

Figure 11-10 Ocean-floor topography. This map, which shows the significant variations in the morphology of the sea floor, was created using satellite-gathered data on the elevation of the seawater surface. Increased gravitational pull produced by masses such as submarine mountain ranges attracts large quantities of seawater, causing upward expansion of the ocean surface (light blue areas); conversely, decreased gravitational pull, produced by depressions such as trenches, lowers the ocean surface (dark blue areas).

Rifting and the Origin of Ocean Basins

As we've learned, new oceans form where an existing continent rifts into two or more smaller continents that then diverge. Rifting begins when a warm current of mantle material, rising within a convection cell, stretches and thins the continental crust above it, causing this surface to break into a three-branched fracture (Fig. 11-11). As the mantle current flows beneath the rifting lithosphere, two of the branches separate further, while the third becomes inactive. This inactive branch forms a linear depression known as an *aulacogen,* or *failed arm.* Many of the world's great river valleys, such as the Mississippi and the Amazon, may comprise failed arms of ancient rifts. Intraplate earthquakes that have occurred in the New Madrid area of Missouri and Tennessee (see Highlight 10-1) may be due to the release of stresses built up along such an ancient fracture.

The two active branches of a rift are marked by high heat flow, normal faulting, frequent shallow earthquakes, and widespread basaltic volcanism. As rifting progresses, the rift valley widens and expands until it reaches an adjacent ocean, which floods into the low-lying graben. At this stage, the flooded graben forms a linear sea, such as the present-day Red Sea and Gulf of Aden. The apparently inactive arm of the rift zone encompassing the Red Sea and the Gulf of Aden is the Great Rift Valley of East Africa (Fig. 11-12).

As the landmasses diverge even further, the rift margins subside. The one-time rift edges become the foundations for continental shelves, accumulating sediment from the eroding horsts above them. By the time rifting ends, the original rift edges are no longer plate margins, nor are they volcanically or seismically active. Instead, they are **passive continental margins.** The passive margins of the Atlantic Ocean (the east coasts of North and South America and the west coasts of Europe and Africa) developed in this way after their initial rifting 180 million years ago. The actual plate margins appear half an ocean away, at the still-active spreading center.

For reasons that we do not understand, sometimes rifting ceases following an initial period of faulting and volcanism. One such aborted rift is believed to have begun in North America about 1.1 billion years ago, as an outpouring of mafic lava reached the surface in the middle of the continent. We can find basalts and gabbros of this age along the north and south shores of Lake Superior in Minnesota and Wisconsin. A linear zone of similar rocks continues further south, buried beneath younger Cambrian rocks in Iowa, Missouri, and the Oklahoma panhandle, forming what may have been the axis of the rift. Had this rifting continued, Midwesterners today might be taking trans-oceanic flights between Milwaukee and Minneapolis.

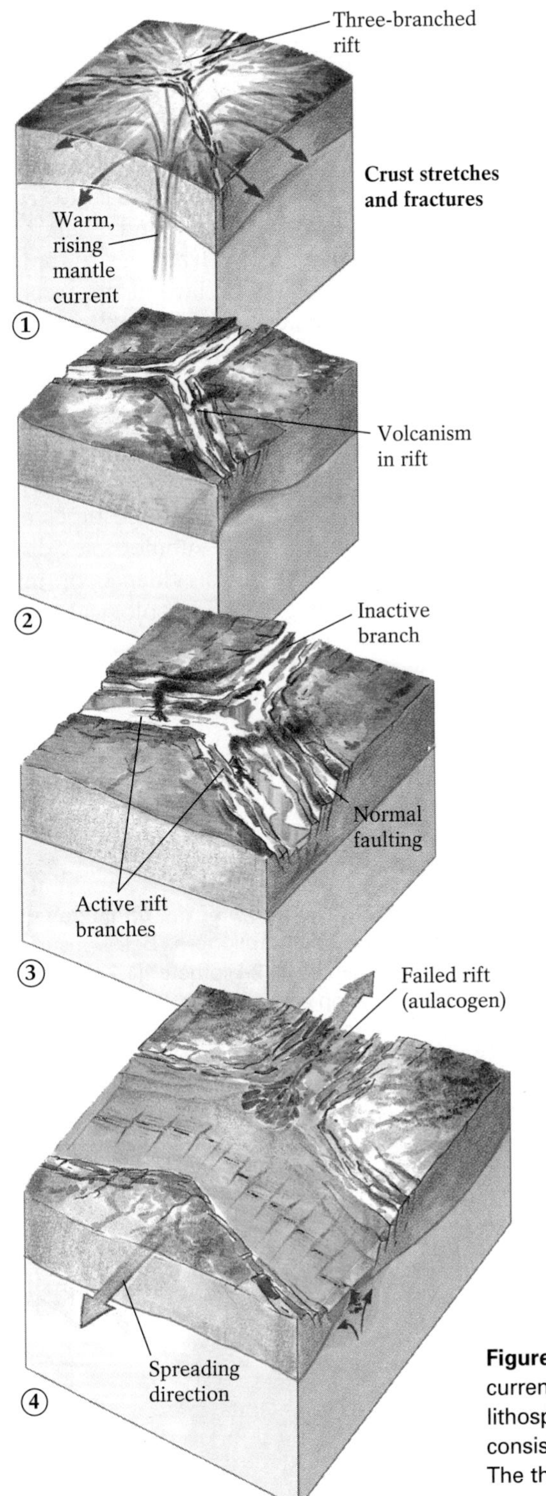

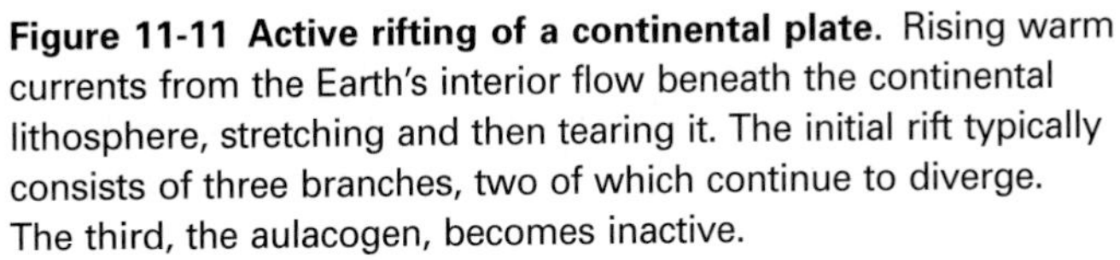
Figure 11-11 Active rifting of a continental plate. Rising warm currents from the Earth's interior flow beneath the continental lithosphere, stretching and then tearing it. The initial rift typically consists of three branches, two of which continue to diverge. The third, the aulacogen, becomes inactive.

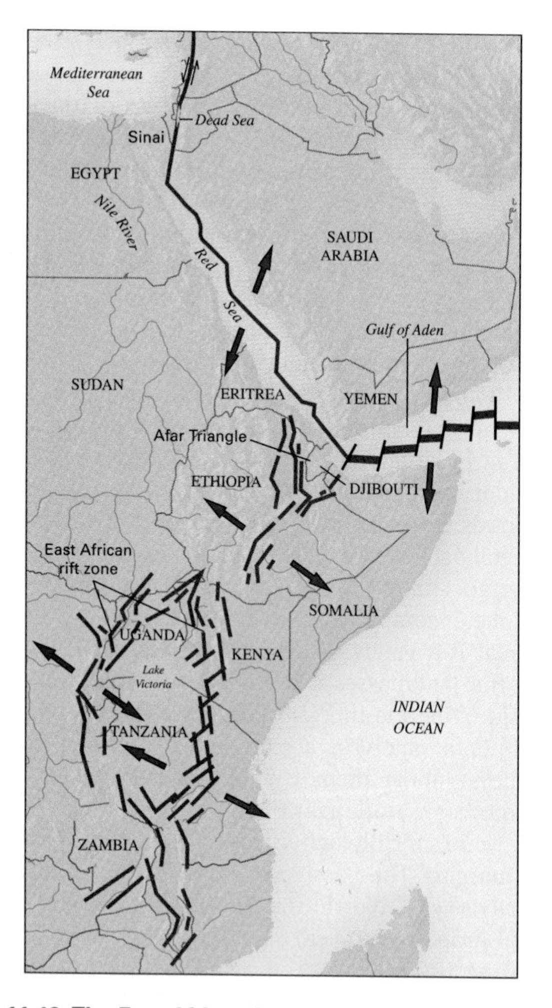

Figure 11-12 The East Africa rift zone. This rift has two actively expanding fractures—the Red Sea and the Gulf of Aden. The valleys and gorges of Kenya and Tanzania are the failed arm, or aulacogen, of the rift.

Divergent Plate Boundaries and the Development of the Ocean Floor

As rifted plates diverge, oceanic lithosphere forms at the spreading center between the two new plate margins. Oceanic lithosphere everywhere is remarkably consistent in structure and thickness, comprising a layered sequence of unconsolidated sediment, basaltic pillow lava, sheeted basalt dikes, gabbro, and peridotite. The rocks may become altered chemically as seawater penetrates the lithosphere's dikes, faults, and fissures, eventually producing serpentinite, a soft, greenish metamorphic rock with a contorted, snake-like appearance. The geological name for the group of rocks that constitute oceanic lithosphere is the **ophiolite suite** (Fig. 11-13).

As an ocean basin grows, an underwater mountain range, or **mid-ocean ridge,** forms from the continuing eruption and accumulation of basaltic lava at the spreading center. Today, mid-ocean ridges stretch continuously for about 65,000 kilometers (40,000 miles) across all major ocean basins. They can span a distance as wide as 1500 kilometers (900 miles), and in places their peaks rise more than 3 kilometers (2 miles) from the ocean bottom. Most ridges are split down the middle by an *axial rift valley,* some of which are deeper than the Grand Canyon and three times as wide as that continental feature. Dives by submersibles directly into these valleys have brought back evidence of remarkable, previously unknown biological and geological processes, as described in Highlight 11-1.

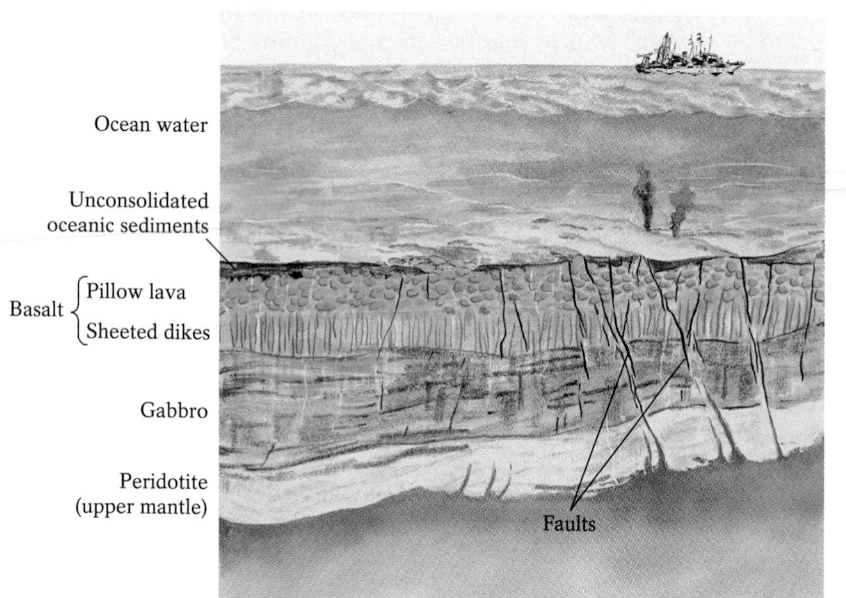

Figure 11-13 The layers of the ophiolite suite, which make up oceanic lithosphere. Below a surface covering of sediments is a 2-kilometer (1.2-mile)-thick layer of basalt, its top characteristically pillowed from its eruption as lava under water. Under the basalt lie 5–6 kilometers (3–4 miles) of gabbro, formed from slow plutonic crystallization and crystal settling. A layer of the ultramafic mantle rock peridotite appears at the base of a typical ocean plate.

Highlight 11-1 The Unseen World of Divergent Zones

Since the early 1970s, oceanographers using deep-sea submersibles have studied the rift valleys of the mid-Atlantic ridge, the Galápagos ridge off the coast of Ecuador and Peru, and the East Pacific rise south of Baja California. They have photographed the eruption of pillow lavas and the chemical interaction of cold seawater and warm basalt. Recent dives into the valleys of the East Pacific rise and the Juan de Fuca ridge (off the coast of Washington state) yielded videos showing *black smokers,* black sulfurous plumes of mineral-laden water rising from vertical chimney-like structures (Fig. 1).

Black smokers develop where seawater seeps down into newly formed oceanic crust, becomes superheated by the underlying magma reservoir, and then dissolves copper, iron, zinc, cobalt, silver, and cadmium out of the warm mafic rocks. These mineral-rich waters, when heated to about 400°C (750°F), rise and erupt at hydrothermal vents in the sea floor. The chimney-like structures from which the black smokers emanate are created when the plumes come in contact with cold seawater, precipitating minerals that encrust the basalt flows around each vent. Similar ore combinations are found within many of the Earth's folded mountains; it is likely that these deposits are remnants of ancient oceanic crust that originated in this manner.

The extremely hot water, volcanic gases such as sulfur dioxide and water vapor, and numerous metallic sulfides that are emitted at hydrothermal vents nourish a complex community of hitherto unknown life forms. Giant clams and exotic tube worms feed on bacteria that thrive in high temperatures, in a world untouched by sunlight (Fig. 2).

Figure 1 Black smokers. These features consist of hot plumes of mineral-rich water vented at volcanically active regions of the sea floor. Accumulations of precipitated minerals form chimney-like structures around the plumes. The "smoke" here consists of hot water and particles of iron, copper, and other sulfide minerals.

Figure 2 Life around black smokers. Bacteria around hydrothermal vents derive energy from heat-generated chemical reactions involving compounds such as hydrogen sulfide (H_2S). More complex creatures, such as these giant tube worms, subsist on these bacteria.

The Earth's mid-ocean ridge system, although continuous, does not form a smooth line; rather, it consists of countless segments interrupted by abrupt perpendicular offsets. These offsets are transform faults where plate segments are sliding past one another. Where adjacent plate segments move in opposite directions, which happens only *between* the offset ridge segments, shearing stress builds up and earthquakes may occur (Fig. 11-14). Beyond the offset ridge segments, adjacent plate segments move in the same direction and earthquakes rarely occur.

Convergence and Subducting Plate Margins

Oceanic plates diverging from mid-ocean ridges may eventually encounter less-dense plates and subduct under them, becoming reabsorbed into the Earth's mantle. Large segments of the ocean floor have been swallowed in this way. As oceanic plates descend, the overriding continental plates become elevated.

The Pacific Example The Pacific is a shrinking ocean, with subduction zones along most of its margins. The spreading ridge of the East Pacific rise no longer appears in the ocean's center, because subduction along the west coasts of the Americas has consumed the eastern part of the Pacific. (In contrast, none of the lithosphere formed at the mid-Atlantic ridge is presently subducting.) The large-scale subduction of Pacific oceanic lithosphere has produced the great mountain ranges of the Andes, Sierra Nevada, Cascades, and Rockies.

Features of Subduction Zones The subduction process produces a wide variety of characteristic features, as illustrated in Figure 11-15. Subduction zones are marked by **ocean trenches,** deep, curvilinear, relatively narrow depressions in the Earth's surface. Some Pacific trenches, particularly in the tectonically active western region, are between 40 and 120 kilometers (25–75 miles) wide and thousands of kilometers long. The deepest is the 11,022-meter (36,161-foot)-deep Marianas trench near the island of Guam. Some trenches, such as the one off the coast of Oregon and Washington, are filled with continentally derived sediments.

As an oceanic plate subducts, rocks of the ophiolite suite and fine-grained, deep-sea sediments, along with coarser, land-derived sediments from the overriding plate, become packed against the inner wall of the trench. Caught in the high-pressure zone between the converging plates, the resulting mass solidifies to become rocky material that is sliced, crushed, and thrust into a chaotic jumble called a *mélange* (Fig. 11-16). Some mélange rocks subduct to depths of 30 kilometers (20 miles) or more, where they metamorphose under the unique combination of low-temperature and high-pressure conditions to form the characteristic subduction-zone metamorphic facies *blueschist.*

The accumulation of mélange rocks in an ocean trench forms a wedge that becomes plastered onto the edge of the overriding plate. This *accretionary wedge* may eventually become uplifted to form a linear range of mountains just inland of the trench; the coast ranges of California, such as those located near Big Sur, originated in this way.

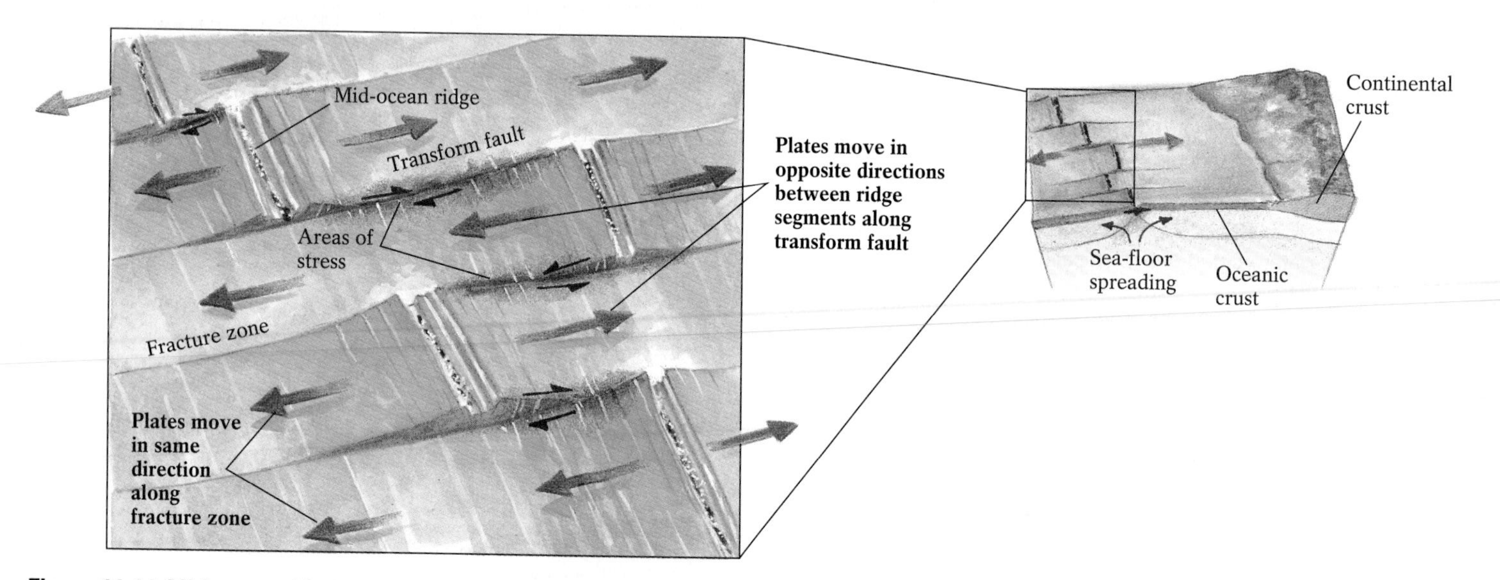

Figure 11-14 Mid-ocean ridge system motion. The direction of plate motion on either side of an oceanic transform boundary is determined by spreading from mid-ocean ridge segments. The plates move in opposite directions only between these segments, so the stresses associated with transform movements build up and earthquakes occur here. Fracture zones are the seismically inactive portions of transform boundaries beyond offset ridge segments, where adjacent plates move in the same direction.

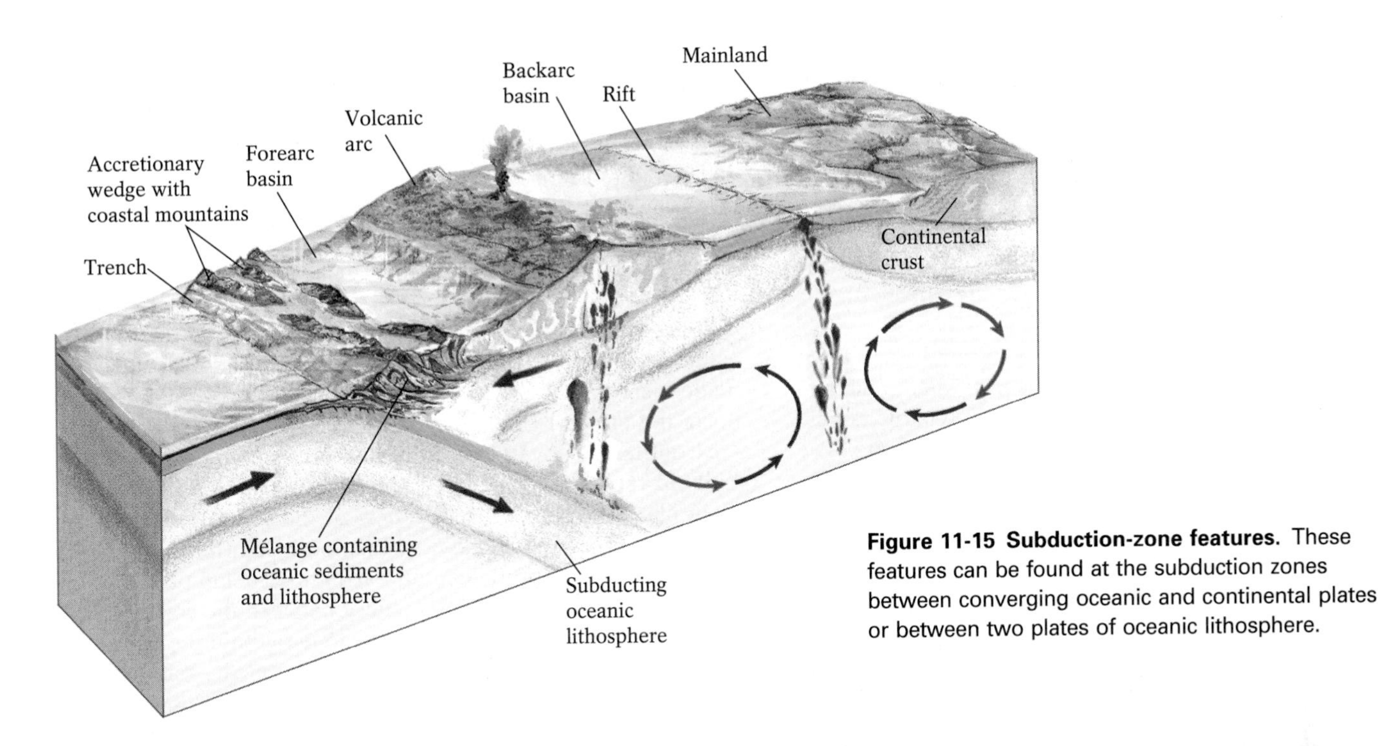

Figure 11-15 Subduction-zone features. These features can be found at the subduction zones between converging oceanic and continental plates or between two plates of oceanic lithosphere.

A **volcanic arc,** a chain of volcanoes fueled by magmas melting and rising from the subducting plate, may form farther inland from the accretionary wedge. Between the accretionary wedge and volcanic arc is a sediment-trapping depression called the *forearc basin,* in which sediment eroded from both the uplifted accretionary wedge and the volcanic arc builds up. On the inland side of a volcanic arc, another sediment-trapping depression, or *backarc basin,* forms. The sediments deposited here are derived both from the eroding volcanic arc and from continental streams flowing toward the sea.

Convergence and Continental Collisions

Given enough time, the entire oceanic portion of any subducting plate will become reabsorbed into the Earth's mantle. At this point, the plate's continental portion (if it has one) reaches the subduction zone and may encounter another continental plate, resulting in continental collision. A continental plate may temporarily subduct, being pulled downward by the subducting oceanic lithosphere. Because continental plates have relatively low density and are buoyant, however, they tend to float back to the surface. The two lightweight continents then compress one another's rocky boundaries, becoming welded together into a single larger block of continental rock. A lofty mountain range marks the site of the collision.

The boundary between collided continents is called a **suture zone.** During a collision, continental crust within a suture zone thickens, because one plate is thrust slightly beneath the other. In addition, this crust thickens as slices faulted from the colliding edges of both plates become swept into fold-and-thrust mountains. Because fragments of an oceanic plate are often trapped between colliding plates and

Figure 11-16 A mélange along the Sonoma County coast in California. These rocks—a jumbled mixture of sea-floor and land-derived materials that have undergone high-pressure/low-temperature metamorphism—formed in the ocean trench associated with the subduction of the Farallon plate under North America.

Highlight 11-2 Convergence and the Birth of the Himalayas

The creation of the Himalaya Mountains, formed from the convergence, collision, and suturing of India to Asia over millions of years, represents a significant event in the geologic history of the Earth (Fig. 1). Among its other effects, the uplift of this mountain chain forever changed the climate of Asia by isolating it from the southern oceans.

The process leading to the Himalayan orogeny began about 180 million years ago, during the Mesozoic Era, when India and Madagascar separated from Pangaea and began to drift rapidly northward. (Soon after, Madagascar rifted away from India.) The Tethys Ocean, which separated the Indian continent from Asia, was being consumed along a subduction zone south of Asia. The drifting India passed over the stationary Reunion hot spot about 65 million years ago, a period marked by the eruption of the extensive Deccan basalt flows. (Today, the Reunion hot spot is about 5000 kilometers [3000 miles] southwest of India in the Indian Ocean.) For the next 30 million years, India moved rapidly northward as subduction along the southern edge of Asia continued.

The collision of India with Asia began some 35 million years ago, when the northern margin of India collided with the Tibetan microcontinent. By about 10 million years ago, the Tethys Ocean had closed completely and the Himalayan orogeny entered its next stage, which continues to the present day. Sediments on the leading edge of the Indian plate and accretionary wedges, volcanic arcs, and batholiths along the southern edge of the Asian plate were folded and thrust-faulted up onto the continents to form the modern Himalayan mountain belt.

Continued convergence of India into the Asian continent results in repeated thrust faulting of the leading edge of the Indian plate, increasing the thickness of the continental crust beneath the emerging Himalayas. This increased thickness is seen today as the highest plateau on Earth, the Tibetan plateau. Several suture zones mark the boundary between the colliding continents and continental fragments trapped within the collisional zone. India is still moving into Asia today, at the rate of about 5 centimeters (2 inches) per year.

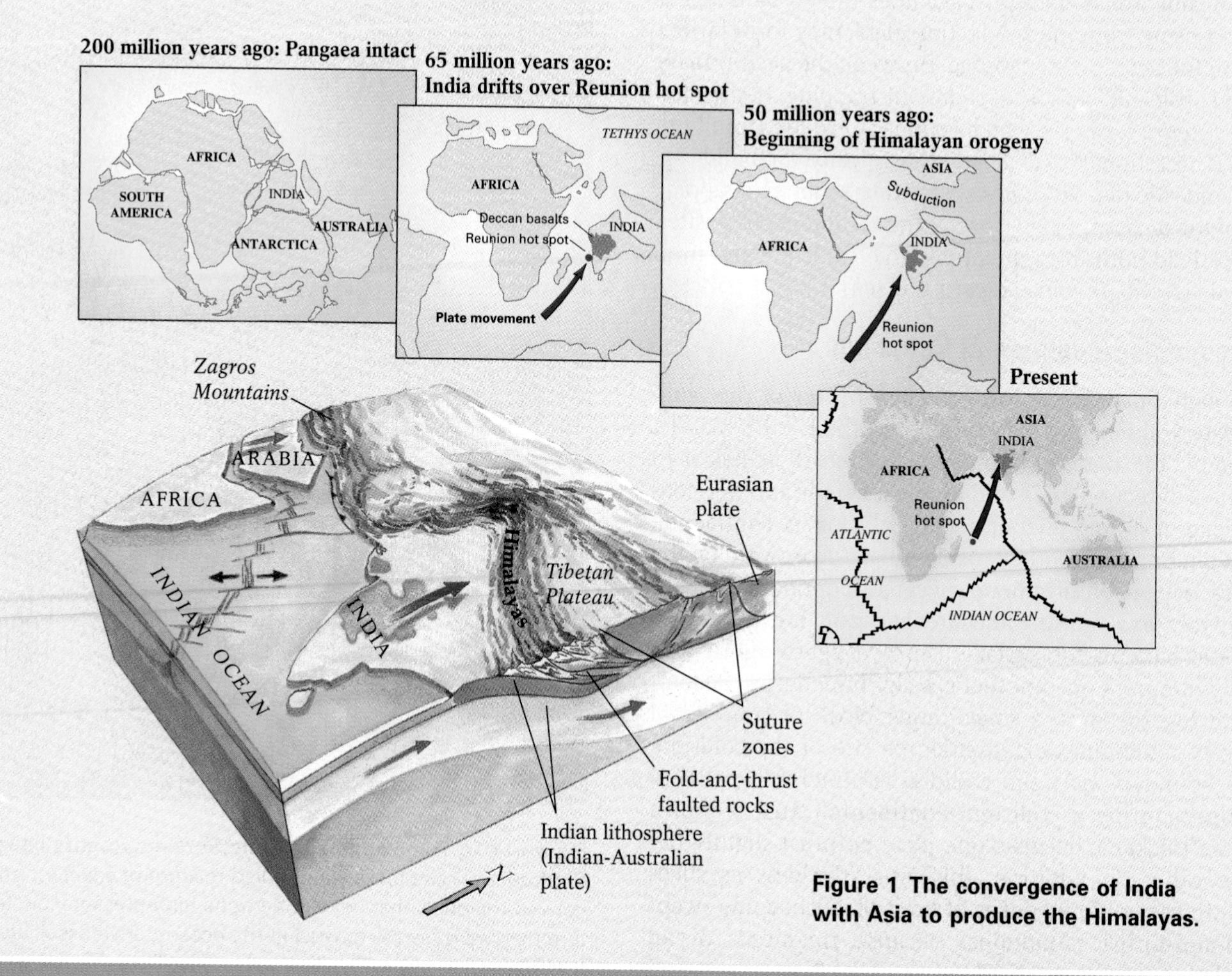

Figure 1 The convergence of India with Asia to produce the Himalayas.

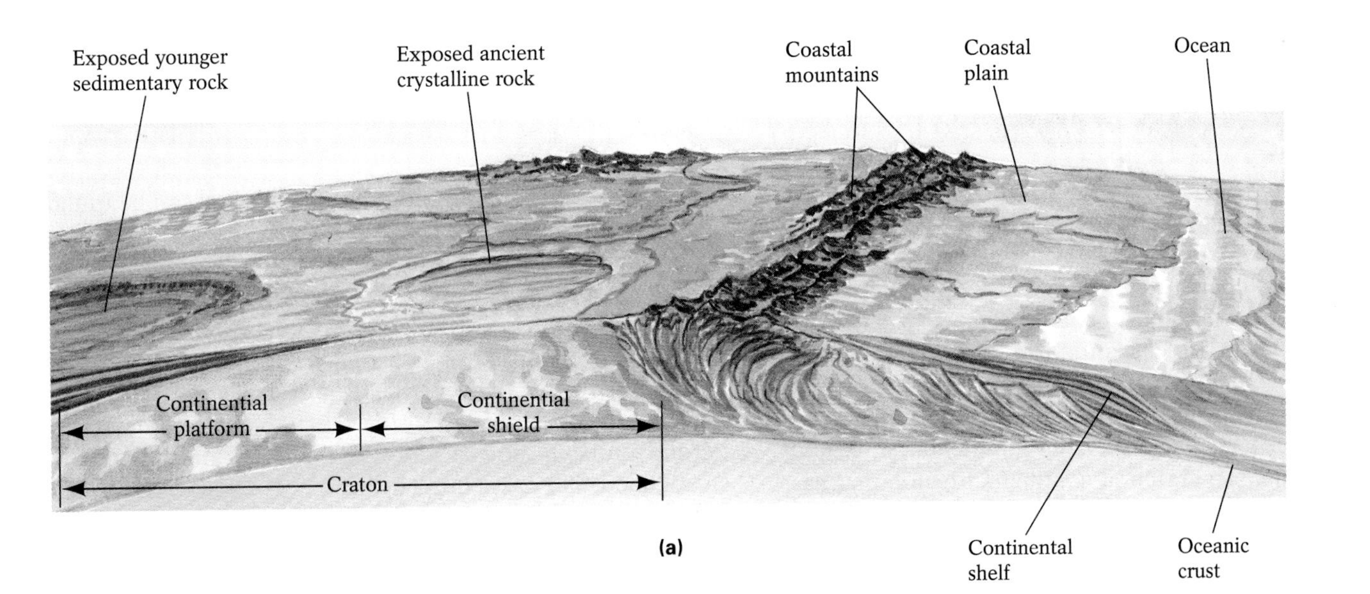

(a)

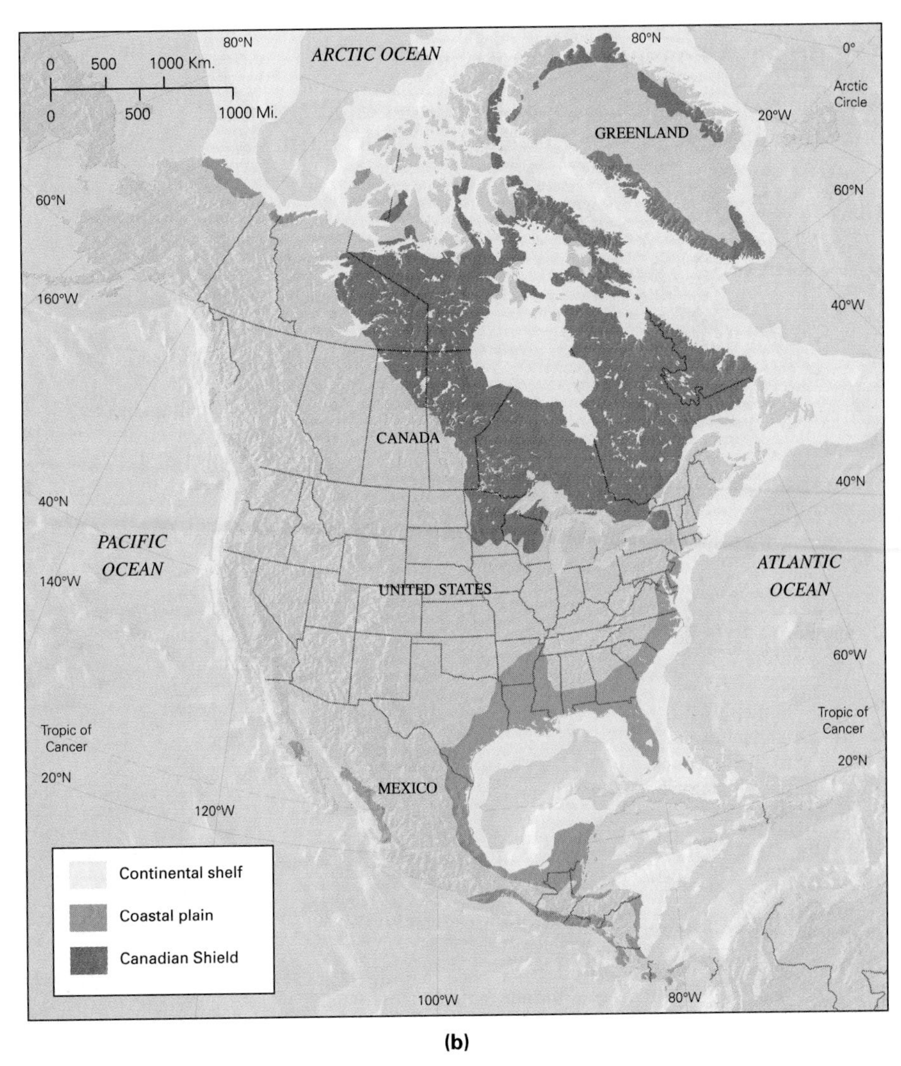

(b)

Figure 11-17 The anatomy of a continent. **(a)** Every continent contains a tectonically stable nucleus, or craton, that consists of one or more ancient, crystalline continental shields and a surrounding platform of sedimentary rock. The outer edge of the continent is typically marked by coastal mountains and plains and a submarine continental shelf. **(b)** The Canadian Shield, North America's ancient crystalline nucleus, is exposed throughout much of Canada as well as northern Minnesota, Wisconsin, and New York.

become attached to the continents, inland mountains rising high above sea level sometimes contain large masses of ophiolite rocks. The complex folded and thrust-faulted mountains in suture zones may also swallow up anything else that lies in the path of the colliding plates. As a result, remnants of stratovolcanoes and felsic batholiths from collided volcanic arcs, or metamorphosed mélanges from the subduction trenches of departed oceanic plates, can sometimes be identified within a mountain's rocky structure.

Continental collisions produced many of the present-day Earth's mountain ranges, including the Alps in southern Europe and the Appalachians in eastern North America. Today, continental plates are actively colliding in only one place: from the northern shore of the eastern Mediterranean, where Africa impinges on southern Europe, across to the Himalayas. The world's highest mountains, the Himalayas of southern Asia, were formed from and continue to grow because of the ongoing continental collision between China and India (see Figure 11-1 and Highlight 11-2). This collision produces frequent earthquakes, including Armenia's devastating tremors in the late 1980s.

The Origin and Shaping of Continents

Every continent has the same basic components (Fig. 11-17). The oldest parts are the **continental shields,** broad areas of exposed crystalline rock in continental interiors that have not changed appreciably for more than a billion years. Every continent contains at least one large shield area. For instance, North America's Canadian Shield extends across much of Canada from Manitoba to the Atlantic coast, dipping into the northern United States from northern Minnesota and Wisconsin to upstate New York's Adirondack Mountains.

Surrounding the continental shield is the **continental platform,** where a veneer of younger sedimentary rock covers the continental shield. Together, the continental shield and platform constitute the **craton,** a continental region that has remained tectonically stable for a vast period of time. At the edges of the craton, near the borders of continents, one finds coastal mountains, coastal plains, and continental shelves.

The Origin of Continental Lithosphere

Because continents do not subduct, continental lithosphere, once formed, becomes a permanent part of the Earth's surface. As a result, rocks as old as 4 billion years can be found on the continents. They provide clues to very early stages in the Earth's history. In the Earth's first few hundred million years of existence, however, no continents existed, nor were there oceans or even an atmosphere like our present one.

About 4 billion years ago, the Earth's newly formed mantle was hotter and less viscous than the present-day mantle and probably flowed more rapidly. Enormous clouds of steam and other gases erupted from it, leading to the beginnings of our present atmosphere and condensing to create the first global ocean. Vigorous mantle convection currents brought great volumes of magma to the surface, and mafic and ultramafic lava spewed from numerous hot spots across the Earth's surface. Some of this lava solidified, forming the first landmasses. These bodies probably consisted of huge basaltic volcanic islands that were too hot, and thus too buoyant, to subduct (Fig. 11-18).

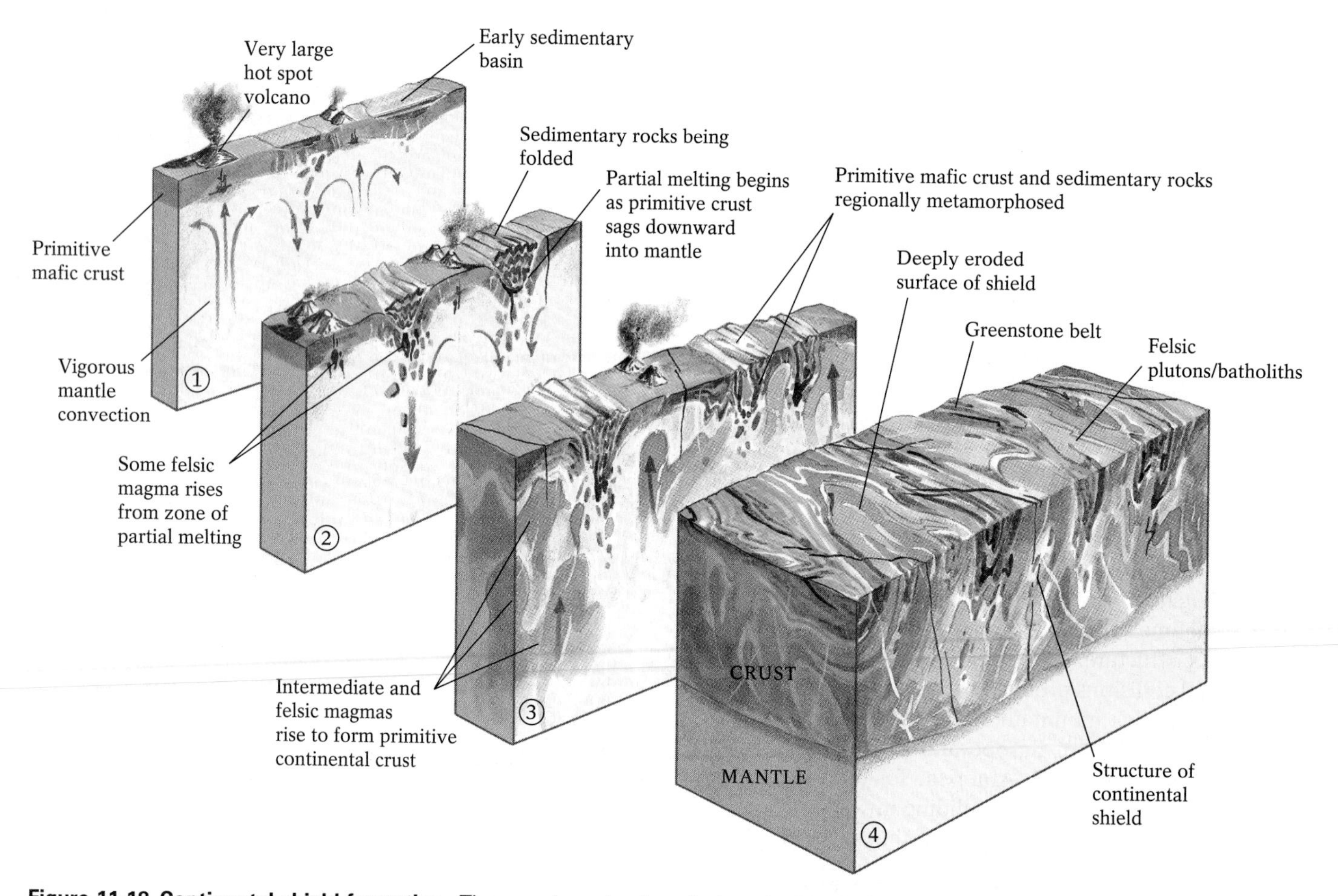

Figure 11-18 Continental shield formation. These regions developed when the Earth's interior was considerably hotter than it is today. Vigorous convection brought warm mantle material to the surface, where it erupted as mafic and ultramafic lavas. As the flows cooled and became denser, they subsided into deeper regions. There they partially melted and differentiated, eventually solidifying as intermediate and felsic plutons. These ancient, low-density plutons are the oldest rocks on Earth and, together with metamorphic greenstone belts, make up continental shields.

As basaltic eruptions continued, the islands grew larger and thickened, and early sedimentary basins developed around them. Some of these islands may have collided to form larger landmasses. The heat and pressure from the collisions metamorphosed both the rocks of the islands and the basin sediments. As older sections of land cooled and became more dense, some parts sagged downward into the hotter interior, folding and dragging the sedimentary basins along with them. The basalts and ultramafic materials partially melted, then became differentiated by density to produce the first intermediate and felsic magmas. Continued partial melting and differentiation of these rock materials eventually created new sets of volcanic islands that consisted largely of low-grade metamorphic basalts called greenstones and felsic plutons of varying compositions. This combination of rock types is found in most ancient continental shields.

After small, lower-density felsic landmasses developed, subduction began to occur at regions where dense oceanic lithosphere and the light continental lithosphere converged. In time, the continental nuclei became enlarged as subduction-zone mélanges accumulated and became attached to continental interiors, and as converging continental plates collided and coalesced. Many of the landmasses subsequently rifted, creating even more centers around which additional subduction and growth took place, which further increased the continental lithosphere. These centers of growth provided the Earth's stable shields, nearly all of which were created during the Precambrian Era. Very little continental material has been added to the Earth in the past 600 million years, probably because the planet's internal heat has diminished and plate tectonic activity has slowed considerably.

The Changing Shape of the Earth's Continents

Bordering North America on both the Atlantic and Pacific coasts are rocks that are no more than 600 million years old. Recently gathered evidence indicates that many of these fault-bounded rock bodies, called **displaced terranes,** originated elsewhere but were transported by plate motion and attached, or *accreted,* to our continent by collisions. Some may have been island arcs that were formed in an ocean basin and then towed to a continental margin by subduction of the intervening ocean plate. Others are probably **microcontinents,** pieces of continental lithosphere broken from larger distant continents by rifting or transform faulting.

Displaced terranes can be distinguished from their surrounding rocks by their different ages, geological structures, stratigraphies, fossil assemblages, and/or magnetic properties. More than 100 such rock bodies of various sizes have been identified along western North America from Alaska to California (Fig. 11-19). Most apparently accreted to the continent within the last 200 million years, when Pangaea rifted apart and the North American plate began to drift westward. The displaced terranes of eastern North America, on the other hand, have a longer and more complex tectonic history. The Appalachian mountain system, for example, is believed to contain terranes acquired during the several stages of the closing of the Atlantic Ocean between 500 and 250 million years ago (see Highlight 9-1). The eastern Canadian provinces and parts of New England contain displaced terranes believed to have accreted to the eastern part of the continent from about 600 to 150 million years ago.

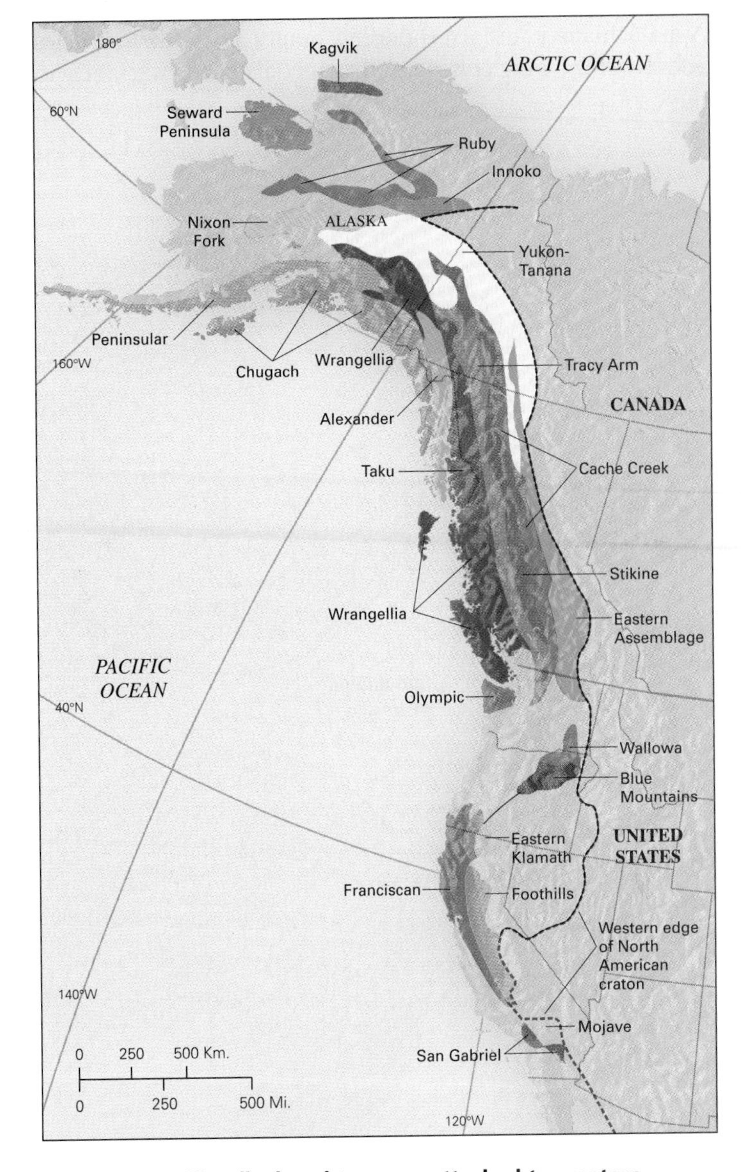

Figure 11-19 The displaced terranes attached to western North America. They are believed to be the remnants of assorted microcontinents, islands, and other unsubductable landmasses that became attached to the continent during subduction of past Pacific plates.

The Formation of Pangaea

Some 500 million years ago, a supercontinent, Gondwana, formed somewhere near the South Pole. Gondwana contained all of the planet's Southern Hemisphere landmasses. At the same time, three northern landmasses formed, each probably separated from the others by a sizable ocean. These independent continents—which would eventually join to create Laurasia, the northern half of Pangaea—included most of what is now North America, northern Europe, and a combination of southern Europe and parts of Africa and Siberia.

The pre-Laurasian landmasses began converging about 420 million years ago. Their collision produced the northern Appalachians and corresponding ranges in the British Isles and Norway, and contributed some displaced terranes to northeastern North America. Siberia collided with northern Europe, completing the formation of Laurasia and producing the Ural Mountains of central Russia. During the next 100 million years, Laurasia and Gondwana collided to form the supercontinent Pangaea (Fig. 11-20). The final collision in the assembly of Pangaea involved three landmasses: Africa, a separate continent that is now southeastern North America, and North America. This collision created the southern Appalachians. By about 225 million years ago, at the close of the Paleozoic Era, Pangaea was a single vast landmass that stretched from pole to pole. Virtually all of the Earth's continental lithosphere remained joined together for the next 50 million years or so, until Pangaea began to rift and water flowed in to form the modern Atlantic, Indian, and Antarctic oceans.

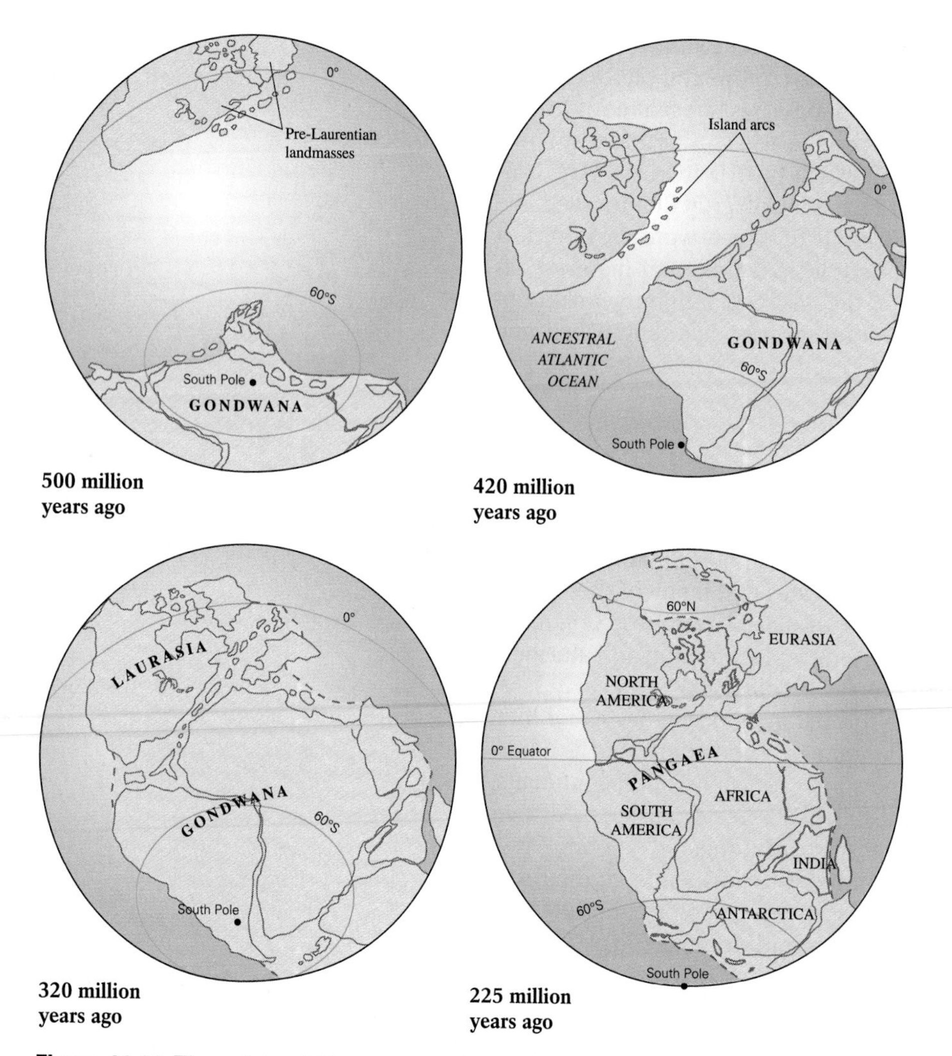

Figure 11-20 The origin of the supercontinent Pangaea.

The Driving Forces of Plate Motion

We have documented conclusively that plates move, and we know a great deal about the plate tectonic history of the Earth; nevertheless, we cannot explain with certainty what drives plate motion. As we saw in Chapter 1, heated material rising from the Earth's mantle by *convection* probably sets plates in motion. Convection is the flow of currents in a fluid caused by variations in their temperature. Hot materials are less dense and therefore rise, while cold materials are more dense and therefore sink. Although slow convection currents in the warm asthenosphere produce movement of the cold, brittle lithosphere directly above it, we do not know whether convection cells within the asthenosphere are solely responsible for plate tectonics. Even our knowledge about the nature of these convection cells remains limited.

Although many geologists accept that convective flow in the mantle serves as an important mechanism of plate movement, some question whether convection cells can exist within the solid lower mantle. If they do, does the mantle flow rapidly enough to drive the Earth's plates at their observed velocities? Advocates of mantle convection continue to question the dimensions of convection cells (Fig. 11-21). Are they shallow, remaining largely confined to the soft, mobile asthenosphere? Do they extend down to 700 kilometers (450 miles), the depth to which subducting plates are known to penetrate? Do they fill the entire mantle, being energized by heat rising from the Earth's liquid outer core? One model proposes that convection cells are stacked in two tiers, with large, deep, slow-moving cells transmitting heat from below to drive the small, shallow, fast-moving cells above them, which carry the Earth's lithospheric plates. An increasing number of hypotheses are casting doubt upon the simple convection-cell model, proposed during the early years of the plate tectonic revolution, which requires cold continental lithospheric plates to move as passengers on warm currents traversing a shallow asthenosphere.

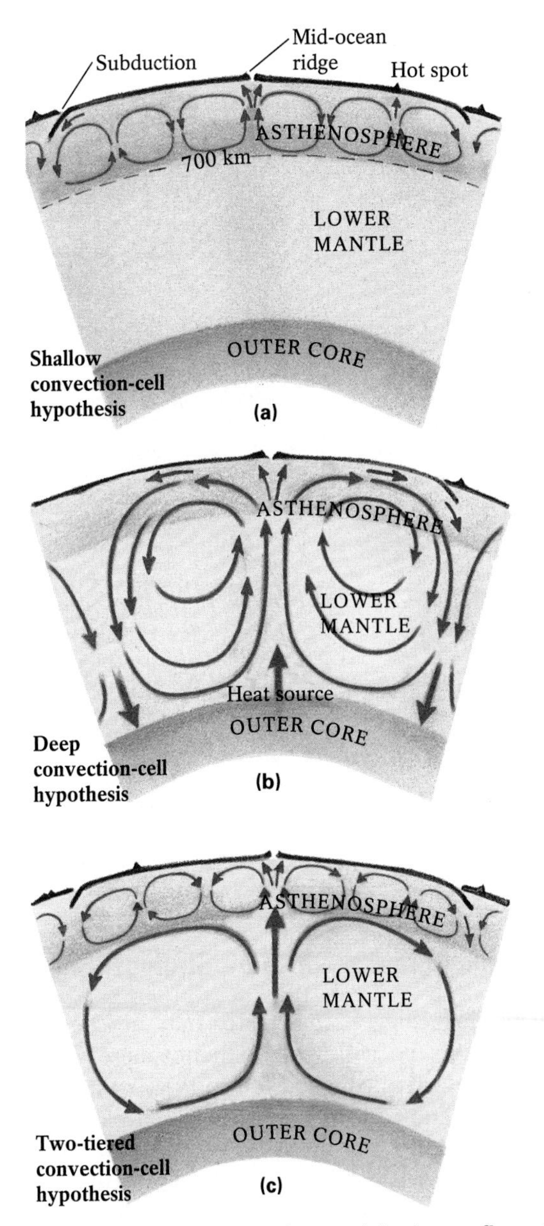

Figure 11-21 Hypotheses proposed to explain the configuration of the Earth's convection cells. (a) The cells may be shallow, confined to the asthenosphere. **(b)** The cells may extend through the entire mantle to the outer core. **(c)** Some hypotheses propose two sets of convection cells that meet at a depth of 670 kilometers (420 miles), the zone within the mantle where seismic waves are known to accelerate in response to changes in the chemistry and structure of mantle materials.

Thermal Plumes—A Possible Alternative to Convection Cells

Some geophysicists have proposed that deep-Earth heat rises through the asthenosphere as scattered vertical columns of warm upwelling mantle material called **thermal plumes,** rather than as distinct convection cells. These narrow plumes, 200 to 400 kilometers (125–250 miles) in diameter, may originate within the asthenosphere or at the mantle–core boundary and then rise beneath both continents and oceans, at plate boundaries as well as plate interiors. Thermal plumes are generally believed to lift up the overlying lithosphere, which becomes domed. They then spread laterally beneath the lithosphere, applying a dragging force to the base of the lithosphere that causes it to rift.

The thermal-plume hypothesis separates divergence from subduction completely. In this view, cooling mantle material descends slowly through the entire mantle—not, as in the convection-cell hypothesis, at individual plate boundaries above the descending arcs of convection cells.

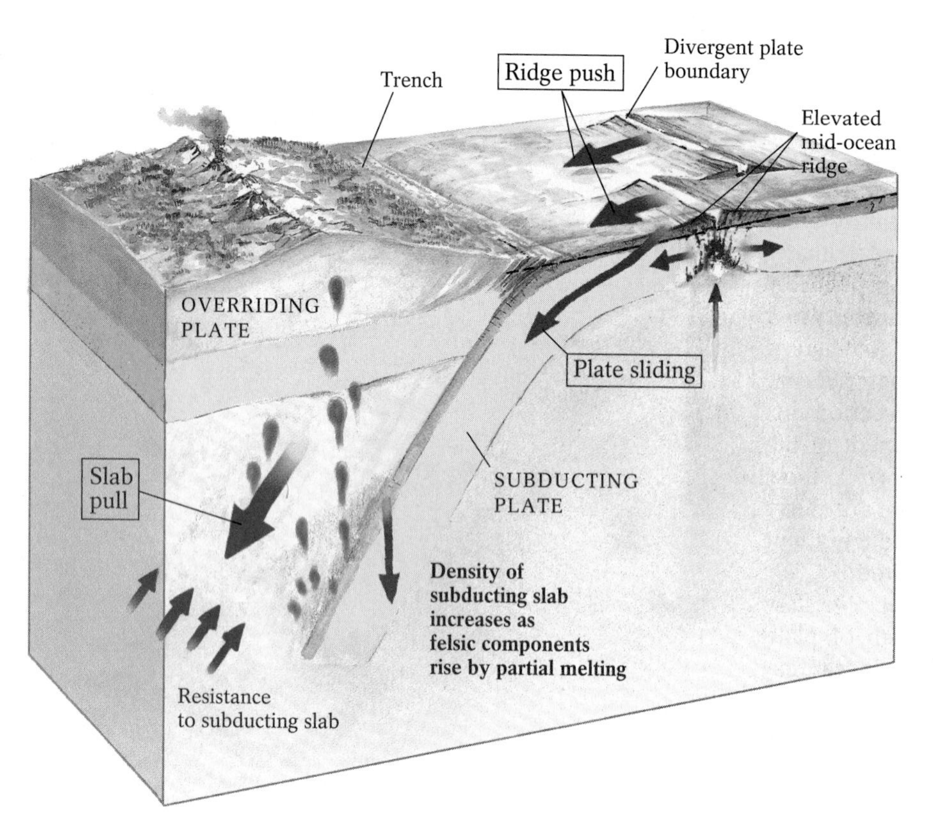

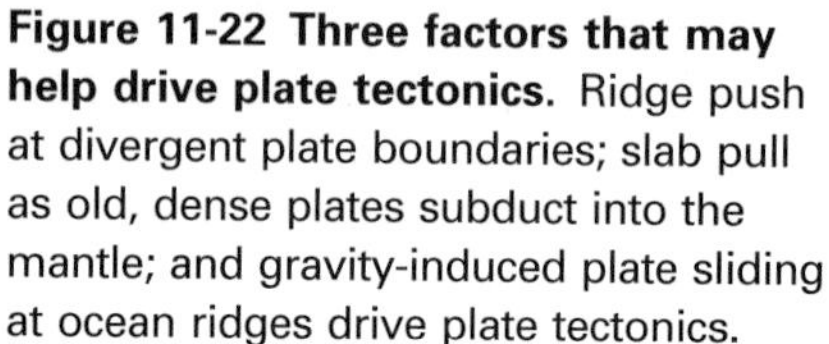

Figure 11-22 Three factors that may help drive plate tectonics. Ridge push at divergent plate boundaries; slab pull as old, dense plates subduct into the mantle; and gravity-induced plate sliding at ocean ridges drive plate tectonics.

Ridge Push, Slab Pull, or Plate Sliding?

Recent findings suggest that plates may not be simply passive hitchhikers on a flowing asthenosphere; instead, they may make an active contribution to their own mobility. Figure 11-22 illustrates three ways in which plates may affect their own motion.

When an ocean grows at a divergent plate boundary, does the rising magma wedge itself between adjoining oceanic plates and actively push them apart? If plates are being pushed, we might expect them to fold up accordion-style, which doesn't happen. In fact, ocean-exploring submersibles diving deep into axial rift valleys have recently identified thousands of tears and fissures within oceanic lithosphere. This evidence strongly suggests that plates are being pulled and stretched at divergent boundaries, not pushed and compressed. Old, cold oceanic lithosphere is denser than warm asthenosphere and therefore sinks. The density of the lower portion of a descending slab also increases as some of its light felsic components melt and rise, thereby concentrating its denser mafic components. The increased density may actively tow oceanic plates into their subduction-zone graves.

Indeed, recent studies show that rates of plate motion are proportionate to the amount of a plate's margin undergoing subduction: The faster-moving plates—typically oceanic plates—are those that are subducting along a significant portion of their margins. This finding supports the hypothesis that gravitational pull on a descending plate contributes to general plate motion. It seems unlikely, however, that this force is powerful enough to influence divergence at the opposite side of the plate, thousands of kilometers away. Moreover, lithospheric plates are so brittle that they would probably break if they were subjected to such large-scale pulling forces for thousands of kilometers. Furthermore, the Atlantic Ocean's floor diverges from the mid-Atlantic ridge at a rate of about 2 centimeters (nearly 1 inch) per year, though no slabs sink along any of its margins.

Another hypothesis suggests that gravity causes oceanic lithosphere to move laterally away from elevated mid-ocean ridges. The continuing eruption of basaltic lava at ocean ridges produces high mountain ranges of still-warm, low-density rocks. Newly formed oceanic lithosphere may be literally sliding down the slope of the uplifted asthenosphere beneath ridge crests. Indeed, mathematical calculations indicate that oceanic plates could slide down even gentler slopes at a rate of several centimeters per year, or roughly the observed rate of divergence for the mid-Atlantic ridge. In the initial stages of rifting, however, plates begin to diverge before an oceanic ridge is created. Hence, gravity alone cannot account for plate motion.

Plate-Driving Mechanisms: A Combined Model

It seems unlikely that a single unified model can explain all plate movements. More likely, the driving mechanisms of plate tectonics combine ridge push, slab pull, gravity-induced sliding, convection cells, and thermal plumes. Some form of convection beneath the lithosphere apparently transports mantle heat upward to drive the mid-ocean plate divergence of Atlantic-style tectonics. Once mantle convection initiates plate motion and oceanic ridge crests develop, gravity may induce the lateral motion of oceanic plates as they slide from topographically high crests over the underlying soft, weak asthenosphere. Gravity also independently enhances plate subduction by returning cold, dense lithosphere to the Earth's interior; this mechanism may be driving Pacific-style tectonics. The motions of continental lithosphere remain the most problematic and warrant further study. In the future, of course, we may learn about even more, as-yet-unknown driving forces.

In the mid-1960s, plate tectonics was an exciting new hypothesis. Since then, it has graduated to a full-fledged theory as Earth scientists have found evidence confirming most of its precepts. Although most scientists now believe that plates exist and move, they continue to explore the oceans and continents intensively in an effort to answer the numerous questions that remain.

Chapter Summary

The movement of the continents, called **continental drift,** was first recognized by the German geophysicist-meteorologist Alfred Wegener in what is considered the first step toward the development of the theory of plate tectonics. Wegener presented several lines of evidence for continental drift—including the complementary outlines of the current continents, similarities between geographically separated animals and rocks, and evidence of changing climates—and modern technology has provided additional support for his theory.

New technology enables Earth scientists to study the results and the mechanisms of plate tectonics. They have several ways to measure the actual velocity of plates: by assessing the patterns of volcanoes that form as plates move over stationary **hot spots,** localized regions where plumes of hot mantle material rise below the base of the lithosphere; by tracking the movement of **marine magnetic anomalies,** alternating stripes of stronger and weaker magnetism along the ocean floor; and by aiming ground-based lasers at reflectors on satellites in fixed orbits.

Our understanding of the Earth's tectonic past comes largely from our growing knowledge of the evolution of the world's ocean floors. Accumulation of geophysical data has helped to clarify the sequence of events that produces ocean basins. Rifting begins when a warm current of mantle material rises under a continental plate, causing it to stretch and thin until it tears. Three radiating rift valleys typically form simultaneously; two of these eventually diverge, creating a new ocean between them, and the third—the failed arm, or aulacogen—commonly becomes inactive. After rifting between two new plates ends, the rift edges become inactive tectonically, forming **passive continental margins.**

The typical rock sequence that develops as new oceanic lithosphere is produced at divergent zones, called the **ophiolite suite,** consists of a layered sequence of unconsolidated sediment, basaltic pillow lava, sheeted basalt dikes, gabbro, and peridotite, sometimes altered chemically by seawater to produce serpentinite. As an ocean basin grows, an underwater mountain range, or **mid-ocean ridge,** develops from the accumulation of basaltic lava at the spreading center.

A diverging oceanic plate gradually cools and becomes more dense, and may eventually sink back into the Earth's interior by subducting beneath a less dense plate far from the mid-ocean ridge. An **ocean trench** develops where a subducting plate flexes downward into the mantle, forming a depression in the Earth's surface. Magma produced by the partial melting of a subducting plate may rise to create a chain of volcanoes, or **volcanic arc,** on the inland side of the trench.

Once an oceanic plate has been completely subducted, the continental blocks on either side collide to form a larger continent. The boundary between them, the **suture zone,** is often marked by a collisional fold-and-thrust mountain range, such as the Himalayas.

The nucleus of a continent consists of ancient bodies of rock called **continental shields;** these bodies are generally surrounded by a **continental platform,** where a veneer of younger sedimentary rocks covers ancient crystalline rocks. Together, the shield and platform constitute the continental **craton,** the tectonically stable portion of a continent. The rocks exposed at the outer edges of the continents are often **displaced terranes,** whose fossils, stratigraphic sequences, rock types, and magnetic signatures suggest that they originated elsewhere. Some displaced terranes are **microcontinents,** pieces of lithosphere broken off from larger continents.

Although abundant evidence confirms the general theory of plate tectonics, new data suggest that plates may not be driven solely by simple convection cells within the mantle. In fact, they may be subject to a combination of forces: convection cells, the pull of gravity at subduction zones and mid-ocean ridges, forceful wedging of rising magma between plates, and **thermal plumes,** rising currents of hot mantle rock.

Key Terms

continental drift (p. 196)
hot spots (p. 199)
marine magnetic anomaly (p. 200)
passive continental margins (p. 203)
ophiolite suite (p. 204)
mid-ocean ridge (p. 204)
ocean trenches (p. 206)
volcanic arc (p. 207)
suture zone (p. 207)
continental shields (p. 209)
continental platform (p. 210)
craton (p. 210)
displaced terranes (p. 211)
microcontinents (p. 211)
thermal plumes (p. 213)

Questions for Review

1. How could you use the distribution of modern and ancient animals to show that continents drift?

2. Why did the scientific community initially reject Wegener's ideas of continental drift? Why were his ideas finally accepted?

3. Describe three methods for determining the velocity of plate motion.

4. What is an aulacogen? How does it form?

5. Describe what happens at a rift margin, from the onset of rifting to the formation of an ocean.

6. Draw a simple sketch of an ophiolite suite (include all of the rock types and structures).

7. Explain why the East Pacific rise, although it is a mid-ocean ridge, is not in the center of the Pacific Ocean.

8. Sketch a convergent boundary between an oceanic plate and a continental plate. Be sure to include all of the important associated landforms.

9. Describe the basic components of a continent. Why are a continent's oldest rocks always in its interior?

10. Briefly describe the fundamental differences between the convection-cell hypothesis and the thermal-plume hypothesis of plate tectonics.

For Further Thought

1. When geologists find ancient glacial deposits in equatorial Africa, they usually interpret them as polar deposits that have drifted from a cold place to a warm place. Formulate another hypothesis to explain this phenomenon.

2. Of the two patterns of marine magnetic anomalies shown below, which shows a faster rate of sea-floor spreading? Explain.

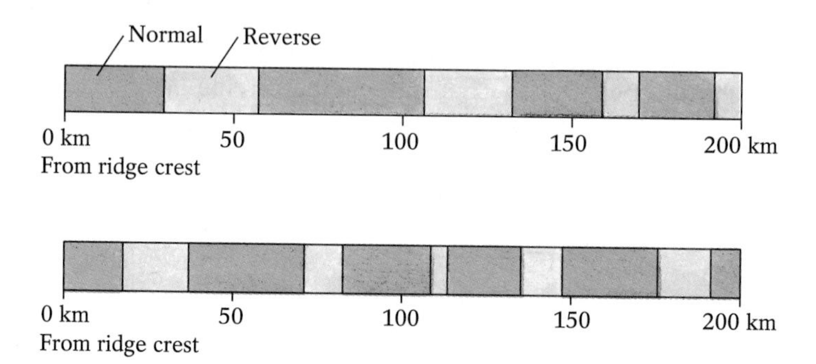

3. What would be some of the major geologic repercussions if a new rift opened between Ohio and Indiana?

4. How would the eastern coast of North America change if the oceanic segments of the plates that make up the Atlantic Ocean basin began to subduct? How would it change if the Atlantic Ocean lithosphere subducted completely?

5. How might the development of a new subduction zone along the East Coast affect plate interactions on the West Coast?

Part 3 Sculpting the Earth's Surface

12

Mass Movement

At 11:37 P.M. on August 17, 1959, an earthquake registering 7.1 on the Richter scale shook thousands of square kilometers in the vicinity of West Yellowstone, Montana. The surface of the ground rolled like sea waves, and the water of lakes and rivers sloshed back and forth. A wall of water rushed across Hebgen Lake and over its dam, sweeping away campgrounds on its way downstream. The quake also triggered a landslide involving more than 80 million tons of rock and weathered regolith, including boulders as large as 9 meters (30 feet) in diameter, as material became dislodged from a steep canyon wall and hurtled downslope at speeds reaching 150 kilometers (90 miles) per hour. As the slide tumbled into Madison Canyon, it compressed and forcefully expelled the air in its path, producing hurricane-force winds that battered the valley. Two-ton cars were blown into the air; one flew more than 10 meters (33 feet) before being dashed against a tree. By the time the slide mass finally came to rest, it had covered the valley floor with 45 meters (150 feet) of bouldery rubble, buried U.S. Highway 287, and taken the lives of 28 campers.

The Madison Canyon disaster is an extreme example of **mass movement,** the process that transports quantities of Earth materials (such as bedrock, loose sediment, and soil) down slopes by the pull of gravity (Fig. 11-1). *Every* slope is susceptible to mass movement. Sometimes the movement is as fast as the Madison Canyon landslide and involves a great mass of material that may travel more than 80 kilometers (50 miles) from its source. More often, the movement is so slow as to be imperceptible and involves just the upper few centimeters of loose soil on a gentle hillside.

Of the 20,000 lives lost as a result of all natural disasters in the United States during the years 1925 to 1975, fewer than 1000 deaths were a direct result of mass movement. Damage associated with mass movement, however, cost a staggering $75 billion, compared with the $20 billion in damage associated with all other natural catastrophes. Thus, although very few of us are likely to perish in a mass-movement event, a strong likelihood exists that we will incur some costs—perhaps because of a cracked house foundation or a living room filled with flowing mud—as a result of this geological process.

Figure 12-1 Mass movement. This massive rockfall occurred in January 1997 along Highway 140, several kilometers east of the Arch Rock entrance to Yosemite National Park, California.

In this chapter, we examine both the underlying and immediate causes of mass movement, the various types of mass movements, and some of the ways their dangers can be prevented.

What Causes Mass Movement?

The principal factor driving all mass movement is gravity, which constantly coaxes materials downslope. Two main factors provide resistance to mass movement: the friction between a slope and the loose material at its surface, and the strength and cohesiveness of the material composing the slope, which prevents it from breaking apart and slipping at its surface. Mass movement occurs when gravity overcomes these factors. The steepness of the slope, the water content of its materials, the amount of vegetative cover, and the slope's history of human and other animal disturbances all influence the mass-movement potential of a slope.

Steepness of Slope

The pull of gravity downslope is proportional to the steepness of the slope. Thus, the steeper a slope, the more likely that material will slide down it. Several natural and artificial processes can steepen slopes and initiate mass movements. For instance, the natural forces of rivers and coastal waves can steepen slopes by undermining them at their base; human activities that can steepen and destabilize slopes include quarrying, road cutting, and dumping excessive amounts of mining waste (Fig. 12-2). Faulting, folding, and tilting of strata can also steepen and destabilize slopes.

Slope Composition

A slope may be composed of any combination of solid bedrock, weathered bedrock, and soil, along with varying quantities of vegetation and water. A slope of solid bedrock tends to be highly stable, even when it is so steep that it forms a vertical cliff. This stability can be greatly reduced, however, if tectonic stresses or weathering (such as frost wedging, root penetration, or chemical dissolution) create cracks or cavities in the rock. Slopes are also less stable when they are parallel to bedding planes or cleavage planes in the rock (Fig. 12-3).

A slope composed of loose, dry particles remains stable only if the friction between its components exceeds the downslope pull of gravity, which increases with the steepness of the slope. Different materials form stable slopes at different angles, depending on the size, shape, and arrangement of their particles. In general, large, flat, angular grains with rough, textured surfaces and a chaotic arrangement create more friction and can form steeper slopes than can smaller grains that are rounded, smooth, and deposited in parallel planes. The maximum angle at which an unconsolidated material can form a stable slope is known as its **angle of repose.** Dry sand, for example, has an angle of repose ranging from 30° to 35°; once a pile of dry sand has attained this slope,

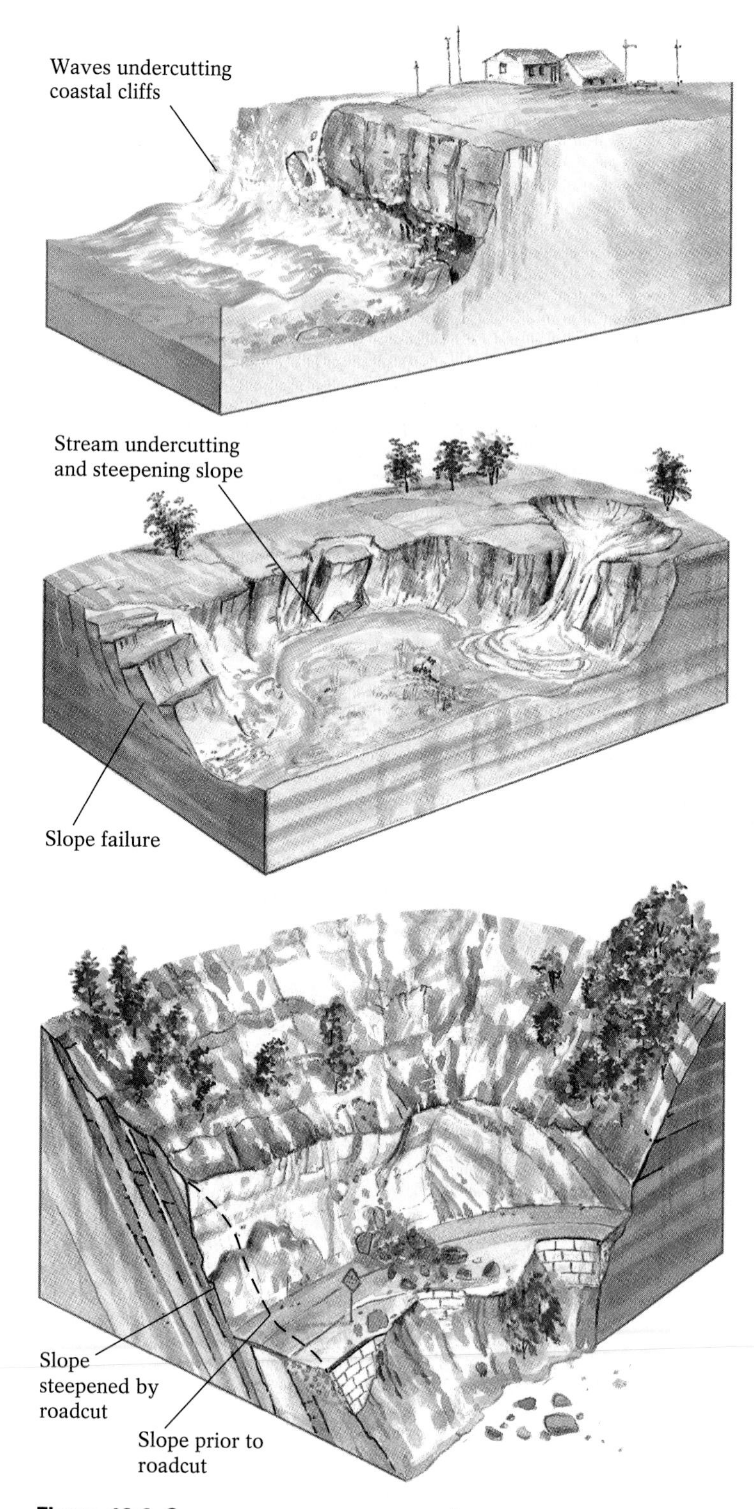

Figure 12-2 Some common processes that oversteepen slopes.

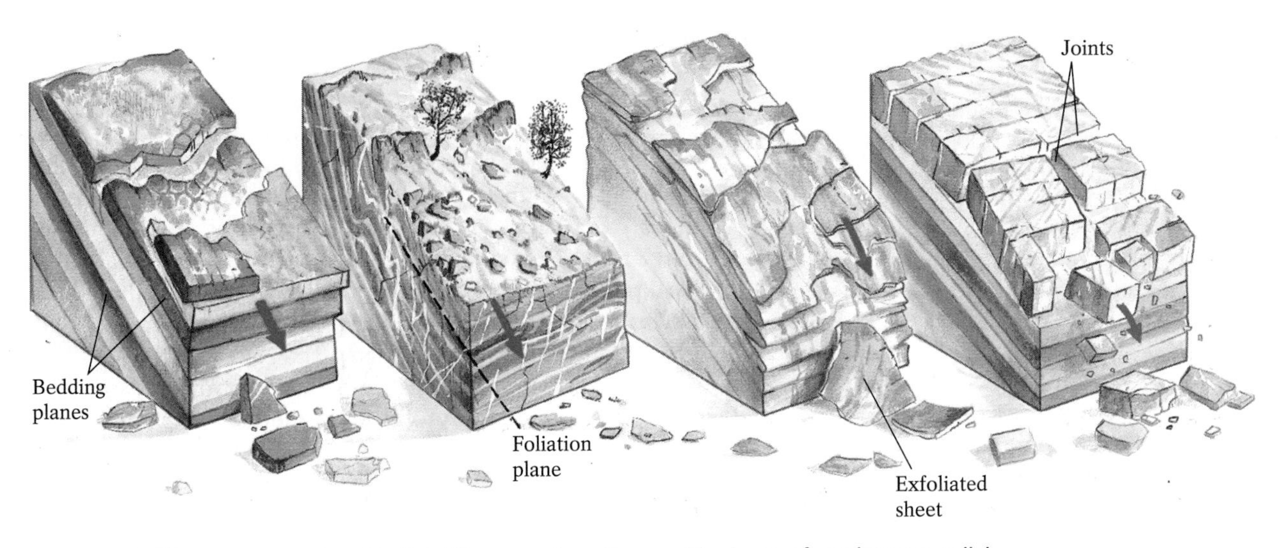

Figure 12-3 Mass-movement potential. Slopes such as these, with planes of weakness parallel to their surface, are especially susceptible to mass movement.

any extra sand added to it would simply cascade down its sides and accumulate at the base of the pile (Fig. 12-4a). In nature, the steepest slopes, greater than 40°, are maintained by large, highly angular boulders. These large boulder piles, called *talus slopes,* often form at the feet of cliffs that have been weathered by frost wedging (Fig. 12-4b).

Vegetation

Vegetation—especially the extensive and deep root networks of large shrubs and trees—binds and stabilizes loose, unconsolidated material. Removal of vegetation by forest fire or clear-cutting for timber or farming purposes allows loose

Figure 12-4 The relationship between angle of repose for unconsolidated materials and particle size and shape. (a) Coarse, angular sand forms a steeper slope than does fine, rounded sand. **(b)** A talus slope, composed of large and irregularly shaped boulders, can form slopes greater than 40°. Photo: Talus slope at Wheeler Peak, Great Basin National Park, Nevada.

material to move downslope, especially shortly after a rainstorm. Several decades ago, farmers in the town of Menton, France, decided to remove olive trees—which have deep, stabilizing roots—from the area's steep slopes so as to plant more profitable shallow-rooted carnations. This horticultural miscalculation contributed to landslides that took 11 lives. Similar widespread tree-cutting in the forests of the Philippines was a primary cause of the tragic landslides that claimed 3400 lives in November 1991.

Water

More than any other factor, water is likely to cause previously stable slopes to fail and slide. Initially, a small amount of water increases the cohesiveness of loose material; damp sand, for example, holds together more effectively than dry sand. Some water also enables the growth of stabilizing vegetation. Excessive water, however, can promote slope failure by reducing friction. It can diminish the friction between surface materials and underlying rocks, between adjacent grains of unconsolidated sediment (for example, turning relatively stable soil into flowing mud), or even between adjacent rock masses that are separated by a plane of weakness (such as a bedding plane, fault, or joint) (Fig. 12-5). Water also promotes slope failure by adding to the weight of slope materials, making them more susceptible to the pull of gravity.

Setting Off a Mass-Movement Event

Before a stable slope becomes unstable and fails, it may develop a fragile balance, or equilibrium, between the forces that tend to drive movement downslope and the forces that tend to resist movement. Some event, either natural or man-made, may then tip the balance, triggering the downslope movement. Natural triggers have included torrential rains, earthquakes, and volcanic eruptions. In 1967, a three-hour downpour in central Brazil triggered hundreds of slope failures, taking more than 1700 lives. The New Madrid, Missouri, earthquakes of 1811–1812 and the eruption of Mount St. Helens in 1980 each produced numerous, damaging mass-movement events.

Human-induced mass movement often results from mismanagement of water. When we overirrigate slopes for farming, install septic fields that leak sewage, divert surface water onto sensitive slopes at construction sites, or over-water sloping lawns, we introduce liquids that destabilize slopes by reducing friction. In one case of inadvertent water mismanagement, a Los Angeles family went on vacation and left its lawn sprinklers turned on. Family members returned to find their hillside lawn and home sitting in the valley below.

Other human actions may trigger mass movements as well. When we clear-cut forested slopes or accidentally set

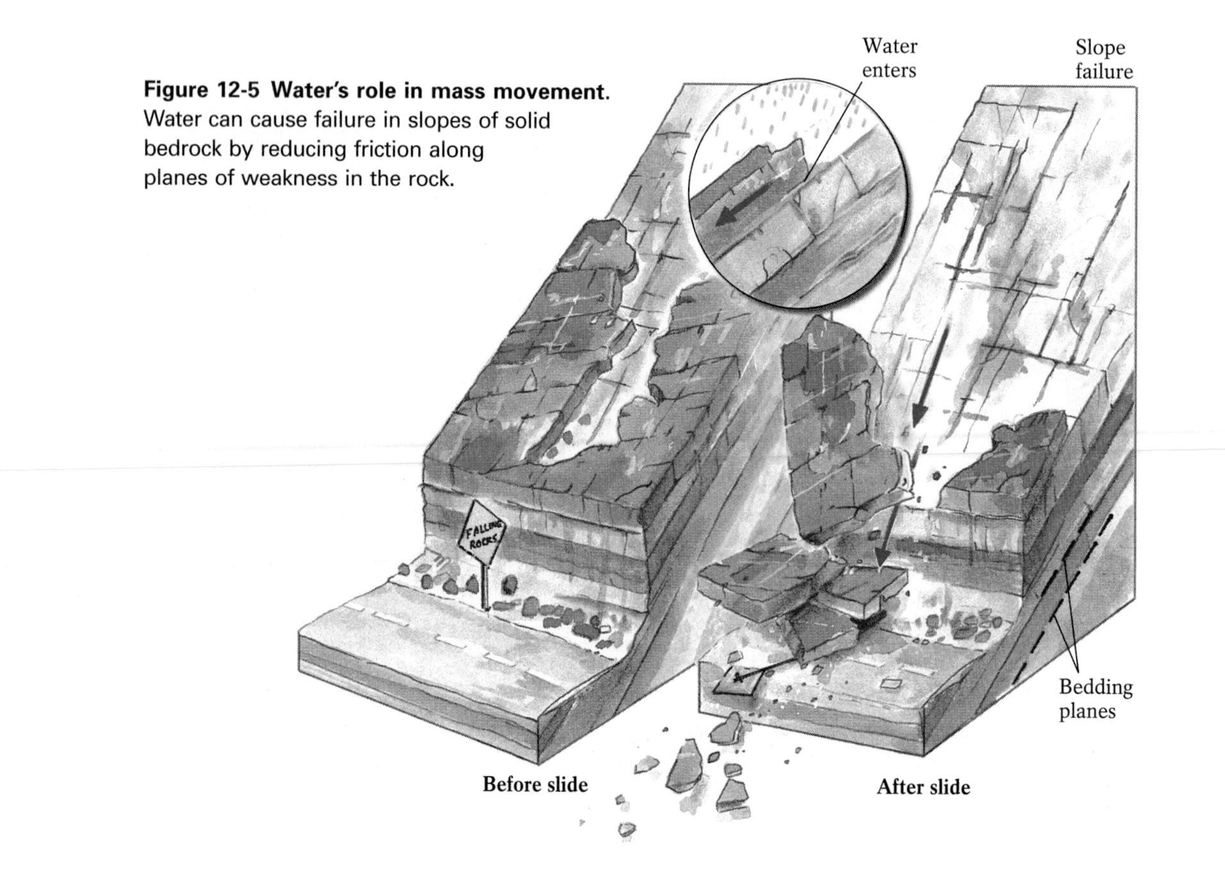

Figure 12-5 Water's role in mass movement. Water can cause failure in slopes of solid bedrock by reducing friction along planes of weakness in the rock.

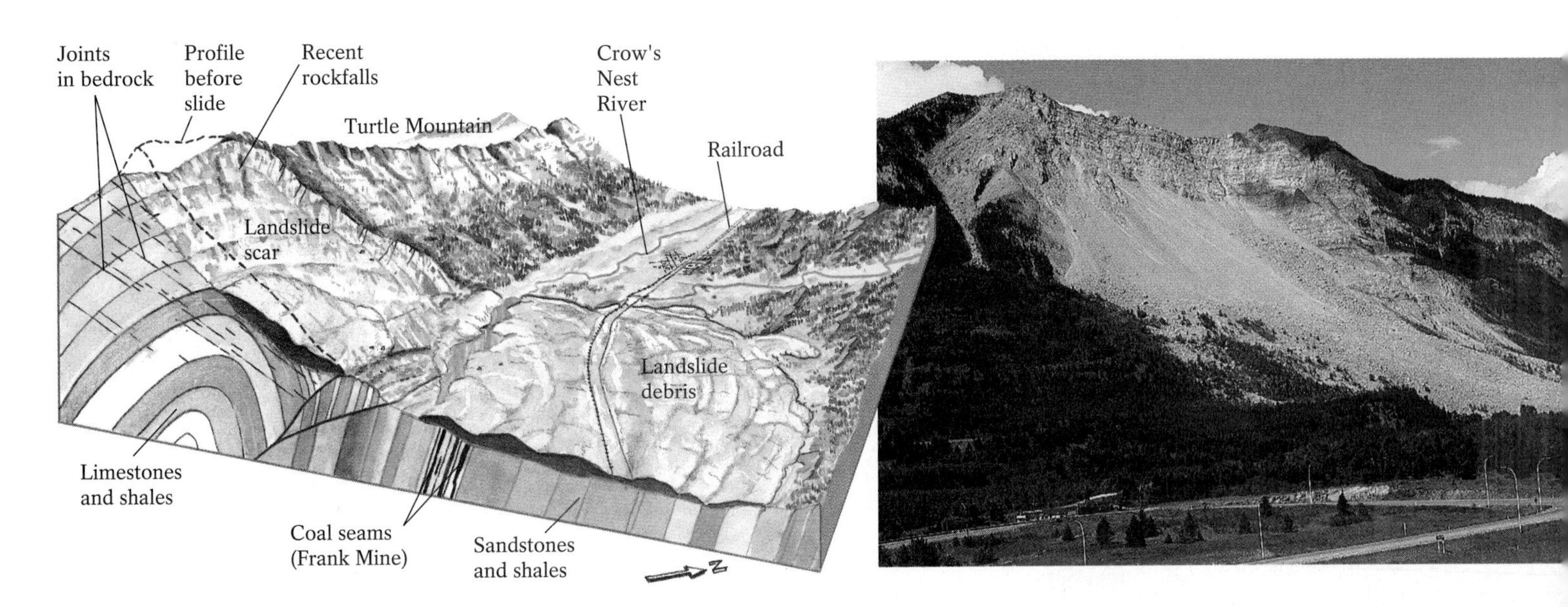

Figure 12-6 The 1903 Turtle Mountain landslide in Frank, Alberta. The disaster began when an enormous mass of limestone broke free and slid downslope to a protruding rock ledge, where it was launched airborne toward the valley below. After a 900-meter (3000-foot) drop, the rock struck the weak shales and coal seams in the valley, shattering into a great avalanche of crushed rock that spread at speeds as high as 100 kilometers (60 miles) per hour to a distance of 3 kilometers (2 miles) across the valley. The mass of crushed rock had such momentum that it actually ascended 120 meters (400 feet) up the opposite side of the valley. Ironically, 16 men working in the mines survived by digging their way out through a soft coal seam. Photo: Turtle Mountain as it looks today, showing evidence of more recent rockfalls within the scar from the 1903 slide.

forest fires, we eliminate the deep, extensive root networks that bind loose materials together. When we cut into the bases of sensitive slopes to clear land for homes or roads—especially if we dump the removed or other material on the top of the slopes—we oversteepen the slopes and jeopardize their equilibrium. Construction of housing developments on hillsides or cliffsides, as is common in California, can initiate mass movement when slopes bearing the extra weight of roads, buildings, and swimming pools later become weakened by heavy rainfall. Where extreme instability exists, even the vibrations of loud sounds—such as passing trains, aircraft sonic booms, and construction blasting—can trigger mass-movement events.

Sometimes human activities, such as mining, can combine with natural factors to increase the probability of mass movement along a slope. One such situation led to the Turtle Mountain landslide of 1903, near the Canadian Rockies town of Frank, Alberta, which claimed 70 lives (Fig. 12-6). The removal of a large volume of coal near the foot of the mountain weakened an already unstable structure containing numerous joints and fractures. The month preceding the slide had been a wet one in the southern Alberta Rockies, and water from the copious snowmelt probably entered and lubricated the fractures in Turtle Mountain, further increasing its potential for mass movement.

Types of Mass Movement

No universally accepted scheme exists for the classification of mass movements. Ask a soil scientist, a geologist, and a civil engineer to classify a mudflow (a rapidly flowing slurry of mud and water), and you'll likely get three different responses. Geologists, however, generally classify mass movements based on the speed and the manner in which the materials move downslope.

Slow Mass Movement

Creep, the slowest type of mass movement, is measured in millimeters or centimeters per year. It occurs virtually everywhere, even on the gentlest slopes, and generally affects unconsolidated materials, such as soil or regolith, whose depths rarely exceed a few meters. Loose material experiencing creep undergoes continuous rearrangement as individual particles become dislodged—for example, by burrowing animals and insects, raindrops, or swaying plants—and respond to the influence of gravity. Because of the friction between this surface material and the underlying slope material, a mass of material undergoing creep tends to move more rapidly at its surface and more slowly at deeper layers, causing implanted

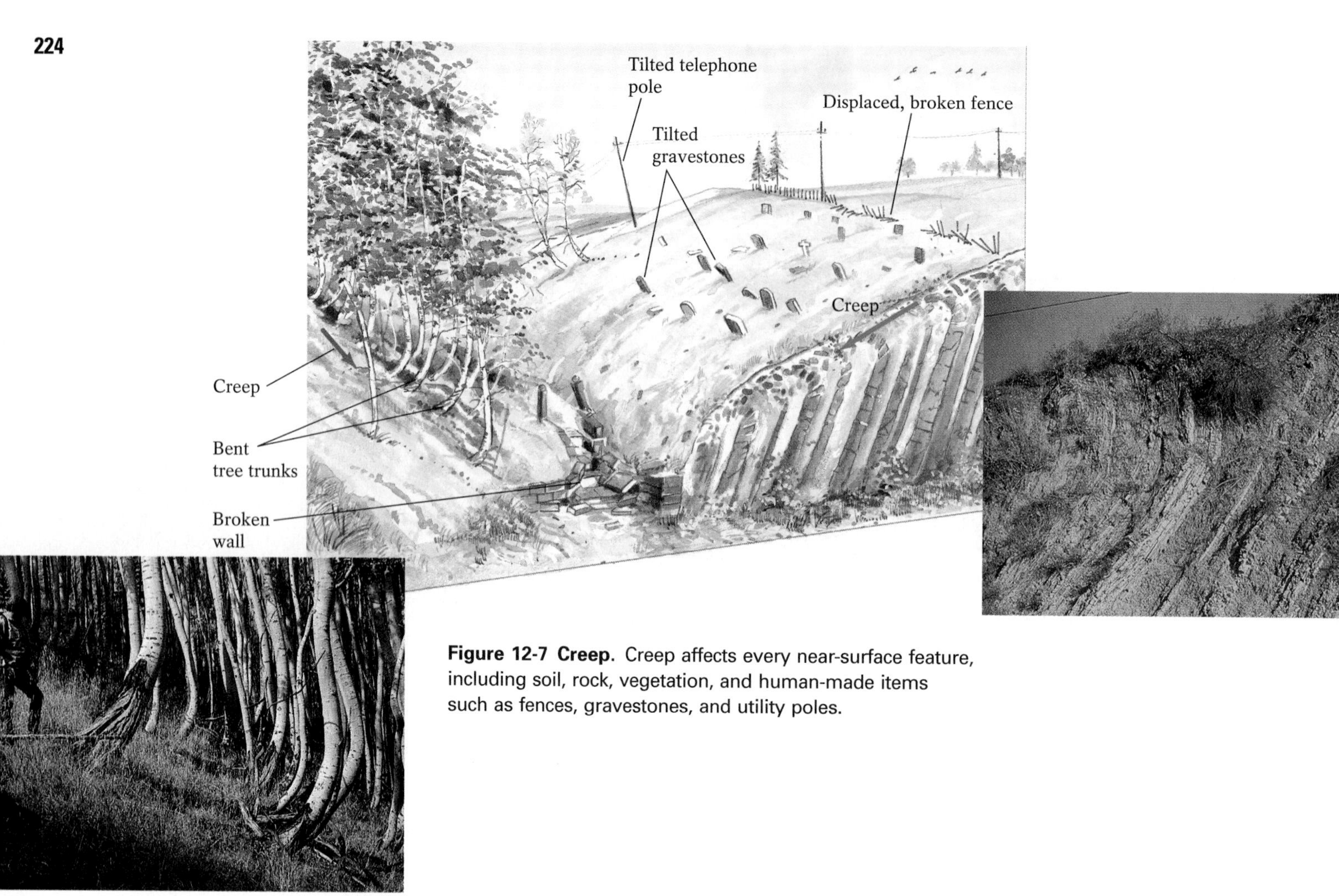

Figure 12-7 Creep. Creep affects every near-surface feature, including soil, rock, vegetation, and human-made items such as fences, gravestones, and utility poles.

objects to slant. A good indication of creep occurs when structures that were originally upright are no longer vertical. Fences, gravestones, and telephone poles inserted on a creeping slope eventually become tilted; trees growing on creeping slopes, because they continually adjust their orientation to absorb maximum sunlight, often have contorted trunks (Fig. 12-7).

In sufficiently cold environments, freezing and thawing can induce creep. Freezing of soil water causes it to expand, and surface materials are lifted perpendicular to the slope. As thawing occurs, gravity pulls the displaced surface particles slightly downslope before they settle back into a stable position (Fig. 12-8). In clay-rich soil, periodic wetting and drying swell and shrink materials and move soil particles in a similar manner. Even on a gentle slope, each of the countless particles in such soils or regolith is gradually but constantly being nudged downhill.

A special variety of creep occurring principally in very cold regions, such as the Alaskan tundra, is **solifluction** ("soil flow"). This comparatively fast form of creep occurs when the warmth of a brief summer season thaws only the upper few meters of frozen soil or regolith. Because the underlying *permafrost* (permanently frozen ground) remains impermeable to water, the thawed soil becomes waterlogged and flows downslope at rates of 5 to 15 centimeters (2–6 inches) per year (Fig. 12-9).

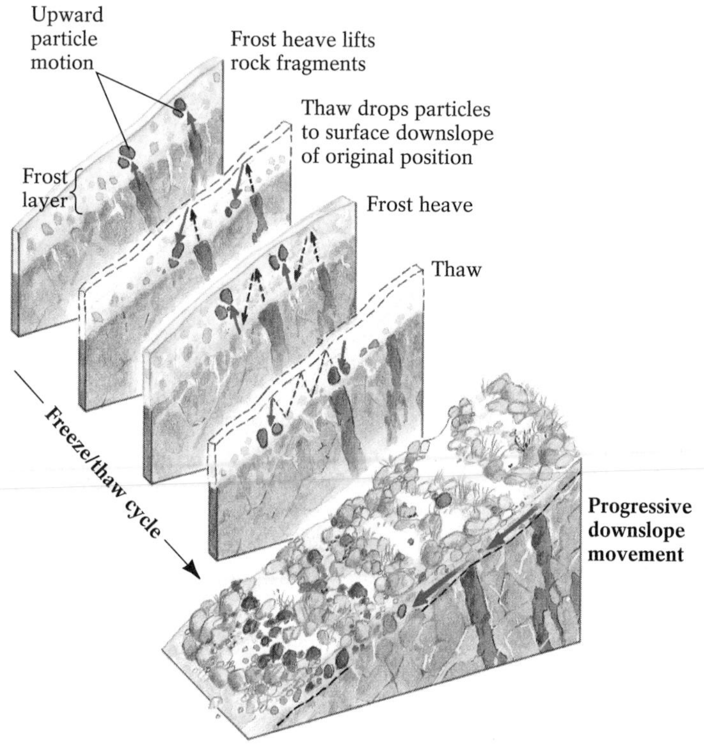

Figure 12-8 Frost-induced creep. Ice that forms beneath soil particles displaces the particles, pushing them toward the surface. Gravity carries the displaced particles a short distance downslope as the frost thaws and the particles sink back down.

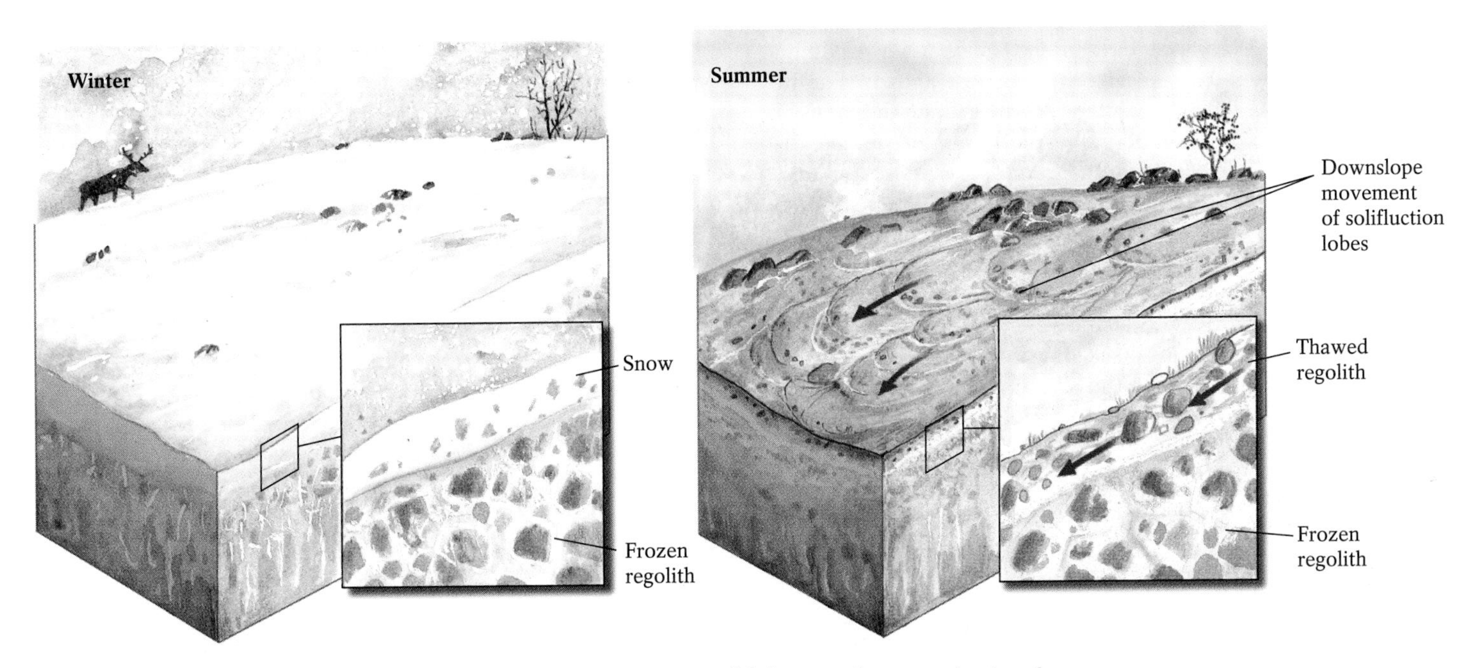

Figure 12-9 Solifluction. Solifluction occurs where the soil and regolith become frozen to depths of hundreds of meters. During the brief summer season, the surface materials thaw out to a depth of only a few meters and flow downslope with the aid of the water released by melting. Solifluction typically creates lobe-shaped masses of slowly moving sediment.

Rapid Mass Movement

Rapid mass movements, rather than occurring at the rate of centimeters per year like creep, can proceed at kilometers per hour or even meters per second. These types of events are further classified according to their different types of motion as falls, slides, slumps, or flows (Fig. 12-10).

Falls A **fall** is the fastest type of rapid mass movement and can be very dangerous. It occurs when rock or sediment breaks free from a steep or vertical slope and plummets through the air to the ground or water below (see Figure 12-1).

Slides A **slide** occurs when a single intact mass of rock, soil, or unconsolidated material detaches and moves downward along a plane of weakness, or *slip plane.* This type of mass movement may involve a small displacement of soil over solid bedrock, or a huge displacement of an entire mountainside (Highlight 12-1). Large slides may be preceded by days, months, or even years of detectable accelerated creep.

The nature of a slip plane varies with the local geology. Unconsolidated sediments often develop slip planes at the boundaries between different types of material, such as between a layer of sand and a layer of clay, or between

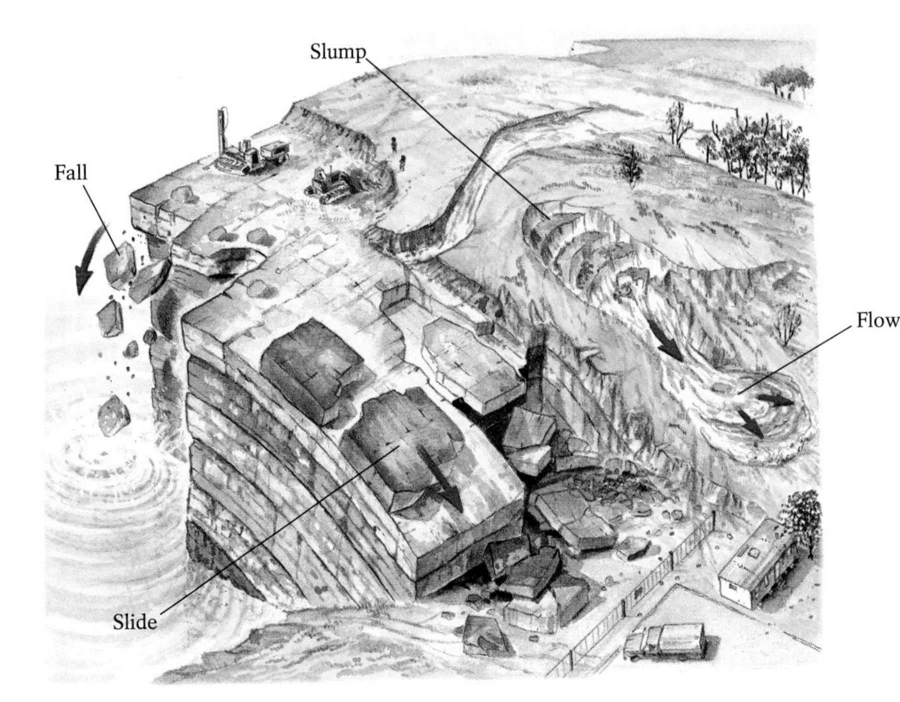

Figure 12-10 Rapid mass movements: falls, slides, slumps, and flows.

Highlight 12-1 The Gros Ventre Slide

On June 23, 1925, just east of the small town of Kelly, Wyoming, near Jackson Hole, the largest known slide in United States history rumbled down the south slope of Sheep Mountain. More than 38 million cubic meters (1.3 billion cubic feet) of sandstone blocks, shale slabs, and loose soil and regolith were transported more than 600 meters (2000 feet) into the valley of the Gros Ventre River. One enormous sandstone mass, traveling intact, brought with it the entire overlying pine forest. The momentum of the mass was so great that it hurtled upslope on the opposite side of the valley more than 100 meters (330 feet) before sliding back down and finally coming to rest on the valley floor. There the slide mass formed a natural dam, 75 meters (250 feet) high, that blocked the Gros Ventre River to create a lake 8 kilometers (5 miles) long and 70 meters (240 feet) deep (Fig. 1 photo). The lake filled so quickly that only 18 hours after the slide, a house 18 meters (60 feet) above the river was floated off its foundation by rising lake water.

Why did this particular mountainside give way? In Sheep Mountain, layers of clay-rich shales alternate with massive beds of limestone and sandstone (Fig. 1). The beds are inclined toward the valley at an angle greater than 20°. Water from the 1925 spring's abundant snowmelt and heavy rainfall could not drain into an impermeable shale layer several meters below the surface, allowing all of the overlying rock to become saturated. The added weight of the water, combined with the moistening of the slip plane above the shale, caused an entire layer of near-surface sandstone to slide.

The area was sparsely populated, and fortunately no lives were lost *during* this monumental event. After heavy May rains two years later, however, the new lake breached the natural dam, cutting a channel 15 meters (50 feet) wide through the slide mass. A 5-meter (16-foot)-high wall of water rushed downhill and washed out the town of Kelly in a furious flood that took the lives of six residents.

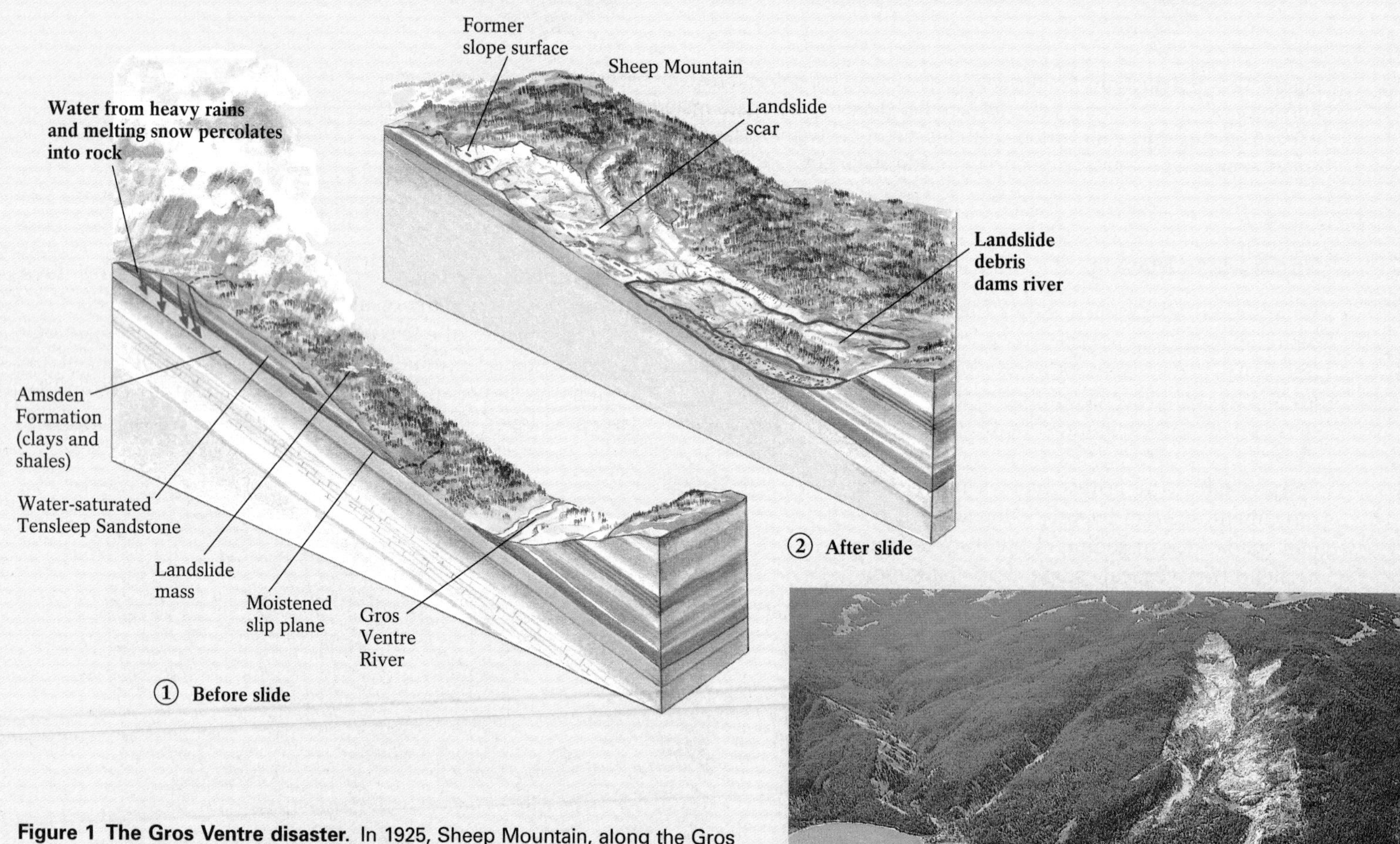

Figure 1 The Gros Ventre disaster. In 1925, Sheep Mountain, along the Gros Ventre River in Wyoming, was the site of one of North America's greatest mass-movement events. Above: The geology of the Sheep Mountain slide. Layers of clay-rich shales alternate with massive beds of limestone and sandstone. Water could not drain into an impermeable shale layer below the overlying sandstone, causing the overlying rock to become saturated and slide downhill. Photo: Landslide debris dams the Gros Ventre River.

unconsolidated sediments and bedrock. In bedrock, slip planes develop where planes of weakness exist within the rock itself (see Figure 12-3). In sedimentary rocks, the slip plane often comprises a bedding plane within a weak rock layer. In plutonic igneous rocks, the slip plane may consist of a large joint produced by exfoliation. In metamorphic rocks, slip is likely to occur along foliation surfaces.

Slumps A slide's slip plane is generally flat. A slide that separates along a concave, crescent-shaped slip plane, moving downward and outward, is called a **slump.** Unlike slides, slumps generally create their own slip planes within a body of unconsolidated material. The steep, exposed cliff face that forms where a slump mass has pulled away from the rest of the slope is known as a **scarp** (Fig. 12-11). Slumps rarely travel far from their point of origin or move at high velocities. The slump block itself, because of the nature of its motion, usually remains intact.

Flows A **flow** occurs when a mixture of rock fragments, soil, and/or sediment moves downslope in the manner of a highly viscous fluid. Flows typically move more swiftly and are more dangerous when they have a high water content. In 1969, for example, southern California received 85 centimeters (35 inches) of rain within a six-week period. The torrential rains completely saturated the clay-rich soils and triggered numerous flows. One hundred lives were lost, 15,000 people were displaced from their mud-covered homes, and $1 billion in property damage was incurred as a result of these flows.

Like other mass movements, flows are classified by their velocity and composition. **Earthflows** are relatively dry masses of clayey or silty regolith that, because of their high viscosity, typically move as slowly as 1 or 2 meters (3–7 feet) per hour (though they sometimes move as rapidly as several meters per minute). Because they flow so slowly, earthflows are rarely life-threatening, but they may badly damage structures in their paths.

A **mudflow** is a swift-flowing slurry of mud and water, the consistency of which can vary from that of wet concrete to that of muddy water (Fig. 12-12). Mudflows are most likely to develop after heavy rain falls on sparsely vegetated slopes with abundant loose regolith. These conditions are commonly found in the canyons and gullies of semi-arid mountains, where loose sediment accumulates between the infrequent storms. As few roots bind the loose materials together, the occasional cloudburst can easily wash large quantities of sediment into arroyos (dry stream channels) and canyons. The saturated sediment then rushes rapidly down the channel.

Occasionally, a partly waterlogged, solid clayey sediment almost instantaneously transforms into a highly fluid mudflow, or **quick clay.** Quick clays develop when ground vibrations—such as those caused by an earthquake, explosion, thunder, or even the passing of heavy vehicles—increases the water pressure between sediment particles. The particles separate from one another, reducing the friction between

Figure 12-11 Slumping and scarps on a cliff in Dorset, England.

Figure 12-12 The tragic mudflow in Sarno, Italy, on May 6, 1998. This mudflow, which took more than 100 lives, formed when water from the heavy spring rainfall mixed with rock and soil on nearby slopes, producing a muddy slurry that overwhelmed the low-lying sections of the village.

Figure 12-13 The Yungay debris avalanche of May 1970. This mass movement, triggered by a major earthquake, buried several towns, taking 30,000 lives. **(a)** Before the avalanche. **(b)** After the avalanche.

them. On November 12, 1955, near the town of Nicolet, Quebec, along the Gulf of St. Lawrence, 165,000 cubic meters (5.8 million cubic feet) of clay liquefied into quick clay that flowed toward the Nicolet River valley. The flow stopped just meters before engulfing the crowded local cathedral. A similar, but much larger and more catastrophic quick-clay flow occurred on May 4, 1971, in the village of Saint-Jean-Vianney, Quebec, where a layer of clay liquefied instantaneously, and 6.9 million cubic meters (240 million cubic feet) of clay buried 40 homes and took 31 lives.

Volcanic eruptions can produce catastrophic mudflows known as *lahars* (see Chapter 4). Lahars occur when hot volcanic ash melts the snow or glacial ice on a mountainside, or when the rainfall that often accompanies an eruption saturates the fresh volcanic ash and soil plus the accumulated ash layers from previous eruptions. A lahar may rush downslope at very high speeds. The 1980 eruption of Mount St. Helens, for example, set off a lahar that traveled between 29 and 55 kilometers (18–35 miles) per hour, sweeping away homes, bridges, and everything else in its path.

Like mudflows, **debris flows** are common in sparsely vegetated mountains in semi-arid climates and are triggered by the sudden introduction of large amounts of water. Debris flows consist of particles that are generally coarser than sand-size grains, and often contain boulders 1 meter (3 feet) or more in diameter. Hence, these materials require a steeper slope than other types of flows to trigger their downward movement. Debris flow velocities range between 2 and 40 kilometers (1–25 miles) per hour.

The swiftest and most dangerous debris flows are **debris avalanches,** which are common in the Appalachians, the Green and White Mountains of New England, and the Cascade and Olympic Mountains of the Pacific Northwest. These mass movements occur on very steep slopes, especially where stabilizing vegetative cover has been removed by fire or logging. Like most flows, they are triggered by heavy rains. During an avalanche, the entire layer of soil and regolith may become detached from the underlying bedrock and rush downslope through narrow valleys. The enormous velocity of debris avalanches, which under certain topographic conditions can even propel debris through the air, accounts for the great damage they can cause. In Yungay, Peru, a powerful earthquake on May 31, 1970, dislodged an avalanche of debris that hurtled downslope at speeds exceeding 200 kilometers (120 miles) per hour (Fig. 12-13). Afterward, grass, flowers, and even tall shrubs on patches of the slope in the avalanche's path were undisturbed, suggesting that the debris flew through the air *above* them.

Avoiding and Preventing Mass-Movement Disasters

Efforts to prevent mass-movement disasters must begin with an assessment of the likelihood of slope failure in a given area. This prediction relies on analysis of both the local geology and the historical records of past mass-movement events. The next step is avoidance of human activities that could potentially contribute to slope failure. Finally, most mass-movement defense plans include an attempt to actively stabilize slopes believed to be at risk.

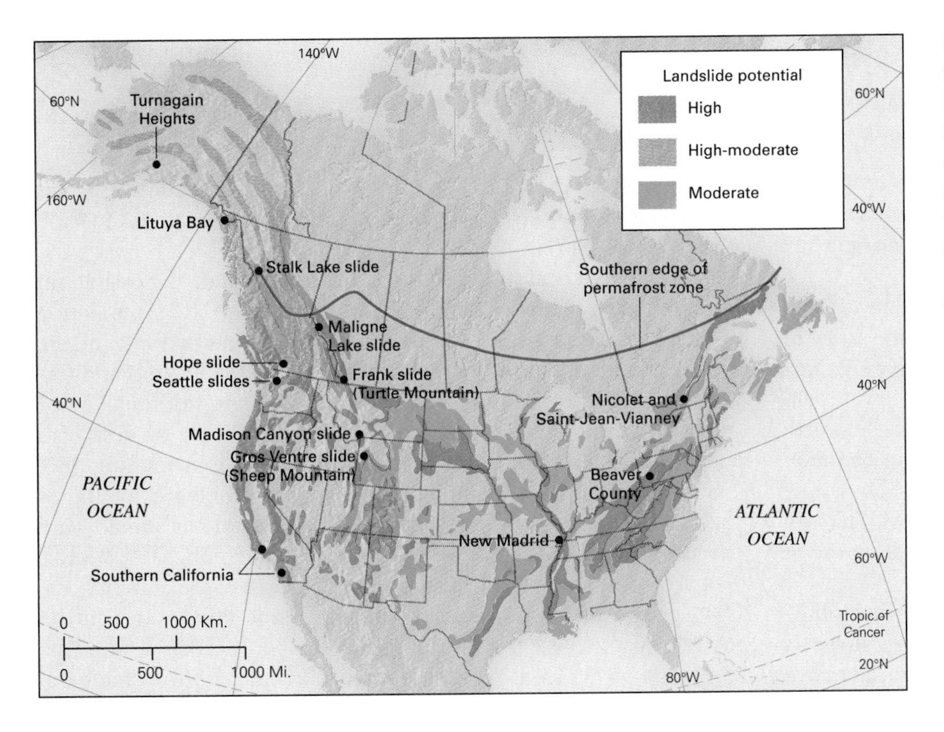

Figure 12-14 A mass-movement hazard map for the United States and Canada. Mass-movement hazards occur more frequently where slopes are steep, such as in western mountains and the Appalachians; where annual rainfall is high, such as in the humid Southeast and misty Northwest; where droughts and human activities destabilize sensitive slopes, such as in southern California; and where repeated freezing and thawing contribute to slope instability, such as in the permafrost regions of Canada.

Predicting Mass Movements

Mass movements may well be the most readily predicted of all geological hazards. Simple eyewitness accounts of past and ongoing mass movements offer the best clues about their future behavior. In the spring of 1935, when the autobahn (German for "highway") was being laid through some clay deposits between Munich, Germany, and Salzburg, Austria, a series of slides caught the German engineers by surprise. Had they heeded the construction crews' observations that the slope "wird lebendig" (was "becoming alive"), they might not have been caught off guard.

The next step in prediction is thorough terrain analysis. Geologists study the composition, layering, and structure of slope materials, determine their water content and drainage properties, measure slope angles, search for field evidence of past mass-movement events (such as landslide scars and jumbles of slide or flow debris), and bury instruments in slopes to monitor the movement and deformation of slope materials. The geological survey offices of both the United States and Canada have issued slide-hazard maps based on such terrain analyses (Fig. 12-14). In addition, the U.S. Office of Emergency Preparedness maintains a list of slide-prone areas.

Although we can never predict exactly when a mass-movement event will occur, we can be alert to the potential dangers of specific slopes. Where valuable property exists, or where mass movement poses a particular threat to safety, we can install expensive landslide-warning devices that signal when a slide or flow is in progress. In the slide-prone southern Rockies, the Denver and Rio Grande Railroad has installed electric slide detectors—wired to respond to the high pressure of an encroaching slide mass—upslope from certain bridges. Contact with a moving mass breaks an electrical circuit, sending a "stop" signal to approaching trains.

Changes in animal behavior may also signal an impending mass-movement event, much as they do for earthquakes. Shortly before the entire Swiss village of Goldau was destroyed in 1806, the town's livestock acted nervously and the beekeeper's bees abandoned their hives. Within hours of these activities, a block of rock 2 kilometers (1.2 miles) long and 300 meters (1000 feet) wide broke loose from a steep valley wall and buried the town and its 457 inhabitants in a massive rockfall.

Avoiding Mass Movements

Given our knowledge of the factors that cause mass movements and our ability to identify potential hazards, it is prudent simply to avoid building on sensitive slopes, especially given that few insurance companies insure homes and property for mass-movement losses. Even after a risk map has been drafted and the potential for slides and flows identified, however, people in susceptible areas are often reluctant to abandon ancestral homelands, commercially or agriculturally valuable properties, and scenic hillsides. Thus avoidance of mass movement is rarely practiced, as the overdevelopment of southern California's steep canyon slopes and unstable oceanfront cliffs attests. Highlight 12-2 examines the importance of mass-movement avoidance to prospective home buyers.

In developing nations, people must often overcultivate unstable slopes to feed the hungry, overgraze already sparse vegetation to maintain their livestock, and deforest slopes to provide wood for dwelling construction and heating. All of these actions promote mass movement. As long as we continue to use and inhabit unstable slopes in these ways, then, *prevention* will remain the most practical way to deal with the threat of mass movement.

Highlight 12-2 How to Choose a Stable Home Site

If you have not already done so, you may eventually wish to purchase your own home. Having studied physical geology, you will want to ensure that your dream house doesn't fall victim to a geological nightmare.

Suppose you're exploring southern California's scenic beachfront locales. You happen upon the mosaic sign for the town of Portuguese Bend and notice that the beautiful ceramic signpost is cracked into two pieces. You check the local real estate listings and find a house priced at $50,000 that should be worth $500,000. Your knowledge of geology, along with your common sense, immediately warns you that something may be wrong here. What else should you look for? How can you tell—in southern California or anywhere else in the world—if you're in mass-movement country?

Examine both the property itself and the entire neighborhood. Do you see any signs of an old mud or debris flow? Is there evidence of a slump scar upslope where a block may have broken away? As you drive through the community and its environs, look for fences that are out of alignment, and for power and telephone lines that seem too slack in some places and too taut in others (Fig. 1).

Next, look carefully at the house itself and, if possible, at neighboring ones. Search for large cracks in the foundation (small cracks may be due to initial drying and settling of the concrete). Doors and windows that stick may indicate that once-linear structural features are now out of line, although poor craftsmanship or high moisture content may also be responsible. A cracked pool lining might explain why a swimming pool doesn't retain water. Finding only one such problem may not indicate danger, but the presence of several problems should send a strong warning signal. If the geology, topography, and hydrology of a home site all raise questions about slope stability, the site may well be prone to progressive slope failure. To confirm your suspicions, try checking newspaper accounts and the records of the local housing authority, contacting the state geological survey, and interviewing the property's neighbors.

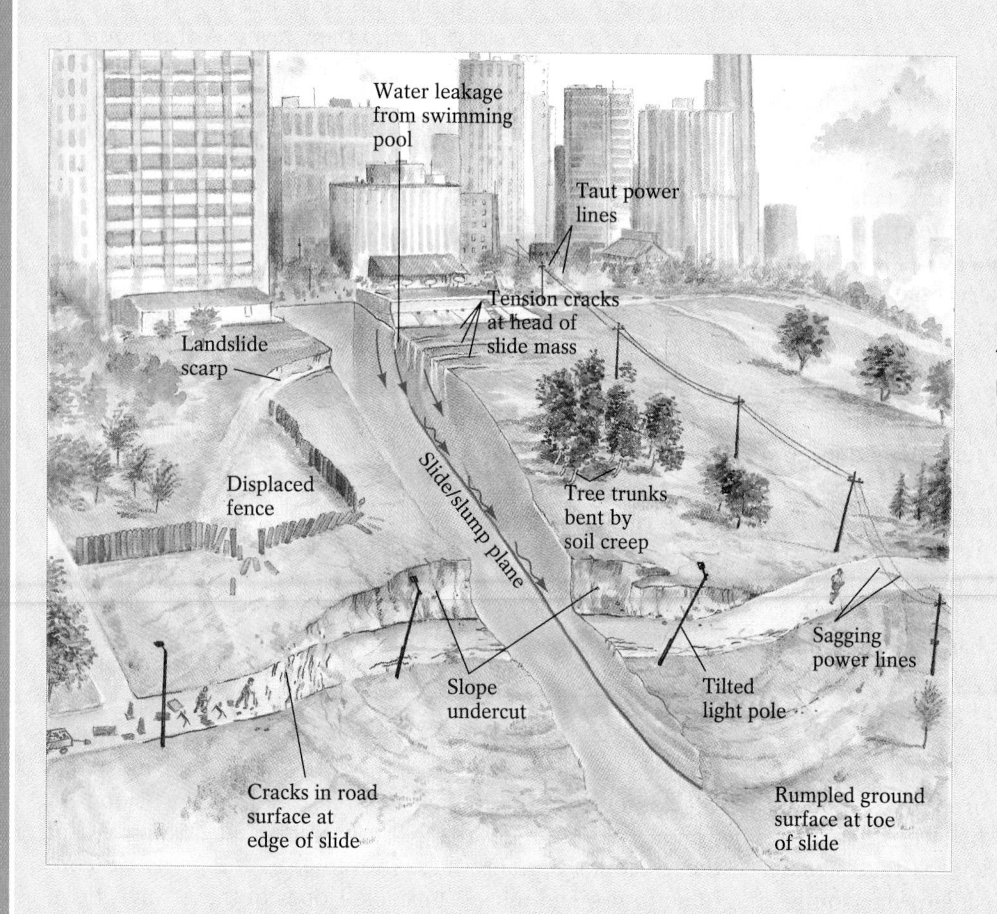

Figure 1 Various signs of past, current, and potential future mass movement in an urban area. If the power lines in a neighborhood are very loose, it suggests that the poles that hold them have moved closer together, as one might expect at the toe of a slide where the slope is bunching up like a rumpled carpet. At the head of the slide, taut lines may indicate that the poles on the slide mass are moving away from those upslope. The poles in the middle of a slide mass may keep their original spacing, because the mass may not be deforming much internally.

(a)

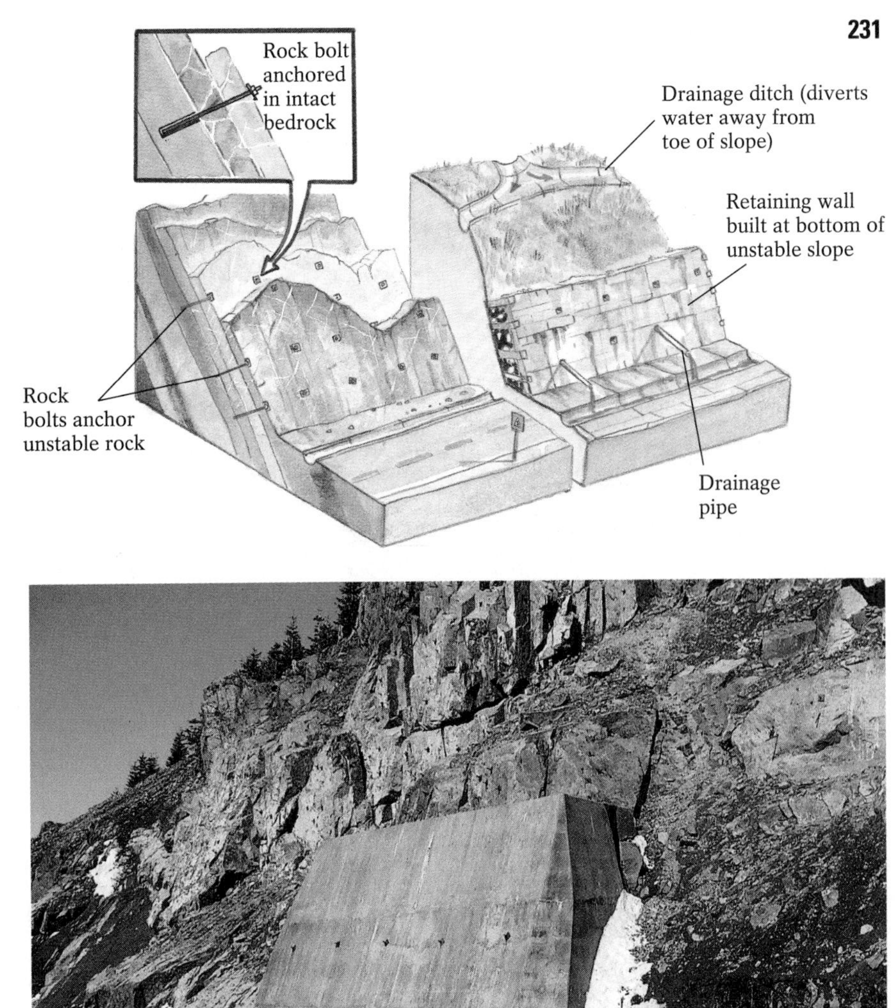

(b)

Figure 12-15 Structural supports to prevent slope failure. (a) Rock bolts stabilize loose-jointed granite on Storm King Mountain, overlooking Storm King Highway in Westchester County, New York. **(b)** A retaining wall supports the ground underlying Route 659, near Blacksburg, Virginia.

Preventing Mass Movements

It is far less expensive and much safer to stabilize a slope before it fails than to clean up after the fact. Unfortunately, the dollars lost to mass movements continue to exceed the money spent to prevent them by a factor of 50.

The first step in developing a prevention plan for any given site is to acquire a clear understanding of its subsurface geology and identify its potential trigger mechanisms. Although stratigraphy is sometimes similar throughout an entire region, the relevant geological features are highly variable in other areas, and tests must therefore be conducted at a number of locations. Once detailed studies have revealed the composition and physical properties of subsurface materials, engineers must determine how best to enhance the forces that resist mass movement or reduce the forces that promote it. Several methods—both nonstructural and structural—can be used, often simultaneously, to increase slope stability.

Nonstructural methods, which do not require large, costly engineering efforts, involve management of vegetation and introduction of soil-strengthening agents. For example, we can plant fast-growing trees and shrubs that will reduce soil moisture and develop extensive, deep root networks to bind nearsurface soils. We can add chemical solutions to soils that bind clay mineral particles, thus increasing slope stability. Cement can even be injected into a slide or flow mass to lend strength and cohesiveness to unconsolidated sediment.

Structural methods, which are very costly, involve constructing retaining structures, actively modifying the terrain, or reducing a slope's water content. Installation of bolts or pins can hold a potential rockslide in place, but this costly procedure is used only in particularly dangerous situations or to protect valuable property (Fig. 12-15). Retaining walls may provide support to the base of a slope; to construct these walls, steel or wooden piles are driven through loose surface materials into underlying stable materials, and the walls are then attached to the pilings.

One way of modifying a slope for safety is to *unload* it by removing the excess weight of buildings and loose soil.

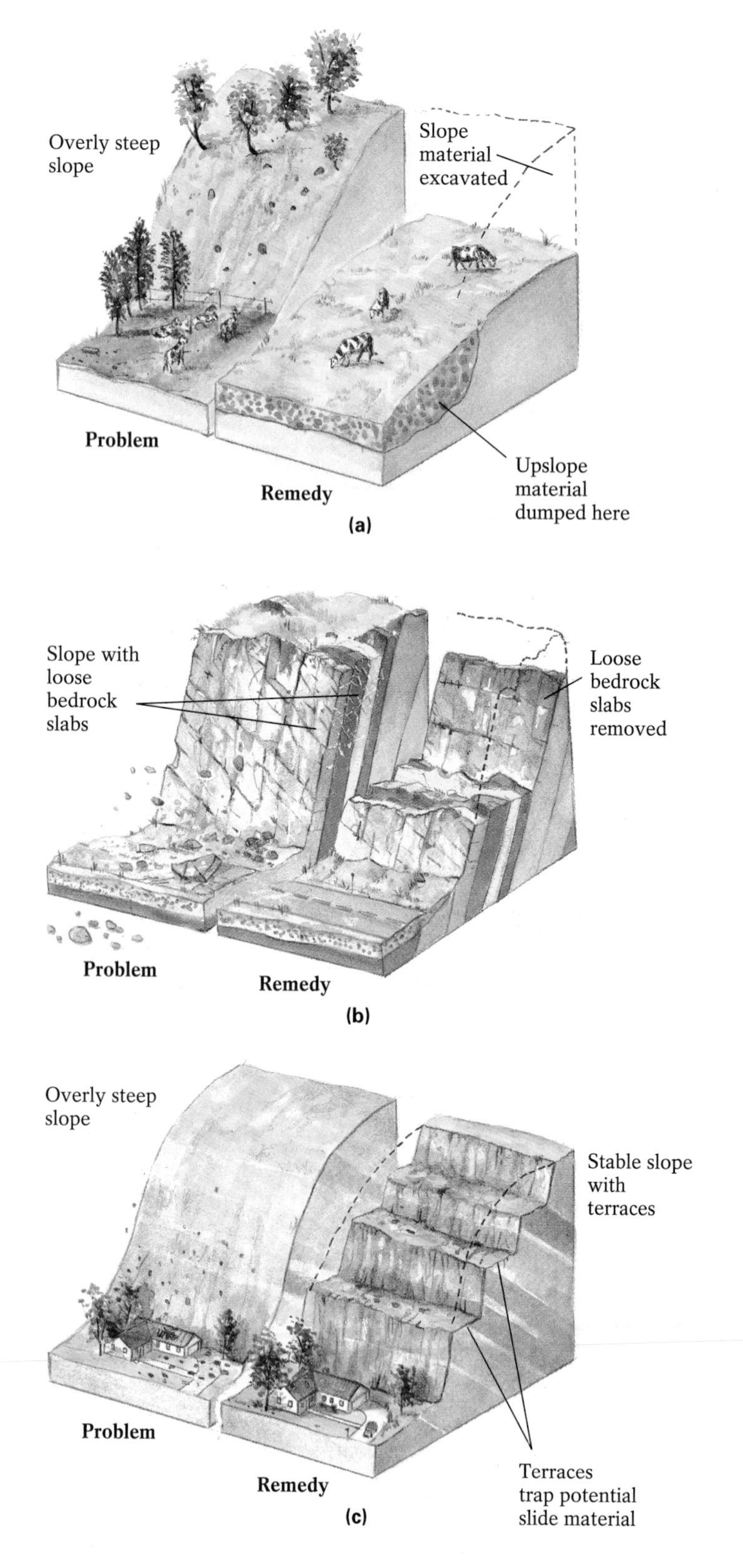

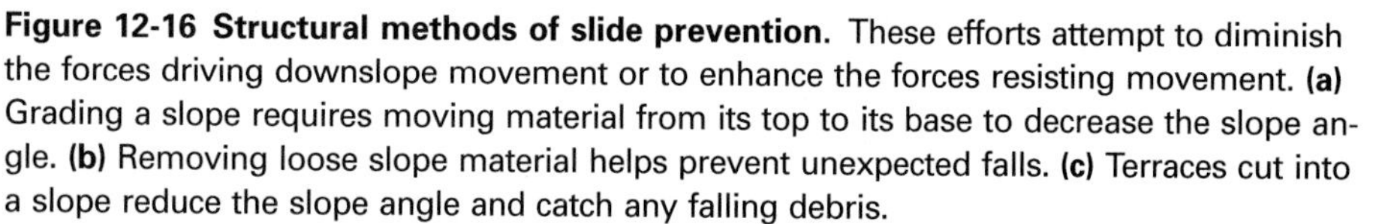
Figure 12-16 Structural methods of slide prevention. These efforts attempt to diminish the forces driving downslope movement or to enhance the forces resisting movement. **(a)** Grading a slope requires moving material from its top to its base to decrease the slope angle. **(b)** Removing loose slope material helps prevent unexpected falls. **(c)** Terraces cut into a slope reduce the slope angle and catch any falling debris.

An overly steep slope can also be *graded* by moving material from the top and spreading it at the toe to reduce the slope angle (Fig. 12-16a). More aggressive approaches involve removing all material overlying a potential slip plane (Fig. 12-16b) or cutting flat terraces into a slope to reduce weight and create areas where small slide masses can safely come to rest (Fig. 12-16c).

Perhaps the best way to reduce the likelihood of mass movement is to reduce the water content of a slope, either by preventing water from entering or by removing the liquid after it has entered. For instance, erecting small earthen barriers to divert water away from a slope or covering a slope with thin plastic sheeting or a layer of concrete can prevent infiltration. In an already-saturated slope, perforated pipes or a layer of coarse permeable gravel can be used to collect and drain off the water. Water can sometimes be removed from the slope by pumping. In Pacific Palisades, California, tunnels were excavated through an unstable slope and hot air circulated through them to evaporate subterranean pore water.

Extraterrestrial Mass Movement

Images from NASA's Apollo, Viking, and other missions reveal that mass movement occurs on two of Earth's nearest neighbors—the Moon and Mars. Although the Moon's gravity is only one-sixth that of Earth's, mass movement is one of the few ways in which its surface has been modified. Because of the nearly total absence of water on the Moon, its mass movement remains limited to dry processes. Lunar mass movement is triggered primarily by meteorites; numerous daily impacts set off avalanches of rock and regolith. After the formation of an impact crater, slumps may occur on the crater's inner walls, which are generally steeper than the angle of repose for loose lunar regolith. Marks left by rolling and sliding boulders may remain unchanged for millions of years on the Moon's surface, which is virtually free of weathering (Fig. 12-17a).

The surface of Mars exhibits a wider range of mass-movement features. Trigger mechanisms for Martian slides and flows include meteorite impacts, "marsquakes," and related faulting. In the past, periodic melting of the planet's

(a)

subsurface permafrost may have released large amounts of water that oversteepened slopes and promoted development of massive slides, slumps, and debris flows. Such melting may have formed rapid mudflows that eroded all or parts of the Valles Marineris, one of Mars' largest canyons (Fig. 12-17b).

In this chapter, we explored the many ways that gravity drives erosion of the Earth's surface. In Chapters 13 through 15, we examine the role played by water—surface stream water, unseen groundwater, and frozen glacier water—in transporting weathered materials and modifying the Earth's landscapes.

Chapter Summary

Mass movement is the process by which masses of Earth materials are transported down slopes because of the pull of gravity. The friction between a slope and the loose material at its surface, as well as the strength and cohesiveness of the material composing the slope, are the principal factors resisting mass movement.

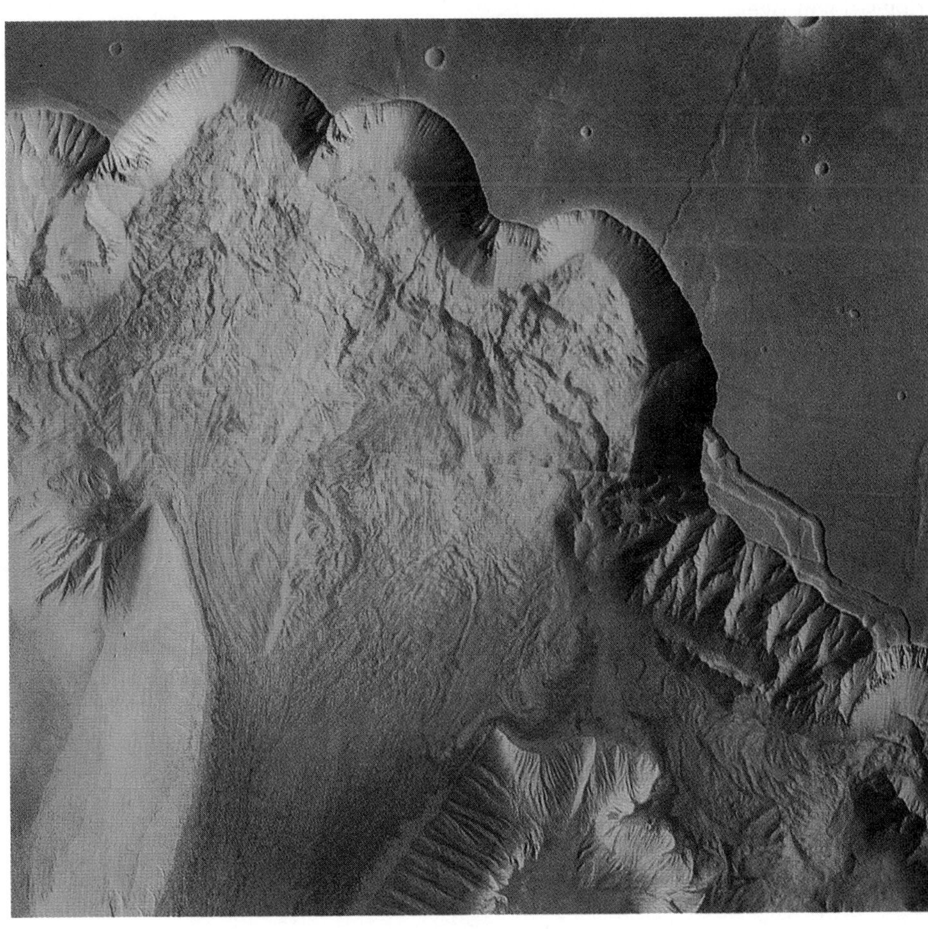

(b)

Figure 12-17 Mass movement on the Moon and Mars. **(a)** A lunar impact crater showing landslide scarps and slumps, photographed by NASA's Apollo 16 mission. (The small hook-shaped mark in the center is a dust particle on the camera lens.) **(b)** Mass movement on the Martian surface can be seen in satellite images of the Valles Marineris, a 2-kilometer (1.2-mile)-deep canyon. Slumps that closely resemble their counterparts on Earth are visible along the margin of the canyon.

The steepness of the slope, the size, shape, and water content of its materials, the slope's vegetative cover, and the slope's history of human-induced and other disturbances help determine whether mass movement will occur. The importance of slope steepness is particularly pronounced for loose unconsolidated material, which is stable only until it reaches its **angle of repose,** the maximum slope angle that loose material can maintain without downslope particle motion. Water contributes to mass movement by reducing the friction between planes of weakness, by reducing the internal friction between loose grains in unconsolidated sediment, and by increasing the weight of slope materials.

A variety of natural and artificial causes can trigger mass movement. Earthquakes, volcanic eruptions, heavy rains, and snowmelt are common natural triggers. Oversteepening and overloading of unstable slopes by dumping or overbuilding are just a few of the ways that human activity triggers mass movement.

We can classify mass-movement processes based on their speed and the manner in which the materials move downslope. Slow mass movement, called **creep,** occurs on virtually all slopes composed of unconsolidated soil or regolith. A special variety of creep occurring principally in cold regions is **solifluction,** the movement of waterlogged soil over frozen ground.

Several types of rapid mass movement exist. In a **fall,** materials become dislodged and plummet from a steep or vertical slope. In a **slide,** rock or regolith detaches along a plane of weakness, or slip plane, and moves as a single intact mass along the slope. A slide that separates along a concave slip plane is a **slump,** which leaves behind a crescent-shaped cliff face, or **scarp.**

A **flow** consists of a slurry of loose material and varying amounts of water. **Earthflows** are relatively dry and quite slow-moving; **mudflows** are faster-moving wet slurries of regolith mixed with water. Occasionally, a partly waterlogged mass of solid clay-rich sediment is instantaneously changed to a **quick clay,** a highly fluid mudflow produced when ground vibrations increase the water pressure between the sediment particles. **Debris flows** consist of particles coarser than sand-size grains; they may contain boulders 1 meter or more in diameter. The most rapidly moving type of flow is a **debris avalanche;** these mass movements occur on very steep slopes and are generally set off by heavy rains.

A slope's potential for mass movement can be predicted from analysis of its stability, evidence of past events, the use of landslide-warning devices, and even changes in animal behavior. Prevention of mass movement may include nonstructural strategies, such as planting vegetation with extensive root systems or adding soil-binding agents to sensitive slopes. Structural approaches include building retaining walls, unloading slopes by removing excess weight from them, grading slopes by moving material from their top to their base, sealing slopes to prevent further water infiltration, and draining water from slopes.

Mass movement also occurs on the Moon and Mars. Lunar mass movement involves dry materials and is triggered by meteorites. Martian mass movement may be caused by periodic melting of its permafrost; it is triggered by meteorite impacts, "marsquakes," and related faulting.

Key Terms

mass movement (p. 219)
angle of repose (p. 220)
creep (p. 223)
solifluction (p. 224)
fall (p. 225)
slide (p. 225)
slump (p. 227)
scarp (p. 227)
flow (p. 227)
earthflows (p. 227)
mudflow (p. 227)
quick clay (p. 227)
debris flows (p. 228)
debris avalanches (p. 228)

Questions for Review

1. List the primary factor that drives mass movements and the primary factors that resist them.

2. What does the "angle of repose" of an unconsolidated sediment describe? What is the general particle size and shape of a sediment that could lie at a steep angle of repose?

3. Describe three ways that water contributes to mass movement.

4. List three different mass-movement trigger mechanisms.

5. Describe how the nature of a slip plane can vary with the local geology.

6. What is the fundamental difference between a flow and a slide?

7. Describe how quick clays change instantaneously from solids to liquids.

8. Describe two structural and two nonstructural ways by which mass movement may be controlled.

9. If you were thinking of moving into a slide-prone region, how would you determine whether your prospective home was located on a stable slope?

10. Draw a simple sketch illustrating how slopes are graded.

For Further Thought

1. How would prospects for mass movement change if a completely clear-cut slope was replanted and became densely reforested? How do you suppose extensive forest fires in Yellowstone National Park have affected the stability of the park's slopes?

2. In which North American states or provinces would you expect to find the most mass movement taking place? Explain your reasoning.

3. Discuss the possible ways that plate tectonic activity could increase the likelihood of mass movement in a given area.

4. Why are there a greater quantity and wider variety of mass-movement events on Earth than on other planets in our solar system?

5. What type of mass movement is shown in the photo to the right? What steps would you take to shore up this material?

13

Streams and Floods

On June 11, 1993, a foot of rain fell in southern Minnesota. Four days later, more than 11 additional inches drenched the same area. Thus began the wettest June, and North America's worst flood, since compilation of weather records began in 1878. Before the wet weather ended two months later, the upper Mississippi River had surged over its banks along an 800-kilometer (500-mile) stretch from St. Paul, Minnesota, to St. Louis, Missouri.

Life throughout a 12-state region was disrupted in countless ways. The swirling floodwaters undermined interstate bridges, washed away roads, and halted all barge traffic along the upper Mississippi, the economic artery of the Midwest. In Hannibal, Missouri, Mark Twain's boyhood home, children caught catfish on streets where they had ridden bicycles only days before. In Davenport, Iowa, ducks swam lazily in the 4 meters (14 feet) of water covering the outfield of Davenport Stadium. Two hundred fifty thousand Iowans in Des Moines went without drinking water for days after floodwaters contaminated the municipal water supply with raw sewage and chemical fertilizers. Rushing, muddy floodwaters—carrying uprooted trees and junked cars—swept away thousands of homes and businesses, displaced more than 50,000 people, took dozens of lives, and caused more than $10 billion in property damage and agricultural losses.

In addition to bringing floods (Fig. 13-1), rivers such as the Mississippi offer countless benefits. For example, they provide a steady supply of water for home and industrial use, transportation, recreation, and irrigation. Rivers can be used to generate clean electric power. Crops grown in the fertile soils near rivers provide nourishment for one-third of the Earth's human population. Historically, the success and prosperity of most major cities have been linked to the size of their rivers.

From the mighty Mississippi River to the smallest creek, any surface water whose flow is normally confined to a channel is called a **stream,** whether it is known locally as a river, creek, brook, or run. (Geologists use the terms *stream* and *river* interchangeably.) Streams are among the most

Figure 13-1 Flood damage from the April 1997 flood in Grand Forks, North Dakota.

common geological features on the Earth's surface, flowing in virtually every geological and geographical setting, with the exception of the Antarctic and large segments of the Arctic. In this chapter, we learn how rivers flow and create distinctive landforms by erosion and deposition. We also see why they flood and discuss efforts to alleviate flooding.

The Hydrologic Cycle

The amount of water in, on, and above the Earth—an estimated 1.36 billion cubic kilometers (326 million cubic miles)—has remained fairly constant for more than a billion years. The majority, about 97.2%, fills the oceans; 2.15% is frozen in icecaps and glaciers; and the remaining 0.65% occurs in lakes, streams, groundwater, and atmosphere (Fig. 13-2). Most of this water originated in the Earth's mantle and was carried to the surface in the form of steams via volcanic eruptions.

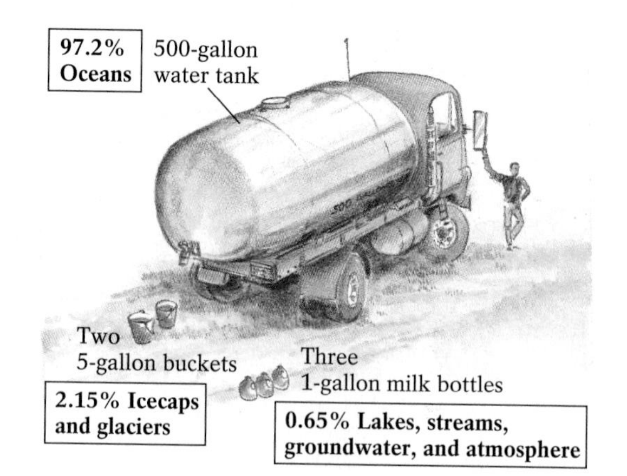

Figure 13-2 Distribution of the Earth's water, by relative volume.

The Earth's water cycles perpetually among its oceans, land, and atmosphere and, via plate subduction and volcanism, between its mantle and its surface. Scientists refer to this phenomenon as the **hydrologic cycle** (Fig. 13-3).

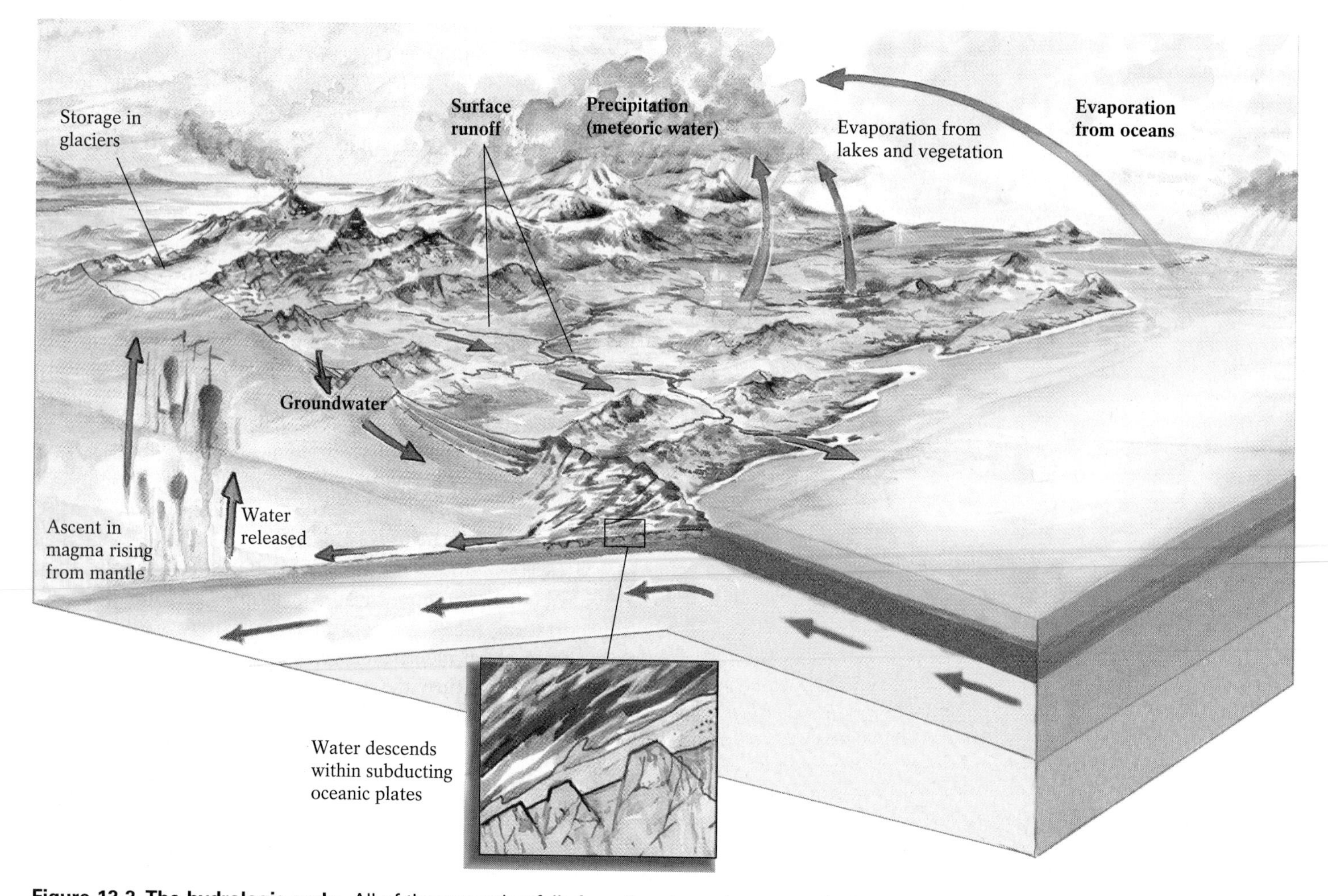

Figure 13-3 The hydrologic cycle. All of the water that falls from the atmosphere onto the Earth's surface eventually enters the vast oceanic reservoir through one or more of the pathways of the cycle.

In the hydrologic cycle, water heated by solar energy evaporates from the Earth's surface, both land and ocean, and enters the atmosphere as water vapor. As the warm, moist air rises, it cools and its water vapor condenses into microscopic water droplets that coalesce to form clouds. If a cloud becomes dense enough to form droplets that are too heavy to remain suspended in the air, the water *precipitates* as rain, snow, sleet, or hail (depending on the temperatures aloft), which falls back into the ocean and onto land.

When water falls on land, one of several things may happen to it. Snow on a high peak may be stored in the snowpack or in a glacier for tens, hundreds, or even thousands of years. Likewise, rain that falls into a lake may remain there for many years. In contrast, water that falls on plants and soil may evaporate back into the atmosphere, run along the surface and into a stream, or be absorbed into the ground, where it might be picked up by plant roots or adhere to soil particles. A small portion of this liquid may infiltrate deeply enough into the soil to become a part of the subsurface *groundwater* (the topic of Chapter 14). From there it may eventually enter a stream and flow once again to the sea, completing the hydrologic cycle.

Streams

The United States boasts an estimated 2 million streams. Each of these flows shares some common characteristics.

Stream Topography

Every stream has a water-collecting area known as a **watershed** or **drainage basin;** this region comprises the total land area from which precipitation (rain and snowmelt) reaches a stream. In North America, watersheds can be smaller than a square kilometer or as large as 3.2 million square kilometers (1.25 million square miles), the area drained by the Mississippi River system. Every drainage basin is bounded by an area of higher topography, called a **drainage divide,** that separates it from adjacent drainage basins. Divides may take the form of low ridges between two small channels or continent-spanning mountain ranges (Fig. 13-4). The Rocky Mountain portion of North America's continental divide, for example, stretches from the Canadian Yukon to New Mexico; rain falling anywhere along this divide flows either westward toward the Pacific Ocean, eastward toward the Atlantic Ocean (via the Gulf of Mexico), or northward toward the Arctic Ocean.

At a drainage divide, rainfall first flows downslope as *overland flow*—broad sheets of unconfined water that are only a fraction of a centimeter thick. These flows represent a stream's *headwaters.* Wherever these headwaters encounter slight surface depressions or changes in surface composition, they erode narrow depressions, tiny channels called *rills.* This spot marks the start of a stream's *transport area,* in which flowing water shapes and travels through progressively larger channels. Downslope, where the rills carry more water, they merge into larger, branching channels; these **tributaries** in turn carry water and sediment to the main stream, or **trunk stream.** The trunk stream traverses the greatest part of the transport area, moving the most water and sediment. At the downstream end of the trunk stream, its *mouth,* the water

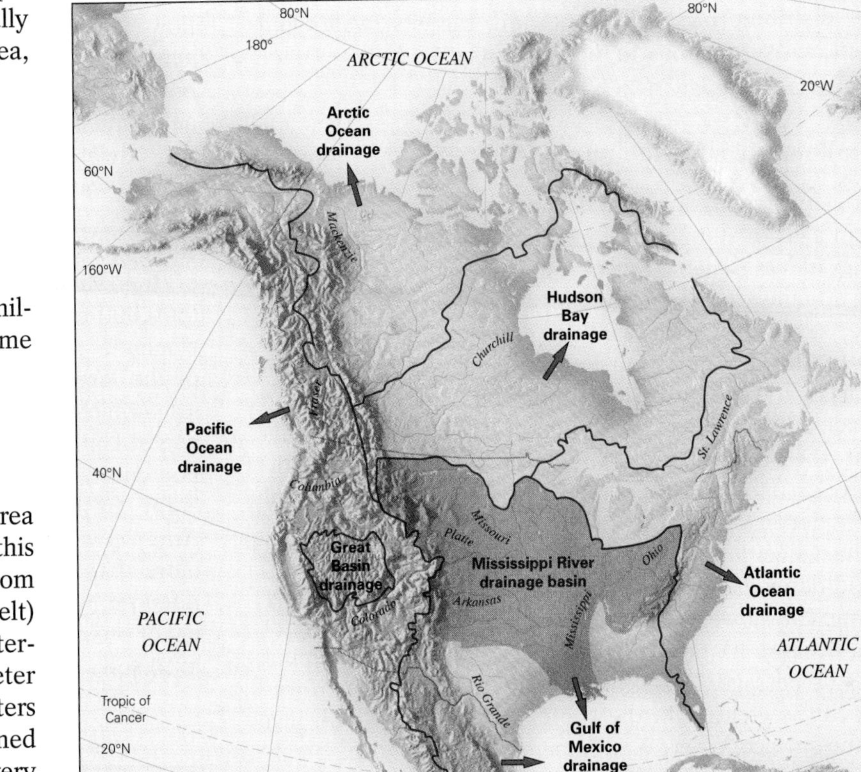

Figure 13-4 North America's drainage divides and major drainage basins. The drainage basin of the Mississippi River encompasses roughly 40% of the continental United States.

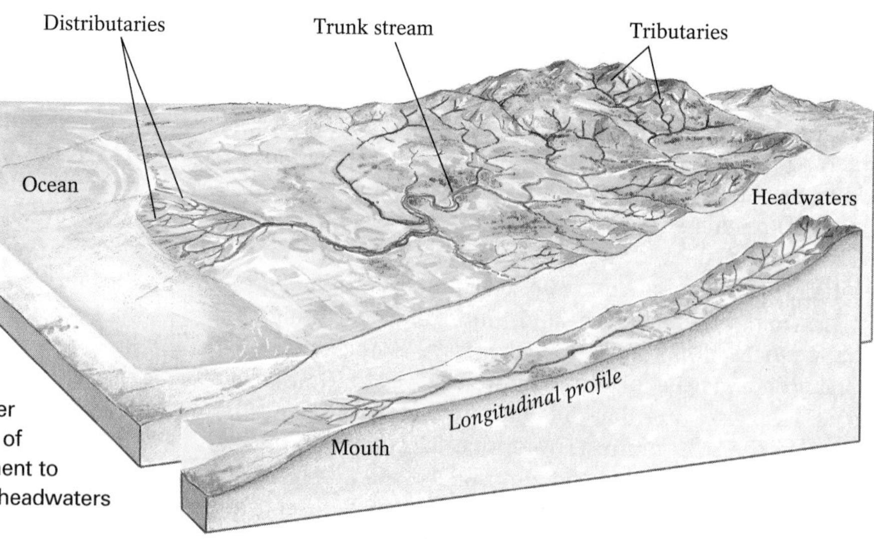

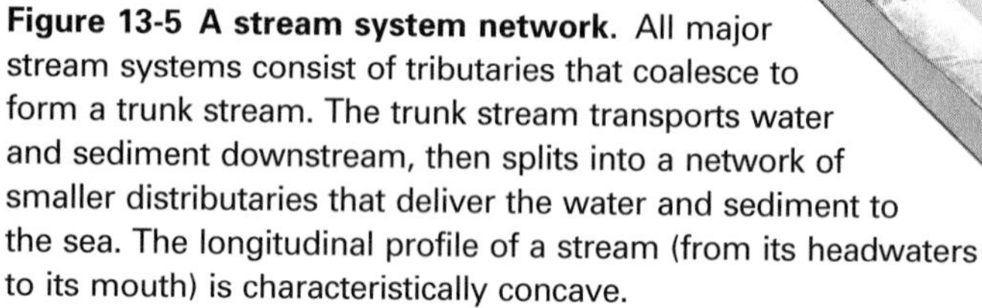
Figure 13-5 A stream system network. All major stream systems consist of tributaries that coalesce to form a trunk stream. The trunk stream transports water and sediment downstream, then splits into a network of smaller distributaries that deliver the water and sediment to the sea. The longitudinal profile of a stream (from its headwaters to its mouth) is characteristically concave.

and sediment load begin to disperse. Numerous small channels, or **distributaries,** may branch off, carrying water and sediment across lowlands to an ocean (Fig. 13-5).

Stream Velocity and Discharge

Stream velocity measures the distance that a stream's water travels in a given amount of time. Slow-moving streams have velocities of less than 0.27 meter (1.0 foot) per second; swift-flowing streams can have velocities exceeding 10 meters (33 feet) per second. In terms that are more familiar, this range is from 0.7 to 22.5 miles per hour. This velocity is driven in large part by the stream's **gradient,** or slope—its vertical drop in elevation over a given horizontal distance. The steeper its gradient, the more rapidly a stream flows. A stream's gradient depends upon the topography over which it flows, but generally decreases from its headwaters to its mouth.

The velocity of the water within a stream is not uniform and depends partly on where in the channel the water is flowing. The slowest water velocity occurs at the sides and bottom, because friction between the water and the channel impedes its passage. In a straight segment of a channel, water velocity is greatest in the center of the stream and just below the surface. Where a stream curves, we find the greatest velocity at the outside of the curve and the lowest velocity at the inside.

Although the overall direction of flow is downstream, within a stream channel the water flows turbulently rather than smoothly, moving in swirls and eddies that reflect the texture of its stream bed. Upstream in the headwaters, friction from the bouldery roughness of the channel bed creates a drag on the water that sends a chaotic streamflow swirling upward, downward, and sideways. Consequently, most of a stream's energy is expended against its bed, and the actual flow velocity remains relatively low. Downstream, velocity actually increases—despite a decrease in gradient—partly because the stream traverses a smoother bed of sand, silt, and clay, which lowers frictional resistance to flow.

Related to stream velocity is stream **discharge,** the volume of water passing a given point per unit of time. The discharge of small tributaries may be in the range of 5 to 10 cubic meters (180–350 cubic feet) per second. The discharge of the Mississippi River is roughly 18,000 cubic meters (600,000 cubic feet) per second. Discharge is dictated principally by the size of a stream's drainage basin. The discharge of the largest river on Earth, the Amazon of South America—which drains an area equivalent to 75% of the continental United States—is about 200,000 cubic meters (7 million cubic feet) per second, or about one-fifth of the Earth's entire freshwater streamflow. (One day's discharge of the Amazon would supply New York City's freshwater needs for more than *5 years!*) In addition, a stream's discharge is influenced by climatic conditions, such as the amount and timing of precipitation and the quantity of snowmelt, and by the local soil's ability to absorb water.

The Geological Work of Streams

In an unchanging environment, a stream's channel would maintain a shape and gradient in which the stream would flow just swiftly enough to transport all sediment supplied to it from the drainage basin, with little net erosion or deposition. A flow in such a state of equilibrium is called a **graded stream.** In reality, however, a stream's environment is always changing, and it must continuously respond to new levels of sediment load and discharge to establish a new graded state.

For example, a sudden increase in a stream's sediment load, such as results from a volcanic mudflow or the collapse of a stream bank, steepens the channel gradient and disturbs the stream's equilibrium. In response to the steeper gradient, the stream flows more swiftly. The faster flow erodes the added sediment until the graded equilibrium is restored (Fig. 13-6a). The streams of central California behaved in this fashion during the Gold Rush days of the nineteenth century, when mining operations dumped millions of tons of debris into them.

Conversely, a graded stream whose normal sediment load suddenly decreases, such as typically occurs downstream from a dam, has a steeper gradient and swifter flow than it needs to carry the reduced load. As a result, the stream quickly erodes additional sediment from its bed, thereby reducing its gradient (Fig. 13-6b). Construction of Hoover Dam, for example, caused the Colorado River to reduce its slope downstream as far south as Yuma, Arizona, located 560 kilometers (350 miles) away from the dam.

These continual adjustments to achieve an equilibrium make stream behavior one of the most dynamic of all geologic processes. Indeed, although streams carry only about one-millionth of the planet's water, they erode, transport, and deposit a massive amount of sediment. For this reason, they are the most important geological agent of surface change.

Stream Erosion

As a stream flows, it cuts downward in its channel toward its **base level,** the lowest level to which the stream can erode. The ultimate base level of most streams is sea level, but streams also may encounter temporary base levels, such as lakes or extremely durable layers of bedrock, that halt their downcutting for considerable lengths of time.

The principal result of this downcutting is the creation and deepening of stream valleys, which grow as overland flow and mass movement remove loosened material from the slopes on either side of a stream channel. The rate at which a valley erodes depends largely on the composition of its rocks and sediments. It could take thousands or millions of years for a stream to cut a valley through a granite batholith, whereas a single powerful flood can carve out a sizable valley in unconsolidated sands.

Valley formation also depends on climate. In temperate and humid regions, chemical weathering contributes to slope erosion by breaking down exposed bedrock; the resulting fragments are more easily removed by overland flow and mass movement. Thus stream valleys in such climates generally have a distinctive V shape, with gently sloping

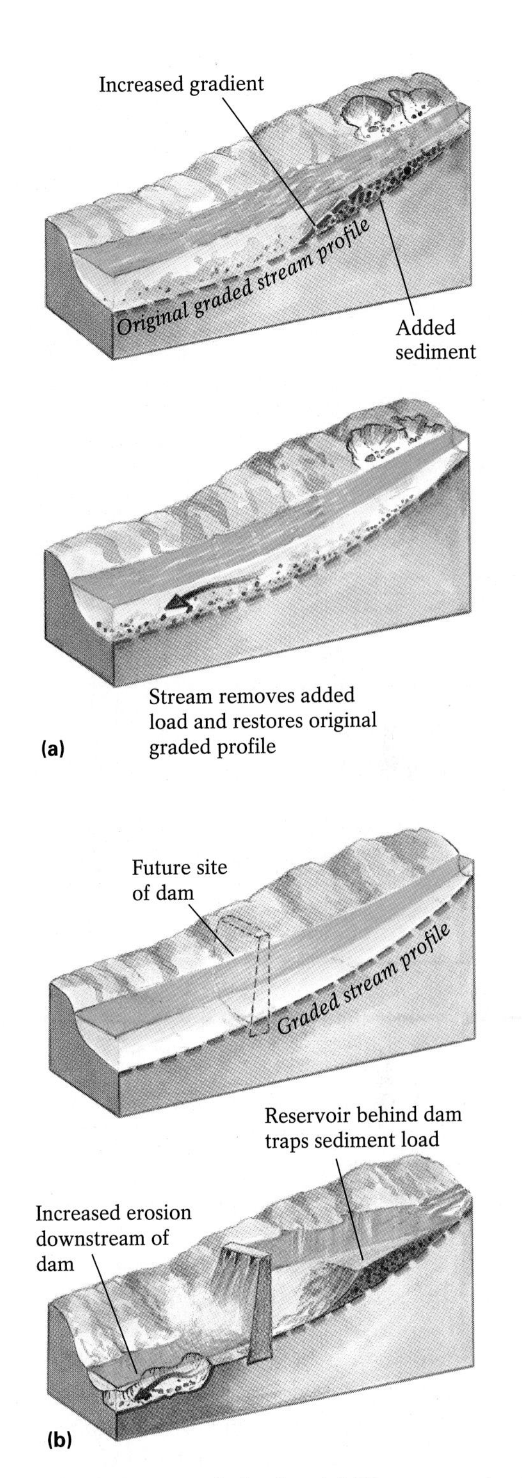

Figure 13-6 Graded stream behavior. (a) When a new load of sediment is added to a graded stream, the stream's gradient is instantaneously increased, causing it to flow faster. The faster-flowing stream erodes the added load and, in time, returns to its original gradient. **(b)** When a stream's sediment load is reduced or eliminated (for example, by being trapped in a new reservoir), the stream erodes more of its bed downstream, thereby decreasing its gradient to form a new graded profile.

Figure 13-7 Stream valley formation. The Yellowstone River in Wyoming cuts down rapidly and vertically through the uplifted rocks of its canyon. Mass movement in the upper portion of the canyon helps produce the valley's characteristic V shape.

walls (Fig. 13-7). In arid regions, on the other hand, only negligible chemical weathering takes place; with little loose sediment to remove, streams tend to produce steep-walled canyons.

Processes of Stream Erosion Streams erode their channels through the processes of abrasion, hydraulic lifting, and dissolution.

Abrasion is the scouring of a stream bed by transported particles. Fine particles suspended in the stream constantly burnish the bed's surface. Some large pebbles rotate in swirling eddies, carving circular depressions called *potholes* into the bedrock; others bounce against the underlying bedrock. Erosion by abrasion is most efficient in swift-flowing, sediment-laden floodwaters.

Hydraulic lifting, erosion by water pressure, occurs when turbulent streamflow through fractures in the bedrock dislodges sediment grains and loosens large chunks of rock. As with abrasion, the most active hydraulic lifting occurs during high-velocity floods; a 1923 flood along the Wasatch Mountain front of central Utah lifted 90-ton boulders and transported them more than 8 kilometers (5 miles) downstream.

Dissolution occurs when a stream flows across and dissolves soluble bedrock such as limestone, dolostone, and evaporites. For instance, the Niagara River, which flows between Lake Erie and Lake Ontario, carries 60 tons of dissolved rock over Niagara Falls every minute.

As noted earlier, a stream's erosive power increases with its velocity, with the increase in the rate of erosion being approximately equal to the square of the increase in velocity. Thus, if a stream's velocity doubles, its erosive power rises by a factor of four. For this reason, a stream's erosive power surges during a flood, when its velocity and discharge increase dramatically.

Drainage Patterns Streams erode their networks of tributary valleys in distinctive drainage patterns (Fig. 13-8), determined largely by the topography, composition, and structure of the affected terrain. When the rocks or sediments that underlie a drainage system are undeformed and share a uniform composition, they tend to erode into a branching *dendritic,* or tree-like, drainage pattern, the most common type of drainage pattern. A *radial* pattern develops on a landscape marked by topographic peaks, such as young volcanic cones or structural domes. A *rectangular* drainage pattern, which resembles a grid of city streets, forms when stream erosion enlarges perpendicular sets of fractures and faults in bedrock. A *trellis* pattern emerges where parallel ridges with erosionally resistant rocks separate narrow valleys of easily eroded rocks. Short, steep tributaries drain down the ridge slopes perpendicular to the main stream valleys, creating a pattern resembling the parallel slats of a garden trellis.

Stream Piracy The opposite sides of a drainage divide often differ in slope. The streams on the steeper slope tend to flow more swiftly and erosively than the ones on the other side, reflecting the difference in gradient. One side of a divide may also be eroded more effectively if it includes less resistant surface rocks, sediments, or soils or if it receives more rain than the other side, giving its streams greater, more erosive discharges. As a result, a stream on one side of the divide may eventually erode a channel through the divide itself and capture the headwaters of streams on the other side, a process colorfully dubbed *stream piracy* (Fig. 13-9). A stream that has lost its headwaters in this fashion, described as *beheaded,* simply stops flowing near the divide, leaving its channel dry.

Channel Patterns Stream channels can be straight, braided, or meandering. Straight channels are rare, but occasionally occur in areas where the channel shape is governed by linear features, such as fractures and faults, in erosion-resistant bedrock. They can also appear where steep slopes compel streams to take the most direct route downslope (usually at their headwaters, where gradients are steepest).

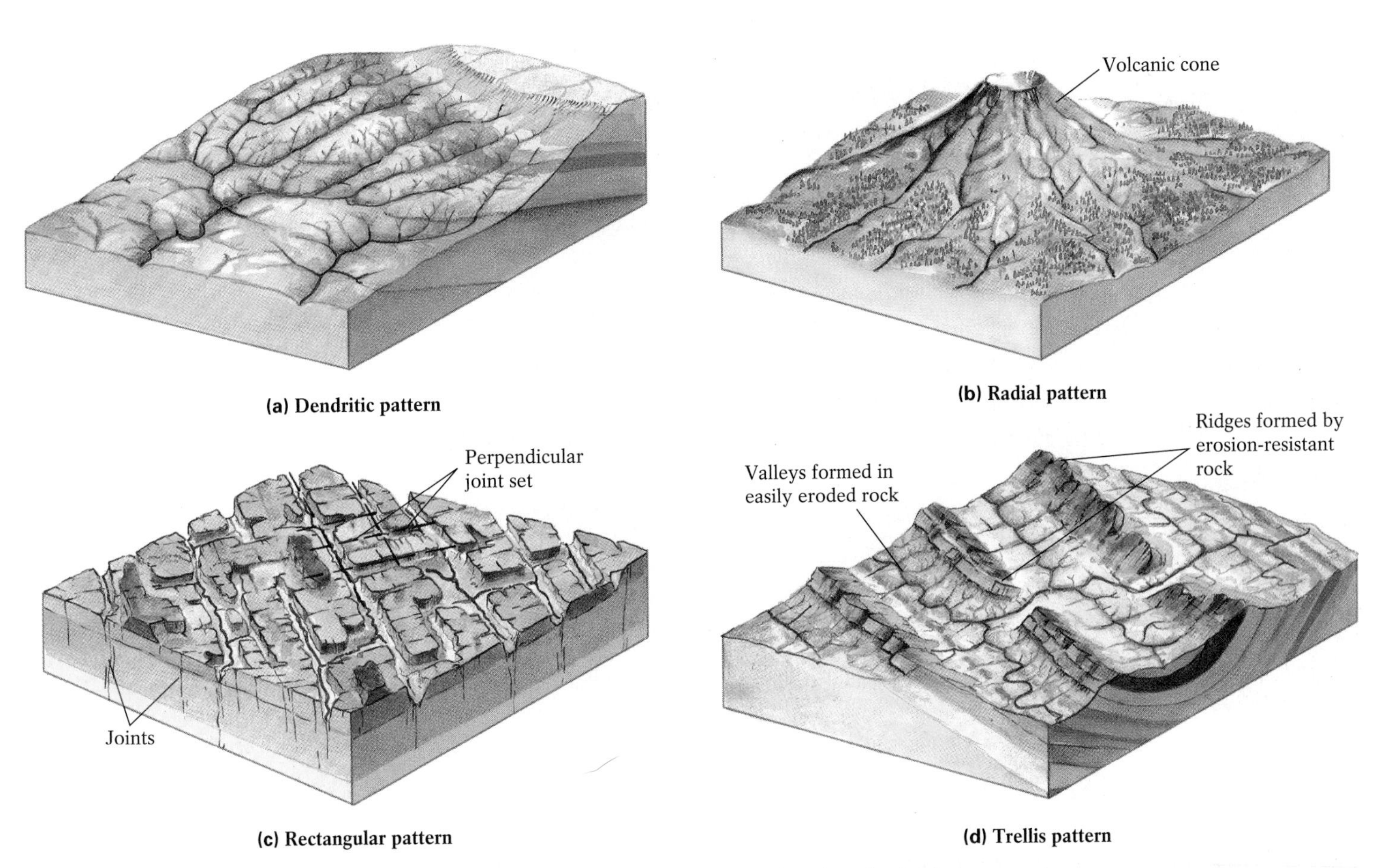

Figure 13-8 Four types of drainage patterns, produced by stream erosion of various bedrock types and under various structural conditions. **(a)** Dendritic drainage patterns often develop on undeformed sedimentary rock of uniform composition. **(b)** Radial drainage patterns usually form on volcanic cones of homogeneous composition. **(c)** Rectangular drainage patterns often appear where bedrock is cut by perpendicular joints. **(d)** The best place to look for trellis drainage is in an area with long parallel folds of sedimentary rock, such as the Ridge and Valley province of the Appalachians.

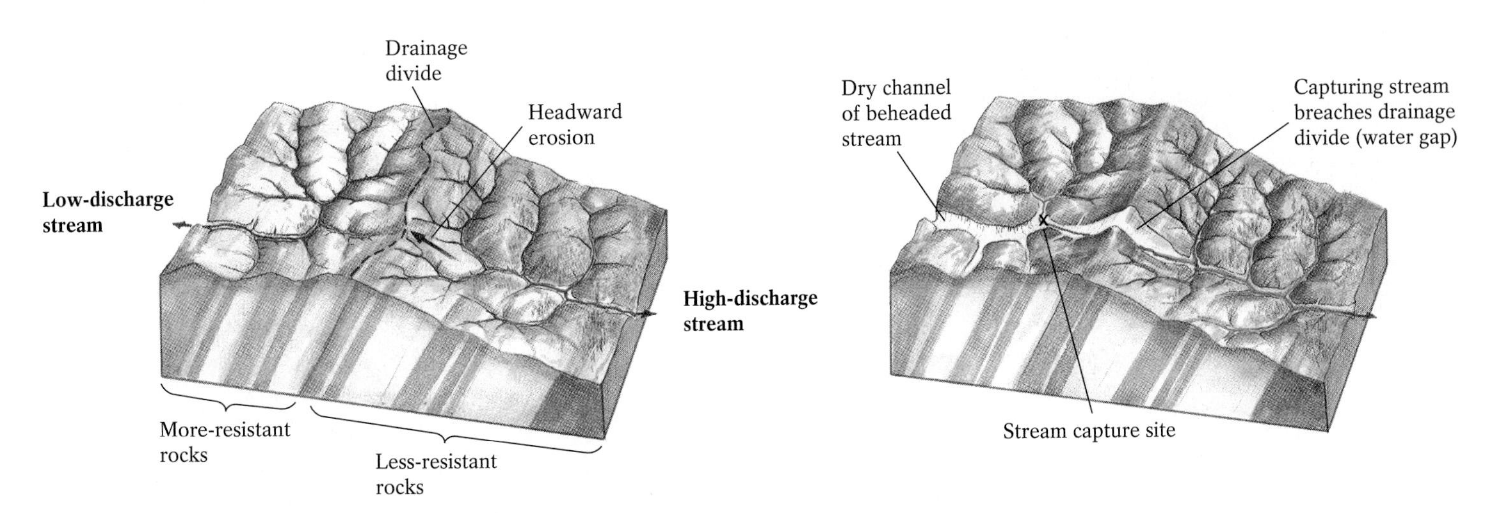

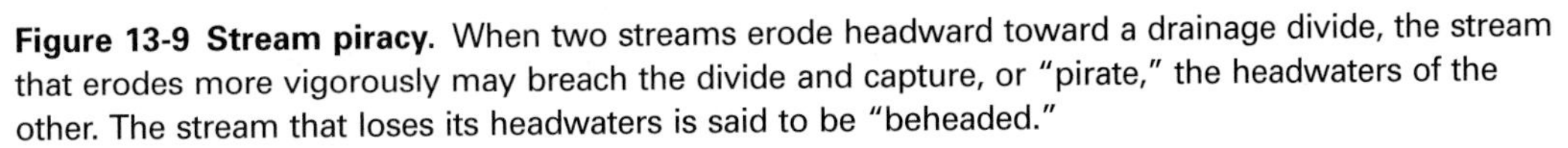
Figure 13-9 Stream piracy. When two streams erode headward toward a drainage divide, the stream that erodes more vigorously may breach the divide and capture, or "pirate," the headwaters of the other. The stream that loses its headwaters is said to be "beheaded."

Figure 13-10 A braided channel of the Toklat River, in Denali National Park, Alaska.

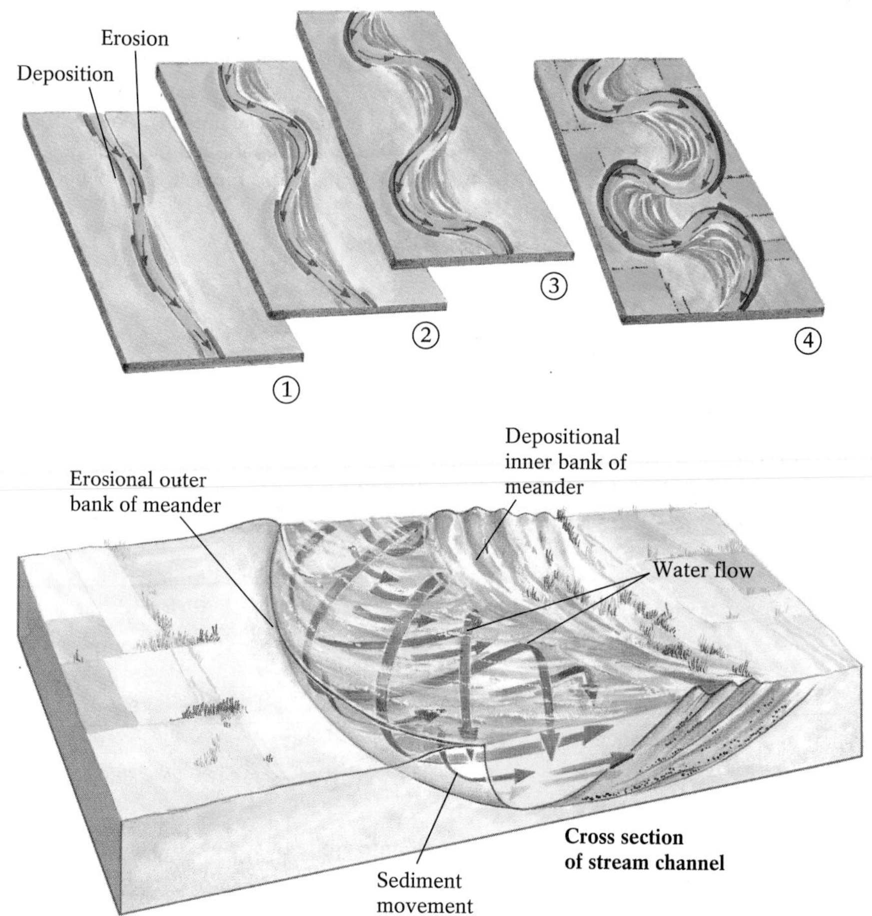

Figure 13-11 The evolution of meandering streams. As a somewhat curving stream flows, the high-velocity water at the outside of each bend erodes the outside bank of the bend. The eroded material is then deposited downstream at the inner bank of the next bend. Meander bends grow increasingly pronounced as progressively more sediment is removed from their outside banks and added to their inside banks. Photo: The meandering Little Bear River in Utah.

Figure 13-12 The evolution of waterfalls and rapids. When the graded profile of a stream is interrupted by a protruding resistant ledge or by faulting (or relative uplift of any kind), the stream at first cascades over the "step" in the landscape. Eventually, the force of the falling water undermines the cliff, causing the falls to retreat upstream. The falls are gradually worn down to rapids, which are in turn eventually completely eroded, returning the stream to its graded state. **(a)** Whitewater falls in Nantahala National Forest, North Carolina; this water drops 125 meters (411 feet) in a distance of 150 meters (500 feet). **(b)** A significantly smaller waterfall in Franconia Notch State Park, White Mountain National Forest, New Hampshire. **(c)** Rapids along the Ellis River in White Mountain National Forest.

Braided streams are networks of converging and diverging streams that are separated by narrow sand and gravel bars (Fig. 13-10). They develop in flows characterized by unusually high sediment load, a turn of events that forces a stream to deposit its excess sediment within its channel bed. These deposits may build up until they break through the water surface, forming islands around which streamflow diverges.

The flat land immediately surrounding a stream channel, which would be submerged if the stream were to overflow its banks, is called its **floodplain.** Within floodplains in relatively flat landscapes, streams tend to form fairly evenly spaced loops, or *meanders.* Hence such streams are called **meandering streams.** Meanders form when the greater flow velocity toward the outsides of the curves erodes sediment from the stream banks there and carries it downstream to be deposited on the inside bank of the next curve, where flow velocity is greatly reduced (Fig. 13-11). In this way, a meandering stream transfers sediments from its erosional outer banks to its depositional inner banks, causing the curves to become exaggerated and occur closer together until they form a series of loops separated only by thin strips of floodplain. During a flood, a stream may cut through one of these separating strips, bypassing an entire loop. The cut-off meander loop then becomes a crescent-shaped body of water called an *oxbow lake.* Subsequent flood deposits may fill oxbow lakes, forming *meander scars.*

Waterfalls and Rapids Waterfalls and rapids occur at sudden drops in topography along a stream's course, usually where erosion has removed softer sections of bedrock, leaving behind more resistant rock as a "step" in the stream's profile, or where faulting has lowered or raised a portion of the stream's profile (Fig. 13-12). Whitewater rapids are often relics

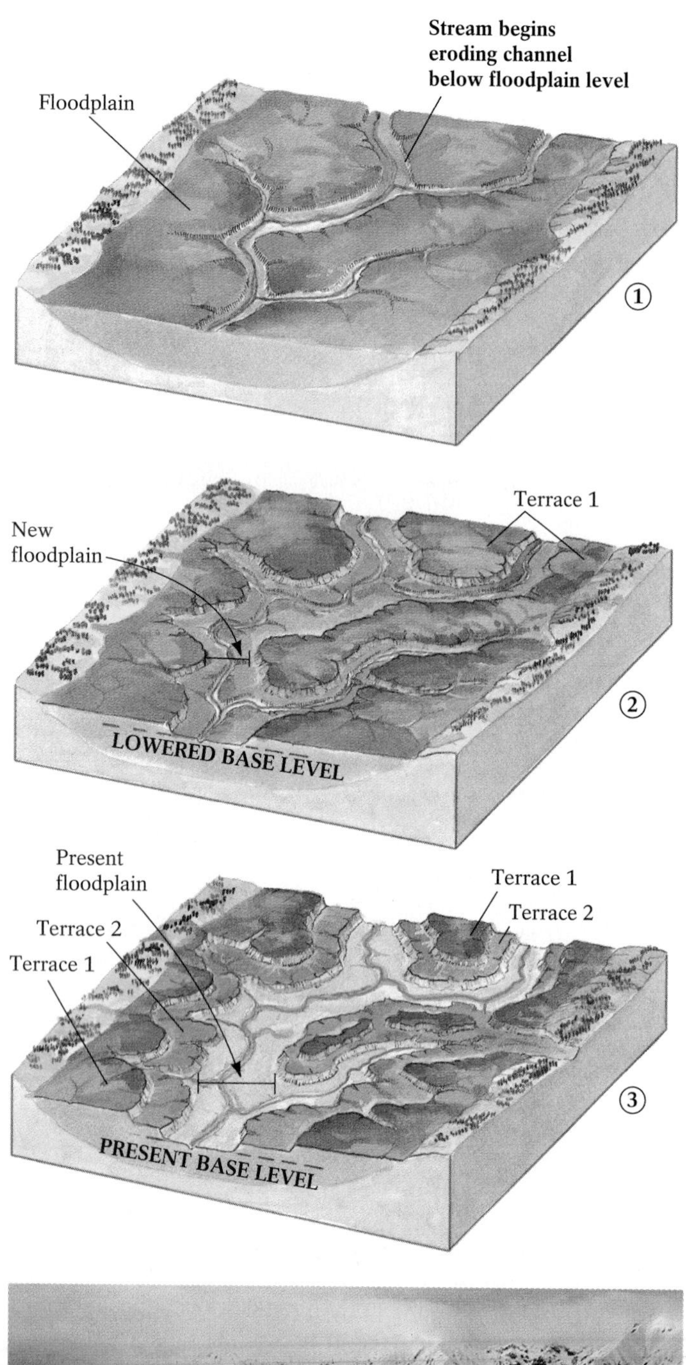

of waterfalls whose "steps" have been eroded away, allowing water to rush turbulently over irregular rocky beds.

Thousands of waterfalls and rapids are scattered across North America, from picture-postcard falls such as Yosemite Falls in California's Sierra Nevada to the countless smaller falls and rapids that dot the springtime streams of the eastern United States. Best known, of course, is Niagara Falls (although Yosemite Falls is nine times as high). Niagara, which is now roughly 55 meters (176 feet) high and 670 meters (2200 feet) wide, has been wearing itself away for 12,000 years. Formed when the last great North American ice sheet retreated north of the Great Lakes region, the falls has since moved southward 11 kilometers (7 miles) from its point of origin at Lake Ontario. Each year billions of liters of water thunder through the 50-kilometer (30-mile)-long gorge of the Niagara River between Lake Erie and Lake Ontario, plunge over the falls, and erode the shale bed below, undercutting the falls. Today, Niagara Falls is retreating slightly less than 0.5 meter (2 feet) per year.

Stream Terraces When a stream's discharge increases over a long period of time (perhaps because of a change to a wetter climate) or when its base level is lowered (perhaps because of a global sea-level change), the flow usually erodes its channel to a lower level. In such a case, the stream forms a new floodplain within the former one. The original floodplain is left high and dry, forming a gently sloping topographic bench known as a *stream terrace.* A terrace usually stands far enough above a stream that it remains untouched by even the most extreme flood. Terraces often appear in sets at progressively lower levels within a single stream valley, recognizable as remnants of a once-continuous floodplain surface (Fig. 13-13).

Stream Transport

The world's streams carry approximately 45 trillion cubic meters of water to the sea every year, along with 9 to 10 billion tons of sediment. The Mississippi River alone delivers more than 1 million tons of sediment *daily* to the Gulf of Mexico. The amount and type of sediment transported by a stream are largely determined by the stream's velocity, the composition and texture of the sediment, and the characteristics of the bedrock that the flow crosses.

Figure 13-13 The creation of stream terraces. **(1)** A terrace forms when a stream erodes its channel below the level of its floodplain, either because its discharge has increased or because its base level has been lowered. **(2, 3)** When a stream has undergone a series of such changes, it may form a series of tiered terraces in a single valley. Photo: Terraces above the Cave River at Arthur's Pass, New Zealand.

The maximum load of sediment that a stream can transport constitutes its **capacity.** Capacity is proportionate to the discharge: The more water flowing in the channel, the greater the volume of sediment that it can transport. The diameter of the largest particle that a stream can transport provides a measure of a stream's **competence.** Competence is proportionate to the square of a stream's velocity: The greater a stream's velocity, the larger the particles that the stream can transport. Thus, when velocity doubles, competence increases by a factor of four. This relationship explains why a stream that ordinarily carries only fine gravel can sweep boulders downstream during a flood. During the 1933 Tehachapi River flood in California, for example, steam locomotives of the Santa Fe railroad were carried hundreds of meters downstream and then buried by tons of transported stream gravel.

Streams transport sediment in different ways, depending on the sediment's particle size (Fig. 13-14). Most of the world's stream-borne sediment—about 7 billion tons annually—travels as **suspended load.** The suspended load consists of very fine solid particles that remain suspended in a stream, because the stream's turbulence and velocity prevent the particles from settling to the bed. Approximately 70% of the Mississippi River's annual load of 500 million tons of sediment is suspended. The Colorado River derives its characteristic red color from suspended particles of red silt and sand eroded from the bedrock of the Grand Canyon region.

Coarse particles that move along the stream bottom form the **bed load.** Unlike its suspended load, which moves constantly, a stream's bed load is transported only when the water's velocity becomes great enough to lift or roll larger particles. The coarse material that makes up bed load includes a wide variety of particle sizes. The sandy portion moves along the bed with a bouncing motion known as *saltation.* Saltating sand grains are lifted from the bed by hydraulic lifting or by collisions with other saltating grains. Once introduced into the streamflow, these grains are carried upward and forward by the stream's turbulence and velocity until the pull of gravity brings them back down to the bed. The larger, heavier cobbles and boulders usually move by *traction,* in which the force of the water slowly rolls them along. The largest boulders can be moved only by major floods. You can often see them settled into the bottom of a stream and covered with algae; little algae would grow if the boulders moved more frequently.

The remainder of the sediment load, which is carried invisibly as dissolved ions in the water, forms the **dissolved load.** The amounts and kinds of dissolved material in a stream depend on several factors, such as climate, vegetation, and bedrock composition, that influence the rate of chemical weathering in the region. In general, higher dissolved loads are found in streams that flow through warm, moist, fairly flat regions filled with lush vegetation and outcrops of soluble bedrock such as limestones and evaporites. Dissolved ions are most abundant in local groundwater, whose slow movement through soluble bedrock promotes dissolution

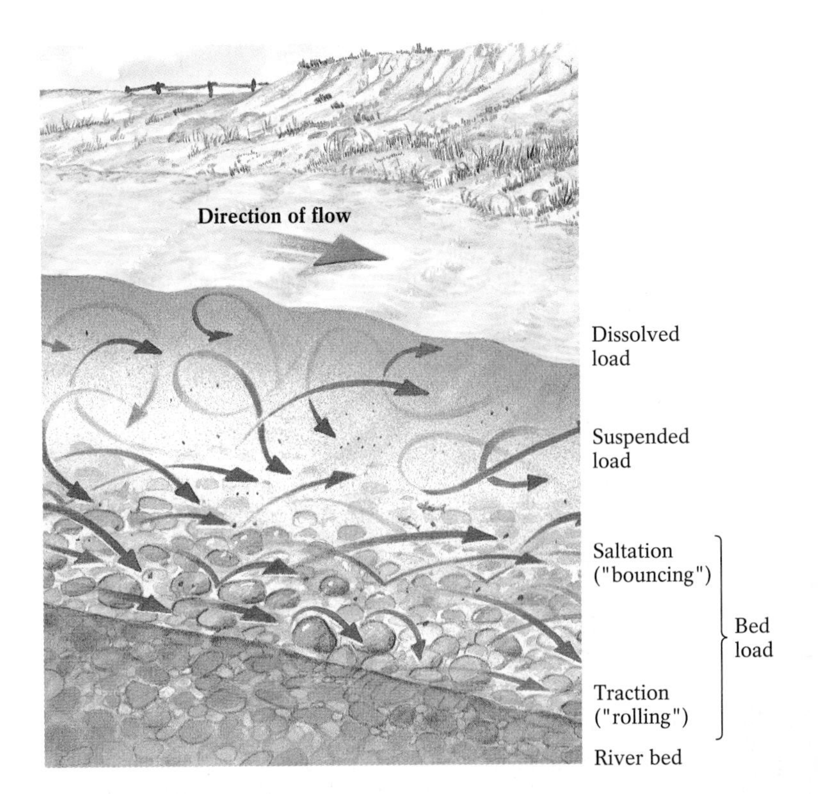

Figure 13-14 Sediment distribution and movement within a stream.

more effectively than contact of a flowing stream with its bed. For streams that derive most of their discharge from groundwater—a type that includes most streams—this bedrock dissolution represents the main source of their dissolved load. Velocity has no effect on a stream's ability to carry dissolved load, although it might affect the rate at which material enters solution.

Stream Deposition

When the downward pull of gravity on stream-borne particles overcomes the effect of stream velocity, the particles are deposited on the stream bed, with the heaviest items settling out first. Deposited stream sediments are described collectively as **alluvium.** As much as 75% of stream alluvium drops into a stream's channel, onto its floodplain, onto flat valley floors at the foot of steep gradients, and into standing bodies of water along its course; the remainder settles where the stream enters the sea.

In-Channel Deposition During normal (nonflooding) flow, a stream with large amounts of coarse bed-load sediment often deposits some of its load in its channel as mid-channel bars. As sand and gravel accumulate, the water becomes diverted around the channel bars and forms a braided stream (see Fig. 13-10).

In-channel deposits may also take the form of **point bars,** which arise when sediments are scoured from the outside banks of meandering streams and deposited on the stream's inner banks (see Fig. 13-11). As a point bar grows, the water flowing over the bar becomes shallower, increasing the friction between the water and the surface of the bar and decreasing the stream velocity even further, thereby promoting even more deposition (Fig. 13-15). Because point bars appear where a meandering stream drops its heaviest particles, gold, platinum, and silver are often concentrated near them. Such gold deposits in the rivers of the Yukon and Alaska beckoned thousands during the gold rushes of the nineteenth century.

Figure 13-15 Growing point bars on the Madison River, Montana.

Floodplain Deposition When a flooding episode begins, stream velocity is increased, and the channeled floodwaters carry a larger-than-normal volume of sediment. When the stream's channel becomes unable to hold all of the water coursing through it, the water flows over its banks. As the water spreads over the relatively flat floodplain, its velocity decreases and it drops its sediment load. Thus, as its name suggests, a floodplain serves as a major repository of flood-borne stream sediment. Figure 13-16 illustrates some typical features associated with floodplains.

During a flood, much of a stream's excess coarse-grained sediment is deposited on top of the stream banks directly adjacent to the channel, forming ridges of sediment called **natural levees**. Natural levees, which grow higher with each flood, tend to be the highest points on a floodplain; they may sometimes act as barriers to prevent lower-volume flows from overflowing the channel banks. Between floods, trees may take root on the levees, stabilizing them and enhancing their usefulness in flood prevention (Fig. 13-17).

After the flooding subsides, the water left standing on the floodplain usually evaporates or slowly infiltrates into the groundwater system. If a floodplain is marked by surface depressions, however, water may remain at the surface as a wetland, or **backswamp.** Backswamps serve as habitats for migrating birds and other wildlife and help to replenish groundwater supplies. In the past, these areas were routinely drained to allow farming on their unusually rich soil or simply to provide land for housing or industry.

Another floodplain feature occurs when growing levees divert the course of river tributaries away from a river, forcing them to flow parallel to the main channel for tens or hundreds of kilometers before they can cross a low spot in the levee and rejoin the trunk stream. Such parallel streams are called *yazoo streams;* they are named for the Yazoo River, which shadows the Mississippi River for 320 kilometers (2000 miles) before finally joining it near Vicksburg, Mississippi.

Deposition at the Foot of Mountains When a stream carrying coarse sediment leaves a narrow, high-gradient mountain valley and flows out onto a wide valley floor, its velocity declines sharply. As a result, the stream deposits its load on the valley floor. Free from the confinement of the narrow valley,

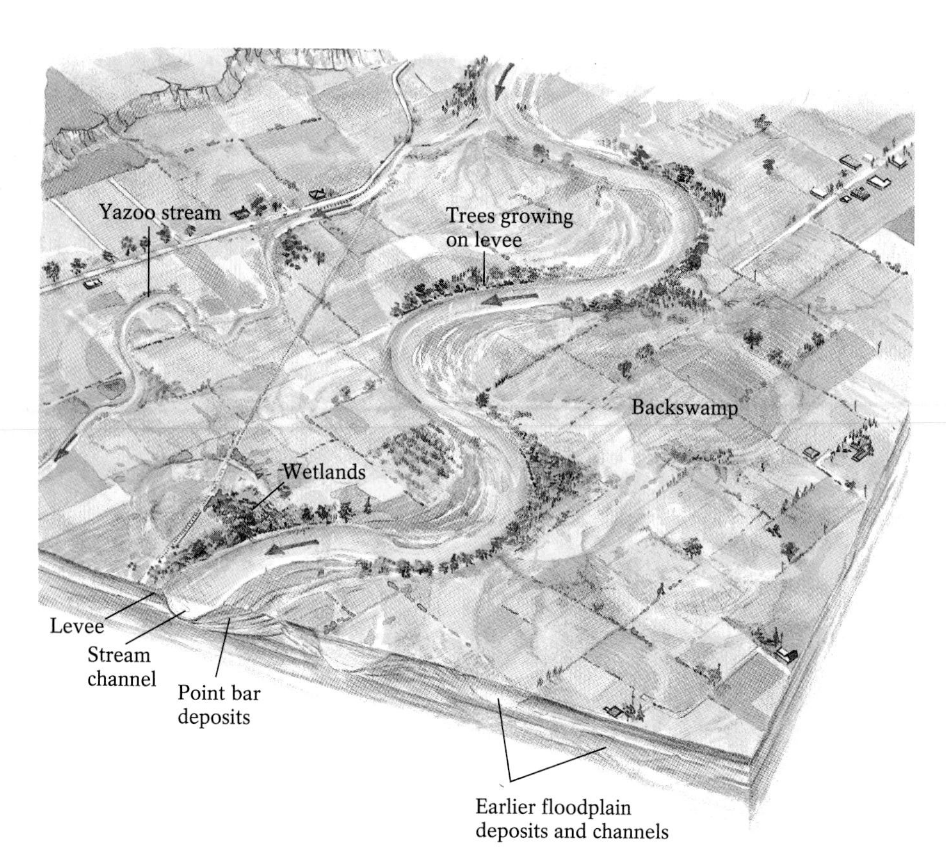

Figure 13-16 Floodplain features. These features are produced when streams overflow their banks and deposit their sediment on the surrounding land.

Figure 13-17 A vegetated levee on the bank of the Mississippi River. Natural levees, composed of coarse sediment, are generally the best-drained parts of the floodplain and consequently are the most capable of supporting trees that cannot tolerate saturated soil conditions.

Figure 13-18 An alluvial fan in Death Valley, California. Alluvial fans form when streams flowing from steep mountain slopes encounter a sharp reduction in slope at the foot of the mountain and drop their sediment loads as fan-shaped deposits.

the stream spreads out across the plain and drops its sediment load in a characteristic triangular shape known as an **alluvial fan** (Fig. 13-18).

Deposition into Standing Water When a stream enters a standing body of water, such as a lake or ocean, its flow decreases abruptly. There the stream deposits its suspended load of fine sand, silt, and clay; this material accumulates in the standing water to form a **delta,** a roughly triangle-shaped alluvial deposit that fans outward from the mouth of a stream (Fig. 13-19). A delta grows outward as long as its stream deposits more sediment than is eroded from the delta by waves and shoreline currents. The Earth's great deltas are found where major rivers—such as the Ganges in Bangladesh, the Indus in Pakistan, the Nile in northern Africa, and the Mississippi in North America—deliver large sediment loads to

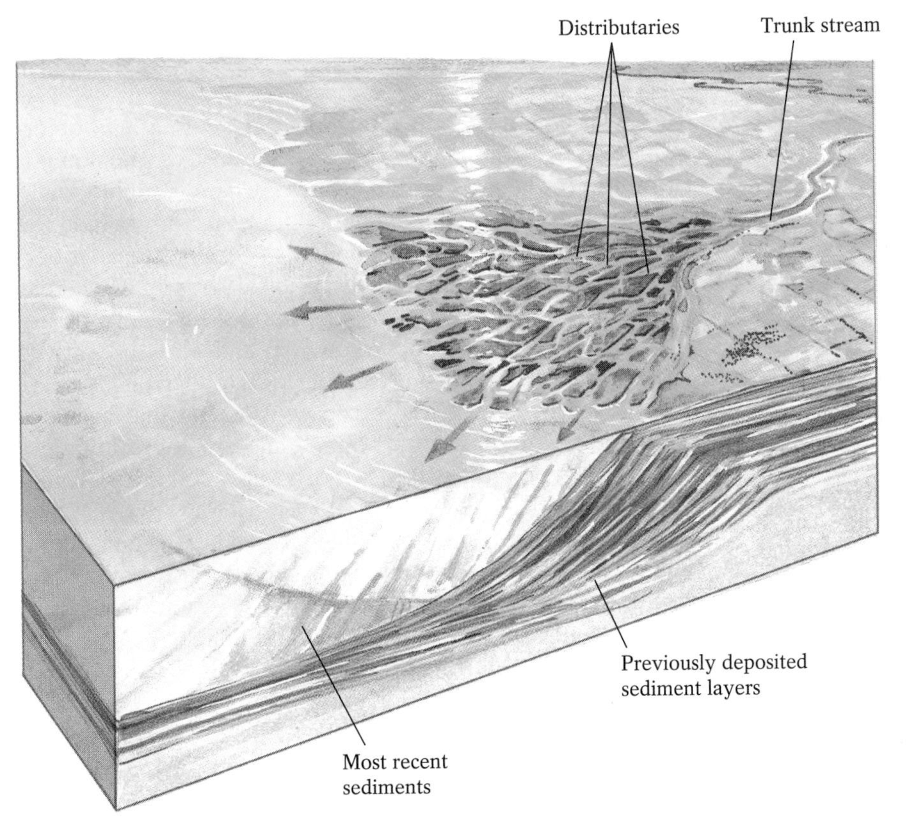

Figure 13-19 The anatomy of a delta.
Photo: The Tengarito River delta, New Zealand.

relatively quiet coastal waters. At the mouths of other large rivers, such as the Amazon in South America and the Niger in Africa, vigorous waves and currents immediately sweep away discharged sediments, and deltas do not form.

As a stream's delta grows, its gradient decreases; consequently, the water flows more slowly and gradually loses the capacity to carry its suspended load. As the stream drops more sediment, its channel becomes clogged. The pent-up flow subsequently branches into a new network of *distributaries.* The channel also becomes shallower, increasing the likelihood that the stream will flood and break through its levees. Freed in this way from its original channel, the stream may eventually discover a more direct route to the sea (with a steeper gradient) and begin to build a new delta at its new outlet. As the new delta grows, the abandoned delta, which lacks a replenishing supply of sediments, begins to erode. Thus the shape and location of a stream's delta change continuously (Highlight 13-1).

Stream Evolution and Plate Tectonics

Our understanding of the evolution of stream systems continues to grow, bolstered by the growing body of knowledge related to plate tectonics. For example, we now know that plate-margin stresses produce much of the local uplift that downcutting streams erode, creating spectacular waterfalls, distinctive terraces, and deeply incised river channels (Fig. 13-20). In addition, tectonic activity often determines the character of regional stream drainage patterns. For example, tectonic stresses cause the faults and fractures that erode to produce a rectangular drainage pattern, whereas folded mountain belts at convergent plate boundaries tend to display trellis drainage. Radial drainage patterns evolve as streams dissect the young volcanoes that are generally found at active plate boundaries.

Figure 13-20 The Goosenecks of the San Juan River, southeastern Utah. These meanders have been deeply incised into the Colorado plateau, which has been tectonically uplifted by interactions between the North American and Pacific plates.

Many large stream systems originate in the folded mountains found at convergent margins; they cross stable mid-plate interiors and then build distributary systems and deltas at passive plate margins. For example, the Amazon River rises in the Andes of western South America, where the South Pacific's Nazca plate subducts beneath the South American plate. The river then flows eastward across the continent for thousands of kilometers before emptying into the Atlantic Ocean. Similar tectonically-produced drainage divides appear in the Alps of southern Europe (which rose from the convergence of the African and Eurasian plates), the Himalayas (produced by convergence of the Indian and Eurasian plates), and the North American Rockies (originating from convergence of the North American and Pacific plates).

Controlling Floods

Many streams overflow their banks every two or three years, causing flooding. Such an event is usually the result of local weather—simply too much rain falls or snow melts, producing too much water for a channel to carry. A variety of factors determine how much precipitation represents "too much."

Factors Promoting Floods

The surface geology and topography of an area help determine whether water runs rapidly off the surface into streams or infiltrates the groundwater system, where it could remain long enough to avert flooding. Important factors include the following issues: whether the bedrock near the surface is permeable, like sandstone, or impermeable, like granite; whether the rock is highly fractured; whether the surface soils are clay-rich types that swell when wet and seal the surface against infiltration; and whether the local topography is steep, flat, or gently sloping. Of course, *any* ground that becomes saturated can no longer absorb any additional precipitation, and the excess water will run off the surface directly into local streams.

Flooding becomes more likely when heavy rains fall on an area where forest fires, droughts, widespread clear-cutting, or urbanization has thinned vegetation. Trees and their root systems tend to keep soils porous and enhance water infiltration, thereby reducing the likelihood of floods. In contrast, roads, buildings, and parking lots prevent surface water from infiltrating the groundwater system, thereby increasing the potential for flooding (Fig. 13-21).

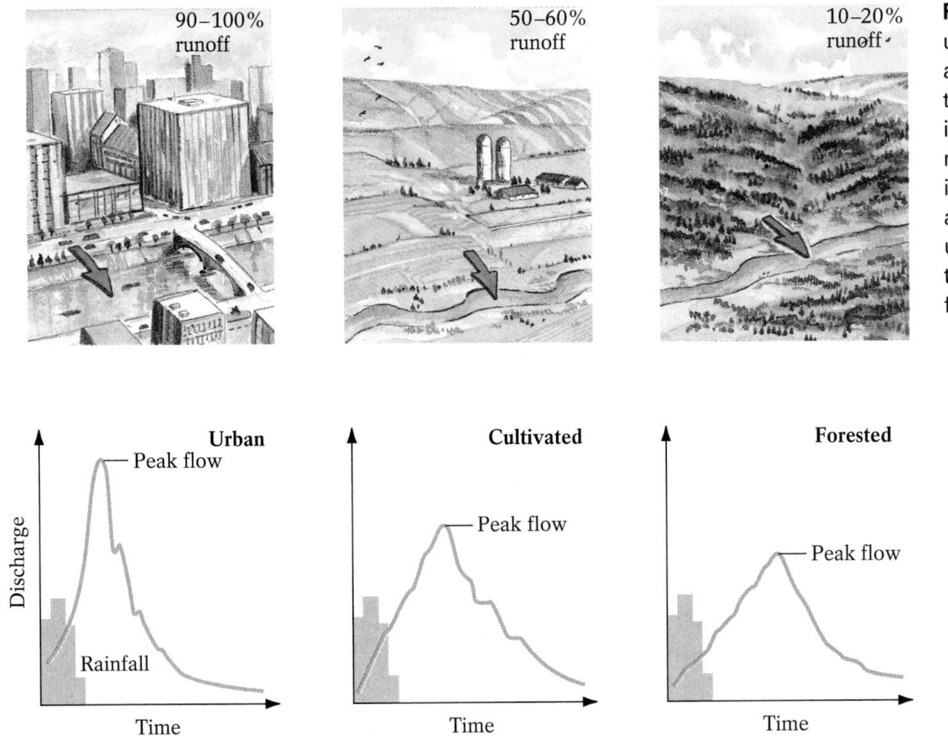

Figure 13-21 Flooding potential. Because urban areas are largely paved over with asphalt and cement, little surface water infiltrates into the groundwater system; in fact, most runs off immediately into local streams. By comparison, much of the precipitation in undeveloped areas is absorbed into the soil. Given the same amount of rainfall, then, stream discharge in urban areas is greater and peaks more quickly than that in undeveloped areas, making the former more prone to flooding.

Floods are usually seasonal, occurring during heavy spring rains and snowmelt in temperate climates and during rainy seasons elsewhere. Most streams have an annual pattern of flooding, to which a region's residents may be compelled to adjust. It is the relatively rare major flooding event, however, that causes the greatest damage. Predicting such floods is necessarily an imprecise art; to a large extent, the irregularities of weather remain a perplexing mystery to us. Our best course is analyze the frequency of past floods and then use these data to determine the statistical probability of a major flood occurring within a given time frame.

Flood Prevention

A number of defensive strategies—structural, nonstructural, and combinations of both—have been employed in an effort to curb the catastrophic effects of floods. The most common stream-containment structures are *artificial levees,* earthen mounds built on the banks of a stream's channel. Artificial levees have been constructed along the Mississippi River to protect croplands since the eighteenth century. The U.S. Army Corps of Engineers has built concrete *flood walls,* which are more expensive, to prevent overflow from the channel at strategic locations, such as along the commercial centers of riverside cities such as St. Louis. *Flood-control dams* are often emplaced upstream from densely populated or economically important areas. These earthen or concrete structures restrict streamflow and temporarily store the water in reservoirs.

Channelization is a structural alteration made to a stream's channel so as to speed the flow of water and thus prevent it from reaching flood height. It may involve clearing obstructions such as fallen trees that slow the stream's flow or widening or deepening a channel by dredging to increase its capacity. More radical channelization, undertaken to increase a stream's gradient and therefore its velocity, usually involves cutting off meanders to straighten a stream. The shorter, straight channel will have a steeper gradient than before, and its increased velocity will transport more water—perhaps rapidly enough to curtail flooding.

Channelization can also have other, more inconvenient, effects on streams. Since 1910, this method has been employed to control the Blackwater River, a tributary of the Missouri River, southeast of Kansas City. The Blackwater had flooded regularly, because it flowed too slowly to accommodate large storm discharges. In an attempt to solve the problem, engineers excavated a straight channel that increased both its gradient and velocity. This approach also caused the subsequently more active river to scour its bed more energetically, however. The river grew from an initial depth of about 4 meters (14 feet) to an average depth of more than 12 meters (39 feet), and from 9 meters (30 feet) wide to more than 71 meters (233 feet) wide. As the river widened, it undermined bridges, which then had to be replaced. Only constant maintenance now keeps the Blackwater River straight.

Structural solutions to flood problems tend to be very expensive and may give a false sense of security to nearby residents. For these reasons, a nonstructural strategy, if

Highlight 13-1 The Mississippi River and Its Delta

The Mississippi River begins its 3750-kilometer (2350-mile) journey as a creek a few meters wide and 10 centimeters (4 inches) deep that trickles from Lake Itasca in the pine forests of northern Minnesota. In its first 2100 kilometers (1300 miles), the upper Mississippi cascades over bedrock falls and rapids wherever it is not dammed or otherwise restrained. At Cairo, Illinois, the Ohio River joins the Mississippi and the two streams form a majestic alluvial valley—the much tamer lower Mississippi valley—that stretches for more than 1600 kilometers (1000 miles) and reaches widths of 48 to 200 kilometers (30–125 miles). By the time the Mississippi empties into the Gulf of Mexico 200 kilometers (120 miles) south of New Orleans, it carries water and sediment from the Missouri, Ohio, and Arkansas rivers, as well as from more than 100,000 other streams, and has deposited fertile silt over more than 770,000 square kilometers (300,000 square miles) of floodplain farmland (and, during the occasional flood, over one or more cities).

The Mississippi began to construct its present-day delta about 9000 years ago. The delta consists of at least seven distinct lobes, each of which was once the river's active delta. Together, these features constitute about 40,000 square kilometers (15,000 square miles) of real estate that has been added to Louisiana. Each inactive lobe was abandoned as the river discovered shorter, steeper routes to the Gulf of Mexico. The Mississippi River *bayous*, the colorful network of lakes and minor streams in southern Louisiana, are abandoned channels crisscrossing inactive lobes. Many more abandoned lobes have eroded away completely. The active portion of today's delta, known as the Balize delta, is 1000 years old (Fig. 1).

The bird's-foot shape of the Balize delta was built by three major distributaries whose gradients are now too gentle to transport sediment effectively. For approximately 100 years, the Mississippi River has been trying to cut a steeper, far shorter (225-kilometer [150-mile]) course along the Atchafalaya River, 100 kilometers (60 miles) to the west of its present channel. This diversion has been prevented only by a gated structure built at the Atchafalaya fork that strictly controls the division of streamflow between the Atchafalaya and the Mississippi. In 1973, a winter of unusually high precipitation produced raging floods in the lower Mississippi valley that might have demolished this structure, allowing the river to finally take the Atchafalaya course and bypass New Orleans, thereby leaving it a sleepy bayou town instead of a center of river commerce. Floodgates at the Atchafalaya fork were opened to divert water from the main channel to Lake Pon-

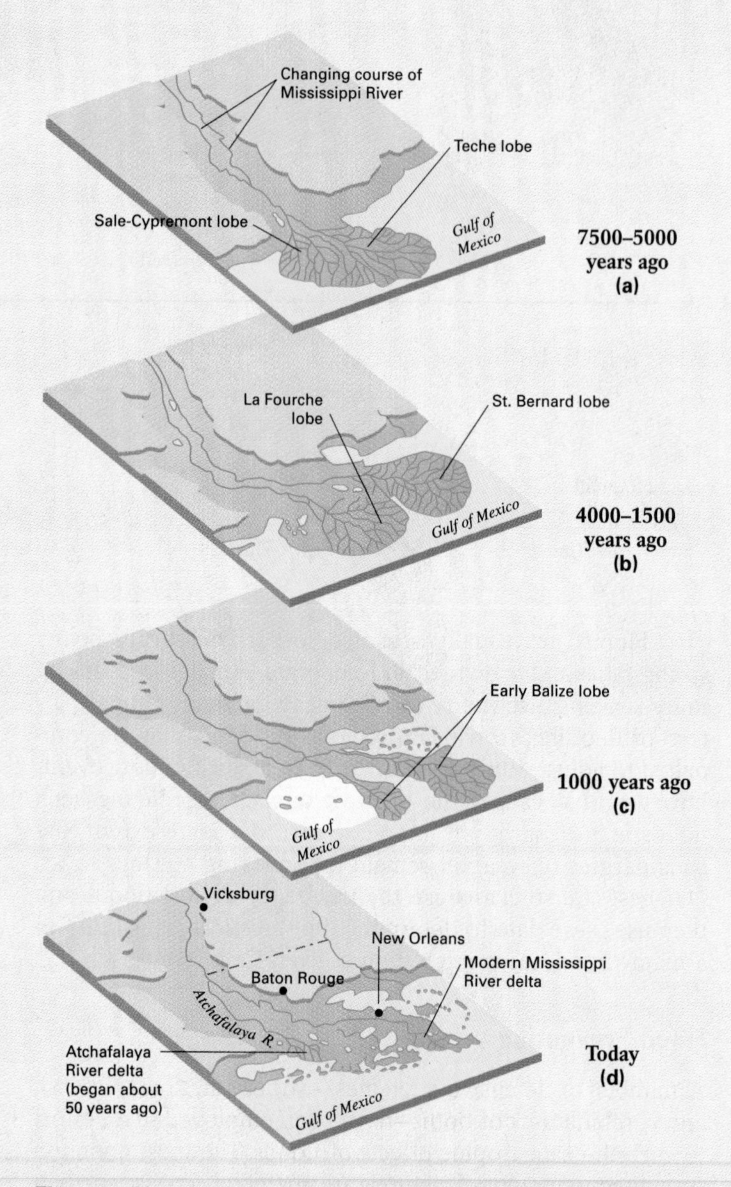

Figure 1 The changing face of the Mississippi River delta in the last 7500 years. During the nineteenth and twentieth centuries, the Atchafalaya River distributary has begun to carry more of the Mississippi's flow; someday it may become the river's main channel. Photo: Today's Mississippi River delta, the active segment of which is centered at Head of Passes, Louisiana.

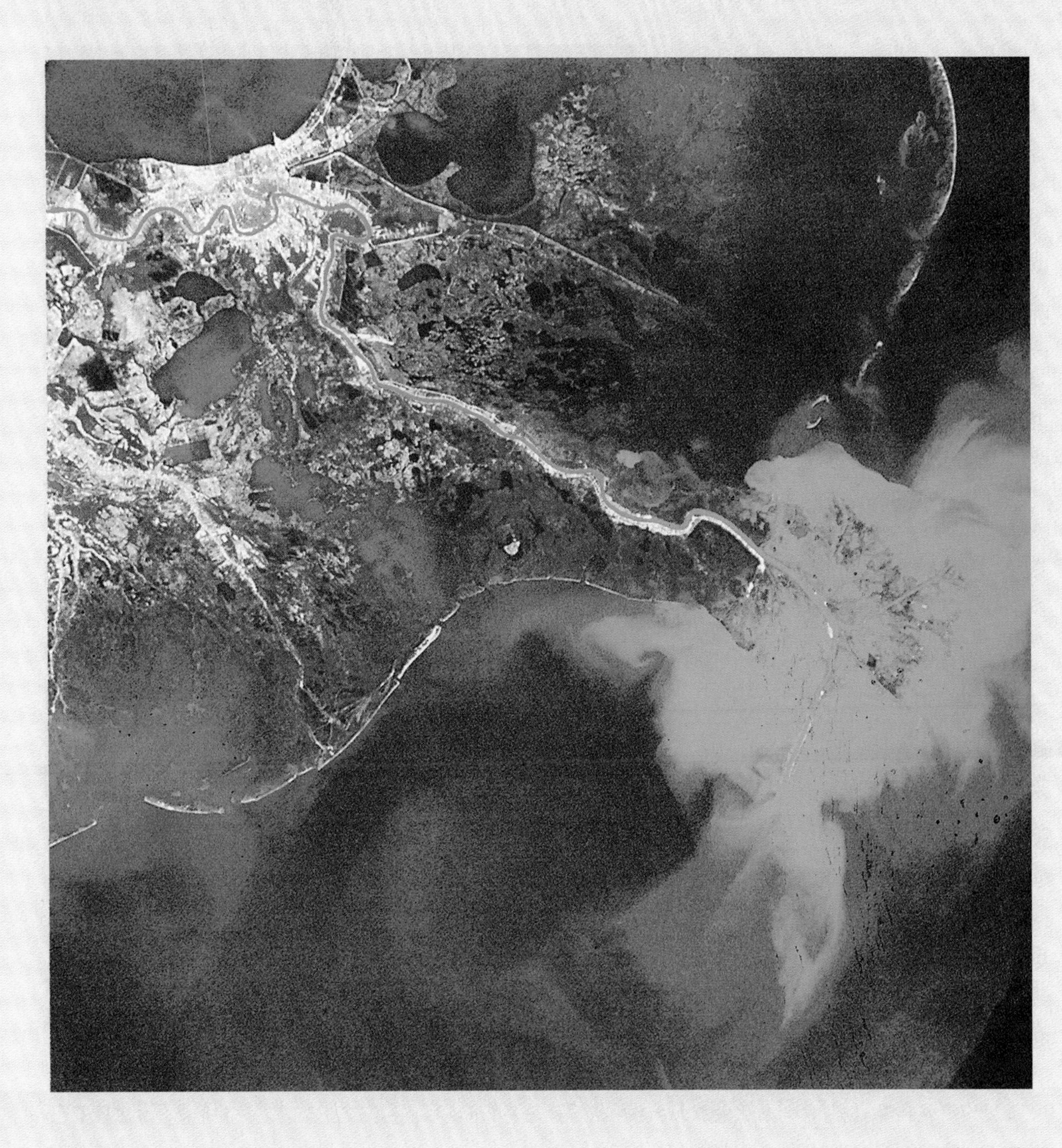

chartrain, thereby preventing the river from overflowing its channel.

Continued dredging to deepen the lower Mississippi channel enough to accommodate the river's flow has thus far saved Baton Rouge and New Orleans from obsolescence. Eventually, however, Mother Nature will probably have her way, and the lower Mississippi will inevitably find its new route to the sea.

Highlight 13-2 Did We Help Cause the Midwestern Flood of 1993?

In the beginning of this chapter, we discussed one of North America's worst floods, which took place along the upper portions of the Mississippi River in the summer of 1993 (Fig. 1). Did human activity—65 years of "managing" the Mississippi River to prevent floods—actually contribute to the magnitude and tragedy of the flood?

Before the last 100 years or so, the Mississippi and its tributaries determined their own boundaries. During periods of high water, the rivers broke through or overflowed natural levees, flooding tens of thousands of square kilometers of the surrounding, largely uninhabited, lands. In the twentieth century, however, millions of people migrated to the region, building cities, towns, and large farms along the rivers' banks. When a flood in 1927 took 214 lives, Congress enacted the first Mississippi River Flood Control Act, assigning the U.S. Army Corps of Engineers the daunting task of confining the river to its channel.

The Corps of Engineers' efforts consist of about 300 dams and reservoirs and thousands of kilometers of artificial levees and concrete flood walls, all designed to prevent the river from spilling onto its natural floodplain. The system also contains numerous pumping stations, spillways, and diversion channels designed to divert water for storage in temporary holding basins. But the events of 1993 showed that such structures simply cannot contain an extraordinary flood. By confining such great discharges to a channel, the artificial retention structures actually caused the swollen rivers to flow more rapidly and violently, thus damaging the very structures designed to restrain them. By denying the river access to its natural floodplains, the structures caused the streams to rise higher than they would have otherwise, ensuring that once they did breach the levees, the floods would cause greater damage. Furthermore, the existence of artificial levees and flood walls had encouraged the growth of cities, towns, and farms closer to the riverbanks than was really safe.

Figure 1 Flooding at Kaskaskia, Illinois.

What does the future hold for the residents of the upper Mississippi valley? Certainly more flooding, but perhaps less human interference with the river's natural behavior. Some communities have proposed that all flood-retention systems be eliminated and that zoning limit future development within the river's floodplain. Others look longingly at St. Louis's 16-meter (52-foot)-high concrete flood wall, which saved that city's downtown business district when the Mississippi reached its record crest at 14.2 meters (47 feet). The debate continues between those who believe we can tame the mighty Mississippi and those who believe we cannot.

feasible, is considered preferable. Nonstructural defenses against flooding depend on identifying high-risk areas, instituting zoning regulations to minimize development in them, and managing resources to minimize the amount of water entering a stream channel at any one time. An important part of any management strategy is the prevention of excessive forest clearing and the prompt suppression of forest fires. This effort helps maintain the vegetation cover, which, as we know, promotes the absorption of water into the groundwater system and reduces stream discharge.

How have we fared in our efforts to prevent floods? Highlight 13-2 focuses on the tragic 1993 flood discussed at the beginning of this chapter, illustrating how nature can confound our most well-intentioned flood-prevention strategies.

Figure 13-22 A 3-mile-wide channel north of the Martian equator. This channel, which was photographed in 1972 by Mariner 9, and other apparently fluvial features on Mars are believed to have formed billions of years ago, when the Martian climate and atmosphere allowed free water to exist and flow at the planet's surface.

Extraterrestrial Stream Activity: Evidence from Mars

Photographs taken by the 1971 Mariner 9 spacecraft and the 1976 Viking probe clearly reveal braided channels, stream-modified islands, dendritic drainage patterns, and catastrophic-flood topography on the surface of Mars. On Earth, flowing water typically produces such features. Yet virtually no free water exists on Mars' surface; all the liquid water at its surface and in its atmosphere would barely fill a small swimming hole. How can we explain the spectacular fluvial (river-related) landforms photographed on Mars?

Although no rain falls on Mars today, during the early life of the solar system water may have accumulated there, just as it did on Earth, from volcanic outgassing of the planet's interior. Early Mars likely had a more significant atmosphere and greater atmospheric pressure, and was almost certainly warmer, than the present-day planet. Thus water could have survived at the surface during Mars' early years. Some of Mars' fluvial-looking features (such as the channel depicted in Figure 13-22) suggest that 3 to 4 billion years ago Mars' surface may have been dotted with lakes and streams nourished by frequent rainstorms.

Some geologists have suggested that Mars' water is now trapped as ice below ground, in the pores of the regolith. Catastrophic events may have released this water from time to time, when the ice was melted by intrusions of magma or by heat generated in meteorite impacts. Most of what appear to be major flood channels are located in Mars' southern volcanic highlands. Although we cannot determine how long ago Mars' catastrophic floods took place, the numerous meteorite-impact craters superimposed on the channels suggest that they may be hundreds of millions of years old.

In Chapter 14, we examine another part of the hydrologic cycle—groundwater. We examine how groundwater accumulates and flows, and how it acts on carbonate bedrock to produce spectacular caves, sinkholes, and other geologic features.

Chapter Summary

All of the water on Earth cycles perpetually among its oceans, land, and atmosphere and, via plate subduction and volcanism, between its mantle and its surface. This never-ending process is called the **hydrologic cycle.**

Any surface water whose flow is confined to a channel is called a **stream.** The area of land from which streams acquire their water is known as a **watershed** or **drainage basin.** Adjacent drainage basins are separated by an area of higher topography called a **drainage divide.** Rainfall initially flows across the slopes of a drainage basin as broad shallow sheets of water, which eventually collects into small channels called rills. Rills combine to form a network of small **tributaries** that feed water into a main or **trunk stream.** The trunk stream, in turn, may eventually split into another network of small channels, or **distributaries,** which empty into an ocean.

Stream velocity is governed principally by the slope, or **gradient,** of the stream bed and by the bed's roughness; friction between the water and the bed may impede flow. Velocity is directly related to stream **discharge,** the volume of water passing a given point per unit of time.

In an unchanging environment, a stream would develop and maintain a shape and gradient in which erosion and deposition would be equal. A stream in such a state of equilibrium is called a **graded stream.** In reality, any change in a stream's environment, such as an increase or decrease in sediment load, is likely to disrupt this equilibrium, prompting the stream to respond by adjusting its gradient toward a new equilibrium state.

A stream cuts downward in its channel toward its **base level,** the lowest level to which it can erode (usually sea level). Stream erosion, combined with mass movement, is responsible for the development of a stream valley's characteristic V shape. A stream erodes its bed by **abrasion, hydraulic lifting** of loose particles, and **dissolution.** The combined effect of surface topography, composition, and structure on stream erosion accounts for the distinctive drainage patterns of tributary networks, stream piracy, waterfalls and rapids, and stream terraces.

Stream channels may be straight, braided, or meandering. Straight streams are relatively rare. **Braided streams** are networks of converging and diverging streams separated by narrow bars of sediment. **Meandering streams** form evenly spaced loops in the surrounding flat land, or **floodplain.**

The maximum volume of sediment that can be transported by a stream—a measure controlled principally by the stream's discharge—is called its **capacity.** The maximum size of its transported particles, which is largely a function of stream velocity, defines a stream's **competence.** The sediment in a stream can travel as fine-grained **suspended load,** bounce or roll along the stream bed as part of its coarser **bed load,** or travel in solution as a **dissolved load.**

Ultimately, the pull of gravity causes transported particles to settle from a stream. These particles, called **alluvium,** may be deposited in a number of locations: within the channel, forming mid-channel bars or **point bars** along its inside curves; on the stream's floodplain, building **natural levees** and **backswamps;** where stream valleys widen at the foot of mountains, building **alluvial fans;** and in bodies of standing water, constructing **deltas.**

Plate tectonics casts light on the evolution of streams. Plate motions may control, among other things, a stream's base level, disruptions to its profile (by faulting and uplift), and its supply of sediment (particularly in volcanic areas).

Floods occur when precipitation cannot infiltrate an impermeable or saturated ground surface and enters a stream so rapidly that the water exceeds the channel capacity. To cope with floods, we can build flood-control structures such as artificial levees, flood walls, and dams, alter channels to allow water to pass through them more swiftly, and institute zoning regulations to minimize development in flood-prone localities.

Photographs of Mars' landscape have revealed several fluvial features, such as braided channels, stream-modified islands, dendritic drainage patterns, and meandering channels. These features suggest that the Martian surface was shaped partially by stream activity early in its history, before the planet's atmosphere thinned and its surface cooled. Today almost no water exists on Mars.

Key Terms

stream (p. 237)
hydrologic cycle (p. 238)
watershed (p. 239)
drainage basin (p. 239)
drainage divide (p. 239)
tributaries (p. 239)
trunk stream (p. 239)
distributaries (p. 240)
gradient (p. 240)
discharge (p. 240)
graded stream (p. 240)
base level (p. 241)
abrasion (p. 242)
hydraulic lifting (p. 242)
dissolution (p. 242)
braided stream (p. 245)
floodplain (p. 245)
meandering stream (p. 245)
capacity (p. 247)
competence (p. 247)
suspended load (p. 247)
bed load (p. 247)
dissolved load (p. 247)
alluvium (p. 247)
point bars (p. 247)
natural levees (p. 248)
backswamp (p. 248)
alluvial fan (p. 249)
delta (p. 249)

Questions for Review

1. Briefly describe the main elements of the hydrologic cycle. List three ways that water can be delayed on land before eventually reaching an ocean.

2 Which factors determine a stream's velocity? Which factors determine a stream's discharge?

3. What do we mean by a graded stream? How does a stream achieve a graded state if its sediment load changes?

4. Describe three ways in which a stream might erode its channel. Under what conditions does erosion take place most rapidly?

5. Sketch three different drainage patterns, and explain how they are influenced by topography.

6. How do straight, braided, and meandering streams differ?

7. List three types of structures produced by stream deposition, and describe the conditions under which they form.

8. How do the sediment particles that are most likely to be transported in suspension differ from those transported by traction? What is the difference between a stream's competence and its capacity?

9. List three ways in which plate tectonics may affect the evolution of a stream profile.

10. What are the principal causes of flooding?

For Further Thought

1. Where would you expect to find a stream that exhibits high competence but low capacity? Where would you expect to find one that exhibits low competence but high capacity?

2. What would be the effect on the Earth's fluvial landscapes if plate tectonic activity suddenly ceased?

3. How would the world's streams respond if climatic warming melted the Antarctic ice sheet?

4. How would the potential for flooding change over a long period of time if all of the dams along the Mississippi River were removed?

5. What kind of deposits do you think underlie the farmland shown in the photo below?

14

Groundwater, Caves, and Karst

For thousands of years, wells and natural springs have supplied clean, abundant groundwater to human communities throughout the world. Even today, the presence of an adequate supply of uncontaminated groundwater can determine whether a region or community will grow and prosper. Pure water has even become a commercially valuable commodity, and both supermarkets and gourmet restaurants sell bottles of ordinary groundwater at exorbitant prices. This source of our trendiest beverage is also responsible for some of the world's most popular tourist attractions, dissolving spectacular caves out of solid bedrock (Fig. 14-1). Most caves form when acidic groundwater, flowing unseen beneath the Earth's surface, gradually enlarges tiny crevices in limestone to create huge, complex cave systems.

Groundwater accounts for 97% of the world's supply of unfrozen fresh water. It provides more than 50% of our drinking water, 40% of our irrigation water, and 25% of water used for industrial purposes. Throughout North America, we are withdrawing groundwater reserves that took thousands of years to accumulate, and supplies (particularly in the Southwest) are being depleted. Between 1955 and 1985, U.S. groundwater use doubled to keep pace with the population growth. In many places, however, groundwater has become contaminated because of improper disposal of wastes and other human activities.

Some of the most urgent issues facing the world's citizens relate to groundwater: where to find it, how to keep it clean, who owns it. In most areas, climate and local geology are capable of providing a continuing supply of cool, refreshing, healthy groundwater for present and future generations—if we learn to properly manage and preserve that reserve.

Figure 14-1 Carlsbad Caverns in New Mexico. This cave was dissolved out of solid limestone bedrock by circulating groundwater.

Groundwater Recharge and Flow

Groundwater *recharge* is the infiltration of water, mostly from precipitation, into the Earth's groundwater systems. After infiltration, groundwater flows through soils and rocks until it

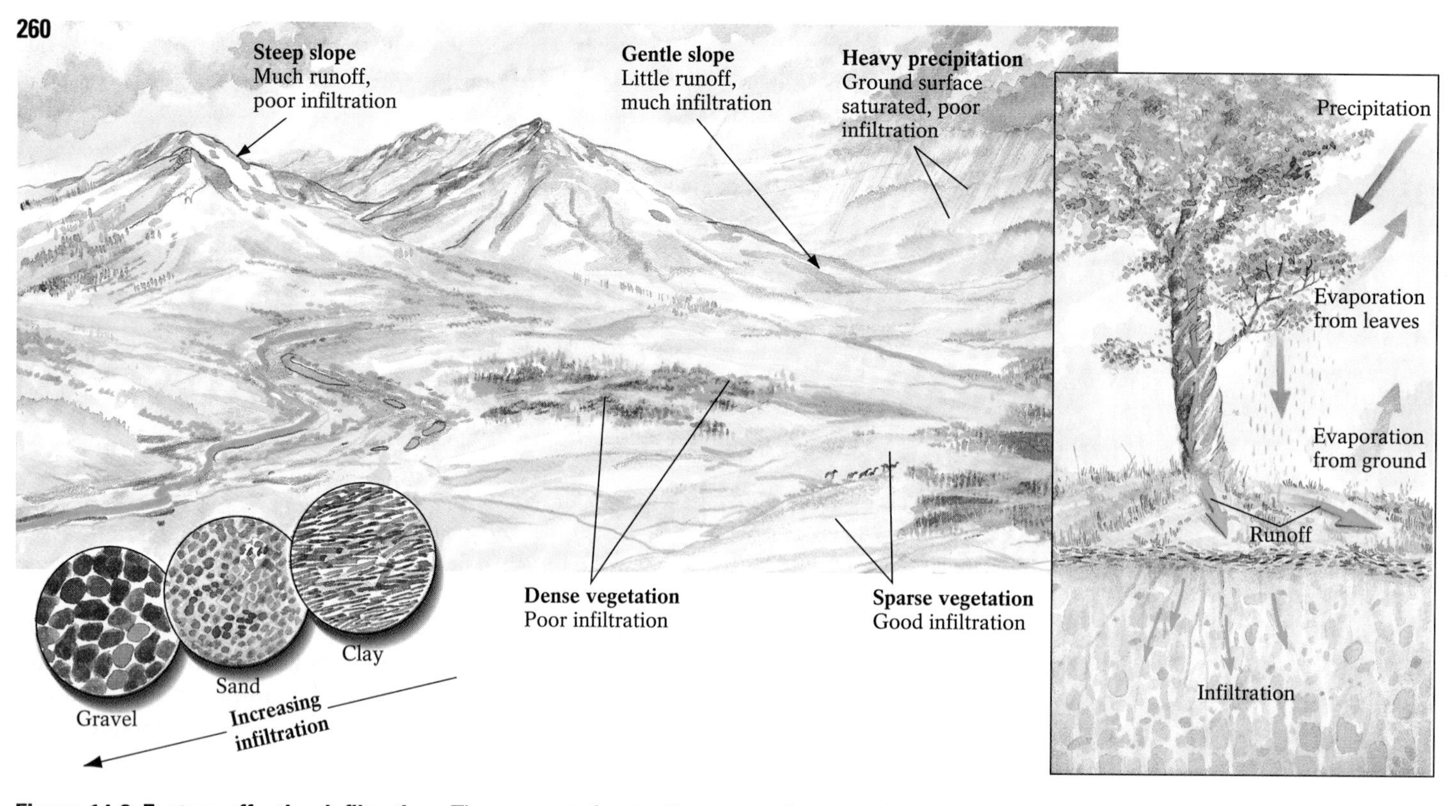

Figure 14-2 Factors affecting infiltration. The amount of water that enters the groundwater system is influenced by soil or bedrock composition, vegetation patterns, surface slope, and climate. Gentle, lightly vegetated slopes composed of permeable materials are most conducive to infiltration.

eventually reemerges, returning to the surface as streams or springs or pooling temporarily in lakes, ponds, or wetlands. The amount of water that infiltrates groundwater systems depends on the local surface materials, topography, vegetation, and precipitation (Fig. 14-2).

Factors Influencing Groundwater Recharge

Extensive infiltration is promoted by abundant, well-connected pore spaces in loose soils and unconsolidated sands and gravels. Exposed bedrock that is fractured or inherently porous, such as coarse sandstone, also allows substantial infiltration by surface water. Clay, on the other hand, consists of closely packed flat particles with only minute pore spaces between them; as a result, it impedes infiltration, as does unfractured crystalline bedrock (such as granite or gneiss).

Surface topography also influences infiltration. Surface water runs slowly down gently sloping terrains, so it has ample time to seep into the ground. On steep slopes, surface water runs rapidly downslope and into nearby streams, so a large percentage of it fails to infiltrate into the groundwater system.

In areas with plentiful vegetation, rain first falls on the leafy canopies of trees, bushes, and grasses, and much of it evaporates before reaching the ground. Most of the rain not intercepted by vegetation lands on the ground and evaporates from the ground surface, runs off as streamflow, or is temporarily stored at the surface in lakes or glaciers. Only a small percentage of the rain actually seeps into the ground to become part of the groundwater system. Plant and tree roots enhance infiltration by opening pathways in the soil into which the water can seep more easily.

Finally, groundwater recharge in any kind of terrain depends on the amount and timing of precipitation over long- and short-term periods. For example, extended droughts may curtail groundwater recharge significantly for several years or more. Shorter-term recharge variations, such as additions to groundwater from spring rains and snowmelt, generally affect the groundwater supply on a seasonal basis. The intensity of precipitation also affects the degree of infiltration. For example, a driving rainstorm packs surface soils and washes fine clay particles into soil pores, clogging the pores and impeding infiltration. A series of drenching storms in close succession increasingly favors runoff over infiltration, because the first storm saturates the surface, preventing water from subsequent storms from finding pathways into soils. Gentle rainfalls for several hours enhance infiltration to the greatest extent.

Movement and Distribution of Groundwater

Water is drawn into the ground by gravity and moves to various depths depending on the properties of the soils, sediments, and rocks that it encounters (Fig. 14-3). When water first moves into the soil, it wets and adheres to the soil particles. Later it may evaporate back into the atmosphere or be absorbed by plants. Once the soil is wet, excess water passes downward through the **zone of aeration,** or unsaturated zone, where the pore spaces in rocks and soils contain both water and air. It continues downward into the **zone of saturation,**

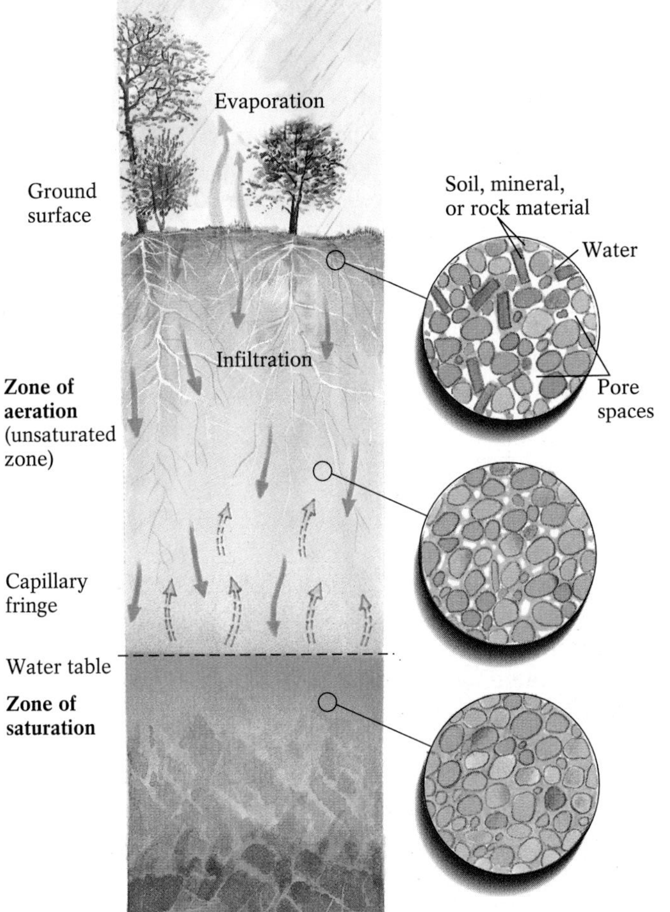

Figure 14-3 The subsurface distribution of water.

where all available pore space is filled with water. The zone of saturation rarely extends beyond 1000 meters (0.6 mile) or so beneath the surface, because most rock below that point contains no pores.

The **water table** is the boundary between the zones of aeration and saturation. The lower part of the aeration zone, which can range from a few tens of centimeters to several meters above the water table, constitutes the *capillary fringe*. In this region, the effect of gravity is partly countered by capillary action within the soil, which causes the water to move upward from the water table into the unsaturated zone. Capillary action results from the attraction of water molecules to mineral surfaces and to other water molecules. To observe a small-scale version of this phenomenon, dip the corner of a paper towel into a glass of water and note how the water rises in the towel against gravity.

Soil Porosity and Permeability Given adequate precipitation, the availability of groundwater is governed by two characteristics of the soil or rock unit into which the water might infiltrate—its porosity and its permeability.

The **porosity** of a soil, rock, or sediment is its volume of pore space, expressed as a percentage of its total volume; this factor determines how much water the material can hold. *Primary* porosity develops when rocks form—for example, it includes the spaces that remain between grains of sand after they have been lithified to sandstone and the vesicles that form near the top of basalt lava flows as they cool (Fig. 14-4). *Secondary* porosity develops after a rock has formed, usually as a result of fracturing or dissolution.

The other important characteristic of a soil or rock unit is its **permeability,** or its ability to allow the passage of a fluid among its pore spaces. The permeability of a rock or sediment reflects the size of its pore spaces and the extent to which they are connected. A material can be porous without being permeable. For example, clay can be highly porous; fresh clayey mud from the Mississippi River's delta south of New Orleans can consist of 80% water. The pore spaces are so small, however, that any water within them tends to

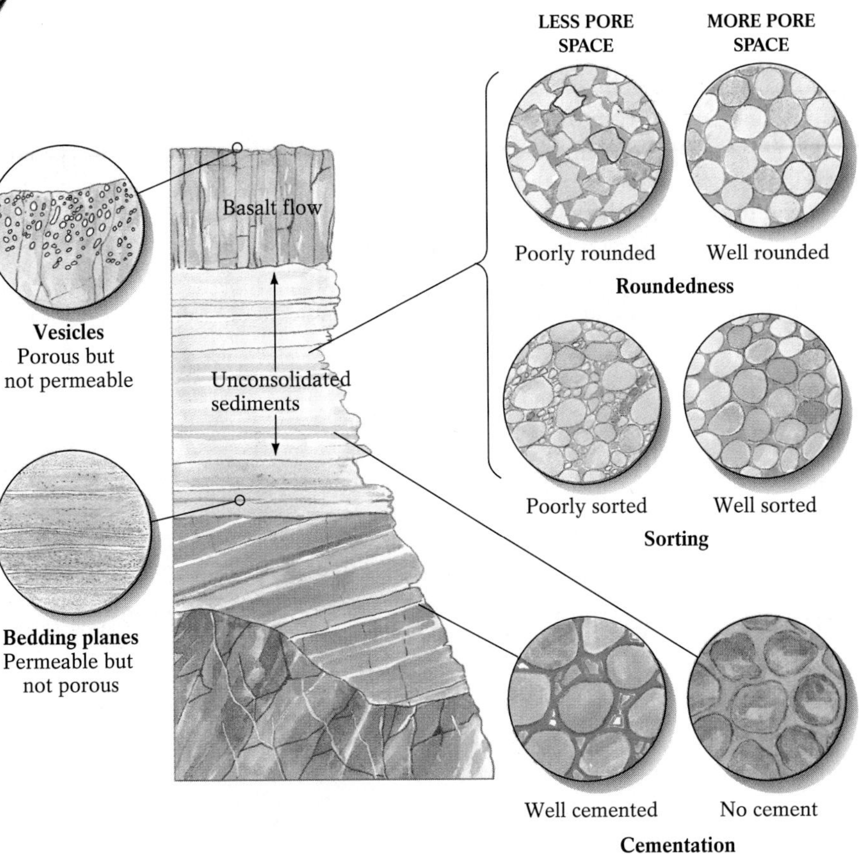

Figure 14-4 Primary porosity in rocks. In sedimentary rocks, primary porosity is determined by such factors as variation in grain shape, sorting, and degree of cementation. The presence of vesicles and bedding planes also increases a rock's primary porosity.

adhere to the clay particles and does not flow among the pore spaces. In addition, because clay consists of minute flat plates, the pore spaces tend to be isolated from each other, so no flow occurs between them. In other cases, a material may be essentially nonporous but still permeable, because it possesses a network of interconnected fractures that can accommodate a modest flow of water. A rock such as basalt, for instance, which often develops a highly interconnected fracture system as it cools, can sometimes serve as a reliable source of groundwater.

Groundwater Flow Groundwater, like water anywhere, flows downward under the influence of gravity. In general, it flows from high spots in the water tables, generally located under hills, to low spots in the water tables, generally found under valleys. Flow direction is also affected by differences in the pressure on the water from the weight of overlying water. Thus, under various circumstances, groundwater under pressure may flow upward against gravity and emerge at the surface as a spring or a seep. Most streams that flow year-round are fed by the upward flow of groundwater.

Several factors influence the rate of groundwater flow, including gravity, pressure, and characteristics (such as porosity and permeability) of the medium through which the liquid flows. To determine the rate of flow of groundwater, geologists may inject an environmentally safe dye into a well and time how long it takes for the dye to arrive at another nearby well. Using this and other methods, geologists have found that groundwater flow is extremely slow, averaging between 0.5 and 1.5 centimeters (0.2–0.6 inch) per day through moderately permeable material, such as poorly sorted sand. The swiftest flow—through well-sorted, coarse, uncemented gravels or highly fractured basalts—can reach 100 meters (330 feet) per day. In comparison, even sluggish streams generally flow at least 2 meters (6.6 feet) per second.

The slow movement of groundwater accounts for its availability for human use. If groundwater flowed as rapidly as rivers, it would not remain stored for long in the ground. Because of this slow flow rate, however, the water that we draw today from a deep well may have fallen as rain thousands of years ago. In the hot, dry parts of Arizona, radiocarbon dating has dated the deep groundwater at more than 10,000 years old; this water precipitated during the last worldwide glaciation, when Arizona's climate was cool and cloudy enough for rainwater to infiltrate the groundwater system instead of evaporating, as much of it does today.

Tapping the Groundwater Reservoir

To determine the depth of an area's water table, we can dig a hole progressively deeper into the ground. The depth at which groundwater begins to seep into the hole—indicating that the surrounding material is saturated with water—marks the height of the local water table. In an area that receives significant rainfall every year, the water table may be only 1 meter (3.3 feet) or so below the surface; in an arid locale, water may not appear for tens of meters below the surface.

The depth configuration of an area's water table often parallels the local topography, although its highs and lows are less pronounced than those of the landscape (Fig. 14-5). Lakes, swamps, and year-round streams are places where the water table intersects the Earth's surface. The depth of the water table is generally established by a long-term balance between recharge (water entering the groundwater system) and discharge (water returning to the surface). Major climatic events, such as storms and droughts, and variations in the amount of water extracted by humans from the ground can raise or lower the water table temporarily. Water table depth

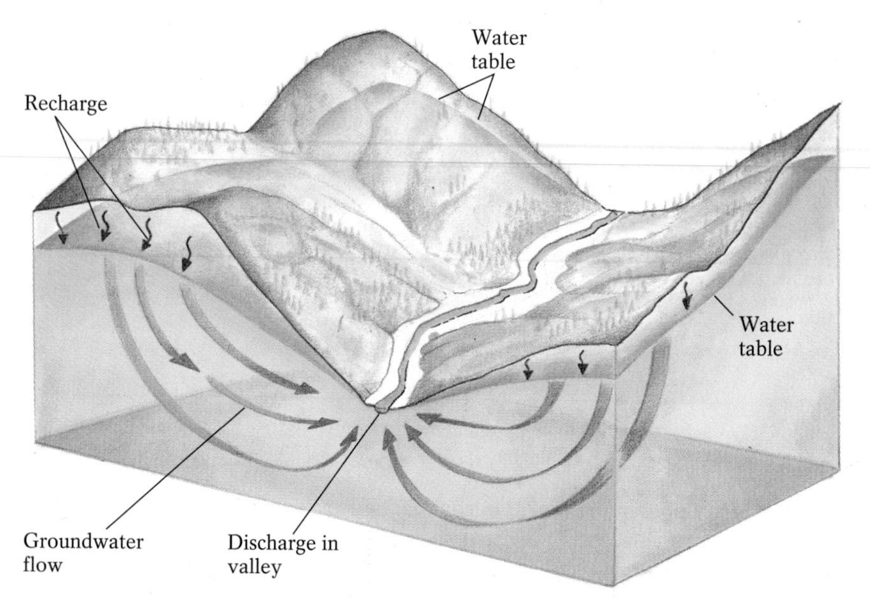

Figure 14-5 Water table configuration and topography. When precipitation falls on an irregular landscape, the slopes receive the most rainfall, because they constitute the largest surface area. The water that infiltrates the hills must then travel through a considerable expanse of soil and rock to reach a stream; it flows slowly, causing it to mound up below the surface of the hills. If groundwater recharge were to cease for a long time—for instance, during a drought—the water table mounds would flatten out in the hilly areas as groundwater flow continued. In most humid areas, however, rainfall occurs frequently enough to replenish the supply beneath the hills and maintain the undulations of the water table.

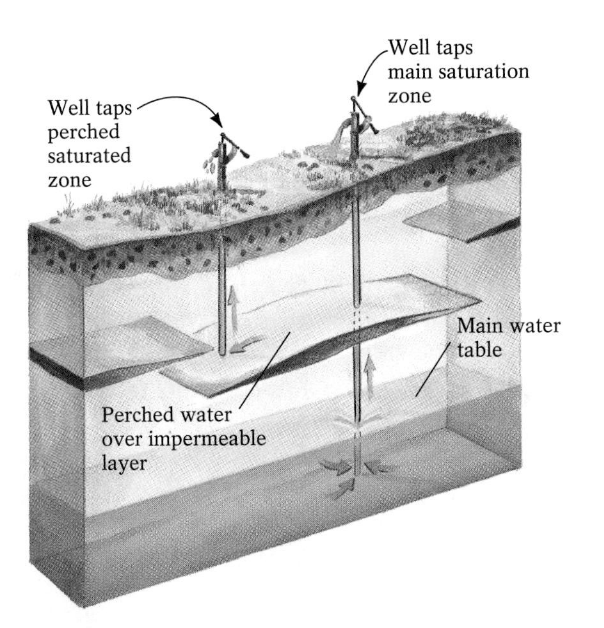

Figure 14-6 Perched water table. Perched water table occurs in the zone of aeration when a local impermeable layer intercepts descending water. To ensure a steady year-round flow of groundwater, wells must be drilled to below the main—not the perched—water table.

also varies seasonally, tending to be higher in winter and lower in summer.

Sometimes a local impermeable layer (such as a layer of clay or shale) may obstruct the downward flow of water through a region's zone of aeration. Water then accumulates above this barrier, saturating part of the overlying zone. In this case, what appears to be the regional water table is actually a *perched water table,* a locally saturated area within the zone of aeration (Fig. 14-6).

Aquifers: Water-Bearing Rock Units

An **aquifer** is a porous and permeable body of geologic material that stores and can transmit significant amounts of groundwater. These rock or soil units therefore serve as humans' source of groundwater. The most productive aquifers are composed of unconsolidated sand and gravel, well-sorted, poorly cemented sandstones, or highly fractured limestones and basalts.

Unconfined Versus Confined Aquifers Water tables are found in *unconfined* aquifers, aquifers that have no overlying impermeable rock or soil. Thus most shallow groundwater comes from unconfined aquifers composed of loose slope material, sands, gravels, and floodplain deposits left by streams and rivers, sands and gravels transported by recent glaciers, and young, fractured lava flows such as those in Hawai'i and the Pacific Northwest (Fig. 14-7). The Ogallala formation, a major unconfined aquifer, is a poorly sorted, generally uncemented mixture of clay, silt, sand, and gravel produced from a combination of these sources; a component of the High Plains regional aquifer system, it underlies much of Nebraska and parts of South Dakota, Wyoming, Colorado, New Mexico, Kansas, Oklahoma, and Texas. The Ogallala supplies 1.2 trillion liters (317 billion gallons) of water per year to irrigate 14 million naturally arid acres. Consequently, these Midwest areas can provide 25% of the United States' feed-grain exports and 40% of the flour and cotton exports.

Confined aquifers are sandwiched between impermeable rock layers called **aquicludes.** Unlike unconfined aquifers,

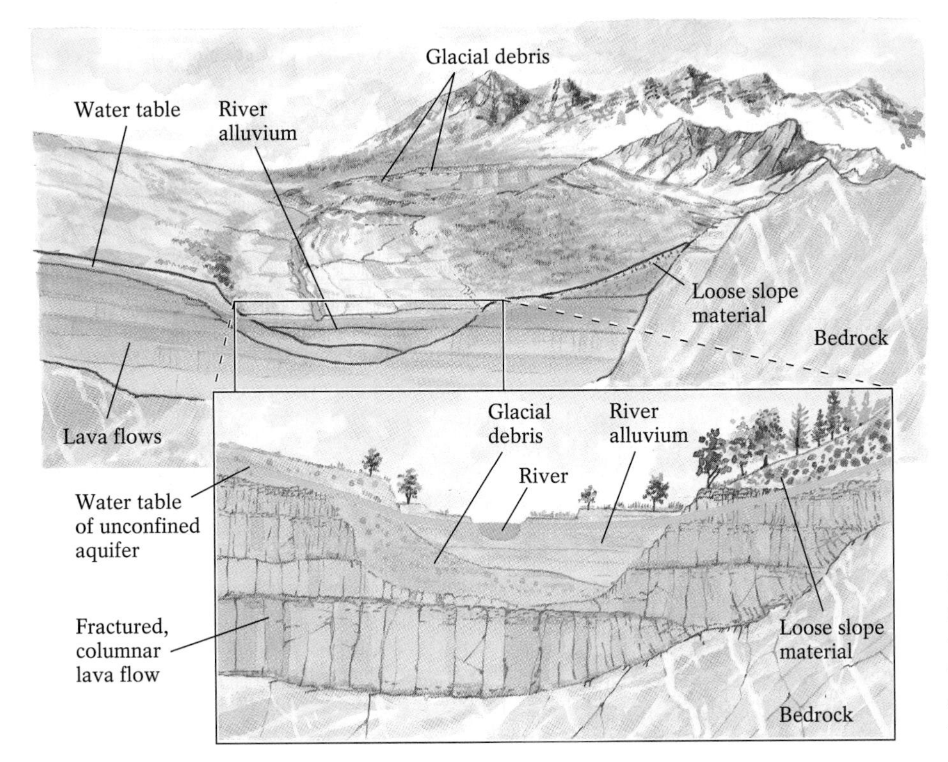

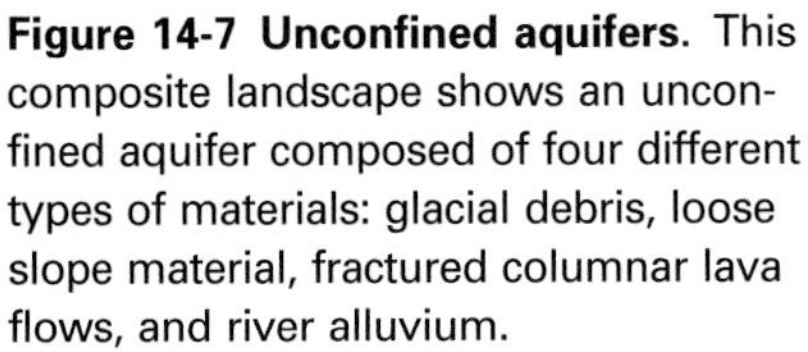

Figure 14-7 Unconfined aquifers. This composite landscape shows an unconfined aquifer composed of four different types of materials: glacial debris, loose slope material, fractured columnar lava flows, and river alluvium.

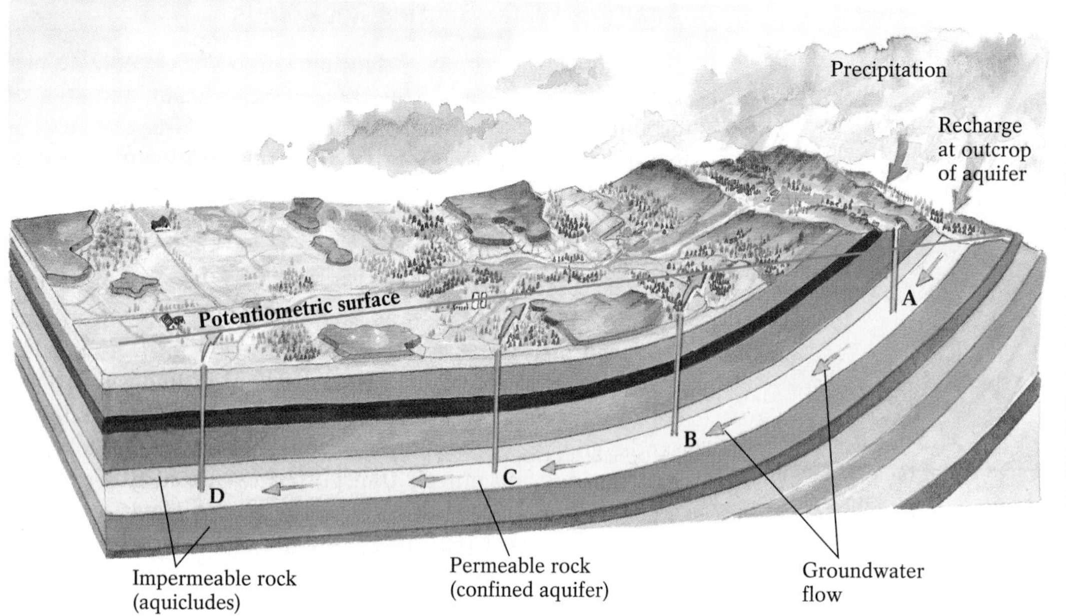

Figure 14-8 Artesian aquifers. High pressure causes groundwater to rise in the wells at points B, C, and D, where the potentiometric surface is significantly higher than the well sites. Water does not flow unaided from well A, because it lies above the potentiometric surface. Photo: Water gushing from a natural artesian well in the Dakota Sandstone aquifer (circa 1910). When this aquifer was first discovered, water gushed freely from wells under unusually high artesian pressure. During the last 75 years, however, more than 10,000 wells have been drilled to tap the Dakota Sandstone aquifer. The extensive withdrawal of water has reduced the pressure within the aquifer to the point that the potentiometric surface is no longer higher than the land surface of the eastern Dakotas. Today, water must be actively pumped in areas where it once flowed forcefully and unassisted.

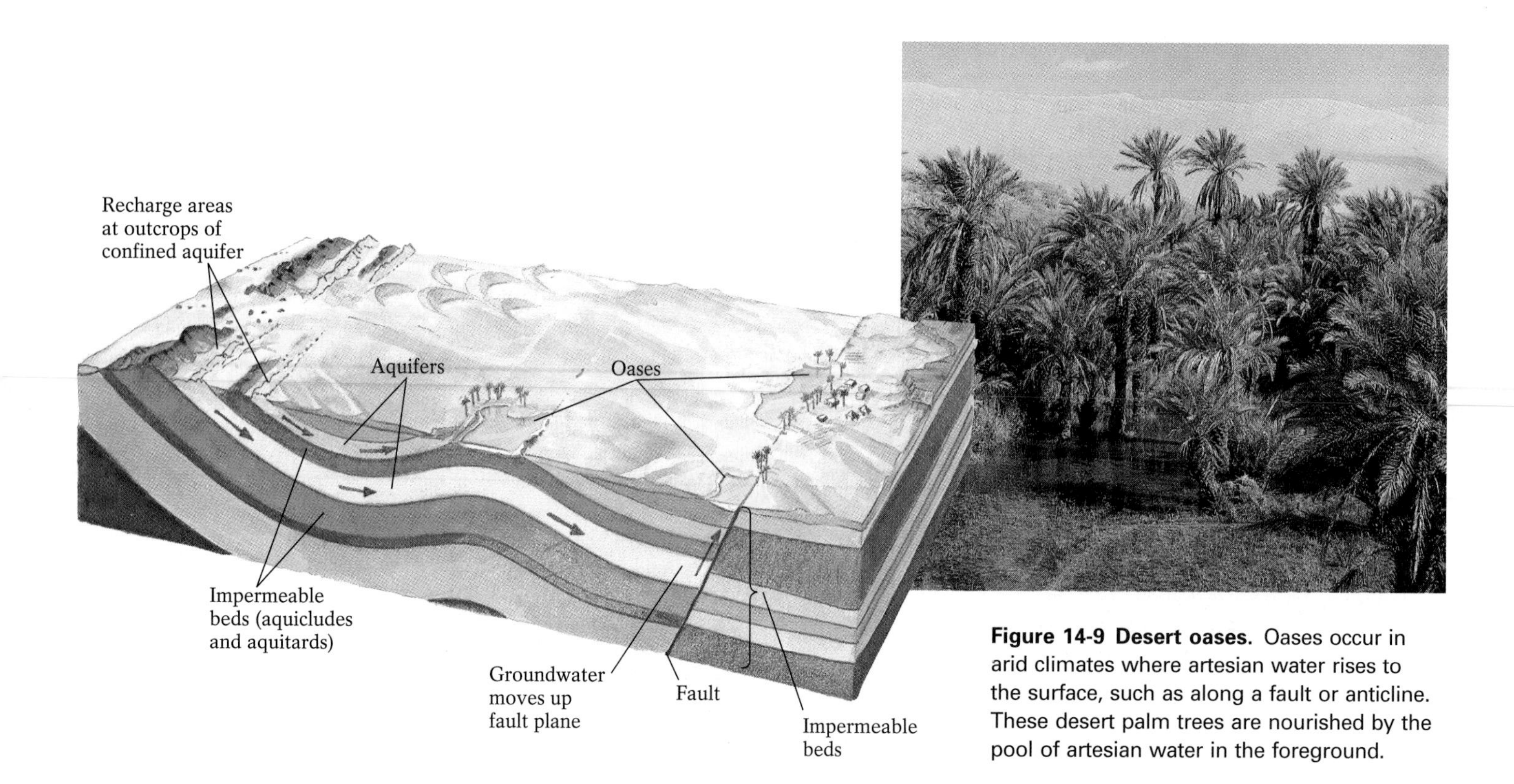

Figure 14-9 Desert oases. Oases occur in arid climates where artesian water rises to the surface, such as along a fault or anticline. These desert palm trees are nourished by the pool of artesian water in the foreground.

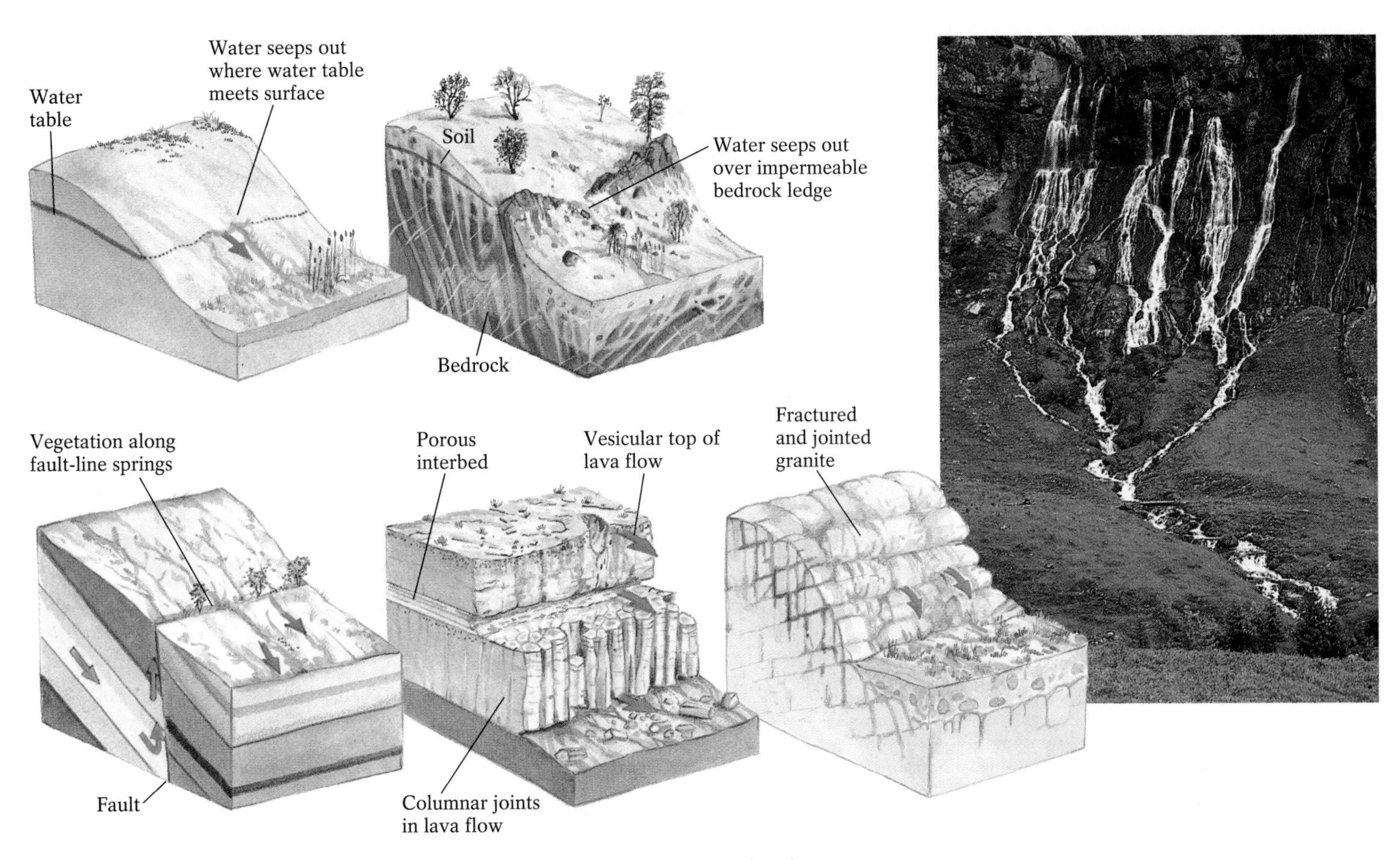

Figure 14-10 Natural springs. Various geological settings in which natural springs occur. Photo: Water gushing from limestone rock in Switzerland.

which receive their recharge largely from local precipitation, the water in confined aquifers may have traveled great distances from its recharge site. The Dakota Sandstone, which is widely tapped for irrigation, is a confined aquifer whose water supply originates in the Black Hills of South Dakota, but then flows at depth beneath the Great Plains of the eastern Dakotas and Nebraska. Similar aquifers can even be found at great depths under deserts—such as in Nevada, eastern Utah, southern Arizona, and parts of southern Montana—where precipitation has entered the groundwater system by infiltrating exposed permeable rocks in mountain ranges both near and far.

Artesian Aquifers Under certain geological conditions, the water in a confined aquifer may rise against the downward pull of gravity and even gush from the ground. This type of release occurs with confined aquifers that are tilted at an angle to the Earth's surface, so that the area of recharge is at a significantly higher elevation than the points of discharge (Fig. 14-8). Because the underlying and overlying aquicludes prevent water from escaping, the weight of the water higher within the aquifer presses on the water below it. When the aquifer is tapped, either naturally (in the form of springs) or artificially (by a well), this pressure drives the water upward. Such **artesian** water tends to rise toward the elevation of the recharge point at the surface, though friction between the water and the surrounding rocks and sediments constrains it at a lower level called the potentiometric surface of the aquifer. When this level is high enough above the Earth's surface, groundwater flows freely from the ground.

Most municipal water supply systems are designed to simulate artesian conditions. Have you ever wondered why the water flows out at a high pressure when you turn on a faucet in your home? The tall water towers in your city or town act like elevated recharge areas, and the municipal water pipes act like confined aquifers. The difference in elevation between the tower and your home causes the water to flow from your faucets under pressure.

Water under artesian pressure can surface in unexpected places. An oasis, for example, is a pool of artesian water (Fig. 14-9). An aquifer that receives its recharge from mountains surrounding a desert may extend hundreds of meters beneath the arid surface. Artesian pressure may then cause water to flow to the surface where the rocks of the aquifer crop out at the surface or where a fault extends down to the aquifer.

Natural Springs and Geysers

Natural springs are sites at which groundwater flows to the surface, usually laterally, and issues freely from the ground. Such springs may develop where groundwater encounters an impermeable bed and is forced to flow around it, or where erosion and other processes have lowered the surface so that it intersects the water table (Fig. 14-10).

Because groundwater remains insulated from fluctuations in air temperature, its temperature tends to approach the mean annual air temperature for the region. A *thermal spring* is a spring whose flow is at least 6°C (11°F) higher than this mean temperature. (A "hot spring" is informally defined as a thermal spring whose water temperature exceeds humans' body temperature.) Most of the more than 1000 thermal springs in North America occur in the West, near the Rocky, Cascade, Olympic, and Sierra Nevada Mountains. The water in these springs is warmed primarily by recent encounters with magmas or with the still-hot rocks of a cooling pluton. After percolating downward for hundreds of meters, the groundwater reaches and is heated by warm igneous rocks—in some places, to the boiling point. As the hot water rises, it usually dissolves large quantities of soluble rock. Upon reaching the surface, the water evaporates, leaving behind the minerals as a deposit called *sinter* (Fig. 14-11). Most eastern hot springs, such as those found in the Appalachians and Ouachitas, result when deeply circulating groundwater that descends hundreds of meters below the Earth's surface, passing through rocks warmed by the Earth's internal heat.

An intermittent surface emission of hot water and steam is called a **geyser** (pronounced "GUY-zer"; from the Icelandic *geysir,* meaning "to rush forth"). Geysers occur where groundwater descends through fractured permeable rock, becomes heated by a shallow magma chamber or a body of young, still-warm igneous rocks, and is then pushed up by steam under great pressure (Fig. 14-12). The pressure of the overlying water column raises the boiling point of the deeper water higher than 100°C (212°F). The superheated deep water expands, pushing some water out of the geyser opening; this release reduces the pressure on the deep water, lowering its boiling point. The deep water instantly flashes to steam, forcefully expelling the water from the entire column. A period of recharging and reheating then takes place, followed by another eruption. The amount of time between eruptions varies with the groundwater supply, the permeability of the rock, and the complexity of the network of passages that delivers the water to the heat source.

The best-known geyser fields appear in areas of current or recent volcanism. For example, volcanism at the divergent zone that bisects Iceland and at the subduction zone under New Zealand produces the famous geysers of those regions. In North America, northern California geysers derive their heat from the subduction that fuels the volcanism of Mount Shasta and Lassen Peak. The hot-spot volcanism of Yellowstone National Park powers the park's numerous geysers. In the past, Yellowstone's Old Faithful geyser—so named because of its formerly remarkable punctuality—shot a jet of steam and boiling water to a height of 50 meters (170 feet) every 65 minutes. Recent earthquake activity in the region, however, has altered the geyser's plumbing, affecting its recharge time. Nowadays Old Faithful erupts *on average* every 79 minutes, but intervals between eruptions may vary between 45 and 105 minutes—far less faithful.

Figure 14-11 Sinter deposited at Mammoth Hot Springs in Yellowstone National Park, Wyoming. Hot groundwater has infiltrated and dissolved carbonate deposits and then precipitated them at the surface to create white cliffs of travertine.

Figure 14-12 Geysers. Geysers develop when groundwater encounters a shallow heat source and erupts at the surface as boiling water and steam. Water pressure at the bottom of the water column inhibits boiling initially. As expansion of the heated bottom water pushes out water near the top of the column, the pressure on the column drops and the water instantaneously turns to steam. Photo: Strokkur geyser in Iceland.

Finding and Managing Groundwater

To access groundwater reservoirs, humans drill wells into the zone of saturation beneath the water table. But how do we decide where to dig? To determine the best places to locate wells, we must rely on our knowledge of the configuration of the local water table.

We can estimate the configuration of the water table by recalling two general principles: the water table is higher beneath hills and lower beneath valleys, and surface water exists where the topography intersects the water table. The locations of hills and valleys, as well as the emplacement of lakes, year-round streams, natural springs, and swamps, thus give us clues about the underlying water table. An area's vegetation may also provide valuable information. Some plant species have extensive, deep root systems for capturing water far below the surface; others have shallow root systems and grow only where abundant water lies close to the surface. Thus such plants indicate the presence of a shallow water table.

We can obtain more detailed knowledge of the water table by drilling a series of test wells and noting the level to which the water rises in each. Because the water in a well dug in an unconfined aquifer rises to the level of the water table, this information can be used to construct a map of the local water table.

Some people hire a water witch, or dowser, to search for water. Highlight 14-1 discusses the merits of this method.

Threats to the Groundwater Supply

As soon as wells are drilled, groundwater withdrawn, and the water table is consequently lowered, issues arise regarding how to manage the water supply and deal with problems related to its use. Especially in densely populated regions or major urban areas, conflicting demands on groundwater can create significant problems. The imperiled Everglades of Florida, where pumping of groundwater has increased to meet the needs of a growing population and from which water has been exported to Florida's farmlands and coastal

Highlight 14-1 Should You Hire a Water Witch?

More than 25,000 water witches are active in the United States today—and a multitude of people swear by them. Water witches find water by walking over an area with a forked stick (or a switch, rod, or wire) in their hands until the stick seems to twist, dip, or jerk uncontrollably downward toward the place where underground water will be found. They claim that the presence of water initiates the behavior of the stick.

Are these people frauds? Probably most are not. Many dowsers don't even charge for their services. But do they really find water? Skeptics suggest that the water witch manipulates the rod by holding it in a delicate balance so that the slightest muscular movement turns it downward. Controlled studies have compared the success rate of water witches with the results obtained by chance and found no significant difference. In one study in Australia, scientists who compared geologists' efforts to find water with those of water witches concluded that the witches caused twice as many dry holes to be dug as the scientists. At Iowa State University, water witches invited to find water along a prescribed course across the campus could not locate the water mains under their feet.

Nevertheless, water witches often do locate water. Geologists believe that most of these successes stem from the fact that, in most humid climates, groundwater can be found almost anywhere you drill. Other successes may reflect the dowser's years of experience at finding water in a specific region and some geological common sense. Over the years, the witches, like most observers, would learn that wells dug in valleys produce water more often than those dug on hilltops, and that certain plants flourish where water lies near the surface. They may even acquire an intuition about groundwater flow and the relation between groundwater and permeable rocks.

Thus, as experienced water witches walk the landscape, they probably process the real information necessary to ascertain the presence of water. Perhaps when they observe the optimal conditions around them, their muscles flex subconsciously and the rod responds.

cities, offers just one example of the types of problems that can follow when groundwater withdrawal exceeds discharge (Fig. 14-13).

Groundwater Depletion When humans withdraw water from a well, the water table in the immediate vicinity is drawn down around the well, forming a **cone of depression** (Fig. 14-14). You have likely observed a similar phenomenon if you have ever sipped a crushed ice drink and seen the liquid disappear around the straw as you drank. Because the rate of groundwater flow is so slow, a cone of depression in a water table does not refill immediately. Especially in highly populated areas containing many wells or areas supporting much irrigation or great industrial activity, cones of depression may intersect, lowering the water table throughout the area. Even a single large well, pumping copious amounts of water for a city water supply or for industrial use, can significantly lower the water table, leaving neighboring wells dry (Fig. 14-15).

If groundwater withdrawal continually exceeds the rate of recharge, the area's water table may be lowered perma-

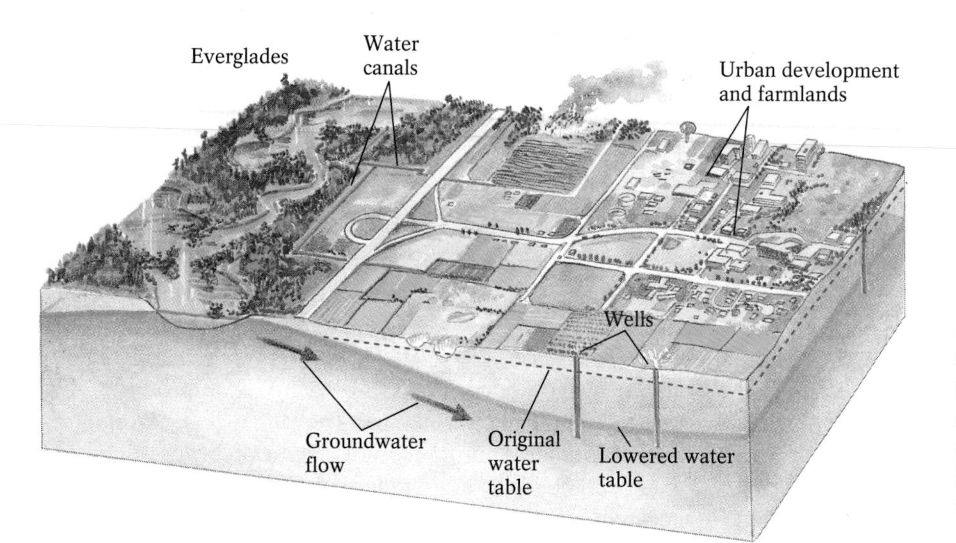

Figure 14-13 Overuse of groundwater. The water table beneath southern Florida once intersected the land surface across a broad area, creating the Everglades. Recently, withdrawal of water from Florida's groundwater reservoir has lowered the water table and caused much of the Everglades to dry out.

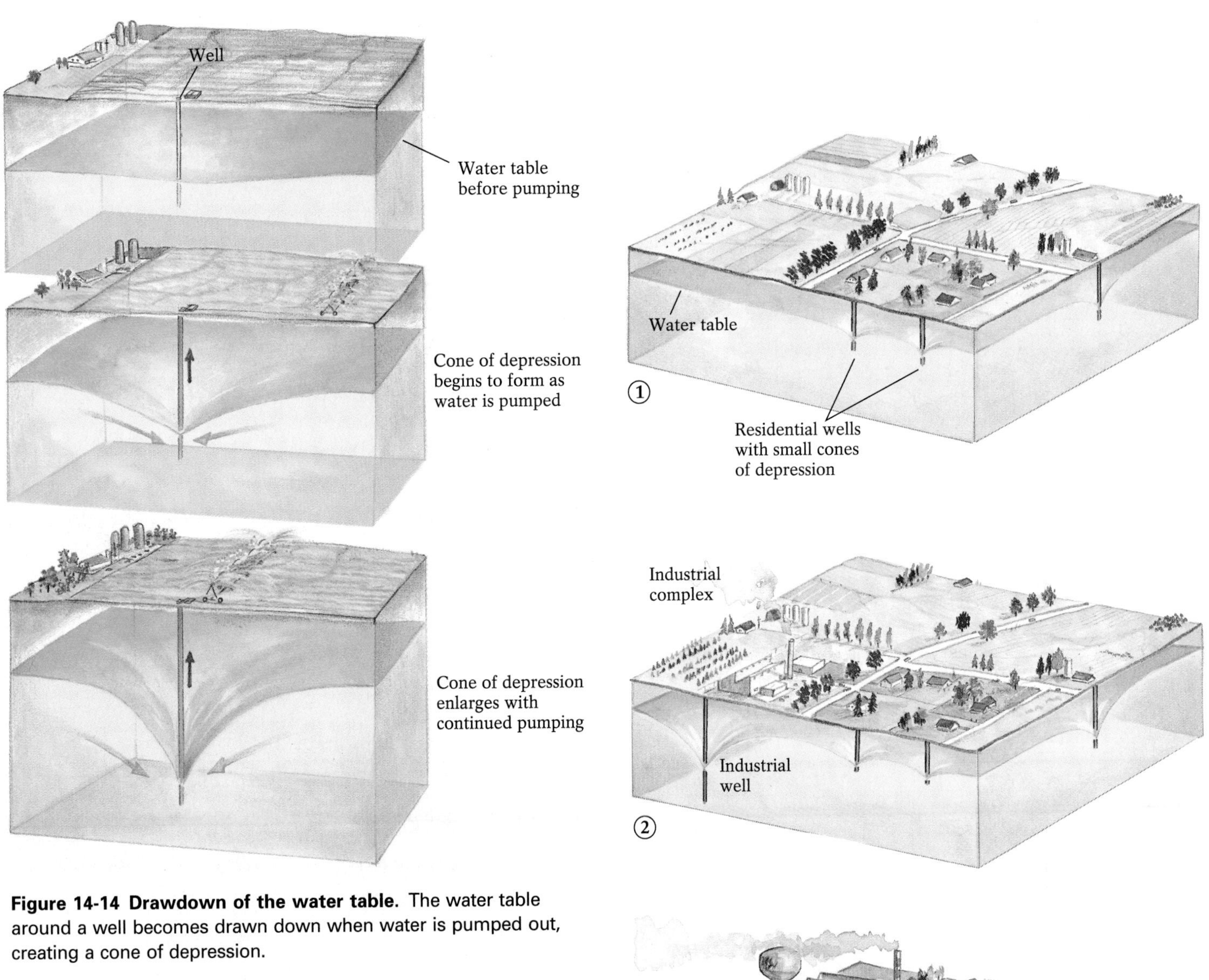

Figure 14-14 Drawdown of the water table. The water table around a well becomes drawn down when water is pumped out, creating a cone of depression.

nently, leaving aquifers largely depleted. Such groundwater "mining" is a particularly serious problem in arid regions such as the American Southwest, where extensive irrigation and enormous losses to evaporation have placed an impossible demand on the groundwater reservoir.

Land Subsidence The presence of groundwater in an aquifer helps support the overlying load of rock, sediment, or soil. When groundwater depletion occurs, less water is present and, as a result, the aquifer may become compressed. This turn of events lowers the surface of the land above the aquifer, a phenomenon known as **subsidence.** Excessive withdrawal of groundwater has led to subsidence in New Orleans (more than 2 meters [7 feet] of subsidence), Las Vegas (more than 1 meter [3 feet]), Mexico City (more than 7 meters

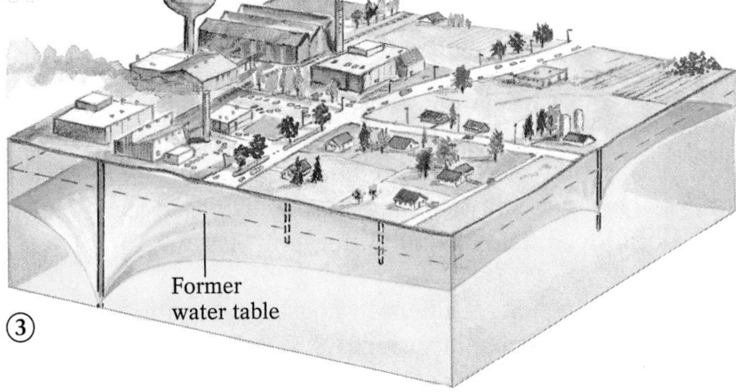

Figure 14-15 The effect of development on a water table. Stage 1: The community consists of farms and some suburban homes having small wells. Stage 2: Industrial development begins, replacing some of the farmland, and groundwater withdrawal increases. Stage 3: A large industrial complex covers much of the farmland, lowering the water table to a level below the depth of the residential wells.

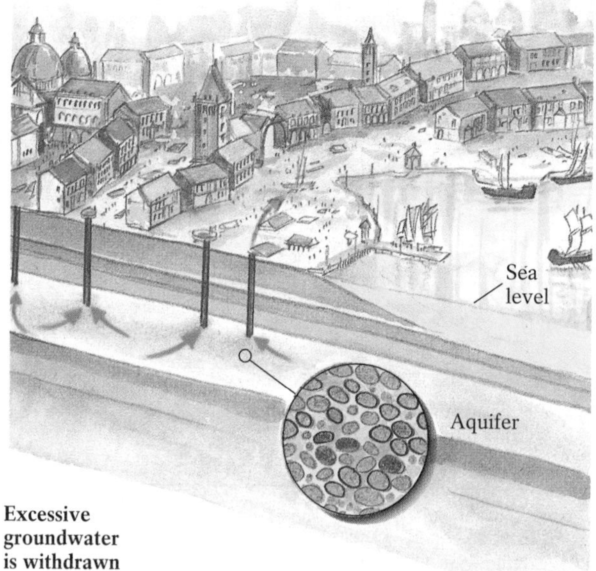

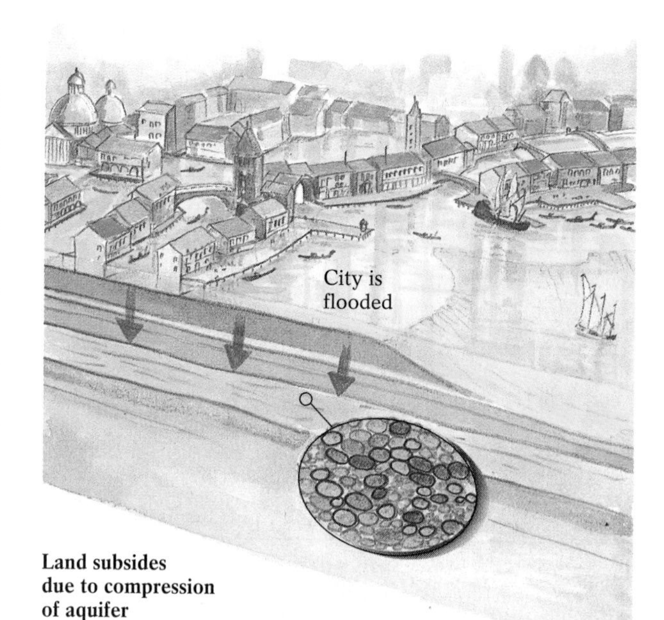

Figure 14-16 Subsidence in coastal areas. Excessive groundwater withdrawal in a coastal city such as Venice, Italy, causes its aquifer to become compressed and the land to sink below sea level. Photo: In Venice today, canals serve as streets.

[23 feet]), and central California (more than 8 meters [26 feet]). The city of Venice, Italy, having subsided more than 3 meters (10 feet), is now plagued by frequent flooding from the Adriatic Sea. Invading saltwater has severely damaged many of the city's historic and architectural treasures (Fig. 14-16).

Saltwater Intrusion Aquifers in coastal communities face a threat from **saltwater intrusion.** As a coastal aquifer is infiltrated by precipitation, creating a freshwater zone of saturation, it may simultaneously be infiltrated by salty marine water from the seaward side of the landmass. Because freshwater is less dense than saltwater and therefore floats above it, the freshwater–saltwater boundary is usually well defined. Indeed, natural recharge provides a constant pool of freshwater above the saltwater. If groundwater depletion significantly lowers the fresh groundwater table, however, the salty groundwater will rise and may completely replace the freshwater. New York City, Long Island, and populous coastal communities along the Gulf of Mexico, the Carolinas, Georgia, Florida, and California have all experienced such problems with saltwater intrusion.

Dealing with Groundwater Depletion

We have several options for addressing the problem of groundwater depletion. The most readily available remedy is water conservation. Particularly during dry spells, common emergency measures may include prohibiting the washing of automobiles and limiting the watering of lawns. More ambitious are projects that supplement inadequate natural recharge to the aquifer facing depletion. For example, engineers might grade a steep slope so that more water infiltrates the ground. In areas of moderate to heavy rainfall, we can construct open *recharge basins* to hold rainwater for gradual

Figure 14-17 A recharge basin/Little League field on Long Island, New York. To counteract high groundwater withdrawal rates, basins such as this one are excavated to collect precipitation and release it into the groundwater system gradually.

release into the groundwater system (Fig. 14-17). In drier regions and areas where water withdrawal consistently exceeds precipitation, water can be imported and actively pumped into the ground. Some communities borrow or buy groundwater from areas that have an abundance of either ground or surface water. For example, Denver receives much of its water through pipelines and tunnels that stretch from the western slopes of the Rockies. Similarly, New York City draws most of its municipal water supply from rural upstate regions through an extensive pipeline and reservoir system.

Another solution to water shortages may involve *desalinization,* the removal of salt from seawater. Distillation of seawater involves solar evaporation or boiling, followed by condensation and collection of the salt-free vapor. Seawater can also be passed through a filter that removes the salts. Although desalinization is costly, it can be a cost-effective option in arid lands with extensive seacoasts. Desalinization plants have operated for some time in Israel and Egypt, and the process may eventually become economically feasible in many other nations.

Maintaining Groundwater Quality

Groundwater naturally contains numerous dissolved ions and, quite often, minerals. Because of this characteristic, groundwater has traditionally been considered beneficial to human health. It can, however, contain harmful contaminants that have been introduced into the groundwater system in a number of ways. For instance, precipitation can infiltrate municipal landfills, incorporate some of the countless chemicals found there (such as household cleansers, paints, and insecticides), and then leak out into an aquifer, introducing these chemicals into the groundwater (Fig. 14-18). The sources yield only a small amount of toxins, however, compared with the vast volumes of agricultural and industrial wastes or the by-products of medical research. Among the most threatening of the latter contaminants, because large quantities are

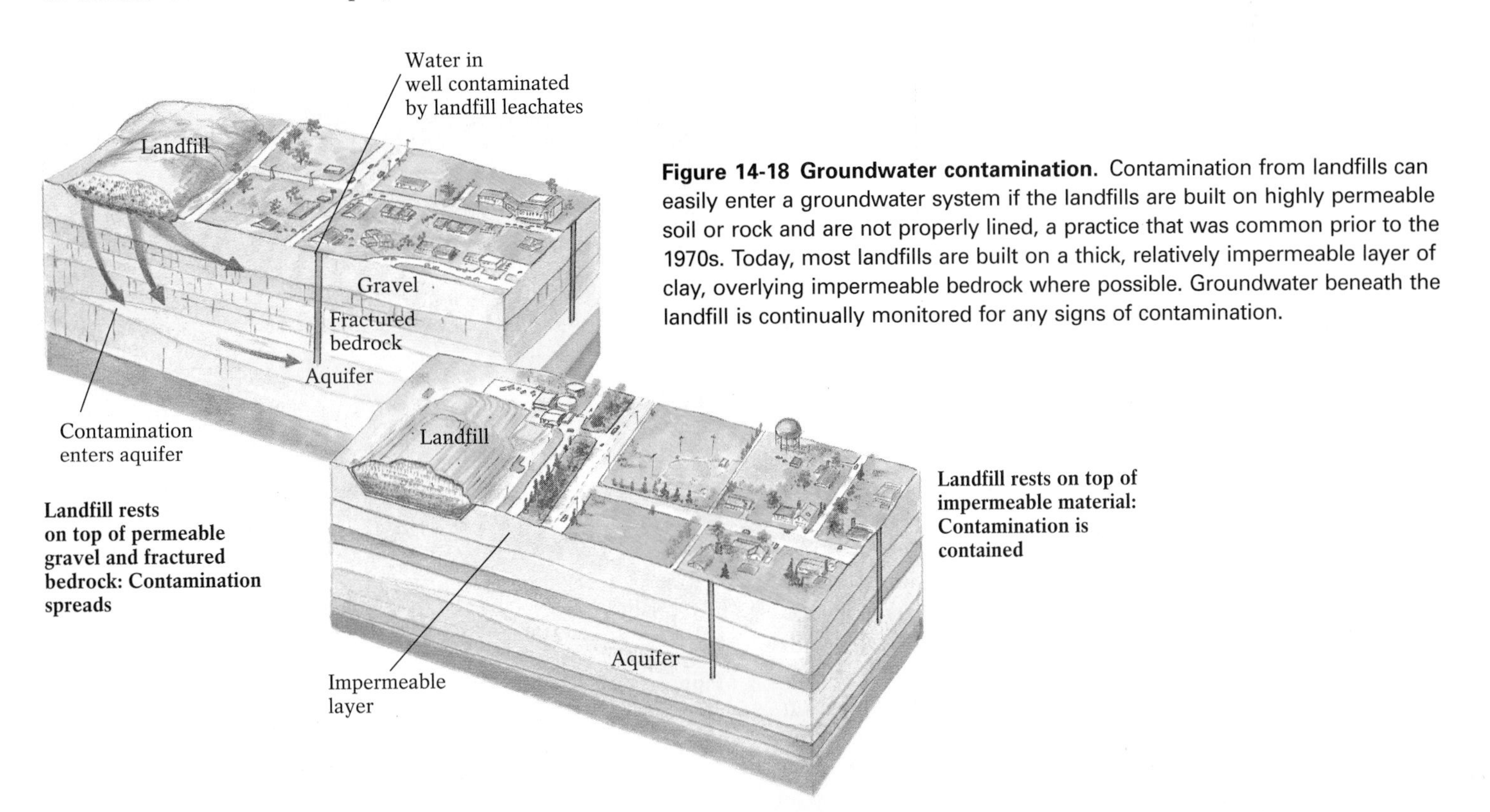

Figure 14-18 Groundwater contamination. Contamination from landfills can easily enter a groundwater system if the landfills are built on highly permeable soil or rock and are not properly lined, a practice that was common prior to the 1970s. Today, most landfills are built on a thick, relatively impermeable layer of clay, overlying impermeable bedrock where possible. Groundwater beneath the landfill is continually monitored for any signs of contamination.

often released, are insecticides and fertilizers, salt and other chemical ice-retardants washed from roadways, carcinogenic industrial by-products (such as PCBs) from factories, biological wastes from cattle feedlots and slaughterhouses, and sewage from overworked septic fields and broken sewer lines.

Once contaminants reach an aquifer, natural processes may act as the aquifer's self-cleaning mechanisms. Some contaminants adhere to clay particles as the water percolates through the soil. Others are completely decomposed by oxidation in the soil. Various microorganisms work to degrade many organic solids. Nevertheless, dissolved chemical pollutants in groundwater generally remain unaffected by these natural cleansing processes, and, because groundwater flows so slowly, it may take many years to flush them out. The problem of groundwater contamination has become serious enough that in the last few decades—beginning in 1970—local, state, and federal governments have enacted myriad laws that provide for monitoring groundwater quality, preventing further contamination, and cleaning up already-contaminated aquifers.

Polluted groundwater systems can be cleansed in several ways—all of them costly, technically demanding, and, given the average rate of groundwater flow, exceedingly slow. The tainted water can be pumped out or freshwater pumped in to dilute the contaminants. Chemicals that either remove or neutralize the contaminants can be introduced into the system, though this treatment is seldom feasible—the rate of groundwater flow is usually too slow to enable the agents to percolate throughout the system. Instead, contaminated groundwater is often pumped to a surface holding pond for treatment and then returned to the groundwater system.

Caves and Karst Topography

Groundwater is not only essential to the well-being of human communities, but also accomplishes a considerable amount of geological "work" as it passes slowly through certain types of rock. One aspect of this geological work is the dissolution of underground soluble rocks to form caves, such as the one shown in Figure 14-1. Groundwater may also dissolve soluble rock at the surface, forming novel landforms that range from towering monoliths to deep depressions in the ground. The features that are created when water dissolves bedrock, either underground or at the surface, are known collectively as **karst.**

Limestone, which is composed primarily of the mineral calcite, is by far the most abundant soluble rock and forms nearly all caves and karst landforms. Groundwater can dissolve limestone because of the groundwater's slight acidity, the result of it having mixed with carbon dioxide first as it passed through the air as precipitation and again as it passed through the soil during infiltration. (The carbon dioxide in soils is produced by respiration of soil organisms and the decomposition of plant and animal remains.) As we learned in Chapter 5, water and carbon dioxide combine to form carbonic acid, which attacks and dissolves the calcite in limestone (Fig. 14-19).

Cave Formation

Caves are natural underground cavities, the most common products of the dissolution of limestone. Almost every state and province in North America boasts some type of cave, and geologists believe that more than half of all caves in North America have not yet been discovered.

The first stage of cave formation occurs as groundwater moves through fractures and between bedding planes in soluble bedrock. This early stage of cave development progresses slowly, because these passages are initially too small to permit significant water circulation. Thousands of years may pass before small amounts of percolating water dissolve enough bedrock to enlarge the cracks to a significant extent. As the cracks become larger, however, more water infiltrates and flows more vigorously, vastly accelerating the rate of dissolution. The growing underground rivulets capture water from narrow cracks; eventually a few primary passages dominate the underground drainage, developing into the main caverns and connecting passageways of the expanding cave system.

When caves first begin to form, dissolution occurs primarily in the zone of saturation, probably at or just below the water table. In that region, resident groundwater combines with inflowing acidic water that is not yet saturated with solutes and can readily dissolve limestone. This period of intense dissolution along fractures and bedding planes below the water table forms the intricate honeycomb pattern of many major cave systems (Fig. 14-20a).

Figure 14-19 Partially dissolved limestone in Indiana. The red-colored soil in the background consists of the oxidized, iron-rich clay that remains after the calcite fraction of limestone has dissolved.

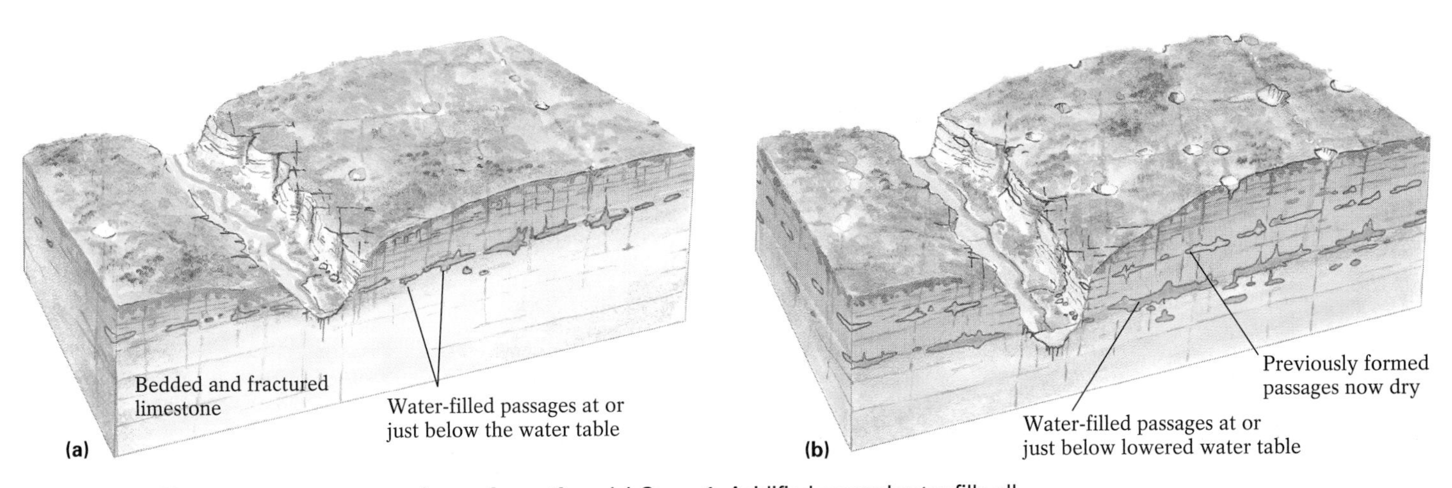

Figure 14-20 The two-stage process of cave formation. (a) Stage 1: Acidified groundwater fills all fractures and bedding planes below the water table and dissolves limestone, forming large caverns and connecting passageways. **(b)** Stage 2: The water table drops below the caverns and passageways, leaving them in an open-air environment.

The second stage of cave formation occurs if the water table drops below its original level, leaving caves high and relatively dry above it. At this point, the cave environment becomes one of open air (Fig. 14-20b). With the water table lowered, erosion by fast-flowing subterranean streams may begin to carve potholes and canyon-like slots into a cave's bedrock floor (Fig. 14-21).

Once a network of cave chambers and connecting passageways has formed and the water table has dropped below the base of the system, water percolating downward from the surface of the ground enters its open-air spaces. This water is generally rich in calcium bicarbonate, the product of carbonic acid reacting with the calcite in limestone. As the water enters the cave, two things can happen:

- In the open-air environment of the cave, some water may evaporate, increasing the concentration of calcium bicarbonate in the remaining water until it becomes saturated.
- Much of the carbon dioxide dissolved in the water may escape into the air, converting the carbonic acid back into carbon dioxide and water.

Both of these processes cause calcium carbonate to precipitate from solution as deposits on cave surfaces. These deposits build up into a variety of formations collectively known as **speleothems.** Speleothems consist largely of the rock *travertine,* the name given to calcium carbonate when it forms cave deposits.

Stalactites are stony travertine structures, resembling icicles, that hang from cave ceilings. They form as water, one drop at a time, enters the roof of a cave through a crack and deposits minute amounts of calcium carbonate. Initially, the center of each growing stalactite is a hollow tube, resembling a soda straw, through which the next drop enters the cave. Eventually precipitated travertine clogs the tube, causing the water to drip down the stalactite's outer surface, where it continues to deposit travertine and creates the irregular icicle shape of a typical stalactite.

Stalagmites are the travertine deposits that accumulate on the floors of caves from water that drips from their ceilings. Thus stalagmites form directly below stalactites. Stalagmites may grow as high as 30 meters (100 feet) and sometimes merge with stalactites to form single speleothems, called *columns,* that reach from the cave's floor to its ceiling.

Speleothems may also assume a variety of other delicate and beautiful shapes. Speleothems that form below a crack in the ceiling of a cave may produce *banded draperies,*

Figure 14-21 Subterranean erosion. An underground river carved this deep channel by dissolution and abrasion.

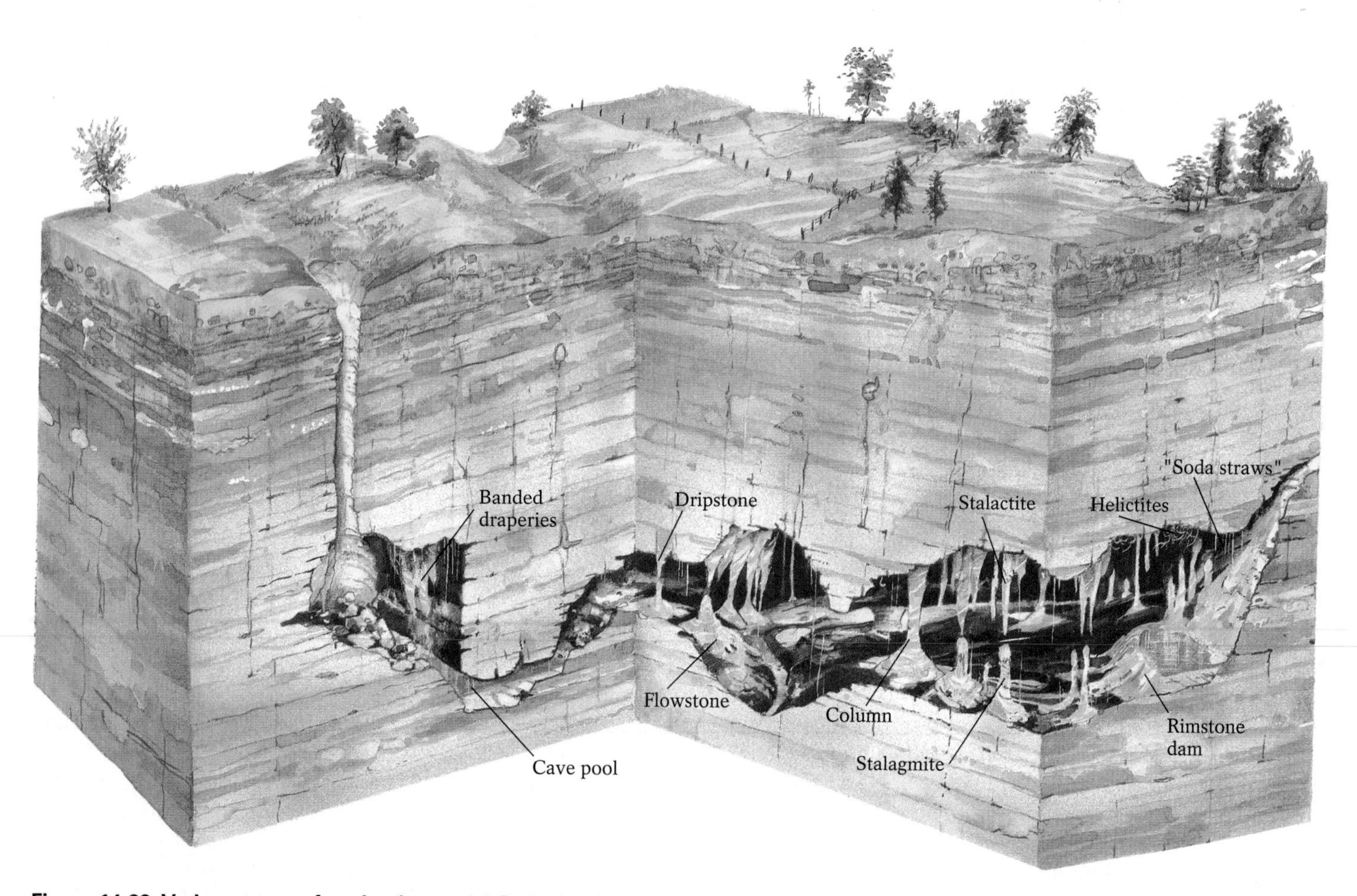

Figure 14-22 Various types of speleothems. **(a)** Stalactites in their early, "soda-straw" stage of formation. **(b)** Stalactites and stalagmites merge to form a column. **(c)** Helictites form by capillary action and can grow in any direction.

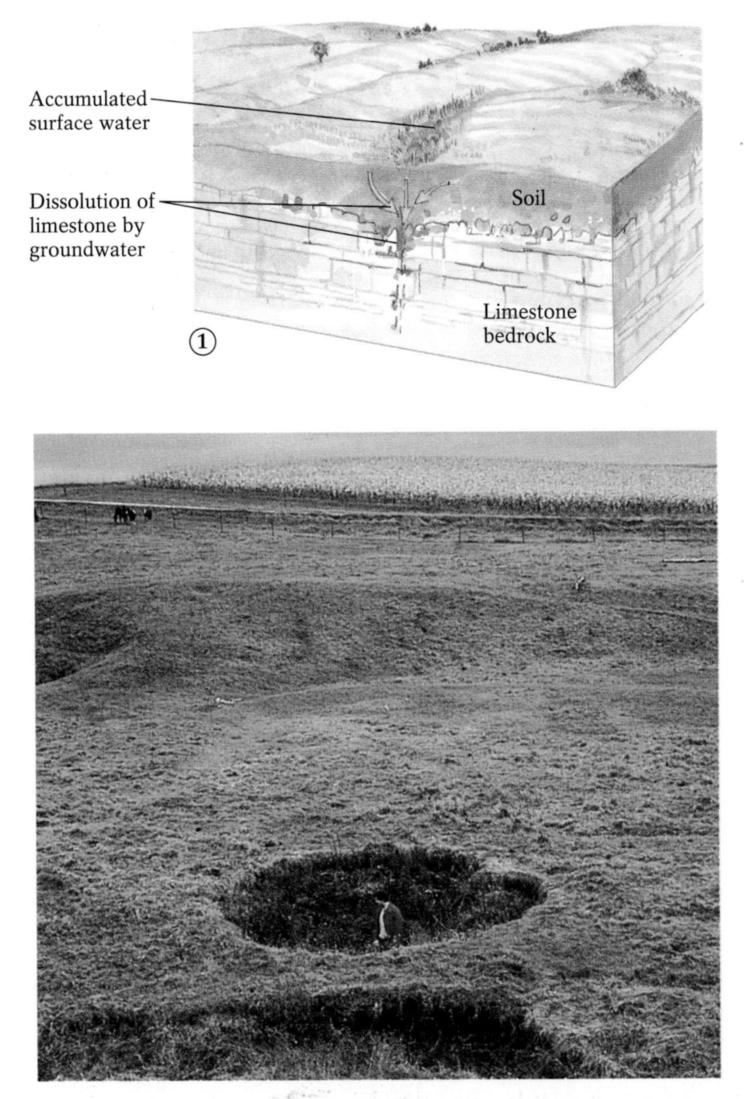

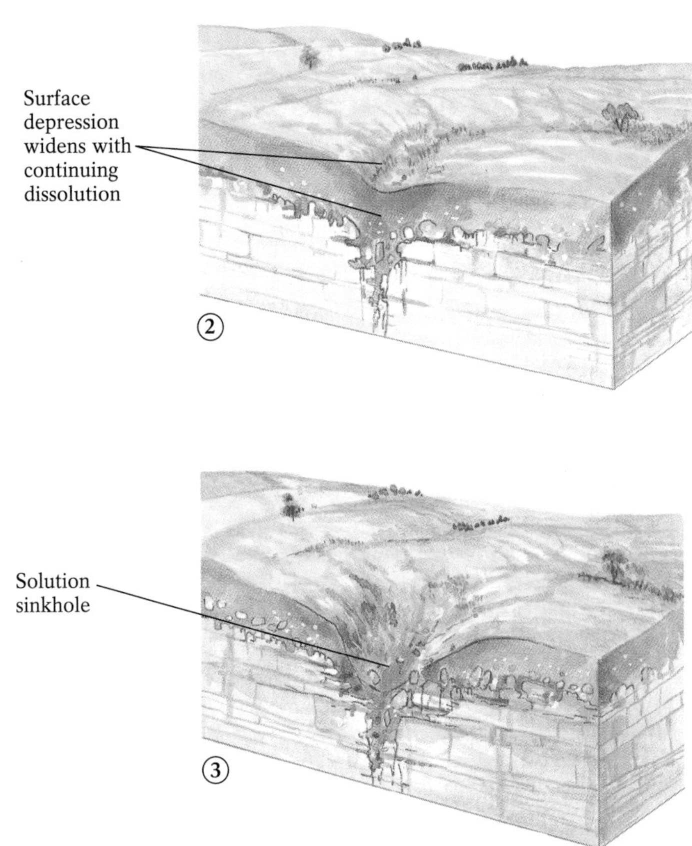

Figure 14-23 The origin of solution sinkholes.
Photo: A sinkhole in Minnesota.

or *drip curtains. Rimstone dams* form when travertine precipitates over obstructions to water outflow from a cave pool. *Helictites* are twig-like structures whose central tubes are so narrow that water passes through them by capillary action. Because capillary action is not dependent on gravity, helictites can grow from any point on a cave ceiling, wall, or floor and in any direction. Figure 14-22 shows some of the great variety of speleothem shapes.

Not all caves have significant speleothem development. Mammoth Cave in Kentucky, despite having largely undecorated walls, has been included in the U.S. national parks system because of its unique character and grand scale (Highlight 14-2).

Karst Topography

Karst topography is the surface expression of karst. It is marked by unique geologic features, principally sinkholes and disappearing streams, that result from the dissolution of soluble bedrock at or near ground level. Other features in karst regions include blind valleys and natural bridges.

Sinkholes are circular surface depressions that appear in most limestone terrains, often occurring in great numbers over broad expanses of highly fractured bedrock. Central Kentucky alone has 60,000 sinkholes; southern Indiana has more than 300,000. Sinkholes form in two different ways: as solution sinkholes or as collapse sinkholes. *Solution sinkholes* are created when acidic groundwater dissolves limestone at or just below the surface. The diameters of these sinkholes widen at the surface over time and become progressively narrower below ground, forming a characteristic funnel shape (Fig. 14-23). Solution sinkholes generally develop on flat or gently sloping landscapes where water tends to persist rather than draining immediately, thereby promoting prolonged contact and extensive dissolution.

Collapse sinkholes are usually deeper than solution sinkholes and typically have steep (sometimes vertical) sides and rocky and irregular floors covered with rubble. They form when the roofs of caves collapse under the weight of overlying rocks and soils, often suddenly and unpredictably. Most collapse sinkholes appear after a lengthy drought or excessive withdrawal of groundwater lowers the regional water

Highlight 14-2 Mammoth Cave

Mammoth Cave is located in moist, temperate, west-central Kentucky between Louisville and Nashville, Tennessee. Its vast labyrinth of connected galleries, some 80 meters (250 feet) high, contains underground lakes, rushing rivers, and spectacular waterfalls. Mammoth Cave was explored intensively centuries ago by native North Americans and more recently by westward-migrating pioneers. In the nineteenth century, the cave—known for its exceedingly long dry tunnels—was considered a natural wonder and rivaled Niagara Falls as the main tourist attraction of eastern North America. In 1972, a connection was found linking the Mammoth Cave system to the adjacent Flint Ridge system. The entire complex, with more than 400 kilometers (250 miles) of interconnected passages, is now recognized as the world's longest surveyed system (Fig. 1).

Mammoth Cave's geological history began some 350 million years ago, when the Ste. Genevieve and St. Louis limestones were deposited on the floor of a shallow sea. Cave formation began in the limestone perhaps 30 million years ago. The most active phase began 2 million years ago, when a massive ice sheet originating in central Canada near Hudson Bay advanced southward toward what is now Kentucky. Although the ice never covered the Mammoth Cave region, it blocked large westward-draining rivers, diverting them into the local stream systems. These enlarged streams cut so deeply through the regional limestone plateau that they drew water down from the area's zone of saturation, lowering the water table and draining the top of the cave system.

Largely because its internal open-air environment is extremely dry, no significant speleothem development has taken place in Mammoth Cave. Most water within the Mammoth–Flint system flows as subterranean streams that have entered the caverns through local surface depressions where bedrock fractures intersect. The bulk of the area is overlain by a moderately impermeable "cap rock" that inhibits infiltration of water. Only in a few places, where this rock has been eroded, does enough water enter to produce speleothems.

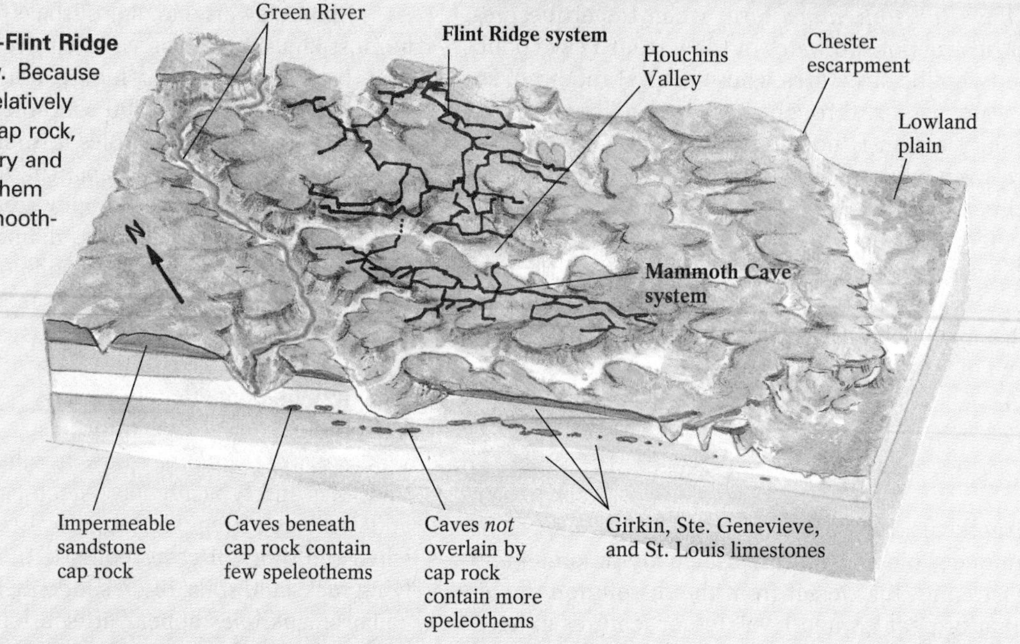

Figure 1 The Mammoth–Flint Ridge cave system in Kentucky. Because most of it is overlain by relatively impermeable sandstone cap rock, Mammoth Cave is fairly dry and undergoes limited speleothem development. Photo: A smooth-walled passageway in Mammoth Cave.

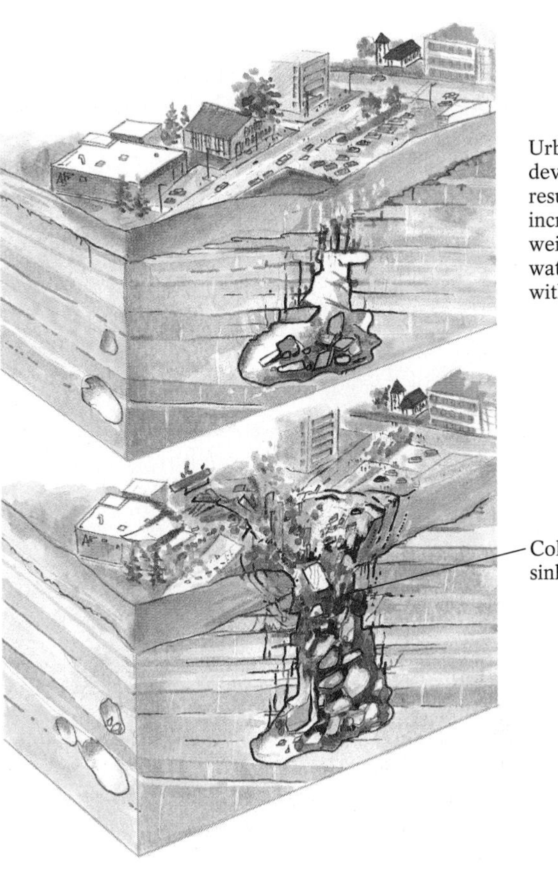

Figure 14-24 The origin of collapse sinkholes. A drop in the water table level, either as a result of drought or from excessive groundwater withdrawal, is a common cause of collapse sinkholes. Photo: A collapse sinkhole in Winter Park, Florida. The sinkhole formed in 1981, when the roof of a large subterranean cavity caved in without warning after a lengthy drought, swallowing a three-bedroom bungalow, a portion of a Porsche dealership, and half a municipal swimming pool.

table, depleting the aquifer of water that might have supported the overlying surface (Fig. 14-24). In recent years, as the population of the southeastern United States has exploded and a great volume of groundwater has been extracted to meet rapidly rising demand, the regional water table has dropped and the surface in many areas has been left unsupported. As a result, thousands of new collapse sinkholes have appeared throughout the Southeast.

Disappearing streams occur where surface streams in karst regions drain into sinkholes and disappear (Fig. 14-25), sometimes descending into well-developed cave systems below. Some disappearing streams eventually return to the surface as natural springs.

Blind valleys are channels that form when adjacent sinkholes, enlarged by continuing dissolution and erosion, coalesce to form larger sinkholes. These features contain water only during periods of heavy rain, when all of the precipitation

Figure 14-25 Disappearing streams. Streams in karst often disappear suddenly down sinkholes that open into caverns and passageways below the surface. In southern Indiana, streams with suggestive names such as Sinking Creek and the Lost River flow underground for more than 10 kilometers (6 miles).

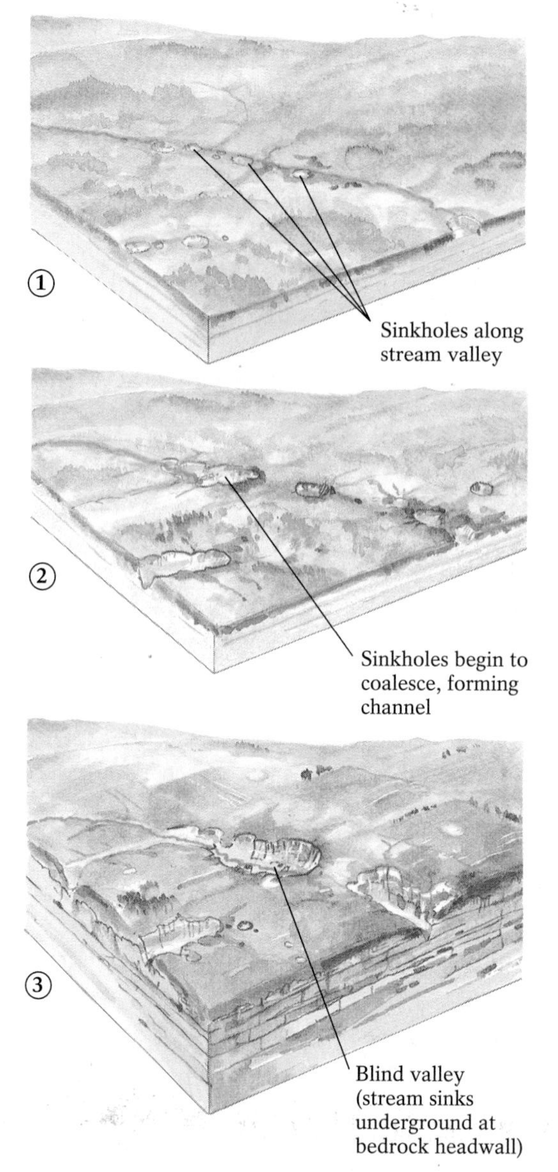

Figure 14-26 Sinkhole expansion. Coalescing sinkholes may eventually form blind valleys and natural bridges. Photo: Natural Bridge, in the Blue Ridge Mountains near Lexington, Virginia, spans Cedar Gorge Creek.

Figure 14-27 Karst towers in the Kwangsi region of southern China. Soluble limestone has been dissolved and carried away in solution, leaving behind the "towers" of less-soluble rock.

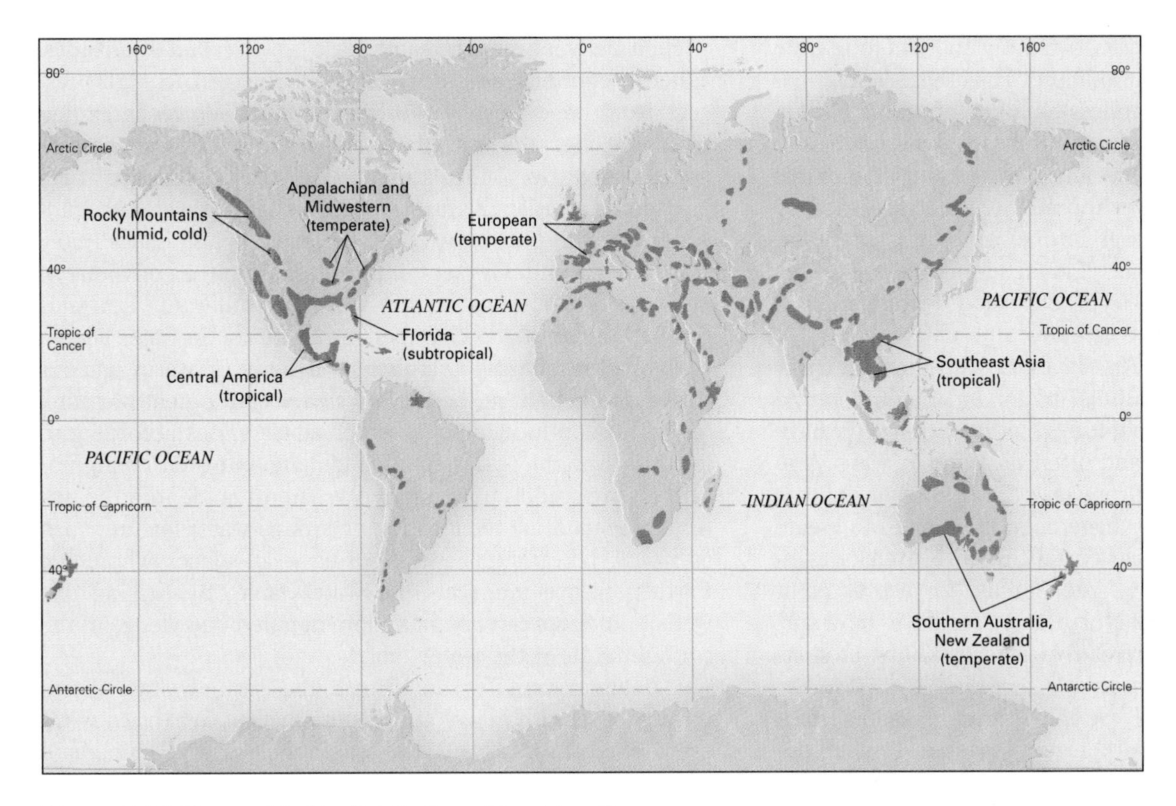

Figure 14-28 The worldwide distribution of karst landforms.

cannot drain into the ground. Unlike regular stream valleys, blind valleys end at bedrock headwalls where the streams seem to disappear, becoming diverted underground.

As a series of neighboring sinkholes expand and coalesce, the surface overlying broad sections of underground stream channels may collapse. Segments of surface material that do not collapse form *natural bridges* over the exposed channel. Perhaps the most famous natural bridge in North America is located in the Blue Ridge Mountains of Virginia, where a remnant of a massive dolomitic limestone has created a bridge approximately 30 meters (100 feet) long (Fig. 14-26).

In tropical climates, abundant rainfall and lush vegetation (which, when it dies, contributes carbon dioxide to the soil) may accelerate limestone dissolution. As a result, karst formations in tropical regions can be particularly dramatic, forming *cockpit karst.* Cockpit karst consists of numerous closely spaced, irregular depressions (cockpits) dissolved out of bedrock and steep-sided conical towers composed of the less soluble, less fractured rock that remains (Fig. 14-27).

Major Karst Sites

Millions of square kilometers of soluble bedrock at or near the Earth's surface have become or are candidates to become karst. Little or no karst is found in arctic regions, where water is often frozen and thus unavailable for dissolution, vegetation is sparse, and permafrost impedes infiltration. In arid regions, low precipitation levels and high evaporation rates similarly discourage dissolution. In temperate regions and humid, mid-latitude areas, however, dissolution proceeds at a relatively rapid pace. The prodigious rainfall and lush vegetation in the humid tropics provide the optimal environment for limestone dissolution, and karst landscapes are highly developed there.

Extensive karst can be found in China, Southeast Asia, Australia, Puerto Rico, Cuba and elsewhere in the Caribbean, the Yucatán peninsula of Mexico, and much of southern Europe. In the United States, nearly 25% of the area encompassed by the contiguous 48 states is karst, including substantial parts of Virginia, Alabama, Florida, Kentucky, Indiana, Iowa, and Missouri. Figure 14-28 shows the worldwide distribution of karst landforms.

Protecting Caves and Karst Environments

Groundwater pollution is an especially troubling problem in karst regions. In nonkarst areas, contaminants in groundwater are often filtered out naturally by soils or by bedrock with primary porosity, such as sandstone. Well-developed karst environments, on the other hand, typically contain numerous sinkholes connected to vast underground networks of caverns and passageways. In such places, water readily enters the ground, moving through it so rapidly that there is little opportunity to filter out contaminants. Thus these landscapes are extremely sensitive to careless handling of waste. For example, when Hidden River Cave in Kentucky was opened to visitors in 1916, it boasted elegant galleries and memorable subterranean boat rides. Over the next few decades, nearby towns disposed of their sewage into several large sinkholes, hopelessly contaminating the regional groundwater system. By the mid-1930s, Hidden River Cave had developed such vile odors that tourism there came to an end.

The majority of North America's most spectacular caves have been designated as national parks and monuments to be preserved for future generations. Others, however, have been treated in mercenary, and often destructive, ways. Amateur collectors have hacked out large quantities of delicate and rare cave crystals. Caves have been dynamited to create more commercially accessible entrances. In Kentucky, gypsum crystals from caves have been mined for use as paint pigments. Nevertheless, most North American caves have survived, largely because their relative inaccessibility protects them from human disturbance.

In Chapters 13 and 14, we have examined the surface and subsurface geological work accomplished as flowing *liquid* water passes through various segments of the hydrologic cycle. In the next chapter, we look at how flowing *frozen* water—as glaciers—has shaped the Earth's surface.

Chapter Summary

Surface water enters the ground wherever topography, geological composition, vegetative cover, and climate permit its infiltration. Directly beneath the surface of the ground is a **zone of aeration,** where pore spaces in rocks and soils are filled with both air and water. Beneath this area lies the **zone of saturation,** in which all available pore spaces are filled with water. The upper surface of the saturated zone is called the **water table.**

Water flows underground when the geological materials through which it travels are sufficiently porous and permeable. **Porosity,** the percentage of a material's pore space relative to its total volume, is a measure of how much water it can hold. **Permeability** measures the ability of a material to transmit water; the presence of connected fractures in solid bedrock and the coarse, well-sorted texture of unconsolidated sediment and soil increase permeability. Groundwater flow is also affected by gravity and by pressure from overlying water.

Aquifers are permeable, water-bearing bodies of geologic materials. Aquifers found at the surface, such as within floodplain deposits and glacial gravels, that are not overlain by impermeable rock are called unconfined aquifers. Confined aquifers, which are found at greater depth, are sandwiched between impermeable rock layers called **aquicludes.** When a confined aquifer's water is subjected to high pressure from large elevation differences between recharge and discharge sites, the water rises above the level of the aquifer and gushes from the ground. Geologists describe such aquifers as **artesian. Natural springs** are sites at which groundwater flows to the surface, usually laterally, and issues freely from the ground. **Geysers,** intermittent surface emissions of hot water and steam, occur where groundwater descends through underground chambers or fractures, becomes heated by a shallow magma chamber or a body of warm igneous rocks, and is then pushed up by steam under great pressure.

The depth of regional water tables varies according to local topography and prevailing climate, but their configuration often parallels topography. We can often identify the water table by noting the location of surface-water features such as rivers, lakes, and natural springs—all places where the Earth's surface intersects the water table. By drilling test wells, geologists can obtain a more detailed knowledge of the configuration of the water table.

Human activity can disturb the groundwater system through overwithdrawal and contamination. When we withdraw water from a well, the water table in the immediate vicinity is drawn down around the well, forming a **cone of depression.** If groundwater withdrawal continually exceeds recharge, the groundwater may become depleted and the local water table may drop significantly and permanently. In some cases, **subsidence** of the land surface may occur as depleted aquifers become compressed. In coastal areas, if the fresh groundwater table is lowered significantly, the salty groundwater will rise and may completely replace the freshwater, a problem known as **saltwater intrusion.**

The features that are created when groundwater dissolves bedrock, either underground or at the surface, are known collectively as **karst.** The most common features of karst are **caves,** natural underground cavities. **Speleothems** consist of travertine formations created when calcium carbonate precipitates on cave ceilings, walls, and floors. **Karst topography** is the surface expression of karst. It is marked by unique geologic features, such as **sinkholes,** circular surface depressions that appear in most limestone terrains, and **disappearing streams,** surface streams that drain into sinkholes and disappear, sometimes descending into well-developed cave systems.

In karst environments, water enters the ground and moves through it so rapidly that contaminants may not be filtered out by soil or porous bedrock. Thus these landscapes are extremely sensitive to careless handling of waste and easily polluted.

Key Terms

zone of aeration (p. 260)
zone of saturation (p. 260)
water table (p. 261)
porosity (p. 261)

permeability (p. 261)
aquifer (p. 263)
aquicludes (p. 263)
artesian (p. 265)
natural springs (p. 265)
geyser (p. 266)
cone of depression (p. 268)
subsidence (p. 269)
saltwater intrusion (p. 270)
karst (p. 272)
caves (p. 272)
speleothems (p. 273)
karst topography (p. 275)
sinkholes (p. 275)
disappearing streams (p. 277)

Questions for Review

1. Discuss three factors that affect groundwater recharge.

2. What is the fundamental difference between the zone of aeration and the zone of saturation? What is a water table?

3. What is the difference between porosity and permeability? What is the difference between primary and secondary porosity?

4. What is an aquifer? Where would you look for an unconfined aquifer? A confined aquifer?

5. Draw a simple sketch showing the conditions necessary to create an artesian groundwater system.

6. Describe three possible results of groundwater depletion. Discuss two ways that communities might solve the problem of dwindling groundwater supplies.

7. Briefly describe how caves form. Where is the water table located during the two stages of cave formation?

8. How do stalactites differ from stalagmites? How does each form?

9. Describe the two ways that sinkholes form.

10. Why is a karst landscape particularly susceptible to groundwater pollution?

For Further Thought

1. Why does groundwater flow so slowly? Why is the water table generally higher beneath hills and lower beneath valleys?

2. How would you solve the groundwater problems facing the Florida Everglades?

3. Suppose you were placed in charge of ensuring the groundwater needs of Los Angeles, California, for the twenty-first century. Name five problems that you would face and the solutions that you would propose.

4. Discuss three reasons why you would not expect to find caves in polar regions.

5. What geological features do you think are shown in the aerial photos below, taken over the same area of the Midwest 43 years apart? Why might these features have grown in number and size between 1937 and 1980?

15

Glaciers and Ice Ages

Figure 15-1 The deeply crevassed Davidson Glacier near Haines, Alaska.

Glaciers provide some of the most awe-inspiring scenery on Earth, drawing millions to Alaska and other high-latitude regions in which they are common (Fig. 15-1). Glaciers also shape the landscapes over which they slowly flow. Without the continental ice sheets of the Pleistocene Epoch, which ended approximately 10,000 years ago, North America would not have its Great Lakes, Niagara Falls, Hudson Bay, Puget Sound, or the 15,000 lakes of Minnesota. There would be no Cape Cod in Massachusetts and no fertile rolling hills in the Midwest and southern Canada. Rivers such as the Missouri and Ohio would drain north to the Arctic and Atlantic Oceans, rather than south to the Mississippi River and the Gulf of Mexico. If Earth had no glaciers today, the shapes of the continents themselves would be substantially different, because sea level would be nearly 70 meters (230 feet) higher than its current level. Landlocked cities such as Memphis, Tennessee, and Sacramento, California, would be seaports, while San Francisco, New York, and many other coastal cities would be mostly under water.

Today, glaciers cover about 10% of the world's land surface. At very high latitudes, such as in the Arctic, Antarctica, and Greenland, these features are common and can exist at any elevation, even at sea level. They occur in mid-latitude regions as well, although only where temperatures are sufficiently cold—at high elevations (at about 2500 to 3000 meters [8000 to 10,000 feet]), such as in the Northern Rockies of Alberta, British Columbia, and Montana.

In this chapter, we discuss what glaciers are, where they exist today, where they appeared in the past, and how they have shaped the landscape. We also consider why only certain periods of the Earth's history have been marked by worldwide glacial expansions, and speculate about when glaciers may again cover vast areas of North America.

Glacier Formation and Growth

A **glacier** is a moving body of ice that forms from the accumulation and compaction of snow. Glaciers flow downslope or outward under the influence of gravity and the pressure of their own weight.

Glacier formation begins with a snowfall and the accumulation of snowflakes (Fig. 15-2). The initial snowpack, like most new snowfalls, tends to be fluffy and of low density. As more snow falls, however, it compresses the underlying snow. Some of the contact points between snow crystals then melt. The resulting water migrates into the pore spaces between snow crystals, refreezes, and binds the snow crystals together. The dense, well-packed snow, called **firn,** created in this way can survive the summer melting season. Repeated melting and refreezing of its interlocking crystals eventually squeezes nearly all air from firn, allowing it to become glacial ice, with a density about the same as that of an ordinary ice cube.

Glaciers are able to form and grow when snow accumulation exceeds the losses from summer melting and sublimation. (*Sublimation* is the process in which a solid changes directly into a gas; in this case, some of the solid snow changes directly into water vapor. This sublimation process accounts for decreases in snow that occur without a "slush" stage.) Cool, cloudy, brief summers that minimize melting probably contribute more to glacier growth than very low winter temperatures; in places where summer warmth completely removes even heavy winter accumulations of snow, such as in Minnesota, glaciers do not occur. The climates that are most likely to produce glaciers are found in higher latitudes and at higher elevations, where it rarely remains very warm for very long. In mountainous regions, glaciers usually form at or above the *snowline,* the lowest topographic level at which we find year-round snow.

The time required for fresh snow to become transformed into glacial ice varies with the rate of snow accumulation and the local climate. In snowy climates with an average annual air temperature close to the melting point of ice (0°C [32°F]), snow may be converted to ice in only a few decades. In extremely cold polar settings, where snowfall is typically sparse and little melting takes place, glacial ice formation may take thousands of years.

Classifying Glaciers

Glaciers are classified broadly by whether the local topography confines them or allows them to move freely (Fig. 15-3). **Alpine glaciers** are confined by surrounding bedrock highlands. Because they are confined, they are relatively small. Three types of alpine glaciers exist: **Cirque glaciers** create and occupy semicircular basins on mountainsides, usually near the heads of valleys; **valley glaciers** develop when growing cirque glaciers flow into preexisting stream valleys; and **icecaps** form at the tops of mountains.

Piedmont ("foot of the mountain") glaciers originate as confined alpine glaciers but then flow onto adjacent lowlands; once unconfined in the lowlands, they can spread radially. Piedmont glaciers that flow to coastlines and into seawater become *tidewater glaciers.*

The only completely unconfined form of glacier is a **continental ice sheet,** an ice mass so large that it blankets much or all of a continent. Today, continental ice sheets cover Greenland and Antarctica. The Antarctic ice sheet actually consists of two ice sheets separated by the Transantarctic Mountains. It is as much as 4.3 kilometers (nearly 3 miles) thick in places and occupies an area about 1.5 times as large as the continental United States. Twenty thousand years ago, such vast ice sheets covered North America to south of the Great Lakes and blanketed western Europe south to Germany and Poland.

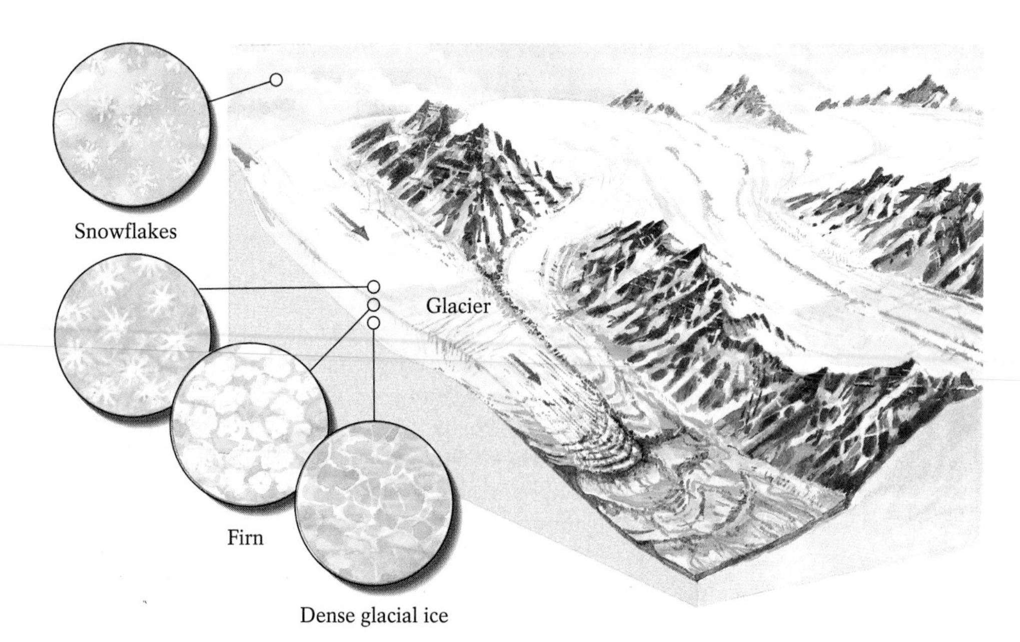

Figure 15-2 The formation of glacial ice from snow. Hexagonal snowflakes that fall on a glacier are gradually changed into rounded crystals. As overlying snow buries the crystals and exerts pressure on them, they become packed into increasingly dense firn and finally glacial ice.

Figure 15-3 Types of glaciers. **(a)** Angel Glacier, a cirque glacier on Mount Edith Cavell, Jasper National Park, Canada. **(b)** A valley glacier in Tongas National Forest, Alaska. **(c)** An icecap in the Sentinel Range, part of the Antarctic continental glacier. **(d)** A tidewater glacier at Kenai Fjords National Park, Alaska.

The Budget of a Glacier

The *budget* of a glacier is the difference between the glacier's annual gain of snow and ice, which occurs in its **zone of accumulation,** and its annual loss of snow and ice, which occurs in its **zone of ablation** (Fig. 15-4). Accumulation occurs principally by snowfall, but is sometimes augmented by avalanches from surrounding slopes. Ablation mainly takes place through melting and sublimation of snow and ice; in the case of glaciers that terminate in bodies of water, ablation may also occur by *calving,* the process in which chunks of ice break off into the water and become *icebergs.* The boundary between the accumulation and ablation zones, which shifts annually in response to that year's climatic conditions, is known as the equilibrium line.

If a glacier's accumulation exceeds its ablation for several consecutive years, its budget is positive and the glacier increases in thickness and area. As a glacier expands in this manner, it advances outward (in the case of unconfined glaciers) or downslope (in the case of confined glaciers); its "leading edge"—the margin that moves forward when the glacier advances—is called the *terminus.* If a glacier's ablation exceeds accumulation for several years, the glacier's budget is negative; the glacier decreases in size and its terminus retreats. If its accumulation is roughly equal to its ablation, the glacier's budget is balanced and its terminus remains stationary.

Glacial Flow

Whether a glacier's terminus is advancing because of accumulation, receding because of ablation, or stationary, ice within the glacier tends to flow forward from the accumulation zone toward the ablation zone. In the accumulation zone, snow becomes compacted into ice and flows down toward the glacier's bed. In the ablation zone, ice is forced to flow upward toward the surface and outward toward the glacier's edges.

Ice flow typically begins when a glacier's thickness exceeds about 60 meters (200 feet). The flow takes place through a combination of two mechanisms: internal defor-

Figure 15-4 The anatomy of a glacier. Every glacier includes an accumulation zone, where more snow and ice are added every year than are lost, and an ablation zone, where more snow and ice are lost than are added annually. An equilibrium line, where the amount of snow and ice added approximately equals the amount lost, forms the boundary between the two zones. In addition to melting and sublimation, a glacier may lose ice by calving if it terminates in a body of water. Photo: Calving of Hubbard Glacier in Wrangell-St. Elias National Park, Alaska.

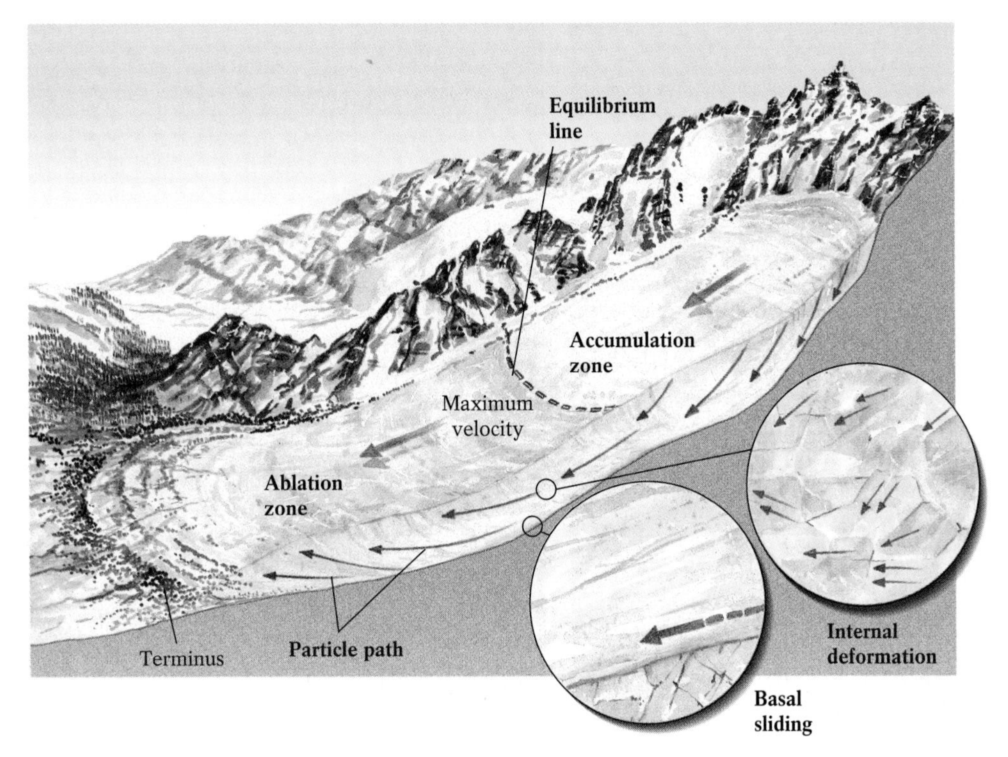

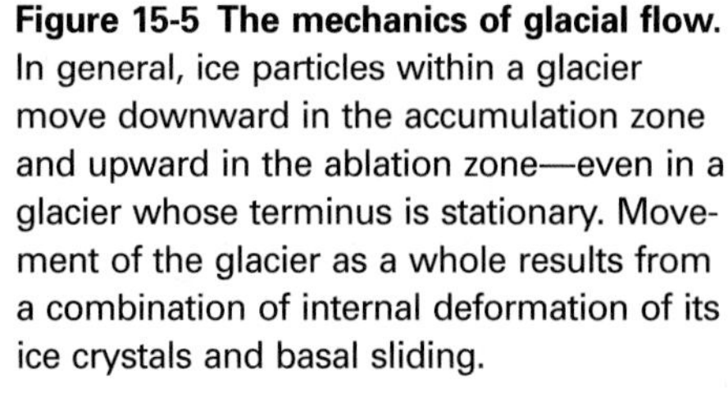
Figure 15-5 The mechanics of glacial flow. In general, ice particles within a glacier move downward in the accumulation zone and upward in the ablation zone—even in a glacier whose terminus is stationary. Movement of the glacier as a whole results from a combination of internal deformation of its ice crystals and basal sliding.

mation and basal sliding. In **internal deformation,** a glacier's ice crystals deform under pressure from overlying ice and snow. The crystals then slip past one another (particularly when the glacier is relatively warm and water lubricates crystal boundaries), or they fracture and move along planes of weakness. All glaciers—even those that remain frozen to their beds—flow to some extent by internal deformation. The extent of the flow depends largely on the temperature of the ice, which may vary from place to place within a glacier. The closer the temperature of ice is to its melting point, the more rapidly it flows.

In **basal sliding,** warmer glaciers, such as those in mid-latitude mountain ranges, thaw at their bases. The resulting film of water enables the glacier to slide along its bed. Glaciers in warm climates are more likely than those in extremely cold climates to move by both basal sliding and internal deformation, and thus they usually flow faster. Figure 15-5 illustrates the factors involved in glacial flow.

The velocity of glacial flow varies widely, depending on both the temperature of the ice and the steepness of the slope on which the glacier lies. Cold, nearly horizontal ice sheets flow slowly, usually creeping along at a rate of only a few meters per year; in contrast, warmer alpine glaciers on steep slopes may flow 300 meters (1000 feet) or more per year. Some glaciers periodically *surge,* or accelerate, usually in brief episodes lasting several months to a few years that are separated by longer periods (10–100 years) of normal flow. Surges apparently occur when a large volume of water accumulates at the base of a glacier, reducing friction and facilitating basal sliding. The most rapid surge known, on the Kutiah Glacier in Northern Pakistan, reached a peak velocity of 110 meters (350 feet) per day and lasted for several months.

The Work of Glaciers

Glacial erosion and deposition have been more effective than almost any other geologic process in shaping the features of North America, northern and central Europe, and Asia, especially in the alpine highlands on these continents. Rapidly flowing glaciers that move largely by basal sliding cause the most erosion, particularly within the zone of accumulation, where ice moves down to the glacier's bed.

Glacial Erosion

Glaciers erode by both abrasion and quarrying. **Glacial abrasion** occurs when rock fragments embedded in the base of a glacier scrape, polish, and cut long striations (scratches) or grooves in the surface of underlying bedrock (Fig. 15-6). Because striations and grooves are usually oriented in the same direction as the ice flow, they may indicate the direction taken by ancient glaciers. **Glacial quarrying** occurs when a glacier lifts masses of bedrock from its bed. This process occurs most often when meltwater at a glacier's base seeps into bedrock cracks and refreezes, dislodging rock fragments by frost wedging (see Chapter 5), and then incorporating those fragments into the glacier.

Abrasion and quarrying together can shape a rock mass into a distinctive asymmetrical form, called a *roche moutonnée* (pronounced "ROSH MOO-ton-nyea"; French for "sheep-like rock") (Fig. 15-7). A glacier typically abrades the up-glacier side of the rock mass (the side from which the glacier advances), but deeply quarries the downglacier side (the side toward which the glacier flows). A roche moutonnée is aligned lengthwise in the direction of glacial flow. Its up-

Figure 15-6 Glacial abrasion in bedrock. **(a)** Glacially striated bedrock in the Blue Mounds State Park in southwestern Minnesota. **(b)** Large glacial grooves in bedrock on Slate Island in Lake Superior, Ontario, Canada.

glacier side consists of a gentle humpbacked slope, whereas the downglacier side is characterized by an abrupt, steep drop. A roche moutonnée's asymmetry provides another indicator of the direction of glacier flow.

Erosional Features of Alpine Glaciation Over time, alpine glacial erosion can sculpt relatively smooth mountain slopes into spectacular rough-hewn peaks and precipitous gorges. The process begins in shady depressions at or above the local snowline, where lack of sunlight and accumulation of wind-drifted snow support the formation of glacial ice and development of cirque glaciers. Quarrying and abrasion by a cirque glacier eventually erode these depressions, making them progressively longer and deeper. Ultimately, the depressions are transformed into deep horseshoe-shaped basins called **cirques.** If the climate warms, a cirque glacier may melt, leaving behind water that forms a cirque lake, or **tarn,** in its basin.

A number of cirque glaciers can form on the same mountain at the same time, carving out a variety of unique alpine landforms (Fig. 15-8). When two cirque glaciers on opposite sides of a mountain erode headward and converge, they form a sharp ridge of erosion-resistant rock at their top

Figure 15-7 Glacial erosion. Glacial abrasion and quarrying combine to shape a roche moutonnée, like this one in Yosemite National Park, California. This distinctive erosional feature is produced when a glacier overrides a bedrock hill, abrading its upglacier side and quarrying its downglacier side. In this photo, the glacier flowed from the right.

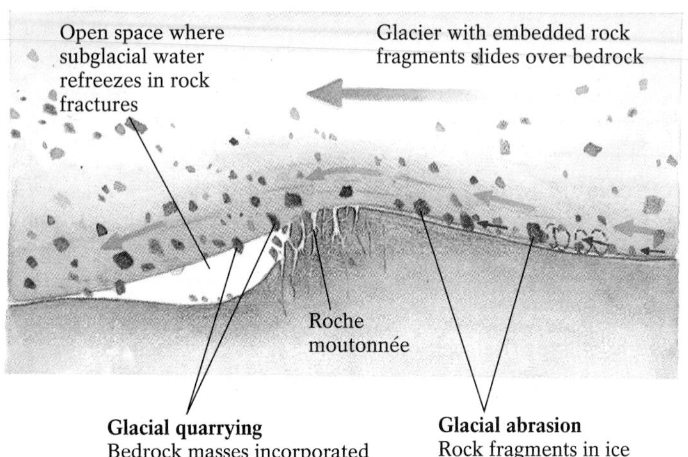

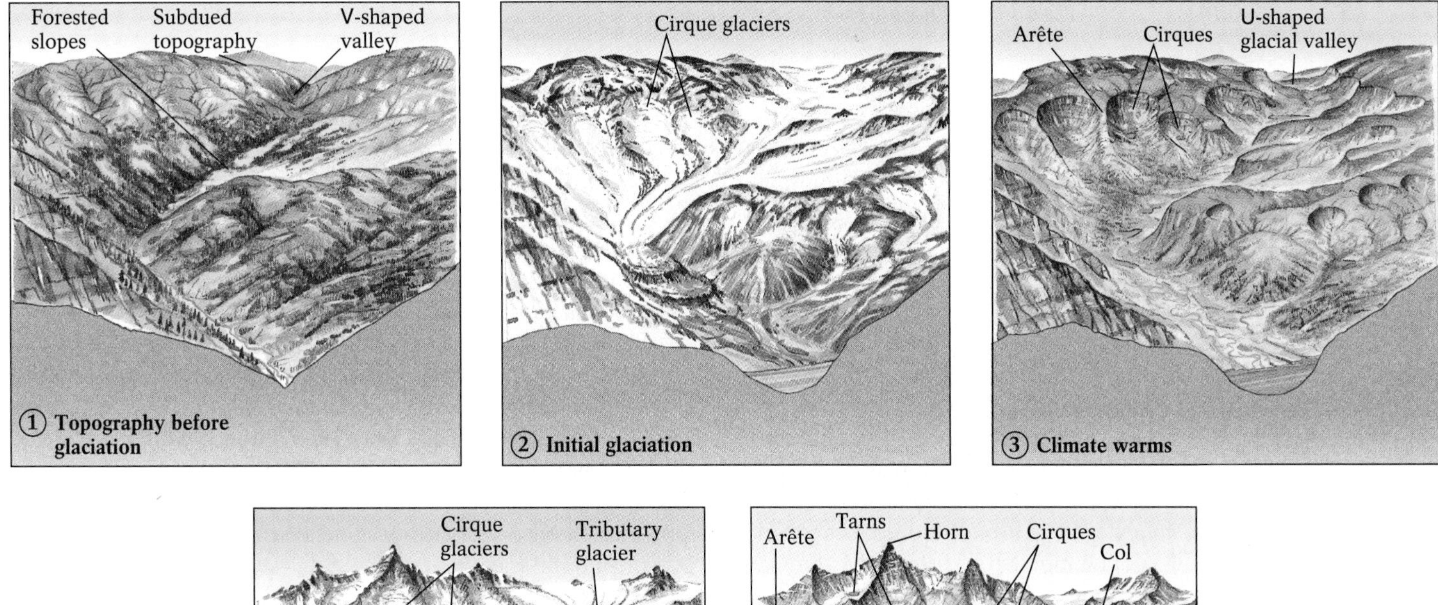

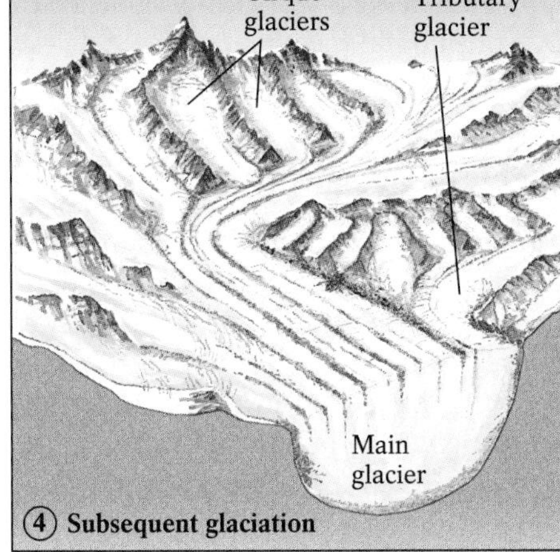

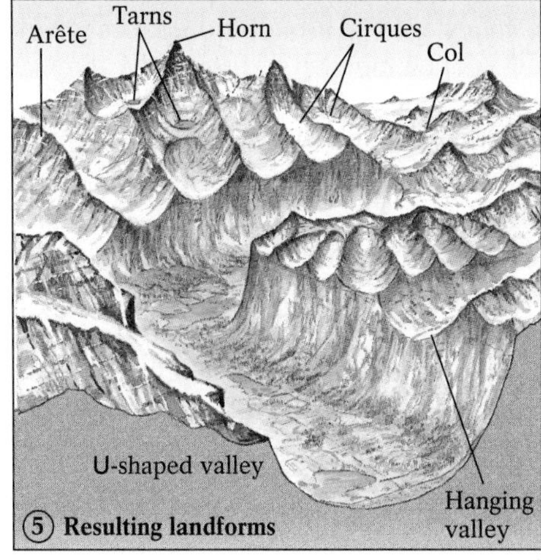

(a)

(b)

(c)

Figure 15-8 The effects of alpine glacial erosion. **(a)** Growth and proliferation of cirques on mountainsides eventually produce horns, arêtes, hanging valleys, and fjords. Another common indicator of glacial erosion is the conversion of V-shaped stream valleys into U-shaped glaciated valleys. **(b)** A U-shaped valley in Tracy Wilderness, southeastern Alaska. **(c)** U-shaped valleys that become flooded with seawater produce fjords, like Bela Bela Fjord in British Columbia, Canada, seen here.

and sides called an **arête.** Continued headward erosion of a cirque can breach part of an arête, producing a mountain pass, or **col.** When three or more cirque glaciers erode a mountain, a steep peak, or **horn,** develops. The Matterhorn in the Swiss Alps is perhaps the best known of these forms, but horns also dot the skyline of Banff and Jasper National Parks in the Canadian Rockies and Glacier National Park in northwestern Montana.

If a cirque glacier's budget is consistently positive, it may spread beyond its basin and flow downslope along a preexisting stream valley. As such a valley glacier moves, it erodes the valley floor and walls, ultimately transforming the narrow, V-shaped stream valley into a broad-floored U-shaped glaciated valley. If seawater submerges a U-shaped valley, a deep, saltwater-filled *fjord* forms.

Small glaciers may develop in tributary stream valleys and eventually join to form a main valley glacier. This main glacier will be thicker and flow more rapidly than its tributaries, eroding its underlying bedrock more rapidly. Eventually, it may undercut the tributary valleys, leaving them perched above the main glacier as *hanging valleys.* If a warm period then ensues, waterfalls will cascade from the hanging valleys; Yosemite Falls in Yosemite National Park, with a vertical drop of 735 meters (2425 feet), is an example of this phenomenon.

Erosional Features of Continental Ice Sheets A continental ice sheet produces erosional landforms on an even larger scale than an alpine glacier does. Through abrasion and quarrying, it can erode entire mountains, producing large-scale asymmetrical landforms called *whalebacks,* similar to but larger than roches moutonnées. As a continental ice sheet flows over stream valleys, it carves monumental U-shaped troughs, which may then become filled with glacier meltwater as the ice sheet retreats. The Finger Lakes of western New York (Fig. 15-9a), North America's five Great Lakes (which are more than 500 meters [1600 feet] deep in places) (Fig. 15-9b), the northern section of Puget Sound in Washington, and Scotland's Loch Ness were all preglacial stream valleys excavated by continental ice sheets.

Glacial Transport and Deposition

The load of debris quarried by a glacier is transported, embedded in the glacial ice, until its deposition. In general, the material is deposited when it reaches the terminus of the glacier or when the glacier's ice begins to melt. Such glacial deposits, or **glacial drift,** are classified into two broad categories: glacial till and meltwater stream deposits.

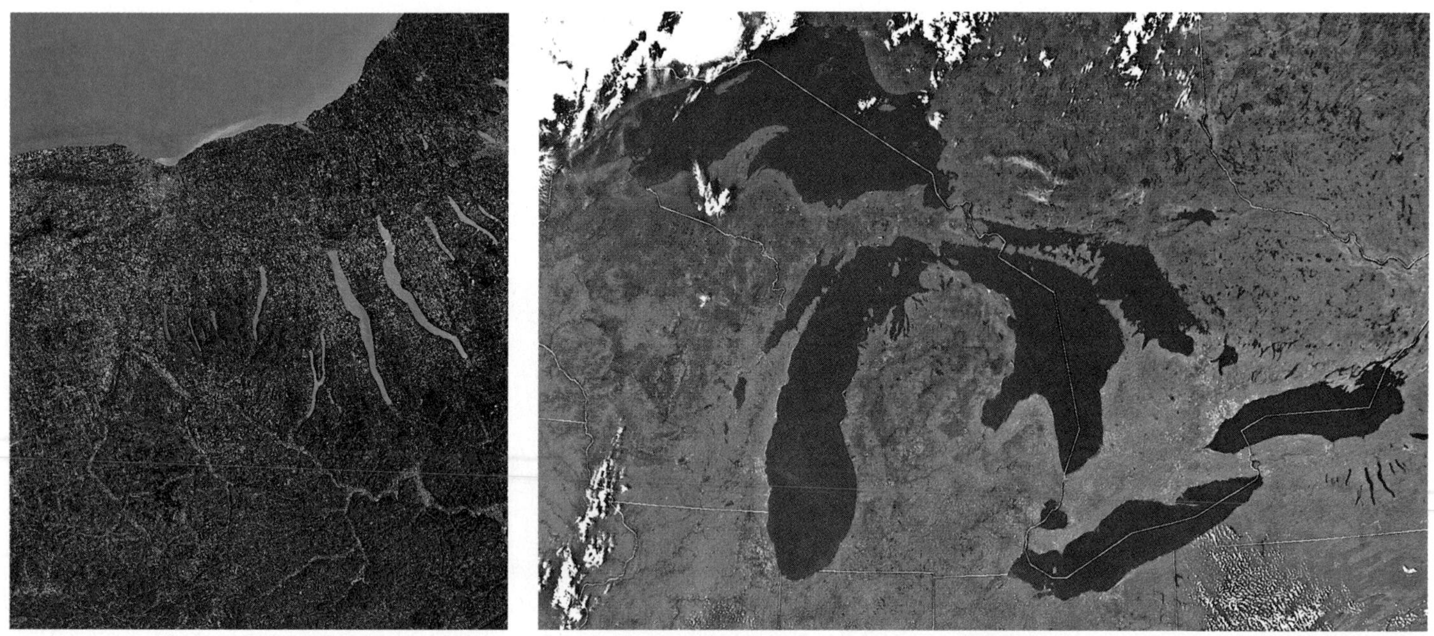

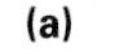

(a) (b)

Figure 15-9 Erosion of preglacial lowlands by continental ice sheets carved out deep basins and valleys. **(a)** During the Pleistocene Epoch, ice accumulating in the Lake Ontario basin (in what is now upstate New York) overflowed and moved south over the landscape, enlarging preexisting stream and river valleys to form deep linear troughs known collectively today as the Finger Lakes. Lake Ontario, itself a glacially carved basin, is north of the Finger Lakes in this satellite image. **(b)** A satellite image showing the five Great Lakes of North America (from left to right, Lakes Superior, Michigan, Huron, Erie, and Ontario). The much smaller Finger Lakes are shown at lower right. *State and international boundaries added for reference.*

Figure 15-10 Large granite erratics on a lake shore in Superior National Forest, Minnesota. The erratics were carried by the Laurentide continental ice sheet about 15,000 years ago, and probably came from the northeast, passing over southwestern Ontario.

Glacial Till **Glacial till** is drift that is deposited directly from glacial ice. Because glaciers can transport a wide range of particle sizes, till is typically poorly sorted, containing everything from fine clays to house-sized boulders. An exceptionally large glacially transported rock that has been eroded from one type of bedrock and deposited on another is known as a **glacial erratic** (Fig. 15-10).

Till deposits generally consist of large rock fragments surrounded by finer-grained sand, silt, and clay. They are deposited almost exclusively in the ablation zone, either by being plastered onto the underlying glacial bed by flowing ice or by sloughing off the glacier's surface as it melts.

Glacial till often forms a **moraine,** a mass of drift that accumulates at the terminus of a glacier. Moraines appear as bands of hills marking the various advances and retreats of a glacier (Fig. 15-11). Even when the terminus of a glacier remains stationary, ice continues to flow from the accumulation zone to the ablation zone; along the way, it erodes and transports sediment to the terminus where the load piles up. The longer the terminus remains stationary, the more till will be deposited in one place and the larger the resulting moraine. If the climate warms and the glacier recedes, a *terminal moraine* will mark the farthest advance of the ice.

If warming is interrupted by a return to glacial conditions, glacial recession will halt and the terminus of the glacier will again remain stationary for a few hundred years. Under these circumstances, the glacier may deposit a *recessional moraine,* upglacier from and usually parallel to its terminal moraine. A glacier that recedes intermittently may deposit several recessional moraines. Terminal and recessional moraines are produced by both continental ice sheets and alpine glaciers.

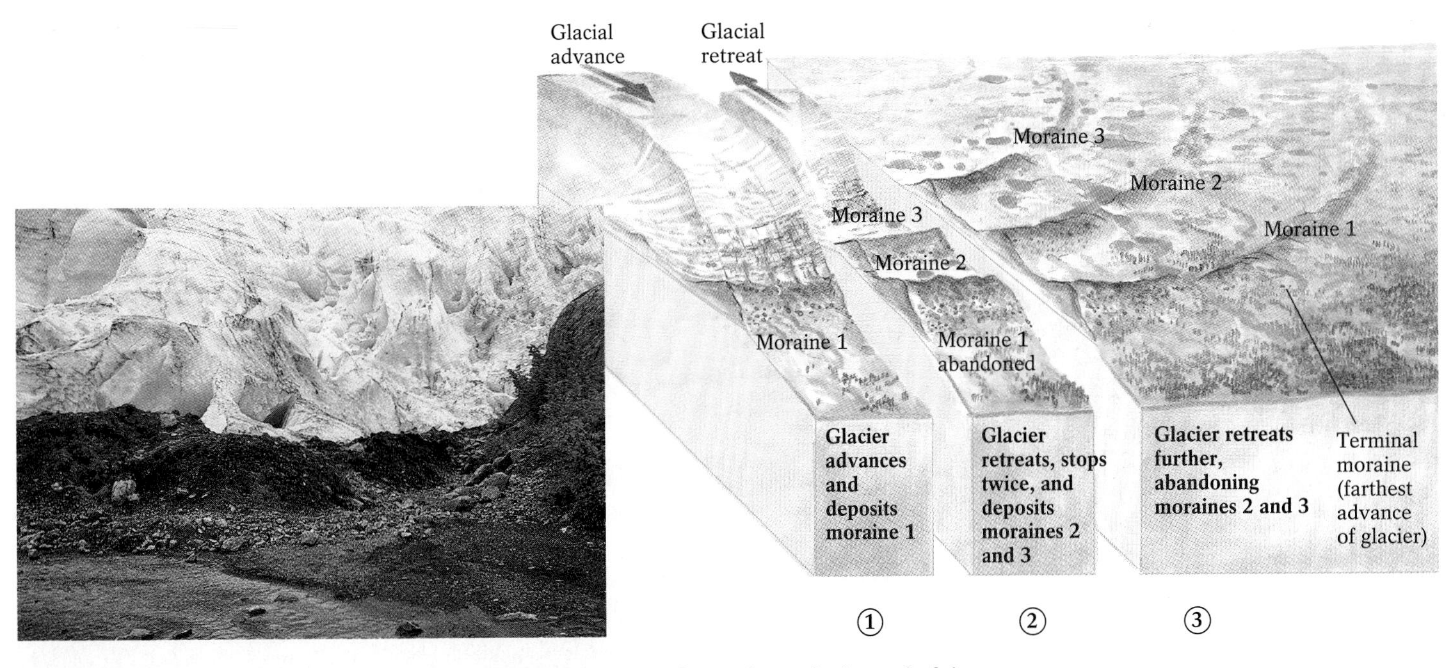

Figure 15-11 Glacial advance and retreat and the deposition of moraines. At the end of the sequence shown, moraine 1 is the glacier's terminal moraine, marking the maximum extent of the glacier. Moraines 2 and 3 are recessional moraines, marking the positions of the glacier's terminus during periods when the glacier's retreat was halted by temporary cooling trends. Photo: A moraine deposited by Exit Glacier at Kenai Fjord in Alaska.

Figure 15-12 The origin of medial moraines. Medial moraines form when adjacent glaciers converge, causing the lateral moraines at their edges to run together. Here several medial moraines are developing along the branches of the Kennicott Glacier in the Wrangell-St. Elias National Park, southeastern Alaska.

Two other types of moraines—lateral and medial—are formed only by alpine glaciers. Because they are topographically confined by their valleys, these glaciers have distinct sides adjacent to the valley walls where they deposit till, forming side or *lateral moraines.* When two valley glaciers merge, their sediment loads are carried down the valley, sandwiched between the ice streams. When these loads are deposited, they remain adjacent to one another, forming a *medial moraine* (Fig. 15-12).

If the climate cools significantly after a glacier has remained stationary for a long period of moraine building, the glacier may advance over previously constructed moraines. It will then reincorporate the moraines' particles and redeposit them as "new" terminal moraines. When ice sheets—thicker and more massive than alpine glaciers—override moraines, they apply enough pressure to reshape the moraines into gently rounded, elongated hills called **drumlins.** Drumlins, which range in height from 5 to 50 meters (17–170 feet), are typically aligned parallel to one another, oriented in the direction of ice flow, and frequently occur in clusters of hundreds or even thousands (Fig. 15-13).

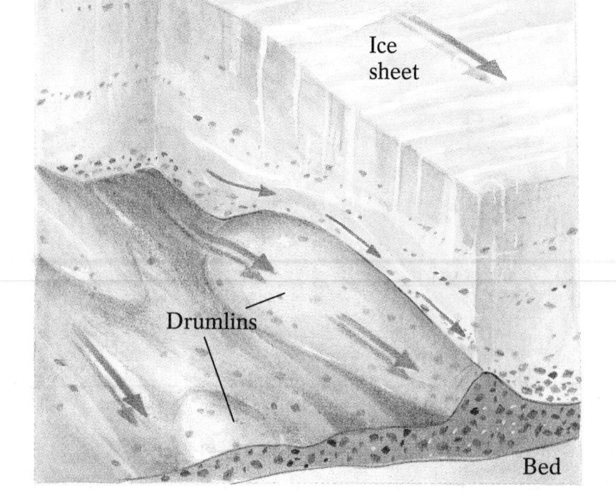

Figure 15-13 The origin of drumlins. Ice sheets passing over preexisting moraines can exert enough pressure to reshape them, forming low oval hills called drumlins. Photo: A drumlin field east of Rochester, New York.

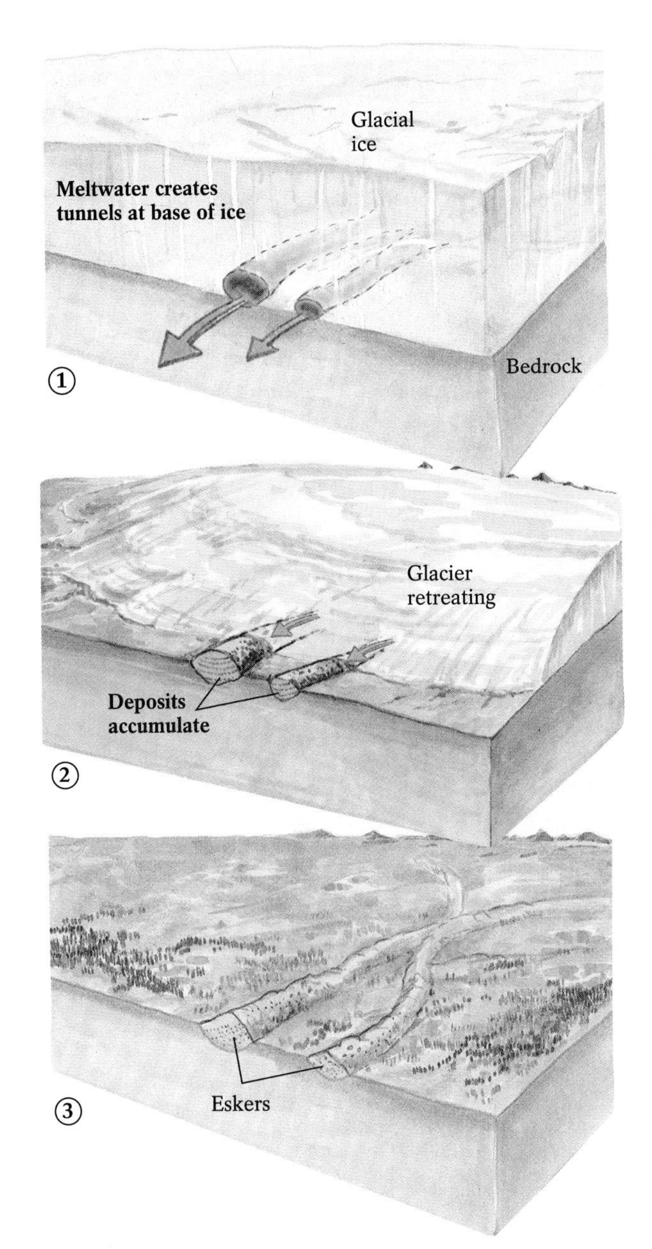

Figure 15-14 The origin of eskers. Eskers typically form under the ablation zone of a glacier, where meltwater erodes curved channels in the bedrock and deposits its load en route. When the glacier retreats, the meltwater flows away or evaporates, leaving meandering ridges of stratified, cross-bedded sand and gravel. Photo: Eskers in Coteau des Prairies, South Dakota.

Meltwater deposits of sand and gravel can also accumulate directly underneath glacial ice, creating sinuous ridges known as **eskers.** Eskers form beneath the ablation zone, when meltwater flowing along a glacier's bed carves S-shaped channels in bedrock. Later, as the glacier diminishes and meltwater flow decreases, sand and gravel are deposited in the channels to form the ridges (Fig. 15-14). Eskers can be seen throughout southern Canada and in much of the northern United States, including Maine, Michigan, Wisconsin, Minnesota, the Dakotas, and eastern Washington. They range in length from less than a kilometer to more than 150 kilometers (100 miles), and can be 30 meters (100 feet) high.

Deposits from Glacial Meltwater Streams At the terminus of a glacier, streams of melted glacial ice flow away from the glacier, depositing a portion of its transported sediment in their path. Such meltwater streams sort and stratify the drift they carry, creating glacial sediments that are dramatically different from till. The most common sediment from glacial meltwater, a mixture of sand and gravel particles, is deposited by braided streams downstream from the glacier as **outwash.** Outwash plains, which are characterized by broad, gently sloping surfaces, accumulate beyond the front of a glacier. Powerful glacial-age winds may erode drying outwash, producing a mass of fine silt, or *loess* (German for "loose"). The fertile plains along the Mississippi River valley are rich in such wind-blown silt deposits all the way to the Gulf of Mexico, even though the southern terminus of North America's Pleistocene ice sheets extended only as far as northern Iowa.

Other Effects of Glaciation

In addition to the local effects of glacial erosion, transport, and deposition, glaciers and glacial conditions can produce a number of other, more far-reaching effects.

Stream Disruption Expanding ice sheets can become continent-wide dams, altering the courses of preexisting streams. The Missouri River and its tributaries, for example, once flowed north to Hudson Bay, until they were blocked and diverted by the North American ice sheet during the Pleistocene Epoch. Now the Missouri River follows a route that formed along the ice sheet's southern terminus, then joins the Mississippi River on its way south to the Gulf of Mexico. Highlight 15-1 discusses a similar drainage disruption, which caused a series of floods.

Highlight 15-1 The Channeled Scablands of Eastern Washington

Today, the Clark Fork River of northwestern Montana is a freely flowing mountain stream. Some 13,000 years ago, however, it became blocked by a broad lobe of ice protruding from the retreating Cordilleran ice sheet, which had covered western Canada to the Canadian Rockies (Fig. 1a). The lobe of ice, known as the Purcell Lobe, flowed through a wide U-shaped valley extending from southern British Columbia to northern Idaho. Along the way, it dammed the Clark Fork's westward flow, creating a lake that became swollen by meltwater from the wasting ice. Known today as Glacial Lake Missoula, this body of water was more than 300 meters (1000 feet) deep and occupied an area about the size of Lake Michigan.

As meltwater continued to feed it, Glacial Lake Missoula rose until it overflowed the ice dam, carving deep channels into the ice. Eventually the dam ruptured, releasing the accumulated waters of Glacial Lake Missoula in a torrent (Fig. 1b). More water flooded eastern Washington than is found in the combined flow of all rivers in the world today. Based on the sizes of deposited boulders, the flood's velocity has been estimated at 50 to 75 kilometers (30–50 miles) per hour.

The effects of this episode are still evident today in the Pacific Northwest (Fig. 2). The flood produced numerous rivers as much as 75 kilometers (50 miles) wide and 250 meters (900 feet) deep, the courses of which are now marked by giant dry waterfalls and massive stream bed ripples. It also excavated enormous channels by tearing hundreds of meters of basaltic lava from the Columbia River plateau, leaving behind thousands of remnant lava *scabs*—masses of basaltic rock that survived the flood intact. The region is thus aptly named the Channeled Scablands.

Geologists speculate that such great floods occurred repeatedly in eastern Washington during the Pleistocene ice age, as glaciers periodically advanced and retreated. We know that the last flood took place about 13,000 years ago, because Mount St. Helens erupted at the same time—ash layers from the eruption are interlayered with the last of the flood deposits and provide geologists with a dating marker. Massive ice sheets and unimaginable floods, plus a fiery volcanic cataclysm, make Washington state of 13,000 years ago a geologist's dream: The evidence left behind gives geologists a storehouse of information about the region's past.

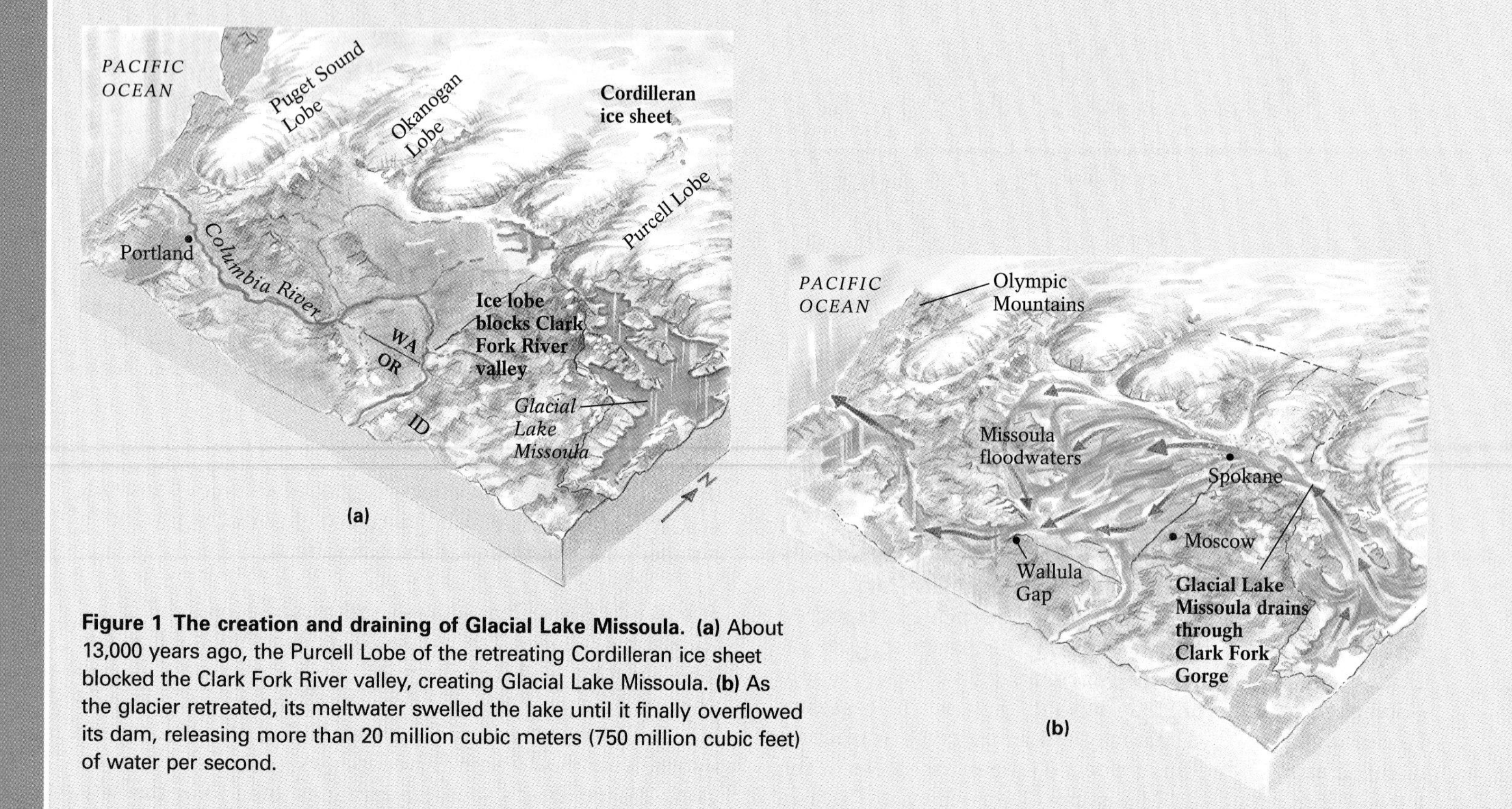

Figure 1 The creation and draining of Glacial Lake Missoula. (a) About 13,000 years ago, the Purcell Lobe of the retreating Cordilleran ice sheet blocked the Clark Fork River valley, creating Glacial Lake Missoula. **(b)** As the glacier retreated, its meltwater swelled the lake until it finally overflowed its dam, releasing more than 20 million cubic meters (750 million cubic feet) of water per second.

(a)

(b)

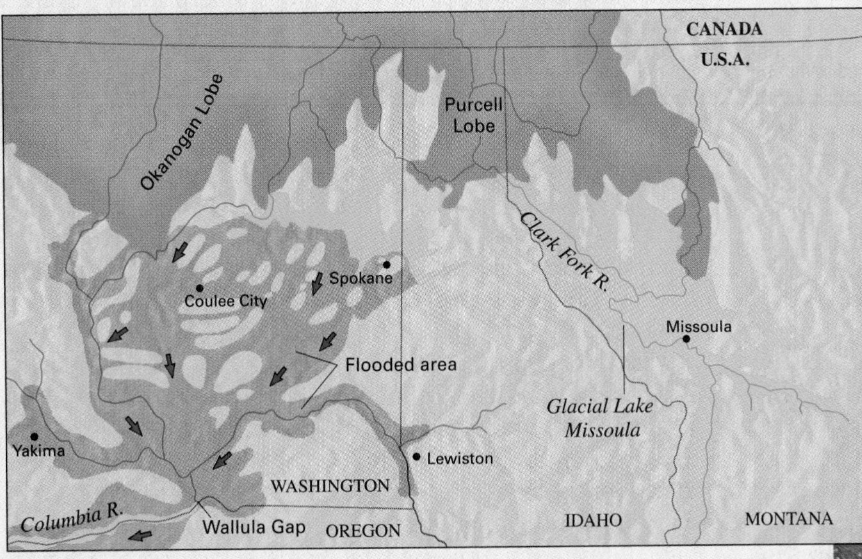

(c)

Figure 2 Effects of the Missoula flooding. **(a)** Massive rivers created by the flooding produced giant bedrock ripple marks, such as these in northwestern Montana. **(b)** This basalt bedrock near Coulee City, Washington, was etched by dry channels and falls where floodwaters once streamed. **(c)** As Glacial Lake Missoula grew, its rising shoreline carved these wave-cut benches seen on the side of Sentinel Mountain at the University of Montana campus.

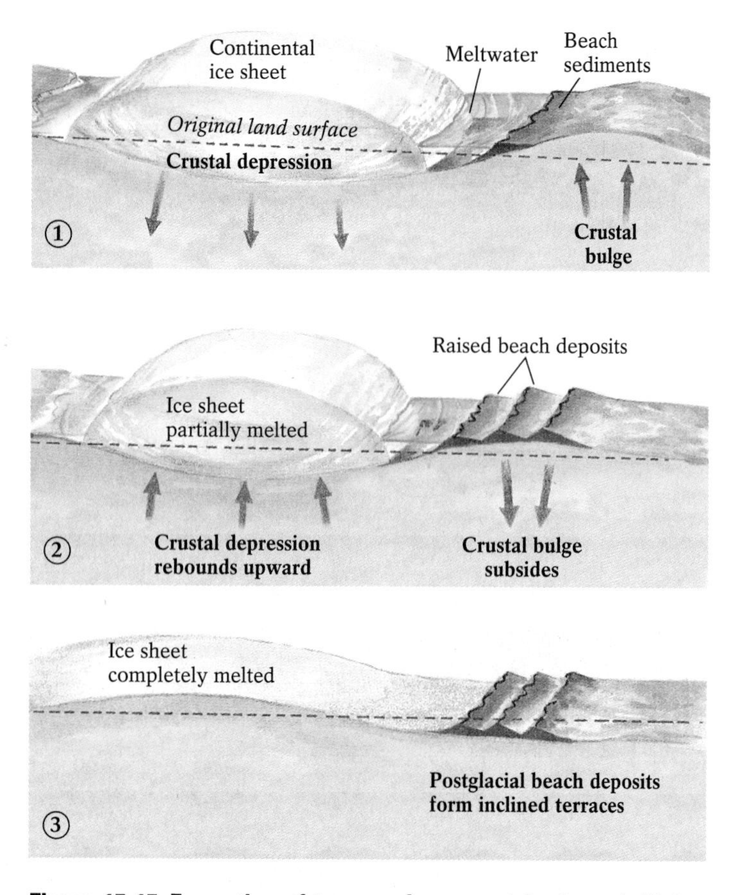

Figure 15-15 Formation of terraces from crustal rebound. Meltwater accumulates in the space between a retreating ice sheet and the crustal bulge at its margin, leaving sediments. As the ice sheet retreats further and the crust slowly returns to its original position, the bulge subsides, leaving the sediment deposits as slightly inclined terraces.

Crustal Depression and Rebound A continental ice sheet weighs so heavily on the Earth's crust that the asthenosphere (the heat-softened portion of the upper mantle) is pushed away from the center of the ice sheet toward its margins, creating a crustal depression under the glacier and topographic bulges along its borders. When the climate warms and the ice sheet melts, the displaced asthenosphere gradually returns to its original position, the depressed crust slowly rebounds, and the marginal bulges disappear. Crustal rebound may lead to the formation of uplifted terraces (Fig. 15-15), examples of which can be seen throughout Scandinavia and around the Hudson Bay region of east-central Canada. The process of crustal rebound, which takes thousands of years because of the extremely high viscosity of the flowing asthenosphere, continues today in those areas.

When crustal depression causes ice sheet margins to bulge and then later subside, the subsiding region may become flooded. The southwestern Netherlands, located near the margin of the former European ice sheet, once stood well above sea level atop just such a bulge. Following the departure of the ice sheet approximately 12,000 years ago, the depressed crust has been rebounding and the related bulge has diminished, causing the southwestern Netherlands to subside—in places to a position below sea level. Since the thirteenth century, only Holland's well-known dikes have kept the North Atlantic from flooding its valuable farmland.

Glacial Effects on Climate The indirect effects of ice sheets sometimes extend even beyond glaciated regions. When air from warmer regions encounters air chilled by ice sheets, the result is cloudy, cool, rainy weather beyond the ice sheet's terminus. During the last period of extensive glaciation, such a humid and cool climate prevailed throughout North America. This climate produced more precipitation than falls today, and less of it evaporated. The remaining water accumulated in landlocked basins, forming *pluvial lakes.* This indirect effect of glaciation reached as far south as Death Valley, California, and the deserts of the American Southwest, covering vast areas of those now-arid lands with shallow lakes. The largest pluvial lake, Lake Bonneville, covered much of Utah, eastern Nevada, and southern Idaho; Utah's Great Salt Lake represents a small remnant of this vast body of water.

Glacial Effects on Sea Levels Because the water in glacial ice ultimately comes from the oceans (by way of the hydrologic cycle), global sea levels drop whenever continental glaciers expand. As a result, during glaciations, much land that was under water becomes exposed, and the coastlines of the continents change. For example, 20,000 years ago, during the last glacial maximum, global sea level was about 130 meters (430 feet) lower than it is today, and North America's Atlantic coastline extended more than 150 kilometers (100 miles) east of what is now New York City onto the continental shelf (Fig. 15-16). Fossilized tree stumps and mastodon bones recovered from this area reveal that the exposed continental shelves were covered by spruce and pine forests and traversed by herds of migrating ice-age mammals.

Lowered sea levels also expose "landbridges," connections between landmasses that are separated by shallow seaways in nonglacial times. Until some 12,000 years ago, such a landbridge linked Siberia in Asia with Alaska in North America; today, the waters of the Bering Strait cover this area. Over this landbridge, humans and giant mammals, such as mammoths and mastodons, migrated into North America, and other animals, such as the camel and the horse, migrated from North America into Asia and Europe.

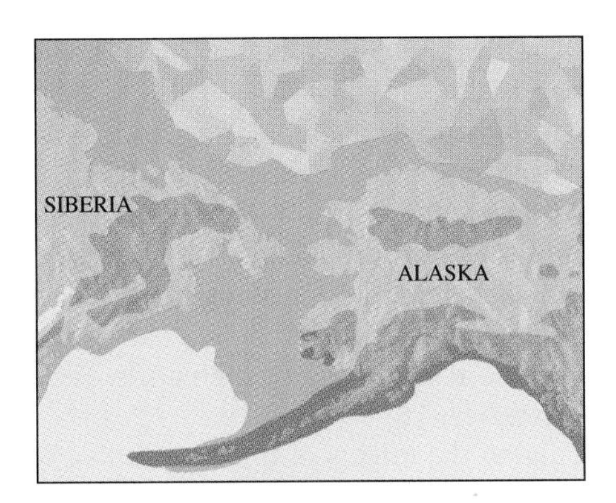

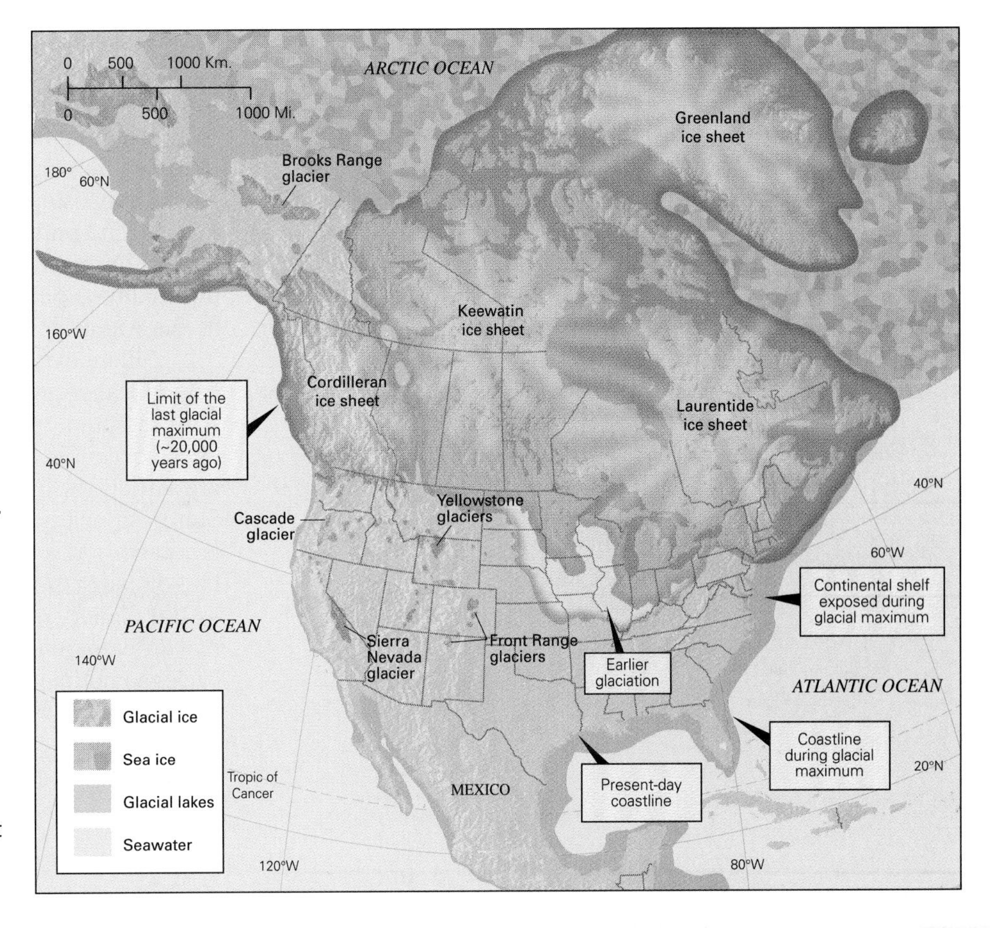

Figure 15-16 Landbridge formation. Twenty thousand years ago, falling sea levels exposed vast areas of the continental shelves. Some of these included strips of land that served as bridges between landmasses. One exposed land strip in particular altered the course of human history in the Western Hemisphere—the strip of land now covered by the shallow Bering Strait, separating Alaska and Siberia. Until about 12,000 years ago, this landbridge permitted human and animal traffic to cross in both directions. Most American archaeologists agree that it provided the principal route by which the first human inhabitants of North America arrived from Asia.

Glaciation and Ice Ages

How have glaciers been able to accomplish so much change on our continents? The answers to this question can be found by examining the occurrence of **ice ages,** the dozen or so periods—each lasting tens of millions of years—during which Earth's climate was substantially cooler than usual and glaciers covered a significant portion of the planet's land surface. During an ice age, climatic fluctuations cause glaciers to alternately grow and advance during *glacial periods,* then thaw and retreat during *interglacial periods.* Earth is currently in the midst of an ice age, called the Quaternary ice age, that has spanned the last 1.6 million years of the Earth's history; because we are now in an interglacial period, however, glaciers today are much less prominent than they were during the last glacial period (10,000 years ago).

What causes ice ages? Many geologists have concluded that ice ages have three prerequisites: sizable landmasses at or near the poles; land surfaces with relatively high elevations; and nearby oceans to provide the moisture that falls as snow. These conditions improve the prospects for an ice age by putting landmasses at latitudes and elevations where the climate is colder and glaciers are most likely to grow. The one phenomenon that can move landmasses to polar regions, raise them to higher elevations, and manipulate the positions of the Earth's ocean basins is, of course, plate tectonics. The most recent ice age began after the landmass we now call Antarctica moved to the South Pole, other major landmasses moved north to the Arctic circle, and, as plates converged, rising mountain ranges placed land at elevations well above the preexisting snowlines.

Plate motion, however, cannot move land to and from the poles or raise and lower land swiftly enough to account for the cycles of repeated glacial advances and interglacial retreats. From ice-core studies, we know that these glacial cycles have lasted for periods of approximately 100,000 years, separated by brief interglacial periods that endured for about 10,000 years. Many glacial geologists believe that glacial fluctuations during an ice age are caused primarily by variations in the Earth's position and orientation relative to the Sun. According to this hypothesis, which was first proposed by Yugoslavian astronomer Milutin Milankovitch around 1930, the total amount of solar radiation the Earth receives may vary only a little, but periodic changes in where and when that radiation is received have profound effects on global climate.

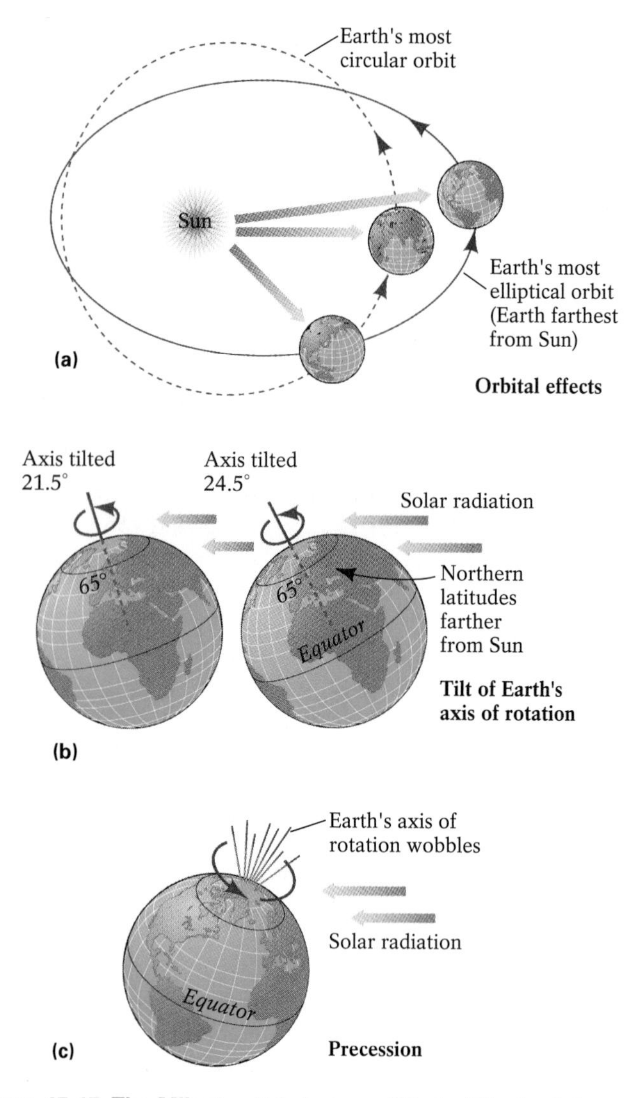

Figure 15-17 The Milankovitch factors. Milutin Milankovitch proposed that three astronomical factors interact to affect the amount of solar radiation striking the Earth's high northern latitudes, possibly triggering the periodic advance and retreat of glaciers around 65°N latitude. The climate at 65°N latitude cools when the Earth's orbit around the Sun is at its most elliptical (especially when the Earth is located farthest from the Sun during the northern summers), when the tilt of the Earth's axis positions the high northern latitudes at a lower angle to incoming solar radiation, and when the "wobble" of the Earth's spin axis carries this crucial latitude out of the path of direct solar radiation. Each of these three factors has a different period of recurrence. When they have coincided in the past, glaciations have occurred.

The Milankovitch hypothesis suggests that the critical region for determining global glacial advances and retreats during ice ages is the vicinity of 65°N latitude, where ice sheets have developed repeatedly during the past several million years. Today, this latitude is largely ice-free. When this region receives less solar radiation and becomes sufficiently cold, and when oceans provide ample moisture to large landmasses such as North America, Scandinavia, and Asia, however, then winter snows falling on these lands can persist through the summer melting season, accumulating from year to year until glaciers form. The region in the vicinity of 65°S latitude, by comparison, has experienced significantly less glacial expansion, because the Southern Hemisphere contains far less land on which glaciers can grow. Once a global glacial expansion begins and the world's climate cools, glaciers do eventually expand in Southern Hemisphere highlands, such as the Andes of South America.

Milankovitch calculated the effects of three periodic astronomical factors to show how they combine to lower the solar radiation levels at 65°N latitude and to explain why this effect occurs at 100,000-year intervals. These three factors are variations in the shape of the Earth's orbit around the Sun from nearly circular to elliptical, variations in the tilt of the Earth's axis toward or away from the Sun, and the wobble, or *precession,* of the Earth's axis and equatorial plane as the planet spins (Fig. 15-17). Each factor follows a cycle of a different length, but, according to Milankovitch, the three cycles sometimes coincide. The result is centuries or even millennia of low summertime radiation at 65°N latitude. The summers are then cool and brief, and alpine glaciers and continental ice sheets expand.

The Earth's Glacial Past

The Earth has experienced numerous ice ages throughout its 4.6-billion-year history.

Pre-Pleistocene Glaciation

The geologic record from the Precambrian, though scattered and fragmentary, yields clear evidence of at least three ice ages during that era. The oldest Precambrian ice age, which occurred about 2.2 billion years ago, can be identified from layers of *tillite,* a sedimentary rock produced by lithification of till. Geologists studying this Precambrian event have discovered signs indicating that during this glaciation, ice may have extended from the Earth's poles to the tropics. Indeed, virtually the entire planet may have been locked in a glacial deep-freeze at this time.

Geologic evidence of global glaciation after the Precambrian is more widespread and better preserved. Layers of tillites and "fossilized" moraines and eskers show that during the Paleozoic Era, about 500 million years ago, an ice sheet covered what is now one of the world's hottest and driest lands, the Sahara of northern Africa. Before plate tectonics moved all the continents to their present locations, the Sahara was located at the South Pole. Ironically, one may now swelter in the merciless African sun while standing within reach of an ancient esker.

Figure 15-18 **Pre-Pleistocene glaciation.** Glacial grooves and tillite in the arid landscape of South Africa suggest that this area was once located near the South Pole, and was cold enough to be glaciated.

The most complete record of any pre-Pleistocene glaciation was left by an extensive continental ice sheet that existed during the Permian Period, about 286 to 245 million years ago. This ice sheet striated the bedrock and deposited extensive tillites in regions that are now Australia, South America, India, and South Africa (Fig. 15-18)—at the time, these areas constituted the supercontinent of Gondwana, located in the vicinity of the South Pole.

There is no evidence of continental glaciation during the Mesozoic Era, the age of the dinosaurs, which lasted from 225 to 65 million years ago. Apparently the Mesozoic was a time of global warming, when the Earth's landmasses had moved from the poles to subtropical and tropical latitudes. In northern Alaska, we find Mesozoic rocks containing remnants of coral reefs, tropical vegetation, and dinosaur fossils. During this time, average global temperatures were as much as 10°C (18°F) warmer than today and few, if any, ice masses appeared on the Earth, even in the planet's polar regions.

Although the most recent ice age is associated with the start of the Pleistocene Epoch 1.6 million years ago, recent evidence suggests that ice-age conditions may have begun to develop as many as 50 million years earlier (Fig. 15-19). Between 50 million and 20 million years ago, world temperatures dropped approximately 5°C (9°F). At roughly the same time, large domes of ice may have begun to grow on Antarctica; striated stones, believed to have been carried northward by floating icebergs, are found today in the South Pacific in the muddy oceanic sediments dating from that period. Gradual global cooling continued, so that by 12 to 10 million years ago ice had formed in the mountains of Alaska. By 3 million years ago, an ice sheet had buried the island of Greenland, and 500,000 years later ice was accumulating on the high plateaus and mountains of North America and Europe. The Quaternary ice age, with its alternating mid-latitude ice-sheet advances and retreats, was set to begin.

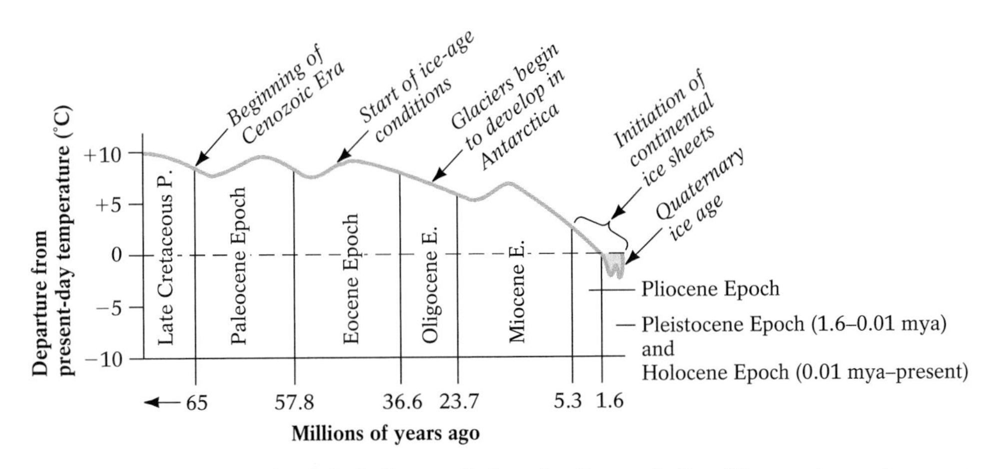

Figure 15-19 **Changes in global climate during the Cenozoic Era.** The most recent (Quaternary) ice age is associated with the start of the Pleistocene 1.6 million years ago, but the Earth had been cooling for about 50 million years prior to that time.

Glaciation in the Pleistocene

By some 1.6 million years ago, Earth had entered an ice age from which it has yet to emerge. The Quaternary ice age (named for the Quaternary Period of the Cenozoic Era) has been marked by as many as 30 episodes of glacial advance and retreat, especially during the Pleistocene Epoch, which ended only 10,000 years ago.

The location of North America's youngest glacial deposits indicates that the continent's last ice sheet extended from the eastern Canadian Arctic south to New York City and west to the foothills of the Rocky Mountains in Montana. This *Laurentide* ice sheet probably originated in Canada west of Hudson Bay and in Labrador and Newfoundland. Another glacial center, the *Cordilleran* ice sheet, formed farther west, between the Canadian Rockies and the Coast Range of British Columbia. It probably began as a network of alpine glaciers that expanded and merged to form an icecap that eventually covered the adjacent valleys and plains. At the time of maximum glacial expansion, a continuous wall of ice stretched 6400 kilometers (4000 miles) across the full breadth of southern Canada and the northern United States (Fig. 15-20).

Between about 13,000 and 12,000 years ago, the world's climate warmed and the ice sheets began to melt. Although the North American ice sheet had retreated well into Canada by 10,000 years ago, it took another 3000 to 4000 years for northern Canada's icecaps to melt completely.

Recent Glacial Events

The last 10,000 years have experienced alternating warmer and cooler periods, but it has never been sufficiently cold, or cold long enough, to reconstitute the mid-latitude ice sheets of the last glacial period. During the most recent cold snap—the Little Ice Age, occurring from about 700 to 150 years ago—winters were colder and snowier than usual and the summers relatively cool and wet. Sustained warming began by 1850 and continues today, although vestiges of pack ice from the Little Ice Age lingered long enough to doom the maiden voyage of the *Titanic* on April 15, 1912. By 1920, virtually no floating ice was observed south of the Arctic circle. Today, most glaciers in North America and Europe are retreating (Fig. 15-21).

Human activities may be contributing to global warming and hastening the melting of the Earth's remaining ice masses. Since the beginning of the Industrial Revolution in the late 1700s, humans' increased use of fossil fuels (coal, oil, and natural gas) has caused the release of enormous quantities of carbon dioxide into the atmosphere. This problem has been exacerbated by accelerating destruction of the vast tropical rainforests, which remove great quantities of carbon

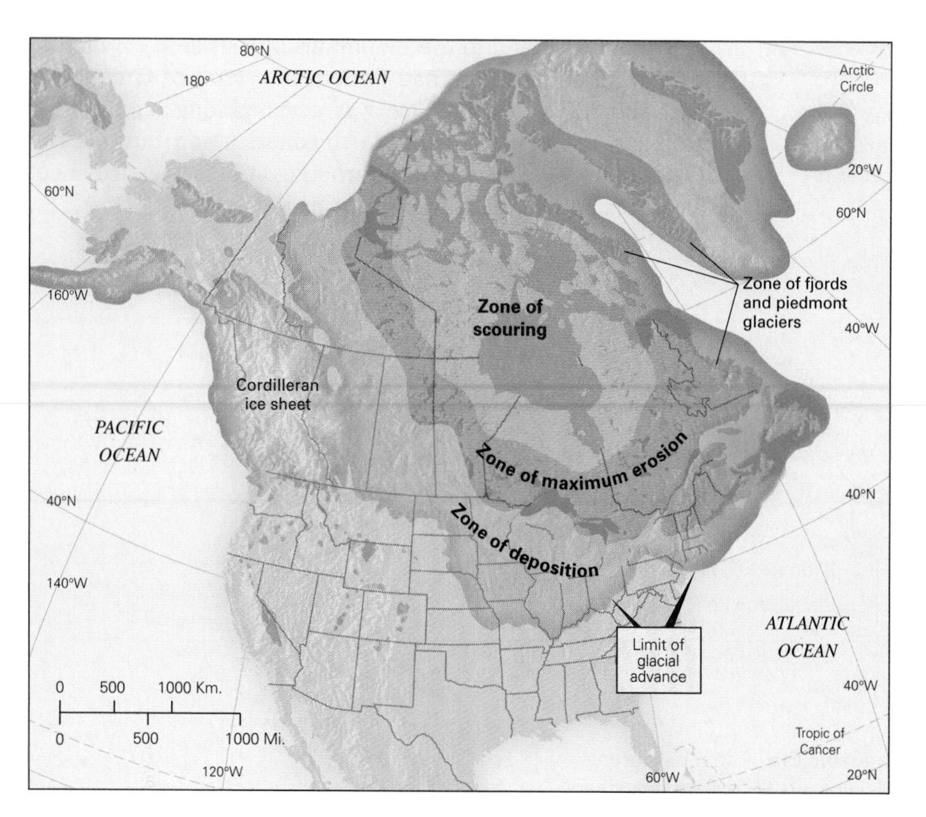

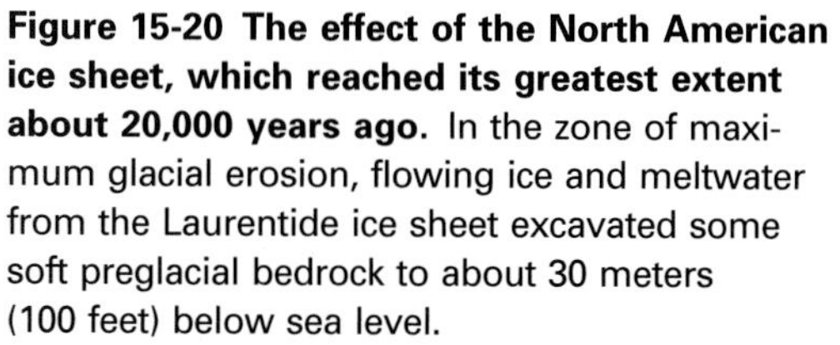
Figure 15-20 The effect of the North American ice sheet, which reached its greatest extent about 20,000 years ago. In the zone of maximum glacial erosion, flowing ice and meltwater from the Laurentide ice sheet excavated some soft preglacial bedrock to about 30 meters (100 feet) below sea level.

Figure 15-21 The retreating Athabasca Glacier in the Columbia Icefield of western Canada. The signposts indicate past positions of the glacier's terminus.

dioxide from the air during photosynthesis. In fact, atmospheric carbon dioxide concentration has risen by 25% since 1850. Because carbon dioxide is a principal *greenhouse gas,* trapping heat within the Earth's atmosphere, global temperatures in the next century could rise by as much as 2° to 4°C (3.5°–7°F) if carbon dioxide production continues at its current rate.

Such a global temperature increase would suffice to melt a significant amount of the Earth's polar ice and to raise global sea level by as much as 50 to 100 centimeters (1–3 feet) by the year 2030. Since 1900, sea level has already risen 20 to 25 centimeters (10 inches). If human activities caused *all* of the world's ice to melt within a few thousand years, global sea level would rise about 70 meters (230 feet) (Fig. 15-22). Under this scenario, the water in New York Harbor would reach to the armpits of the Statue of Liberty, and most of the Atlantic Coast from New Jersey to the Carolinas would be submerged. The entire state of Florida would be covered by the Atlantic, much of Alabama, Mississippi, and Louisiana would be consumed by the Gulf of Mexico, and much of interior California would be engulfed by the Pacific.

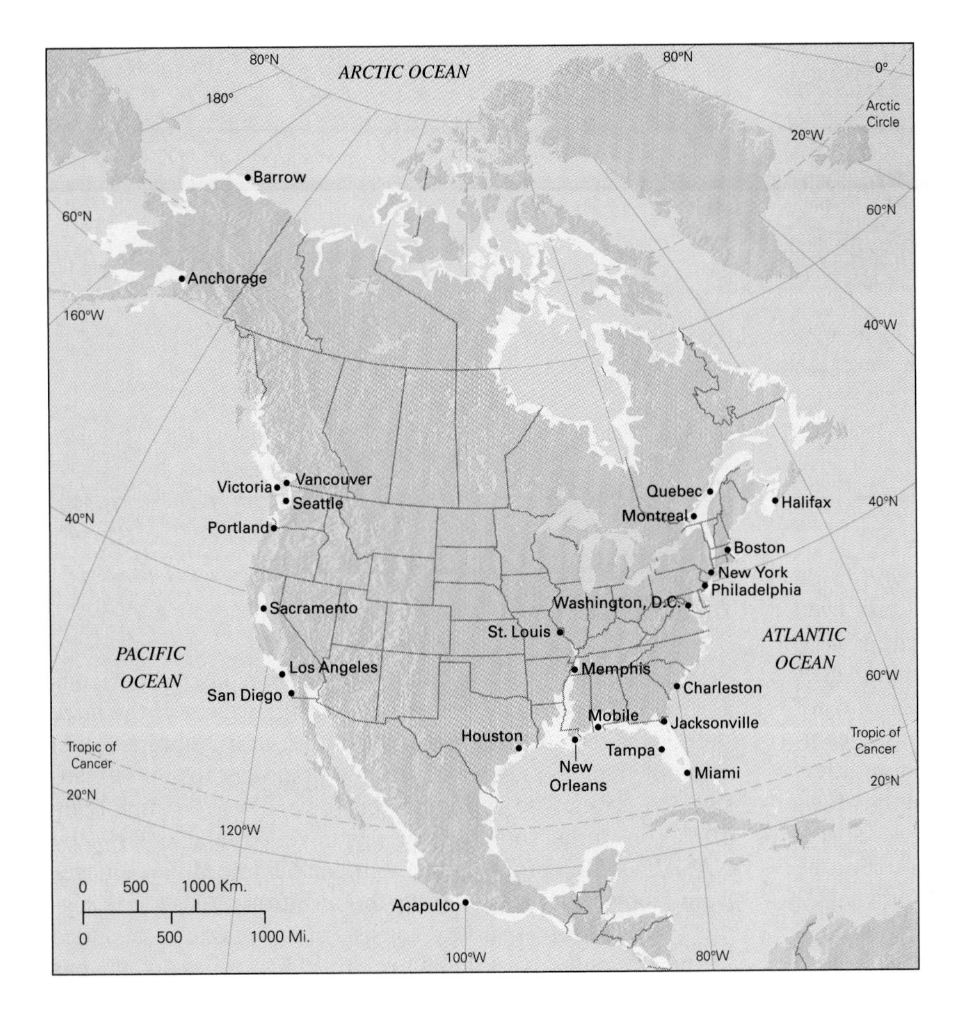

Figure 15-22 The theoretical effect on North America if all ice on Earth today were to melt completely. The global sea level would rise about 70 meters (230 feet), submerging most coastal areas and even some inland areas.

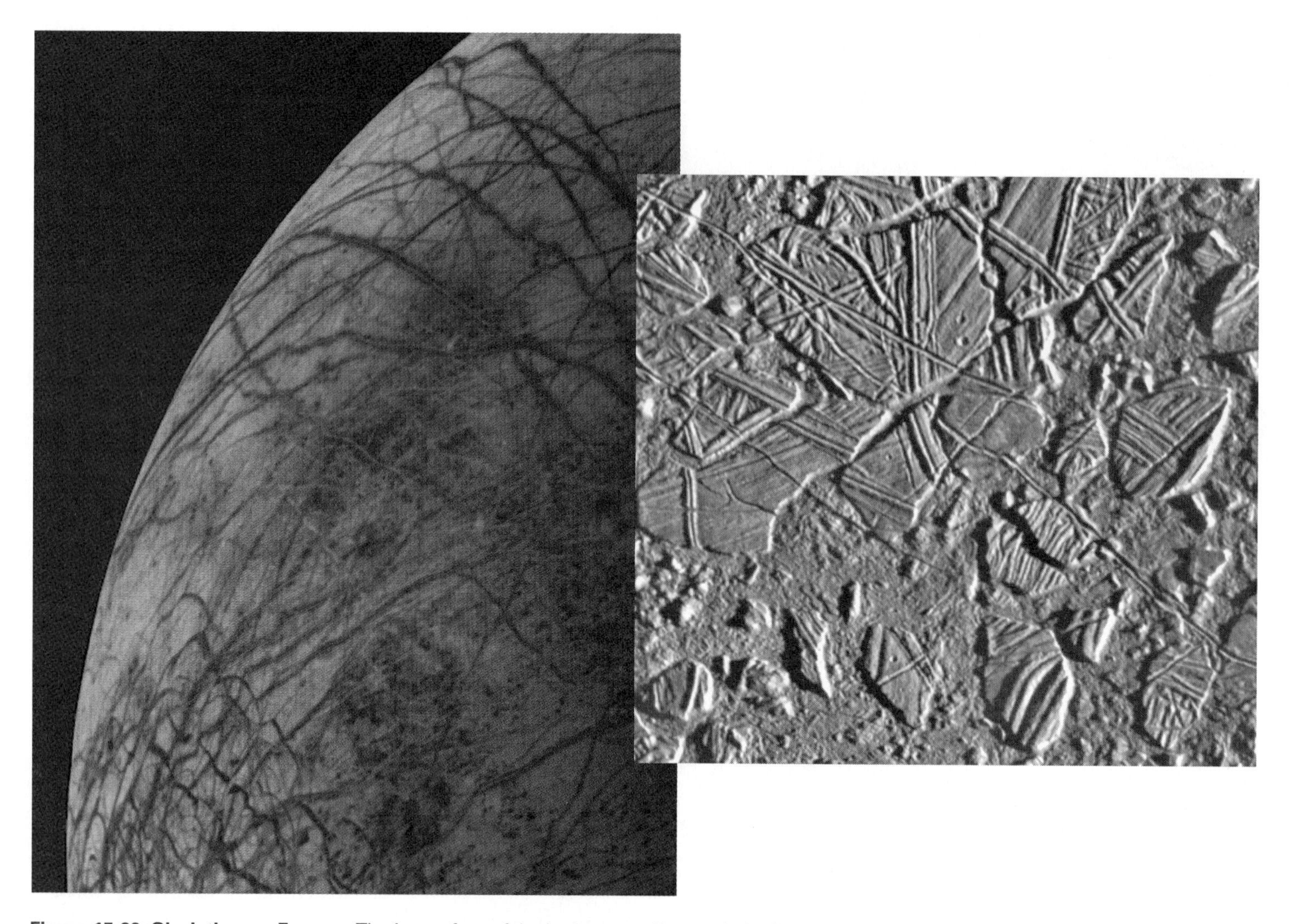

Figure 15-23 Glaciation on Europa. The icy surface of Jupiter's moon, Europa, is broken into solid plates of ice that appear to rift and diverge in ways similar to the Earth's tectonic plates. Photo from the Galileo probe, 1996.

The Future of Our Current Ice Age

How can we predict when the next period of worldwide glacial expansion or melting will occur, when we can't even predict with certainty whether tomorrow's ballgame will be rained out? The Pleistocene record suggests that interglacials generally span a period lasting 8000 to 12,000 years and are followed by 10,000 to 20,000 years of slow intermittent cooling before glacial conditions return. We are now in an interglacial that began about 10,000 years ago, so it may be drawing to a close. The Milankovitch factors predict that ice sheets will again cover vast areas of North America and Europe about 23,000 years from today. Whether human-produced atmospheric levels of carbon dioxide will alter this pattern remains to be seen.

Glaciation on Jupiter's Europa and Mars

Several other planets in our solar system also experience glaciation. Recent satellite surveys of the surfaces of the large outer planets—Jupiter, Saturn, Uranus, and Neptune—and some of their moons have revealed polar icecaps and extensive ice cover. The most exciting new discovery involving glaciation in our solar system was unveiled in June 1996 by NASA's Galileo probe, which sent back extraordinary images from its orbit around Jupiter's moon Europa. As seen in Figure 15-23, Europa's surface consists of an icy crust—believed to be roughly 10 kilometers (6 miles) thick—that has

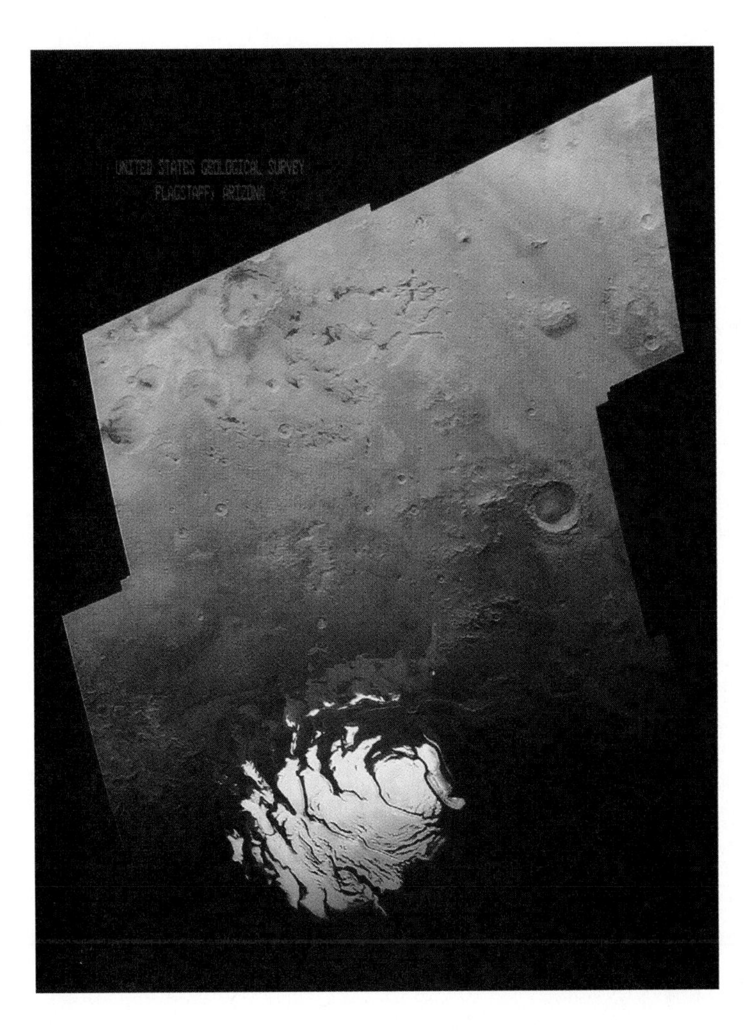

Figure 15-24 An icecap near the south pole of Mars. This photograph was taken by the Viking Orbiter 2 during Mars' southern summer in 1972.

broken into plates reminiscent of terrestrial ice floes. Some scientists believe that these icy plates float on a subsurface ocean of water or slushy ice that may "erupt" as ice "volcanoes" through fissures in the solid crust.

Closer to home, Mars exhibits white patches near its poles that appear to grow and shrink seasonally, much as the Earth's polar ice does (Fig. 15-24). Mars' polar ice is probably only a few meters thick, and most of it may consist of frozen carbon dioxide (dry ice), the main component of its atmosphere. In the Martian summer, the northern hemisphere icecap recedes rapidly at first, then more slowly. The rapid shrinkage may be caused by sublimation of frozen carbon dioxide at the edges of the icecap, where it is thinnest. Any ice that lasts through the summer is probably frozen water, which sublimates imperceptibly slowly in Mars' atmosphere.

Satellite photographs of Mars reveal troughs and ridges in the polar regions that resemble the glaciated valleys and large moraines found on Earth. The terrain around the icecaps consists of alternating horizontal layers of light and dark materials, which may be composed of stratified outwash and loess. As many as 50 layers, each of which is 15 to 35 meters (50–120 feet) in thickness, have been counted at a single site. Nevertheless, confirmation of the planet's complex history of multiple glaciations will probably have to await a visit to Mars by a glacial geologist.

Chapter Summary

Glaciers are bodies of ice, formed by the compaction of snow, that flow downslope or outward under the influence of gravity. Glaciers form when snow accumulates, becomes well-packed **firn,** and then refreezes to ice. They tend to arise in climates marked by moist snowy winters and, especially, cool cloudy summers that minimize melting.

Glaciers are classified on the basis of their topographic setting. Relatively small **alpine glaciers**—which include **cirque glaciers** (which erode and occupy basins in mountainsides), **valley glaciers** (which flow into preexisting stream valleys), and **icecaps** (which form on mountaintops)—are confined by surrounding mountains. **Continental ice sheets,** which are unconstrained in this way, cover vast areas of land.

The budget of a glacier is the difference between the glacier's annual gain of snow and ice, which occurs in its **zone of accumulation,** and its annual loss, which occurs in its **zone of ablation.** If its budget is positive, the glacier will grow; if it is negative, the glacier will recede.

Glacial flow occurs by a combination of **internal deformation,** in which individual ice crystals fracture or slip past one another, and **basal sliding,** in which the entire glacier slides across a thin film of meltwater at its base. Regardless of whether the terminus is growing, stationary, or receding, the ice within a glacier always flows toward the zone of ablation.

Glaciers erode a landscape by both abrasion and quarrying. **Glacial abrasion** occurs when rock particles embedded in a glacier's base begin to make linear scratches, or striations, and grooves in the underlying bedrock surface.

In **glacial quarrying,** the glacier breaks off and removes large masses of bedrock. Alpine glaciers erode smooth, unglaciated slopes to produce horseshoe-shaped basins, or **cirques,** in the mountainsides. When a cirque glacier melts, a cirque lake, or **tarn,** forms in its basin. The growth of two or more cirques on a mountain can erode the intervening rock into several features: sharp pointy peaks, or **horns;** long serrated ridges, or **arêtes;** mountain passes through the ridges, or **cols.** When alpine glaciers flow into preexisting stream valleys, they erode a large quantity of bedrock and sediment, converting V-shaped stream valleys into U-shaped glaciated valleys.

There are many types of glacial deposits, or **glacial drift.** Drift consisting of unsorted, unstratified sediments deposited directly from ice makes up **glacial till.** Tills sometimes contain **glacial erratics,** exceptionally large, glacially transported rocks that have been eroded from one type of bedrock and deposited on another. Glacial till often accumulates at a glacier's terminus as a **moraine.** Moraines remain as hills of sediment after a glacier retreats, marking its various advances and retreats. When continental ice sheets override moraines, they reshape the moraines into gently rounded, elongated hills called **drumlins.**

Drift consisting of sorted, stratified sediments deposited by meltwater downstream of a glacier is **outwash.** Wind erosion of drying outwash produces a mass of fine silt, called loess. Meltwater deposits of sand and gravel can also accumulate directly underneath glacial ice, forming sinuous ridges known as **eskers.**

In addition to creating erosional and depositional landforms, glaciation can have more far-reaching effects. For example, glaciers may become continent-wide dams, diverting the courses of streams, or their weight may produce huge crustal depressions and associated uplifts. They may change the climate of surrounding regions, increasing precipitation and producing pluvial lakes. Extensive glaciation can even cause global sea levels to drop, thereby altering continental coastlines.

Plate tectonics has periodically moved the Earth's continents toward the poles and raised mountains above their preexisting snowlines, thus making their climate colder and preparing the Earth for **ice ages.** Various astronomical factors—the shape of the Earth's orbit around the Sun, the tilt of its axis, and the wobble of its equatorial plane as the planet rotates on its axis—moderate the amount of solar radiation that reaches the Earth at 65°N latitude. When these factors interact to cause worldwide cooling, global glaciation is likely to occur.

Ice ages have occurred in both the Precambrian and Paleozoic Eras and during the Pleistocene Epoch of the Cenozoic Era. The last extensive glaciation occurred some 20,000 years ago, and the global climate has generally been warming ever since. Human activity—especially activities that release excessive amounts of carbon dioxide into the atmosphere—may increase global warming and affect the course of glaciation in the future. Nevertheless, the gradually changing position of the continents and the inevitable variations in solar radiation may still combine to produce another worldwide glacial expansion in about 20,000 to 25,000 years.

Recent satellite surveys of the surfaces of the large outer planets — Jupiter, Saturn, Uranus, and Neptune — and some of their moons have revealed polar icecaps and extensive ice cover. Jupiter's moon Europa may possess a crust of ice that is broken into plates that float on a subsurface ocean of water or slushy ice. Closer to home, icecaps on Mars, composed largely of frozen carbon dioxide, grow and shrink seasonally. Mars also exhibits numerous glacially derived landscape features.

Key Terms

glacier (p. 283)
firn (p. 284)
alpine glaciers (p. 284)
cirque glaciers (p. 284)
valley glaciers (p. 284)
icecaps (p. 284)
continental ice sheet (p. 284)
zone of accumulation (p. 286)
zone of ablation (p. 286)
internal deformation (p. 287)
basal sliding (p. 287)
glacial abrasion (p. 287)
glacial quarrying (p. 287)
cirques (p. 288)
tarn (p. 288)
arête (p. 290)
col (p. 290)
horn (p. 290)
glacial drift (p. 290)
glacial till (p. 291)
glacial erratic (p. 291)
moraine (p. 291)
drumlins (p. 292)
outwash (p. 293)
eskers (p. 293)
ice ages (p. 297)

Questions for Review

1. What are three types of topographically confined glaciers?

2. Draw a simple diagram of a confined glacier showing its zone of accumulation and its zone of ablation. Show the flow path of a hypothetical ice particle within the glacier.

3. What is the budget of a glacier? Describe the behavior of a glacier's terminus if its budget is positive, negative, and balanced.

4. Explain how the two glacial flow mechanisms, basal sliding and internal deformation, work.

5. Distinguish between glacial abrasion and glacial quarrying. Describe three distinctive features or landforms produced by these erosional processes.

6. List five prominent erosional features produced by alpine glaciers.

7. Describe the difference between terminal, recessional, lateral, and medial moraines.

8. Describe the difference in origin and texture between till and outwash.

9. What are three major effects of glaciers that may act on areas far removed from the glaciers?

10. What factors are believed to cause ice ages?

For Further Thought

1. In the Northern Hemisphere, the lateral moraines of alpine glaciers are often larger on the south-facing sides of east–west oriented valleys. Why?

2. Some eskers actually "climb" up and over topographic ridges, in seeming defiance of gravity. How can the water that deposits eskers flow uphill?

3. If latitude 65°N is crucial in terms of the Milankovitch theory, why isn't latitude 65°S equally crucial?

4. Speculate about what might happen to the world's climate and glaciers if the Antarctic ice sheet surged into the surrounding oceans, and vast areas of the ocean were displaced by floating ice.

5. Identify the "mystery glacial landform" in the photograph at the right.

Deserts and Wind Action

Unlike any landscape described earlier in this book, Death Valley National Monument, located in southeastern California, displays the wind-swept sands and austere barrenness characteristic of an arid landscape (Fig. 16-1). The valley owes its unique appearance to its extreme lack of water. Without appreciable water, little chemical weathering can take place—soils are thin, dry, and crumbly, and winds readily sweep loose particles into dunes or sandblast exposed rock surfaces with them. These characteristics are typical of **deserts,** regions that receive very little annual rainfall and are generally sparsely vegetated.

Every major continent, although surrounded by water, contains at least one extensive dry region. Deserts account for as much as one-third of the Earth's land surface—more area than is occupied by any other terrestrial environment. Relatively few deserts, however, resemble the popular image of endless tracts of drifting sand. In North Africa's Sahara, the world's largest desert, only 10% of the surface is sand-covered. Even the Arabian Desert, Earth's sandiest, is only 30% sand-covered. Desert climates are *arid.* That is, they are characterized by dryness rather than by temperature, and thus they can be found in both cold and hot regions. Such diverse landscapes as ice-bound Antarctica, the fog-shrouded coasts of Peru and Chile, and the near-continuous 8000-kilometer (5000-mile) stretch of land across northern Africa and the Arabian peninsula to southern Iran are all considered deserts, based on their extreme lack of surface water.

The term "desert" is misleading in its implication that the land is literally deserted, devoid of life. In fact, hot deserts are home to some of the Earth's hardiest plants and animals (Fig. 16-2), which have developed special adaptations that enable them to survive in extremely dry conditions. Some hot-desert plants produce seeds that can endure 50 years of drought. Some have small, waxy leaves that minimize water loss to evaporation. Most possess thick, spongy stems that store water from the occasional cloudburst and produce deep root systems through which to tap groundwater supplies. Some desert plants resemble dead twigs for months or even years on end, until the occasional downpour arouses them into a brief but memorable bloom.

Figure 16-1 Wind-swept sand dunes at Mesquite Flats, Death Valley National Monument, California.

(a) (b) (c)

Figure 16-2 Some desert life forms. (a) Desert shrubs and flowers in southwestern Colorado. **(b)** A roadrunner in the Sonoran Desert. **(c)** A thorny devil lizard (*Moloch horridus*) from Rainbow Valley, Alice Springs, Australia.

Animals found in hot deserts include insects, reptiles, birds, and mammals. Some birds fly hundreds of kilometers to find water, which they then carry back to their young in absorbent abdominal feathers. Desert rodents may live their entire lives without a single drink of water, instead absorbing the moisture they need by consuming desert plants. Most desert animals are nocturnal, avoiding activity during the hottest and driest times of day and venturing out only at night, after temperatures have dropped considerably.

In this chapter, we examine the processes that shape a desert's unique landscape. We see how water's brief, intermittent appearances actually contribute significantly to the evolution of desert landforms. In addition, we discuss the role of wind, another major agent of surface change in deserts. Finally, we consider how human activity contributes to the expansion of deserts.

Characteristics of Deserts

How do we determine if an area is dry enough to be called a desert? One common guideline defines a desert as any region that receives less than 250 millimeters (10 inches) of precipitation annually, an amount that in most areas is insufficient to sustain crops without irrigation. Many deserts, however, receive far less precipitation. The Sahara, the world's driest region, receives an average of only 0.4 millimeter (0.016 inch) of precipitation annually. In fact, the Sahara may have *no* measurable precipitation for years; its annual average may be attributable to a few storms occurring over the course of a decade.

A more precise method of identifying deserts relies on an **aridity index,** a ratio of a region's potential annual evaporation to its recorded average annual precipitation. In an area with an aridity index of 1.0, the amount of annual precipitation equals the area's potential for evaporation. Such an area would have a humid climate. An area with an aridity index greater than 4.0—where the potential for evaporation is at least four times greater than the annual precipitation—is classified as a desert. Two of the driest spots on Earth, the eastern Sahara of Africa and the Atacama Desert of Peru, have an aridity index of 200; such areas are described as *hyper-arid.* An area with an aridity index between 1.5 and 4.0 is classified as *semi-arid;* semi-arid regions, such as the Great Plains east of the Colorado Rockies, are capable of supporting a greater diversity of life than are deserts.

Types of Deserts

Deserts may be cold, temperate, or hot. Some cool deserts actually support significant vegetation despite their aridity, because low temperatures retard evaporation and maximize the amount of moisture available for plant growth. In one such desert in northern Scandinavia, for example, 250 millimeters (10 inches) of annual precipitation is sufficient to support a dense forest. In contrast, no forests can survive in southern Nevada, which receives the same amount of annual precipitation, because high temperatures there hasten evaporation.

Although deserts may occur in any climatic zone, most—and the best known—are in hot places. Where are the hottest places on Earth? At El Azizia, Egypt, a desolate outpost in the Libyan Desert of northeastern Africa, the air temperature rose to a blistering 58°C (136°F) in the shade on September 13, 1922. The North American record, 56°C (133°F), occurred in Death Valley, California, in July 1913. In such places, where no surface moisture is available to evaporate, humidity remains low and few clouds filter daytime sunlight or retain the heat that radiates at night from sun-heated rock surfaces. Consequently, days become even hotter, and nighttime temperatures may plummet by as much as 40°C (72°F) from daytime highs.

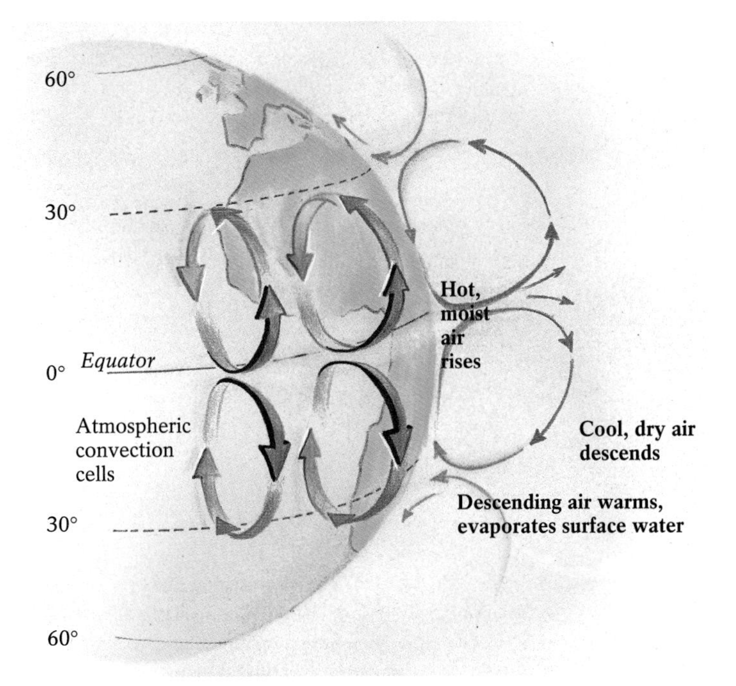

Figure 16-3 Creation of a subtropical desert. Atmospheric convection cells form as warm, moist equatorial air rises, cools, and releases its moisture as tropical rain. When it reaches about 10 kilometers (6 miles) above the surface, it spreads northward and southward from the equator and descends in the subtropical latitudes 20° to 30° north and south of the equator. There the already-dry air warms and causes surface moisture to evaporate rapidly, producing desert conditions on land.

Subtropical Deserts Arid conditions may exist for a variety of reasons, and thus deserts can be found at all latitudes. Nevertheless, most deserts are located in the subtropical belts between latitudes 20° and 30° on either side of the equator. At these latitudes, global atmospheric circulation patterns cause already-dry air to descend to the Earth's surface, thereby warming it significantly. Because warmer air can hold more moisture than cooler air, almost all surface water in these regions evaporates (Fig. 16-3). These air circulation patterns are largely responsible for the two subtropical belts of desert lands shown in Figure 16-4. Just north of the Tropic of Cancer in the Northern Hemisphere lie the Sahara, the Arabian Desert of the Middle East, and the parched landscapes of Mexico and the American Southwest. Straddling the Tropic of Capricorn in the Southern Hemisphere are the Kalahari and Namib Deserts of southwestern Africa, the Atacama Desert of Peru and Chile, and the interior deserts of the Australian outback.

Rain-Shadow Deserts Many smaller deserts form on the *leeward* (away from the wind) side of coastal mountain ranges. As moist air rises up the windward side of a mountain range, it cools and loses much of its capacity to hold water. As a result, most of this moisture precipitates on the windward mountain slopes. The leeward slopes and the regions beyond them remain dry, creating **rain-shadow deserts.** The desert lands of Nevada, Arizona, and eastern California, found leeward of the

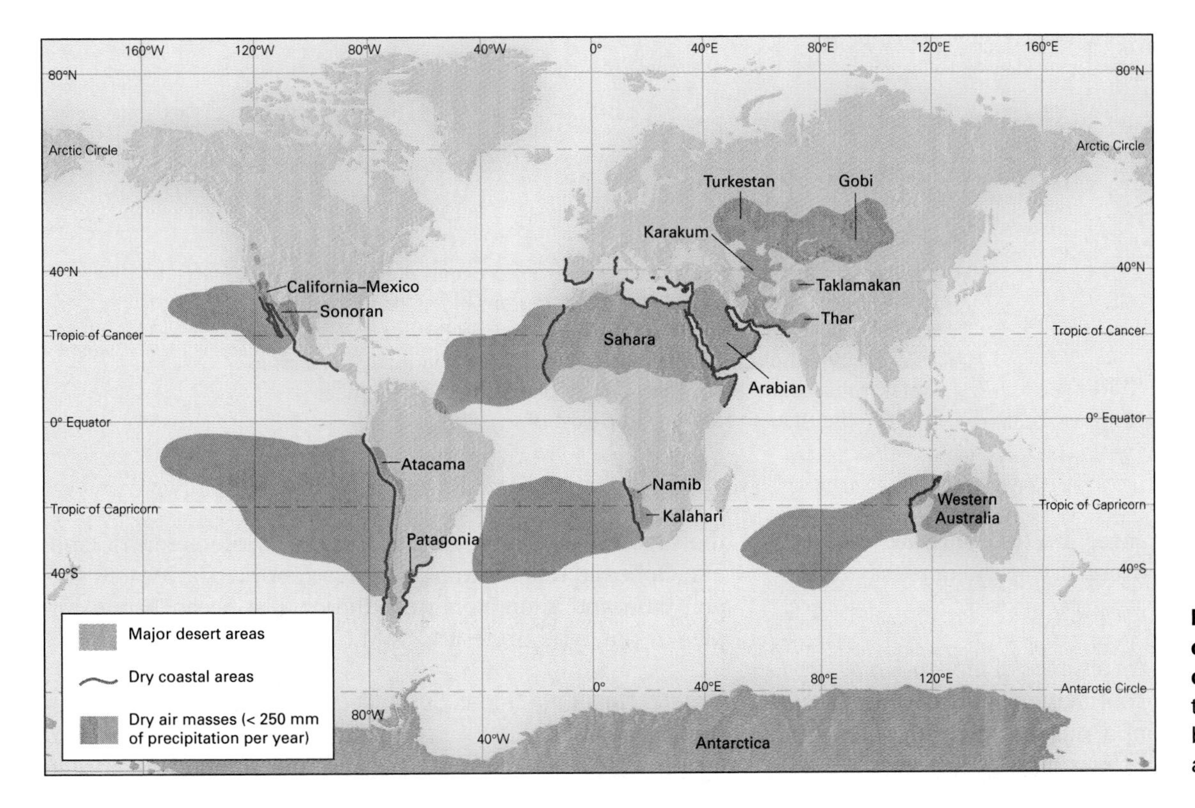

Figure 16-4 The worldwide distribution of the Earth's deserts. Most deserts lie in the two subtropical belts between 20° and 30° north and south of the equator.

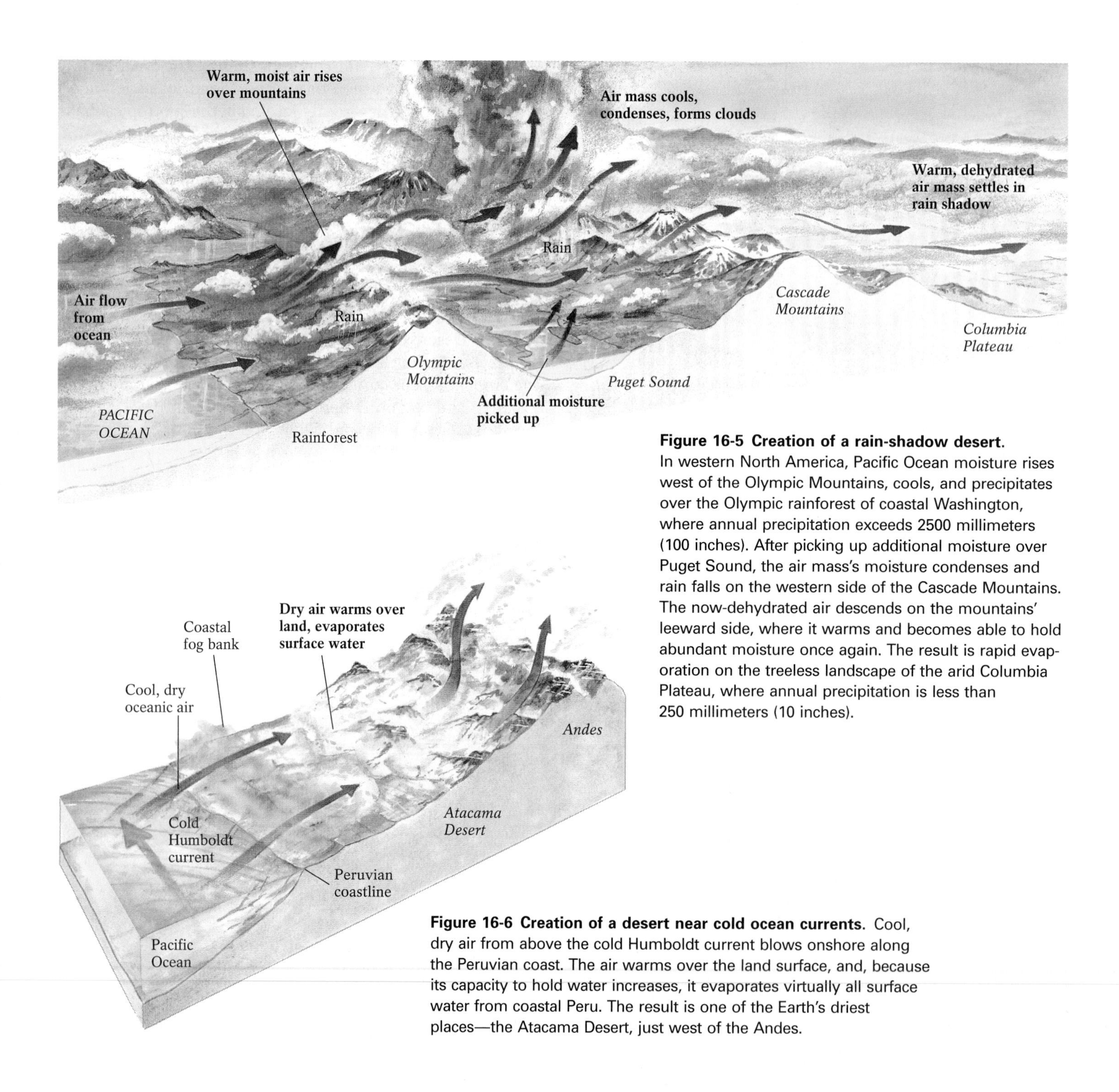

Figure 16-5 Creation of a rain-shadow desert. In western North America, Pacific Ocean moisture rises west of the Olympic Mountains, cools, and precipitates over the Olympic rainforest of coastal Washington, where annual precipitation exceeds 2500 millimeters (100 inches). After picking up additional moisture over Puget Sound, the air mass's moisture condenses and rain falls on the western side of the Cascade Mountains. The now-dehydrated air descends on the mountains' leeward side, where it warms and becomes able to hold abundant moisture once again. The result is rapid evaporation on the treeless landscape of the arid Columbia Plateau, where annual precipitation is less than 250 millimeters (10 inches).

Figure 16-6 Creation of a desert near cold ocean currents. Cool, dry air from above the cold Humboldt current blows onshore along the Peruvian coast. The air warms over the land surface, and, because its capacity to hold water increases, it evaporates virtually all surface water from coastal Peru. The result is one of the Earth's driest places—the Atacama Desert, just west of the Andes.

Sierra Nevada mountain range, are rain-shadow deserts; Pacific air masses have been wrung dry by the time they reach them (Fig. 16-5).

Continental Interior Deserts An absence of moisture may create deserts in the interiors of continents, simply because these regions lie too far away from a major source of water. Air masses that reach these areas have already lost virtually all their moisture after crossing thousands of kilometers of land. The Gobi and Taklamakan deserts in central Asia, located several thousand kilometers from the nearest ocean, are examples of this type of desert.

Deserts Near Cold Ocean Currents Some deserts in warm subtropical regions occur close to the ocean but in areas adjacent to cold ocean currents. Because the air above the ocean currents is cold, it contains little water. As this dry air moves over land, it warms, enabling it to hold more water; as a result, the air evaporates almost all of the surface water in these coastal desert regions. The Atacama Desert of Peru and Chile, for example, owes its aridity in large part to the Pacific Ocean's cold Humboldt current, which flows along the coast of western South America (Fig. 16-6). The Namib and Kalahari Deserts, on the southwest coast of Africa, are also examples of this type of desert.

Polar Deserts Some places are both extremely arid and numbingly cold. Because they receive little solar radiation, such high-latitude locations as northern Greenland, Arctic Canada, northern Alaska, and Antarctica have temperatures that remain below freezing year-round, even during their brief summers when the sun never sets. At the same time, global atmospheric circulation brings only cold dry air to these lands, thus creating desert conditions.

Weathering in Deserts

Features such as steep angular cliffs, sharp-edged stones, and relatively thin soils may give the impression that little or no weathering takes place in deserts. Mechanical-weathering processes are at work, however; indeed, these forces produce most of a desert's loose sandy regolith. As discussed in Chapter 5, extreme daily temperature fluctuations cause rock surfaces to expand and contract repeatedly, prompting mineral grains to flake off the surface and join the desert's supply of shifting sands. At the same time, ongoing evaporation of any available water precipitates salt crystals within rock fractures. As these crystals grow, they act like minute crowbars, forcing adjacent mineral grains apart.

Because water is necessary for chemical reactions, it was long believed that the lack of water in deserts prevented chemical weathering from occurring there. Cold desert nights, however, promote condensation of dew on rocks, and some deserts experience a brief rainy season. Any moisture present, along with organic acids produced by desert vegetation, may support a small amount of chemical weathering. Even hyper-arid regions may contain enough moisture to oxidize iron-rich silicates, making possible the colorful hues of many desert rocks (Fig. 16-7).

The Work of Water in Deserts

Although desert landscapes appear vastly different from the topography of other environments, they are nevertheless shaped by many of the same processes that act on virtually all land surfaces. Despite conditions of extreme

Figure 16-7 Chemical weathering in deserts. Although oxidation, like other forms of chemical weathering, proceeds slowly in deserts, it has created colorful rocks such as these in the Chinle formation in Utah. Oxidation of iron-bearing silicates in the area's sandstones produces red-orange iron oxide, which gives these rocks their characteristic hues.

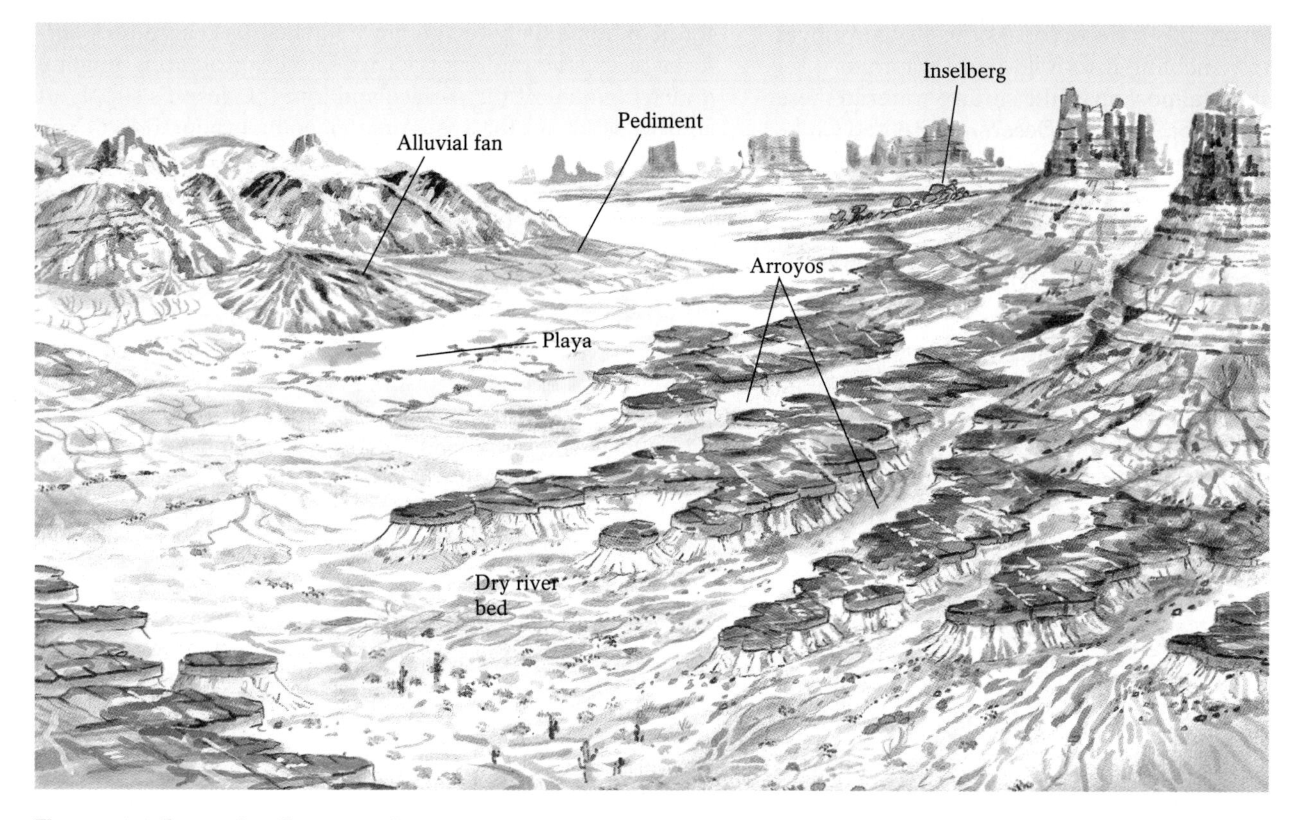

Figure 16-8 Desert landforms produced by water. Intermittent surface-water flow in deserts produces arroyos and pediments. Although the arroyo pictured here is dry, it fills rapidly with rushing water during a desert cloudburst. Deposition from desert streams creates alluvial fans and playas.

aridity, water serves as the primary sculptor of desert landforms. Brief occasional cloudbursts cause rapid erosion and subsequent deposition that create the desert landscape (Fig. 16-8).

Stream Erosion in Deserts

In arid regions, only a minimal network of soil-binding plant roots exists to anchor sediment particles, so flowing water can easily remove these grains. Moreover, infrequent desert storms produce immediate and intense surface runoff that is not slowed by vegetation, as it would be in grasslands or forests. The rapidly moving water of short-lived desert streams can, over thousands of years, erode numerous **arroyos,** stream channels that remain dry most of the year. The high-velocity streams that occasionally rush through arroyos, which arise from brief but intense desert storms, excavate their channels vertically, forming long, narrow, steep-sided canyons.

When desert streams remove sediment from a mountain, the resulting erosional surface forms a **pediment** (Latin for "foot"), a gently sloping, platform-like structure that extends from the base of the mountain. Pediments enlarge slowly as mountains recede by continuing stream erosion, sometimes continuing to grow until nearly all the mountain has dwindled away. These features are usually thinly covered with loose debris that is being transported downslope. After advanced pediment development, part of the mountain may remain as a steep-sided knob of resistant bedrock called an **inselberg** (German for "island mountain").

Stream Deposition in Deserts

Whenever erosion takes place, deposition is sure to follow. Unlike the process observed in moist, temperate regions, deposition in arid regions usually occurs relatively close to the erosion site. Because surface water evaporates or infiltrates so quickly in deserts, sediments tend to be dropped before they have been transported very far, producing the characteristic depositional features of arid regions.

During the occasional desert torrent, surface runoff flows rapidly at first, typically being confined to narrow arroyos with steep walls. At the outlet of the arroyos, where the canyon walls end, the unconfined water slows, spreads out, and soon evaporates into the air or infiltrates into the soil. In the process, the sediment loads washed from the surrounding steep mountain slopes are deposited, forming great alluvial fans that slope gently to the desert floor (see Figure 13-18).

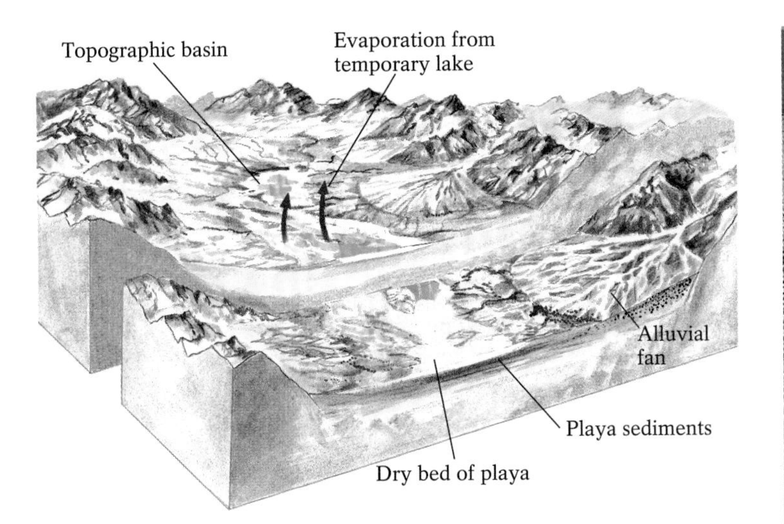

Figure 16-9 Playas. A playa is a dry basin in the desert floor from which a temporary lake has evaporated. Photo: Evaporite deposits in Devil's Golf Course, a playa in Death Valley, California. Ninety meters (295 feet) below sea level, this playa posed a significant obstacle to westward migration in the late nineteenth century.

During brief periods of higher-than-average precipitation in a desert's interior, water drains toward topographic lows—that is, closed basins—where it may collect as a temporary lake. This type of lake may evaporate in a few days or weeks, leaving behind a **playa,** a dry lake bed on the desert floor (Fig. 16-9). A playa's dry bed typically contains a mixture of fine-grained, clastic sediments and precipitated salts initially dissolved from soluble rock, many of which have utility as industrial chemicals. Some such salts have been mined from the playas of the American Southwest for more than 100 years. Death Valley, for example, has long served as a source of sodium borate, or borax, which is used in many pottery glazes and household cleansers.

If the water flowing into a playa does not encounter and dissolve soluble rock, the resulting playa may contain only fine-grained, clay-rich sediment. Such a surface can be baked so hard by the desert sun that it can be used as a natural landing strip for aircraft. For example, Rogers Dry Lake, on Edwards Air Force Base in the Mojave Desert, north of Los Angeles, is an ideal landing site for NASA's space shuttles.

When precipitation and runoff from surrounding mountains exceed evaporation and infiltration on the desert floor over an extended time, a large lake may form that can last for years or centuries, even in the ultradry desert air. Most such desert lakes become extremely saline (salt-rich) as evaporation continually concentrates dissolved salts. Large saline water bodies of this type include the Great Salt Lake in Utah, the Middle East's Dead Sea, and Lake Chad, a 22,000-square-kilometer (8500-square-mile) body of water in the southern Sahara.

The Work of Winds in Deserts

Winds, a second major force in shaping deserts, are air currents set in motion by heat-induced changes in air pressure. Deserts are subject to winds both from global atmospheric circulation patterns and from local pressure differentials caused by daytime heating and nighttime cooling of the desert floor. During the day, heated surface air rises to create a low-pressure zone; in the evening, relatively cooler and higher-pressure surface air rushes in to replace it, often at a velocity of 100 kilometers (60 miles) per hour or more.

Erosion by Wind

When you walk along a beach on a blustery day and feel the sting of the blowing sand against your legs, you are experiencing the power of wind erosion. After water, wind is the second most effective agent of surface change in deserts. By itself, wind cannot erode solid bedrock in the same way as flowing water or ice, but it can remove and transport large quantities of loose, unconsolidated material. Desert landscapes are more vulnerable to wind erosion than most other geographical settings, because they are covered with loose materials, and have little vegetation that can potentially anchor the soil and hold its particles down.

Deflation Wind gusts blowing across a dry, treeless desert lift sand- and silt-sized particles from the surface, but leave behind larger and heavier pebbles and cobbles. Some clay-sized particles also remain, because they adhere to the surface.

Figure 16-10 A blowout caused by deflation in Sand Hills State Park, Texas.

The removal of large quantities of loose material by wind is known as **deflation.** Deflation excavates distinct depressions in a desert floor and leaves a coarse pavement of large particles.

Deflation usually lowers the landscape slowly, by only a few tens of centimeters per thousand years. In extraordinary cases, however, such as occurred during North America's Dust Bowl of the 1930s, as much as 1 meter (3.3 feet) of fine-grained topsoil can be swept away in just a few years. Land surfaces may deflate either over broad expanses or in local areas only. Small lowered regions, known as **blowouts,** form where surface vegetation has been previously disturbed by animal trampling, overgrazing, range fires, drought, or human activity. Blowouts are often little more than 3 meters (10 feet) across and 1 meter deep. Because water helps make loose deflatable material more cohesive, these depressions are seldom deeper than the driest uppermost layers of regolith. Thousands of blowouts dimple the semi-arid Great Plains of North America from Texas to Saskatchewan (Fig. 16-10).

Deflation basins, large blowouts, occur where local bedrock is particularly soft or where faulting has produced a broad area of crushed bedrock. Sizable deflation basins are shaped by the winds that blow across the western Great

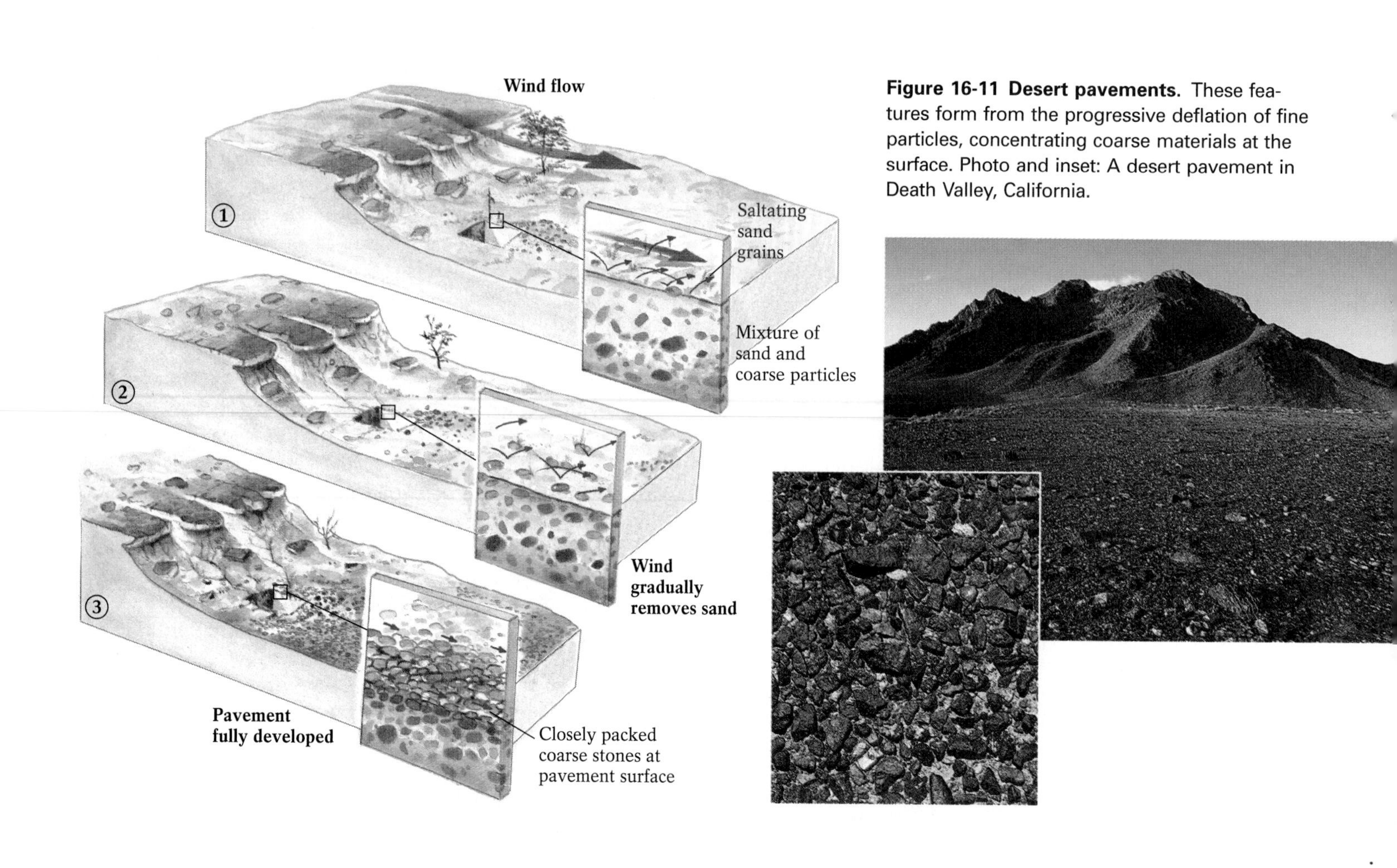

Figure 16-11 Desert pavements. These features form from the progressive deflation of fine particles, concentrating coarse materials at the surface. Photo and inset: A desert pavement in Death Valley, California.

Figure 16-12 Remnants of wind abrasion. In desert areas, such as this one in Goblin Valley, Utah, it is common to see balanced rocks perched precariously on narrow pedestals that have been cut by wind abrasion.

Plains. Near Laramie in southeastern Wyoming, soft fine-grained bedrock has been deflated to produce Big Hollow, a depression 15 kilometers (9 miles) long, 5 kilometers (3 miles) wide, and 30 to 50 meters (100–160 feet) deep.

A **desert pavement** comprises a surface of closely packed stones left behind by deflation (Fig. 16-11). It develops rapidly at first, because surface particles of sand and silt are readily carried off. The underlying layer of fine-grained materials usually remains protected under the coarser material, however, and further development therefore proceeds more slowly. A mature desert pavement, consisting of one or two layers of pebbles thick and with virtually no exposed fine regolith, may take hundreds or thousands of years to develop. When a mature desert pavement is disturbed (for example, by an off-road vehicle), fine regolith becomes exposed once more and deflation resumes until the protective pavement is restored.

Abrasion **Wind abrasion** occurs when a wind-borne supply of eroded particles is hurled against a surface and sandblasts it. In a Saharan sandstorm, for example, several hours of wind abrasion can strip the finish from a Jeep and pit the windshield until it is no longer transparent.

Wind abrasion produces several distinctive desert features, such as the rock pedestal shown in Figure 16-12. It may shape the exteriors of stones into flat sharp-edged faces; such **ventifacts** (Latin for "wind-made") are commonly strewn across desert floors (Fig. 16-13). Ventifacts' characteristic polished faces with pitted "orange-peel" textures result from prolonged wind abrasion on the side of the ventifact that faces the prevailing wind direction.

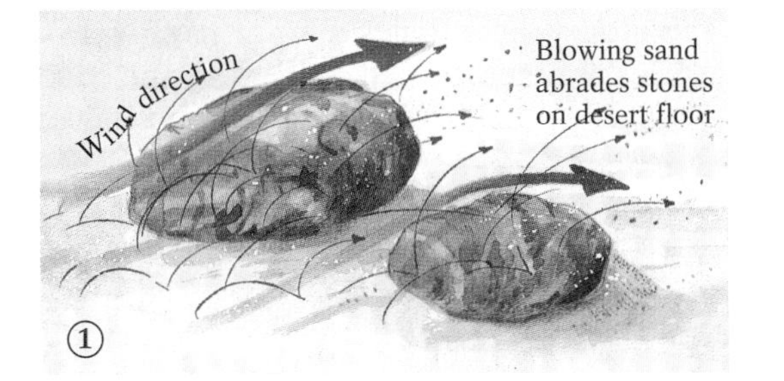

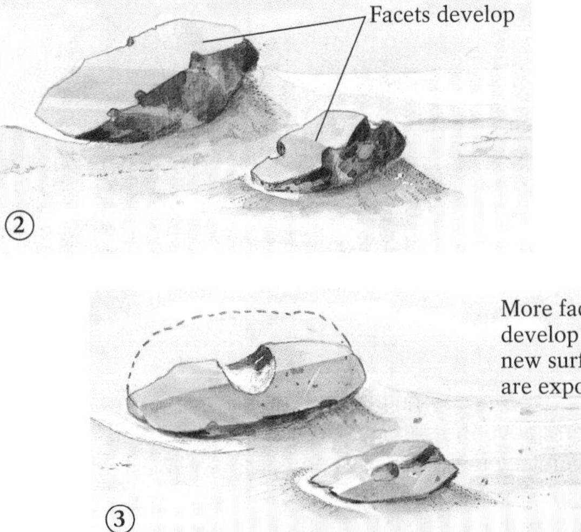

Figure 16-13 The origin of ventifacts. Ventifacts form when wind blowing predominantly from one direction abrades desert-floor stones, creating flat surfaces and sharp edges. As the wind changes direction or the stones shift position, exposing other surfaces to wind abrasion, more facets are produced on the newly exposed surfaces. Photo: A wind-etched ventifact.

Figure 16-14 **Yardangs from the Namib Desert, Namibia, and (inset) White Desert, Egypt.**

Wind abrasion can also produce large-scale, streamlined desert ridges called **yardangs** (from the Turkish *yar,* meaning "steep bank"). Yardangs most commonly form where strong one-directional winds abrade soft sedimentary bedrock layers, leaving behind more resistant layers. The resulting landform, which somewhat resembles the inverted hull of a sailboat (Fig. 16-14), is surrounded by wind-abraded depressions. Yardangs are usually wider on the windward side and taper leeward. They may be as much as tens of kilometers long and tens of meters high.

Transportation by Wind

In deserts, wind-transported particles roll, or saltate, along the surface much like the sand grains near the bottom of a stream. They may also travel several to tens of meters above the land's surface, suspended by the wind's turbulence. Unlike rivers or glaciers, wind has an extremely low viscosity and cannot easily lift a loose particle directly from the surface. Furthermore, vegetation and coarse sediment on the ground interrupt the air's passage just above the surface, creating a layer of virtually "dead air." To move any particle in this layer, wind must first reach a velocity great enough to begin rolling some of the particles along the surface of the ground. Once these particles start to move, they collide with other particles and propel these grains upward (Fig. 16-15). The airborne particles are then carried forward by the wind in curved paths until gravity pulls them down. The movement

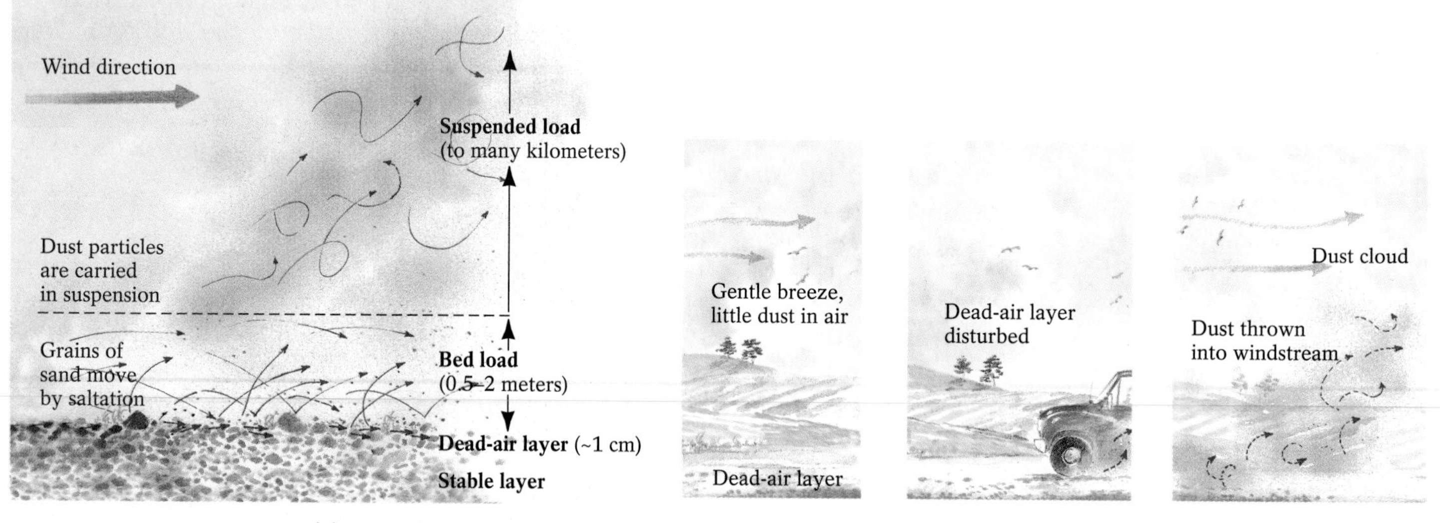

Figure 16-15 Transport of wind-borne sediment. (a) Like stream-borne particles, wind-borne particles may travel by rolling or saltating along the surface or by being carried in suspension above the surface. Greater velocity is needed to transport particles in wind, however, because air has a relatively low viscosity and because ground-level obstructions slow air movement, creating a dead-air layer. **(b)** On a relatively windy day, little dust rises from a dry country road, because the particles do not extend above the dead-air layer; thus they remain relatively unaffected, even by strong wind gusts. When a passing truck or a tractor disturbs the dead-air layer, it throws dust up above the zone of motionless air and into the windstream, and a thick cloud of dust rises behind the vehicle's tires. The dust blows away in the wind long after the truck and its turbulence are gone.

Highlight 16-1 America's Dust Bowl

In the early 1930s, strong winds, blowing across a drought-parched landscape, stripped the plowed wheat fields of their rich topsoil from the Texas Panhandle to the prairie provinces of Canada (Alberta, Saskatchewan, and Manitoba). Southeastern Colorado, western Kansas and Nebraska, and the Texas and Oklahoma panhandles were particularly hard-hit. Towering columns of swirling dust, called *black rollers,* darkened the sky at noon (Fig. 1). In May 1934, powerful winds lasting for 36 hours swept dust from the Great Plains into a dense black cloud that cast a 2000-kilometer (1200-mile)-long shadow across the eastern half of the continent. Prairie dust fell in upstate New York as "black rain" in autumn and in the mountains of Vermont as "black snow" in winter.

The dust in the Great Plains was sometimes so dense that it could penetrate the fabric of any garment and bury entire fields of crops. Many people and farm animals died from suffocation or "dust pneumonia," a malady akin to miner's silicosis (a condition brought on by the inhalation of mine dust). As one eyewitness reported:

> These storms were like rolling black smoke. We had to keep the lights on all day. We went to school with the headlights on and with dust masks on. I saw a woman who thought the world was coming to an end. She dropped down to her knees in the middle of Main Street in Amarillo and prayed out loud: "Dear Lord! Please give them a second chance."

A combination of natural and human factors produced the Dust Bowl. Much of the Great Plains had been overcultivated. Settlers had plowed the thick tough prairie sod to plant great fields of wheat until, by 1929, more than 100 million acres were under cultivation. For years, rainfall was abundant and harvests were plentiful. Unfortunately, removal of the protective grass cover left the land vulnerable to wind deflation when drought came. In the early 1930s, precipitation decreased to less than 50 centimeters (20 inches) per year. Crops failed everywhere. With no roots from trees or crops to anchor it, hundreds of millions of tons of rich topsoil blew away. Thousands of farmers, their crops blighted and their soils depleted, abandoned their lands and moved on. The westward migration of "Okies" to California was immortalized in John Steinbeck's novel, *The Grapes of Wrath.* Songwriter Woody Guthrie captured the bleak outlook of the times in "So Long, It's Been Good to Know You," written on April 14, 1935, the date of one of the decade's worst dust storms.

Figure 1 Black rollers. Ominous clouds of dust obscured the midday sun during the Dust Bowl years of the 1930s. This photo was taken in April 1935, in Mills, New Mexico.

By 1939, a few wet years, along with state and federal attempts to improve farming practices, had ended the Dust Bowl tragedy. The prairie states and provinces became habitable and cultivatable again. Although damaging droughts occurred again between 1950 and 1980, extensive irrigation and crop rotation ensured that there was less threat of a recurrent major problem. Residents of the plains have learned from the past and now adjust to naturally recurring dry periods.

of large saltating grains is quickly overcome by gravity and they soon fall back to the surface, striking other grains and, in turn, setting them in motion.

In addition to mobilizing other bed-load particles, saltating grains may set into motion smaller particles, which winds then carry aloft as a suspended load. Most of the suspended load in an airstream consists of dust, relatively flat particles only 0.15 millimeter (0.006 inch) or less in diameter. Because of their flat shape and large surface area relative to their weight, dust particles lifted above the dead-air layer by turbulent winds can be swept into the upper atmosphere, thousands of meters high, where they may remain for several years and travel over thousands of kilometers. In North America in the 1930s, dust from the drought-stricken fields of the Great Plains of Texas and Oklahoma fell on fresh snow in New England. In March 1935, dust from eastern Colorado was carried by sustained winds traveling at 80 kilometers (50 miles) to upstate New York, 3000 kilometers (2000 miles) away. Highlight 16-1 further describes the tragedy of America's Dust Bowl.

Figure 16-16 Deposition of wind's bed load. The dune field of the Great Sand Dunes National Monument continues to grow as loose material eroded from the arid lands to the west is carried by westerly winds and dropped against the Sangre de Cristo Mountains of southern Colorado.

Figure 16-17 Complexes of dunes in the Gobi Desert in western China.

Deposition by Wind

If wind can excavate blowouts and deflation basins and darken the sky with a veil of dust, what happens when it stops blowing? Airborne particles begin to fall, with the largest, heaviest saltating grains dropping closest to their source, and the flattest, lightest particles of dust settling farthest downwind. In this way, wind sorts its deposits. The appearance and location of wind's depositional landforms are determined by the size and amount of sediment, the constancy and direction of the wind, the presence of stabilizing vegetation, and the type of sediment load (bed load versus suspension).

Bed-Load Deposition: Sand Dunes The most notable landform created by wind deposition is the **dune,** a wind-built mound or ridge of sand (Fig. 16-16). Sand dunes form in both arid and humid climates, wherever we find a sufficient supply of sand that is initially not stabilized by vegetation and strong winds that blow relatively constantly. Dunes are found not only in deserts, but also along the sandy shores of oceans, seas, and large lakes (such as Lake Michigan), and near large, dry, sandy floodplains. Dunes typically form where an obstacle—anything from a small picket fence to a mountain range—interrupts the flow of saltating sand or where the wind slows to a speed at which it can no longer carry sediment. They also appear where a narrow obstacle, such as a clump of vegetation or a large rock, forces the windstream to diverge and go around it. As a dune grows, it becomes an obstacle itself, enabling the dune to become even larger.

Dunes reach their height limit when the wind speed is insufficient to carry a load of sand to the top of the dune and deposit it. Most dunes are 10 to 25 meters (35–80 feet) high. Those of Saudi Arabia and China exceed 200 meters (660 feet), but these landforms are actually complexes of dunes deposited one atop another, not individual dunes (Fig. 16-17). The sand-sized particles that make up dunes most often consist of weathering-resistant quartz grains, but there are exceptions. Gypsum grains, eroded from evaporite deposits on the slopes of the San Andres Mountains to the west, form the dunes at White Sands, New Mexico, for example; the dunes on the island of Bermuda are made of the broken shells of marine organisms.

Most dunes are asymmetrical, sloping gently (about 10°) on the windward side and more steeply (about 34°) on the leeward side, at the angle of repose of dry sand (discussed in Chapter 12). As sand grains saltate up the gentle windward slope and reach the dune crest, they spill over into the leeward wind shadow. Here the decreased wind velocity quickly deposits the particles on the steep leeward slope, or *slip face.*

As sand saltates up the windward slope and is deposited on the leeward side, a sand dune migrates, traveling downwind (Fig. 16-18). Small dunes tend to migrate faster and farther, because less sand must be moved. The rate of migration ranges from a few meters per year for large dunes in vast sandy deserts with gentle variable winds, to hundreds of meters per year for small dunes on bare rocky desert floors with strong one-directional winds. Coastal dunes may even migrate seaward and extend a coastline offshore by several kilometers.

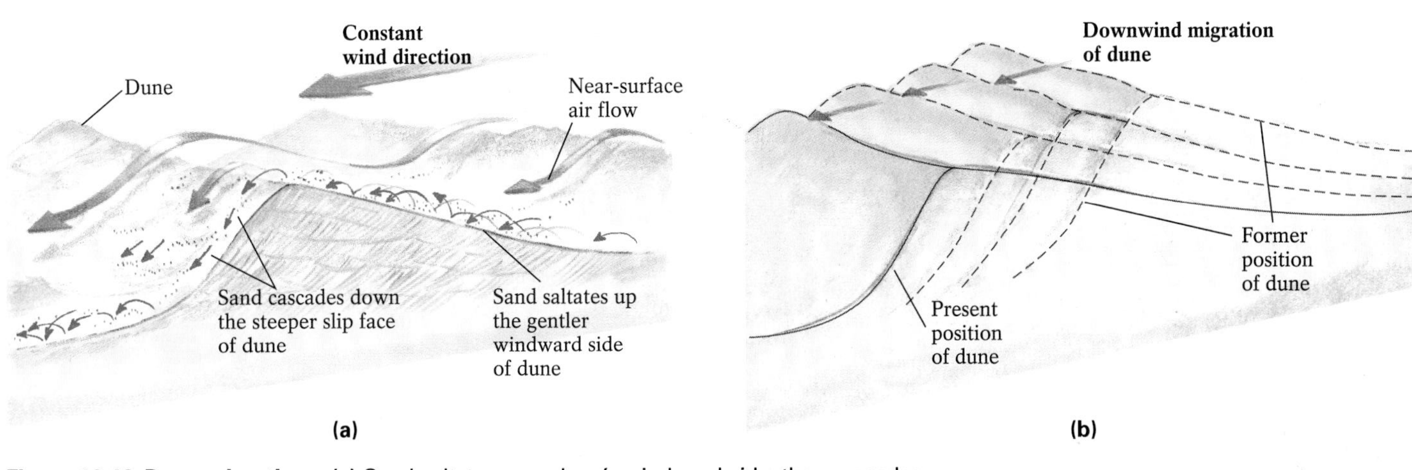

Figure 16-18 Dune migration. **(a)** Sand saltates up a dune's windward side, then cascades down its slip face. **(b)** This progressive transport of sand from the windward to the leeward side of dunes causes downwind migration of the dunes.

Dune Shapes Dune shapes reflect the local conditions, type of sand, degree of aridity, nature of the prevailing winds, and type and amount of vegetation present. **Transverse dunes** consist of a series of parallel ridges that typically occur in arid and semi-arid regions marked by plentiful sand, relatively constant wind direction, and scarce vegetation. These dunes form perpendicular to the prevailing wind direction and have a gentle windward slope and steep leeward slope (Fig. 16-19). In the Sahara, these landforms can be as large as 100 kilometers (60 miles) long, 100 to 200 meters (330–660 feet) high, and 1 to 3 kilometers (0.6–2 miles) wide. Transverse dunes can also develop along the shores of oceans

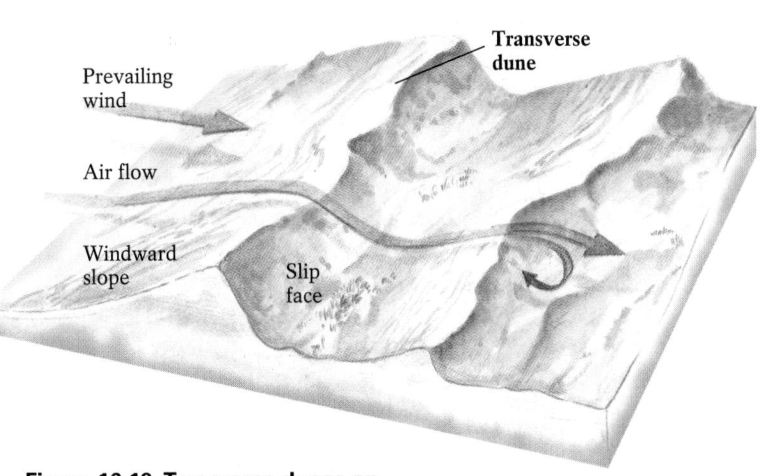

Figure 16-19 Transverse dunes on Mesquite Flat, Death Valley, California.

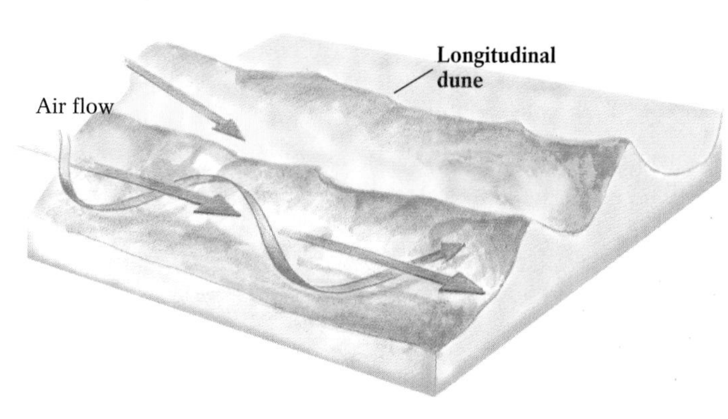

Figure 16-20 Longitudinal dunes at Stovepipe Wells, Death Valley National Monument, California.

and large lakes, where strong onshore winds shape the abundant sand. Such dunes dot the southeastern shore of Lake Michigan at the Indiana Dunes National Lakeshore in Indiana and Warren Dunes State Park in Michigan.

Longitudinal dunes are also parallel ridges; unlike transverse dunes, however, they are oriented parallel to the prevailing wind direction (Fig. 16-20). Longitudinal dunes form when sand supply is moderate and wind direction varies within a narrow range. Small ones may be only 60 meters (200 feet) long and 3 to 5 meters (10–16 feet) high. In the Libyan and Arabian Deserts, where strong winds blowing from several directions converge over the dune crest to create one sinuous form, longitudinal dunes can reach 100 kilometers (60 miles) long and 100 meters (330 feet) high.

Barchan (pronounced "BAR-kane") **dunes** are crescent-shaped ridges that form perpendicular to the prevailing wind as sand begins to accumulate around small patches of desert vegetation (Fig. 16-21). Barchan dunes develop in arid

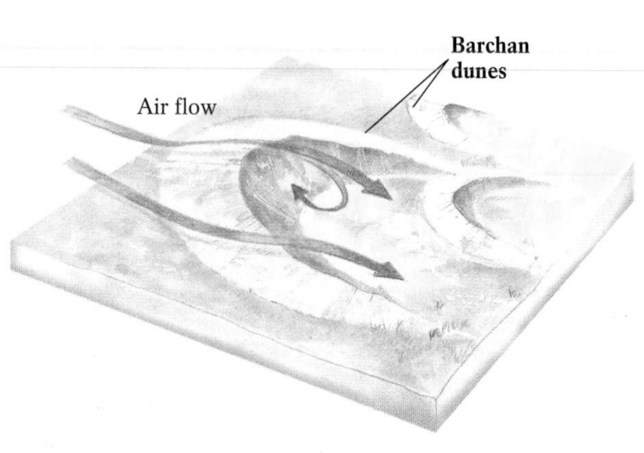

Figure 16-21 Barchan dunes in the Baja Desert of Baja California.

regions on flat, hard ground where little sand is available and wind direction remains constant. As they grow, these landforms become thicker and higher in their centers, where air flow is impeded most and more sand is deposited. Because their horns (the points of the crescents) are thinner than their centers, the horns migrate downwind more rapidly. The resulting barchan and its characteristic sharply pointed horns are therefore extended in the downwind direction.

Parabolic dunes are horseshoe-shaped, differing from barchans in that their horns point upwind (Fig. 16-22). They commonly form along sandy ocean and lake shores. Parabolic dunes develop when transverse dunes are exposed to accelerated wind deflation, especially after the removal of some vegetation. A small deflation hollow forms on the transverse dune's windward side, and the wind-excavated sand piles up downwind. The horns, a remnant of the original transverse dune, usually remain covered by vegetation and anchored in place; the rest of the parabola continues to migrate downwind, forming a horseshoe shape that can become quite elongated.

Star dunes, the most complex of the dune types, form when winds blow from three or more principal directions or when wind direction shifts constantly (Fig. 16-23). They tend to grow vertically to a high central point and may have three or four arms radiating from the center. Continued variability of wind direction causes the position of star dunes to remain relatively fixed.

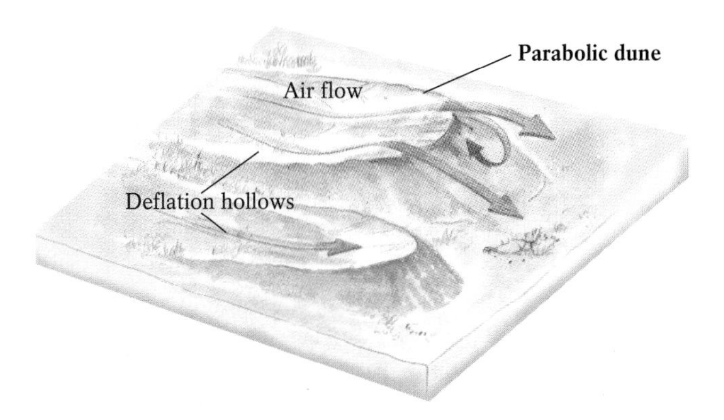

Figure 16-22 Parabolic dunes.

Suspended-Load Deposition: Loess **Loess** is a yellow-brown or ocher-colored wind-borne silt deposit that blankets hills, slopes, and valleys evenly, in a manner suggesting that it literally fell from the sky. Loess is always deposited downwind

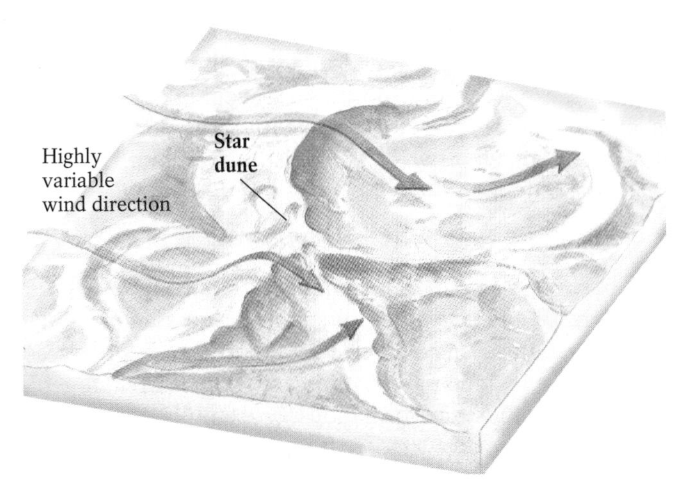

Figure 16-23 Star dunes (in the foreground), Death Valley National Monument, California.

Figure 16-24 Loess (windblown silt) deposits east of the Missouri River, Monona County, western Iowa.

of a plentiful supply of silt (Fig. 16-24). Although drying glacial outwash is often the source of the silt (see Chapter 15), some loess may originate in nonglacial environments. For example, the fertile loess plains west of the Mississippi River in eastern Kansas and Nebraska apparently were eroded by wind from the ancient dust-storm deposits that make up the sand hills of northwestern Nebraska. Similarly, loess now lying on the Palouse Hills of southeastern Washington was lifted by westerly winds from the drying floodplain of the mighty Columbia River to the west.

Northern China is home to the world's largest loess deposits. Their phenomenal 300-meter (1000-foot) thickness derives from the nearly endless supply of silt in the vast Gobi Desert to the west. This airborne dust eventually settles and washes into the Huang Ho (Yellow River) and Yellow Sea, giving these bodies of water their distinctive color.

Reconstructing Paleowind Directions

Winds blowing in the past, or *paleowinds,* have left traces of their intensity and prevailing direction in loess and in dune deposits that have since turned to stone. These preserved traces can tell us much about the geologic past. When geologists study a sandstone formation, for example, they search for clues to determine whether it was deposited by wind or water. Excellent sorting, cross-bedded sands displaying the angle of repose of dry sand, the occasional discovery of sharply faceted ventifacts, the presence of sand grains polished and pitted by mid-air collisions, and the absence of ultralight mica flakes (which are easily swept away in suspension) suggest a wind-deposited sandstone. After confirming a wind-related origin for the rock, geologists can use any of several wind-generated features to reconstruct ancient wind directions. For example, they may examine the shape of yardangs, the position of ventifact faces, the asymmetry of dunes, and the thickness and texture of loess deposits to hypothesize about the conditions during the rock's formation.

Sand dunes rarely survive intact in sedimentary rocks; instead, they are usually buried under other dunes that migrate over them. Although the gentle windward slope and the steep leeward slope of a dune may vanish, the inclined bedding of its original leeward slope may be preserved under the sands of advancing dunes, creating dune-type cross-bedding. The leeward slope beds are inclined at the angle of repose of dry sand (30°–34°), indicating the paleowind direction that prevailed when the bedding layer formed (Fig. 16-25).

Geologists can also analyze certain properties of loess to estimate past wind directions. A loess layer is generally thicker near its source and thins downwind, where its particles are finer. The 10-meter (33-foot)-thick layers of loess that occupy the eastern banks of southward-flowing rivers in the Midwest thin to less than a few centimeters in the East. Hence, when they were deposited 12,000 years ago, the prevailing winds must have been westerly, as they are today.

Figure 16-25 Sand dune preservation. Numerous ancient, lithified sand dunes are found in national parks in the arid Southwest. The Navajo Sandstone, for example, crops out throughout Zion National Park in southwestern Utah. This memento dates from 180 million years ago, when what is now Utah was on the continent's west coast (before the lithosphere on which California is located was added to the continent). We can be sure that the Navajo Sandstone originated on land, and therefore was deposited by wind, because the formation contains dinosaur footprints. From the orientation of the cross-beds, we can reconstruct the northerly and westerly paleowinds that formed the migrating coastal sand dunes that later became this cross-bedded outcrop.

Desertification

Desertification is the invasion of desert conditions into formerly nondesert areas. Common symptoms of desertification include a significant lowering of the water table, a marked reduction in surface-water supply, increased salinity in natural waters and soils, progressive destruction of vegetation, and an accelerated rate of erosion.

Current Desertification Trends

Northern and western Africa have been experiencing rapid desertification for 2000 years. About 8000 years ago, well after the end of the last major worldwide glaciation, the Namib Desert of southwestern Africa was a lush savannah that supported advanced Stone Age societies. It remained a fertile grassland, teeming with wildlife, for the next 6000 years. Today, however, the region is a vast desert.

Two factors were involved in turning the northern African savannah and grassland into the modern Sahara. One—drought—is natural and unavoidable; the other—overpopulation and land mismanagement—is human-induced. Drought or overpopulation, or both, can start the process of desertification. When a land's inhabitants cannot produce enough to provide for their needs, they tend to overgraze their cattle, plant crops without replenishing the soil, cut down trees for shelter and fuel, and draw more water from springs and other resources than is naturally resupplied. Gradually soils become depleted of their nutrients, and the removed trees and their roots do not grow back. Without vegetation, soils cannot hold water or prevent wind erosion. Precipitation from the occasional cloudbursts washes away the loose topsoil and, in hot climates, the rainfall evaporates before it can enter the groundwater supply. Eventually the parched land is left with virtually no productive capacity. The process of desertification, once begun, tends to be self-perpetuating. Left with no source of sustenance, the land's inhabitants must migrate elsewhere in search of food, water, and shelter.

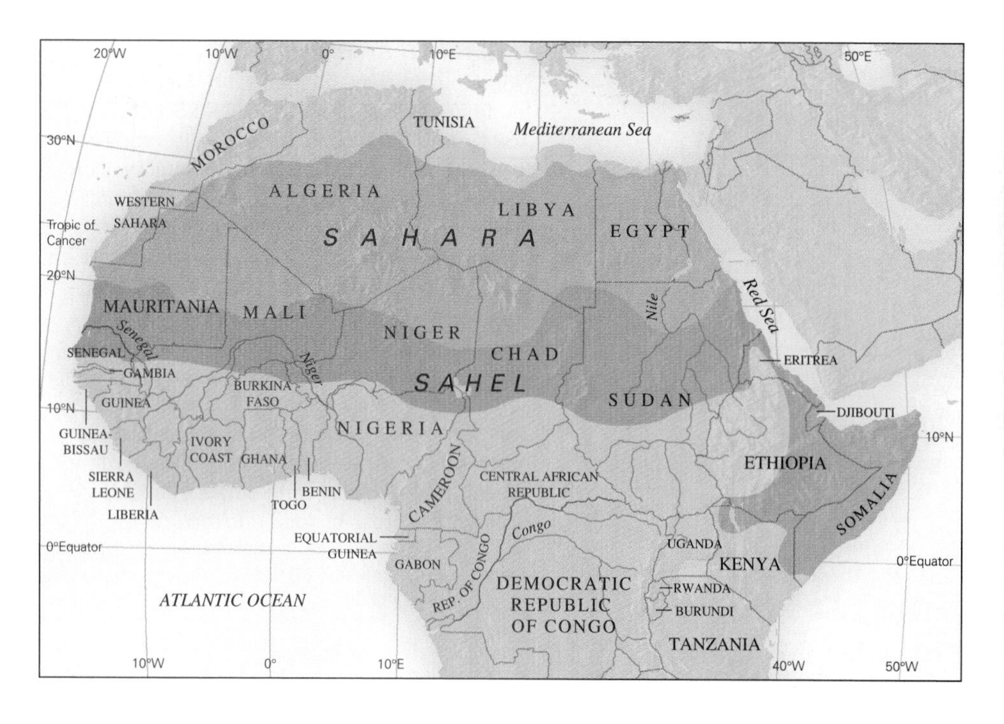

Figure 16-26 Skeletons of a cattle herd in the sun-baked lands of the Sahel, which underwent rapid desertification from 1970 through the early 1990s. (left) Map of the Sahel.

The Sahel, the region of Africa located immediately south of the Sahara, has been particularly hard-hit by desertification in recent decades (Fig 16-26). After decades of overgrazing and overfarming had depleted the soil, the worst drought of the century struck in the early 1970s. In 1970, no rain fell at all. Large herds of domestic cattle, sheep, goats, and camels demolished the existing grasses; the topsoils, bereft of any binding roots, were easily eroded by wind. By the time of the next year's rain, the ground was baked so hard as to be impermeable, and surface runoff accelerated soil erosion. Animals died by the millions. The starving human population migrated to the region's larger cities, which tripled in size as refugee camps sprang up around them. Worldwide relief efforts were "too little, too late" to prevent the deaths of hundreds of thousands of people from starvation, malnutrition, and associated diseases. The drought continued into the 1980s, ending only in the mid-1990s.

As much as 35% of the world's land, now barely sustaining a population of 850 million people, consists of semi-arid margins of already-arid lands. In the next few decades, drought and overpopulation could potentially convert them from dry but habitable grasslands into deserts unable to sustain significant human populations. As many as 70 nations face this threat; about half of them are in Africa, where the Sahara is advancing southward by as much as 50 kilometers (30 miles) per year. Indeed, choking dust storms and newly formed dune fields already threaten some northern and western cities in this region.

In the United States, overpopulation and urban growth, overgrazing, and excessive groundwater withdrawal, particularly in the arid and semi-arid Sunbelt states, have already increased soil salinity and accelerated erosion. Today, about 27% of the United States' nondesert land is facing encroaching desertification.

Reversing Desertification

Efforts to reverse desertification focus on controlling wind erosion and on managing water more effectively. Trees serve as a windbreak to halt shifting sands, thereby helping to retain both soil cover and groundwater. To stabilize migrating dunes, we can plant fast-growing trees such as poplars, which have deep, sand-binding roots. In northeastern China, an entire forest, 500 by 800 kilometers (300 by 500 miles), was planted upwind of 90,000 acres of prime farmland. Dunes in China have even been leveled by hand and covered with topsoil to create productive new farmland.

In some extremely parched places, inventive technologies have been developed to reclaim recently desertified lands and to make portions of ancient deserts more habitable for human populations. In the Middle East, for example, systems of wells, channels, and collecting pools are used to collect and store infrequent storm runoff underground, which protects the water from evaporation. Throughout the region, farmers apply a specially designed plastic mulch to their farmlands or cover their fields with plastic sheeting, punctured for plant stems, to retain irrigation moisture. Computers monitor and regulate water flow through pipes and canals, delivering water where it is needed at times when evaporation is minimal. In Israel, "drip" agriculture—using perforated garden hoses that snake amid plantings—delivers water constantly to individual plants, literally drop by drop. Water is even drawn from the Mediterranean Sea and desalinized for agricultural use. In the American Southwest, billions of liters of Colorado River water have been channeled into the Sonoran Desert of southern Arizona, providing sufficient irrigation to produce food for hundreds of thousands of people (Fig. 16-27).

These methods are but a few of the many techniques that can enable arid lands to support a human population. Unfortunately, such techniques remain unaffordable to many of the countries that need them the most.

Wind Action on Mars

The Earth is not the only windswept planet in the solar system. Because all of the planets are heated unevenly by the Sun's rays, those that have atmospheres also experience wind action. Mars is an especially windy planet.

The Mariner 9 space probe went into orbit around Mars on November 13, 1971, after a five-month journey from Earth. In a disappointing development, its first photographs depicted a planet-wide dust storm. For three months, dust that rose as high as 6 kilometers (4 miles) obscured the entire Martian surface. (On Earth, where the force of gravity is stronger and the air is more than 100 times as dense, it would have taken winds in excess of 160 kilometers [100 miles] per hour to raise dust to that height.) Mars' global dust storms apparently start when summer comes to its southern hemisphere. As the edges of the planet's frozen carbon dioxide polar cap rapidly sublimate, large temperature differences arise between the still-solid polar cap remnants and the warming surrounding landscape. These temperature variations produce differential air pressures. The resulting high winds create dust storms that spread northward, engulfing the planet's entire surface for months.

After the dust finally settled in 1972, Mariner 9 was able to transmit exciting photographic images of the Martian surface. A dune field, 60 by 30 kilometers (40 by 20 miles), was sighted on the floor of a large crater (Fig. 16-28). Viking

Figure 16-27 Circles of irrigated land in the Yuma Valley of Arizona's Sonoran Desert. Irrigation is largely responsible for the agricultural productivity and rapid population growth of this region.

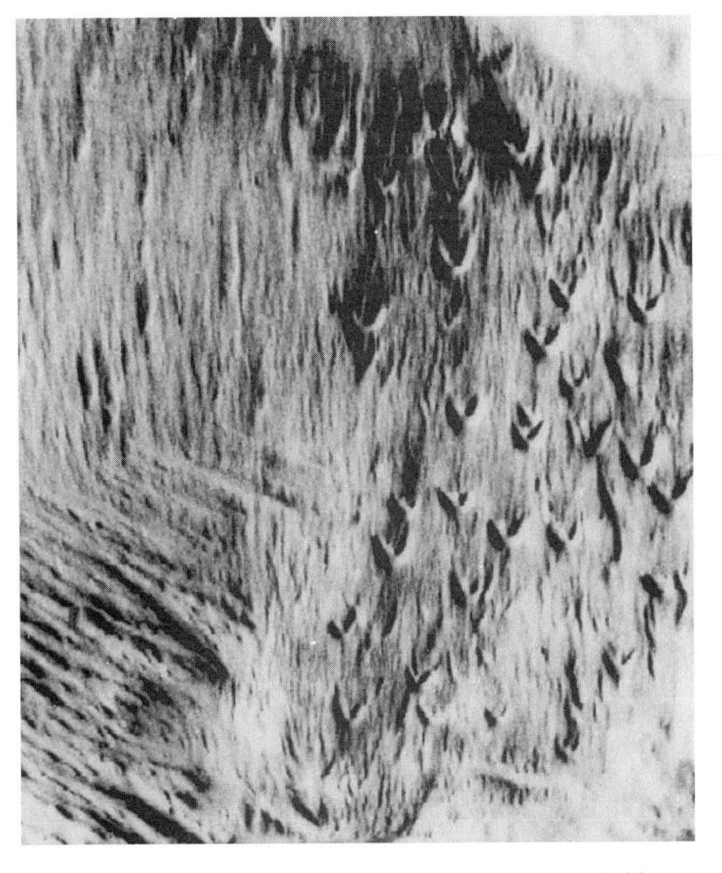

Figure 16-28 Martian sand dunes. These crescent-shaped ridges on Mars are remnants of barchan dunes produced by the planet's strong surface winds.

probes landed on the Martian surface a few years later, on July 20 and September 3, 1976. The landers descended onto loess-covered plains, where they photographed interspersed ventifacts, barchan and longitudinal dunes, yardangs, and other wind-generated landforms.

Chapter Summary

Deserts are generally defined as dry regions, typically with sparse vegetation, that receive less than 250 millimeters (10 inches) of precipitation annually. More specifically, a desert is a region with an **aridity index,** or ratio of potential annual evaporation to average annual precipitation, greater than 4.0. Five principal types of deserts exist: (1) the subtropical deserts found between about 20° to 30° latitude both north and south of the equator, where dry air descends toward the surface as part of the global atmospheric circulation pattern; (2) deserts occurring on the dry leeward side of major coastal mountain ranges, known as **rain-shadow deserts;** (3) interior deserts, located in the remote centers of continents far from sources of water; (4) coastal deserts appearing near cold ocean currents, which are formed by dry onshore winds; and (5) polar deserts, which are both extremely cold and dry.

Water is the primary sculptor of desert landforms. One such landform is **arroyo,** a channel that carries a stream during periods of high discharge but remains dry most of the year. A desert mountain that is receding due to stream erosion may develop a gently inclined surface at its base called a **pediment.** After advanced pediment development, part of the mountain may persist as a resistant bedrock knob, or **inselberg.**

In deserts, surface water evaporates or infiltrates rapidly, forming distinctive depositional features. Alluvial fans accumulate where a slope ends and the desert floor begins. A **playa** is a dry lake bed that develops when a desert-floor lake evaporates.

Wind, which arises because of differences in air temperature and pressure, is the second most effective erosive agent in deserts. Wind erodes by **deflation,** removing finer particles from the surface. Small depressions in the ground caused by deflation are called **blowouts; deflation basins** are large blowouts. The layer of coarser particles left behind by deflation constitutes **desert pavement.**

In **wind abrasion,** wind-carried particles are hurled against rock surfaces, effectively sandblasting the rock. The windward surfaces of individual rocks and boulders may be beveled by wind abrasion to form **ventifacts.** Wind abrasion also produces streamlined ridges of resistant rock known as **yardangs.**

When wind reaches sufficient velocity, desert soil grains begin rolling and collide with other grains, initiating saltation. Saltating grains make up a windstream's coarse bed load; finer particles are carried aloft as suspended load. When its transport energy decreases, the wind deposits the bed load, creating **dunes,** hills of loose wind-borne sand. Five principal types of dunes exist: (1) **transverse dunes** are parallel ridges that develop perpendicular to the prevailing wind direction; (2) **longitudinal dunes** are parallel ridges that develop parallel to the prevailing wind direction; (3) crescent-shaped **barchan dunes,** whose horns point downwind, lie perpendicular to prevailing winds; (4) horseshoe-shaped **parabolic dunes,** whose horns point upwind, also lie perpendicular to prevailing winds; and (5) **star dunes** grow vertically to form a high central point with three or four radiating arms. Silt-sized particles carried as the wind's suspended load are deposited as blanket-like layers of **loess.**

Drought and overpopulation can cause **desertification,** the invasion of desert conditions into formerly nondesert areas. The symptoms of desertification include significant lowering of the water table, marked reduction in surface-water supply, increased salinity in natural waters and soils, progressive destruction of native vegetation, and accelerated rates of erosion. Desertification currently threatens as many as 70 nations, about half of them in Africa.

Mars, a particularly windy planet, experiences planet-wide seasonal dust storms. NASA's Mariner and Viking probes have photographed loess-covered plains, ventifacts, dunes, and other wind-generated landforms on its surface.

Key Terms

deserts (p. 307)
aridity index (p. 308)
rain-shadow deserts (p. 309)
arroyos (p. 312)
pediment (p. 312)
inselberg (p. 312)
playa (p. 313)
deflation (p. 314)
blowouts (p. 314)
deflation basins (p. 314)
desert pavement (p. 315)
wind abrasion (p. 315)
ventifacts (p. 315)
yardangs (p. 316)
dune (p. 318)
transverse dunes (p. 319)
longitudinal dunes (p. 320)
barchan dunes (p. 320)
parabolic dunes (p. 321)
star dunes (p. 321)
loess (p. 321)
desertification (p. 323)

Questions for Review

1. Briefly describe three different types of deserts and the conditions that contribute to their aridity.

2. Draw a simple sketch to illustrate the rain-shadow effect.

3. Cite evidence that both chemical and mechanical weathering occur in arid regions.

4. List three landforms that are created through the work of water in arid regions.

5. What is a playa? Why are playas generally a rich source of precipitated salts?

6. How does erosion by deflation differ from erosion by abrasion? Briefly describe one product of deflation.

7. What are ventifacts and yardangs? How can we tell the direction of the wind that formed them?

8. Sketch the basic shape of a transverse dune, viewed from the side. What is the essential difference between a barchan dune and a parabolic dune?

9. What is loess? How can it help geologists reconstruct paleowind directions?

10. What are two causes of desertification? How do they produce their effects?

For Further Thought

1. Speculate about how changes in the configuration of the Earth's continents through plate tectonic activity might increase or decrease the total area of arid regions.

2. What would happen to the distribution of deserts if global warming were to have a significant effect on the Earth's climate?

3. Determine the prevailing wind direction of the landscape in the photograph at the right.

4. Describe how global wind patterns may have changed through Earth's history. (*Hint:* The Earth's rate of rotation has been slowing throughout its history.) Speculate about the future of the Earth's global wind patterns.

5. What actions would you favor in combating the global trend toward increased desertification?

17

Shores and Coastal Processes

North Americans have a passion for vacationing near shores and coasts, and more than half of us live within 80 kilometers (50 miles) of the Atlantic or Pacific Ocean or near one of the Great Lakes. Shores and coasts, at once scenic and educational, are wonderful places to observe natural processes—particularly the action of waves, tides, and nearshore currents.

All shorelines change constantly through natural processes. Sometimes those processes act rapidly and dramatically. Waves from a powerful storm, such as those shown in Figure 17-1, may produce immediate alterations to the shoreline. On January 2, 1987, for example, waves from a powerful winter storm gouged more than 20 meters (65 feet) of dunes and beaches from Nauset Beach, at the bend in the "elbow" of Cape Cod along the Atlantic coast of Massachusetts. The beach, a 20-kilometer (12-mile)-long pile of sand that had been built and shaped by waves and tides during the past 4000 years, had sheltered the bayside town of Chatham and its fishing fleets for centuries. On that night, however, Nauset was breached by 6-meter (20-foot)-high waves that also swept away nearly a kilometer of the cape's offshore islands, the barrier that had protected Chatham from the Atlantic's waves and storms.

The entire land region bordering a body of water is called a **coast**. Coasts extend inland until they encounter a different geographical setting, such as a mountain range or a high plateau. A **shoreline** is the boundary where a body of water meets the adjacent dry land. A *shore*, the strip of coast closest to a sea or lake, often includes a sandy strip of land, or a *beach*. In this chapter, we examine the variety of processes that shape and change our coasts—the waves, currents, and tides that erode and deposit coastal materials. We also look at several different types of coasts and consider how human activity sometimes affects the evolution of coasts.

Figure 17-1 Crashing surf along the central Oregon coast.

Waves, Currents, and Tides

Moving water is the great agent of geologic change at the Earth's coasts. Water can be set in motion by the wind, which produces most waves and currents, and by the combined effects of the gravitational pull of the Moon and Sun and the rotation of the Earth, which alternately raise and lower water surfaces, producing tides.

Waves and Currents

All waves—be they earthquake, sound, water, or any other kind—transport energy. The ultimate source of the energy transported by water waves is solar radiation, which heats the atmosphere more in some regions (such as near the equator) than in other regions (such as near the poles). This disparity in heating creates zones of low atmospheric pressure (more heated) and high atmospheric pressure (less heated). Air flows from areas of high pressure to areas of low pressure, creating wind.

When wind crosses the surface water of an ocean or lake, friction between the air and the water causes waves, in which the water alternately rises to form *crests* and falls to form *troughs* (Fig. 17-2). A wave's *height* is the vertical distance between its crest and trough. In the open ocean, waves commonly measure 2 to 5 meters (7–18 feet) high; during hurricanes, however, wave heights can exceed 30 meters (100 feet). Wave *length* is the distance between two adjacent waves, measured from crest to crest or trough to trough. Ocean waves typically occur 40 to 600 meters (135–2000 feet) apart. The time required for one wave length to pass a stationary point is called the wave's *period*. Ocean waves commonly have a period of a few seconds. Wave *velocity*, the speed at which an individual wave travels, is typically 30 to 90 kilometers (20–60 miles) per hour in mid-ocean.

The Movement of Waves Wave characteristics such as height, length, and velocity are determined by four factors: wind speed, wind duration, constancy of wind direction, and the distance that the wind travels across the water surface, or *fetch*. A strong wind that blows for a long time from the same direction across a large body of water produces a closely spaced series of large, fast-moving waves. Pacific Ocean waves tend to be very large, in part because the Pacific is so huge and its landmasses so widely spaced. Similarly, the band of the global ocean system that lies south of the southern tip of South America is virtually uninterrupted by any landmasses and generates some of the world's largest waves, commonly exceeding 15 meters (50 feet) in height. The largest wind-generated wave ever recorded, sighted in the northern Pacific in 1933 by the captain of the USS *Ramapo*, was 34 meters (115 feet) high. It was generated by persistent gale-force winds.

Although mid-ocean waves move outward from a wind source, only the wave *form* moves significantly outward; the water within the wave follows a rolling circular path, rising and falling as the wave passes but moving only a short distance from its original position. This **oscillatory motion** decreases with depth. Below a depth equal to about one-half the wave length (defined earlier as the distance between two successive crests), called the *wave base*, it is virtually absent. Water below the wave base remains undisturbed by waves passing above. In deep water, surface waves have no effect on the sea floor; in shallow water, however, a wave base may intersect the sea floor, disturbing its loose sediments.

After a journey of perhaps thousands of kilometers, waves cross a continental shelf and approach a coast. As the shelf becomes progressively shallower landward, the sea-floor depth decreases to the wave base and eventually interrupts the water's oscillatory motion (Fig. 17-3). A set of waves whose length is 100 meters (330 feet), for example, begins to drag against the sea floor at a depth of 50 meters (165

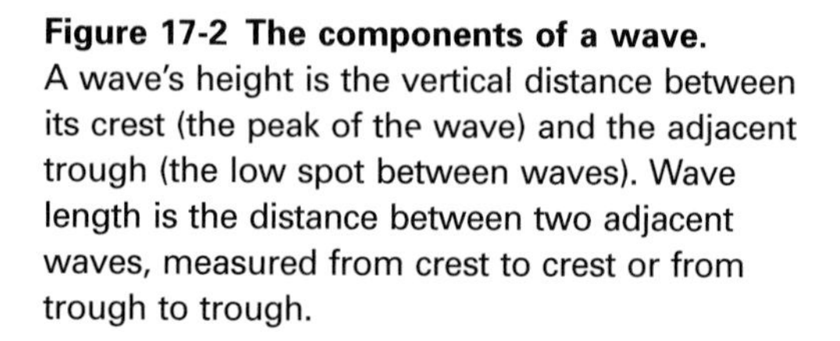

Figure 17-2 The components of a wave. A wave's height is the vertical distance between its crest (the peak of the wave) and the adjacent trough (the low spot between waves). Wave length is the distance between two adjacent waves, measured from crest to crest or from trough to trough.

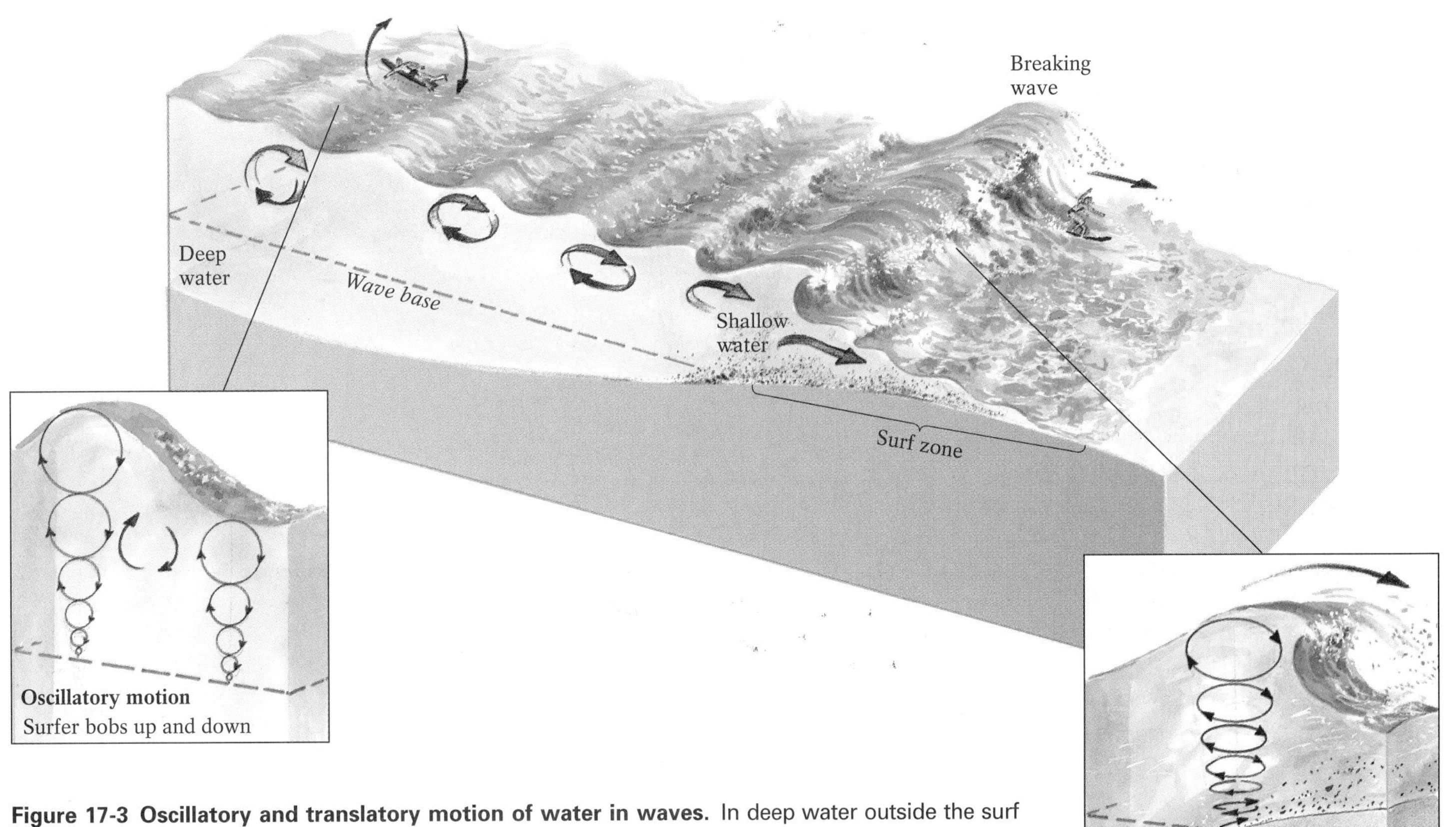

Figure 17-3 Oscillatory and translatory motion of water in waves. In deep water outside the surf zone, water moves in vertical circles but does not move landward; its motion is oscillatory. In the surf zone, however, the water is shallow. The wave base has touched the sea floor, causing the waves to slow and break; translatory motion carries the water (as well as loose sand particles) shoreward.

feet). When a wave touches bottom, its orbital motion becomes restricted, causing its velocity to decrease. It spends some of its energy moving loose sediment on the sea floor. The seaward waves, which are still in deep water, begin to catch up with the slower waves. As a result, the waves bunch up, their lengths and periods decrease, and their heights increase. As a wave continues into the near-shore shallows, its crest moves more rapidly than its bottom, which continues to drag on the sea floor. Eventually, the crest overruns the rest of the wave, and the wave falls over, forming a "breaker." The offshore shallow area where waves break is known as the *breaker zone*. The water in a breaking wave moves landward as *surf*, low, foamy waves. In surf, the water itself—not just the wave form—actually moves forward, thus displaying **translatory motion**.

The water that hurtles up the beach after a wave breaks is known as *swash*. The water that returns to the sea is called *backwash*. Swash and backwash commonly roll and push sand back and forth along the shore, eroding some from coastal rocks and depositing some sand inland that was once located farther out at sea.

Wave Refraction and Coastal Currents Waves generally strike a coast at an angle, with part of a wave entering shallower water sooner than the rest. When the first-arriving part touches bottom, it loses energy and slows down, and its crest breaks. The rest of the wave, still in deeper water, continues to move at a high speed and pivots around the slow, shallow-water segment, much in the same way that a marching band turns a corner (by having the marchers closest to the corner take small steps while their comrades farthest from the corner take large, quick steps). This **wave refraction,** or bending, causes the last-arriving portion of the wave to be positioned nearly parallel to the coast before it breaks. Waves that first approach the beach at a 50° to 60° angle are typically refracted to less than a 5° angle (Fig. 17-4).

As each refracted wave breaks and strikes the coast, its surf pushes the swash ahead of it up the beach at a small angle. Under the force of gravity, the water then returns to the sea as backwash—not at an angle, but perpendicular to the shoreline. This zig-zag motion of swash and backwash propels the water down the coast as a **longshore current.** Longshore currents flow nearly parallel to the shoreline,

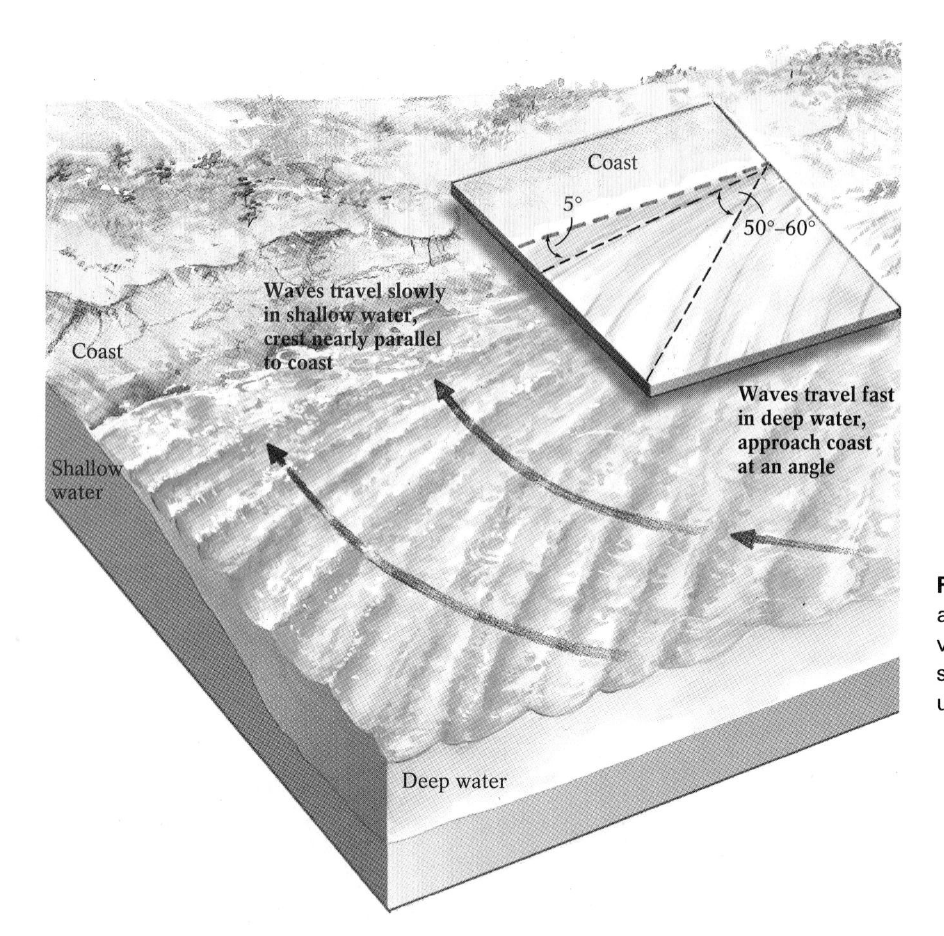

Figure 17-4 Wave refraction. Virtually all waves approach the coast at an angle, but the slower velocity of the part of the wave that reaches shallow water first bends the incoming wave until it is almost parallel to the coastline.

Figure 17-5 Longshore currents. The swash and backwash of breaking waves combine to produce a longshore current, which travels parallel to the shoreline and transports sediment along the coast. If you have ever swum in the ocean and been unable to immediately locate the place where you entered the water, you were probably carried down the beach by the longshore current.

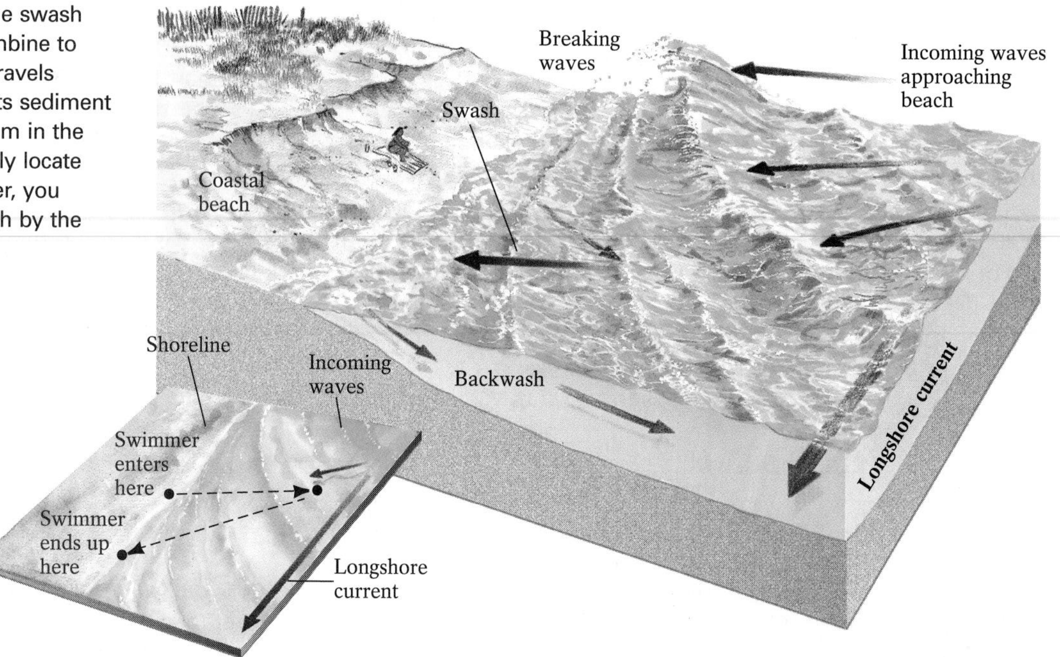

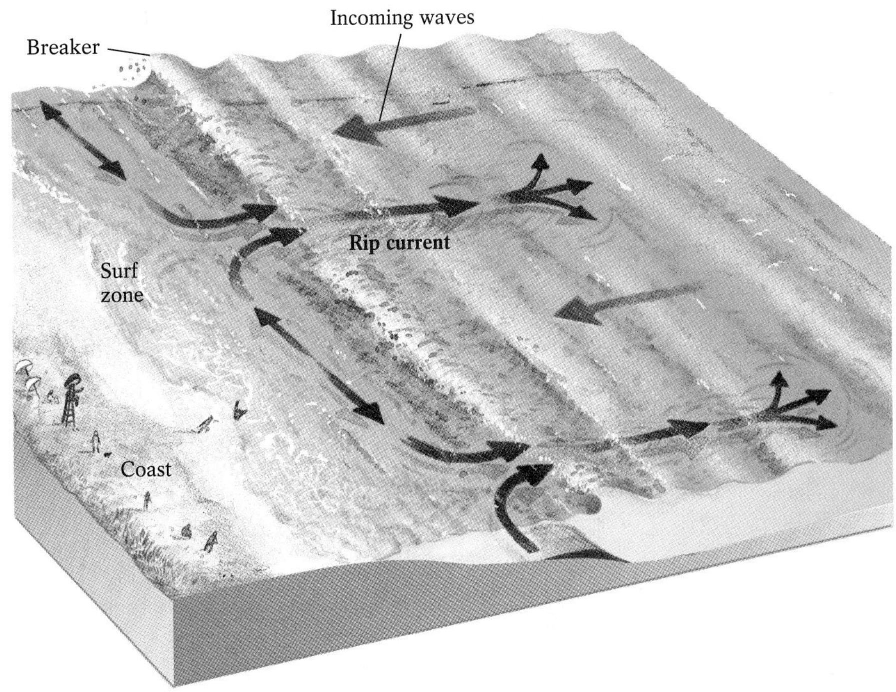

Figure 17-6 Rip currents. A rip current occurs when backwash builds up in the surf zone and then flows seaward through the incoming waves. An ocean swimmer who is unable to return to shore because of a strong seaward rip current should swim parallel to the coast for a short distance to escape it.

carrying sediment—and sometimes swimmers—down the coast (Fig. 17-5).

Using fluorescent dye or colored grains of sand, geologists and oceanographers have monitored longshore currents and learned that their velocity typically ranges between 0.25 and 1.0 meter (1–3 feet) per second. The velocity can, however, reach several meters per second and be powerful enough to lift loose bottom sediment into suspension and transport it down the coast for great distances. Some of the sand on the Outer Banks of North Carolina, for instance, originated on the rocky coast of Maine, 1500 kilometers (1000 miles) to the north.

Rip currents, known to swimmers as "undertow," flow straight out to sea, moving water and sediment *perpendicular* to the shoreline (Fig. 17-6). Rip currents occur when water that has accumulated in the surf zone moves seaward. They usually interfere with incoming waves, causing them to break before reaching the beach or even preventing them from breaking altogether. Swimmers caught in powerful rip currents can bypass the current by swimming parallel to the shore for a few tens of meters; surfers, however, may choose to ride rip currents seaward for a swift, effortless return to the breaker zone.

Tides

Tides are the daily rises and falls of the surfaces of oceans and large lakes that move shorelines alternately landward (during high tides) and seaward (during low tides). They result from the gravitational pull of the Moon and the Sun and from the rotation of the Earth. Because the Moon is much closer to the Earth than is the Sun, the Moon exerts the dominant gravitational force. The portion of the Earth's oceans that is facing the Moon at any given time is therefore pulled moonward into a bulge, or *high tide.*

As the Earth rotates on its axis, its coastal regions experience two high tides each day: one when they are on the side of the Earth facing the Moon, and a second when they are on the opposite side, away from the Moon. The opposite-side tidal bulge occurs because the spinning of the Earth forces the water outward, away from the Earth, at the same time that the Moon's gravitational pull, which would pull the water back inward, is at a minimum because of the Moon's increased distance from the far side of the planet. As the Earth rotates, the tidal bulges remain in line with the Moon, thus appearing to move around the Earth's oceans (Fig. 17-7).

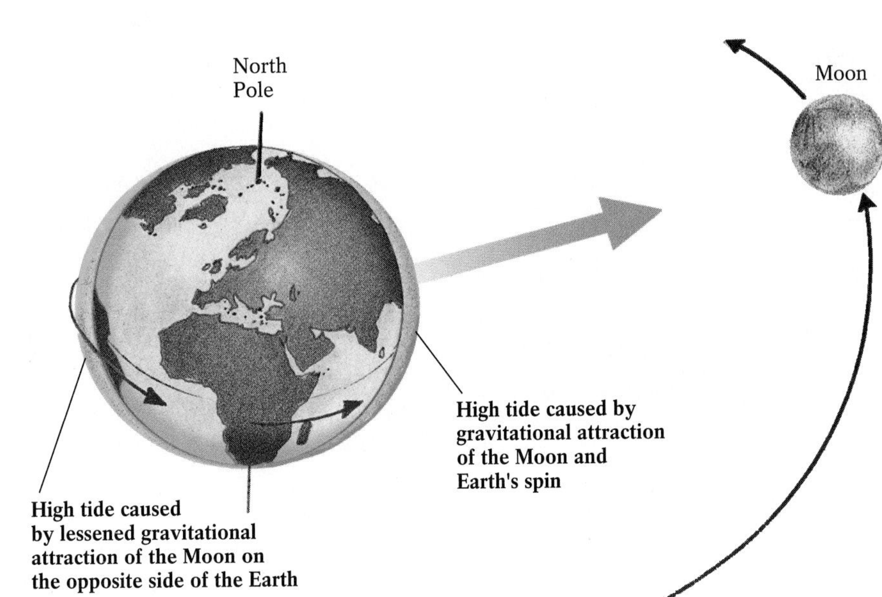

Figure 17-7 Tidal bulges. The Earth's high tides occur where the gravitational attraction of the Moon is strongest (where the Moon is closest) and where it is weakest (where the Moon is farthest away). Thus the tides are influenced by the rotation of the Earth.

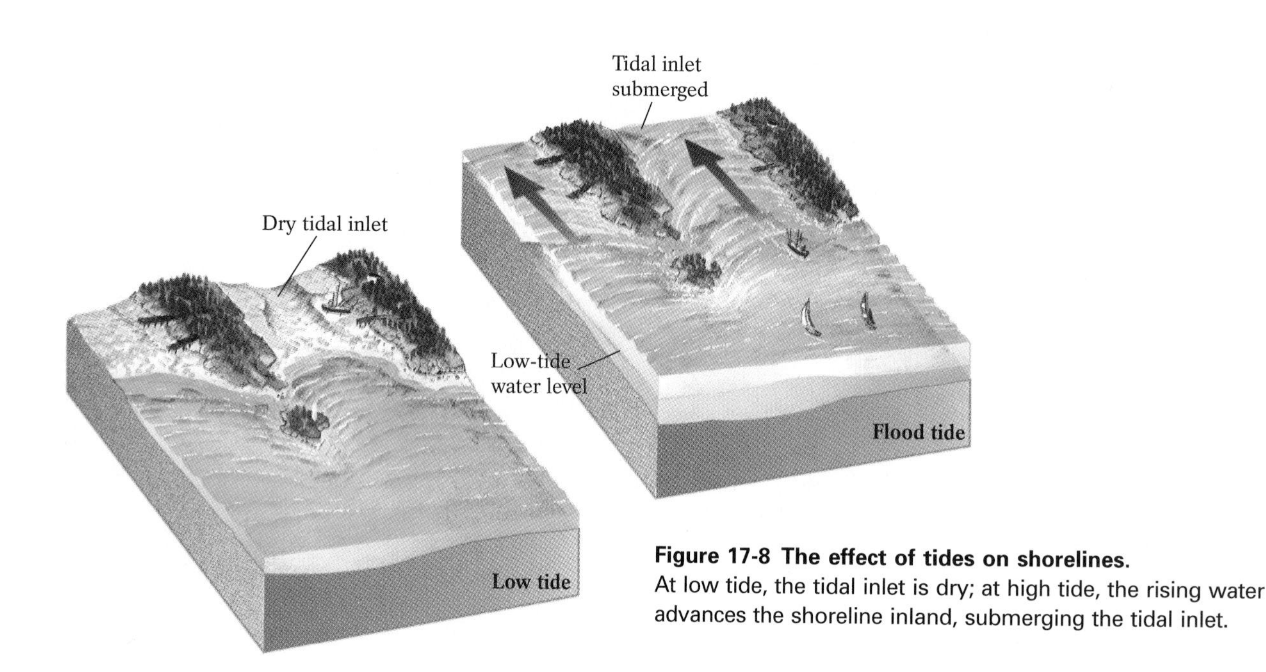

Figure 17-8 The effect of tides on shorelines. At low tide, the tidal inlet is dry; at high tide, the rising water advances the shoreline inland, submerging the tidal inlet.

In addition to the two high tides, two *low tides* occur simultaneously on opposite sides of the Earth. These tides are located halfway between the Earth's closest and farthest points from the Moon, where both the gravitational pull of the Moon and the effect of the Earth's rotation are minimal.

This simple model begins to break down, however, when we stand at an ocean beach with the Moon directly overhead, but note that the tide is out (that is, experience a low tide). Many factors affect the behavior of ocean tides, including the manner in which the shapes of the surrounding landmasses interfere with tidal movements. Each ocean basin acts independently, with its tidal bulge traveling around the basin like a circulating wave. Given the velocity of the Earth's rotation, tidal bulges in oceans cannot move swiftly enough to always keep high tides directly beneath the Moon. In fact, the tides would have to speed along at about 1600

Figure 17-9 The tidal range at the Bay of Fundy, between the Canadian Maritime provinces of New Brunswick and Nova Scotia. **(a)** Eroded columns of rock at high tide. **(b)** The same columns at low tide.

kilometers (1000 miles) per hour to keep pace with the Earth's rotation. Thus a time lag inevitably separates the Moon's position overhead and the location of the high-tide bulge. While a high tide is always occurring somewhere in the basin, it does not necessarily take place at the precise spot where the Moon is directly overhead.

A rising tide elevates the water surface of an ocean, advancing the shoreline landward; a falling tide lowers the water surface of an ocean, causing the shoreline to recede seaward (Fig. 17-8). The vertical difference between local low and high tides is an area's *tidal range.* Tidal ranges vary according to the irregularity of the coastline and the size of the body of water. Rising tides funneled into restricted bays and estuaries, for example, cause a large volume of additional water to pile up in a small area; in this situation, the tidal range is usually quite large. The tidal range at Seattle, Washington, in the relatively narrow constriction of Puget Sound, is from 3 to 4 meters (10–13 feet). The Bay of Fundy, on the eastern coast of Canada, has an even larger tidal range, reaching 20 meters (65 feet) (Fig. 17-9). When the coastline is relatively straight, the additional water carried by a rising tide spreads over a large area, moderating its effect. The resulting tidal range is usually small—as little as 0.5 meter (1.5 feet)—as observed in the Pacific Ocean along Hawai'ian beaches. Lakes and other inland water bodies—especially large ones, such as the Great Lakes—also experience high and low tides, but these bodies usually are too small to have a sizable tidal range.

Tidal fluctuations themselves do not produce major geological effects. When storms strike during a high tide, however, waves penetrate much farther inland than they would normally. In such a case, they may flood beachfront communities and erode beach cliffs ordinarily protected by wide beaches.

Processes That Shape Coasts

The most significant factor shaping shorelines and coasts is the constant battering of waves. Wherever wave energy is concentrated, rock and sediment are eroded; where wave energy is dissipated, the eroded material is deposited. The combination of coastal erosion and deposition tends to straighten an irregular coastline—for example, by removing rock and sediment from projecting coastal cliffs and depositing them in quiet coastal inlets.

Coastal Erosion

An average 14,000 waves strike the exposed rocks and beaches at a given coast every day, some with sufficient force to trigger nearby seismographs. Waves erode principally by forcing water and air under high pressure into rock crevices (Fig. 17-10). A small, 2-meter (6.5-foot)-high wave, for example, exerts nearly 15 metric tons of pressure per square

Figure 17-10 Coastal erosion mechanisms. Crashing surf erodes solid bedrock by forcing water and air into crevices in the rock. Photo: Waves crashing against bedrock at Schoodic Point, Acadia National Park, Maine.

Figure 17-11 Lighthouse at Morris Island, South Carolina. Coastal beach erosion has left this formerly land-based structure partially submerged and stranded off the coast.

meter (more than 2000 pounds per square foot) of rock or sediment surface. Although this force is applied for only a fraction of a second, it is repeated every six or so seconds. Over time, it may dislodge large masses of bedrock or remove a great volume of loose sediment.

Wave-induced erosion is affected not only by wave size and energy, but also by the erodibility of local rocks or sediments. This erodibility depends on the strength of the exposed rock or sediment as well as the extent to which it is fractured. Softer, more fractured bedrock is more readily eroded than harder, less fractured rocks, whether the erosion occurs through the work of streams, winds, or waves. After waves loosen and remove rocks, the rocks may remain in the breaker zone and be hurled repeatedly against the coastline, thus accelerating its erosion.

The slope of the local sea floor also influences the amount of wave energy that strikes a given coast. When the slope is gradual and the water remains shallow for a considerable distance from the shoreline, waves tend to break farther offshore. In such locations, wave bases intersect the sea floor and break well before they reach the shore, and wave energy is dissipated in churning up the sea-floor sediments. When the slope is steep, incoming waves may never touch bottom—instead, they strike the coast with full force. This situation is particularly common where a coastline is irregular and contains **headlands**, cliffs that project into deep water.

Major Coastal Erosion Sites Each year, erosion along about 85% of California's 1100 kilometers (700 miles) of Pacific coastline claims an average of 15 to 75 centimeters (0.5–2.5 feet) of land. Erosion is also claiming vast tracts of land along the Gulf Coast of Texas and Louisiana. In a recent nine-month period, more than 3 meters (10 feet) of the coastline of Chambers County, Texas, was lost to waves and currents in coastal Galveston Bay, southeast of Houston. Along the East Coast, the beaches of North Carolina have been worn back by as much as 20 meters (65 feet) during the past decade, largely by storms like July 1996's Hurricane Bertha. The landmark lighthouse at Cape Hatteras, which was built more than 1500 meters (5000 feet) inland in 1879, now stands, imperiled, at the shore. Because of the beauty and historic value of the lighthouse, efforts are now being made to move it to a safer place. A similar light on Morris Island near Charleston, South Carolina, which once stood on dry land, is now about 500 meters (1700 feet) offshore and surrounded by water (Fig. 17-11)

Large inland lakes are susceptible to coastal erosion as well. In North America, the eastern shores of the Great Lakes, which are often buffeted by powerful late-autumn storm waves, have suffered from rapid coastal erosion in recent years. Highlight 17-1 focuses on the ramifications of such erosion along the eastern shores of Lake Michigan.

Landforms Produced by Coastal Erosion The great energy of waves that strike headlands rapidly wears down their cliffs into a series of distinctive erosional features (Fig. 17-12). Initially, a *wave-cut notch* forms at the base of a cliff. Continued wave action enlarges the notch until it undercuts the cliff, removing the foundation of the overlying rock masses and allowing them to fall into the surf. After enough rock has fallen, the remaining cliff base becomes a platform called a **wave-cut bench.** This bench becomes enlarged as the surf, laden with rock debris, further abrades the cliff face. Eventually, the cliff retreats so far that it no longer remains in the breaker zone. The wide bench then protects the cliff from further erosion to some extent, except during large storms. The bench may also accelerate headland erosion, by enhancing the refraction of incoming waves against the headland's flanks.

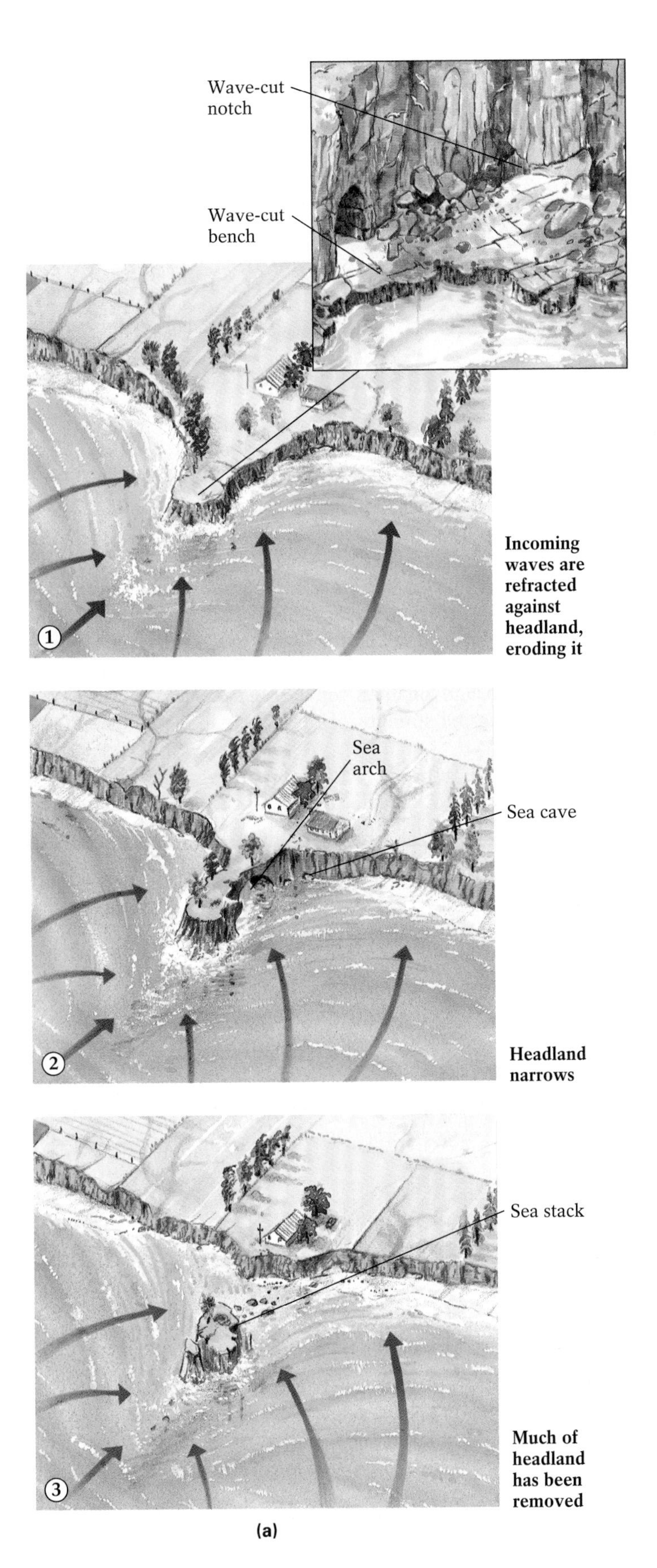

(a)

(b)

(c)

Figure 17-12 The evolution of erosional coastal landforms. **(a)** Wave fronts are refracted against the flanks of headlands, forming wave-cut notches and benches. As the headland narrows, the notches erode to form sea caves. Two sea caves, eroding at both sides of a headland, form a sea arch. Finally, the arch collapses, isolating a sea stack. **(b)** Sea caves on Cape Kildare, Prince Edward Island, Canada. **(c)** A sea arch at Land's End, England.

As waves approach a wave-cut bench, the wave front encounters the shallow bottom and slows down, breaking over the bench. The waves in the adjacent deeper-water bays pass the headland without touching bottom. They can therefore travel more rapidly, refracting toward the flanks of the exposed headland and concentrating their energy there. This action produces a series of notable coastal landforms, which have an especially dramatic appearance when the headland cliffs contain rocks of differing resistance to wave erosion. At first, battering waves erode the cliff rock to form **sea caves.** Further wave action can excavate caves on both sides of the headland until they join, forming a **sea arch.** Over time, continued wave erosion may undermine the supporting foundation of the arch until it finally collapses. Only an isolated remnant of the original headland, a **sea stack,** may then remain.

Coastal Transport and Deposition

Deposition by coastal waves occurs for much the same reason that deposition by wind and rivers does: When the energy needed for its transport drops below a critical level, sediment can no longer be carried. Wave energy may be lost or interrupted for several reasons:

Highlight 17-1 *Lake Michigan's Eroding Shoreline*

Thousands of Michiganders live in towns from Benton Harbor to Ludington on the eastern shore of Lake Michigan, and thousands more from nearby Chicago vacation there. For the past two decades, residents and visitors have watched as erosion has removed lakefront beaches and caused the lake's eastern bluffs to recede (Fig. 1). This beach and cliff erosion has taken place unusually rapidly for two reasons: The area is composed largely of unconsolidated glacial deposits, and powerful winter storm winds push large waves against the lake's eastern shore. Compounding the problem, recent mild winters have reduced the lake ice that protects the bluffs from winter storm waves. Also, several years of unusually high precipitation have produced record high lake levels, so that water now covers part of the shore. In 1964, when the water level of the Great Lakes was 1.5 to 2.0 meters (5–7.5 feet) lower than it is today, wide beaches received and dissipated much of the wave energy that now strikes the bluffs directly. The few relatively dry years that have occurred did not lower lake levels sufficiently to reduce the rate of erosion. Hence the bluffs continue to recede, shrinking about 0.4 meter (1.5 feet) annually. It would take about five years of drought to restore the lakes to their pre-1964 levels.

Human activity has hastened erosion of Lake Michigan's beaches and cliffs. For instance, overdevelopment—building too many homes and other structures along the lakeshore—has removed much of the protective vegetation from bluff tops and eliminated vegetation that stabilizes the sands of energy-absorbing dunes. Meanwhile, disposal of wastewater and sewage in septic fields and on bluff-top farmlands has saturated cliff slopes, reducing the strength of their sediments and promoting mass movement.

What does the future hold for Lake Michigan's scenic bluffs? The warmer, wetter climate of the past two decades may continue; in fact, the drier period of the past century may have been an anomaly. If so, then the lake has merely been returning to its normal level, and accelerated bluff erosion and loss of shoreline property may be long-term phenomena along Lake Michigan's shore.

Figure 1 Erosion at Lake Michigan. Waves generated by westerly winds easily erode the loose glacial sediments of this coastal bluff at St. Joseph, Michigan, on the eastern shore of Lake Michigan.

- Wind velocity and wave force vary seasonally.
- Water depth increases abruptly.
- Waves, refracted from headlands, are channeled into bays.
- A barrier—whether natural or artificial—prevents waves from reaching the shore, interrupting the longshore current.

When waves deliver more sediment to a shore than the amount removed or redistributed by near-shore currents, the excess sediment becomes deposited, most often as a beach.

A **beach** is defined as a dynamic, relatively narrow segment of coast that is washed by waves or tides and covered with sediments of various sizes and compositions. It may consist of only sand, or it may contain coarse gravel or even cobbles. Beach sediment is typically white (from quartz grains) or beige (from shell fragments), but may even be black (from mafic volcanic fragments). More than 90% of beach sediment—an estimated 15 billion cubic meters per year—originates from inland and upland sources and is delivered by coastal streams. A smaller amount derives from erosion of headlands and beach cliffs and an even smaller amount from the offshore surf zone. A beach is naturally maintained when the amount of sediment received by the shore balances the amount exported by longshore currents.

A beach's boundaries stretch from the low-tide line landward, ending where topography changes—for example, at a sea cliff or sand-dune field, or where permanent vegetation begins. A typical beach, illustrated in Figure 17-13, includes a *foreshore,* which extends from the low-tide line to the high-tide line, and a *backshore,* which stretches from the high-tide line to a sea cliff or vegetation line. The steepest part of the foreshore comprises the *beach face,* which receives the swash of breaking waves. The backshore, which swash reaches only during major storms, may contain a *berm,* a horizontal bench or landward-sloping mound of sediment deposited by storm waves. Some beaches contain several parallel berms from different storms, and others have none, depending on the extent of storm activity.

Longshore Transport As we've learned, when waves break at an oblique angle to the shore, the zig-zag motion of their swash and backwash combine to create a longshore current that transports sediment down the coast parallel to the shore. The longshore current moves the sediment both on the beach face, as **beach drift**, and within the surf zone, as **longshore drift.** Normally, sand is carried offshore more effectively in the surf zone than on the beach face. In a storm, however, a grain on a beach may travel down-current as far as 1000 meters (3300 feet). Longshore drift can transport enormous quantities of sand. For example, more than 1.5 million tons of sand per year pass by the California coast at Oxnard, and 750,000 tons of sand per year, including some eroded from granitic cliffs on the Maine coast, pass by Sandy Hook, New Jersey.

Landforms Produced by Coastal Deposition When a longshore current suddenly encounters deeper water, such as where the shoreline is broken by the entrance to a bay, the current is interrupted. Incoming waves do not touch bottom in such regions; because they do not break, they fail to produce the swash and backwash that drive the current. All the sediment being carried by the longshore current is deposited at this point as a **spit,** a finger-like ridge of sand that extends from

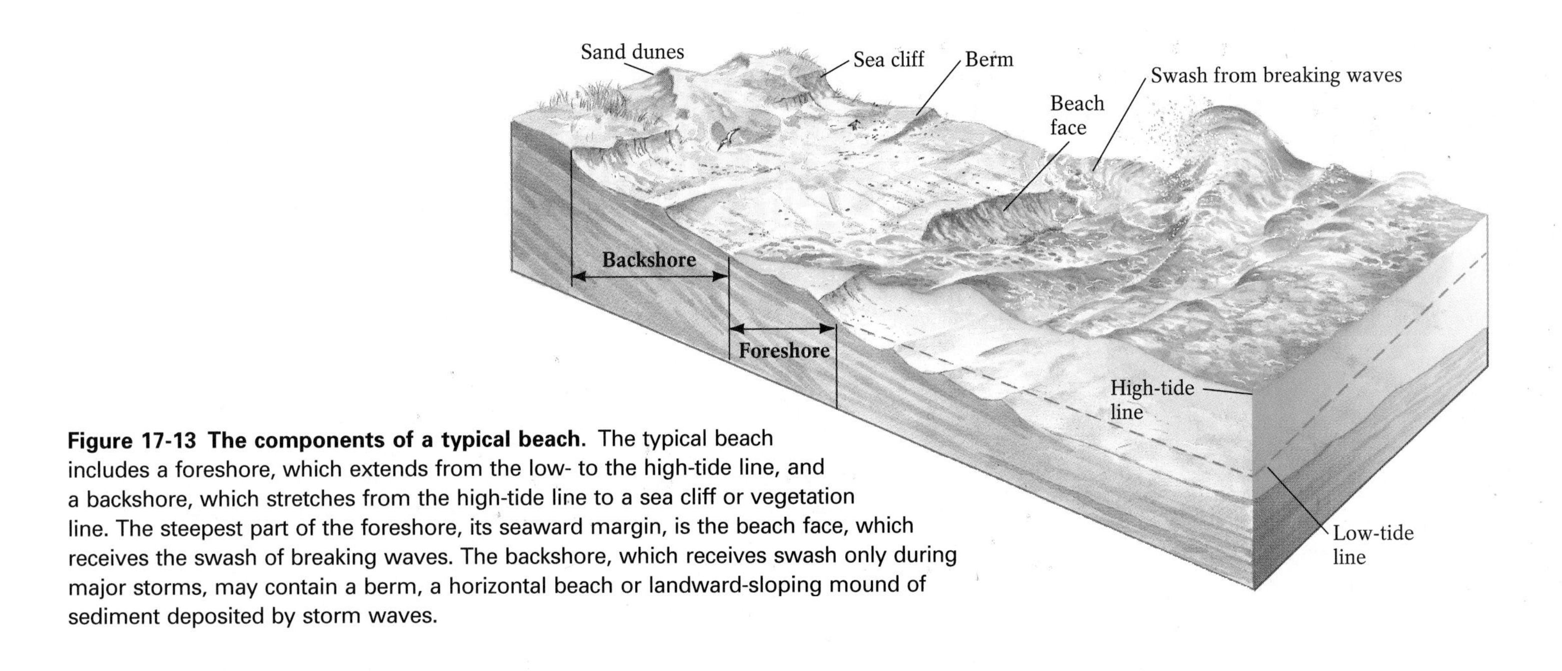

Figure 17-13 The components of a typical beach. The typical beach includes a foreshore, which extends from the low- to the high-tide line, and a backshore, which stretches from the high-tide line to a sea cliff or vegetation line. The steepest part of the foreshore, its seaward margin, is the beach face, which receives the swash of breaking waves. The backshore, which receives swash only during major storms, may contain a berm, a horizontal beach or landward-sloping mound of sediment deposited by storm waves.

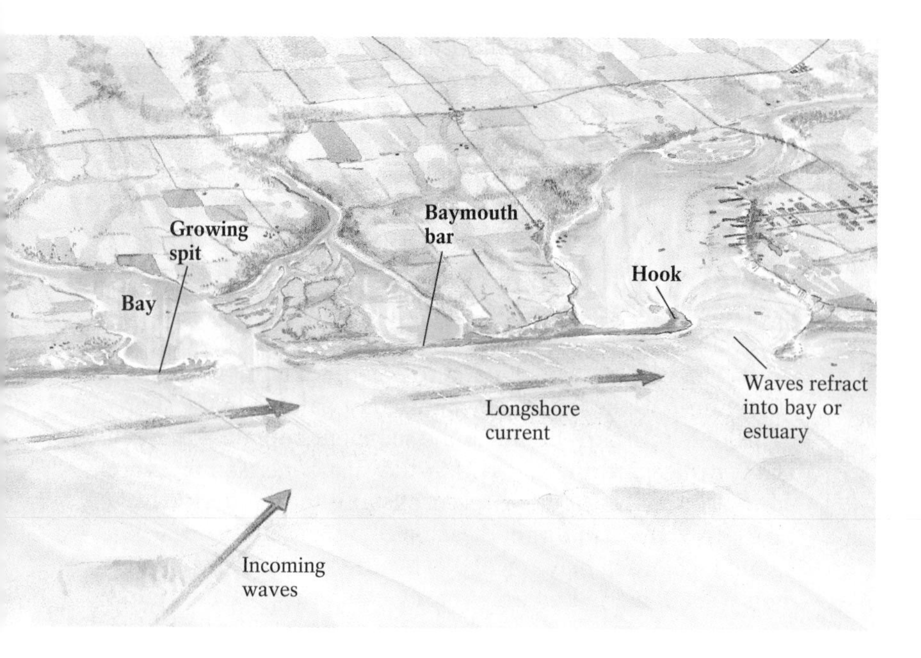

(a)

(b)

Figure 17-14 Deposition of spits, hooks, and baymouth bars. These features typically develop when the sediment being carried by longshore transport settles out where water becomes deeper, such as at the mouth of a bay. **(a)** A spit at Cape Henlopen, Delaware. **(b)** Baymouth bars on the south shore of Martha's Vineyard, Massachusetts.

Figure 17-15 Deposition of a tombolo landward of a sea stack. The sea stack absorbs the wave energy, interrupting the longshore current and causing it to drop its load of sand. Photo: Deposition of tombolo in the lee of a large sea stack, Big Sur, California.

the coast into the open water of the bay entrance. As more sediment accumulates, the spit can grow by tens of meters per year; the growth depends on the supply of sand and the intensity of wave energy. Where incoming waves and the resulting longshore currents approaching a bay are particularly strong, a growing spit becomes curved, forming a **hook.** A spit may ultimately become a **baymouth bar** if it grows completely across a bay entrance (Fig. 17-14).

Coastal deposition also occurs when the waves that drive the longshore current find their path to the shore blocked. A sea stack, for example, may intercept incoming waves, thus interrupting the longshore current. In this case, waves break on the seaward surface of the stack, but quiet water prevails on its landward side. Wave-borne sand is deposited in the wave-free zone on the landward side of the stack, because no wave energy is available to carry it along the coast. The sediment accumulates as a **tombolo,** a sandy landform that grows from the mainland to a stack (Fig. 17-15). Likewise, longshore currents are reduced where the continental shelf has a gentle slope. Waves touch bottom, break, and dissipate much of their energy farther offshore. In such a case, the waves do not remove the sediment loads delivered by streams to the beaches, and the beaches tend to widen.

Human-Induced Coastal Deposition Some human-made structures can disrupt the natural balance between the amount of sediment delivered to the shore and the amount removed by the longshore current, causing beaches to grow in some places and shrink in others. **Breakwaters,** walls designed to intercept incoming waves, are built to create quiet, wave-free zones that curtail coastal erosion and protect boats in harborages. About 70 years ago, one such breakwater was constructed offshore of Santa Barbara, California, to protect private boats from being battered by Pacific waves. Predictably, the structure interrupted the longshore current, causing sediment to be deposited seaward of the beach (Fig. 17-16). After 30 years, the beach had widened by hundreds of meters, clogging the harborage.

Groins are shore-protection structures that jut out perpendicular to the shoreline so as to interrupt longshore drift and trap sand, thereby restoring an eroding beach. They are typically built where a wide sandy beach, such as Miami Beach, is vital to a community's economic life. **Jetties** are structures, typically built in pairs, that extend the banks of a stream channel or tidal outlet beyond the coastline. They direct and confine channel flow to keep channel sediment moving, thus preventing sediment from filling the channel. Both groins and jetties broaden upcurrent beaches by trapping sand there (Fig. 17-17). The drawback to these types of beach-enhancing structures is that their interruption of the longshore current prevents sediment from reaching and replenishing downcurrent beaches, which subsequently become eroded.

(a) (b)

Figure 17-16 **(a)** Santa Barbara Harbor in 1931. **(b)** Santa Barbara Harbor in 1977. The breakwater at Santa Barbara was built in the 1920s to provide safe harborage for pleasure boats. It interrupted the longshore current, causing sand to be deposited behind it and on the seaward edge of the beach. This action extended the beach but closed off the passage from the marina to the sea. Today, sand must be regularly dredged from behind the breakwater and pumped down-current to keep the passage open.

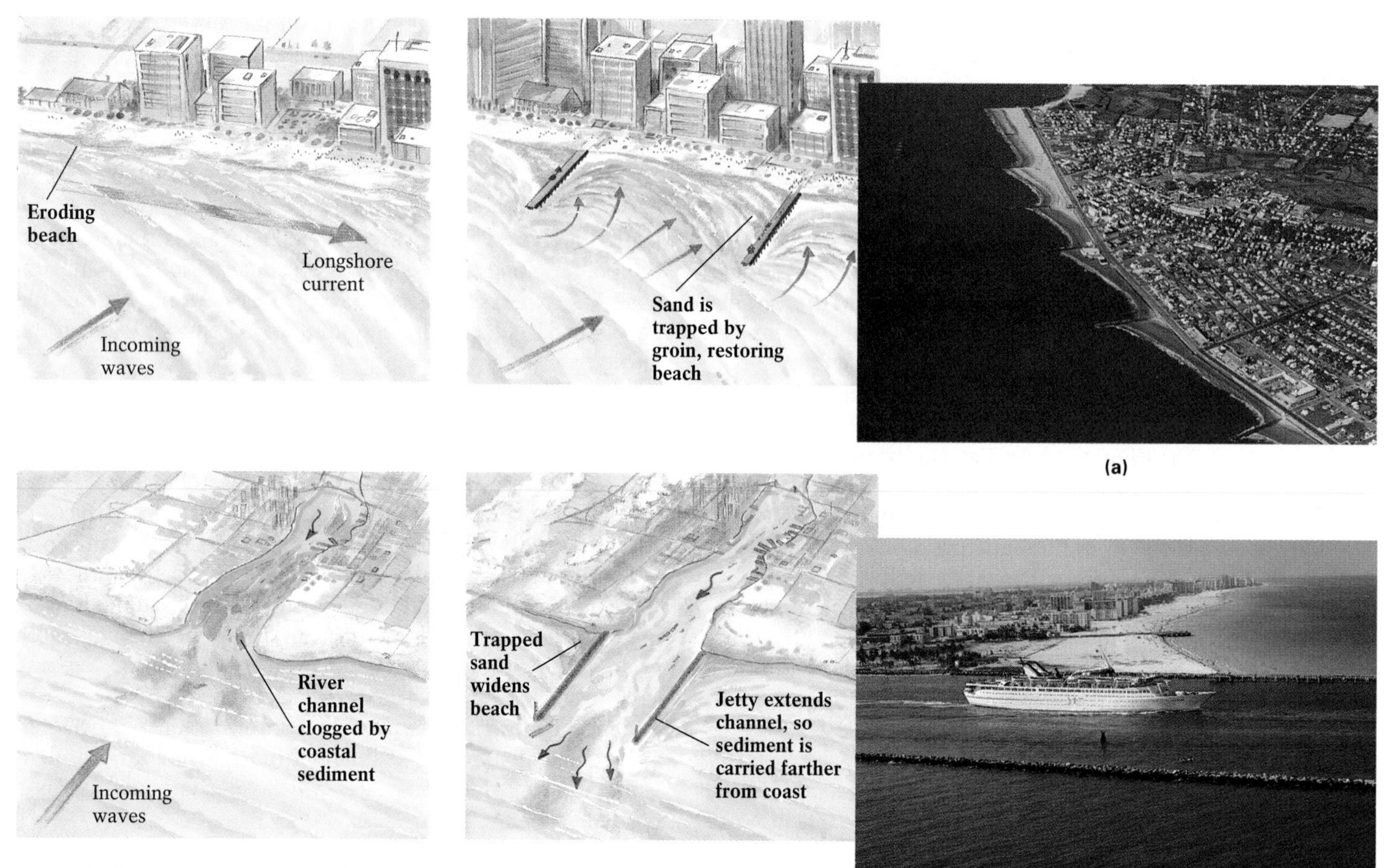

Figure 17-17 Groins and jetties. By interrupting longshore currents, groins and jetties cause longshore drift to accumulate on their upcurrent side, broadening beaches there. **(a)** Groins off Cape May, New Jersey. **(b)** A cruise ship in a jetty in Miami Beach, Florida.

Types of Coasts

Every coast is shaped by a combination of erosional, depositional, and other processes that is distinctive to its geological setting.

Primary coasts are shaped by nonmarine processes (Fig. 17-18). Glacial erosion, for example, produced such primary coasts as the fjords of Alaska and British Columbia; glacial deposition of terminal and recessional moraines shaped the northeast Atlantic from Cape Cod, Massachusetts, to the southwestern edge of Long Island, New York. Stream deposition produces primary coasts such as the Gulf Coast of Louisiana, where stream deltas extend into the marine environment. Primary coasts can also be shaped by the interaction of biological processes with geological processes, such as at carbonate-reef coasts, or by fluctuating sea levels.

Secondary coasts are shaped predominantly by ongoing marine erosion or deposition, such as the processes described in the previous section. Secondary erosional coasts contain cliffed headlands, wave-cut terraces, and an assortment of sea caves, stacks, and arches. Secondary depositional coasts include beaches, spits, hooks, and tombolos. Some coastal features, such as offshore barrier islands, may form as the result of either primary or secondary coastal processes.

Barrier Islands

Barrier islands are nearly continuous ridges of sand, located parallel to the main coast but separated from it by a bay or lagoon. These types of islands can rise as much as 6 meters (20 feet) above sea level, and are 10 to 100 kilometers (6–60 miles) long and 1 to 5 kilometers (0.6–3 miles) wide. They are usually composed of a relatively narrow beach, perhaps 50 meters (165 feet) wide, and a broader zone of inland dunes that makes up most of the island.

Barrier islands are the most common North American coastal feature, lining the East Coast for 1300 kilometers (850

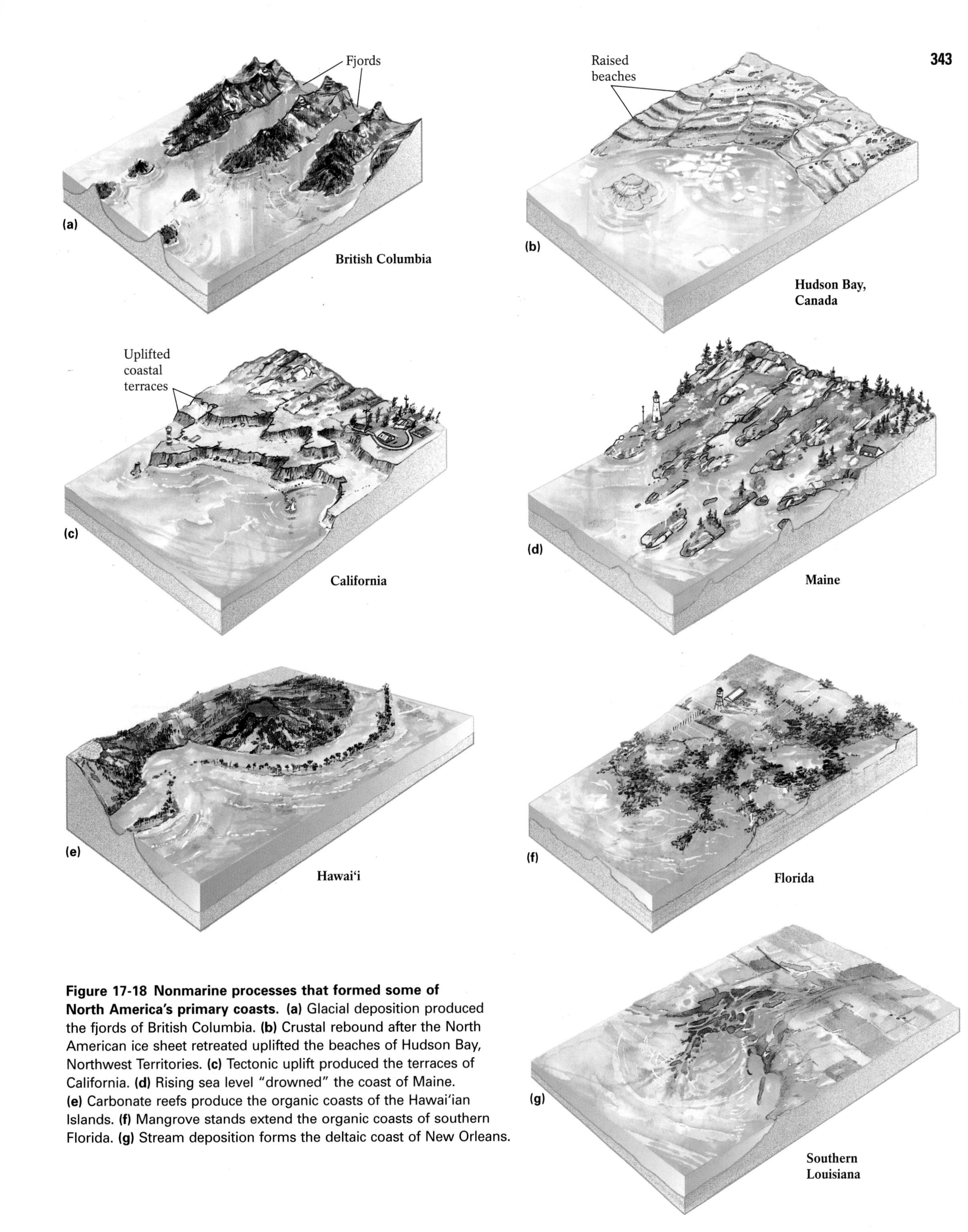

Figure 17-18 Nonmarine processes that formed some of North America's primary coasts. **(a)** Glacial deposition produced the fjords of British Columbia. **(b)** Crustal rebound after the North American ice sheet retreated uplifted the beaches of Hudson Bay, Northwest Territories. **(c)** Tectonic uplift produced the terraces of California. **(d)** Rising sea level "drowned" the coast of Maine. **(e)** Carbonate reefs produce the organic coasts of the Hawai'ian Islands. **(f)** Mangrove stands extend the organic coasts of southern Florida. **(g)** Stream deposition forms the deltaic coast of New Orleans.

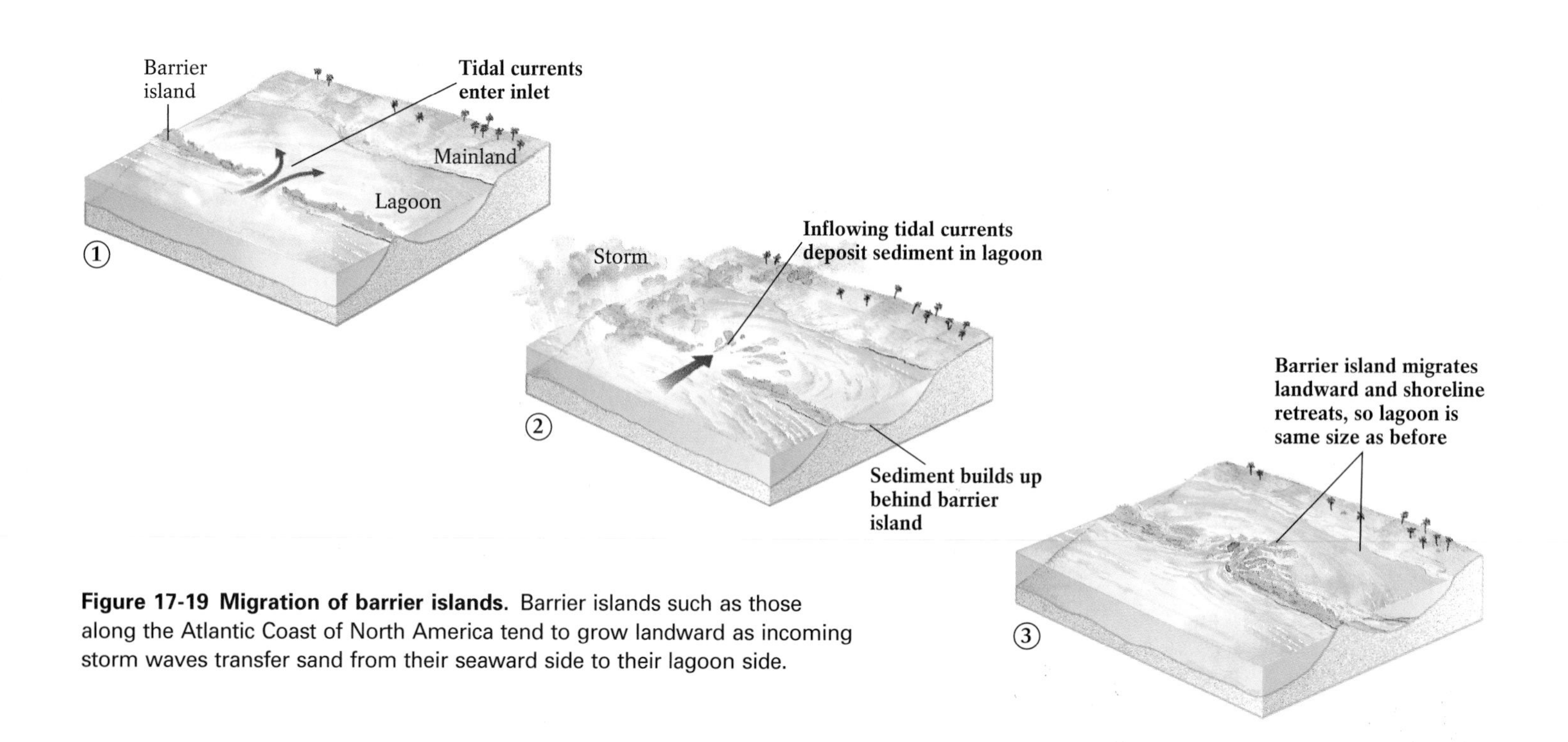

Figure 17-19 Migration of barrier islands. Barrier islands such as those along the Atlantic Coast of North America tend to grow landward as incoming storm waves transfer sand from their seaward side to their lagoon side.

miles) from eastern Long Island, New York, south to Florida, and then continuing for another 1300 kilometers along the Gulf Coast to eastern Texas. This string of 295 islands is interrupted only by occasional tidal inlets or by the flow of major streams such as the Hudson River at New York Harbor or the Delaware River at Delaware Bay.

Some barrier islands are thought to have formed when elongated spits were breached by tidal currents or powerful storm waves to create separate islands. Others probably originated when waves breaking offshore above a broad, gently sloping continental shelf moved sand from the bottom and deposited it as a sand bar, which gradually grew into an island. Still other barrier islands may be remnants of a sand-dune system that bordered the continent during the last period of worldwide glacial expansion, when sea level was as much as 140 meters (460 feet) lower than it is today. As the glaciers melted, the rising sea surrounded and isolated the dunes, forming a primary coast.

Whatever their origin, once barrier islands begin to form, erosional and depositional processes combine to shape and move them. Longshore drift gradually redistributes their sand down-current, causing the islands to migrate parallel to the coast. At the same time, onshore winds, inflowing tidal currents, and storm waves transfer sand from the seaward side of these islands to the lagoon side, causing the islands to migrate landward (Fig. 17-19). The landward migration rate for barrier islands along the Atlantic Coast ranges from 0.5 to 2 meters (1.5–6.5 feet) per year, but increases considerably during extraordinary storms. The moving island rarely overruns the mainland coast, because the shoreline behind the island retreats at the same time.

North America's long chain of barrier islands is immeasurably valuable as the first line of defense against storm waves in the hurricane-prone Southeast and the Gulf Coast. Ironically, however, more than one-quarter of the United States' barrier islands have been transformed into vacation resorts, overbuilt with costly condominiums, hotels, casinos, and private residences. Such development inevitably eradicates much of the islands' protective frontal dunes, hastening the processes of erosion and island migration. As a result, the great financial investments made in developing such properties often pale when compared with the price of protecting them against natural island migration. The protective role played by barrier islands, the processes that shape them, and the costs of human interference with natural coastal processes are especially clear on North Carolina's Outer Banks, as discussed in Highlight 17-2.

Organic Coasts

Some coasts—and particularly those associated with mangrove swamps or coral reefs—form when erosion and deposition act in conjunction with biological processes. Mangrove trees, which live in standing tidal water in tropical climates, grow from an extensive web of long roots that rise above the water. This root system dissipates much of the energy of waves that enter the mangrove swamp, creating a quiet-water environment. It also traps fine sediment, creating *tidal mud flats.* Hardy mangrove seedlings take root in the growing mud flats, enabling the mangrove forest to grow seaward and thus extend the coast outward. In North America, mangrove swamps are expanding along the southern tip of the Florida mainland

Highlight 17-2 North Carolina's Outer Banks

The Outer Banks, stretching along the eastern coast of North Carolina, offer a wonderful outdoor laboratory for assessing the effects of human interference with an efficient natural system. This string of barrier islands includes both a developed coast, the Cape Hatteras National Seashore, and an undeveloped coast, the pristine Cape Lookout National Seashore (Fig. 1).

The interior of Cape Hatteras has been stabilized for more than 50 years by a 10-meter (35-foot)-high, human-made line of dunes that protects State Highway 12 from flooding by winter storms and periodic hurricanes. Built beyond the reach of damaging salt spray and wind- and water-borne sand, the artificial dunes are lushly vegetated. During storms, however, the dunes deflect wave energy, redirecting it toward the seaward beaches and increasing erosion there. Since the dunes were built, erosion has narrowed the island's 200-meter (about 650-foot)-wide beaches to only 30 meters (about 100 feet).

The inland side of the Outer Banks along Pamlico Sound is also being eroded, largely because the high artificial dunes keep away the sand that would otherwise wash over lower natural dunes. This measure interrupts the natural tendency of a barrier island to migrate landward.

Disruption of the natural barrier-island system along the Hatteras section has forced North Carolina to compensate by replenishing its beaches with imported sand, an expensive program that has so far failed to halt erosion. If erosion of that segment of the Outer Banks continues at its present rate, little of Cape Hatteras may be left in the future.

At Cape Lookout, on the other hand, the banks are naturally adjusted to the area's periodic storms. Broad beaches and low dunes absorb erosive energy from storm waves. In the natural system, the vegetation growing along salt marshes has evolved to be somewhat resistant to salt spray and flooding. It traps landward-migrating sand, replenishing the natural interior dunes. When a powerful storm hits, the island grows and migrates landward, instead of eroding.

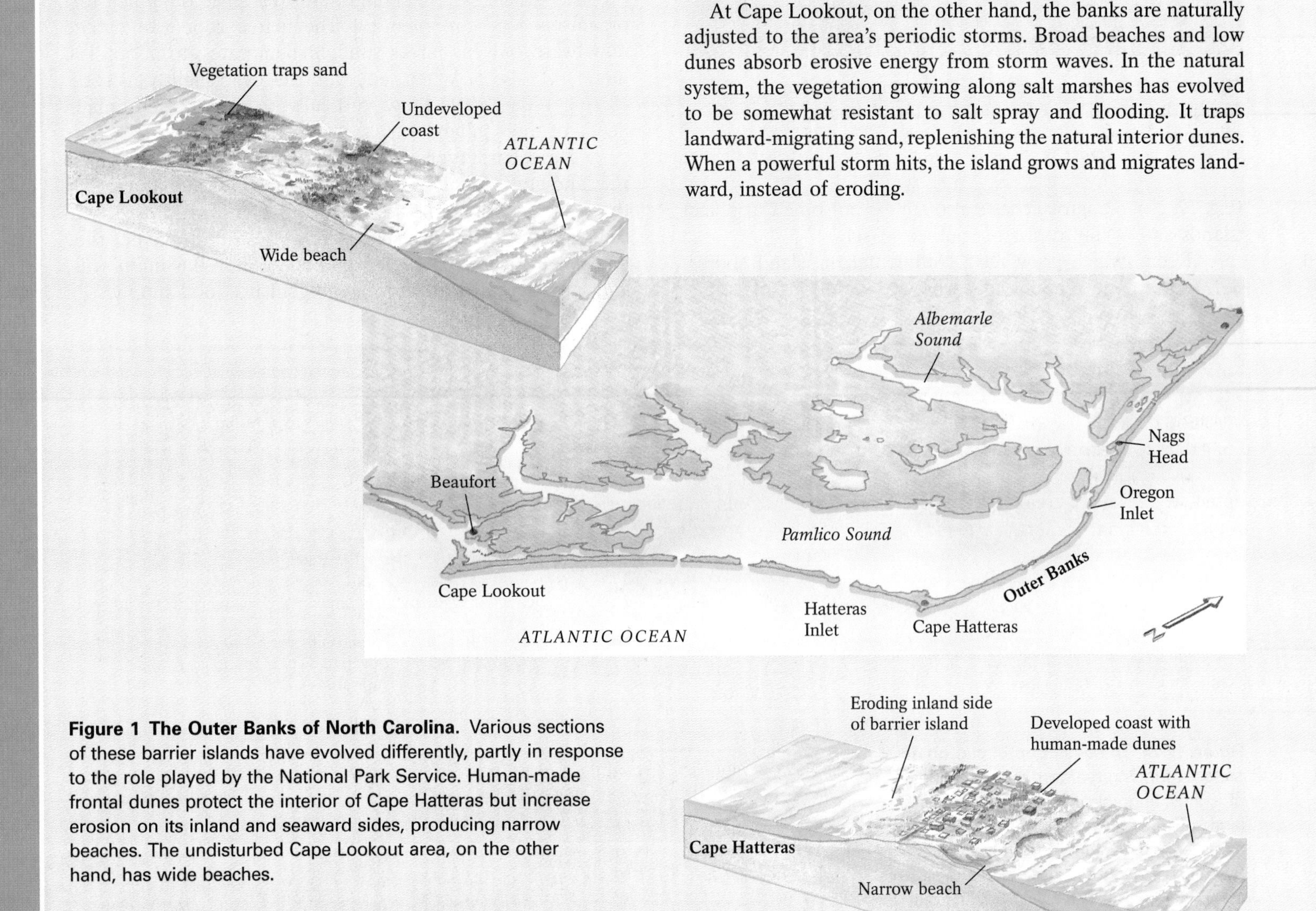

Figure 1 The Outer Banks of North Carolina. Various sections of these barrier islands have evolved differently, partly in response to the role played by the National Park Service. Human-made frontal dunes protect the interior of Cape Hatteras but increase erosion on its inland and seaward sides, producing narrow beaches. The undisturbed Cape Lookout area, on the other hand, has wide beaches.

Figure 17-20 A typical mangrove coast in the Florida Everglades. Mangrove coasts form as the protruding roots of mangrove trees trap sediment.

(Fig. 17-20); mangrove coasts also appear on some Caribbean islands and in the tropics of southeast Asia.

Reefs, which grow near continental or island shores, derive from the remains of carbonate secretions of corals, algae, and sponges. Because reef-building organisms flourish only in regions with warm near-surface water, most reefs are found in the tropics, within the geographic band from about latitude 30°N to 25°S. Clear water is a necessary condition for their formation, because most reef organisms are filter-feeders, taking in a large quantity of water and filtering out microscopic plankton to eat. These creatures could not survive if they ingested the suspended particles in turbid water. Also, reef organisms require moderately salty water, so reefs are unlikely to form where a large influx of freshwater dilutes the local seawater or where evaporation concentrates the seawater, making it too salty.

The active portion of a reef is generally located near the sea surface, where sunlight and the algae on which reef organisms feed are plentiful. The active zone can extend downward only as far as sunlight penetrates, to about 100 meters (330 feet). When sea level rises even slightly, as is occurring today, reefs grow vertically to satisfy the sunlight needs of their inhabitants.

Three basic types of carbonate reefs exist: fringing, barrier, and atoll (Fig. 17-21). **Fringing reefs** rest directly against the coast of a landmass, such as along the margins of volcanic islands in the Caribbean and the South Pacific. They are usually from 0.5 to 1.0 kilometer (0.3–0.6 mile) wide and grow seaward toward their organisms' food supply.

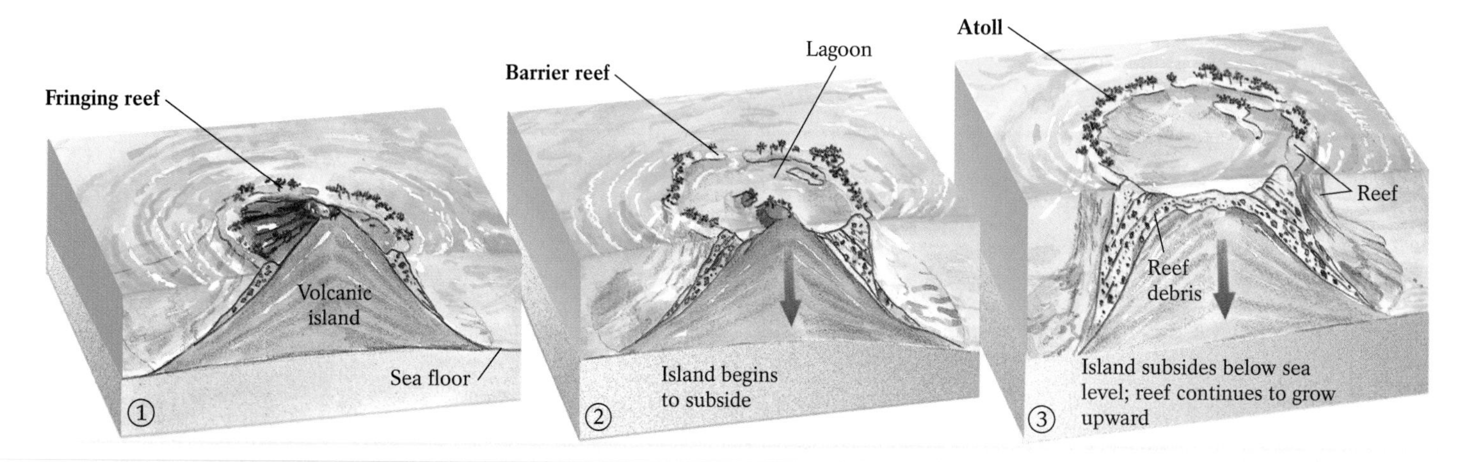

(a)

(b)

Figure 17-21 The evolution of carbonate reefs. A fringing reef forms on the side of a subsiding volcanic island. As the island subsides, the fringing reef becomes a barrier reef, separated from the volcano by a circular lagoon. After the island sinks completely below sea level, only the reef remains visible—as an atoll—because it continues to grow toward the water's surface. **(a)** Great Barrier Reef, Australia. **(b)** Wake Island, a coral atoll in the Pacific Ocean.

Barrier reefs are built on the local continental shelf, and a wide lagoon separates them from the mainland. Incoming waves break against the reef, which protects the mainland coast from wave erosion. Some barrier reefs originate as a fringing reef around a subsiding landmass, such as a volcanic island. As the island subsides and much of its area descends below sea level, the reef grows vertically, keeping the surface of its active zone always within a few meters of sea level.

Atolls, the most common type of reefs, are circular structures that extend from very great depth to the sea surface and enclose relatively shallow lagoons. They consist of barrier reefs that once surrounded oceanic islands that subsequently subsided completely below sea level.

Coastal Evolution and Plate Tectonics

The different characteristics of North America's East, West, and Gulf coasts are largely attributable to their distinctive plate tectonic settings, although local differences also occur because of such factors as past glaciation, sea-level fluctuations, and even human activity. The plate setting strongly determines the steepness of the local continental shelf and, consequently, the behavior of waves and the prospects for coastal erosion and deposition. Every major type of plate boundary develops its own distinct coastal style.

Divergent plate boundaries that have recently rifted, such as those found in the Gulf of California between Baja California and the Mexican mainland, have steep continental shelves that are vulnerable to head-on attack by waves. As a result, these regions feature recently formed erosional landforms, such as wave-cut benches and sea stacks. Passive continental margins, which long ago diverged from an active spreading center and subsided, are characterized by broad, gently sloping continental shelves. Incoming waves break offshore on the shallow shelf, creating such depositional features as spits, beaches, and barrier islands. Most of the East Coast of North America shows coastal evolution at a passive plate margin.

Along convergent plate boundaries, coasts generally have narrow, steep continental shelves, and tectonic uplift continuously produces sea cliffs along the shoreline. Without offshore shallows to dissipate wave energy, incoming waves strike the coast directly, forming such erosional features as sea stacks and arches and wave-cut benches. The Northwest Coast of North America exhibits coastal evolution at a convergent plate margin.

The East Coast of North America

North America's East Coast is bounded by a broad, tectonically quiet continental shelf that is slowly subsiding under the weight of continent-derived, river-transported sediment. In addition, the entire coast is slowly submerging because of the global sea-level rise of the past century. Variations in the East's coastal topography primarily reflect the region's many types of bedrock and local differences in its glacial history. Glacial erosion carved deep fjords and impressive U-shaped valleys into the soft sedimentary rocks along the northeastern coast of Canada and on the eastern coasts of Baffin Island and the Labrador Peninsula. To the south, glacial erosion of exposed granite plutons produced the low, rocky coasts of the Canadian Maritime provinces and of Maine and New Hampshire. As the local ice sheets melted, crustal rebound lifted (and continues to lift) these coasts and their continental shelves, enabling oncoming waves to shape these coasts by erosion.

The coast from Boston to Long Island, New York, a zone of glacial deposition, consists primarily of unconsolidated sediments. Vulnerable to attack by crashing surf, these deposits have been eroded and then carried by longshore transport, only to be redeposited farther down the coast as spits, baymouth bars, barrier islands, and beaches (Fig. 17-22). South of the glacial terminus in New York City to the Florida Keys, the East Coast boasts a number of different features: in the Middle Atlantic states, large bays such as Delaware Bay, Chesapeake Bay, and Pamlico Sound, which formed from drowned river valleys that were flooded by rising sea levels at the end of the last great glaciation; barrier islands and inland lagoons and marshes that parallel the coastline for more than 1000 kilometers (about 600 miles); and the reefs of tropical southern Florida.

Figure 17-22 East Coast shorelines. Depositional shorelines, such as this one at Curritick Sound, Virginia, are characterized by wide sandy beaches and offshore islands.

Figure 17-23 West Coast shorelines. The stepped terraces at Palos Verdes Hills in southern California are typical of a West Coast shoreline uplifted by plate convergence. Each terrace is a wave-cut bench formed when that terrace was at sea level.

The West Coast of North America

The West Coast from Alaska through California completely differs from the East Coast of North America. Along much of its length, converging plates and tectonic uplift have produced a steep, narrow continental shelf and rising coastal mountains. The stretch of coast from Alaska to northern Washington was extensively glaciated during recent ice expansions. Consequently, it contains primary coasts with major westward-draining fjords and U-shaped valleys. South of the glacial terminus, from southern Washington through Oregon and into southern California, wave erosion has cut into the uplifted terraces, shaping headlands into a near-continuous chain of cliffs and rugged offshore islands (Fig. 17-23). Eroded sediment is deposited as narrow beaches in protected bays.

Protecting Coasts and Coastal Environments

Human efforts to prevent coastal erosion usually involve building structures, such as breakwaters, that deflect or weaken waves to prevent them from striking the coast with great force. *Riprap,* a heap of angular boulders piled along a shoreline, reduces coastal erosion by absorbing the brunt of wave energy (Fig. 17-24a). Eventually—after several years or even decades—breaking waves remove the riprap, but riprap is relatively easy and inexpensive to replace. *Seawalls* are sturdy, longer-lasting structures, generally built along a shoreline, that repel waves seaward to divert some of their energy (Fig. 17-24b). They are, however, expensive to build and maintain.

Human-made sand dunes can provide a first line of defense against the encroaching sea by absorbing the energy of waves. Unfortunately, they tend to migrate with the wind unless they are stabilized in some way. On the Outer Banks of North Carolina, for example, officials have planted hardy vegetation whose roots can hold the soil against wind erosion, and sand has been added to create a higher and continuous dune line. On the southern shore of New York's Long Island, driftwood and old tires are piled up in some places to fortify the area's natural dunes.

Instead of trying to prevent erosion, some communities have attempted to restore shrinking beaches by replacing the sand lost to erosion. The sand for such beach nourishment is typically dredged from nearby lagoons, inland sand dunes, and offshore sand bars. In one successful beach nourishment project, along Mississippi's Gulf Coast at Biloxi and Gulfport, sand was imported from offshore sites to create the world's longest artificial beach (30 kilometers [20 miles] long).

Individual human beings have little power to control the geological forces that affect our coastlines. Our governments, however, can often protect citizens from the consequences of inevitable coastal change. A few states offer

(a) (b)

Figure 17-24 Beach-protection structures. (a) Riprap at Carlsbad, California, about 30 kilometers (20 miles) north of San Diego. **(b)** Seawall along the Gulf Coast of Louisiana.

low-interest loans to property owners who are willing to relocate their ocean-front homes to less vulnerable sites. Twenty-nine of the 30 coastal and Great Lakes states operate coastal-zone management programs. Many states prohibit construction of permanent coastal structures that would be vulnerable to swamping by rising sea levels. In addition, they may ban construction of barriers that interfere with natural longshore currents.

Environmental organizations can also prevent damage from coastal evolution, sometimes by purchasing environmentally sensitive coastal property to preserve it and protect it. For example, the Nature Conservancy of Washington, D.C., has recently bought long parcels of the barrier islands off Virginia to prevent future overdevelopment of these fragile coastal environments.

Chapter Summary

Shorelines are the boundaries between bodies of water and dry land. Landward of ocean shorelines are **coasts,** which extend inland until they encounter a different geographical setting. Coasts are shaped by the sediments delivered to them, primarily through the work of streams, and by the waves that erode and redeposit those sediments.

Waves arise because of friction between wind and the sea surface. Wind speed, duration, direction, and the extent of uninterrupted water surface across which the wind is blowing determine the height, velocity, and wavelength of the subsequent waves. A wave moving across deep, open ocean exhibits **oscillatory motion,** in which water particles move in a rolling circular path. When a wave reaches shallow water, the sea floor slows the base of the wave, the wave breaks, and its water moves with **translatory motion** toward shore as surf.

When a wave approaches the coast from an angle, the first part of the wave front to encounter shallow water slows, causing the rest of the wave front to swing around until it lies nearly parallel to the coastline; this process is called **wave refraction.** Nevertheless, virtually all waves strike the coast at some small angle; their water moves up the beach at that angle, as swash, but takes a perpendicular path, as backwash, back to the ocean. This zig-zag motion causes the water to travel down the coast as a **longshore current.** Longshore currents transport most coastal sediment. **Rip currents** occur when water that has accumulated in the surf zone moves seaward.

Tides are the daily rises and falls of the ocean surface. They are primarily caused by the gravitational pull on the ocean surface by the Moon and from the force created as the Earth spins on its axis.

Most coastal erosion occurs when waves strike a coast with sufficient energy to remove loose materials. When the water immediately offshore is deep, the waves may crash with their full force against coastal **headlands,** cliffs that jut seaward. Erosion of a headland can produce a **wave-cut bench,** a shelf of rock that remains after waves have removed a cliff.

When waves refract against the side of a headland, they may erode **sea caves.** If sea caves on opposite sides of a headland join, the remaining rock forms a **sea arch.** Continued wave erosion can cause the foundation of a sea arch to collapse, leaving only an isolated remnant, a **sea stack.**

A **beach** is a sandy strip of land at the edge of a coast. Sediment transported along the beach by swash and backwash moves by **beach drift;** sediment transported offshore within the surf zone moves by **longshore drift.** Landforms produced from the deposition of this sediment include **spits** (finger-like ridges of sand that extend from the coast into a bay entrance), **hooks** (spits that have grown to form a curved end), and **baymouth bars** (areas where spits have grown completely across a bay entrance). A **tombolo** is a sandy strip of land that stretches from the mainland to a sea stack. Human-built structures that intercept waves or longshore currents, causing deposition, include the following: **breakwaters,** walls designed to intercept incoming waves; **groins,** structures that jut out perpendicular to the shoreline to interrupt longshore drift and trap sand; and **jetties,** structures that extend the banks of a stream channel or tidal outlet to prevent sediment from filling the channel.

Primary coasts are generated by nonmarine processes such as glaciation, stream deposition, or organic processes. **Secondary coasts** are products of marine erosion and deposition. **Barrier islands** are nearly continuous ridges of sand, located parallel to the main coast but separated from it by a bay or lagoon; they may be either secondary or primary coasts. In the latter case, they may consist of the remnants of a drowned, ancient sand-dune system.

Organic coasts include those formed by mangrove tree roots in swamps and those formed by carbonate reefs. Carbonate reefs, which are commonly found along the margins of tropical volcanic islands, include **fringing reefs** (built directly against the coast of a landmass), **barrier reefs** (built on the local continental shelf and separated from the mainland by a wide lagoon), and **atolls** (circular structures that extend from a very great depth to the sea surface and enclose relatively shallow lagoons).

Plate tectonic settings influence the nature of coastlines. Recently diverged plate margins and convergent plate boundaries have steep continental shelves, and their coastlines include erosional landforms such as wave-cut benches and sea stacks. Passive continental margins, on the other hand, have broad continental shelves, and their coastlines typically contain depositional features such as beaches and spits.

Human efforts to prevent coastal erosion usually involve building structures, such as breakwaters, that deflect or weaken waves, or piling riprap or constructing human-made sand dunes along a shoreline to absorb the brunt of wave energy. State governments may prohibit development along vulnerable coastlines, and may offer low-interest loans to property owners who are willing to relocate their ocean-front homes to less vulnerable sites.

Key Terms

coast (p. 329)
shoreline (p. 329)
oscillatory motion (p. 330)
translatory motion (p. 331)
wave refraction (p. 331)
longshore currents (p. 331)
rip currents (p. 333)
tides (p. 333)
headlands (p. 336)
wave-cut bench (p. 336)
sea caves (p. 338)
sea arch (p. 338)
sea stack (p. 338)
beach (p. 339)
beach drift (p. 339)
longshore drift (p. 339)
spit (p. 339)
hook (p. 341)
baymouth bar (p. 341)
tombolo (p. 341)
breakwaters (p. 341)
groins (p. 341)
jetties (p. 341)
primary coasts (p. 342)
secondary coasts (p. 342)
barrier islands (p. 342)
fringing reefs (p. 346)
barrier reefs (p. 347)
atolls (p. 347)

Questions for Review

1. Draw a simple diagram showing wave crests, wave troughs, wave height, and wave length.

2. Briefly describe why waves "break." How does water movement change in the breaker zone?

3. Why are most wave fronts nearly parallel to the shoreline on arrival?

4. How does a longshore current develop?

5. Explain why most of the Earth's coasts experience two high tides and two low tides each day.

6. What three principal factors control the rate of coastal erosion?

7. Draw a sketch showing the main components of a beach. Using arrows, indicate the source of most beach sediment and the methods by which it is most commonly removed.

8. Describe how the following depositional coastal landforms develop: spit, hook, baymouth bar, tombolo.

9. Discuss how breakwaters, groins, and jetties interrupt longshore transport and affect beach size.

10. What is the difference between a primary coast and a secondary coast? Explain how barrier islands may originate as either primary or secondary coasts.

For Further Thought

1. Why do you suppose surfing is more popular on the western than on the eastern coast of North America? Speculate about the prospects for surfing on the western coast of Europe or Africa.

2. Suppose you acquired some valuable beachfront property in Oregon. How would you protect this land from coastal erosion? (*Note:* You have unlimited financial resources.) In your plan, try to minimize the negative secondary effects that follow most cases of human interference with natural coastal systems.

3. What would be the effect on California's beaches if all dams were removed from its coast-bound streams?

4. How would the East Coast of North America be affected if widespread subduction resumed there?

5. How might the coasts of North America change if Earth entered another period of worldwide glacial expansion?

18

Human Use of the Earth's Resources

Until recently, the Earth's resources, such as those being mined in Figure 18-1, were believed to be unlimited. Yet, today, we face serious shortages of many essential materials. For example, scientists believe that the world's recoverable supply of crude oil, from which we obtain gasoline, may last only another 50 to 100 years at the current rate of use. How have we managed to exhaust our stores so quickly? Can we compensate for these losses?

To find the answers to these questions, we must understand the growth rate of the world's population, the amount of natural resources used by each individual, and the search for alternative resources. In the United States alone, each person directly or indirectly uses about 10,000 kilograms (22,000 pounds) of raw materials each year; most of this amount consists of stone and cement for the construction of roads and buildings, but it also includes about 500 kilograms (1100 pounds) of steel, 25 kilograms (55 pounds) of aluminum, and 200 kilograms (440 pounds) of industrial salt (mostly for cold-weather road maintenance). Each American also uses nearly 3800 liters (1000 gallons) of oil per year. Collectively, Americans account for approximately 30% of world oil consumption. The United States, which possesses only 6% of the world's population, uses nearly 30% of its minerals, metals, and energy. One American uses as many as 30 times as much material and energy as a person in an emerging nation.

Resource consumption worldwide is rising at an accelerating rate as world population increases (it is now approaching 6 billion—three times what it was in 1920—and expected to double by 2040) and people everywhere strive to achieve the benefits of technological development. Unless we identify new supplies of depleted resources or find substitutes for them, and until we manage industrial development in ways that limit resource depletion, impending shortages will force people everywhere to change their ways of life.

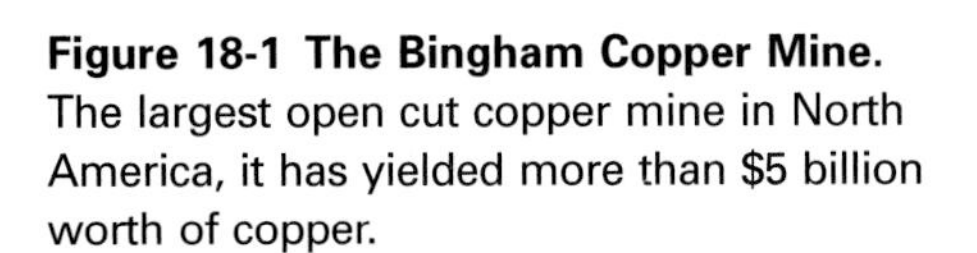

Figure 18-1 The Bingham Copper Mine. The largest open cut copper mine in North America, it has yielded more than $5 billion worth of copper.

Reserves and Resources

Reserves are natural resources that have been discovered and can be exploited profitably with existing technology and under prevailing economic conditions. We know the locations of reserves, and we have the means to extract them; most importantly, their economic value in the marketplace exceeds the cost of their extraction. An **ore** is a mineral deposit that can be mined for a profit; this is an economic, not a geological, term.

Resources, in contrast, are deposits that we know or believe to exist, but that are not exploitable today, whether for technological, economic, or political reasons. Estimates of resources beneath the Earth's surface are derived from exploratory drilling, geophysical modeling, and extrapolation from known reserves. World oil reserves, for example, are estimated at 700 billion barrels (a *barrel* is a volume equaling 159 liters or 42 U.S. gallons), whereas world oil resources are thought to total 2 trillion barrels.

A low-value resource may become reclassified as a reserve if new technology develops to exploit it more cheaply, if economies improve, or if political events allow easier access to it. On the other hand, if new technologies render a resource obsolete, economies languish, or the political climate prevents access to deposits, a once-profitable reserve may be transformed into a relatively valueless resource.

Some natural resources are *renewable,* meaning they are naturally replenished over short time spans (such as trees) or available continuously (such as sunlight). *Nonrenewable* natural resources form so slowly that they are typically consumed much more quickly than nature can replenish them; they include fossil fuels, such as coal, oil, and natural gas, and metals, such as iron, aluminum, gold, silver, and copper. Some resources, such as copper and aluminum, can be recycled for reuse. Others may be either renewable or nonrenewable, depending on how we use them. Soil, for example, is a renewable resource if farmers follow sound agricultural practices. It is nonrenewable when it is depleted of its nutrients by overplanting or when it erodes because of overgrazing, overplanting, or deforestation.

To maintain our standard of living and, at the same time, protect the global environment, we must all become more knowledgeable about the Earth's energy and mineral resources. To use dwindling resources wisely, we need to understand how they formed, how their use will affect the environment, and how long the known supplies are likely to last. This chapter addresses these and other crucial issues in regard to fossil fuels, alternative energy sources, and both metallic and nonmetallic minerals.

Fossil Fuels

Worldwide, we get relatively little of our energy from renewable sources such as solar energy, wind power, and power from streams. Instead, we obtain most of our energy from nonrenewable **fossil fuels,** which are derived from the organic remains of past life. The principal fossil fuels are oil, natural gas, and coal. Oil and natural gas are by now well past their peak production period. Extraction of the world's reserves of crude oil peaked around 1975, and at the current rate of use these reserves will be virtually exhausted within the next century or so. Nevertheless, for economic and technological reasons, the nations of the world continue to draw nearly 95% of their total energy from a dwindling supply of fossil fuels that are, in practical terms, nonrenewable (at least on a human time scale).

Petroleum

The most common and versatile of the fossil fuels is **petroleum,** a group of gaseous, liquid, and semi-solid substances composed chiefly of **hydrocarbons,** molecules consisting entirely of hydrogen and carbon. Typically pumped from the ground as dark, viscous *crude oil,* petroleum is refined—that is, separated into its various hydrocarbon components—to produce propane for camp stoves, motor oil and gasoline for cars, tar and asphalt for roads, and natural gas and heating oil to warm our homes and workplaces. Petroleum also provides the major ingredients for plastics, synthetic fibers, dyes, cosmetics, explosives, certain medicines, certain fertilizers, and records, tapes, and compact discs.

Humans have used petroleum for thousands of years. Some 4500 years ago, Babylonians collected crude oil bubbling from natural pools to make glue for attaching metal projectile points to spears. In ancient Iraq, oil seeping from rocks in the valleys of the Tigris and Euphrates Rivers was used in mortar for setting bricks, in grout for securing tiles in ancient mosaics, and for waterproofing boats.

Modern use of petroleum began in 1816, when gas extracted from coal was employed to fuel Baltimore's gaslights. Combustible gas was discovered in 1821, when a water well in Fredonia, New York, was accidentally ignited, producing a spectacular flame. Wooden pipes were installed to carry the gas to 66 gaslights in downtown Fredonia. Commercial use of oil began in 1847, when a merchant in Pittsburgh bottled and sold natural "rock oil" as a lubricant for machines in the home and workplace. In 1852, Canadian chemists, using oil from what is now Oil Springs, Ontario, refined kerosene from rock oil for use in home lamps, which soon eliminated much of the candle and whale-oil industries. The oil-well industry was born on Sunday, August 27, 1859, when Edwin Drake of Titusville, Pennsylvania, pumped oil from the first true oil well (Fig. 18-2).

The Origin of Petroleum The creation of petroleum generally begins in marine basins in tropical environments, where a rich diversity of microscopic plants and animals exists (Fig. 18-3). When the organisms die, they start to decay by oxidation; eventually the oxygen in the bottom waters of the environment becomes depleted, and decay ceases. Layers

Figure 18-2 The discovery of a new fuel. The first commercial oil well was developed in 1859 in Titusville, Pennsylvania, by Edwin Drake (at right). Drake's well, which was 21.2 meters (70 feet) deep, yielded 35 barrels of oil per day.

of sediment and additional organic material may then bury the organic remains, retarding their subsequent decay. As sediments accumulate, pressure and geothermal heat convert the organic molecules to **kerogen,** a waxy solid organic material.

Kerogen is converted to various liquid and gaseous hydrocarbons at temperatures between 50° and 100°C (100°–200°F) and at a depth of 2 to 4 kilometers (1.2–2.5 miles). At the start of this process, kerogen's large, complex molecules form highly viscous hydrocarbons, such as those found in tar, petroleum jelly, and paraffin wax. With increasing heat, these molecules break down to form smaller, simpler, less viscous ones, such as those found in diesel oil, kerosene, and gasoline. At temperatures exceeding 100°C (200°F), liquid petroleum is converted to various natural gases, ranging from those with relatively complex molecules, such as butane, to the simplest, lightest natural gases—propane, ethane, and methane. At temperatures of approximately 200°C (400°F) and at depths of 7 kilometers (4 miles) or more, methane, the lightest gas, breaks down completely, and the rocks no longer contain hydrocarbons.

The rocks in which petroleum forms are called **source rocks;** they typically consist of shales and siltstones lithified from fine-grained, organic-rich muds. Oil and gas are rarely found in source rocks, because most liquid and gaseous hydrocarbons are readily expelled from their compacting source muds. These materials tend to migrate upward into adjacent permeable **reservoir rocks,** such as well-sorted sandstones and highly fractured or porous limestones, and continue on this path until they are trapped by impermeable rock.

Geologic activity can also destroy such oil traps. Uplift and erosion can remove the trapping cap rock, allowing oil or gas to escape at the surface. A new fault, or the extension of an old one, can breach an oil trap and allow its oil to seep

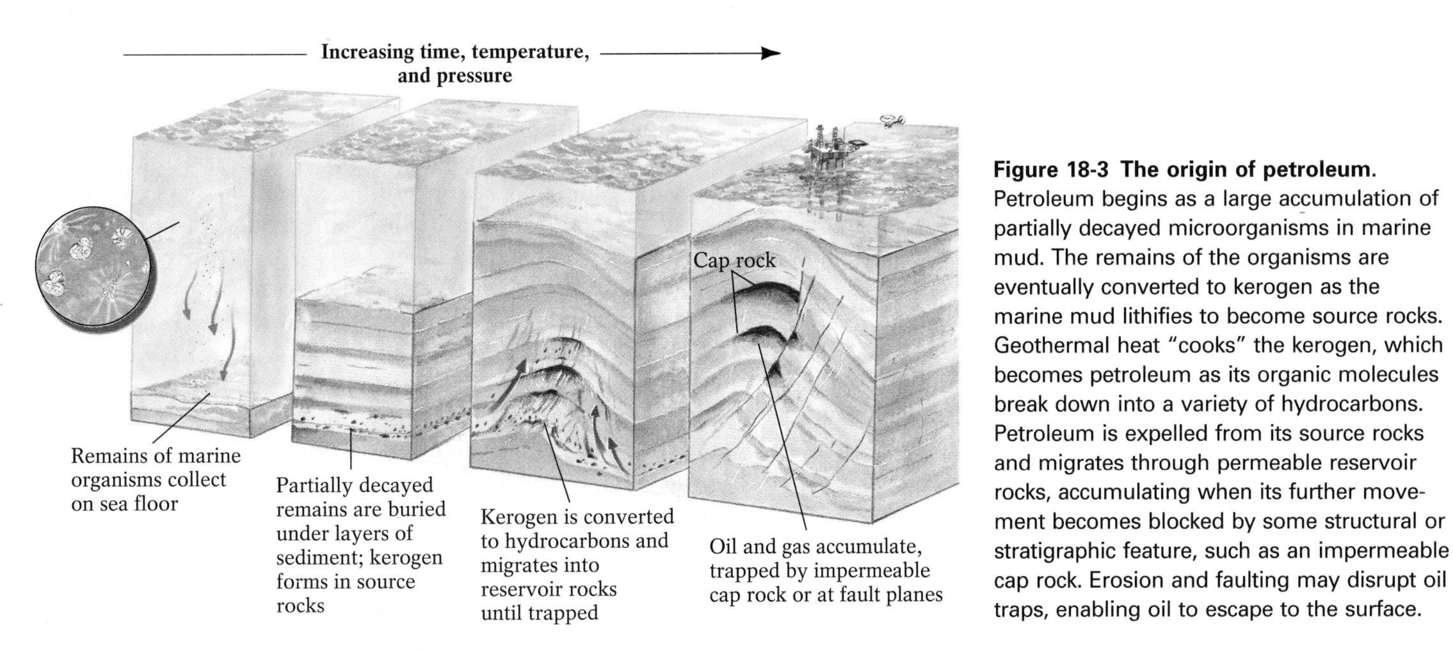

Figure 18-3 The origin of petroleum. Petroleum begins as a large accumulation of partially decayed microorganisms in marine mud. The remains of the organisms are eventually converted to kerogen as the marine mud lithifies to become source rocks. Geothermal heat "cooks" the kerogen, which becomes petroleum as its organic molecules break down into a variety of hydrocarbons. Petroleum is expelled from its source rocks and migrates through permeable reservoir rocks, accumulating when its further movement becomes blocked by some structural or stratigraphic feature, such as an impermeable cap rock. Erosion and faulting may disrupt oil traps, enabling oil to escape to the surface.

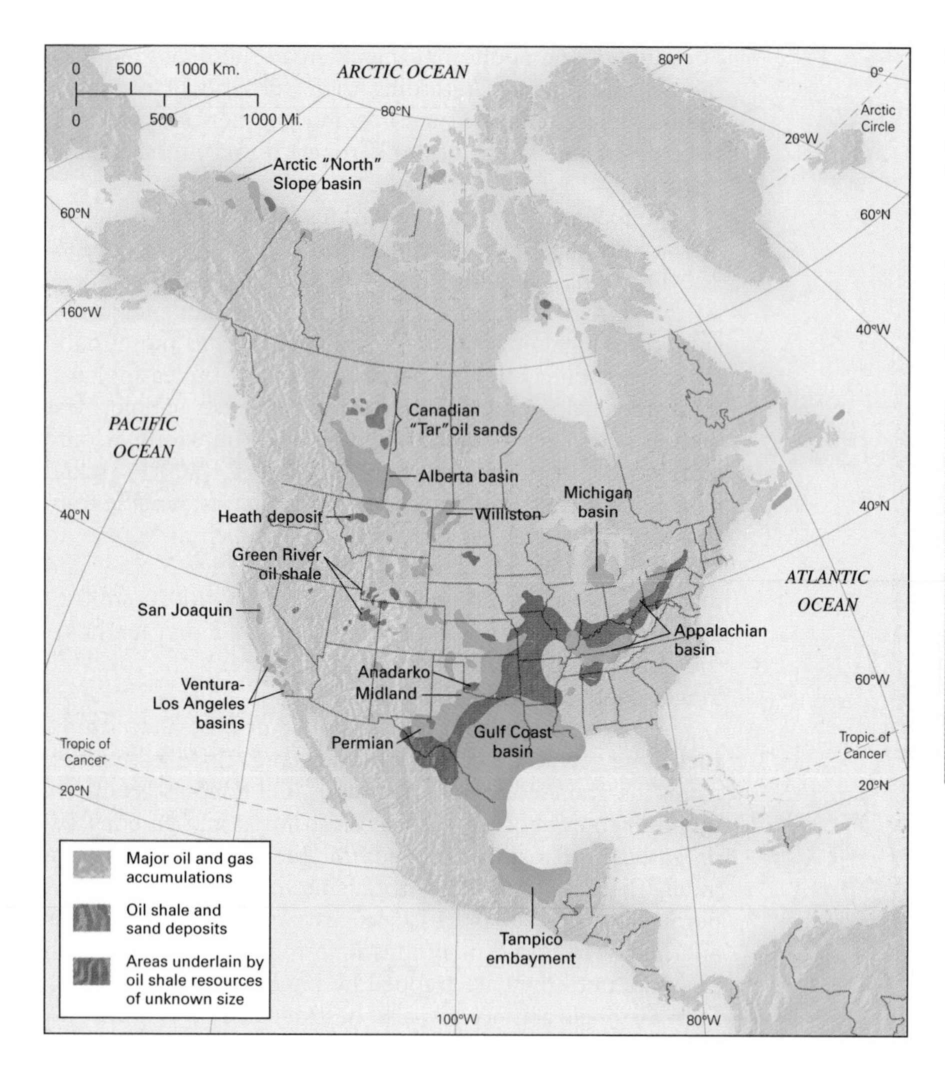

Figure 18-4 North American oil, gas, and oil shale and sand deposits, most of which are in the United States. The world's greatest body of exploitable oil shale, the Green River Shale of western Wyoming and northern Colorado and Utah, originated as a deposit from a freshwater lake that occupied about 40,000 square kilometers (15,000 square miles) 50 million years ago. In some places, this rock formation is 650 meters (2100 feet) thick. Photo: A hand specimen of oil shale from the Green River formation.

out. As a result of these processes, much of the oil and gas formed before about 65 million years ago has long since leaked from its traps and evaporated at the surface.

Geologists estimate that less than 0.1% of all marine organic matter buried at the sea floor eventually becomes trapped as usable petroleum. Some settings may lack adequate heat to convert organic matter to kerogen and petroleum. In others, deposits may experience high temperatures but not reside at great enough depth, enabling shallow-forming hydrocarbons to escape without becoming trapped. The conditions required to produce, trap, and retain hydrocarbons are rarely found together, which explains why most marine rocks are petroleum-free. We do not know how long it takes for oil and gas to form. No known petroleum sources are less than 1 to 2 million years old, so the process must take at least that much time.

Oil Shale and Oil Sand **Oil shale,** a black-to-brown clastic sedimentary rock consisting of a mixture of waxy kerogen and fine mineral grains, is a common source rock. Because it was never buried deeply enough to raise its temperature to that required to convert kerogen to oil, oil shale does not actually contain liquid oil. By crushing and heating the rock to more than 500°C (900°F), however, we can break down the molecules that compose the kerogen and extract various petroleum products.

Based on current and anticipated increased rates of usage, the estimated 2 to 5 trillion barrels of shale oil underlying the United States (which has about two-thirds of the world's oil shale) could meet America's petroleum needs for more than 500 years. According to conservative estimates, this country's largely untapped oil shales (Fig. 18-4) contain more than ten times the known oil reserves of the Middle East. Nevertheless, economic, technological, and environmental concerns keep us from using this oil to heat our homes and fuel our cars. Current world prices for crude oil are not high enough to economically justify the cost of developing this resource.

Oil sand, a mixture of unconsolidated sand and clay that contains a semi-solid, tar-like hydrocarbon called **bitumen,** poses similar development challenges. Bitumen can be refined to produce gasoline, fuel oil, or other commercially viable hydrocarbon products, but it adheres strongly to the mineral grains in oil sand. As a result, it does not flow, nor can it be pumped from the ground. Oil sand must be mined, then heated with hot water and steam to release the bitumen in a more fluid state. The world's largest oil sand deposit, which is located about 400 kilometers (250 miles) north of Edmonton in Alberta, Canada, has been producing crude oil from oil sand since 1967. Today it produces about 200,000 barrels per day, about 16% of Canada's oil. This deposit is commercially exploitable because it lies close enough to the surface to be extracted profitably by surface mining (a less costly process than underground mining, but one that produces enormous piles of residual oil sand that can contaminate local surface and groundwater). For most oil sand deposits, however, the extraction plus processing costs exceed the potential value of the energy obtained.

Coal and Peat

Coal, the Earth's most abundant fossil fuel, is a combustible organic sedimentary rock (see Chapter 6) that forms from the highly compressed remains of land plants. It contains the energy stored in living plants via photosynthesis, the process through which sunlight, water, and carbon dioxide produce the materials for plant growth. When coal burns, it releases the energy that was stored in plants millions of years ago.

Native Americans used coal thousands of years ago to fire pottery; a thousand years ago, Europeans mined it to heat homes and fuel the fires of smelting industries. Cheap, plentiful coal powered the newly invented steam engine in the Industrial Revolution of nineteenth-century Europe and North America. By 1900, coal supplied 90% of the United States' energy demand for industry and domestic heating.

The use of coal has declined during the last 40 years, although it still provides about 23% of U.S. energy needs and as much as 55% of the electricity generated in the nation. Coal's relative decline largely reflected the abundance of oil and natural gas, which could be economically extracted with fewer problems. Coal can be difficult and sometimes dangerous to mine, and it is costly to process. When burned, it also pollutes the air more than the other fossil fuels do. As humans deplete its replacements, however, coal will likely rebound as our principal fossil fuel, especially in electricity-producing power plants. New technology makes it economically feasible to convert coal to liquid and gaseous fuels; when used in these forms, existing world resources could meet energy needs for hundreds of years. The United States has more than 30% of the world's currently accessible coal, though what may be the world's largest coal field has recently been discovered in Antarctica.

The Origin of Coal and Peat The luxuriant plant growth that produced the world's coal deposits likely flourished in tropical or semitropical swamps that became completely covered by the growth of more vegetation or by overlying sediments. The cumulative weight of overlying deposits squeezed water from the porous mass of partially decomposed vegetation. As plant remains became more deeply buried, increasing pressure, geothermal heat, and bacterial reactions removed more water as well as the organic gases, such as carbon dioxide and methane, that were produced by bacterial activity and oxidation.

These processes create a variety of fossil fuels categorized by their increasing carbon and decreasing gas and water content. Low pressure produces **peat**, a soft, brown mass of compressed, largely undecomposed, and still recognizable plant structures; it consists of only about 50% carbon. In rural Ireland, England, and elsewhere in Europe, peat is dried and then burned as fuel for home heating and other domestic uses.

Bacterial action continues in peat, eventually decomposing it enough to form a kerogen-rich vegetative muck that becomes compressed as **lignite.** This soft, brown coal consists of approximately 70% carbon (plus 20% water and 10% oxygen); its higher carbon content makes lignite a more efficient heat source than peat. Deep burial of lignite and the accompanying rise in pressure and geothermal heat eventually convert this material to lustrous black **bituminous coal** having a carbon content of 80% to 93%. This form of coal produces more heat, with much less smoke, than either peat or lignite.

With even higher pressure and heat, lignite and bituminous coal may be metamorphosed to **anthracite,** a hard, jet-black coal containing 93% to 98% carbon and very little gas. Anthracite burns with an extremely hot flame and very little smoke. Unlike lignite and bituminous coal, which are widely distributed, anthracite exists only in low-grade metamorphic zones in mountainous regions—notably the Appalachians of northeastern Pennsylvania (Fig. 18-5).

Coal and Peat Locations Peat, lignite, and bituminous coal typically occur as distinct seams or beds in sequences of detrital sedimentary rocks, particularly those that accumulated in warm, moist coastal environments. A series of alternating detrital sediments and coal beds typically indicates the past occurrence of cycles of rising and falling sea level and their corresponding periods of land submergence (and thus detrital sedimentation) and land emergence (and ensuing swamp development).

Much of North America's vast coal deposits formed during ancient episodes of high worldwide sea level, when shallow seas invaded the continent's interior. The coal of Utah, Montana, and the Dakotas, for example, formed at the swampy margins of such seas. Coal deposits have also formed on the vegetated floodplains of intermountain rivers. For instance, the coal of Wyoming's Powder River basin occurs as layers between floodplain sediments. In addition, coal deposits may accumulate at continental margins, where wide continental shelves are flooded periodically by fluctuating sea levels. The coal of the Pennsylvania and West Virginia Appalachians formed in such an environment.

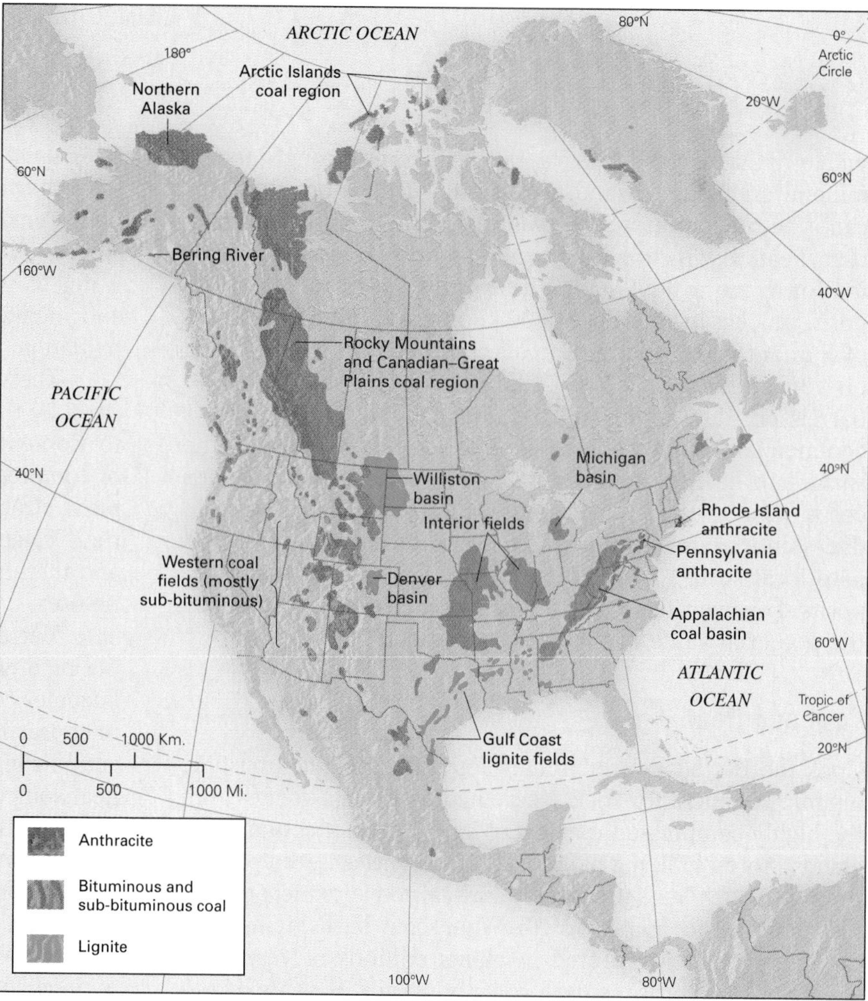

Figure 18-5 Major coal deposits of North America. Photo: A coal seam in Alaska.

Peat and coal have probably been accumulating continuously since land plants first appeared 450 million years ago. Today, coal is being formed in the Great Dismal Swamp of coastal Virginia and North Carolina, where a 2-meter (7-foot)-thick layer of vegetation has accumulated over a 5000-square-kilometer (2000-square-mile) area during the last 5000 years. The next major rise in sea level that buries these organic deposits with detrital sediment will halt their further decomposition and eventually provide the pressure and heat needed to convert them into peat and coal.

Fossil Fuels and the Environment

Widespread use of coal, oil shale, and other fossil fuels can produce a variety of environmentally damaging side effects. In recent years, concerns about acid rain, global warming, and massive marine oil spills have led to increased public awareness and inspired legislation intended to reduce or prevent such effects.

Acid Rain Burning fossil fuels releases secondary materials into the atmosphere, reducing the quality of the air that we breathe. When coal is burned, for example, coal ash—fragments of noncombustible silicates and toxic metals—enters the atmosphere. Burning sulfur- and nitrogen-rich hydrocarbons releases sulfur and nitrogen oxides, which then combine with water in the atmosphere to form sulfuric acid and nitric acid, the principal human-related source of **acid rain** (Fig. 18-6). Acid rain may damage forests and crops, kill aquatic life in lakes, and accelerate the destructive weathering of many human-made structures. The most pronounced damage from acid rain occurs downwind of major coal-burning industrial regions.

In some settings, the environmental damage from acid rain may be lessened by the area's geology; calcium-rich soils or exposed carbonate bedrock (limestone or dolostone), for example, may react with the acids and neutralize them before they can do much harm. Where calcium-poor soils overlie granitic bedrock, as in the lake and forest country of the Canadian Shield area of the northern Great Lakes and adjacent parts of southeastern Canada, acid rain damage to the ecosystem can be severe. Eastern Canada receives acid rain produced from emissions from the coal-burning smokestacks of the heavy steel and automotive industries of the American Midwest. In addition to installing scrubbing devices that remove much of the sulfur and nitrogen from these emissions, attempts to neutralize this threat involve low-flying aircraft that spread limestone dust across the landscape.

Sulfur dioxide and nitrogen oxides produced

Pollutants carried into atmosphere

Reaction with water vapor in atmosphere produces sulfuric acid and nitric acid

Pollutants deposited nearby

- Corrosion of buildings, stones, etc.
- Acidification of lakes, rivers, and other bodies of water

Acid rain/snow

- Forests damaged
- Fish in lakes killed
- Vital nutrients leached from soils

Figure 18-6 The origin of acid rain. Acid rain has two major causes: (1) the release of sulfur dioxide into the atmosphere from burning of sulfur-rich coal, where it combines with water to form sulfuric acid; and (2) the emission of nitrogen oxides mostly from automobile exhaust, which combine with water to produce nitric acid. Once these pollutants enter the atmosphere, their effects can spread to nearby and even distant environments.

Global Warming Burning all types of fossil fuels increases the volume of carbon dioxide in the atmosphere. Measurements made from 1958 to 1984 at the top of Hawai'i's Mauna Loa, far from industrial pollution sources and large population centers, showed a 9% increase in global atmospheric carbon dioxide. Atmospheric carbon dioxide absorbs and traps heat from the Earth's surface, creating what is known as the greenhouse effect and warming the planet. Some climatologists predict that if atmospheric carbon dioxide levels continue to rise at their present rate, atmospheric temperatures could rise 1.5° to 4.5°C (2.5°–8°F) by the middle of the twenty-first century. This phenomenon would cause droughts in some regions, increase desertification in others, and significantly reduce the agricultural potential of today's marginally cultivatable lands. It would also promote melting of a substantial portion of the Earth's ice masses, which would cause sea levels to rise worldwide, perhaps by 1 to 2 meters (3.3–6.6 feet) per 100 years, for the next 200 to 500 years. A rise of 2 to 10 meters (6.5–33 feet) would accelerate coastal erosion and submerge a substantial amount of coastal land, including most major port cities on every continent (as discussed in Chapter 15).

Marine Oil Spills Occasionally, an oil tanker becomes damaged and leaks its cargo of crude oil into the sea. Extraordinary environmental damage resulted from one such marine oil spill in March 1989, when the supertanker *Exxon Valdez* went aground in Alaska's Prince William Sound. More than 10.2 million gallons of crude oil flowed from the ship's cracked hull into the sea, much of which washed up on the shore in layers tens of centimeters thick. The oil killed thousands of birds and marine mammals (Fig. 18-7), halted herring and

Figure 18-7 The disastrous effects of oil spills. One problem associated with oil exploration and development is the potential for oil spills, which can cause large-scale environmental damage to oceans and coastal habitats. The negative impact on marine life and habitats can be profound, as dramatized by the oil-covered wildlife found following the 1989 breakup of the supertanker *Exxon Valdez* in Alaska's Prince William Sound. The effects of this spill are still being felt today.

salmon fishing during the peak season, and coated hundreds of kilometers of Alaska's coastline.

Oil spills in calm seas can usually be confined within floating barriers placed around the oil, allowing workers to skim the oil from the surface. In Alaska, unfortunately, inclement weather produced rough seas that breached the floating barriers; the Exxon Company recovered only some 500,000 gallons of its oil. Efforts to ignite and burn the spilled oil failed as well, although some of the lightest hydrocarbons evaporated and some were consumed by oil-eating bacteria. Ultimately, millions of gallons washed onshore, necessitating an enormous cleanup operation. Workers soaked up standing pools of oil with peat moss, wood shavings, and even chicken feathers. High-powered steam hoses were used to remove some of the oil coating; workers even scrubbed rocks individually by hand. After several months and some $4 billion in cleanup costs, more than 85% of the spilled oil was gone. The remainder consisted of thick asphalt clumps that could not be scrubbed, did not evaporate in sunlight, and would not decompose by bacterial action. These clumps fouled breeding grounds and other wildlife habitats on land, as well as major fishing grounds.

Alternative Energy Resources

As fossil fuel reserves dwindle and environmental damage related to their use increases, governments and industries are seeking alternative ways to meet their populations' growing energy needs. Most existing alternative energy sources, such as solar and wind power, are renewable. One major alternative, nuclear energy, relies on uranium, a nonrenewable natural resource.

Renewable Alternative Energy Sources

Renewable alternative energy resources are those resources that can be used virtually without depletion or that are replenished over a relatively short time span. They include geothermal, hydroelectric, tidal, solar, and wind energy and the energy produced by burning such renewable organic materials as trees and agricultural waste.

Geothermal Energy Reykjavik, the capital of Iceland, is relatively pollution-free, because it has a clean, inexpensive

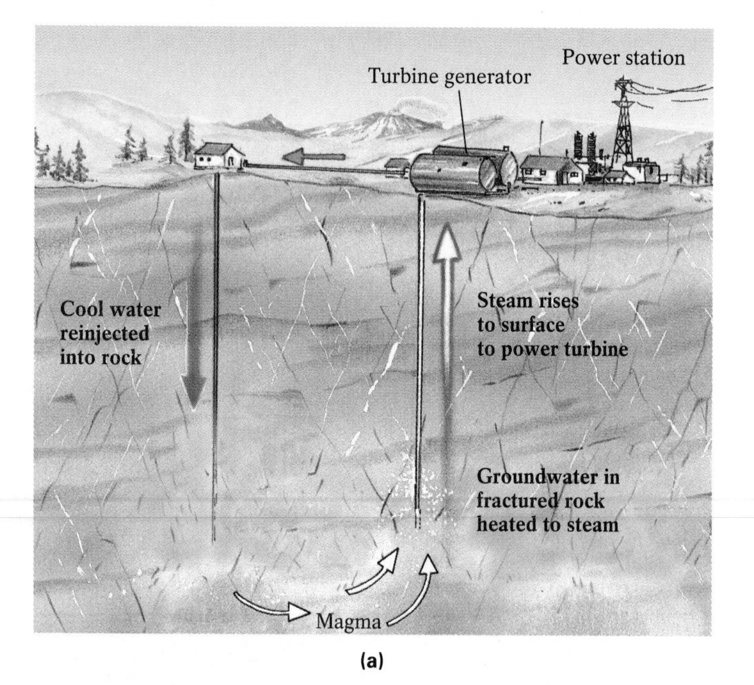

(a)

(b)

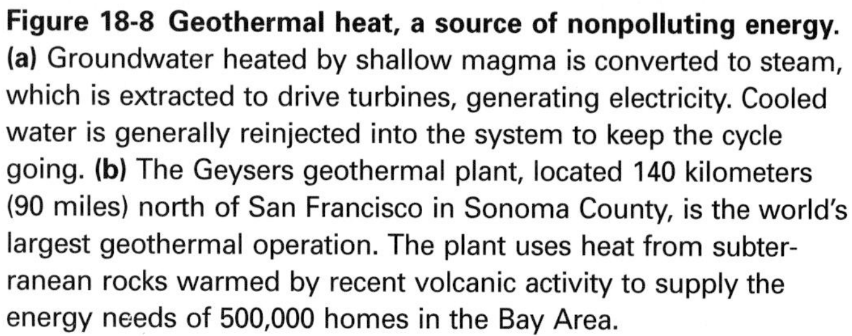

Figure 18-8 Geothermal heat, a source of nonpolluting energy. **(a)** Groundwater heated by shallow magma is converted to steam, which is extracted to drive turbines, generating electricity. Cooled water is generally reinjected into the system to keep the cycle going. **(b)** The Geysers geothermal plant, located 140 kilometers (90 miles) north of San Francisco in Sonoma County, is the world's largest geothermal operation. The plant uses heat from subterranean rocks warmed by recent volcanic activity to supply the energy needs of 500,000 homes in the Bay Area.

source of energy: Heat from shallow, hot rock and magma pooled beneath the surface is used to convert groundwater to hot water and steam. The hot water is circulated through pipes and radiators to heat homes and municipal buildings; the steam drives electric generators (Fig. 18-8).

Since human beings first tapped such *geothermal heat* as an energy source in Larderello, Italy, in 1904, about 20 countries have used it, including the United States. More nations would take advantage of this relatively inexpensive, nonpolluting energy if they could. Unlike oil, coal, and natural gas, however, geothermal energy cannot be transported and must be used close to its source. Every nation using geothermal energy is located on a currently or recently active plate margin or near an intraplate hot spot, where magmatic heat has not yet dissipated.

Hydroelectric Energy For many centuries, humans have used falling water as an energy source to mill flour, saw logs, and power numerous machines. Today, hydroelectric facilities employ falling water to produce electricity. To generate hydroelectric power, a high-discharge stream is impounded by a dam to create enough vertical drop to rotate the blades of large turbines.

Hydroelectric power is widely available; since 1983, nearly one-third of all new electricity-generating plants built in the United States have been hydroelectric installations. At present, however, the United States generates only about 15% of its electricity in this way. The Federal Power Commission estimates that if every sizable river in the United States were dammed for this purpose, hydroelectric power could supply 50% of our total electricity needs. Global hydroelectric development lags even further; only 6% of the world's hydroelectric potential is being used, and in South America and Africa, where its potential is greatest, only 1% has been developed. Canada, at the other extreme, obtains 75% of its electricity from this clean resource.

Although hydroelectric power is nonpolluting, some environmental problems are associated with its use. Dams can disrupt the local ecological balance by altering or destroying wildlife habitats; they may also impede natural erosion processes—their reservoirs eventually fill with sediment that would otherwise replenish coastal beaches. Decisions to build dams must therefore balance their environmental costs against their energy yield.

Tidal Power In coastal areas with a high tidal range—the difference in the water surface level between high and low tide—energy from rising and falling water levels can be harnessed by building a dam across a narrow bay or inlet. The dam's gates are then opened during rising tides and closed to trap the water at its maximum height. During low tide, the elevated water is channeled seaward through turbines connected to electrical generators, producing renewable, pollution-free energy (Fig. 18-9).

Tidal power production, however, requires a minimum tidal range of 8 meters (26 feet), and it disturbs the ecology of the surrounding coastal habitats. As yet no tidal-power facilities in North America have been constructed, although Passamaquoddy Bay in northeastern Maine, with a tidal range of 15 meters (50 feet), is a strong candidate for future

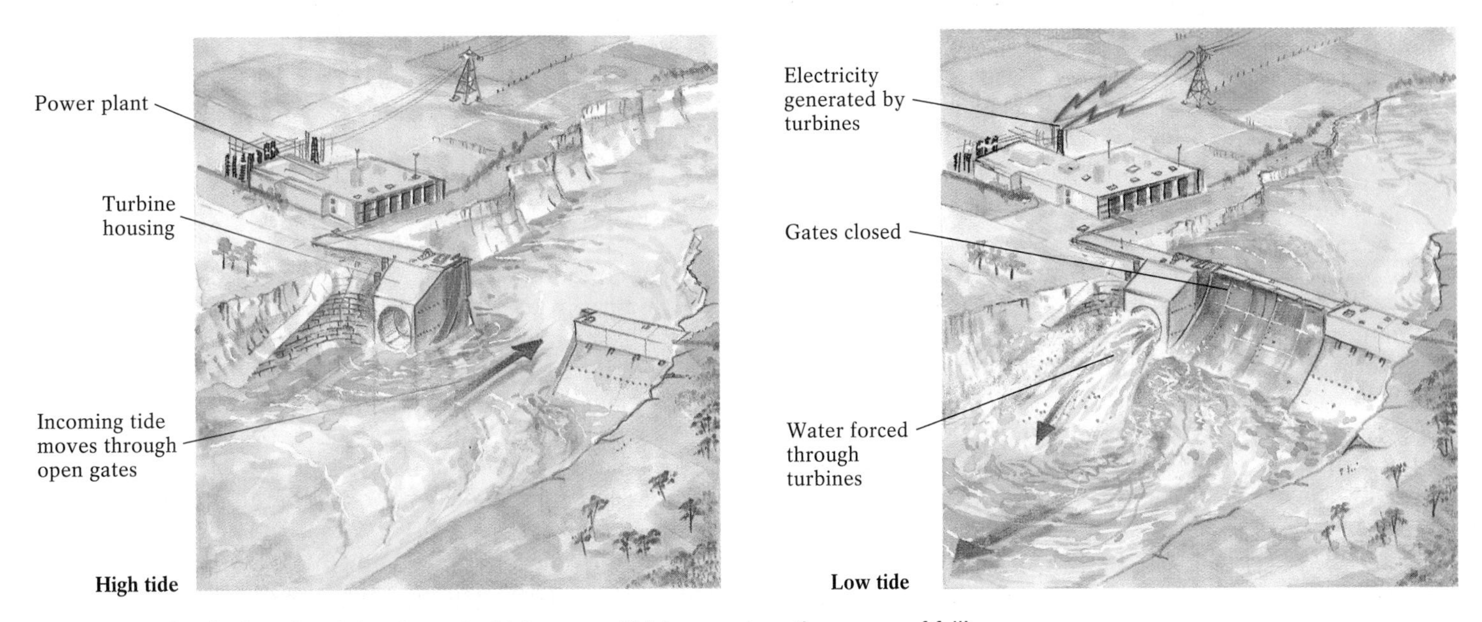

Figure 18-9 Producing electricity through tidal power. Tidal power taps the energy of falling water by trapping water at high tide and then releasing it seaward through electricity-producing turbines at low tide.

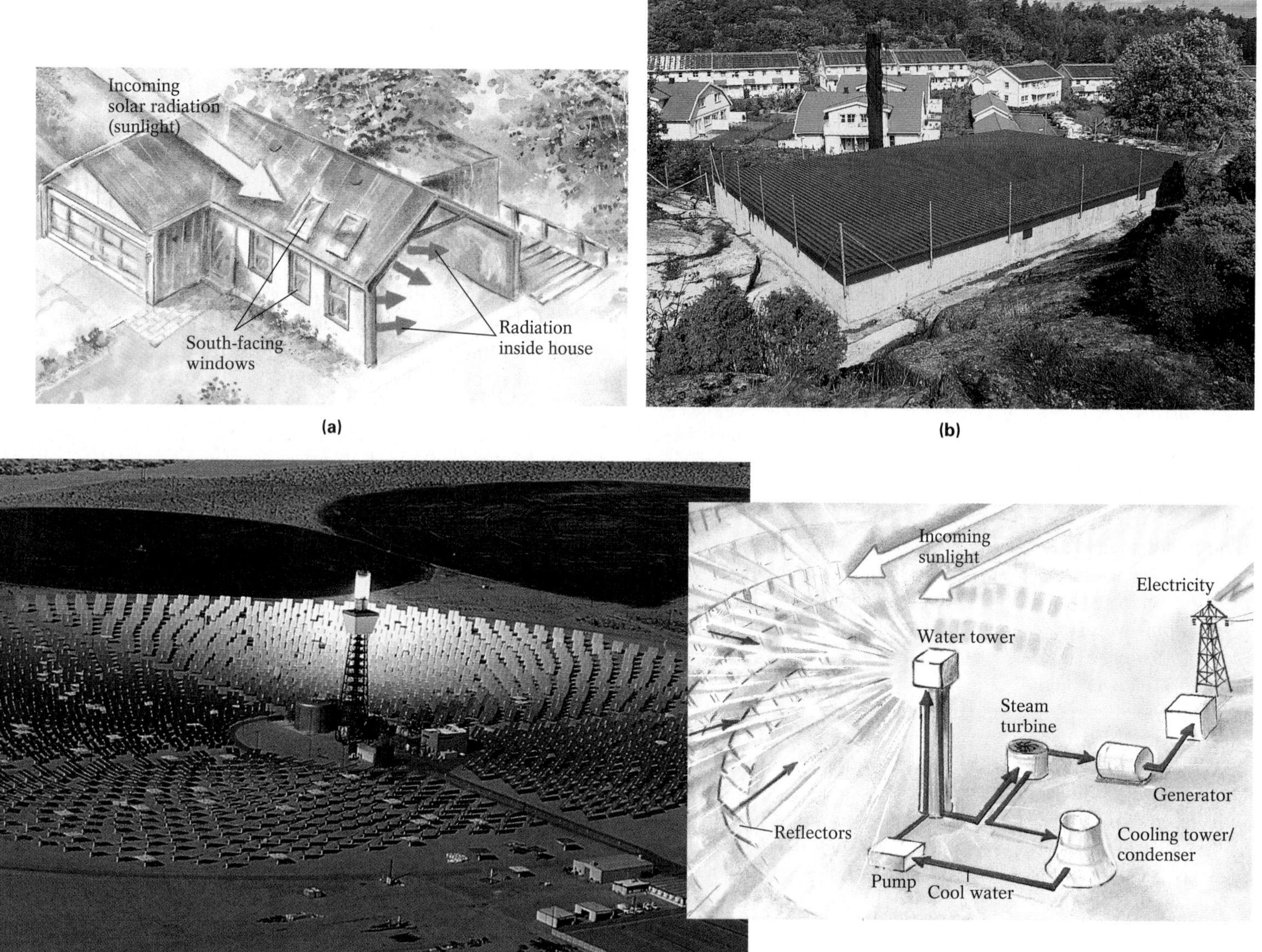

Figure 18-10 Using solar power. **(a)** Solar energy passively heats a single dwelling. **(b)** Water-filled panels on rooftops provide hot water and space heating, in a type of active solar heating. **(c)** This facility near sunny Dagget, California, uses solar energy to generate electricity. Reflecting mirrors focus on a water tower, concentrating the Sun's energy and converting the water to turbine-driving steam.

development, as is the Bay of Fundy in New Brunswick, Canada, with a tidal range of 20 meters (65 feet). Maximum development of the United States' potential tidal power would provide only 1% of the nation's total electric needs, although it could become a significant supplement to other energy sources in some localities. Worldwide potential for this energy source, only slightly better, is about 2% of the total electricity demand.

Solar Energy Solar-powered pocket calculators and wristwatches take advantage of an energy source that requires no expensive drilling or destructive strip mining, cannot be monopolized by unfriendly political regimes, and produces no hazardous wastes or air pollution. The Sun, which is expected to shine for another 5 billion years or so, is a totally renewable and easily accessible energy source. Solar power can be used to heat buildings and living spaces and to generate electricity—energy needs that together account for two-thirds of North America's energy consumption.

Solar heating can be either passive or active. Passive solar heating distributes the heat naturally by radiation, conduction, and convection. At mid-northern latitudes, the simplest way to heat spaces passively is to construct buildings with windows facing south. Sunlight passes through the window glass and heats objects within the room; their heat then radiates to warm the air (Fig. 18-10a). Such an architectural design, coupled with efficient insulation, sharply reduces both air pollution and the cost of heating with fossil fuels.

Active solar heating works via water-filled, roof-mounted panels with black linings that absorb maximum sunlight. The solar-heated water is circulated throughout the building for space heating or directly to the building's hot-water system (Fig. 18-10b). Solar panels are most productive in mild, sunny climates such as Florida, Texas, California, and the Southwest, where they can provide as much as 90% of a building's heating needs. Even in colder regions, such as northeastern North America, where solar panels are less productive, they can significantly reduce the need for other energy sources.

Solar energy can also be used to generate electricity. In one such system, an array of many mirrors reflects sunlight onto a large water tower; the water is heated to create steam, which drives turbines attached to electrical generators (Fig. 18-10c). Electricity can be generated by solar energy more efficiently in some regions than in others. Unfortunately, the technology to maximize its effectiveness at low cost is not yet widely available. Satisfying the current U.S. electricity needs would require a system of collecting mirrors that would occupy approximately 25,000 square kilometers—about one-tenth the size of the state of Nevada—which is clearly not feasible. As a result, substantial utilization of solar energy for the generation of electricity probably remains decades away.

Wind Power Like falling water, the tidal motion, and sunlight, wind power is a clean, renewable, nonpolluting energy source that has long been used by humans. The picturesque windmills of the Netherlands and those of rural midwestern North America have pumped groundwater and powered sawmills and flour mills for centuries. Wind power, however, is rarely cost-effective on a large scale: Only in a few sparsely populated mountain passes, such as Altamont Pass, east of San Francisco (Fig. 18-11), do winds blow constantly, forcefully, and from a consistent direction—all requirements for the practical application of wind power.

Figure 18-11 Wind power. The productivity of this "wind farm" at Altamont Pass, California—where 2000 wind turbines are connected to electrical generators—is made possible by the sustained westerly winds that buffet California's Sierra Nevada.

Biomass Fuels derived from plants and animals are known collectively as **biomass fuels.** In developing countries, as much as 35% of the energy used for cooking and heating comes from burning two common biomass fuels, wood and animal dung. These types of fuels also include grain alcohol (used as an additive to gasoline), methane gas that rises from decaying garbage in landfills, combustible urban trash, and plant waste from crops such as sugarcane, peanuts, and corn. The most widely used biomass fuel is wood, which today heats about 10% of North America's homes, more than are heated by electricity from nuclear power plants.

Biomass fuel is a renewable resource. Although trees grow slowly, continuous planting and harvesting of these plants can produce a steady supply. Unlike most other renewable energy resources, however, biomass fuels can create air pollution and desertification problems when used on a wide scale or implemented poorly. As with oil and coal, burning of this resource introduces noxious gases and particles into the air, reducing air quality and increasing global warming. Moreover, in arid regions, overreliance on scrub and trees for energy removes the root systems that help retain water and soil; this trend contributes to desertification (discussed in Chapter 16), which consequently eliminates animals that provide dung for fuel.

Nuclear Energy: A Nonrenewable Alternative

In the mid-twentieth century, physicists harnessed the energy released from the decaying nuclei of radioactive isotopes (nuclear fission). Today, they are attempting to tap the energy produced when atomic nuclei become fused together (nuclear fusion). These energy sources may provide a twenty-first-century solution to the inevitable exhaustion of fossil fuels.

Nuclear Fission When the nuclei of the radioactive isotope uranium-235 are bombarded with neutrons, they split into several lighter elements, releasing part of the binding energy of the atom's nucleus, additional neutrons, and an enormous quantity of heat. The released neutrons, in turn, bombard other uranium nuclei, setting off a chain reaction (Fig. 18-12), known as **nuclear fission.** In commercial nuclear reactors, the heat from controlled nuclear fission reactions is used to convert water to steam, driving turbines and thereby producing electricity. To prevent the chain reactions from getting out of control, workers move neutron-absorbing control rods in and out of the target uranium fuel to regulate the rate of reactions, and the reactor itself is lined with layers of a neutron-absorbing material such as graphite.

Uranium, a relatively uncommon element in the Earth's crust, becomes concentrated in fluids of granitic magmas during the late stages of cooling. Because it is highly soluble in water, uranium readily dissolves when igneous rocks are weathered. Groundwater then transports this element into permeable sediments and sedimentary rocks, where it bonds to the surfaces of clay particles and organic matter. These concentrations serve as the source of most uranium, which is mined from ancient stream sands and gravels. In North America, uranium deposits are found in the Mesozoic stream gravels of Colorado, New Mexico, Wyoming, and Texas, often associated with fossilized plant and wood remains. Uranium-235 accounts for only 0.7% of naturally occurring uranium, whereas its nonfissionable isotope U-238 makes up 99.3%. At its current rate of use, recoverable reserves of U-235 may be sufficient to power the world's 575 nuclear power plants for only another 30 years or so.

To address the issue of dwindling U-235 supplies, nuclear physicists are experimenting with *breeder reactors,* which produce fissionable plutonium-239 from the abundant nonfissionable U-238. In these reactors, uranium-238 is placed together with a small amount of U-235. The U-235 decays spontaneously, and its neutrons bombard the U-238 nuclei, converting them to Pu-239. Fission of the manufactured plutonium atoms produces a surplus of neutrons that can then be used to create, or "breed," more fissionable material (Pu-239) than they consume (U-235).

Currently, few breeder reactors are on-line, primarily because Pu-239 is a weapons-grade nuclear material. Funding for the Clinch River breeder reactor in Tennessee was canceled in the mid-1980s because of public concern about the proliferation of nuclear weapons. Today, the Super Phoenix reactor in France is the world's largest operating breeder facility.

In 1974, the U.S. Geological Survey predicted that the United States would obtain 60% of its electricity from nuclear plants by 2000; in 1993, this figure was about 16% and dropping. Meanwhile, Western Europe and Japan, both

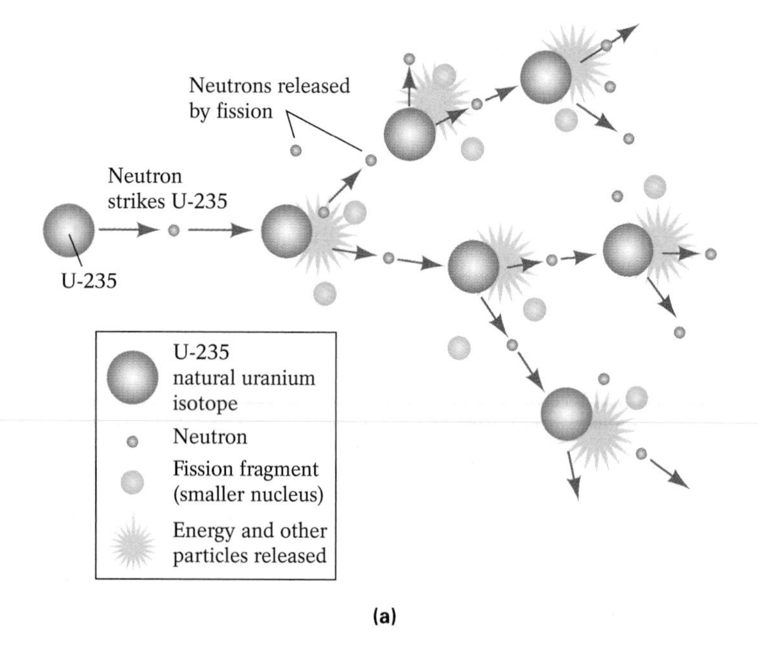

Figure 18-12 Nuclear fission. (a) A neutron released by the natural decay of a U-235 nucleus initiates a chain reaction. **(b)** Uranium fuel is contained within 12 meters (40 feet) of water at the Indian Point 2 nuclear power plant in Buchanan, New York.

of which have little indigenous fossil fuel, have been expanding their nuclear-energy facilities; France, for example, derives 65% of its electricity from its many reactors. A number of reasons explain the United States' reluctance to develop nuclear energy. In addition to concerns about the theft of weapons-grade fuels, technological problems have arisen involving reactor safety and radioactive-waste disposal, and several other economic, political, and psychological problems are associated with the growth of nuclear energy.

In 1979, an instrument malfunction led plant operators at Three Mile Island, Pennsylvania, to conclude that too much water was flushing through the reactor's cooling system. They responded by reducing water flow, leaving the reactor core uncovered for several hours and allowing it to overheat to critical temperatures. Although the reactor core suffered damage, little measurable radiation escaped. In 1986, technicians, working at a poorly designed nuclear facility at Chernobyl, U.S.S.R., accidentally allowed a runaway chain reaction to develop. Two small explosions blew the roof off the building, showering the immediate area with radioactive material. Eastern Europe and Scandinavia were covered with a cloud of radioactive steam. Numerous cases of radiation sickness and cancer developed in Chernobyl's immediate vicinity, and agriculture and commerce were disrupted over a wide area.

In recent years, the safety of nuclear reactors has increased to the point that the risks associated with them are exceedingly small. Because radioactive nuclear waste is so toxic that it must be isolated from all life for thousands of years, however, safe disposal of this material continues to pose the industry's greatest technological problem. Spent fuel rods, internal machinery from reactor cores, and the waste products of nuclear-fuel processing must be disposed of in ways that prevent any leakage to the atmosphere or groundwater system.

Nuclear Fusion **Nuclear fusion** occurs when extremely high pressure and temperature cause the atomic nuclei of light elements to fuse together to form heavier atoms. Like fission, this process releases an enormous amount of heat energy that can be used to convert water to steam, driving turbines that produce electricity. Unlike with fission, the principal fuel of fusion is hydrogen, one of the most abundant elements; likewise, rather than toxic radioactive waste, fusion produces helium, a harmless inert gas.

Fusion's potential is enormous: The energy that could be generated by the hydrogen in 1 cubic kilometer (0.24 cubic mile) of seawater exceeds the total energy stored in the world's oil reserves. Sustained small-scale hydrogen fusion is now possible. In recent experiments, heavy isotopes of hydrogen (tritium, ^{3}H, and deuterium, ^{2}H), were compressed in a powerful magnetic field and then heated by intense laser bursts to about 50 million°C, producing the heavier element helium (He) and an enormous amount of heat. As yet, nuclear fusion is not economically feasible as a practical energy source, because the amount of energy needed to achieve fusion still exceeds the amount it produces. Much further study and expensive experimentation will be required before this process will become commercially viable.

Mineral Resources

Minerals, as we saw in Chapter 2, are naturally occurring inorganic solids that consist of specific proportions of chemical elements, whose atoms are arranged in a systematic internal pattern. Of the 3000 or so known minerals, only a few dozen are economically valuable. These substances include both metallic minerals and nonmetallic rocks and minerals.

Metals

The first known human use of metals, which took place about 9000 years ago in what is now Turkey, occurred when people hammered naturally pure copper into amulets, tools, and weapons. By 6000 years ago, metallurgists in Europe and Asia Minor were separating metals from their host rocks by *smelting;* that is, they heated rocks to the melting points of their incorporated metals and then collected the molten metal for further processing. The earliest smelting operations extracted copper from copper sulfides such as chalcocite and chalcopyrite. By 5000 years ago, lead, tin, and zinc, as well as copper, were being smelted and combined in their molten state. The resulting *alloys,* or metal mixtures, were harder than their component metals and maintained sharper points and edges. Bronze, a mixture of copper and tin, was a common alloy that dominated the tool and weapon industries that flourished between 5000 and 2650 years ago; its widespread use characterized the archaeological period known as the Bronze Age. By about 2650 years ago, metallurgists had learned to smelt and work iron, which quickly replaced bronze as the principal metal used in the manufacture of tools, weapons, and armor, thereby ushering in the Iron Age. The demand for metals has grown ever since.

In general, metals combine readily either with oxygen, to form oxides, or with sulfur, to form sulfides. Common oxides include those of tin, uranium, iron, and aluminum; common sulfides include those of zinc, lead, iron, copper, and molybdenum. *Native* metals, such as gold, platinum, and silver, do not combine with other elements.

Most metals are scattered thinly through the Earth's crust. Their extraction becomes economically feasible only where various rock-forming processes have gathered them into mineable concentrations.

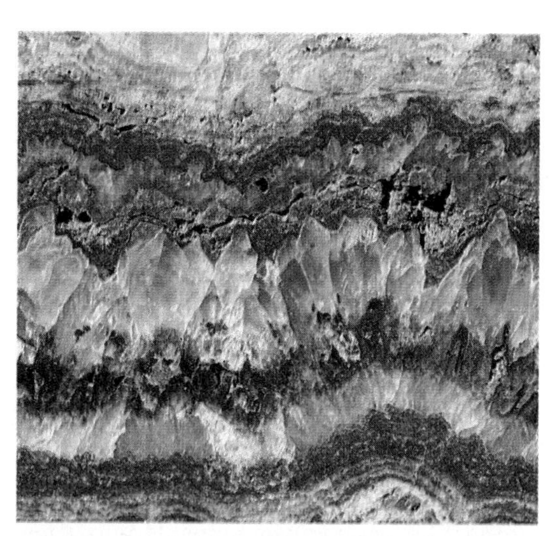

(a)

(b)

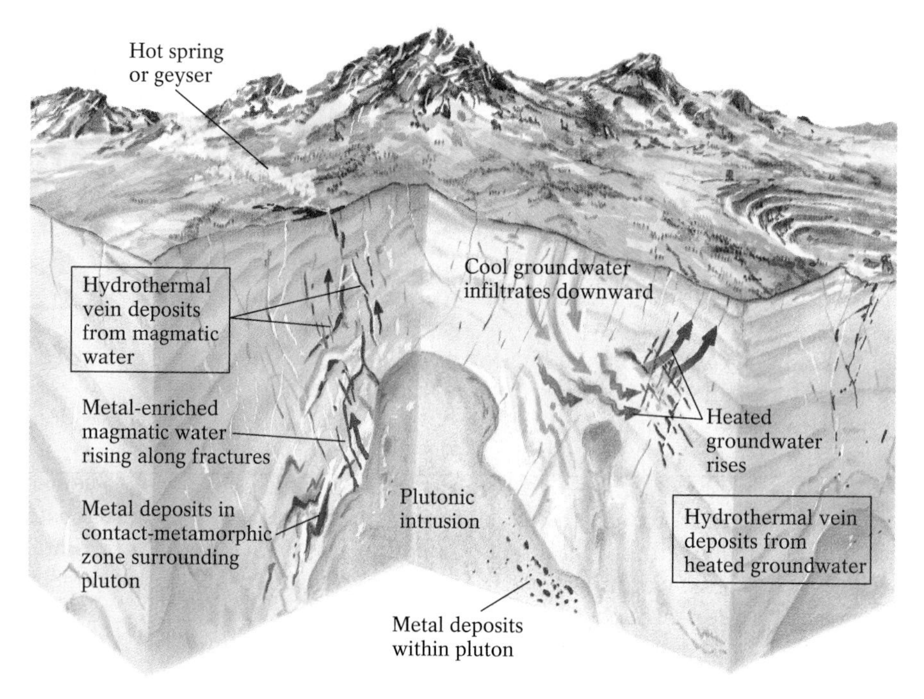

Figure 18-13 Hydrothermal processes that concentrate valuable metals. Metals may precipitate from magmatic water or from circulating groundwater or seawater heated by the magma. **(a)** These veins of gold and silver ore formed when hot solutions of the metals infiltrated cracks around a cooling magma. **(b)** These massive sulfide deposits in Quebec formed at a mid-ocean ridge when seawater heated by basaltic magma emerged at hot springs and precipitated large quantities of metals on the ocean floor.

Processes That Concentrate Metals Metals are often found concentrated in **hydrothermal deposits,** formed by the precipitation of metallic ions from hot, ion-rich water (Fig. 18-13). A common source of this metal-rich, hot water is a cooling magma that contains metallic ions left over after most silicate minerals have crystallized (see Chapter 3). Another source is deep-circulating groundwater that has been heated by close contact with shallow magmas or young, warm, plutonic rock; as this hot water rises, it dissolves metals from the pluton and surrounding rocks. The resulting solutions may contain copper, lead, zinc, gold, silver, platinum, and uranium. The hot water typically infiltrates faults, fractures, and bedding planes in surrounding rocks, where it cools and deposits its solutes as metal-rich mineral *veins*.

Massive hydrothermal sulfide deposits form where cold seawater enters the fractures associated with divergent zones, then becomes heated by shallow basaltic magma. As the water rises through the fractures, it dissolves some of the trace metals in the basaltic mid-ocean ridge; it also picks up the sulfurous gases typical of basaltic eruptions. The resulting hydrothermal solution contains sulfides of copper, zinc, manganese, lead, and iron; these materials eventually precipitate on the sea floor.

The processes by which minerals crystallize from magma can also produce metal deposits. In *gravity settling,* dense, early-crystallizing minerals sink to the bottom of the magma chamber, where they accumulate in layers. In *filter pressing,* tectonic forces compress a magma chamber and force the still-liquid portion of the magma into fractures in adjacent rocks, causing large crystals to become concentrated against the wall of the chamber. Gravity settling and filter pressing of mafic and ultramafic magmas have produced valuable ore bodies composed of iron, chromium, titanium, and nickel.

When hot, ion-rich fluids move through rock bodies, their heat can produce contact-metamorphic mineral alterations in the host rock. These changes often result in metallic ores. For example, when the fluids enter impure limestones

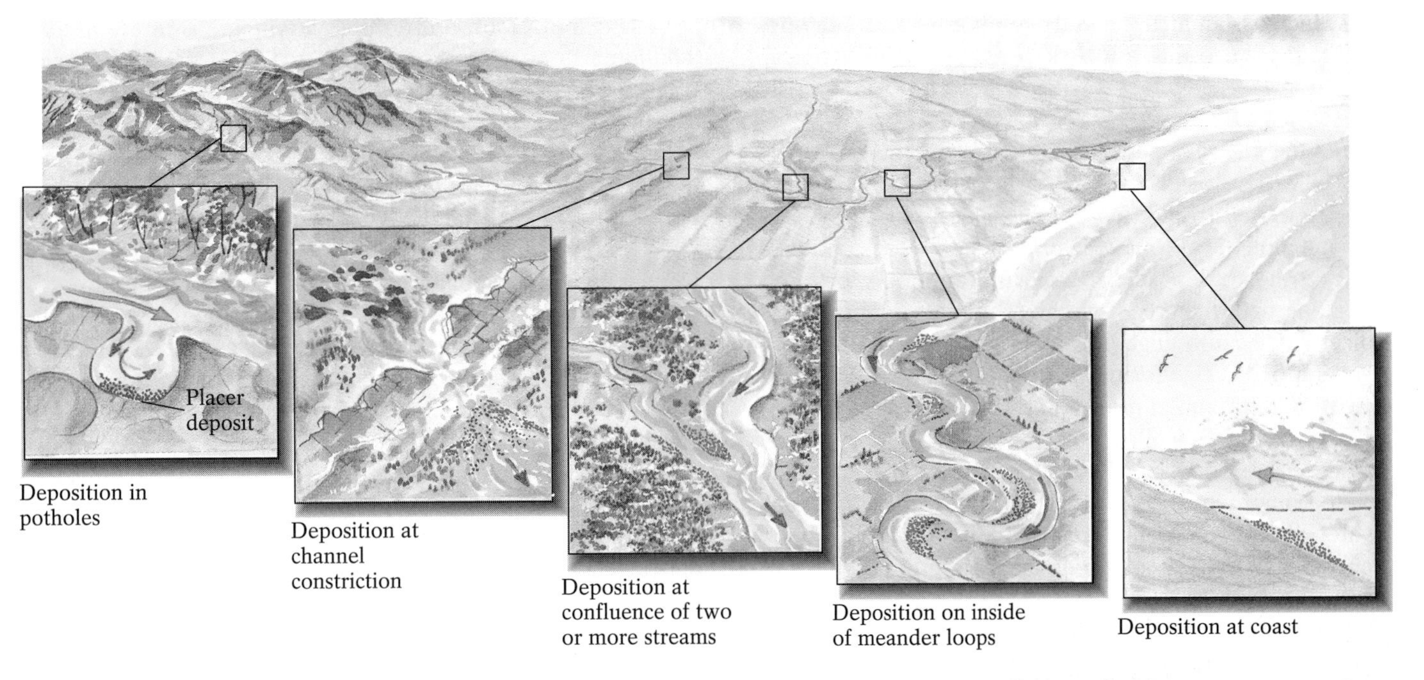

Figure 18-14 Placer deposits. These mineral concentrations develop wherever the velocity of flowing water is significantly reduced, such as at potholes in a stream bed, downstream from a constriction in a channel, at the confluence of two or more streams, at the inside of a meander bend, and at coasts, where dense minerals settle from the water and are reworked by wave action. **(a)** Placer deposits of emeralds being mined in Colombia. **(b)** Emerald merchants in Colombia show their wares.

(a)

(b)

and dolostones containing aluminum-rich clay, chemical reactions in the contact zone may produce corundum. Other metallic deposits produced by contact metamorphism include iron, zinc, lead, and copper.

Sedimentary processes produce a range of valuable metal deposits. Water-borne heavy materials such as gold, platinum, and tin become sorted during transport based on their density and durability, concentrating as **placer deposits** wherever flowing water slows (Fig. 18-14). For instance, nuggets of gold eroded from hydrothermal vein deposits in California's Sierra Nevada batholiths have been carried down the range's numerous streams for hundreds of thousands of years, some to be eventually found by lucky prospectors panning for gold in that region. Extremely durable minerals such as diamonds, sapphires, rubies, and emeralds form placer deposits as well, because they resist abrasion and can survive long journeys that might wear away softer minerals.

Metals at or near the Earth's surface may become concentrated when other minerals around them weather and dissolve, leaving behind a metallic residue, or when the metals themselves are dissolved by acidic groundwater and then precipitated in concentrated form at the water table. These *secondary enrichment* processes take place only in an environment characterized by a warm climate, abundant atmospheric water, permeable bedrock (to promote groundwater flow), and a soluble matrix surrounding valuable materials. The major ore produced by secondary enrichment is bauxite, an aluminum oxide that serves as the most abundant source of aluminum. Bauxite is found in tropical locations, such as Jamaica, as well as in nontropical regions, such as France, Australia, and Arkansas, that were tropical in the past.

Metals and the Environment The mining of metals is associated with a variety of environmental problems. Most major deposits are mined from open pits by operations that begin by removing millions of tons of *overburden,* the rock and regolith that cover ore deposits. These waste materials, or *mine tailings,* along with the great volume of waste from ore mills, are typically piled into huge hills and often left without covering vegetation. In this form, they are highly susceptible to rapid erosion and mass movement. In addition, surface-water runoff from waste piles may clog regional streams with silt or pollute those flows with toxic dissolved metals or acid. In the mine pits themselves, ore rocks may come in contact with water, thereby contaminating it with sulfides. Sulfides oxidize to form sulfuric acid, producing acid ground and surface water.

These and other problems have prompted the U.S. Congress to pass legislation requiring monitoring and control of mine-water discharge. State and federal regulations also require land reclamation after a mining operation ends. The mine operator must isolate the sulfides from the groundwater system, restore the topography, and replant vegetation to prevent erosion.

Nonmetals

The search for gold, silver, and platinum has unleashed veritable stampedes of prospectors and altered the histories of entire states and nations. Although we have not seen any "gravel rush" or "phosphate rush" matching the gold rushes of nineteenth-century America, such nonmetals are at least as useful and valuable as precious metals. Nonmetal resources range from the building materials derived from common rocks to the natural mineral fertilizers used to increase our food supply.

Nonmetal Building Materials The combined economic value of natural building materials is exceeded only by that of petroleum. Limestone, the most widely used building material, provides crushed stone for roadbeds, cut stone for monuments and other stately buildings, and the key ingredient in Portland cement, a staple of the construction industry. Other types of stone are quarried for specific building purposes. For example, slate is used for roofing, flooring, fireplaces, and patios; slabs of polished granite, diorite, gneiss, and other attractive coarse-grained rocks are employed in the facades of office and government buildings; retaining walls that support unstable slopes are constructed from blocks of basalt; and tons of crushed limestone, granite, marble, and schist lie beneath highways as roadbed fill.

Gypsum, a soft and soluble nonmetal, is rarely used as a building stone. Heating gypsum to 177°C (351°F), however, drives off about 75% of its water, leaving behind a powdery substance known as plaster of Paris (named for the gypsum quarries near Paris that produce high-quality plaster). Plaster of Paris is used in the manufacture of plaster board (wallboard) and in such fields as art, dentistry, and orthopedics.

Mixing sand and gravel with cement forms concrete, which is used principally in building foundations and roads. A single kilometer of a four-lane highway, for example, requires about 40 tons of gravel for its concrete. Common sources of sand and gravel include river channel and bar deposits, coastal offshore bars, beach deposits, sand dunes, and glacial eskers and outwash (Fig. 18-15).

Clay minerals are end products of the chemical weathering of feldspars. As a result, they remain relatively stable in the Earth's surface weathering environment. When wet and plastic, clays can be shaped into a variety of building materials, such as decorative terra cotta bricks, tiles, and pipes, that harden when fired.

Nonmetals in Agriculture and Industry Soon after the year 2000, unless we curb world population growth, the number of people on Earth will exceed 7 billion. Feeding so many humans using soil that has, in many areas, already been depleted of nutrients by overcultivation will require a vast supply of chemical fertilizers to hasten plant growth and increase the rate of sugar and starch production.

Figure 18-15 A gravel operation in glacial outwash in Ontario, Canada.

Phosphorus and potassium are two major components of agricultural fertilizers. Phosphorus is derived primarily from the mineral apatite, which is found in certain marine sedimentary rocks, in guano deposits (bird and bat droppings) in caves, and in some calcium-rich igneous rocks. In North America, most phosphorus comes from ancient sedimentary deposits in Montana and Wyoming and from recently deposited sediments along coastal North Carolina and Florida. Potassium is obtained primarily from sylvite, an evaporite deposited in shallow marine basins during periods of climatic warming. Large sylvite deposits exist in New Mexico, Utah, Colorado, Montana, and Saskatchewan.

Among the most common nonmetal resources is pure quartz sand, which can be melted and then cooled rapidly to produce glass. Ancient Egyptians were the first to manufacture glass from melted sand, 4000 years ago. Today, 30 million tons of sand are quarried each year in North America from the purest quartz arenite deposits, such as the St. Peter sandstone of Illinois and Missouri, which is 95% quartz. Slight impurities in the sand subsequently impart color to the glass.

Another nonmetal resource with multiple industrial applications is *asbestos,* a generic term for a variety of magnesium-rich white, gray, and green fibrous silicates, including the minerals serpentine and chrysotile. This nonflammable material is used primarily in fireproof clothing, in electrical insulation, and in the linings of automobile brakes. Its use in the construction industry has ended, however, because inhalation of asbestos dust has been linked to several lung and digestive-tract diseases. Asbestiform silicates are produced primarily by metamorphism of olivines and pyroxenes in ultramafic mantle rocks of continental collision zones. The high-grade metamorphic rocks of the northern Appalachians of Vermont and Quebec contain asbestos reserves.

Future Use of Natural Resources

The United States has about 50 billion barrels of known oil reserves recoverable by current methods and may have an additional, as-yet-undiscovered, 35 billion barrels. The world's oil reserves total approximately 700 billion barrels. At the current worldwide petroleum use rate of nearly 21 billion barrels per year, these reserves will last only 35 years or so. Although shortages and price increases may eventually reduce the rate of oil consumption, somewhat extending the life of current supplies, ways must be found to extend the finite crude oil supply or to develop alternative energy sources before the reserves run out during the twenty-first century.

Large new petroleum finds are unlikely; virtually all potential oil-producing rock formations have already been explored. One way to maximize our oil reserves, however, is to extract as much as possible from known reservoir rocks.

Figure 18-16 Nodules of manganese oxide. These nodules, which are found abundantly on the ocean floor, may become a source of valuable minerals in the near future.

Currently economical methods of oil extraction remove only 20% to 30% of the oil within the rocks. As oil shortages increase, it may become more economical to use already-available enhanced recovery methods that can reclaim as much as 50% of the oil. Extraction of more than that percentage is virtually impossible, because the oil tends to adhere tightly to the rock particles.

Other technological initiatives seek alternatives to petroleum for transportation energy. These options include the liquefication and gasification of coal as fuel, and the development of practical electric automobiles, solar-powered buses, and trains propelled by powerful electromagnets.

The supply of mineral resources remains adequate, although some are located in remote places. Valuable deposits of important metals and nonmetals are still being discovered. Indeed, major ore bodies have recently been unearthed in Chile, Australia, and Siberia. Extensive outcrops rich in zinc, lead, copper, nickel, cobalt, and uranium have also been found in Antarctica, along with a 120-kilometer (70-mile)-long, 100-meter (300-feet)-thick deposit of iron, which is large enough to meet the world's iron demand for 200 years.

Soon we may be able to mine the vast mineral deposits on the sea floor, which in some places is covered by billions of nodules of manganese oxide, some approaching the size of bowling balls. These nodules (Fig. 18-16), which contain lesser amounts of iron, nickel, copper, zinc, and cobalt, accumulate where deep-sea sedimentation occurs slowly enough to allow them to grow without being buried. Collectively, they may represent the Earth's largest mineral deposit.

Most mineral resources can be recycled. Steel from "tin" cans, automobiles, and old bridges, mercury from discarded thermometers, copper from electrical wire, and platinum from the catalytic converters of abandoned automobiles can all be reclaimed for reuse. Recycling offers several benefits: It reduces the volume of waste requiring disposal, causes less land area to be disturbed by new mining operations, and minimizes the energy needed to mine and refine new ores. Scrap aluminum recycling, for example, uses only one-twentieth the energy needed to mine and process an equivalent amount of new aluminum from bauxite (Fig. 18-17).

Everything we have, or can make, comes from the Earth. Now that you understand the processes that built, moved, and shaped the continents and ocean floors, and the vast amount of time over which the Earth and its resources developed, you can appreciate why we must act wisely in managing the materials that make our modern lives possible. If we fail in this task, most assuredly, little will remain for future generations. Industry, local governments, and the world community of nations must cooperate to ensure that the search for, development of, and use of the planet's resources do not squander those precious resources or irreversibly damage our shared environment. With the knowledge you have acquired from your geology training, you are now prepared to contribute to the ongoing debate about the need to balance resource development and environmental protection. By all means, use your training and make yourself be heard.

Figure 18-17 Recycling as a method of preserving resources. Beverage cans are processed into high-density bales in a recycling plant in New Jersey. Their aluminum can then be recycled.

Chapter Summary

Reserves are quantities of natural resources that have been discovered and can be exploited profitably with current technology. An **ore** is a mineral deposit that can be mined for a profit; this classification is based on economics, rather than geology. **Resources** are deposits that we believe to exist, but that are not exploitable today because of various technological, economic, or political issues. Renewable natural resources, such as trees and solar energy, are either replenished naturally over a short time span or can be used continuously without being depleted. Nonrenewable resources, such as fossil fuels, form so slowly that they are consumed much more quickly than nature can replenish them.

Fossil fuels are energy sources derived from the organic remains of past life. The most common and versatile of the fossil fuels is **petroleum,** a group of gaseous, liquid, and semi-solid substances composed chiefly of **hydrocarbons,** molecules consisting entirely of hydrogen and carbon. Today, most of our energy needs are met by two petroleum fuels, oil and natural gas. These fuels typically form in marine basins where abundant microscopic organisms live, die, and eventually become buried by younger sediments. Pressure and geothermal heat transform the organic remains first into **kerogen,** a solid waxy material, and finally into oil and natural gas. These petroleum products form in **source rocks,** typically shales and siltstones formed from marine muds, then migrate into permeable **reservoir rocks. Oil shale** is a common source rock containing kerogen. **Oil sand** is unconsolidated clay or sand that contains a semi-solid, tar-like hydrocarbon called **bitumen.** Both kerogen and bitumen can be refined to produce gasoline, fuel oil, and other commercially useful products.

Coal is a sedimentary rock consisting of the combustible, highly compressed remains of land plants. It begins to form with the accumulation of **peat,** a soft, brown mass of compressed, largely undecomposed plant materials. Bacterial action within the sediment gradually breaks the vegetation down into an organic muck. Over time, pressure from the weight of overlying deposits transforms peat into **lignite,** and finally into **bituminous coal.** Metamorphic conditions may then transform lignite and bituminous coal to **anthracite.**

Burning sulfur- and nitrogen-rich hydrocarbons releases sulfur and nitrogen oxides, which may combine with water in the atmosphere to form sulfuric acid and nitric acid, the components of **acid rain.** Other environmental concerns associated with the use of fossil fuels include global warming and massive marine oil spills.

As fossil fuel supplies dwindle, we face a pressing need to discover or further develop alternative energy sources. Such options may include geothermal energy, hydroelectric power, tidal power, solar power, wind power, energy from the burning of **biomass fuels,** and energy released by the splitting of uranium atoms (**nuclear fission**). **Nuclear fusion** occurs when extremely high pressure and temperature cause the atomic nuclei of ultralight elements to fuse together to form heavier atoms; it is also a potential source of energy.

In addition to fossil fuels, the other principal group of resources is minerals, including both metals and nonmetals. Metals—such as iron, copper, aluminum, gold, and silver—are concentrated into ore deposits by igneous, metamorphic, and sedimentary processes. **Hydrothermal deposits** form by the precipitation of metallic ions from hot, ion-rich water. **Placer deposits,** containing materials such as gold, platinum, and tin, form when streams carry the materials and then selectively deposit them according to their density or durability.

Nonmetals are widely used as building materials (such as sand and gravel for concrete, limestone for cement, gypsum for plaster of Paris, clay for bricks) and agricultural fertilizers (such as phosphorus and potassium). Among the nonmetals commonly used in industry are pure quartz sand, used for glass production, and asbestos, used for fireproof clothing and electrical insulation.

At the current worldwide rate of use, our petroleum supplies will last only 35 years or so. New technologies must therefore be developed that will allow us to use these supplies more efficiently, or even to replace them with alternative sources of energy. The worldwide supply of mineral resources remains adequate, with valuable deposits of important metals and nonmetals still being discovered.

Key Terms

reserves (p. 353)
ore (p. 354)
resources (p. 354)
fossil fuels (p. 354)
petroleum (p. 354)
hydrocarbons (p. 354)
kerogen (p. 355)
source rocks (p. 355)
reservoir rocks (p. 355)
oil shale (p. 356)
oil sand (p. 356)
bitumen (p. 356)
peat (p. 357)
lignite (p. 357)
bituminous coal (p. 357)
anthracite (p. 357)
acid rain (p. 358)
biomass fuels (p. 363)
nuclear fission (p. 364)
nuclear fusion (p. 365)
hydrothermal deposits (p. 366)
placer deposits (p. 367)

Questions for Review

1. How do resources differ from reserves? Under what circumstances might a resource become a reserve, and vice versa?

2. What are fossil fuels and why are they considered nonrenewable resources? List three kinds of fossil fuels.

3. What is the key physical property of a petroleum reservoir rock? Why are oil and gas found more often in reservoir rocks than in source rocks?

4. Discuss two reasons why petroleum is not more abundant, given that it has had hundreds of millions of years to accumulate.

5. How do lignite, bituminous coal, and anthracite form from peat? What are the advantages of using the later-forming types of coal rather than the earlier-forming ones?

6. Briefly describe three environmental hazards commonly associated with the use of fossil fuels.

7. List four alternative energy sources and briefly explain how they are used.

8. Discuss two geological processes by which metals may become concentrated in rocks.

9. Name three types of nonmetallic mineral resources and describe their uses.

10. List four benefits gained from recycling mineral resources.

For Further Thought

1. Using your general knowledge of the geology of North America, explain why aquatic life in the lakes of Yosemite National Park, California, would be more susceptible to the effects of acid rainfall than aquatic life in the lakes of Florida and Indiana.

2. Referring to the illustration in Figure 18-14, in which areas would you search for oil? Gold? Sand and gravel?

3. Speculate on why copper mining in the United States has been so drastically reduced in recent years. Under what circumstances might U.S. mining companies resume extensive copper mining?

4. Propose a U.S. energy plan for the year 2050. Which energy sources would you develop? Why?

5. Suppose the Earth's natural resources become completely exhausted and it is decided to prospect on neighboring planets. On which planet might we find extensive iron, chromium, and platinum deposits? On which might we find sand and gravel? Could any of our planetary neighbors provide us with limestone or gypsum?

19

A Brief History of Earth and Its Life Forms

Assembling the puzzle that is Earth's history depends on both geological and fossil evidence. From geological evidence, geologists can reconstruct the physical stage upon which life forms developed over time. The study of fossil evidence, known as paleontology, reveals how these life forms evolved.

One of the most astonishing paleontological finds of this century occurred at the end of summer in 1909 in the Canadian Rockies. Charles D. Walcott, head of the Smithsonian Institute, was fossil hunting along Burgess Ridge near the town of Field, British Columbia. Splitting open several blocks of shale, Walcott and his party found numerous fossils consisting of the impressions of a remarkable number of soft-bodied organisms. In the following years, and continuing even today, thousands of fossil specimens have been extracted from the Burgess Shale. The extraordinary part of this find is that it was almost like viewing a completely intact scene of life in the Cambrian Period (early in the Paleozoic Era), some 530 million years ago (Fig. 19-1). One of the greatest discoveries of middle Cambrian life, the Burgess Shale brought into focus an otherwise obscure part of the history of life on Earth.

Fossils can be found in many parts of the world, including many localities in the United States. Grade school students commonly bring in sharks' teeth or small clamlike fossils for their teachers to examine and identify. For a number of reasons, however, even the most fossil-rich localities possess only a limited variety of fossil types. First, most living organisms are restricted to specific types of habitats and do not exist everywhere in the world at any one time. For example, most corals flourish only in warm shallow marine environments. Second, because such a narrow set of conditions supports fossilization, many species are probably completely absent from the planet's fossil record (see Highlight 8-1). Furthermore, most fossils represent only the hard parts—such as the shells, bones, exoskeletons, or teeth—of the original organisms. Paleontologists must use all of their knowledge and experience to develop even a partial picture of the long-extinct ecosystem in which such organisms might have lived. By combining this information with that derived from same-aged fossils found in many different locations,

Figure 19-1 Revelations from the fossil record. A diorama of organisms found in the Burgess Shale, one of the most important paleontological discoveries of this century.

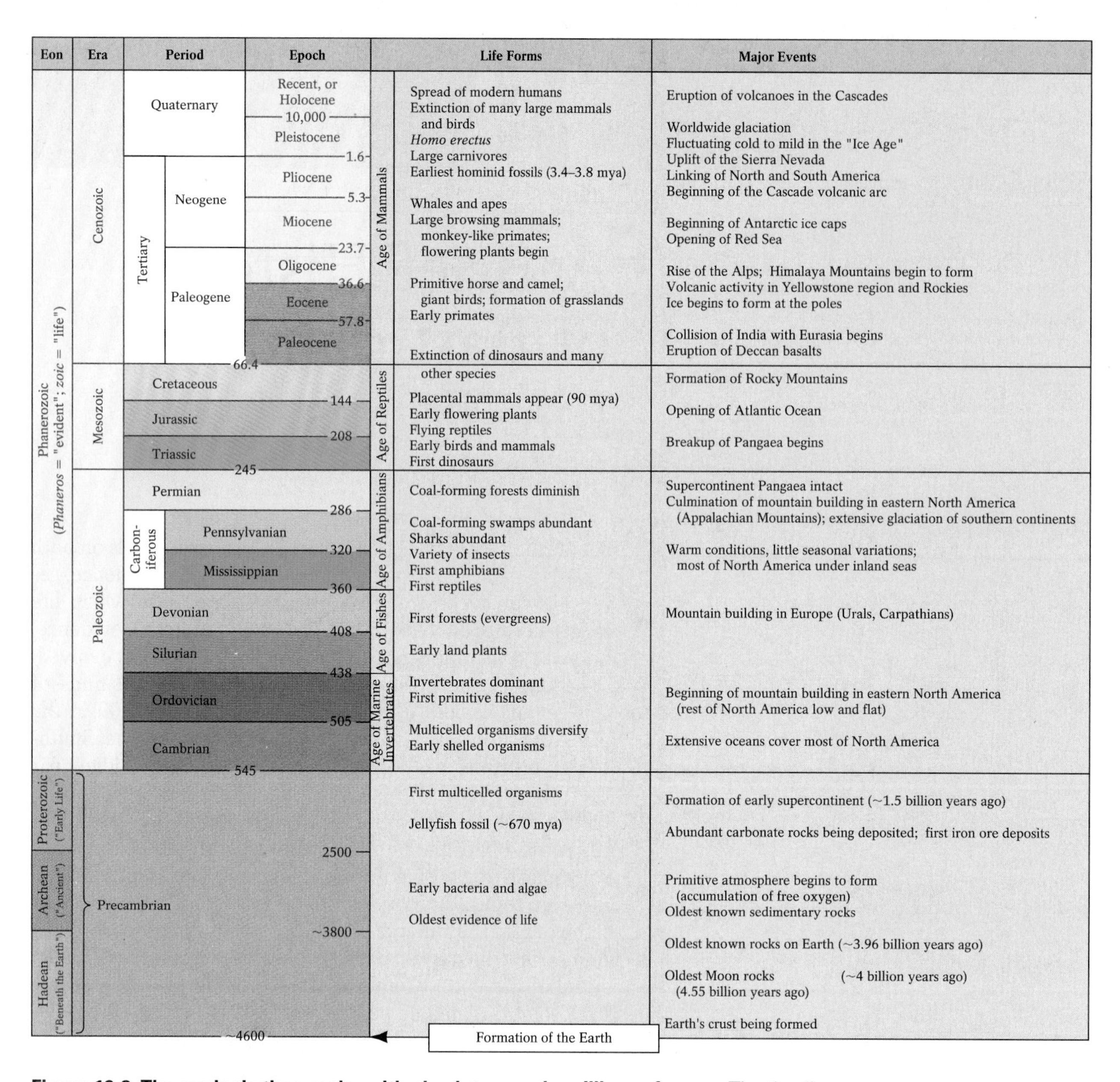

Eon	Era	Period	Epoch	Boundary (mya)	Age	Life Forms	Major Events
Phanerozoic (*Phaneros* = "evident"; *zoic* = "life")	Cenozoic	Quaternary	Recent, or Holocene		Age of Mammals	Spread of modern humans Extinction of many large mammals and birds	Eruption of volcanoes in the Cascades
				10,000			
			Pleistocene			*Homo erectus* Large carnivores	Worldwide glaciation Fluctuating cold to mild in the "Ice Age" Uplift of the Sierra Nevada
				1.6			
		Tertiary — Neogene	Pliocene			Earliest hominid fossils (3.4–3.8 mya)	Linking of North and South America Beginning of the Cascade volcanic arc
				5.3			
			Miocene			Whales and apes Large browsing mammals; monkey-like primates; flowering plants begin	Beginning of Antarctic ice caps Opening of Red Sea
				23.7			
		Tertiary — Paleogene	Oligocene				Rise of the Alps; Himalaya Mountains begin to form
				36.6			
			Eocene			Primitive horse and camel; giant birds; formation of grasslands Early primates	Volcanic activity in Yellowstone region and Rockies Ice begins to form at the poles
				57.8			
			Paleocene				Collision of India with Eurasia begins Eruption of Deccan basalts
				66.4		Extinction of dinosaurs and many other species	
	Mesozoic	Cretaceous			Age of Reptiles		Formation of Rocky Mountains
				144		Placental mammals appear (90 mya) Early flowering plants	
		Jurassic				Flying reptiles	Opening of Atlantic Ocean
				208			
		Triassic				Early birds and mammals First dinosaurs	Breakup of Pangaea begins
				245			
	Paleozoic	Permian			Age of Amphibians	Coal-forming forests diminish	Supercontinent Pangaea intact Culmination of mountain building in eastern North America (Appalachian Mountains); extensive glaciation of southern continents
				286			
		Carboniferous — Pennsylvanian				Coal-forming swamps abundant Sharks abundant	
				320			
		Carboniferous — Mississippian				Variety of insects First amphibians First reptiles	Warm conditions, little seasonal variations; most of North America under inland seas
				360			
		Devonian			Age of Fishes	First forests (evergreens)	Mountain building in Europe (Urals, Carpathians)
				408			
		Silurian				Early land plants	
				438			
		Ordovician			Age of Marine Invertebrates	Invertebrates dominant First primitive fishes	Beginning of mountain building in eastern North America (rest of North America low and flat)
				505			
		Cambrian				Multicelled organisms diversify Early shelled organisms	Extensive oceans cover most of North America
				545			
Proterozoic ("Early Life")	Precambrian					First multicelled organisms Jellyfish fossil (~670 mya)	Formation of early supercontinent (~1.5 billion years ago) Abundant carbonate rocks being deposited; first iron ore deposits
				2500			
Archean ("Ancient")						Early bacteria and algae Oldest evidence of life	Primitive atmosphere begins to form (accumulation of free oxygen) Oldest known sedimentary rocks
				~3800			
Hadean ("Beneath the Earth")							Oldest known rocks on Earth (~3.96 billion years ago) Oldest Moon rocks (4.55 billion years ago) (~4 billion years ago) Earth's crust being formed
				~4600		Formation of the Earth	

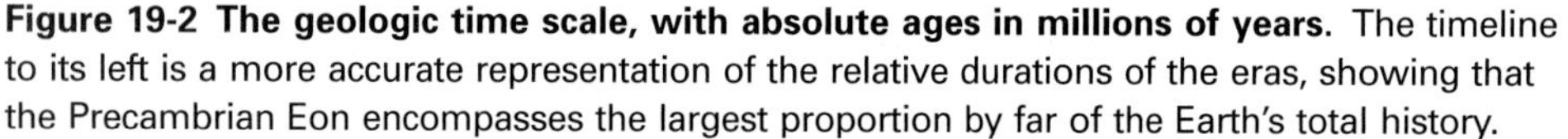

Figure 19-2 The geologic time scale, with absolute ages in millions of years. The timeline to its left is a more accurate representation of the relative durations of the eras, showing that the Precambrian Eon encompasses the largest proportion by far of the Earth's total history.

paleontologists strive to create a wider, more accurate picture of the regional or even world environment that existed at the time. Because of the sparse nature of the evidence, their work is comparable to that of anthropologists who try to re-create an entire human civilization based on only a few shards of pottery.

The Geologic Time Scale

The **geologic time scale** (Fig. 19-2) organizes all of Earth history into blocks of time during which important events occurred. Although many geologic events coincide with the divisions of the geologic time scale, most units of the scale are determined primarily by biological—not geological—developments. The basic divisions of the geologic time scale were established during the nineteenth and early twentieth centuries by geologists and paleontologists who identified changing *fossil assemblages*—groupings of fossil organisms that, when found together within sedimentary strata, indicate the relative age of the strata—and then applied the principles of superposition, faunal succession, and cross-cutting relationships (discussed in Chapter 8) to establish their sequence. Within the last century, radiometric dating has enabled scientists to assign absolute-age ranges to these divisions.

The largest time spans shown on the geologic time scale are **eons,** which designate the major developments of the Earth. The earliest of these eons is the **Hadean Eon,** a time from which no fossils and virtually no rocks remain. The ear-

Figure 19-3 **An outcrop of Acasta Gneiss, in the Northwest Territories of Canada.** Grains of the mineral zircon that have been extracted from this gneiss have been dated at 3.96 billion years, making them some of the few surviving remnants of the Hadean Eon.

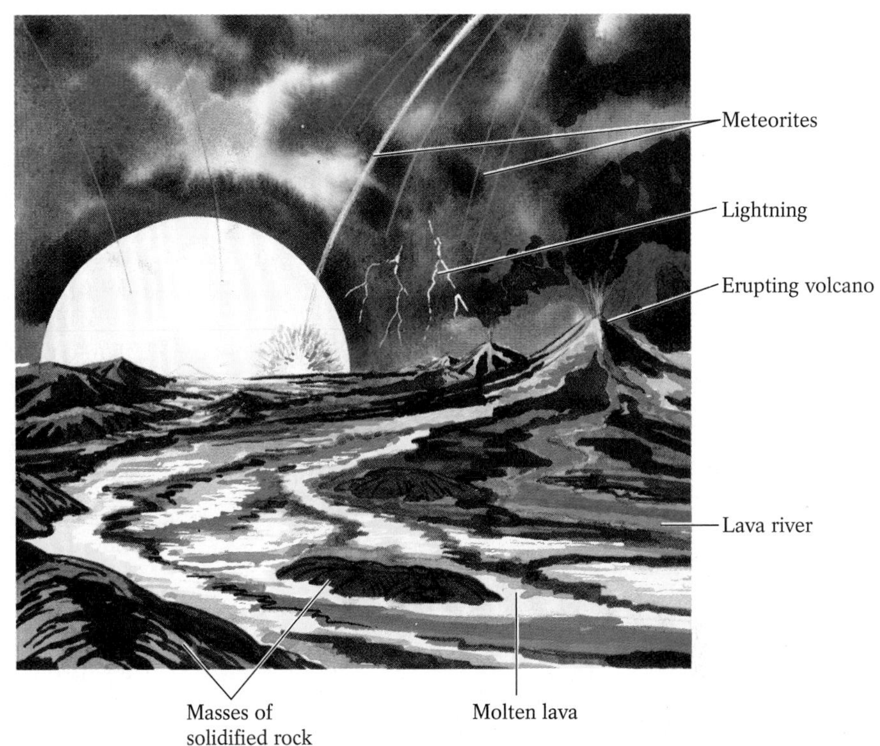

Figure 19-4 The molten Earth. The surface of the Earth during the Hadean Eon is believed to have been composed of a sea of gurgling molten lava in which only a few isolated islands and masses of mafic and ultramafic rock began to form. Constant meteoritic bombardment, residual heat from the Earth's formation, and heat from radioactive decay kept the surface inhospitably hot for many millions of years.

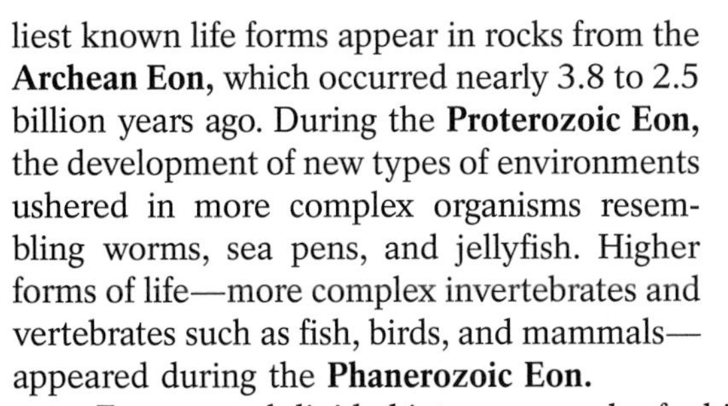

liest known life forms appear in rocks from the **Archean Eon,** which occurred nearly 3.8 to 2.5 billion years ago. During the **Proterozoic Eon,** the development of new types of environments ushered in more complex organisms resembling worms, sea pens, and jellyfish. Higher forms of life—more complex invertebrates and vertebrates such as fish, birds, and mammals—appeared during the **Phanerozoic Eon.**

Eons are subdivided into **eras,** each of which is defined by its dominant life forms; the Mesozoic Era of the Phanerozoic Eon, for instance, is the "Age of Reptiles." Eras are subdivided into **periods,** which record less dramatic distinctions. Periods, in turn, are subdivided into **epochs,** based on still more subtle biological criteria and some nonbiological criteria. Each of these successive subdivisions marks progressively smaller units of time. The lengths of time covered vary, however, even for subdivisions at the same level, because the key characteristics and geological conditions that define them do not share a uniform duration.

Precambrian History

Evidence of life on Earth before 570 million years ago (the beginning of the Cambrian Period) is sparse. As a result, even though it represents 88% of all geologic time, Precambrian history is divided into only three major divisions, while the last 12% of all geologic time is the most detailed portion of the geologic time scale. The three major divisions of Precambrian history are the Hadean, Archean, and Proterozoic eons.

The Hadean Eon

The Hadean ("beneath the Earth") Eon began with the formation of the Earth approximately 4.6 billion years ago and continued until some 3.77 billion years ago, when the first simple bacteria appeared. Because virtually none of the rocks or geologic formations from that time period still exists, we know little about this eon (Fig. 19-3). The Hadean is thought to have been marked by extensive meteoric bombardment and by the differentiation of the interior of the Earth into layers of contrasting densities (discussed in Chapter 1). The crust of the Earth began to form during the Hadean, beginning with the development of many mafic and ultramafic microcontinents and volcanic islands that later coalesced as the nuclei of our present-day continents (see Chapter 11).

During the Hadean Eon, there were no animals, plants, or even bacteria. No photosynthesis took place, the atmosphere did not contain any oxygen (but rather mostly nitrogen, carbon dioxide, and water vapor), and no oceans existed. There was insufficient landmass for plate tectonics to have begun. The small bodies of land that did appear on the planet, within the sea of molten lava at the Earth's surface, were hot, desolate, and completely barren (Fig. 19-4).

The Archean Eon

The Archean ("ancient") Eon stretched from about 3.77 to 2.5 billion years ago, encompassing an interval of almost 1.3 billion years. The beginning of the Archean Eon was marked by the first appearance of life—early single-celled bacteria called cyanobacteria, or blue-green algae (Fig. 19-5).

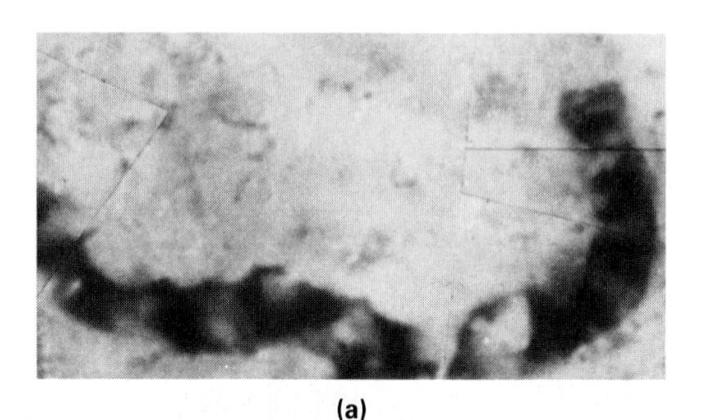

(a)

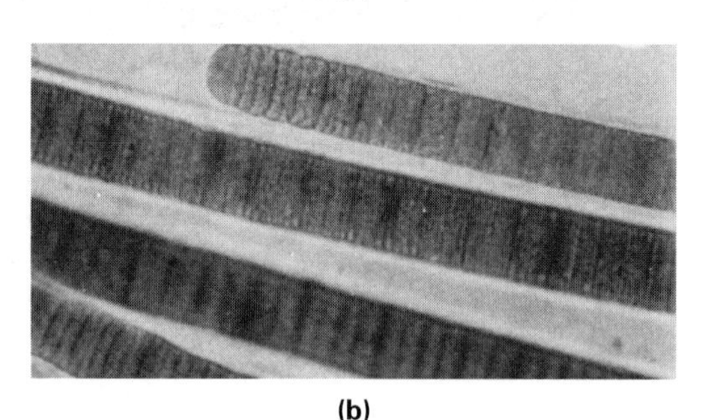

(b)

Figure 19-5 Early life forms. The first organisms on Earth were probably simple one-celled blue-green algae. **(a)** A fossilized Archean alga (about 3.5 billion years old). **(b)** *Lyngbya,* a modern species of blue-green alga.

The oldest fossil evidence of blue-green algae—organisms so simple in structure that they are **prokaryotic** (pronounced "PRO-kar-ee-AT-ic"), or lacking a nucleus—was discovered in 1980 in northwestern Australia, in a layer of sedimentary chert dating from 3.77 billion years ago. Other findings include numerous specimens of Archean-age **stromatolites,** layered reef structures that develop as sediments become entrapped within colonies of blue-green algae (Fig. 19-6).

Early in the Archean Eon, atmospheric water vapor began to condense to form oceans. Landmasses resembled those found in the Hadean Eon—no large continents, but many microcontinents and volcanic islands. The latter part of the Archean, however, saw the initial development of the continental cratons. Microcontinents and volcanic islands began to collide and become sutured together, creating small continental collision zones and subduction zones. This early form of plate tectonics resulted in metamorphism of the predominant ultramafic and mafic volcanic rocks, a few felsic volcanic rocks, and deep-sea sedimentary rocks. In fact, almost all surviving rocks of Archean age are metamorphic, consisting primarily of granite-gneiss complexes and secondarily of the low-grade metamorphic basalt greenstone (Fig. 19-7). Archean limestones and sandstones also exist, but they are rare, since no large continents—and hence, no large continental shelves to receive sediment—were in existence during the early Archean Eon.

The Proterozoic Eon

The Archean Eon was followed by the Proterozoic ("early life") Eon, which lasted from 2.5 billion to 570 million years ago. This time period spanned almost 2 billion years, or 42% of Earth's history.

(a)

(b)

Figure 19-6 Stromatolites. Fossilized algal reef structures are the most prevalent evidence of life prior to the Cambrian Period. **(a)** Archean-age stromatolites. **(b)** Modern stromatolites in Shark Bay, Western Australia.

Figure 19-7 Archean rocks of the Canadian Shield area of North America. The granite-gneiss complexes are metamorphic remnants of Archean volcanic islands. The greenstone belts are metamorphosed basalts and sediments that were deposited in basins between the volcanic islands.

Proterozoic Life Forms Life during the early part of the Proterozoic still consisted of simple, non-nucleated, single-celled bacteria. Some 1.4 billion years ago, however, larger and more complex organisms appeared that were composed of nucleus-containing, or **eukaryotic** (pronounced "YOU-kar-ee-AT-ic"), cells (Fig. 19-8). The internal structure and metabolic processes of eukaryotic cells were tremendously more complex than those of prokaryotic cells. Eukaryotes were also much more diversified, leading scientists to speculate that these organisms were the distant ancestors of all higher forms of life. Because no predators existed that might eat either the prokaryotic blue-green algae or the eukaryotic algae that now coexisted with them, the Proterozoic seas were probably well saturated with these organisms. As a result, photosynthesis by these organisms increased the atmospheric oxygen content from a meager 1% to an estimated 10% during the Proterozoic Eon. The introduction of these massive amounts of oxygen into the atmosphere made the emergence of higher life forms possible. It was probably not until about 400 million years ago, during the Devonian Period of the Paleo-

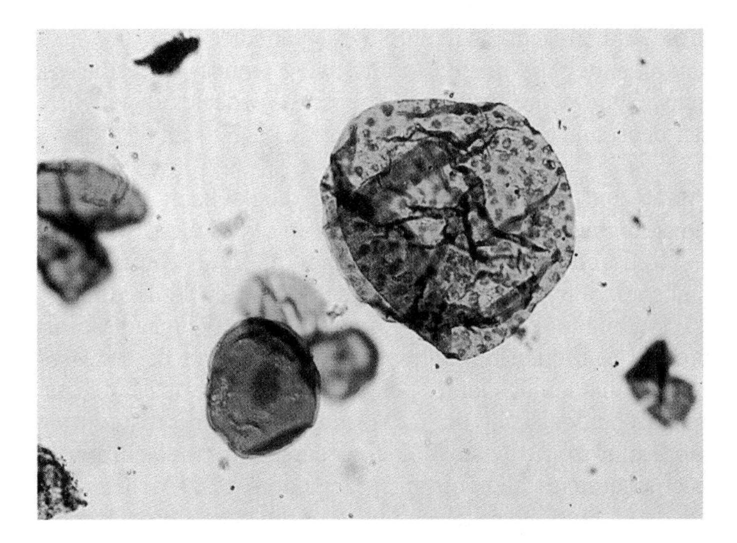

Figure 19-8 A eukaryotic bacterium from the Proterozoic, about 850 million years old. Far more complex than the non-nucleated prokaryotes, the single-celled eukaryotic organisms of the Proterozoic Eon represented a major advance in the Earth's life forms.

Figure 19-9 The Ediacaran fauna, probably Earth's first multicellular life forms. (a) *Charniodiscus arboreus,* an Ediacaran sea pen. **(b)** Jellyfish-like *Mawsonites spriggi.* **(c)** *Spriggina floundersi,* an early segmented worm.

zoic Era, that continued photosynthesis produced enough oxygen to attain the present level of 21% of the atmosphere, however.

Approximately 800 million years ago, but perhaps as long as 1.4 billion years ago, tremendously more complex multicellular algae appeared on Earth. Eventually, very late in the Proterozoic, the first soft-bodied marine animals emerged, called **Ediacaran fauna** (named for the Ediacara Hills of southern Australia, where geologists have found thousands of fossil impressions of these animals). Because only impressions of these animals exist, paleontologists cannot definitively state whether they are related to any living groups. Nevertheless, many believe the fossils represent early species of jellyfish, sea pens, and various types of worms (Fig. 19-9).

Proterozoic Plate Tectonics During the Archean Eon, collisions among small microcontinents and volcanic islands had resulted in the first subduction zones. As these events forced mafic and ultramafic rocks into the interior of the Earth, partial melting took place that produced less-mafic magmas. This process continued throughout the Archean and the Proterozoic, so that a much larger percentage of the Earth's landmass had become felsic by the end of the Proterozoic. Also during the Proterozoic, the small cratons created by the collision of microcontinents and volcanic islands during the Archean collided and themselves formed most of the large cratons we see today. The most active period of crustal development took place between 3.0 and 2.5 billion years ago. During this time, the bulk of the North American craton was formed, as well as the cratons of most of the other present-day continents.

Although the evidence is not conclusive, from about 1.2 to 1.0 billion years ago (around the same time that the first multicellular organisms appeared), almost all of the Earth's landmasses may have been sutured together into three or four supercontinents, and possibly into one large supercontinent—a Proterozoic version of Pangaea (Fig. 19-10). Finally, about 800 million years ago, almost at the end of the Proterozoic Eon, the supercontinents began to rift, and the resulting continental landmasses began to drift apart.

Evidence in the form of ancient glacial deposits indicates that the Earth experienced periods of extensive glaciation, or ice ages, during the Proterozoic Eon. The first, which covered extensive portions of the cratons that later became North America, South Africa, and Australia, probably began during the late Archean Eon and lasted until about 2.3 billion years ago. The second, which occurred during the late Proterozoic (from 900 to 600 million years ago), consisted of at least four separate glacial advances that, at one time or another, covered extensive areas of each of the continents.

The Paleozoic Era

Thanks to a far greater body of evidence, our knowledge of the most recent eon, the Phanerozoic, is much more detailed than our knowledge of the Precambrian eons. The Phanerozoic ("evident life") Eon began 570 million years ago with the appearance of organisms with fossilizable hard shells,

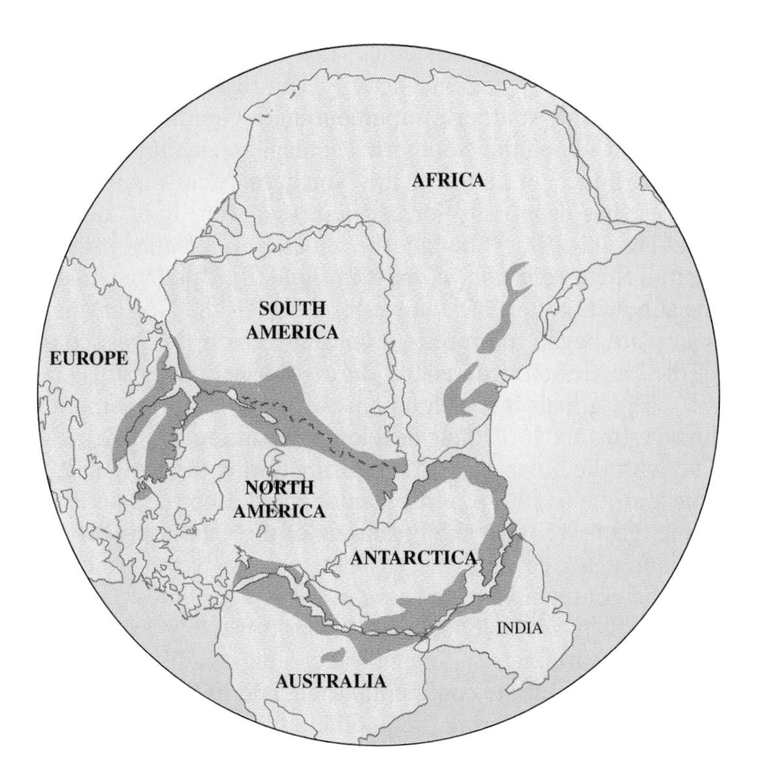

Figure 19-10 Early Pangaea. Evidence indicates that an early version of the supercontinent Pangaea may have existed during part of the Proterozoic Eon. Numerous orogenic episodes have since changed the shape of many of the continents, so the exact arrangement of this early Pangaea remains unknown. (Pink areas indicate rocks of similar age at the edges of modern continents, where the continents may have been linked in the past.)

prominent external spines, and internal skeletons; this eon continues today. It is divided into three eras: the Paleozoic, Mesozoic, and Cenozoic.

The **Paleozoic** ("ancient life") **Era** designates the time period from 570 million to 245 million years ago—a span of 325 million years, accounting for 7% of Earth's history. Geologically, this era was characterized by intense mountain building, intense erosion, *transgressions* and *regressions* (advances and retreats) of continent-wide shallow inland seas, rearrangement of continental landmasses, and the opening and closing of ocean basins. These events provided a multitude of new environments for the planet's diverse organisms. Biologically, the era was dominated by marine invertebrates, but it also saw the appearance of higher animals—fish (late Cambrian), amphibians (Devonian), and reptiles (early Pennsylvanian)—and land plants (middle Silurian).

The Early Paleozoic

The early Paleozoic Era encompassed the Cambrian, Ordovician, and Silurian periods.

Biological and Geological Characteristics The Cambrian Period (and thus the Phanerozoic Eon) began with the sudden widespread appearance of numerous marine organisms with hard parts. These animals included the now-extinct trilobites, various clam-like organisms, snail-like organisms, sponges, and organisms related to starfish and sand dollars (Fig. 19-11).

Figure 19-11 Life in the shallow seas of the Devonian Period. This diorama shows many of the life forms that may have been ancestors to some of our present-day organisms. Many, including the trilobite, were arthropods, related to today's insects and to crustaceans such as lobsters and crabs. Crinoids are related to today's starfish, sand dollars, and sea urchins.

At this time, North America lay within the equatorial regions, rotated nearly 90° clockwise from its present orientation so that the equator ran through present-day Canada toward the present-day middle-southern states. The resulting warmer climates undoubtedly played a central role in the appearance of the many new life forms that emerged during this time.

Mass extinctions also occurred in the early Paleozoic Era. At the end of the Ordovician Period, more than 100 families of marine invertebrates ceased to exist. Some geologists suspect that a late Ordovician ice age may have been the cause, but hard evidence is elusive and the exact reason for this mass extinction remains unknown—a statement that is equally true of the many other mass extinctions that have occurred in a nearly cyclical pattern throughout geologic time.

Twice during the early Paleozoic, North America was almost completely covered by shallow oceans, now referred to as the *Sauk* and *Tippecanoe* seas. Today, throughout much of the United States (in Missouri, Kentucky, Kansas, and Tennessee, for example) one can see tremendous expanses of marine limestone that attest to previous occupation by these and other ancient inland seas (Fig. 19-12). Sediment deposited as these seas transgressed over the continent also appears in the walls of the Grand Canyon in the form of sandstone, shale, and limestone.

Early Paleozoic Plate Tectonics The early Paleozoic was a time of significant tectonic activity. Throughout much of the Paleozoic, the southern supercontinent Gondwana, which contained all of the Southern Hemisphere landmasses, remained intact; in contrast, the Northern Hemisphere landmasses that eventually formed Laurasia were quite mobile (see Fig. 11-20). Although the mountains of the western United States did not yet exist in the early Paleozoic, ongoing subduction beneath the western edge of the North American plate began to produce a volcanic arc off the west coast of the North American craton during the last part of this era.

The Appalachian Mountains of eastern North America underwent much of their development during the Paleozoic Era, with the northern portion arising during the early Paleozoic and the southern portion emerging during the late Paleozoic (Fig. 19-13). At the beginning of the Cambrian Period, a shallow sea covered the eastern coast of North America, and the continental craton was overlain with limestone. The region was tectonically inactive, just as it is today. Sandstone, shale, and carbonate sediments were being steadily deposited on the craton, on the continental shelf, and in the deeper water off the coast. During the middle Ordovician Period, however, the Proto-Atlantic (*Iapetus*) Ocean began to close, with subduction zones forming on both the American and European sides of the ocean. This action deformed the eastern coast of North America, uplifting parts of the continental mar-

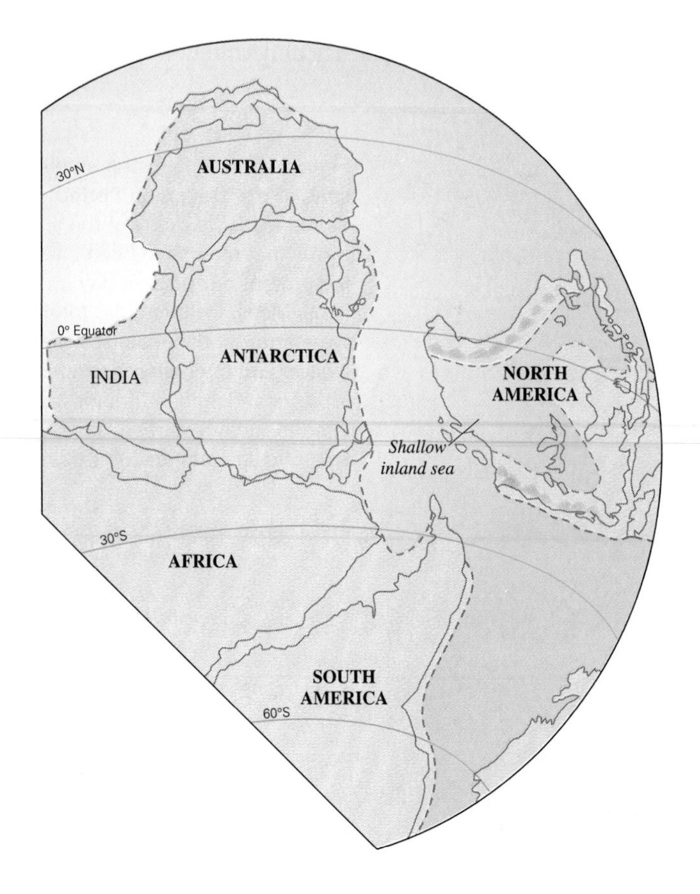

Figure 19-12 The reach of inland seas. Several times in geologic history, vast shallow inland seas, teeming with a tremendous variety of life forms, covered much of the North American continent. The limestones and dolomites of mid-America are rife with fossils from these ancient marine environments. Photo: Layers of limestone rock in Cowley County, Kansas.

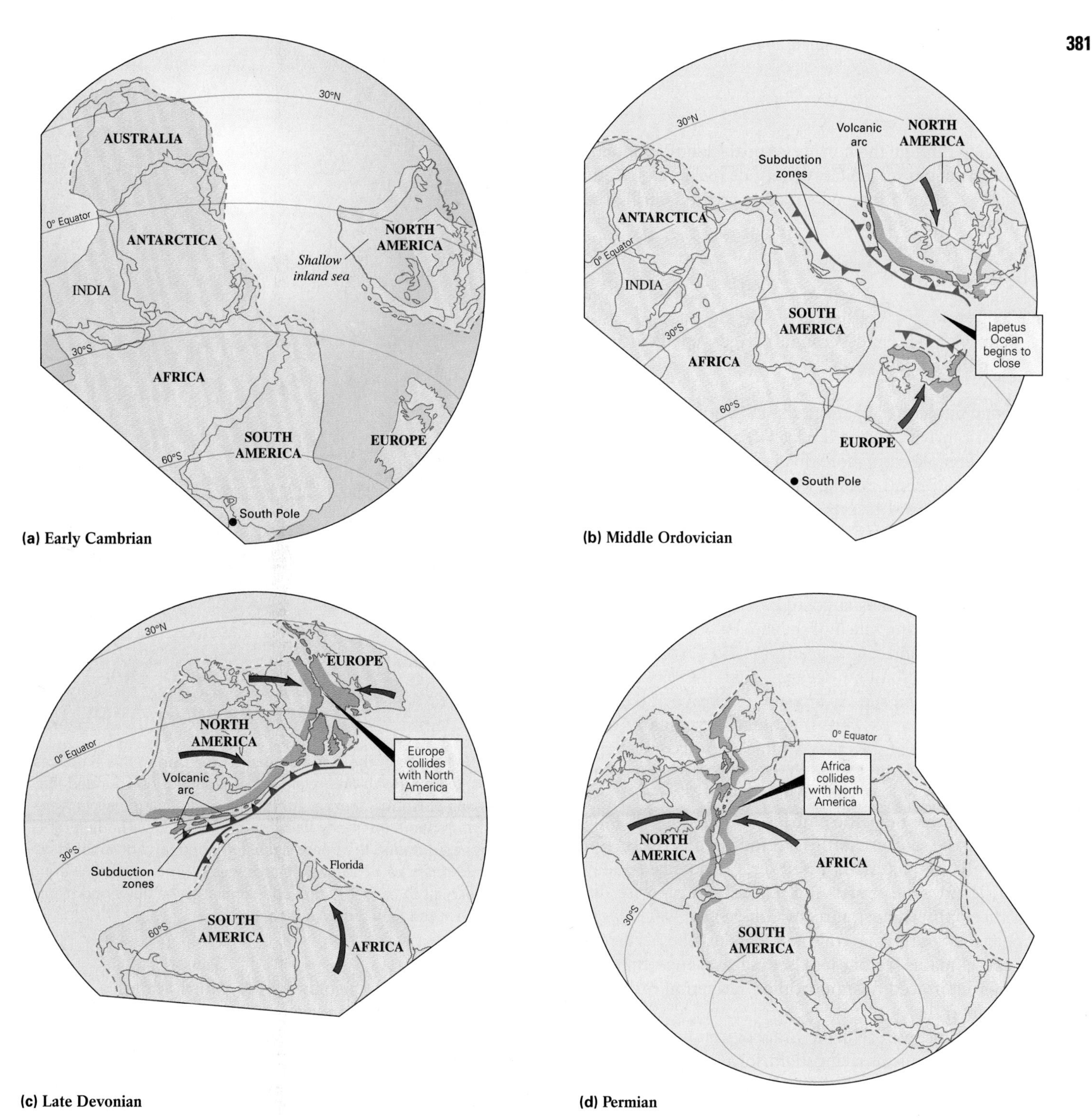

Figure 19-13 Formation of the Appalachian Mountains. The northern portions of the Appalachian Mountains were formed from the collision of North America and Europe during the late Devonian Period of the Paleozoic Era. Later, during the Permian Period, the southern portions were formed by the collision of North America and northern Africa.

gin and downwarping other parts to form sedimentary basins. In addition, a volcanic arc formed off the eastern coast of North America (much as another volcanic arc—Japan—later formed off the coast of the Asian continent). Volcanism also took place on the eastern side of the ocean, creating volcanic mountains in western Europe. Erosion of these volcanic mountains and uplifted regions produced sediment that began to fill basins at the margins of both continents. Finally, during the late Devonian Period of the late Paleozoic Era, the Proto-Atlantic Ocean closed completely between North America and Europe. The resulting continental collision merged all of these sediments, volcanic and intrusive igneous materials, and both continental and oceanic crustal material, much of which became metamorphosed and highly faulted by the collision process. Driving through the northern Appalachian Mountains today, one can see an incredible variety of rocks that attests to the complex composition and structure created during the formation of these mountains.

The Late Paleozoic

The late Paleozoic Era, consisting of the Devonian, Mississippian, Pennsylvanian, and Permian periods, spanned from 408 to 245 million years ago.

Biological and Geological Characteristics While invertebrate marine life still thrived during the late Paleozoic, life on land became more complex and more diverse (Fig. 19-14). The Devonian Period (408–360 million years ago) saw the appearance of the first winged insects, as well as the first amphibians and the first extensive forests. The first sharks, which differed but little from the sharks of today, appeared in the oceans. During the Pennsylvanian Period (320–286 million years ago), the first reptiles appeared.

The climate during much of the late Paleozoic was generally warm and humid. Once again, much of North America was covered by shallow oceans—first the *Kaskaskia,* and then the *Absaroka* sea. As a result, extensive and abundantly vegetated swamps were prevalent, particularly during the Mississippian and Pennsylvanian periods, which are collectively known as the Carboniferous Period. Almost all of our present-day coal resources are derived from these swamps, which grew near the shifting margins of the great inland seas of the Carboniferous Period. Other regions of the world were characterized by deserts, evidenced today by ancient desert sand dunes and extensive evaporite deposits. In the southernmost parts of Gondwana, which were located near the South Pole, continental glaciation occurred. All of these climatic factors left behind evidence that Alfred Wegener later used to support his continental drift theory (see Chapter 11).

At the end of the Paleozoic Era, environmental conditions for Earth's organisms changed drastically. Extensive tectonically induced mountain building and volcanism brought about continental uplift and regression of the inland seas. Seasonal changes became more pronounced, arid regions became more extensive, and climates became much more severe. As a result, Earth saw a mass extinction of both marine and terrestrial organisms, including jawed fishes and the trilobites that had introduced the Paleozoic Era. Corals and other sea life, amphibians, reptiles, and many groups of plants were greatly diminished. This period of mass extinction provided a clear demarcation between the end of the Paleozoic Era and the beginning of the Mesozoic Era.

Late Paleozoic Plate Tectonics With Gondwana already existing as a single landmass, and North America and northern Europe having become sutured together during the early Paleozoic, the late Paleozoic witnessed the final assembly of Pangaea during the Permian Period. North America collided with Africa and with Gondwana (in the area that is now South America), an event that completed the formation of the southern Appalachian Mountains.

Figure 19-14 Life in the late Paleozoic Era. During this time, vast swamps, teeming with both plant and animal life, surrounded the inland seas of North America. Amphibians and flying insects were abundant, as were the many varieties of marine life. The first reptiles appeared during the Pennsylvanian Period, and diversified during the Permian. Late Paleozoic plants were primitive nonflowering species similar to modern ferns and conifers—the flowering plants so dominant today had not yet appeared.

On the western side of North America, the subduction beneath the North American plate of its neighbor, the Farallon plate, had formed a volcanic island arc off the coast during the last part of the Early Paleozoic. Now, from the Devonian through the early Mississippian periods, in a tectonic episode called the Antler orogeny, further subduction caused the volcanic islands to collide with the craton; this process resulted in extensive faulting and crushing of the sedimentary basin that lay between the islands and the mainland, and it produced a mountain chain called the Antler Mountains (see Fig. 19-17). The Antler Mountains have since been worn away by erosion; indeed, many of the mountains now present in western North America were not born until the Mesozoic Era. (The majestic Rocky Mountains, which are geologically still very young, were not born until the Cenozoic Era, only 66 million years ago.)

Figure 19-15 A new conception of dinosaurs. More bird-like than reptile-like, many dinosaurs are believed to have attended their young in huge dinosaur nests. Groups of fossil nesting sites have been found in the western United States. Photo: Geologist John Horner with a dinosaur hatchling and eggs found at Egg Mountain in Montana.

The Mesozoic Era

The **Mesozoic** ("middle life") **Era,** which gave rise to the dinosaurs, began 245 million years ago and lasted until the extinction of the dinosaurs 66 million years ago (see Highlight 1-1). Thus the "Age of Reptiles" lasted 179 million years, or for almost 4% of Earth's history. (By comparison, humans have been in existence during at most only 0.02% of Earth's history.)

Mesozoic Life Forms

In recent years, paleontologists have been reassessing their perceptions of dinosaurs, the dominant animals in the Mesozoic Era. The term *dinosaur* means "terrible lizard," and in the past dinosaurs were considered simply as large reptiles. More recent evidence, however, indicates that at least one of the two orders of dinosaurs may have been more bird-like than reptile-like. These *Ornithischian* dinosaurs may have been warm-blooded and possessed a bird-like pelvic structure. Dinosaur nesting sites indicate that adult dinosaurs may have cared for newly hatched baby dinosaurs while they remained in their nests—a behavior similar to that seen in modern birds (Fig. 19-15).

The rise and fall of the dinosaurs was not the only distinction of the Mesozoic Era. The first mammals appeared in the late Triassic. Mammal skull fragments, jaws, and teeth—many with reptile-like features—indicate that mammals throughout the Mesozoic were probably small, shrew-like nocturnal creatures. Mammals did not gain any dominance until the Cenozoic Era, however, after the dinosaurs had become extinct. Plants began to diversify tremendously during the early Mesozoic Era, giving rise to an abundance of pines, cedars, and firs, as well as the first flowering plants. Birds also first appeared during this era, perhaps developing from a line of Ornithischian dinosaurs during the Jurassic Period. *Archaeopteryx,* often considered to be the first bird, was about the size of a chicken and was basically a reptile with feathers.

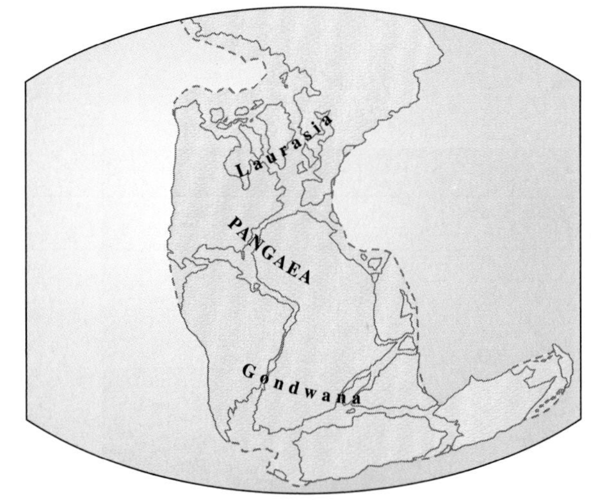

(a) Intact Pangaea

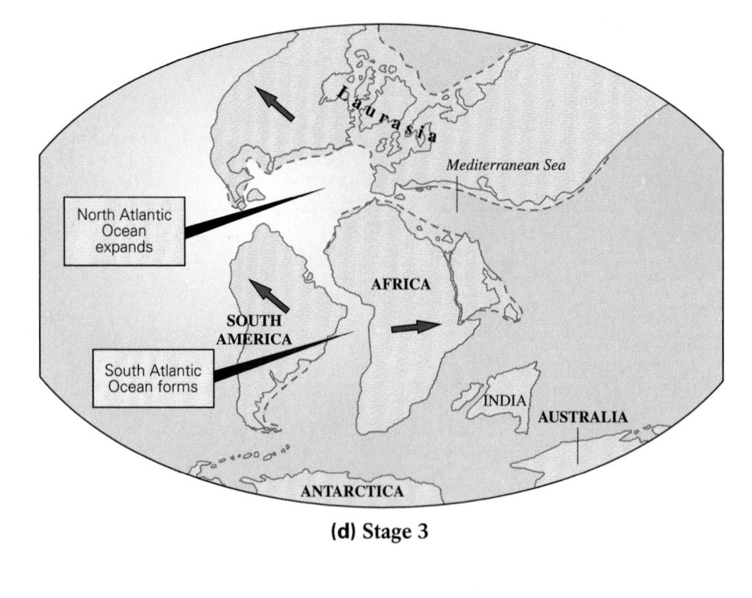

(d) Stage 3

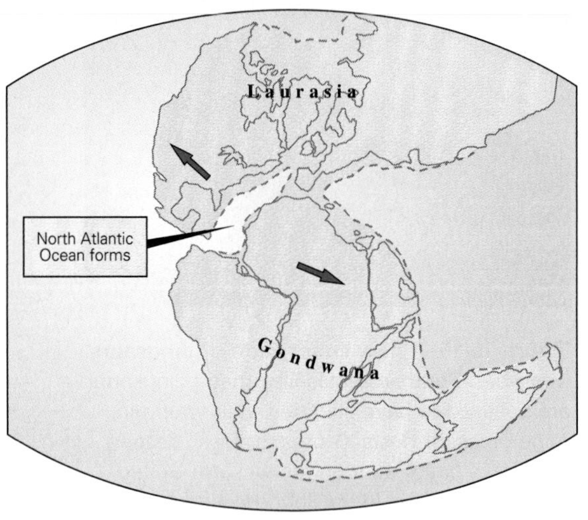

(b) Stage 1

(e) Stage 4

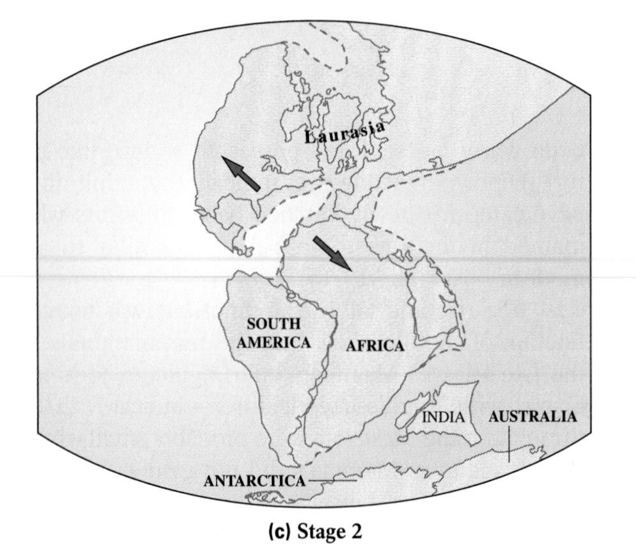

(c) Stage 2

Figure 19-16 The four stages of the breakup of Pangaea. **(1)** Rifting and separation of Laurasia from Gondwana. **(2)** Breakup of the Gondwana continents. **(3)** Separation of South America from Africa and formation of the Mediterranean Sea. **(4)** Separation of Greenland from Europe and North America.

Mesozoic Plate Tectonics

During the Triassic Period, Pangaea began to break apart, an event that greatly altered oceanic and atmospheric circulation patterns and led to a worldwide change in climate, allowing for the establishment of many new and varied environments. The breakup of Pangaea occurred in four stages (Fig. 19-16). The first stage involved the rifting and separation of Laurasia from Gondwana and the opening of the Atlantic Ocean; North America became separated first from Africa (leaving a small portion of Africa stuck to North America, seen today as part of South Carolina and Georgia) and then from South America, forming the Gulf of Mexico. The second stage involved the breakup of some of the Gondwana continents; Australia and Antarctica remained together and moved away from South America and Africa, which also stayed together. India became a separate landmass. The third stage began with the separation of South America from Africa during the late Jurassic. At the same time, the Eurasian plate was pivoting so that the *Tethys Sea* began to close, forming the modern Mediterranean Sea. Australia and Antarctica separated at this time, and Greenland began to separate from Europe. The fourth

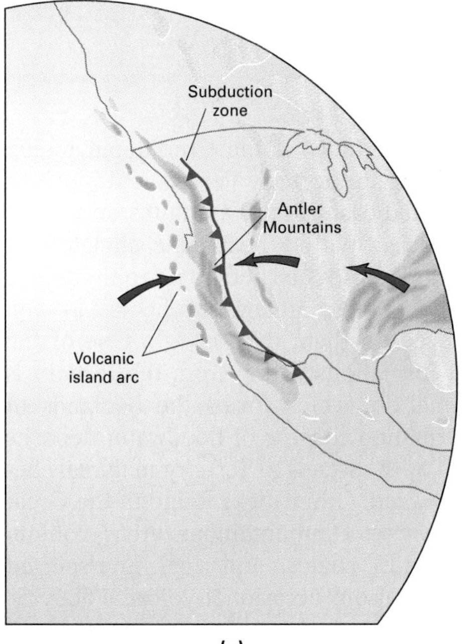

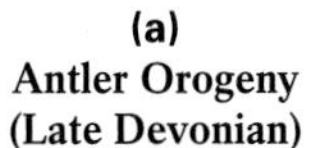

(a)
Antler Orogeny
(Late Devonian)

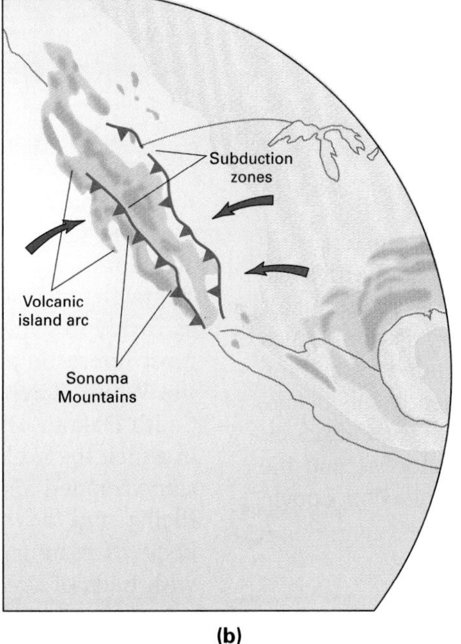

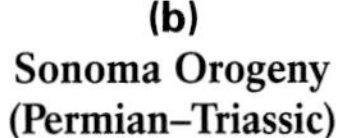

(b)
Sonoma Orogeny
(Permian–Triassic)

(c)
Nevadan Orogeny
(Jurassic–Cretaceous)

Figure 19-17 The Cordilleran Mountains. The development of these mountains of western North America took place during five separate orogenies: **(a)** the Antler, during the late Devonian; **(b)** the Sonoma, spanning the Permian–Triassic boundary; **(c)** the Nevadan, spanning the Jurassic–Cretaceous boundary; **(d)** the Sevier, during the Cretaceous; and **(e)** the Laramide, spanning the Cretaceous–Tertiary boundary.

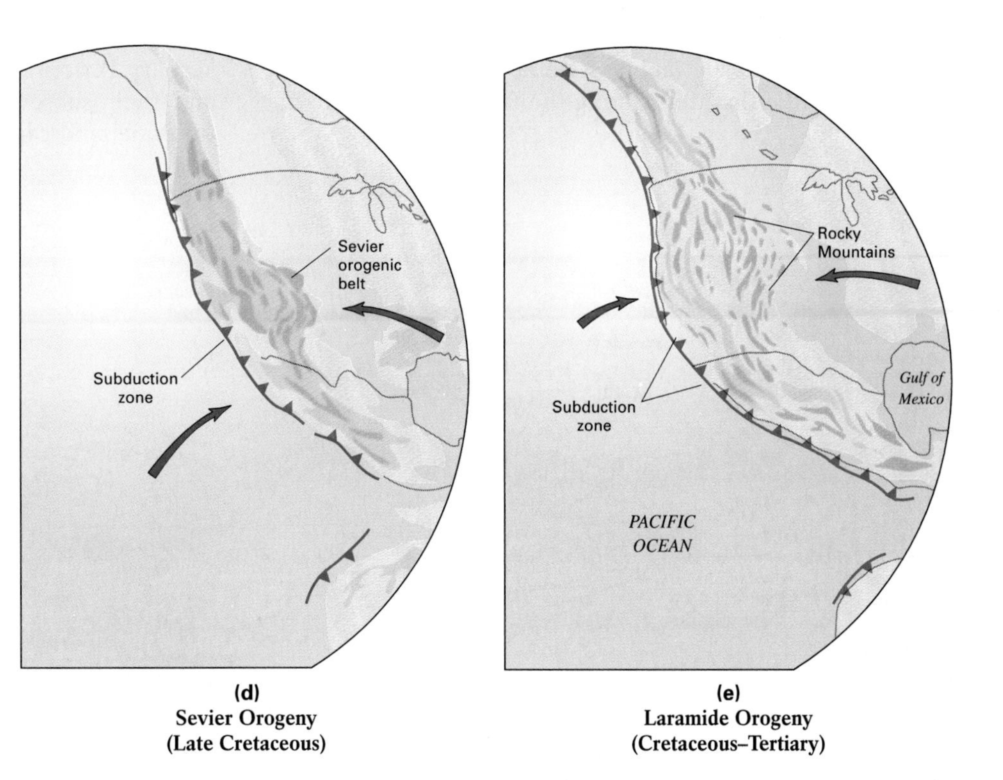

(d)
Sevier Orogeny
(Late Cretaceous)

(e)
Laramide Orogeny
(Cretaceous–Tertiary)

stage, the complete separation of Greenland from both Europe and North America, did not occur until the Cenozoic Era.

In the western and now-mountainous portion of North America (the Cordilleran region), almost no tectonic activity had taken place since the Antler orogeny of the Devonian and early Mississippian periods. Then, approximately 245 million years ago, at the Paleozoic–Mesozoic boundary, North America again collided with an eastward moving island arc, producing the Sonoma Mountains. The island arc accreted to the continent, forming a new continental shelf that gradually became covered by a blanket of sandstones, shales, and limestones. As the sea regressed, the continental shelf became exposed to erosion, resulting in the spectacular landscapes we now see in the Painted Desert area of Arizona.

During the late Triassic Period, a subduction zone remained along the western margin of North America; this zone continued to govern the tectonics of the western part of the continent until the middle of the Cenozoic Era. Subduction beneath the North American plate caused countless numbers of volcanic arcs and microcontinents to accrete to the North American continent, forming as much as 70% of the Cordilleran region (see Fig. 11-19). In addition to these displaced terranes, and beginning with the Antler orogeny during the Devonian Period, a total of five separate major orogenic events over a period of almost 350 million years combined to produce the complex mountainous structures of the western United States (Fig. 19-17). Today's Rocky Mountains, relatively young and still rugged, resulted from the final episode, the Laramide orogeny, which took place during the Tertiary Period.

The Cenozoic Era

The era in which we are presently living, the **Cenozoic** ("recent life") **Era,** represents only the latest 1.4% of Earth's history, yet it began a full 66 million years ago. The Cenozoic Era is divided into two periods. The Tertiary Period, which began at the end of the Mesozoic Era and ended 1.6 million years ago, includes the Paleocene, Eocene, Oligocene, Miocene, and Pliocene epochs. The Quaternary Period includes the two most recent epochs, the Pleistocene and the Holocene; we are currently living in the Holocene Epoch.

The Tertiary Period

Biological and Geological Characteristics It was during the relatively brief Tertiary Period, almost the last bit of geologic time, that many of the great landforms and geographic features with which we are familiar emerged. Furthermore, many of the plants and animals we now consider common appeared for the first time during the Tertiary—the "Age of Mammals."

In the West, the mountains of the Cordilleran region continued to erode during the Tertiary Period, and much of the material was deposited east of the mountains on what is now the Great Plains area. With a hospitable climate and abundant vegetation, the prairies became home to thousands of grazing animals such as deer, camels, and horses, as well as carnivores that preyed on them (Fig. 19-18). One of the most interesting rock formations dating from the Tertiary is the White River Formation, which forms the badlands of South Dakota. This formation consists of floodwater deposits in which the skeletons of thousands of Tertiary mammals became trapped and preserved. Other rocks found in the Great Plains and near the western mountainous areas contain beds of nonmarine sands, shales, and coal interspersed with beds of volcanic ash from occasional volcanic activity.

Tertiary Plate Tectonics Subduction of the Pacific plate under the North American plate, as well as subduction of the smaller Farallon plate, which for a time lay between the two, produced tremendous regional uplifts in the west during the Tertiary Period. The Rocky Mountains were greatly uplifted, and their subsequent erosion produced some of the picturesque landscapes we see today in that region. Uplift and

Figure 19-18 Blossoming of animal life in the Miocene Epoch of the Tertiary Period. This epoch witnessed vast herds of animals that roamed the plains, most of which eventually became extinct in North America. The animals included early species of camels, horses, and deer, pig-like peccaries, and elephant-like mastodons. Meat-eating species such as the saber-toothed cat also flourished at this time.

faulting produced the Basin and Range Province (Arizona, Nevada, New Mexico, and southern California) and the rugged Grand Tetons of Wyoming. In addition, the Colorado Plateau was pushed up at this time. The increased gradient of the Colorado River accelerated its erosive power, allowing it to fashion the magnificent Grand Canyon and other spectacular scenery. The Sierra Nevada Mountains, which had formed in the Nevadan orogeny during the Jurassic Period, became further uplifted during the Pliocene through the Pleistocene, exposing them to greater erosional processes. Increased erosion by both streams and glaciers cut the Sierra Nevadas into their present impressive forms.

During the Eocene Epoch, the leading edges of the Farallon plate disappeared beneath the North American plate, enabling part of the Pacific plate to butt up against North America. The Pacific plate, however, was moving in a northwesterly direction relative to the North American plate. Rather than subducting beneath the North American plate, this movement created a transform fault, now known as the San Andreas fault. The Columbia Plateau, located in Oregon, Washington, and Idaho, was created by a tremendous outpouring of basaltic lavas that erupted from deep intraplate fissures in the late Tertiary and Quaternary. Approximately 4 million years ago, a volcanic mountain chain, the Cascades, formed west of the Columbia Plateau. The Cascades are the site of Mount St. Helens, Mount Rainier, and about a dozen other still-active volcanos. The mountain chain was created by subduction of the relatively small Juan de Fuca plate; like the Cocos plate that is subducting today beneath Mexico, it is actually a remnant of the more extensive Farallon plate that once subducted beneath the entire west coast of North America (Fig. 19-19).

East of the North American plate, the Atlantic Ocean continued to widen as new oceanic crust was generated at the mid-ocean ridge. Little orogenic activity took place along North America's eastern regions, though broad gentle uplifts occurred along the Atlantic coastal plain. Late in the Tertiary Period, one of these uplifts hoisted an extensive region of marine limestone above sea level; today this region is the state of Florida.

Along the Gulf Coast, frequent marine transgressions extended the shallow *Tejas Sea* inland up what is now the Mississippi River basin, at one point extending as far north as southern Illinois. For much of the time, rapid deposition of deltaic sands occurred in the Gulf. At other times, shale layers were deposited, becoming interlaced with the sands. The many natural oil and gas traps found today in the Gulf of Mexico were produced by this interfingering of shale source rocks and sandstone reservoir rocks (see Chapter 18).

Prior to the Miocene Epoch of the Tertiary, Antarctica had basked in a semitropical climate. Beginning in the Miocene, however, the worldwide climate began to change, and snow and ice began to accumulate on that continent, transforming it into the frozen desert we see today. In Africa,

Figure 19-19 The Trinity Alps, part of the Cascade Range in northern California. The Cascades, which extend from northern California to British Columbia, were created by subduction of a neighboring plate under the western edge of North America about 4 million years ago.

the Great Rift Valley—where evidence of the earliest human beings would later be found—began to form. Throughout the Tertiary and continuing today, the Eurasian plate collided with the African and Indian plates to form the Alps, Himalayas, and several other lofty mountain ranges in that part of the world.

Finally, about 2 million years ago, global climate began to change drastically. The Pleistocene ice ages were about to begin.

The Quaternary Period

The Quaternary Period is divided into two epochs, the Pleistocene and the Holocene. The Pleistocene, which lasted from 1.6 million to about 10,000 years ago, was the "Ice Age," noted primarily for its alternating episodes of glacial advance and retreat (discussed in Chapter 15). The Holocene is the epoch in which we are now living.

Almost all of the history of the human race lies within the Quaternary Period. Although fossils of the earliest members of the human family (*Australopithecus afarensis*) have been dated at 3.4 to 3.8 million years, placing them near the end of the Tertiary Period, the first "true" humans (*Homo erectus*) appear in the fossil record at about 750,000 years ago, well within the Pleistocene. The skeleton of *Homo erectus* closely resembled that of modern humans, albeit with a substantially smaller cranial cavity and a more massive face that was shaped differently from ours. Evidence shows that *Homo erectus* employed tools, used and controlled fire for warmth and for cooking, and constructed communities of wood and stone shelters.

Figure 19-20 The birth of modern human beings. **(a)** The Cro-Magnon skull, at right, is identical to skull types found today, whereas the Neanderthal skull, at left, is noticeably larger, with heavy browridges and a protruding jaw. **(b)** This famous panorama of bulls, horses, and other creatures painted on a cave wall in Lascaux, France, some 15,000 years ago is evidence that the Cro-Magnon had a highly developed culture.

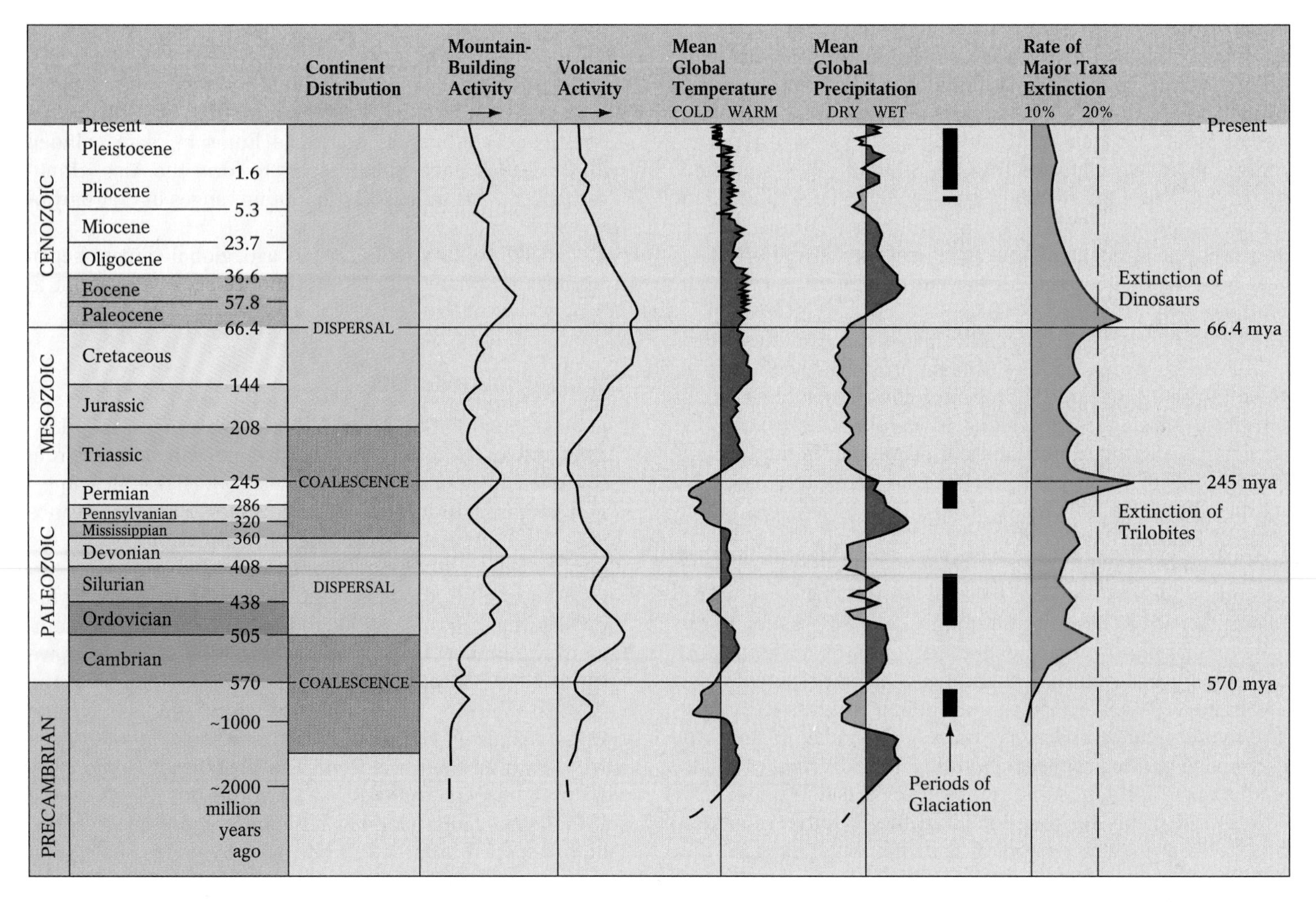

Figure 19-21 Composite chart showing all of the major geological and biological events in the 4.6-billion-year history of the Earth.

The Neanderthals, the humans we generally think of as "cavemen," first appeared in the middle Pleistocene in Europe. They are considered the first *Homo sapiens.* Contrary to the once-popular view, the Neanderthals were not dim-witted ape-like people. They had a true Stone Age culture—using a variety of stone tools and weapons—and left complex burial sites, evidence of a highly developed society. To survive in the harsh climate of an ice age, they frequently lived in caves or in stone shelters.

The first modern human beings, the Cro-Magnon, appeared in Europe about 34,000 years ago (Fig. 19-20). They may have either replaced the Neanderthals, possibly through warfare, or simply assimilated them through interbreeding. Like the Neanderthals, the Cro-Magnon frequently lived in caves; complex cave paintings found in France and other parts of the world have been attributed as their handiwork. These nomadic hunters were highly skilled, making use of bows and arrows, and may have been at least partially responsible for the extinction of many ice-age mammals, such as mammoths, mastodons, and saber-toothed cats. The Cro-Magnon people were in every way modern human beings. Indeed, their skull types are still found today among northern and western European and north African people.

Putting It All Together

It is extremely difficult to assimilate the tremendous volume of information that has been collected on Earth's long geologic past. By studying composites of the information, such as that shown in Figure 19-21, however, we can begin to fathom the wonderful changes that have taken place on our planet. New evidence constantly comes to light that alters portions of this picture. Clues are put together in new patterns, new fossils are found and dated, new technology is utilized in an ever-widening range of geologic applications, and new work fills in gaps in local, regional, and global geologic histories. The more we learn, the more we realize that the amazingly complex puzzle that is the Earth's history will probably never be completed.

Chapter Summary

The **geologic time scale** organizes all of Earth's history into blocks of time based primarily on important biological developments. The largest time spans, **eons,** designate the major developments. They are, from earliest to latest, the **Hadean, Archean, Proterozoic,** and **Phanerozoic.** Eons are divided into **eras,** which are subdivided into **periods,** which are further divided into **epochs.** Each level of these divisions is based on progressively more subtle biological criteria and some nonbiological criteria.

The Hadean Eon comprised almost the first billion years of Earth's history, from 4.6 to 3.77 billion years ago. It was characterized by extensive meteoric bombardment and by the differentiation of the Earth's interior into layers. During the Hadean, the crust of the Earth began to form, beginning with microcontinents and volcanic islands. There were no oceans, no oxygen in the atmosphere, and no life forms.

The Archean Eon encompassed the next 1.3 billion years of Earth's history. The beginning of the Archean is marked by the first appearance of life—simple **prokaryotic** (non-nucleated), single-celled bacteria called blue-green algae. Fossilized reefs built by these algae are called **stromatolites.** During the Archean, many of the microcontinents and volcanic islands collided in an early form of plate tectonics, producing a variety of metamorphic rocks.

The Proterozoic Eon began 2.5 billion years ago and ended 570 million years ago, accounting for 42% of the Earth's history. During the Proterozoic, the first **eukaryotic** (nucleated) organisms appeared. Photosynthesis by both prokaryotic and eukaryotic organisms was probably responsible for increasing atmospheric oxygen levels significantly. Later in the Proterozoic Eon, multicellular algae and then soft-bodied marine animals, called **Ediacaran fauna,** emerged. By the end of the Proterozoic, all of Earth's landmasses—which included the cratons of most present-day continents—had assembled into three or four supercontinents, and possibly into one large supercontinent. Two separate ice ages occurred during the Proterozoic, one at the beginning and one at the end, about 2 billion years later.

The Phanerozoic Eon is divided into the **Paleozoic, Mesozoic,** and **Cenozoic** eras. The Paleozoic began with the sudden appearance of many new varieties of marine invertebrates, including trilobites and many others. Later during the Paleozoic, the first insects, amphibians, reptiles, and forests emerged. Two mass extinctions occurred during this era, one at the end of the early Paleozoic and another at the end of the era. Geologically, the era was characterized by mobile Northern Hemisphere landmasses, which collided to produce many mountains, including much of the Appalachians. Transgressions and regressions of continent-wide inland shallow seas during the Paleozoic led to extensive continental deposition of sediment, and vast swamplands at the margins of the seas produced the plants that later became the coal resources we use today. Toward the end of the Paleozoic, the supercontinent Pangaea was assembled.

The Mesozoic Era, from 245 to 66 million years ago, was dominated by reptiles such as the dinosaurs, but also saw the emergence of mammals, flowering plants, and birds. During this era, Pangaea broke apart in four separate stages, beginning with the rifting of Laurasia from Gondwana and culminating with the final rifting of Greenland from both Europe and North America. Much of the Cordilleran region of western North America was assembled through accretion of several microcontinents and volcanic arcs. Five separate orogenic episodes built the mountains we now see in the American West.

The Cenozoic Era is divided into the Tertiary Period, which ended 1.6 million years ago, and the Quaternary Period, which continues today. The Tertiary was the "Age of Mammals," producing many of the animals and plants with which we are familiar today. Many modern geographical features developed into their present forms during the Tertiary, including in North America the Great Plains, the Rocky Mountains, the Basin and Range Province, the Colorado Plateau, the Sierra Nevadas, the Columbia Plateau, the Cascade Mountains, and the state of Florida. Elsewhere in the world, the Great Rift Valley, the Alps, and the Himalayas began to form.

The Quaternary Period is divided into two epochs. The Pleistocene, lasting from 1.6 million to about 10,000 years ago, is most notable for its ice ages, which were characterized by repeated episodes of glacial advance and retreat. The Holocene is the epoch in which we now live. Fossils of the first true humans, *Homo erectus,* appeared approximately 750,000 years ago. The first *Homo sapiens,* the Neanderthals, appeared in the middle Pleistocene. The first modern humans, the Cro-Magnon, appeared about 34,000 years ago.

Key Terms

geologic time scale (p. 374)
eons (p. 374)
Hadean Eon (p. 374)
Archean Eon (p. 375)
Proterozoic Eon (p. 375)
Phanerozoic Eon (p. 375)
eras (p. 375)
periods (p. 375)
epochs (p. 375)
prokaryotic (p. 376)
stromatolites (p. 376)
eukaryotic (p. 377)
Ediacaran fauna (p. 378)
Paleozoic Era (p. 379)
Mesozoic Era (p. 383)
Cenozoic Era (p. 386)

Questions for Review

1. Why was the discovery of the Burgess Shale fossils of special significance to paleontologists?

2. List the divisions and subdivisions of the geologic time scale. On what basis were these divisions made?

3. Why are almost all rocks of Archean age metamorphic?

4. When and how did our atmosphere attain its present level of oxygen?

5. Construct a timeline, spanning from 4.6 billion years ago to the present. Divide the timeline into the four eons of the geologic time scale, maintaining approximately correct proportionate lengths for each eon. Indicate on the line the appearance of each of the major biological groups mentioned in this chapter.

6. In what era did the Appalachian Mountains begin to form? What plate tectonic event was the main force behind their development?

7. What is believed to be the reason for the mass extinction that occurred at the end of the Paleozoic Era?

8. What significant evolutionary events divide the three eras of the Phanerozoic Eon?

9. Describe the four stages of the breakup of Pangaea.

10. What were the main geologic events of the Cenozoic that formed many of the geographic features found in North America today?

For Further Thought

1. The Earth has existed for 4.6 billion years—an almost incomprehensible span of time. As a quick project, find some object you can use to visualize this number. For example, you might count the dots in a ceiling tile and calculate how large a ceiling would be required to contain 4.6 billion dots, or you might count the number of grass blades in a square foot of lawn and calculate the size of a lawn required to contain 4.6 billion blades of grass.

2. Why is the variety of fossil types at any one fossil locality usually quite limited?

3. Why does the composition of the rocks of the North American craton differ from those of the eastern and western margins of the continent?

4. Several times during geologic history, shallow seas covered much of North America. Speculate on events that might have caused these tremendous fluctuations in sea level.

5. How might plate tectonic interactions have affected the survival of species throughout geologic time?

Appendix A

Conversion Factors for English and Metric Units

Length	1 centimeter	=	0.3937 inch
	1 inch	=	2.54 centimeters
	1 meter	=	3.2808 feet
	1 foot	=	0.3048 meter
	1 yard	=	0.9144 meter
	1 kilometer	=	0.6214 mile (statute)
	1 kilometer	=	3281 feet
	1 mile (statute)	=	1.6093 kilometer
Velocity	1 kilometer/hour	=	0.2778 meter/second
	1 mile/hour	=	0.4471 meter/second
Area	1 square centimeter	=	0.16 square inch
	1 square inch	=	6.45 square centimeters
	1 square meter	=	10.76 square feet
	1 square meter	=	1.20 square yard
	1 square foot	=	0.093 square meter
	1 square kilometer	=	0.386 square mile
	1 square mile	=	2.59 square kilometers
	1 acre (U.S.)	=	4840 square yards
Volume	1 cubic centimeter	=	0.06 cubic inch
	1 cubic inch	=	16.39 cubic centimeters
	1 cubic meter	=	35.31 cubic feet
	1 cubic foot	=	0.028 cubic meter
	1 cubic meter	=	1.31 cubic yard
	1 cubic yard	=	0.76 cubic meter
	1 liter	=	1000 cubic centimeters
	1 liter	=	1.06 quart (U.S. liquid)
	1 gallon (U.S. liquid)	=	3.79 liters
Mass	1 gram	=	0.035 ounce
	1 ounce	=	28.35 grams
	1 kilogram	=	2.205 pounds
	1 pound	=	0.45 kilogram
Pressure	1 kilogram/square centimeter	=	0.97 atmosphere
	1 kilogram/square centimeter	=	14.22 pounds/square inch
	1 kilogram/square centimeter	=	0.98 bar
	1 bar	=	0.99 atmosphere
Temperature	°F (degrees Fahrenheit)	=	°C(9/5) + 32
	°C (degrees Celsius)	=	(°F – 32)(5/9)

Appendix B

A Statistical Portrait of Planet Earth

Surface Areas

Landmasses	150,142,300 kilometers² (57,970,000 miles²)
Oceans and Seas	362,032,000 kilometers² (138,781,000 miles²)
Entire Earth	512,175,090 kilometers² (197,751,500 miles²)

Distribution of Water, by Volume

Oceans and Seas	1.37×10^9 kilometers³ (3.3×10^8 miles³)
Glaciers	2.5×10^7 kilometers³ (7×10^6 miles³)
Groundwater	8.4×10^6 kilometers³ (2×10^6 miles³)
Lakes	1.25×10^5 kilometers³ (3×10^4 miles³)
Rivers	1.25×10^3 kilometers³ (3×10^2 miles³)

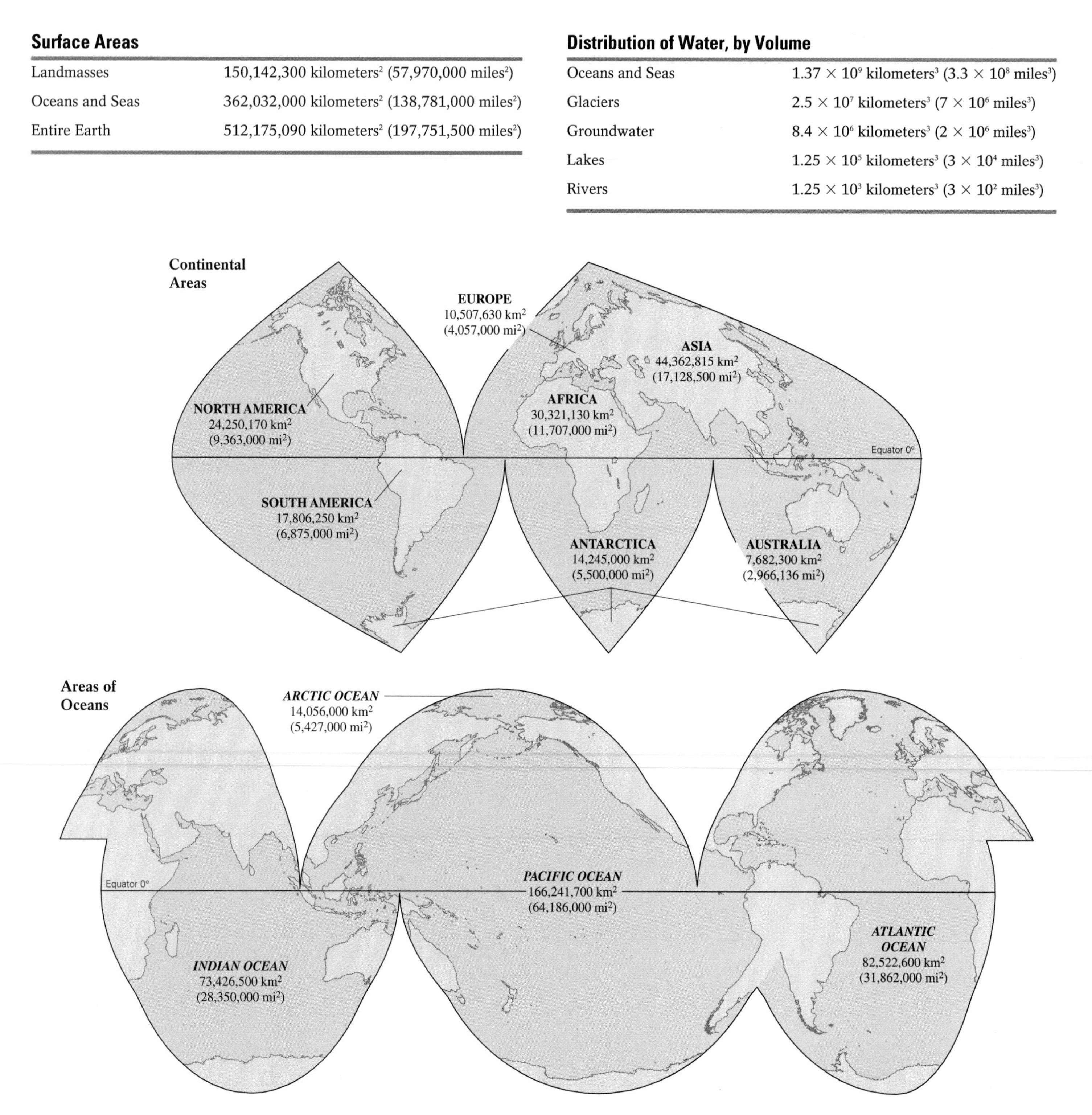

Elevations, Depths, and Distances

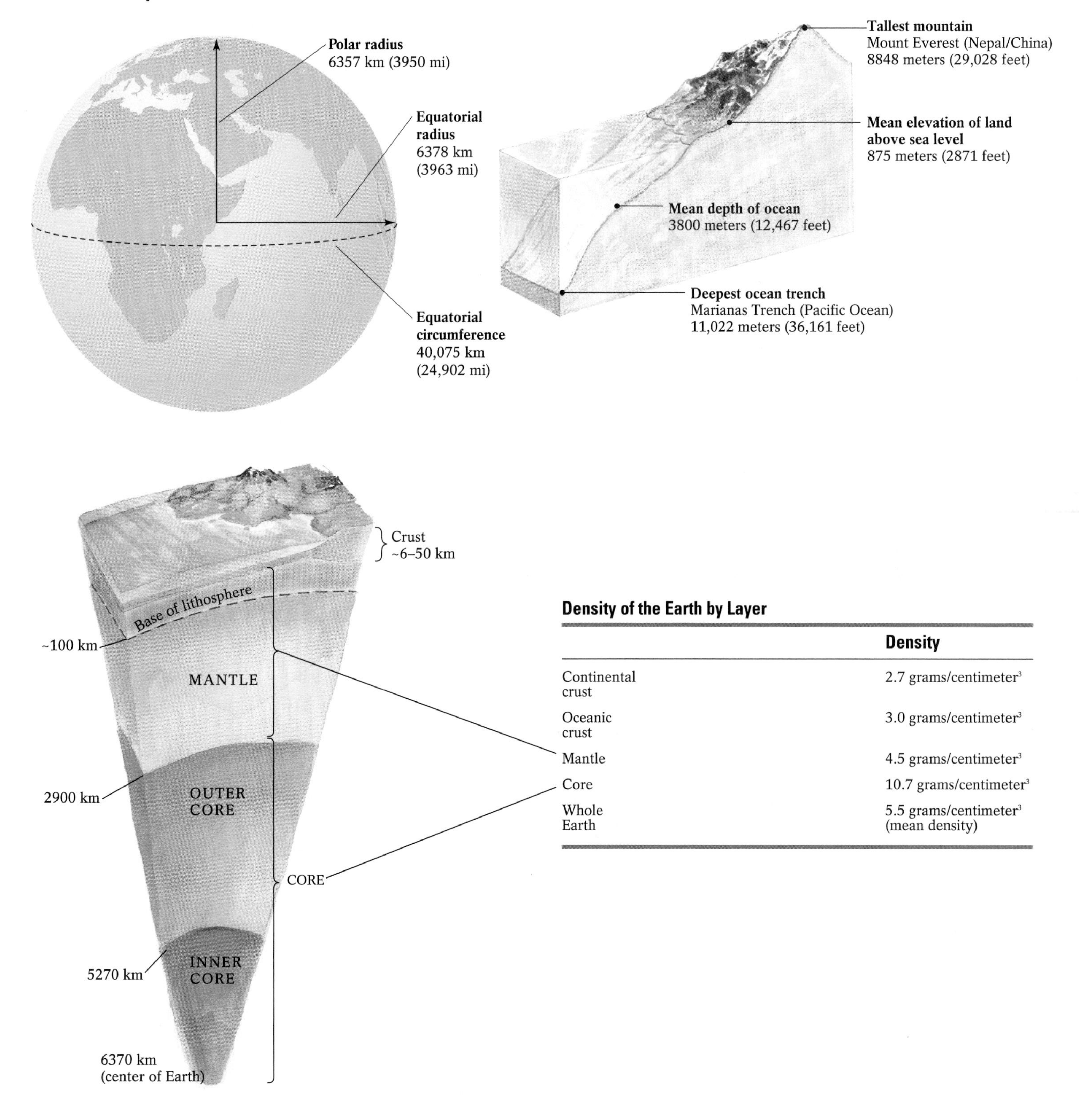

Density of the Earth by Layer

	Density
Continental crust	2.7 grams/centimeter^3
Oceanic crust	3.0 grams/centimeter^3
Mantle	4.5 grams/centimeter^3
Core	10.7 grams/centimeter^3
Whole Earth	5.5 grams/centimeter^3 (mean density)

Appendix C

Reading Topographic and Geologic Maps

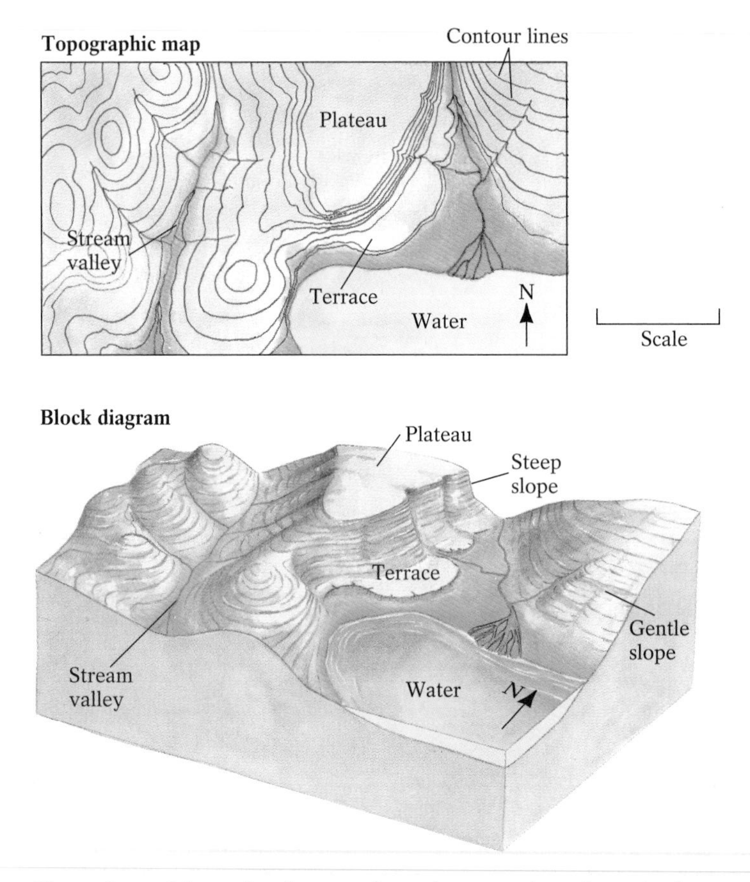

Throughout this textbook, maps have been used to show the locations of and relationships among various features. Geologists use maps for soil management, flood control, environmental planning, finding such resources as ores and groundwater, and determining optimal locations for fuel pipelines, highways, recreational areas, and the like.

Most maps show all or part of the Earth's surface, drawn to scale. A map's scale is usually shown at the bottom of the map. It may feature miles, kilometers, or any other convenient unit of measurement. Most maps from the United States Geological Survey are drawn to a scale of 1:24,000, meaning that 1 centimeter = 240 meters or 1 inch = 2000 feet; the latter would be read as "one inch on this map is equal to two thousand feet in real-world distance." The U.S. Geological Survey has been phasing in metric topographic maps using a scale of 1:100,000.

Most maps show two dimensions, marked by longitude (vertical) and latitude (horizontal) lines. A third dimension—elevation or depth—can be shown on *topographic maps,* which show relief using contour lines that mark specific heights above or depths below sea level. Every point on a contour line is at the same elevation. Every contour line closes upon itself—delineating the border of a discrete area—although its entire length may not be visible within the margins of a given map. A map's contour interval, or the distance between adjacent contour lines, is critical to both the usefulness and the appearance of the map: Contour lines that occur close together represent a steep slope; those that are farther apart represent a more gentle one. In an area of low relief, intervals designating elevation differences of only 5 or 10 feet are appropriate, whereas terrain with great relief may require intervals of 50 feet or more. On a given map, all contour intervals are the same.

In the past, accurate map-making depended predominantly upon actual measurements made on site by surveyors, geologists, and others. The advent of aerial photography, and then of space-based satellite photography, has made accurate standardized revisions possible.

The U.S. Geological Survey uses color to indicate specific features:

black and/or red solid lines	:	major roads
brown lines	:	contour lines
black	:	human-made structures, names
light red lines	:	town limits
blue	:	water features
green	:	wooded areas
white	:	open fields, deserts, and other nonvegetated areas

Dotted or dashed lines are used for temporary features (that is, solid blue represents a lake, whereas a dashed blue line marks a seasonal stream). Any special symbols used on a map are explained in its legend, which appears with the scale in the bottom margin.

A region's underlying geology can be represented on a *geologic map.* Symbols representing various types of rock have been standardized; such commonly accepted rock symbols have been used throughout this book. Standardized colors are used to show rock ages. The key to a geologic map's labeling, colors, and symbols usually appears at its side margins.

Columbus Quadrangle, New Jersey

Scale 1:24,000
Contour interval 10 feet

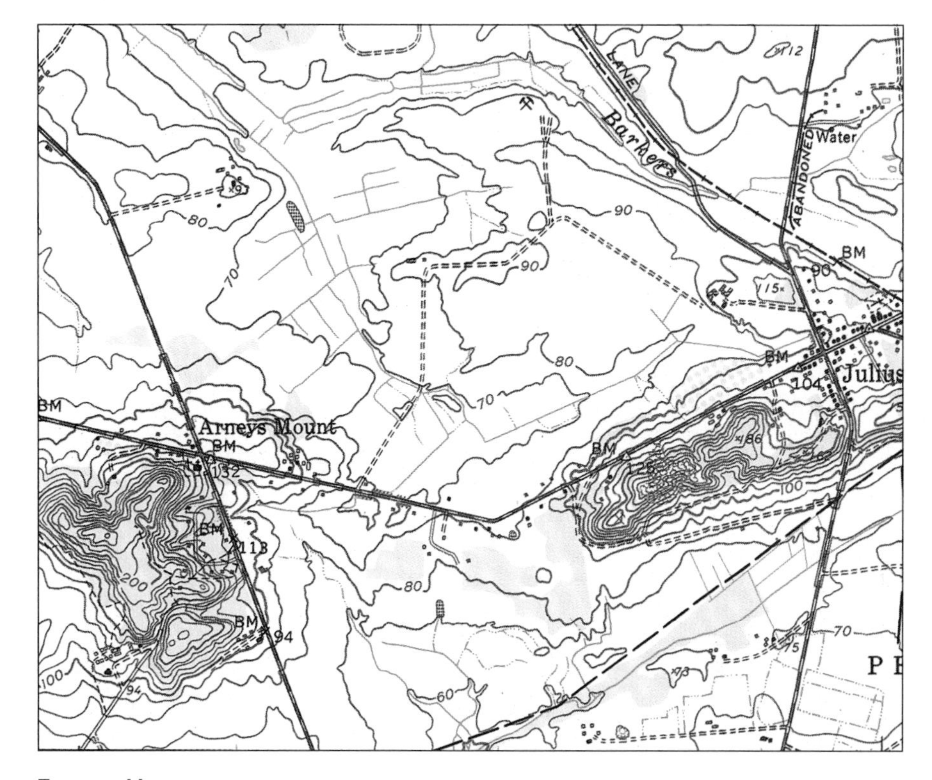

Topographic map

Geologic map

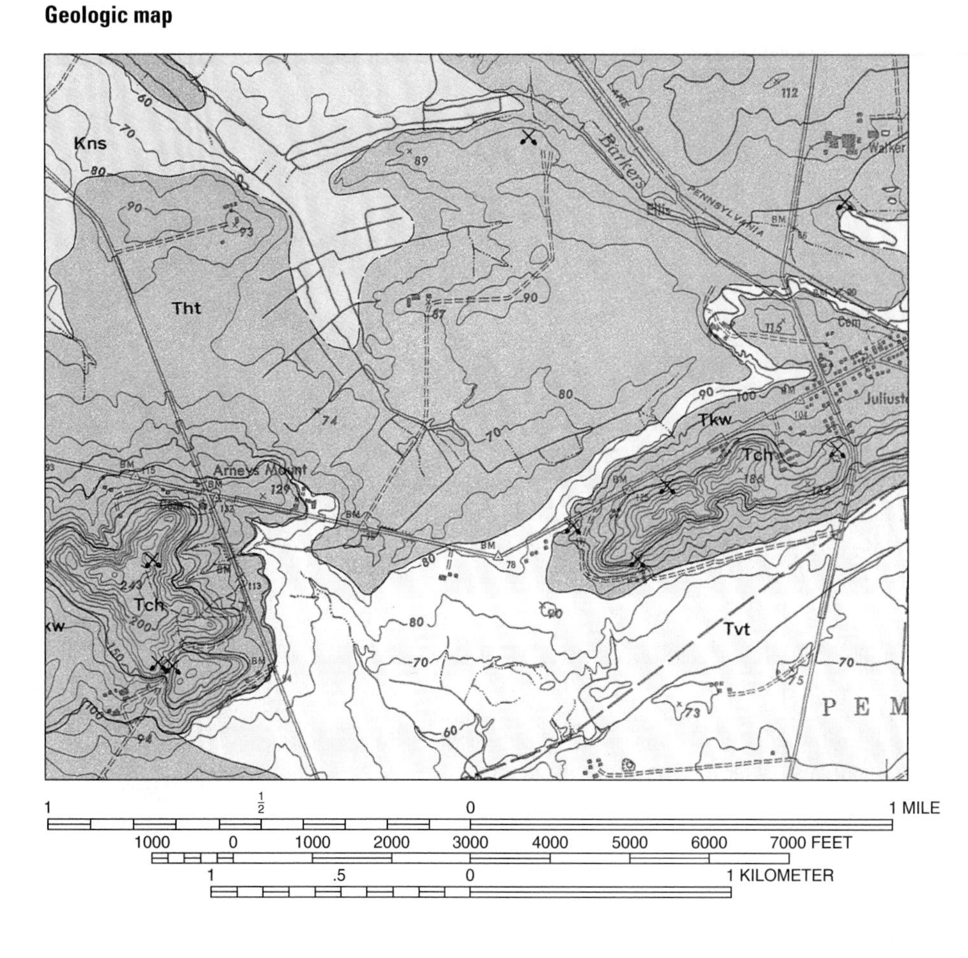

Tertiary Period Deposits

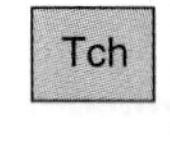

Cohansey sand
Sand, quartz, light-gray to yellow-brown, medium- to coarse-grained, pebbly, ilmenitic, micaceous, stratified

Kirkwood formation
Sand, quartz, light-gray to tan, fine- to very fine-grained, clayey, micaceous, ilmenitic, kaolinitic, sparingly lignitic, massive-bedded

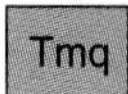

Manasquan formation
Sand, quartz, dark green-gray, medium- to coarse-grained, glauconitic, clayey

Vincentown formation
Upper member—calcarenite, quartz and glauconite, dusky-yellow to pale-olive, clayey; lower member—sand, quartz, dark-gray, poorly sorted, fine to coarse, clayey, glauconitic, entire formation very fossiliferous

Tht
Hornerstown sand
Sand, glauconite, dusky-green, medium- to coarse-grained, clayey, massive-bedded

Cretaceous Period Deposits

Red Bank sand
Lower member—sand, glauconite, dark grayish-black, coarse-grained, very clayey, micaceous, lignitic, massive-bedded (upper member not present)

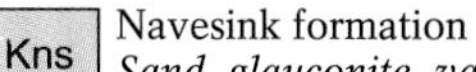
Kns
Navesink formation
Sand, glauconite, varying amounts of quartz, greenish-black to brown, medium- to coarse-grained, clayey

Kml
Mount Laurel sand
Sand, quartz, reddish-brown to green-gray, poorly sorted, fine- to coarse-grained, glauconitic, massive-bedded

Appendix D

Mineral-Identification Charts

Most Common Rock-Forming Minerals

			MINERAL OR GROUP NAME	COMPOSITION/ VARIETIES	CRYSTAL FORM/OTHER DIAGNOSTIC FEATURES	CLEAVAGE/ FRACTURE	USUAL COLOR/ LUSTER	STREAK	HARDNESS
Light-colored; abundant in all rock types	SILICATES	Framework	Feldspar	Potassium (orthoclase) feldspar ($KAlSi_3O_8$)	Coarse crystals or fine grains	Good cleavage in two directions at 90°; cleavage surfaces not striated	White to gray or pink, with pearly luster	White	6
				Sodium, calcium (plagioclase) feldspar Albite: $NaAlSi_3O_8$ Anorthite: $Ca_2Al_2Si_3O_8$		Good cleavage in two directions at 90°; cleavage surfaces striated	White to gray, sometimes green or yellowish	White	
			Quartz	SiO_2	Six-sided crystals; individual or in masses	No cleavage; conchoidal fracture	Colorless or slightly smoky gray, pink or yellow	White	7
		Sheet	Mica	Muscovite $KAl_3Si_3O_{10}(OH)_2$	Thin, disc-shaped crystals	One perfect cleavage plane; splits into very thin sheets	Colorless or slightly gray, green, or brown, with vitreous luster	White	2–2½
				Biotite $K(Mg, Fe)_3AlSi_3O_{10}(OH)_2$	Irregular foliated masses	One perfect cleavage plane; splits into thin sheets	Black, brown, or green, with vitreous luster	White or gray	2½–3
				Chlorite $(Mg, Fe)_5(Al, Fe)_2Si_3O_{10}(OH)_8$			Yellowish, brown, green, or white	White or colorless	2–2½
Dark-colored; abundant in metamorphic and igneous rocks		Double Chain	Amphibole	Actinolite $Ca_2(Mg, Fe)_5Si_8O_{22}(OH)_2$	Long, six-sided crystals, fibrous or in aggregates	Two good cleavage planes at 56° and 124°	Pale to dark green or black, with vitreous luster. Pure actinolite white.	Pale green or white	5–6
				Hornblende $(Ca, Na)_{2\text{-}3}(Mg, Fe, Al)_5Si_6(Si, Al)_2O_{22}(OH)_2$					
		Single Chain	Pyroxene	Augite $Ca(Mg, Fe, Al)(Al, Si)O_6$	Short, four- or six-sided crystals	Two good cleavage planes at about 90°	Light to dark green, with vitreous luster	Pale green	5–6
				Diopside $CaMg(Si_2O_6)$			White to light green, with vitreous luster	White or pale green	
				Orthopyroxene $MgSiO_3$			Pale gray, green, brown, or yellow, with vitreous luster	White	
		Single Tetrahedra	Olivine	$(Mg, Fe)_2SiO_4$	Small grains and granular masses	No cleavage; conchoidal fracture	Grayish green or brown, with vitreous, glassy luster	White	6½–7
			Garnet	$(Ca_3, Mg_3, Fe_3, Al_2)n(SiO_4)_3$	12- or 24-sided crystals	No cleavage; conchoidal fracture	Deep red, with vitreous to resinous luster	White	6½–7½
Light-colored; abundant in sedimentary rocks	CARBONATES		Calcite	$CaCO_3$	Fine to coarsely crystalline. Effervesces rapidly in HCl.	Three oblique cleavage planes, forming rhombohedral cleavage pieces	White or gray, with pearly luster	White	3
			Dolomite	$CaMg(CO_3)_2$	Fine to coarsely crystalline. Effervesces slowly in HCl when powdered.		Colorless, white, or pink, and may be tinted by impurities, with pearly luster	White to pale gray	3½–4
	CLAY MINERALS (Hydrous alumino-silicates)		Kaolinite	$Al_2Si_2O_5(OH)_4$	Very fine grains; found as bedded masses in soils and sedimentary rocks; earthy odor	Earthy fracture	White to buff, or tinted gray by impurities	White, off-white, or colorless	1½–2½
			Illite	$K_{0.8}Al_2(Si_{3.2}Al_{0.8})O_{10}(OH)_2$					
			Smectite	$Na_{0.3}Al_2(Si_{3.7}Al_{0.3})O_{10}(OH)_2$					

Accessory or Less-Abundant Rock-Forming Minerals

		MINERAL OR GROUP NAME	COMPOSITION/ VARIETIES	CRYSTAL FORM/OTHER DIAGNOSTIC FEATURES	CLEAVAGE/ FRACTURE	USUAL COLOR/ LUSTER	STREAK	HARDNESS
Light-colored; common in sedimentary rocks	SULFATES	Gypsum	$CaSO_4 \cdot 2H_2O$	Tabular crystals in fine-to-granular masses	One perfect cleavage plane, forming thin sheets; also two other good cleavage planes	Colorless to white, with vitreous luster	White	1–2½
	SULFATES	Anhydrite	$CaSO_4$	Granular masses	Three good to perfect cleavage planes at 90°	White, gray, or blue-gray, with pearly to vitreous luster	White	3–3½
		Halite	NaCl	Perfect cubic crystals, soluble in water. Tastes salty.	Three excellent cleavage planes at 90°	White or gray, with pearly luster	White	2½
Light-colored; mainly in metamorphic and igneous rocks	ALUMINO-SILICATES	Kyanite	Al_2SiO_5	Long, bladed, or tabular crystals	One perfect cleavage plane, parallel to length of crystals	White to light blue, with vitreous luster	White	5 along cleavage plane, 6½–7 across cleavage plane
	ALUMINO-SILICATES	Sillimanite		Long, slender crystals or fibrous masses		White to gray, with vitreous luster	White	6–7
	ALUMINO-SILICATES	Andalusite		Coarse, nearly square crystals	Irregular fracture	Red, reddish brown, or green, with vitreous luster	White	5–6
		Serpentine	$Mg_6Si_4O_{10}(OH)_8$	Fibrous or platy masses	Splintery fracture	Light to dark green or brownish yellow, with pearly luster	White	4–6
		Talc	$Mg_3Si_4O_{10}(OH)_2$	Foliated masses. Feels soapy.	Perfect in one direction, forming thin flakes	White to pale green, with pearly or greasy luster	White	1–1½
		Corundum	Al_2O_3	Short, six-sided crystals	Irregular fracture	Usually brown, pink, or blue, with adamantine luster	None	9
		Fluorite	CaF_2	Octahedral or cubic crystals	Cleaves easily	White, yellow, green, or purple, with vitreous luster	White	4
Dark-colored; common in metamorphic rocks	SILICATES	Epidote (paired tetrahedra)	$Ca_2(Al,Fe)Al_2Si_3O_{12}(OH)$	Usually granular masses; also slender prisms	One good cleavage direction, one poor	Yellow to dark green, with vitreous luster	White or gray	6–7
	SILICATES	Staurolite (single tetrahedra)	$Fe_2Al_9Si_4O_{22}(O,OH)_2$	Short crystals, some cross-shaped	One poor cleavage direction	Brown or reddish brown to black, with vitreous luster	Off-white to white	7
		Graphite	C	Scaley, foliated masses. Feels greasy.	One direction of cleavage	Steel gray to black, with vitreous or pearly luster	Gray or black	1–2
Dark-colored; common in all rock types		Apatite	$Ca_5(PO_4)_3(OH,F,Cl)$	Granular masses	Poor cleavage	Green, brown, or red, with adamantine or greasy luster	White	5
		Magnetite	Fe_3O_4	Granular masses	Uneven fracture	Black, with metallic luster	Black	5½
		Hematite	Fe_2O_3	Granular masses	Uneven fracture	Brown-red to black, with earthy, dull, to metallic luster	Brick red	5½
		Limonite	$2Fe_2O_3 \cdot 3H_2O$	Earthy masses	Uneven fracture	Yellowish brown to black	Brownish yellow	5–5½
Metallic luster; common in all rock types	SULFIDES	Pyrite	FeS_2	Cubic crystals or granular masses	Uneven fracture	Pale brass yellow, with metallic luster	Greenish black	6–6½
	SULFIDES	Galena	PbS	Cubic crystals or granular masses	Three perfect cleavage planes at 90°	Silver gray with metallic luster	Gray	2½
	SULFIDES	Sphalerite	ZnS	Granular masses	Six perfect cleavage planes at 60°	White to green, brown, or black, with resinous to submetallic luster	Reddish brown to yellow brown	3½–4
	SULFIDES	Chalcopyrite	$CuFeS_2$	Granular masses	Irregular fracture	Brass yellow	Greenish black	3½–4

Minerals and Elements of Industrial or Economic Importance

MINERAL OR GROUP NAME	COMPOSITION	USUAL COLOR/ LUSTER	STREAK	HARDNESS	OTHER PROPERTIES/ COMMENTS	ORIGIN
Asbestos (Chrysotile)	$Mg_3Si_2O_5(OH)_4$	White to pale green, with pearly luster	White	1–2½	Flexible, nonflammable fibers	A variety of serpentines; found mostly in metamorphic rock
Bauxite	$Al(OH)_3$	Reddish to brown, with dull luster	Pale reddish brown	1½–3½	Found in earthy, clay-like masses. Principal source of commercial aluminum.	Weathering of many rock types
Chalcopyrite	$CuFeS_2$	Yellow, with metallic luster	Greenish black	3½–4½	Uneven fracture; softer than pyrite. Iridescent tarnish. Most common ore of copper.	Hydrothermal veins and porphyry copper deposits
Chromite	$FeCr_2O_4$	Black, with metallic or submetallic luster	Dark brown	5½	Massive; granular; compact. Most common ore of chromium. Used in making steel.	Ultramafic igneous rocks
Copper (native)	Cu	Red, with metallic luster	Red	2½–3	Malleable and ductile. Tarnishes easily.	Mafic igneous rock (basaltic lavas); also in oxidized ore deposits
Galena	PbS	Silver-gray, with metallic luster	Gray	2½	Perfect cubic cleavage. Most important ore of lead; also commonly contains silver.	Hydrothermal veins
Gold (native)	Au	Yellow, with metallic luster	Yellow	2½–3	Malleable and ductile	Sedimentary placer deposits; hydrothermal veins
Hematite	Fe_2O_3	Brown-red to black, with earthy, dull, or metallic luster	Dark red	5½–6½	Granular, massive. Most important source of iron.	Found in all types of rocks. Commonly in contact-metamorphic aureoles around mafic igneous sills.
Magnetite	Fe_3O_4	Black, with metallic luster	Black	5½	Uneven fracture. An important source of iron. Strongly magnetic.	Found in all types of rocks
Platinum (native)	Pt	Steel-gray or silver-white, with metallic luster	Off-white, metallic	4–4½	Malleable and ductile. Occasionally magnetic.	Mafic igneous rocks and sedimentary placer deposits
Silver (native)	Ag	Silver-white, with metallic luster	Silver-white	2½–3	Malleable and ductile. Tarnishes to dull gray or black.	Mostly in hydrothermal veins; also in oxidized ore deposits
Sphalerite	ZnS	Shades of brown and red, with resinous to adamantine luster	Reddish brown	3½–4	Perfect cleavage in six directions at 120°. Most important ore of zinc.	Hydrothermal veins

Precious and Semi-precious Gems*

MINERAL OR GROUP NAME	COMPOSITION	USUAL COLOR/ LUSTER	STREAK	HARDNESS	OTHER PROPERTIES	ORIGIN
Beryl (aquamarine, emerald)	$Be_3Al_2(SiO_3)_6$	Blue, green, yellow, or pink, with vitreous luster	White	7½–8	Uneven fracture; hexagonal crystals	Cavities in granites and pegmatites, and schist
Corundum (ruby, sapphire)	Al_2O_3	Gray, red (ruby), blue (sapphire), with adamantine luster	None	9	Short, six-sided crystals; irregular, occasionally cleavage-like fracture ("parting")	Metamorphic rocks; some igneous rocks
Diamond	C	Colorless or with pale tints, with adamantine luster	None	10	Octahedral crystals	Peridotite; kimberlite; sedimentary placer deposits
(S) Garnet	$(Ca_3, Mg_3, Fe_3, Al_2)n(SiO_4)_3$	Deep red, with vitreous to resinous luster	White	6½–7½	No cleavage; 12- or 24-sided crystals	Contact-metamorphic and regionally metamorphosed rocks, and sedimentary placer deposits
(S) Jadeite (jade)	$NaAl(Si_2O_6)$	Green, with vitreous luster	White or pale green	6½–7	Compact fibrous aggregates	High-pressure metamorphic rocks
Olivine (peridot)	$(MgFe)_2SiO_4$	Light to dark green, with vitreous, glassy luster	White	6½–7	Uneven fracture, often in granular masses	Basalt, peridotite
(S) Opal	$SiO_2 \cdot nH_2O$	White with various other colors, with vitreous, pearly luster	White	5½–6½	Conchoidal fracture; amorphous; tinged with various colors in bands	Low-temperature hot springs; weathered near-surface deposits
(S) Quartz (includes amethyst, citrine, agate, onyx, bloodstone, jasper, etc.)	SiO_2	Colorless, white, or tinted by impurities, with luster depending on variety	White	7	No cleavage; six-sided crystals	Origin specific to gem variety. (Quartz found in all rock types except ultramafic igneous rocks.)
Topaz	$Al_2SiO_4(F, OH)_2$	Colorless, white, or pale pink or blue, with vitreous luster	White	8	Cleavage in one direction; conchoidal fracture	Pegmatite, granite, rhyolite
Tourmaline	$(Na, Ca)(Li, Mg, Al)$-$(Al, Fe, Mn)_6(BO_3)_3$-$(Si_6O_{18})(OH_4)$	Black, brown, green, or pink, with vitreous luster	White to gray	7–7½	Poor cleavage; uneven fracture; striated crystals	Metamorphic rocks; pegmatite; granite
(S) Turquoise	$CuAl_6(PO_4)_4$-$(OH)_8 \cdot 5H_2O$	Blue-green, with waxy luster	Blue-green or white	5–6	Massive	Hydrothermal veins
Zircon	$Zr(SiO_4)$	Colorless, gray, green, pink, or light blue, with adamantine luster	White	7½	Poor cleavage	Felsic igneous rocks and sedimentary placer deposits

*Gems are classified as semi-precious *(S)* if they are more accessible than precious gems and/or their properties are somewhat less valued.

Appendix E

How to Identify Some Common Rocks

Rocks are aggregates of minerals, and some are aggregates of organic materials. They are generally classified according to whether they are of igneous, sedimentary, or metamorphic origin, and then on the basis of their physical properties—primarily texture and mineral composition.

Igneous Rocks

Igneous rocks are formed when a magma cools and crystallizes. The chemical composition of a magma determines the minerals and rocks that will be formed when it cools. The rate at which a magma cools determines the size of the resulting crystals, which in turn determines the texture of the resulting rocks. Both composition and texture are used to classify igneous rocks.

Color is used only broadly as a diagnostic property for igneous rocks. The descriptions *light, dark,* and *intermediate* are generally agreed on. But many igneous rocks comprise two different-colored components; some blend various grays and pinks. Color, then, can help reduce the possible identities of a rock, but in most cases, color is not the defining property.

I. *Phaneritic*, or coarsely crystallized rocks: Composed of individual crystals of about 1 to 5 millimeters in length

A. If the rock contains quartz, it is probably **granite.**

B. If the rock contains no quartz, and from 30% to 60% feldspar, it is **diorite.**

C. If the rock contains no quartz and less than 30% feldspar, it is **gabbro.**

D. If the rock contains neither quartz nor feldspar, it is likely to contain ferromagnesian minerals and be ultramafic **peridotite** (consisting largely of olivine or pyroxene).

II. *Aphanitic* rocks: Individual crystals are fine-grained or microscopic

A. If quartz crystals can be identified in the rock, it is **rhyolite;** if the rock is light gray, white, or light green but too fine-grained to determine whether quartz is present, it is probably rhyolite.

B. If the rock contains no visible quartz crystals but does have approximately equal proportions of white or gray feldspar and ferromagnesian minerals, it is **andesite.**

C. If ferromagnesian minerals can be identified in the rock, it is **basalt;** if the rock is dark gray or black, and too fine-grained to identify any crystals, it is probably a basalt.

III. *Porphyritic* rocks: Contain macroscopic crystals (crystals that can be seen without the aid of a microscope) embedded in a matrix of smaller macroscopic or microscopic crystals. Use the descriptions above for aphanitic rocks to identify porphyritic rhyolites, basalts, and so on.

IV. *Glassy* rocks: Rocks look like solid glass. The most common is **obsidian,** which is rhyolitic in composition.

Sedimentary Rocks

Sedimentary rocks are made up of particles derived either from preexisting rocks or from organic debris. They are classified on the basis of their texture and mineral composition.

In identifying sedimentary rocks, the first step is to determine whether the rock contains carbonate minerals. This step is done by applying a drop of dilute hydrochloric acid to the rock surface. (*Note:* HCl must be used under close supervision to avoid burning one's skin or clothing.)

I. Rocks that do not effervesce (fizz), even when powdered—but may effervesce in some places, such as the fine cement between grains. Such rocks contain little or no carbonate minerals.

A. Rocks with **clastic** texture: Composed of grains in a cement matrix

1. If the grains are more than 2 millimeters in diameter . . .
   a. and are angular, the rock is **sedimentary breccia.**
   b. and are rounded, the rock is **conglomerate.**
2. If the grains are from 1/16 to 2 millimeters in diameter, and the rock feels gritty, it is **sandstone.**
   a. If more than 90% of the grains are quartz, the rock is **quartz sandstone.**
   b. If more than 25% of the grains are feldspar, the rock is **arkose.**
   c. If more than 25% of the grains are fine fragments of shale, slate, basalt, and the like, the rock is **lithic sandstone.**
   d. If more than 15% of the rock is fine-grained matrix material, the rock is **graywacke.**
3. If the rock is fine-grained, feels smooth, . . .
   a. and the grains are visible with a hand lens, the rock is **siltstone.**
   b. and the grains are invisible even with a hand lens, and the rock is . . .
      i. laminated, or layered, it is **shale.**
      ii. unlayered, it is **mudstone.**

B. Rocks with **crystalline** texture: Composed of microscopic, interlocking crystals

1. If the rock dissolves in water, it is **rock salt.**
2. If its crystals are fine to coarse, and have a hardness of 2 on the Mohs scale, the rock is **gypsum.**

C. Rocks with indeterminate texture

1. If the rock is smooth, very fine-grained, and fractures conchoidally, it is **chert;** if it is dark in color, it is **flint.**
2. If the rock is black or dark brown and breaks easily, soiling the fingers, it is **coal.**

II. Rock that effervesces strongly is **limestone.** Limestone may be clastic or crystalline in texture.

A. If the rock is clastic in texture and contains fossils, it is **bioclastic limestone.**

1. If it is composed of whole, recognizable fossils, it is **coquina.**
2. If it is very fine-grained, light-colored, and powdery, it is **chalk.**

B. If the rock is clastic and composed of small spheres, it is **oolitic limestone.**

C. If the rock is coarsely crystalline and contains different-colored layers, it is **travertine.**

III. Rock that effervesces only when hammered to a powder is **dolostone.** Dolostone may be crystalline, or less commonly, clastic in texture.

Metamorphic Rocks

Metamorphic rocks are rocks that have had their composition or structure changed by intense heat or pressure. Two factors determine the nature of a metamorphic rock: the composition of the parent rocks and the combination of metamorphic factors that acted upon them. Different circumstances can produce different textures; thus texture is the primary basis for metamorphic rock identification.

I. Is the rock *foliated*? If so, identify the type of foliation and, if possible, the rock's mineral content.

A. If the rock is fine-grained and readily splits into sheets, it is **slate,** and will have an earthy luster.

B. If the rock is slate-like but has a silken luster, it is **phyllite.**

C. If the rock contains flat or needle-like mineral crystals that are virtually parallel to one another, it is **schist.** A schist that primarily contains mica is a mica schist; there are, similarly, garnet mica schists, hornblende schists, and talc schists. A schist that contains serpentine is **serpentinite.**

D. If the rock consists of separate layers of light and dark minerals, it is **gneiss.** Light-colored layers consist of feldspars and perhaps quartz; dark-colored layers are probably biotite, amphibole, or pyroxene.

II. Is the rock nonfoliated? If so . . .

A. Is quartz its primary constituent? If so, the quartz grains will be interlocking and the rock will be hard enough to scratch glass. The rock is **quartzite.**

B. Does it consist of interlocking, coarse crystals of calcite or dolomite? If so, it is **marble.**

C. Does it consist primarily of dark grains too fine to be seen unaided? If so, it is probably **hornfels;** it may also contain a few larger crystals of less common minerals.

Appendix F

Sources of Geologic Literature, Photographs, and Maps

American Geological Institute (AGI)
4220 King Street
Alexandria, VA 22302-1502
http://www.agiweb.org
or
AGI Publications Center
P.O. Box 205
Annapolis Junction, MD 20701
http://www.agiweb.org/pubserv.html

Issues the monthly "Bibliography and Index of Geology," which includes worldwide references and contains listings by author and subject.

Canada Centre for Remote Sensing
588 Booth Street
Ottawa, Ontario K1A-0Y7
CANADA
http://www.ccrs.nrcan.gc.ca/

Source for obtaining aerial photographs and satellite imagery of Canada.

Earth Science Information Center
U.S. Geological Survey
507 National Center
Reston, VA 20192
http://mapping.usgs.gov/esic/esic.html

Publishes topographic and other scale maps of the United States. Indexes listing available map scales are available for all the states.

Earthquake Engineering Research Institute
499 14th Street, Suite 320
Oakland, CA 94612-1934
http://www.eeri.org

Provides a valuable field guide for learning about and studying earthquakes.

Federal Emergency Management Agency
P.O. Box 2012
Jessup, MD 20794-2012
ATTN: Publications
http://www.fema.gov

Provides a number of publications aimed at helping citizens prepare for geologic disasters, including "Are you Ready? Your Guide to Disaster Preparedness Checklist." Also issues a general "FEMA Publications Catalog."

Geological Society of America
3300 Penrose Place
P.O. Box 9140
Boulder, CO 80301-9140
http://www.geosociety.org

Publishes a large and diverse number of papers and journals concerning the geology and mineral resources of North America.

National Air Photo Library
615 Booth Street
Room 180
Ottawa, Ontario K1A-0E9
CANADA
http://napl.ccm.nrcan.gc.ca/collection.html

Source for obtaining aerial photographs and satellite imagery of Canada.

National Climatic Data Center (NCDC)
Federal Building
151 Patton Avenue
Asheville, NC 28801-5001
http://www.ncdc.noaa.gov

Issues a monthly publication of climatological data, such as temperature and precipitation statistics for any given state or region. The NCDC maintains up-to-date weather records for the entire United States.

National Geophysical and Solar–Terrestrial Data Center (part of the National Oceanic and Atmospheric Administration)
325 Broadway
Boulder, CO 80303
http://www.noaa.gov

Maintains worldwide computer files of earthquakes recorded by seismographs and historic earthquake data.

National Oceanic and Atmospheric Administration (NOAA)
Outreach Unit
1305 East West Highway
Rm. 1W204
Silver Spring, MD 20910
http://www.noaa.gov

This federal agency maintains a large collection of aerial photographs of the coastal regions of the United States. A detailed index is available.

State geological surveys

All states maintain a geological survey that produces technical reports, maps, and other publications at the state or county level.

U.S. Department of Agriculture
Aerial Photography Field Office
2222 West 2300 South
Salt Lake City, UT 84119-2020
http://www.fsa.usda.gov/dam/APFO/airfto.htm

Can provide aerial photographs of most of the United States. A variety of scales is available.

U.S. Geological Survey (USGS)
Information Services
Box 25286
Building 810
Denver Federal Center
Denver, CO 80225
http://www.usgs.gov

Produces a large number of maps, reports, circulars, professional papers, bulletins, water resources publications, and so on. Circular number 777 of the USGS provides an annually updated guide to obtaining information on the earth sciences.

U.S. Geological Survey (USGS)
EROS Data Center
47914 252nd Street
Sioux Falls, SD 57198-0001
http://edcwww.cr.usgs.gov

Provides computer listings of most satellite imagery, high-altitude aerial photography, and photographs obtained during the Apollo, Skylab, and Gemini space missions. Worldwide coverage and indexes are available.

Appendix G

Careers in the Geosciences

Careers in the geosciences are numerous and varied. Several have been alluded to throughout this book. Employers in these fields include the energy industry, which hires about half of all professional geoscientists; the mining industry; federal and state governments; consulting firms, especially in environmental issues and hydrogeology; and academia, which employs more than 10% of all geoscientists.

Each of these broad fields, of course, involves a number of subdisciplines: the energy industry, for example, requires expertise in sedimentology and stratigraphy; mining careers require a background in economic geology, petrology, mineralogy, crystallography, or structural geology. The following is a list of some of the occupations in which people with an interest in the geosciences are employed:

Economic geologists conduct field investigations to determine the locations and economic viability of mineral deposits; they also investigate the genesis of mineral deposits. Mining companies typically employ economic geologists.

Engineering geologists investigate the geologic factors that affect human-made structures such as bridges, dams, and buildings, and the geologic effects of mass wasting.

Environmental geologists work in assessing, solving, and preventing problems associated with the pollution of soil, bedrock, and groundwater. For example, they assist in the selection of suitable locations for municipal- and hazardous-waste facilities such as landfills and waste-storage facilities, and may also help design such facilities.

Geochemists investigate the nature and distribution of chemical elements in the geologic environment.

Geochronologists determine the age of geologic materials, helping to reconstruct the geologic history of the Earth.

Geomorphologists study landforms, the rates and intensity of processes that created them, and the relationship of these landforms to the underlying geologic structures and climates that existed during their evolution.

Geophysicists attempt to determine the internal structure and properties of the Earth. They may focus on specific factors such as seismic waves, geomagnetism, or gravity.

Glaciologists investigate the physical and chemical properties of glacial masses. They also study the development, movement, and decay of glaciers and ice sheets and their deposits.

Hydrogeologists study the production, distribution, movement, and quality of groundwater in the Earth's crust. Many hydrogeologists work in the environmental industry, in cooperation with environmental geologists.

Hydrologists study the distribution, movement, and quality of surface bodies of water.

Marine geologists investigate the topography and sediments of the world's oceans. Marine geologists may work closely with petroleum geologists in off-shore exploration projects. They often work closely with oceanographers.

Mineralogists study the formation, composition, and properties—both physical and chemical—of minerals.

Paleontologists collect fossils and determine their age, reconstruct past environments, and regionally correlate rocks by determining the evolutionary sequence of fossil assemblages found in the rocks. Many paleontologists work in the oil industry.

Petroleum geologists work in the oil industry, and are involved with the exploration and production of petroleum and natural gas. They may also research unconventional sources such as oil shales and sands.

Petrologists study the mineralogical relationships of rocks, and specialize in determining the genesis of rocks; they may focus on either sedimentary, igneous, or metamorphic petrology.

Planetary geologists study the planets and their satellites to better understand the evolution of the solar system. Most planetary geologists are employed by universities or advanced research organizations such as NASA (National Aeronautics and Space Administration).

Sedimentologists study the formation of sedimentary rocks by assessing the processes of transportation, erosion, and deposition. Many sedimentologists work in the petroleum industry.

Stratigraphers decipher the sequence of rocks using the principle of superposition. They study the time and space relationships of rock sequences in an effort to determine the geologic history of an area.

Structural geologists investigate the phenomena and structures produced by the deformation of the Earth's crust, such as faults and folds. Most of the data used in structural geology are collected during detailed field work.

An academic background that includes courses in math and other sciences as well as in the specific geological fields mentioned here would be practical preparation for a geoscience career.

Glossary

ablation zone See *zone of ablation.*

abrasion A form of *mechanical weathering* that occurs when loose fragments or particles of rocks and minerals that are being transported, as by water or air, collide with each other or scrape the surfaces of stationary rocks.

absolute dating The fixing of a geological structure or event in time, as by counting tree rings. See also *relative dating.*

accumulation zone See *zone of accumulation.*

acid rain Rain that contains such acidic compounds as sulfuric acid and nitric acid, which are produced by the combination of atmospheric water with oxides released when *hydrocarbons* are burned. Acid rain is widely considered responsible for damaging forests, crops, and human-made structures, and for killing aquatic life.

aeration zone See *zone of aeration.*

alluvial fan A triangular deposit of *sediment* left by a stream that has lost velocity upon entering a broad, relatively flat valley.

alluvium A deposit of *sediment* left by a stream in the stream's channel or on its *floodplain.*

alpine glacier A mountain *glacier* that is confined by highlands.

andesite The dark, aphanitic, *extrusive rock* that has a *silica* content of about 60% and is the second most abundant volcanic rock. Andesites are found in large quantities in the Andes Mountains.

andesite line The geographic boundary between the *basalts* and *gabbros* of the Pacific Ocean basin and the *andesites* at the subductive margins of the surrounding continents.

angle of repose The maximum angle at which a pile of unconsolidated material can remain stable.

anthracite A hard, jet-black coal that develops from *lignite* and *bituminous coal* through *metamorphism,* has a carbon content of 92% to 98%, and contains little or no gas. Anthracite burns with an extremely hot, blue flame and very little smoke, but it is difficult to ignite and both difficult and dangerous to mine.

anticline A convex *fold* in rock, the central part of which contains the oldest section of rock. See also *syncline.*

aquiclude An impermeable body of rock that may absorb water slowly but does not transmit it.

aquifer A *permeable* body of rock or *regolith* that both stores and transports groundwater.

Archean Eon The portion of Earth history that followed the *Hadean Eon* and preceded the *Proterozoic Eon,* beginning about 3.77 billion years ago and ending about 2.5 billion years ago, and is distinguished by the first appearance of simple single-celled life forms.

arête A sharp ridge of erosion-resistant rock formed between adjacent *cirque glaciers.*

aridity index The ratio of a region's potential annual evaporation, as determined by its receipt of solar radiation, to its average annual precipitation.

arroyo A small, deep, usually dry channel eroded by a short-lived or intermittent desert stream.

artesian Of, being, or concerning an *aquifer,* in which water rises to the surface due to pressure from overlying water.

asthenosphere A layer of heat-softened but solid, mobile rock comprising the lower part of the upper *mantle* from about 100 to 350 kilometers beneath the Earth's surface. See also *lithosphere.*

atoll A circular reef that encloses a relatively shallow lagoon and extends from a very great depth to the sea surface. An atoll forms when an oceanic island ringed by a *barrier reef* sinks below sea level.

atom The smallest particle that retains all the chemical properties of a given *element.*

atomic mass 1. The sum of *protons* and *neutrons* in an atom's nucleus. 2. The combined mass of all particles in a given atom.

atomic number The number of *protons* in the nucleus of a given atom. *Elements* are distinguished from each other by their atomic numbers.

aureole A section of rock that surrounds an intrusion and shows the effects of *contact metamorphism.*

backswamp A wetland area formed when flood water fills surface depressions in a stream's *floodplain.*

barchan dune A crescent-shaped *dune* that forms around a small patch of vegetation, lies perpendicular to the prevailing wind direction, and has a gentle, convex windward slope and a steep, concave leeward slope. Barchan dunes typically form in arid, inland deserts with stable wind direction and relatively little sand.

barrier island A ridge of sand that runs parallel to the main coast but is separated from it by a bay or lagoon. Barrier islands range from 10 to 100 kilometers in length and from 2 to 5 kilometers in width. A barrier island may be as high as 6 meters above sea level.

barrier reef A long, narrow reef that runs parallel to the main coast but is separated from it by a wide lagoon.

basal sliding The process by which a *glacier* undergoes thawing at its base, producing a film of water along which the glacier then flows. Basal sliding primarily affects glaciers in warm climates or mid-latitude mountain ranges.

basalt The dark, dense, aphanitic, *extrusive rock* that has a silica content of 40% to 50% and makes up most of the ocean floor. Basalt is the most abundant volcanic rock in the Earth's crust.

base level The lowest level to which a *stream* can erode the channel through which it flows, generally equal to the prevailing global sea level.

basin A round or oval depression in the Earth's surface, containing the youngest section of rock in its lowest, central part. See also *dome.*

batholith A massive discordant *pluton* with a surface area greater than 100 square kilometers, typically having a depth of about 30 kilometers. Batholiths are generally found in elongated mountain ranges after the country rock above them has eroded.

baymouth bar A narrow ridge of sand (a *spit*) that stretches completely across the mouth of a bay. (Also called *bay bar* and *bay barrier.*)

beach The part of a *coast* that is washed by waves or tides, which cover it with *sediments* of various sizes and composition, such as sand or pebbles.

beach drift 1. The process by which a *longshore current* moves *sediments* along a beach face. 2. The sediments so moved. Beach drift typically consists of sand, gravel, shell fragments, and pebbles. See also *longshore drift.*

bedding The division of *sediment* or *sedimentary rock* into parallel layers (beds) that can be distinguished from each other by such features as chemical composition and grain size.

bed load A body of coarse particles that tend to move along the bottom of a *stream.*

biogenic chert A type of *sedimentary rock* composed of silica-based organic debris such as the shells and skeletons of small marine organisms. Unlike *inorganic chert,* which forms in the shape of nodules, organic chert typically forms in layers.

biogenic limestone A type of *sedimentary rock* composed mostly of calcite (calcium carbonate), formed from the shells and skeletons of small marine organisms. See also *inorganic limestone.*

biomass fuel A renewable fuel derived from a living organism or the by-product of a living organism. Biomass fuels include wood, dung, methane gas, and grain alcohol.

bitumen Any of a group of solid and semi-solid *hydrocarbons* that can be converted into liquid form by heating. Bitumens can be refined to produce such commercial products as gasoline, fuel oil, and asphalt.

bituminous coal A shiny black coal that develops from deeply buried *lignite* through heat and pressure, and that has a carbon content of 80% to 93%, which makes it a more efficient heating fuel than lignite.

blowout A small area of land that has been somewhat lowered due to *deflation,* usually where stabilizing surface vegetation has been previously disturbed.

body wave A type of *seismic wave* that transmits energy from an earthquake's *focus* through the Earth's interior in all directions. See also *surface wave.*

bond To combine, by means of chemical reaction, with another atom to form a *compound.* When an atom bonds with another, it either loses, gains, or shares electrons with the other atom.

Bowen's reaction series The sequence of *igneous rocks* formed from a mafic *magma,* assuming mineral crystals that have already formed continue to react with the liquid magma and so evolve into new minerals, thereby creating the next rock in the sequence.

braided stream A network of converging and diverging *streams* separated from each other by narrow strips of sand and gravel.

breakwater A wall built seaward of a coast to intercept incoming waves and so protect a harbor or shore. Breakwaters are typically built parallel to the coast.

breccia A *clastic rock* composed of particles more than 2 millimeters in diameter and marked by the angularity of its component grains and rock fragments.

brittle failure The rupturing of rock, a type of permanent deformation caused by great stress under relatively low temperature and pressure conditions. See also *plastic deformation.*

burial metamorphism A form of *regional metamorphism* that acts on rocks covered by 5 to 10 kilometers of rock or sediment, caused by heat from the Earth's interior and *lithostatic pressure.*

caldera A vast depression at the top of a *volcanic cone,* formed when an eruption substantially empties the reservoir of *magma* beneath the cone's summit. Eventually the summit collapses inward, creating a caldera. A caldera may be more than 15 kilometers in diameter and more than 1000 meters deep.

capacity The ability of a given *stream* to carry *sediment,* measured as the maximum quantity it can transport past a given point on the channel bank in a given amount of time. See also *competence.*

carbon-14 dating A form of *radiometric dating* that relies on the 5730-year half-life of radioactive carbon-14, which decays into nitrogen-14, to determine the age of rocks in which carbon-14 is present. Carbon-14 dating is used for rocks from 100 to 100,000 years old.

carbonate One of several minerals containing one central carbon atom with strong *covalent bonds* to three oxygen atoms and typically having *ionic bonds* to one or more positive ions.

catastrophism The hypothesis that a series of immense, brief, worldwide upheavals changed the Earth's crust greatly and can account for the development of mountains, valleys, and other features of the Earth. See also *uniformitarianism.*

cave An opening beneath the surface of the Earth, a *karst* feature generally produced by *dissolution* of carbonate bedrock.

cementation The process by which loose sediment grains are bound together by *precipitated* minerals originally dissolved during the *chemical weathering* of preexisting rocks.

Cenozoic Era The latest era of the *Phanerozoic Eon,* following the *Mesozoic Era* and continuing to the present time, and marked by the presence of a wide variety of mammals, including the first hominids. It is divided into two periods, the Tertiary and the Quaternary.

chemical sediment *Sediment* that is composed of previously dissolved minerals that have either precipitated from water or been extracted from water by living organisms and deposited when the organisms died or discarded their shells.

chemical weathering The process by which chemical reactions alter the chemical composition of rocks and minerals that are unstable at the Earth's surface and convert them into more stable substances; *weathering* that changes the chemical makeup of a rock or mineral. See also *mechanical weathering.*

cinder cone A *pyroclastic cone* composed primarily of cinders.

cirque A deep, semi-circular basin eroded out of a mountain by an *alpine glacier.*

cirque glacier A small *alpine glacier* that forms inside a *cirque,* typically near the head of a valley.

clastic Being or pertaining to a *sedimentary rock* composed primarily from fragments of preexisting rocks or fossils.

coal A member of a group of easily combustible, organic *sedimentary rocks* composed mostly of plant remains and containing a high proportion of carbon.

coast The area of dry land that borders on a body of water.

col A mountain pass eroded out of a ridge by an *alpine glacier.*

compaction The process by which the volume or thickness of sediment is reduced due to pressure from overlying layers of sediment.

competence The ability of a given *stream* to carry *sediment,* measured as the diameter of the largest particle that the stream can transport. See also *capacity.*

composite cone See *stratovolcano.*

compound Two or more *elements* bonded together in specific, constant proportions. A compound typically has physical characteristics different from those of its constituent elements.

compression *Stress* that reduces the volume or length of a rock, as that produced by the *convergence* of plate margins.

cone of depression An area in a *water table* along which water has descended into a well to replace water drawn out, leaving a gap shaped like an inverted cone.

confining pressure See *lithostatic pressure.*

conglomerate A *clastic rock* composed of particles more than 2 millimeters in diameter and marked by the roundness of its component grains and rock fragments.

contact metamorphism *Metamorphism* that is caused by heat from a magmatic intrusion.

continental collision The *convergence* of two continental plates, resulting in the formation of mountain ranges.

continental drift The hypothesis, proposed by Alfred Wegener, that today's continents broke off from a single supercontinent and then plowed through the ocean floors into their present positions. This explanation of the shapes and locations of Earth's current continents evolved into the theory of *plate tectonics.*

continental ice sheet An unconfined *glacier* that covers much or all of a continent.

continental platform That portion of a continent where the *continental shield* is covered with a veneer of younger rock. The continental shield and the continental platform together constitute a continent's tectonically stable *craton.*

continental shield A broad area of exposed crystalline rock in a continental interior that is the oldest part of the continent (e.g., the Canadian Shield in North America). The continental shield together with the *continental platform* constitute a continent's tectonically stable *craton.*

convection cell The cyclical movement of material in the *asthenosphere* that causes the plates of the *lithosphere* to move. Heated material becomes less dense and rises toward the solid lithosphere, through which it cannot rise further and therefore begins to move horizontally, dragging the lithosphere along with it and pushing forward the cooler, denser material in its path. The cooler material eventually sinks down lower into the mantle, becoming heated there and rising up again, continuing the cycle. See also *plate tectonics.*

convergence The coming together of two lithospheric plates. Convergence causes *subduction* when one or both plates is oceanic, and mountain formation when both plates are continental. See also *divergence.*

core The innermost layer of the Earth, consisting primarily of pure metals such as iron and nickel. The core is the densest layer of the Earth, and is divided into the outer core, which is believed to be liquid, and the inner core, which is believed to be solid. See also *crust* and *mantle.*

correlation The process of determining that two or more geographically distant rocks or rock strata originated in the same time period.

covalent bond The combination of two or more atoms by sharing electrons so as to achieve chemical stability. Atoms that form covalent bonds generally have outer energy levels containing three, four, or five electrons. Covalent bonds are generally stronger than other bonds.

crater See *volcanic crater.*

craton That region of a continent, consisting of the *continental shield* and the *continental platform,* that has been tectonically stable for a vast period of time and contains its oldest rocks.

creep The slowest form of *mass movement,* measured in millimeters or centimeters per year. It affects unconsolidated materials such as soil or *regolith,* the particles of which are continuously rearranged by gravity on virtually any slope.

cross-bed A bed made up of particles dropped from a moving current, as of wind or water, and marked by a downward slope that indicates the direction of the current that deposited them.

cross-cutting relationships See *principle of cross-cutting relationships.*

crust The outermost layer of the Earth, consisting of relatively low-density rocks. See also *core* and *mantle.*

crystal A mineral in which the systematic internal arrangement of *atoms* is outwardly reflected as a latticework of repeated three-dimensional units that form a geometric solid with a surface consisting of symmetrical planes.

crystal structure 1. The geometric pattern created by the systematic internal arrangement of atoms in a mineral. 2. The systematic internal arrangement of atoms in a mineral. See also *crystal.*

daughter isotope An *isotope* that forms from the radioactive decay of a *parent isotope.* A daughter isotope may or may not be of the same element as its parent. If the daughter isotope is radioactive, it will eventually become the parent isotope of a new daughter isotope. The last daughter isotope to form from this process will be stable and nonradioactive.

debris avalanche The sudden, extremely rapid *mass movement* downward of entire layers of *regolith* along very steep slopes. Debris avalanches are generally caused by heavy rains.

debris flow 1. The rapid, downward *mass movement* of particles coarser than sand, often including boulders 1 meter or more in diameter, at a rate ranging from 2 to 40 kilometers per hour. Debris flows occur along fairly steep slopes. 2. The material that descends in such a flow.

deflation The process by which wind erodes a surface by picking up and transporting loose rock particles.

deflation basin A large-scale *blowout,* occurring where local bedrock is particularly soft or has been extensively crushed by faulting.

delta An *alluvial fan* having its apex at the mouth of a *stream.*

desert A region with an average annual rainfall of 10 inches or less and sparse vegetation, typically having thin, dry, and crumbly soil. A desert has an *aridity index* greater than 4.0.

desertification The process through which a desert takes over a formerly nondesert area. When a region begins to undergo desertification, the new conditions typically include a significantly lowered *water table,* a reduced supply of surface water, increased salinity in natural waters and soils, progressive destruction of native vegetation, and an accelerated rate of erosion.

desert pavement A closely packed layer of rock fragments concentrated in a layer along the Earth's surface by the *deflation* of finer particles.

detrital sediment *Sediment* that is composed of transported solid fragments of preexisting igneous, sedimentary, or metamorphic rocks.

dike A discordant tabular *pluton* that is substantially wider than it is thick. Dikes are often steeply inclined or nearly vertical. See also *sill.*

dilatancy The expansion of a rock's volume caused by *stress* and deformation.

diorite Any of a group of dark, phaneritic, *intrusive rocks* that are the *plutonic* equivalents of *andesite.*

dip The angle formed by the inclined plane of a geological structure and the horizontal plane of the Earth's surface.

dip-slip fault A *fault* in which two sections of rock have moved apart vertically, parallel to the *dip* of the fault plane.

directed pressure Force exerted on a rock along one plane, flattening the rock in that plane and lengthening it in the perpendicular plane.

disappearing stream A surface *stream* that drains rapidly and completely into a *sinkhole.*

discharge The volume of a *stream's* water that passes a given point per unit of time.

displaced terrane A body of foreign rock that has been carried from elsewhere by plate motion and attached to a continent's coast by collision.

dissolution A form of *chemical weathering* in which water molecules, sometimes in combination with acid or another compound in the environment, attract and remove oppositely charged ions or ion groups from a mineral or rock.

dissolved load A body of sediment carried by a *stream* in the form of *ions* that have dissolved in the water.

distributary One of a network of a small *streams* carrying water and sediment from a *trunk stream* into an ocean.

divergence The process by which two lithospheric plates separated by *rifting* move farther apart, with soft mantle rock rising between them and forming new oceanic *lithosphere.* See also *convergence.*

dolostone A *sedimentary rock* composed primarily of dolomite, a mineral made up of calcium, magnesium, carbon, and oxygen. Dolostone is thought to form when magnesium ions replace some of the calcium ions in limestone, to which dolostone is similar in both appearance and chemical structure.

dome A round or oval bulge on the Earth's surface, containing the oldest section of rock in its raised, central part. See also *basin.*

drainage basin The area from which water flows into a *stream.* Also called a *watershed.*

drainage divide An area of raised, dry land separating two adjacent *drainage basins.*

drumlin A long, spoon-shaped hill that develops when pressure from an overriding *glacier* reshapes a *moraine.* Drumlins range in height from 5 to 50 meters and in length from 400 to 2000 meters. They slope down in the direction of the ice flow.

dune A usually asymmetrical mound or ridge of sand that has been transported and deposited by water or wind. Dunes form in both arid and humid climates.

dynamothermal metamorphism A form of *regional metamorphism* that acts on rocks caught between two *converging* plates and is initially caused by *directed pressure* from the plates, which causes some of the rocks to rise and others to sink, sometimes by tens of kilometers. The rocks that fall then experience further dynamothermal metamorphism, this time caused by heat from the Earth's interior and *lithostatic pressure* from overlying rocks.

earthflow 1. The *flow* of a dry, highly viscous mass of clay-like or silty *regolith,* typically moving at a rate of 1 or 2 meters per hour. 2. The material that descends in such a flow.

earthquake A movement within the Earth's *crust* or *mantle,* caused by the sudden rupture or repositioning of underground rocks as they release *stress.*

Ediacaran fauna The Earth's first soft-bodied marine animals—believed from fossil impressions to be related to modern jellyfish, sea pens, and various types of worms—which appeared late in the *Proterozoic Eon.*

elastic deformation A temporary change in the shape or volume of a rock, caused by less stress than that which produces *brittle failure* or *plastic deformation.*

electron A negatively charged particle that orbits rapidly around the *nucleus* of an *atom.* See also *proton.*

element A form of matter that cannot be broken down into a chemically simpler form by heating, cooling, or chemical reactions. There are 106 known elements, 92 of them natural and 14 synthetic. Elements are represented by one- or two-letter abbreviations. See also *atom, atomic number.*

energy level The path of a given electron's orbit around a nucleus, marked by a constant distance from the nucleus.

eons The largest time spans shown on the geologic time scale, designating the major developments (mostly biological) in Earth history. See also *Hadean Eon, Archean Eon, Proterozoic Eon,* and *Phanerozoic Eon.*

epicenter The point on the Earth's surface that is located directly above the *focus* of an *earthquake.*

epochs The time spans that subdivide *periods* in the *geologic time scale.*

eras The second-largest time spans shown on the *geologic time scale* (subdividing the *eons*), defined by their dominant life forms. See also *Paleozoic Era, Mesozoic Era,* and *Cenozoic Era.*

erosion The process by which particles of rock and soil are loosened, as by *weathering,* and then transported elsewhere, as by wind, water, ice, or gravity.

esker A ridge of *sediment* that forms under a glacier's *zone of ablation,* made up of sand and gravel deposited by meltwater. An esker may be less than 100 meters or more than 500 kilometers long, and may be anywhere from 3 to over 300 meters high.

eukaryotic Having a cell *nucleus.* Describes the more complex life forms that evolved from the Earth's first simple *prokaryotic* organisms as well as all higher organisms that evolved subsequently.

evaporite An inorganic *chemical sediment* that *precipitates* when the salty water in which it had dissolved evaporates.

extrusive rock An *igneous rock* formed from *lava* that has flowed out onto the Earth's surface, characterized by rapid solidification and grains that are so small as to be barely visible to the naked eye.

fall The fastest form of *mass movement,* occurring when rock or sediment breaks off from a steep or vertical slope and descends at a rate of 9.8 meters per second. A fall can be extremely dangerous.

fault A fracture dividing a rock into two sections that have visibly moved relative to each other.

fault-block mountain A mountain containing tall *horsts* interspersed with much lower *grabens* and bounded on at least one side by a high-angle *normal fault.*

faunal succession See *principle of faunal succession.*

firn Firmly packed snow that has survived a summer melting season. Firn has a density of about 0.4 grams per cubic centimeter. Ultimately, firn turns into glacial ice.

floodplain The flat land that surrounds a *stream* and becomes submerged when the stream overflows its banks.

flow A type of *mass movement* involving a mixture of solid, mostly unconsolidated particles that moves downslope like a viscous liquid although it may be relatively dry. The higher the water content of a flow, the more swiftly it moves.

focus (plural **foci**) The precise point within the Earth's *crust* or *mantle* where rocks begin to rupture or move in an *earthquake.*

fold A bend that develops in an initially horizontal layer of rock, usually caused by *plastic deformation.* Folds occur most frequently in *sedimentary rocks.*

fold-and-thrust mountains Large, complex mountain systems resulting from the collision of continental plates. Typically, these rocks consist of marine sediments that have been intensely *folded, thrust-faulted,* and, in places, intruded and metamorphosed by large *plutons.* The Alps, the Appalachians, the Carpathians, the Himalayas, and the Urals are all fold-and-thrust mountains.

foliation The arrangement of a set of minerals in parallel, sheet-like layers that lie perpendicular to the flattened plane of a rock. Occurs in *metamorphic rocks* on which *directed pressure* has been exerted.

fossil A remnant, an imprint, or a trace of an ancient organism, preserved in the Earth's crust.

fossil fuel A nonrenewable energy source, such as oil, gas, or coal, that derives from the organic remains of past life. Fossil fuels consist primarily of *hydrocarbons.*

fractional crystallization The process by which a *magma* produces crystals that then separate from the original magma, so that the chemical composition of the magma changes with each generation of crystals, producing *igneous rocks* of different compositions. The *silica* content of the magma becomes proportionately higher after each crystallization.

fringing reef A reef that forms against or near an island or continental *coast* and grows seaward, sloping sharply toward the sea floor. Fringing reefs usually range from 0.5 to 1.0 or more kilometers in width.

frost wedging A form of *mechanical weathering* caused by the freezing of water that has entered a pore or crack in a rock. The water expands as it freezes, widening the cracks or pores and often loosening or dislodging rock fragments.

gabbro Any of a group of dark, dense, phaneritic, *intrusive rocks* that are the *plutonic* equivalent to *basalt.*

geochronology The study of the relationship between the history of the Earth and time.

geologic time scale The division of all of Earth history into blocks of time distinguished by geologic and evolutionary events, ordered sequentially and arranged into *eons* made up of *eras,* which are in turn made up of *periods,* which are in turn made up of *epochs.*

geology The scientific study of the Earth, its origins and evolution, the materials that make it up, and the processes that act on it.

geyser A *natural spring* marked by the intermittent escape of hot water and steam.

glacial abrasion The process by which a *glacier* erodes the underlying bedrock through contact between the bedrock and rock fragments embedded in the base of the glacier. See also *glacial quarrying.*

glacial drift A load of rock material transported and deposited by a *glacier.* Glacial drift is usually deposited when the glacier begins to melt.

glacial erratic A rock or rock fragment transported by a *glacier* and deposited on bedrock of different composition. Glacial erratics range from a few millimeters to several yards in diameter.

glacial quarrying The process by which a *glacier* erodes the underlying bedrock by loosening and ultimately detaching blocks of rock from the bedrock and attaching them instead to the glacier, which then bears the rock fragments away. See also *glacial abrasion.*

glacial till Drift that is deposited directly from glacial ice and therefore not *sorted.* Also called *till.* See also *glacial drift.*

glacier A moving body of ice that forms on land from the accumulation and compaction of snow, and that flows downslope or outward due to gravity and the pressure of its own weight.

gneiss A coarse-grained, foliated *metamorphic rock* marked by bands of light-colored minerals such as quartz and feldspar that alternate with bands of dark-colored minerals. This alternation develops through *metamorphic differentiation.*

graben A block of rock that lies between two *faults* and has moved downward to form a depression between the two adjacent fault blocks. See also *horst.*

graded bed A *bed* formed by the deposition of sediment in relatively still water, marked by the presence of particles that vary in size, density, and shape. The particles settle in a gradual slope with the coarsest particles at the bottom and the finest at the top.

graded stream A stream maintaining an equilibrium between the processes of erosion and deposition, and therefore between aggradation and degradation.

gradient The vertical drop in a *stream's* elevation over a given horizontal distance, expressed as an angle.

granite A pink-colored, felsic, *plutonic rock* that contains potassium and usually sodium feldspars, and has a quartz content of about 10%. Granite is commonly found on continents but virtually absent from the ocean basins.

groin A structure that juts out into a body of water perpendicular to the *shoreline* and is built to restore an eroding beach by intercepting *longshore drift* and trapping sand.

Hadean Eon The earliest portion of Earth history, which began with the formation of the Earth approximately 4.6 billion years ago and ended at the beginning of the *Archean Eon* about 3.77 billion years ago, when the first simple single-celled organisms appeared.

half-life The time necessary for half of the atoms of a *parent isotope* to decay into the *daughter isotope.*

headland A cliff that projects out from a *coast* into deep water.

historical geology The study of the origin and evolution of the Earth and all of its life forms and geologic structures.

hook A *spit* that curves sharply at its coastal end due to a particularly strong *longshore current.*

horn A high mountain peak that forms when the walls of three or more *cirques* intersect.

hornfels A hard, dark-colored, dense *metamorphic rock* that forms from the *intrusion* of magma into shale or *basalt.*

horst A block of rock that lies between two *faults* and has moved upward relative to the two adjacent fault blocks. See also *graben.*

hot spot An area in the upper *mantle,* ranging from 100 to 200 km in width, from which magma rises in a plume to form *volcanoes.* A hot spot may endure for ten million years or more.

hydraulic lifting The *erosion* of a stream bed by water pressure.

hydrocarbon A molecule that is made up entirely of hydrogen and carbon.

hydrologic cycle The perpetual movement of water among the mantle, oceans, land, and atmosphere of the Earth.

hydrolysis A form of *chemical weathering* in which ions from water replace equivalently charged ions from a mineral, especially a silicate.

hydrothermal deposit A mineral deposit formed by the precipitation of metallic ions from water ranging in temperature from 50° to 700°C.

hypothesis A tentative explanation of a given set of data that is expected to remain valid after future observation and experimentation. See also *theory.*

ice age A period during which the Earth is substantially cooler than usual and a significant portion of its land surface is covered by *glaciers.* Ice ages generally last tens of millions of years.

icecap An *alpine glacier* that covers the peak of a mountain.

igneous rock A *rock* made from molten (melted) or partly molten material that has cooled and solidified.

inclusions See *principle of inclusions.*

index fossil The *fossil* of an organism known to have existed for a relatively short period of time, used to date the rock in which it is found.

index mineral See *metamorphic index mineral.*

inorganic chert A type of *sedimentary rock* composed of microscopic silica crystals and formed by *precipitating* directly from silica-rich water. Typically found in the shape of nodules, often within bodies of limestone or *dolostone.* See also *biogenic chert.*

inorganic limestone A type of *sedimentary rock* composed mostly of calcite (calcium carbonate) and formed by *precipitating* directly from water. See also *biogenic limestone.*

inselberg A steep ridge or hill left in an otherwise flat, typically desert plain where a mountain has eroded.

internal deformation The rearrangement of the planes within ice crystals, due to pressure from overlying ice and snow, that causes the downward or outward flow of a *glacier.*

intrusive rock An *igneous rock* formed by the entrance of *magma* into preexisting rock.

ion An *atom* that has lost or gained one or more *electrons,* thereby becoming electrically charged.

ionic bond The combination of an atom that has a strong tendency to lose electrons with an atom that has a strong tendency to gain electrons, such that the former transfers one or more electrons to the latter and each achieves chemical stability. The resulting *compound* is electrically neutral.

isotope One of two or more forms of a single element; the atoms of each isotope have the same number of protons but different numbers of neutrons in their nuclei. Thus, isotopes have the same *atomic number* but differ in *atomic mass.*

jetty A structure that is built to extend the banks of a stream channel or tidal outlet beyond the coastline to direct the flow of a stream or tide and keep the sediment moving so that it cannot build up and fill the channel. Jetties are typically built in parallel pairs along both banks of the channel. Jetties that are built perpendicular to a *coast* tend to interrupt *longshore drift* and thus widen *beaches.*

karst The features that are produced, either at the Earth's surface or underground (such as *caves*), when groundwater dissolves bedrock (usually limestone). See also *karst topography.*

karst topography The surface expression of *karst,* characterized principally by *sinkholes* and *disappearing streams.*

kerogen A solid, waxy, organic substance that forms when pressure and heat from the Earth act on the remains of plants and animals. Kerogen converts to various liquid and gaseous *hydrocarbons* at a depth of 7 or more kilometers and a temperature between 50° and 100°C.

laccolith A large *concordant pluton* that is shaped like a dome or a mushroom. Laccoliths tend to form at relatively shallow depths and are typically composed of granite. The country rock above them often erodes away completely.

lahar A flow of pyroclastic material mixed with water. A lahar is often produced when a snow-capped volcano erupts and hot pyroclastics melt a large amount of snow or ice.

lava *Magma* that comes to the Earth's surface through a *volcano* or fissure.

lignite A soft, brownish coal that develops from *peat* through bacterial action, is rich in *kerogen,* and has a carbon content of 70%, which makes it a more efficient heating fuel than peat.

liquefaction The conversion of moderately cohesive, unconsolidated *sediment* into a fluid, water-saturated mass.

lithification The conversion of loose *sediment* into solid *sedimentary rock.*

lithosphere A layer of solid, brittle rock comprising the outer 100 kilometers of the Earth, encompassing both the crust and the outermost part of the upper *mantle.* See also *asthenosphere.*

lithostatic pressure The force exerted on a rock buried deep within the Earth by overlying rocks. Because lithostatic pressure is exerted equally from all sides of a rock, it compresses the rock into a smaller, denser form without altering the rock's shape.

loess A blanket of silt that is produced by the erosion of glacial outwash and transported by wind. Much loess found in the Mississippi Valley, China, and Europe is believed to have been deposited during the *Pleistocene Epoch.*

longitudinal dune One of a series of long, narrow *dunes* lying parallel both to each other and to the prevailing wind direction.

longshore current An ocean current that flows close and almost parallel to the *shoreline,* caused by the combination of swash and backwash.

longshore drift 1. The process by which a *longshore current* moves *sediments* within a surf zone. 2. The sediments so moved. Longshore drift typically consists of sand, gravel, shell fragments, and pebbles. See also *beach drift.*

lopolith A large, saucer-shaped concordant *pluton* produced when dense mafic magma depresses the country rock below it.

magma Molten (melted) rock that forms naturally within the Earth. Magma may be either a liquid or a fluid mixture of liquid, crystals, and dissolved gases.

mantle The middle layer of the Earth, lying just below the *crust* and consisting of relatively dense rocks. The mantle is divided into two sections, the upper mantle and the lower mantle; the lower mantle has greater density than the upper mantle. See also *core* and *crust.*

marble A coarse-grained, nonfoliated *metamorphic rock* derived from limestone or *dolostone.*

marine magnetic anomaly An irregularity in magnetic strength along the ocean floor that reflects *sea-floor spreading* during periods of magnetic reversal.

mass movement The process by which such Earth materials as bedrock, loose *sediment,* and *soil* are transported down slopes by gravity.

meandering stream A *stream* that traverses relatively flat land in fairly evenly spaced loops and separated from each other by narrow strips of *floodplain.*

mechanical exfoliation A form of *mechanical weathering* in which successive layers of a large *plutonic rock* break loose and fall when the erosion of overlying material permits the rock to expand upward. The thin slabs of rock that break off fall parallel to the exposed surface of the rock, creating the long, broad steps that can be found on many mountains.

mechanical weathering The process by which a rock or mineral is broken down into smaller fragments without altering its chemical makeup; *weathering* that affects only physical characteristics. See also *chemical weathering.*

Mercalli intensity scale A scale designed to measure the degree of intensity of *earthquakes,* ranging from I for the lowest intensity to XII for the highest. The classifications are based on human perceptions.

Mesozoic Era The intermediate era of the *Phanerozoic Eon,* following the *Paleozoic Era* and preceding the *Cenozoic Era,* and marked by the dominance of marine and terrestrial reptiles and the appearance of birds, mammals, and flowering plants.

metallic bonding The act or process by which two or more atoms of electron-donating elements pack so closely together that some of their electrons begin to wander among the nuclei rather than orbiting the nucleus of a single atom. Metallic bonding is responsible for the distinctive properties of metals.

metamorphic differentiation The process by which minerals from a chemically uniform rock separate from each other during *metamorphism* and form individual layers within a new *metamorphic rock.*

metamorphic facies 1. A group of minerals customarily found together in *metamorphic rocks* and indicating a particular set of temperature and pressure conditions at which metamorphism occurred. 2. A set of *metamorphic rocks* characterized by the presence of such a group of minerals.

metamorphic grade A measure used to identify the degree to which a *metamorphic rock* has changed from its parent rock. A metamorphic grade provides some indication of the circumstances under which the metamorphism took place.

metamorphic index mineral One of a set of minerals found in *metamorphic rocks* and used as indicators of the temperature and pressure conditions at which the metamorphism occurred. A metamorphic index mineral is stable only within a narrow range of temperatures and pressures and the metamorphism that produces it must take place within that range.

metamorphic rock A *rock* that has undergone chemical or structural changes. Heat, pressure, or a chemical reaction may cause such changes.

metamorphism The process by which conditions within the Earth alter the mineral content, chemical composition, and structure of solid rock without melting it. *Igneous, sedimentary,* and *metamorphic rocks* may all undergo metamorphism.

microcontinent A section of continental *lithosphere* that has broken off from a larger, distant continent, as by *rifting.*

mid-ocean ridge An underwater mountain range that develops between the margins of diverging plates, produced by accumulation of mantle basalt that continues to erupt after plate *rifting.*

migmatite A rock that incorporates both *metamorphic* and *igneous* materials.

mineral A naturally occurring, usually inorganic, solid consisting of either a single element or a *compound,* and having a definite chemical composition and a systematic internal arrangement of atoms.

mineraloid A naturally occurring, usually inorganic, solid consisting of either a single element or a *compound,* and having a definite chemical composition but lacking a systemic internal arrangement of atoms. See also *mineral.*

mineral zone An area of rock throughout which a given *metamorphic index mineral* is found, presumed to have undergone metamorphism under uniform temperature and pressure conditions.

Moho (abbreviation for Mohorovičić) The seismic discontinuity between the base of the Earth's *crust* and the top of the *mantle. P waves* passing through the Moho change their velocity by approximately 1 kilometer per second, with the higher velocity occurring in the mantle and the lower in the crust.

moraine A single, large mass of *glacial till* that accumulates, typically at the edge of a glacier.

mudcrack A fracture that develops at the top of a layer of fine-grained, muddy sediment when it is exposed to the air and as a result dries out and contracts.

mudflow The rapid flow of typically fine-grained *regolith* mixed with water. There may be as much as 60% water in a mudflow.

mudstones Fine-grained *sedimentary rocks* composed largely of silt, clay minerals, and mica that have settled out of still water. These include shale and siltstone.

native elements *Elements* that do not combine with others in nature. Minerals composed of native elements, such as gold, silver, and diamond, contain only those elements.

natural levee One of a pair of ridges of *sediment* deposited along both banks of a *stream* during successive floods.

natural spring A place where groundwater flows to the surface and issues freely from the ground.

neutron A particle that is found in the *nucleus* of an *atom,* has a mass approximately equal to that of a *proton,* and has no electric charge.

normal fault A *dip-slip fault* marked by a generally steep *dip* along which the hanging wall has moved downward relative to the footwall.

nuclear fission The division of the *nuclei* of *isotopes* of certain heavy *elements,* such as uranium and plutonium, effected by bombardment with *neutrons.* Nuclear fission causes the release of energy, additional neutrons, and an enormous quantity of heat.

nuclear fusion The combination of the *nuclei* of certain extremely light *elements,* especially hydrogen, effected by the application of high temperature and pressure. Nuclear fusion causes the release of an enormous amount of heat energy, comparable to that released by *nuclear fission.*

nucleus (plural **nuclei**) 1. The central part of an *atom,* containing most of the atom's mass and having a positive charge due to the presence of *protons.* 2. The membrane-bound structure found in the cells of *eukaryotic* organisms that contains their genetic information, in the form of DNA.

nuée ardente A sometimes glowing cloud of gas and *pyroclastics* erupted from a *volcano* and moving swiftly down its slopes. Also called a *pyroclastic flow.*

ocean trench A deep, linear, relatively narrow depression in the sea floor, formed by the *subduction* of oceanic plates.

oil sand A mixture of unconsolidated sand and clay that contains *bitumen,* a semi-solid *hydrocarbon.*

oil shale A brown or black clastic *source rock* containing *kerogen.*

ophiolite suite The group of *sediments, sedimentary rocks,* and mafic and ultramafic *igneous rocks* that make up the oceanic *lithosphere.*

ore A mineral deposit that can be mined for a profit.

original horizontality See *principle of original horizontality.*

orogenesis Mountain formation, as caused by *volcanism, subduction, plate divergence, folding,* or the movement of fault blocks. Also called *orogeny.*

oscillatory motion The circular movement of water up and down, with little or no change in position, as a wave passes.

outwash A load of *sediment,* consisting of sand and gravel, that is deposited by meltwater in front of a *glacier.*

oxidation The process whereby a mineral's ions combine with oxygen ions. A mineral that is exposed to air may undergo oxidation as a form of *chemical weathering.*

oxide One of several minerals containing negative oxygen ions bonded to one or more positive metallic ions.

Paleozoic Era The earliest era of the *Phanerozoic Eon,* marked by the presence of marine invertebrates, fish, amphibians, insects, and land plants.

parabolic dune A horseshoe-shaped *dune* having a concave windward slope and a convex leeward slope. Parabolic dunes tend to form along sandy ocean and lake shores. They may also develop from *transverse dunes* through *deflation.*

parent isotope A radioactive *isotope* that changes into a different isotope when its nucleus decays. See also *daughter isotope.*

parent material The source from which a given soil is chiefly derived, generally consisting of bedrock or *sediment.*

partial melting The incomplete melting of a rock composed of minerals with differing melting points. When partial melting occurs, the minerals with higher melting points remain solid while the minerals whose melting points have been reached turn into *magma.*

passive continental margin A border that lies between continental and oceanic *lithosphere,* but is not a plate margin. It is marked by lack of seismic and volcanic activity.

peat A soft, brown mass of compressed, partially decomposed vegetation that forms in a water-saturated environment and has a carbon content of 50%. Dried peat can be burned as fuel.

pediment A broad surface at the base of a receding mountain. The pediment develops when running water erodes most of the mass of the mountain.

peridotite An *igneous rock* composed primarily of the iron-magnesium *silicate* olivine and having a silica content of less than 40%.

periods The time spans that subdivide *eras* in the *geologic time scale,* and are themselves divided into *epochs.*

permeability The capability of a given substance to allow the passage of a fluid. Permeability depends upon the size of and the degree of connection among a substance's pores.

petroleum The most common and versatile fossil fuel, comprising a group of naturally occurring substances made up of *hydrocarbons.* These substances may be gaseous, liquid, or semi-solid.

Phanerozoic Eon The eon that started 570 million years ago, when numerous fossils of sea shells began to be formed, and that continues to the present time.

phyllite A foliated *metamorphic rock* that develops from *slate* and is marked by a silky sheen and medium grain size.

placer deposit A deposit of heavy or durable minerals, such as gold or diamonds, typically found where the flow of water abruptly slows.

plastic deformation A permanent change in the shape or volume of a rock caused by great *stress* under higher temperatures and pressures than that which produce *brittle failure.*

plate tectonics The theory that the Earth's *lithosphere* consists of large, rigid plates that move horizontally in response to the flow of the *asthenosphere* beneath them, and that interactions among the plates at their borders cause most major geologic activity, including the creation of oceans, continents, mountains, volcanoes, and earthquakes.

playa A dry lake basin found in a desert.

pluton An *intrusive rock,* as distinguished from the preexisting country rock that surrounds it.

plutonic rock See *intrusive rock.*

point bar A low ridge of *sediment* that forms along the inner bank of a *meandering stream.*

polymorph A mineral that is identical to another mineral in chemical composition but differs from it in *crystal structure.*

porosity The percentage of a soil, rock, or sediment's volume that is made up of pores.

primary coast A *coast* formed primarily by nonmarine processes, such as *glacial erosion,* sea-level fluctuations, or biological processes. See also *secondary coast.*

principle of cross-cutting relationships The scientific law stating that a *pluton* is always younger than the rock that surrounds it.

principle of faunal succession The scientific law stating that the organisms of Earth have evolved in a definite order over time, and that this is reflected in the fossil record.

principle of inclusions The scientific law stating that rock fragments contained within a larger body of rock are always older than the surrounding body of rock.

principle of original horizontality The scientific law stating that *sediments* setting out from bodies of water are deposited horizontally or nearly horizontally in layers that lie parallel or nearly parallel to the Earth's surface.

principle of superposition The scientific law stating that in any unaltered sequence of rock strata, each stratum is younger than the one beneath it and older than the one above it, so that the youngest stratum will be at the top of the sequence and the oldest at the bottom.

principle of uniformitarianism The scientific law stating that the geological processes taking place in the present operated similarly in the past and can therefore be used to explain past geologic events.

prokaryotic Lacking a cell *nucleus.* Describes the first simple one-celled organisms to appear on Earth as well as many modern bacteria. See also *eukaryotic.*

Proterozoic Eon The portion of Earth history that followed the *Archean Eon* and preceded the *Phanerozoic Eon,* beginning about 2.5 billion years ago and ending about 570 million years ago, and is distinguished by the development of the first complex, multicellular organisms.

proton A positively charged particle that is found in the *nucleus* of an *atom* and has a mass approximately 1836 times that of an *electron.*

P wave (abbreviation for **primary wave**) A *body wave* that causes the *compression* of rocks when its energy acts upon them. When the P wave moves past a rock, the rock expands beyond its original volume, only to be compressed again by the next P wave. P waves are the fastest of all *seismic waves.* See also *S wave.*

pyroclastic cone A usually steep, conic *volcano* composed almost entirely of an accumulation of loose pyroclastic material. Pyroclastic cones are usually less than 450 meters high. Because no *lava* binds the *pyroclastics,* pyroclastic cones erode easily.

pyroclastic eruption A volcanic eruption of viscous, gas-rich magma. Pyroclastic eruptions tend to produce a great deal of solid volcanic fragments rather than fluid *lava.*

pyroclastic flow A rapid, extremely hot, downward stream of *pyroclastics,* air, gases, and ash ejected from an erupting *volcano.* A pyroclastic flow may be as hot as 800°C or more and may move at speeds higher than 150 kilometers per hour.

pyroclastics (used only in the plural) Particles and chunks of *igneous rock* ejected from a volcanic vent during an eruption.

quake See *earthquake.*

quartzite An extremely durable, nonfoliated *metamorphic rock* derived from pure *sandstone* and consisting primarily of quartz.

quick clay A partly waterlogged solid clay sediment that almost instantaneously becomes a highly fluid *mudflow,* usually when ground vibrations increase the water pressure between its particles, causing them to separate and reducing the friction between them.

radiometric dating The process of using relative proportions of *parent* to *daughter isotopes* in radioactive decay to determine the age of a given rock or rock stratum.

rain-shadow desert A desert that forms when moist air on the windward side of a mountain rises and cools, causing precipitation there and leaving the leeward side of the mountain dry.

recrystallization The process by which unstable minerals in buried sediment are transformed into stable ones.

regional metamorphism *Metamorphism* that affects rocks over vast geographic areas stretching for thousands of square kilometers.

regolith The unconsolidated products of *mechanical* and *chemical weathering* that cover almost all of the Earth's land surface; composed of *soil, sediment,* and fragments of underlying bedrock.

relative dating The fixing of a geologic structure or event in a chronological sequence relative to other geologic structures or events. See also *absolute dating.*

reserve A known *resource* that can be exploited for profit with available technology under existing political and economic conditions.

reservoir rock A permeable rock containing oil or gas.

resource A mineral or fuel deposit, known or not yet discovered, that is not currently available for human exploitation.

reverse fault A *dip-slip fault* marked by a hanging wall that has moved upward relative to the footwall. Reverse faults are often caused by the *convergence* of lithospheric plates.

rhyolite Any of a group of felsic *igneous rocks* that are the *extrusive* equivalents of *granite.*

Richter scale A logarithmic scale that measures the amount of energy released during an *earthquake* on the basis of the amplitude of the highest peak recorded on a seismogram. Each unit increase in the Richter scale represents a 10-fold increase in the amplitude recorded on the seismogram and a 30-fold increase in energy released by the earthquake.

rifting The tearing apart of a *plate* to form a depression in the Earth's *crust* and often eventually separating the plate into two or more smaller plates.

rip current A strong, rapid, and brief current that flows out to sea, moving perpendicular to the *shoreline.*

ripple marks A pattern of wavy lines formed along the top of a bed by wind, water currents, or waves.

rock A naturally formed aggregate of usually inorganic materials from within the Earth.

rock-forming mineral One of the twenty or so minerals contained in the rocks that compose the Earth's crust and mantle.

saltwater intrusion Infiltration of a coastal *aquifer* by salty marine water, which may occur if the aquifer is depleted of fresh water from overuse or lack of *precipitation.*

sandstone A *clastic rock* composed of particles that range in diameter from 1/16 millimeter to 2 millimeters. Sandstones make up about 25% of all sedimentary rocks.

saturation zone See *zone of saturation.*

scarp The steep cliff face that is formed by a *slump.*

schist A coarse-grained, strongly foliated *metamorphic rock* that develops from *phyllite* and splits easily into flat, parallel slabs.

scientific law 1. A natural phenomenon that has been proven to occur invariably whenever certain conditions are met. 2. A formal statement describing such a phenomenon and the conditions under which it occurs. Also called *law.*

scientific method The technique that involves gathering all available data on a subject, forming a *hypothesis* to explain the data, conducting experiments to test the hypothesis, and modifying or confirming the hypothesis as necessary to account for the experimental results.

sea arch A span of rock produced when *sea caves* on both sides of a coastal cliff, or *headland,* erode toward one another until they are joined.

sea cave A cave eroded out of a coastal cliff, or *headland,* by the battering of waves refracted against its flanks. (See *wave refraction.*)

sea-floor spreading The formation and growth of oceans that occurs following *rifting* and is characterized by eruptions along *mid-ocean ridges,* forming new oceanic *lithosphere,* and expanding ocean basins. See also *divergence.*

sea stack A steep, isolated island of rock, separated from a *headland* by the action of waves, as when the overhanging section of a *sea arch* is eroded.

secondary coast A *coast* shaped primarily by ongoing marine erosion or deposition, such as by sea currents and waves. See also *primary coast.*

sediment A collection of transported fragments or precipitated materials that accumulate, typically in loose layers, as of sand or mud.

sedimentary environment The continental, oceanic, or coastal surroundings in which sediment accumulates.

sedimentary facies 1. A set of characteristics that distinguish a given section of sedimentary rock from nearby sections. Such characteristics include mineral content, grain size, shape, and density. 2. A section of sedimentary rock so characterized.

sedimentary rock A *rock* made from the consolidation of solid fragments, as of other rocks or organic remains, or by *precipitation* of minerals from solution.

sedimentary structure A physical characteristic of a *detrital sediment* that reflects the conditions under which the sediment was deposited.

seiche The movement of waves back and forth in an enclosed or partially enclosed basin, such as a lake or bay, due to *seismic waves* during an *earthquake.*

seismic gap A locked *fault* segment that has not experienced seismic activity for a long time. Because *stress* tends to accumulate in seismic gaps, they often become the sites of major *earthquakes.*

seismic wave One of a series of progressive disturbances that reverberate through the Earth to transmit the energy released from an *earthquake.*

shearing stress *Stress* that slices rocks into parallel blocks that slide in opposite directions along their adjacent sides. Shearing stress may be caused by transform motion.

shield volcano A low, broad, gently sloping, dome-shaped structure that forms over time as repeated eruptions eject *basaltic lava* through one or more vents and the lava solidifies in approximately the same volume all around.

shoreline The boundary between a body of water and dry land.

silicate One of several rock-forming minerals that contain silicon, oxygen, and usually one or more other common elements.

silicon-oxygen tetrahedron A four-sided geometric form created by the tight bonding of four oxygen atoms to each other, and also to a single silicon atom that lies in the middle of the form.

sill A concordant tabular *pluton* that is substantially wider than it is thick. Sills form within a few kilometers of the Earth's surface. See also *dike.*

sinkhole A circular, often funnel-shaped depression in the ground that forms when soluble rocks dissolve.

slate A fine-grained, foliated *metamorphic rock* that develops from shale and tends to break into thin, flat sheets.

slide The *mass movement* of a single, intact mass of rock, soil, or unconsolidated material along a weak plane, such as a *fault,* fracture, or *bedding* plane. A slide may involve as little as a minor displacement of soil or as much as the displacement of an entire mountainside.

slump 1. A downward and outward *slide* occurring along a concave slip plane. 2. The material that breaks off in such a slide.

soil The top few meters of *regolith,* containing both mineral and organic matter.

soil horizon A layer of soil that can be distinguished from the surrounding soil by such features as chemical composition, color, and texture.

soil profile A vertical strip of soil stretching from the surface down to the bedrock and including all of the successive *soil horizons* in a given location.

soil taxonomy A soil classification system that categorizes soils based on such attributes as their physical characteristics, chemistry, origin, and relative age.

solifluction A form of *creep* in which soil flows downslope at 0.5 to 15 centimeters per year. Solifluction occurs in relatively cold regions when the brief warmth of summer thaws only the upper meter or two of *regolith,* which becomes waterlogged because the underlying ground remains frozen and therefore the water cannot drain down into it.

sorting The process by which a given transport medium separates out certain particles, as on the basis of size, shape, or density.

source rock A rock in which *hydrocarbons* originate.

speleothem A mineral deposit of calcium carbonate that precipitates from solution in a *cave.*

spheroidal weathering The process by which *chemical weathering,* especially by water, decomposes the angles and edges of a rock or boulder, leaving a rounded form from which concentric layers are then stripped away as the weathering continues.

spit A narrow, finger-like ridge of sand that extends from land into the open water of a coastal bay, deposited by a *longshore current* interrupted by deeper water.

star dune A *dune* with three or four arms radiating from its usually higher center so that it resembles a star in shape. Star dunes form when winds blow from three or four directions, or when the wind direction shifts frequently.

stratification See *bedding.*

stratovolcano A cone-shaped *volcano* built from alternating layers of *pyroclastics* and viscous *andesitic lava.* Stratovolcanoes tend to be very large and steep.

strain The change in the shape or volume of a rock that results from *stress.*

stream A body of water found on the Earth's surface and confined to a narrow topographic depression, or channel, down which it flows and transports rock particles, sediment, and dissolved particles. Rivers, creeks, brooks, and runs are all streams.

stress The force acting on a rock or another solid to deform it, measured in kilograms per square centimeter or pounds per square inch.

strike 1. The horizontal line marking the intersection between the inclined plane of a solid geological structure and the Earth's surface. 2. The compass direction of this line, measured in degrees from true north.

strike-slip fault A *fault* in which two sections of rock have moved horizontally in opposite directions, parallel to the line of the fracture that divided them. Strike-slip faults are caused by *shearing stress.*

stromatolites Layered reef-like structures created by the accumulation of sediments in colonies of blue-green algae. Fossilized stromatolites are the most common evidence of life during the *Archean Eon.*

subduction The sinking of an oceanic *plate* edge as a result of *convergence* with a plate of lesser density. Subduction often causes *earthquakes* and creates *volcano* chains.

subsidence The lowering of the Earth's surface, caused by such factors as compaction, a decrease in groundwater, or the pumping of oil.

sulfate One of several minerals containing positive sulfur ions bonded to negative oxygen ions.

sulfide One of several minerals containing negative sulfur ions bonded to one or more positive metallic ions.

superposition See *principle of superposition.*

surface wave One of a series of *seismic waves* that transmits energy from an earthquake's *epicenter* along the Earth's surface. See also *body wave.*

suspended load A body of fine, solid particles, typically of sand, clay, and silt, that travels with stream water without coming into contact with the stream bed.

suture zone An area where two continental plates have joined together through *continental collision.* Suture zones are marked by extremely high mountain ranges, such as the Himalayas and the Alps.

S wave (abbreviation for **secondary wave**) A *body wave* that causes the rocks along which it passes to move up and down perpendicular to the direction of its own movement. See also *P wave.*

syncline A concave *fold,* the central part of which contains the youngest section of rock. See also *anticline.*

tarn A deep, typically circular lake that forms when a *cirque glacier* melts.

tension *Stress* that stretches or extends rocks, so that they become thinner vertically and longer laterally. Tension may be caused by *divergence* or *rifting.*

tephra (plural noun) Pyroclastic materials that fly from an erupting volcano through the air before cooling, and range in size from fine dust to massive blocks.

theory A comprehensive explanation of a given set of data that has been repeatedly confirmed by observation and experimentation and has gained general acceptance within the scientific community but has not yet been decisively proven. See also *hypothesis* and *scientific law.*

thermal expansion A form of *mechanical weathering* in which heat causes a mineral's crystal structure to enlarge.

thermal plume A vertical column of upwelling *mantle* material, 100 to 250 kilometers in diameter, that rises from beneath a continent or ocean and can be perceived at the Earth's surface as a *hot spot.* Thermal plumes carry enough energy to move a plate, and they may be found both at plate boundaries and plate interiors.

thrust fault A *reverse fault* marked by a *dip* of 45° or less.

tide 1. The cycle of alternate rising and falling of the surface of an ocean or large lake, caused by the gravitational pull of the Sun and especially Moon in interaction with the Earth's rotation. Tides occur on a regular basis, twice every day on most of the Earth. 2. A single rise or fall within this cycle.

till See *glacial till.*

tombolo A sandy coastal landform that grows from a mainland to a *sea stack,* produced when waves are intercepted by the stack and deposit their sediment load on its landward side.

topography The set of physical features, such as mountains, valleys, and the shapes of landforms, that characterizes a given landscape.

transform motion The movement of two adjacent lithospheric plates in opposite directions along a parallel line at their common edge. Transform motion often causes *earthquakes.*

translatory motion Motion of water in which the water itself actually moves.

transverse dune One of a series of *dunes* having an especially steep slip face and a gentle windward slope and standing perpendicular to the prevailing wind direction and parallel to each other. Transverse dunes typically form in arid and semi-arid regions with plentiful sand, stable wind direction, and scarce vegetation. A transverse dune may be as much as 100 kilometers long, 200 meters high, and 3 kilometers wide.

tributary A *stream* that supplies water and sediment to a larger main stream.

trunk stream A main *stream* into which *tributaries* carry water and sediment.

tsunami (plural **tsunami**) A vast sea wave caused by the sudden dropping or rising of a section of the sea floor following an *earthquake.* Tsunami may be as much as 30 meters high and 200 kilometers long, may move as fast as 250 kilometers an hour, and may continue to occur for as long as a few days.

unconformity A boundary separating two or more rock layers of markedly different ages, marking a gap in the geologic record.

uniformitarianism The hypothesis that current geologic processes, such as the slow erosion of a coast under the impact of waves, have been occurring in a similar manner throughout the Earth's history and that these processes can account for past geologic events. See also *catastrophism* and *principle of uniformitarianism.*

upwarped mountain A mountain consisting of a broad area of the Earth's *crust* that has moved gently upward without much apparent deformation, and usually containing *sedimentary, igneous,* and *metamorphic rocks.*

valley glacier An *alpine glacier* that flows through a preexisting stream valley.

van der Waals bond A relatively weak kind of intermolecular bond that forms when one side of a molecule develops a slight negative charge because a number of *electrons* have temporarily moved to that side of the molecule, and this negative charge attracts the *nuclei* of the *atoms* of a neighboring molecule, while the side of the molecule with fewer electrons develops a slight positive charge that attracts the electrons of the atoms of neighboring molecules.

ventifact A stone whose sides have been flattened and sharpened by *wind abrasion.* Ventifacts are commonly found strewn across a *desert* floor.

volcanic arc A chain of *volcanoes* fueled by magma that rises from an underlying *subducting* plate.

volcanic cone A cone-shaped mountain that forms around a vent from the debris of *pyroclastics* and *lava* ejected by numerous eruptions over time.

volcanic crater A steep, bowl-shaped depression surrounding a vent. A volcanic crater forms when the walls of a vent collapse inward following an eruption.

volcanic dome A bulb-shaped solid that forms over a vent when *lava* so viscous that it cannot flow out of the *volcanic crater* cools and hardens. When a volcanic dome forms, it traps the volcano's gases beneath it. They either escape along a side vent of the volcano or build pressure that causes another eruption and shatters the volcanic dome.

volcanic rock See *extrusive rock.*

volcanism The set of geological processes that result in the expulsion of *lava, pyroclastics,* and gases at the Earth's surface.

volcano The solid structure created when lava, gases, and hot particles escape to the Earth's surface through vents. Volcanoes are usually conical. A volcano is "active" when it is erupting or has erupted recently. Volcanoes that have not erupted recently but are considered likely to erupt in the future are said to be "dormant." A volcano that has not erupted for a long time and is not expected to erupt in the future is "extinct."

watershed See *drainage basin.*

water table The surface that lies between the *zone of aeration* and the underlying *zone of saturation.*

wave-cut bench A relatively level surface formed when waves erode the base of a cliff, causing the overlying rock to fall into the surf. A wave-cut bench stands above the water and extends seaward from what remains of the cliff.

wave refraction The process by which a wave approaching the shore changes direction due to slowing of those parts of the wave which enter shallow water first, causing a sharp decrease in the angle at which the wave approaches until the wave is almost parallel to the coast.

weathering The process by which exposure to atmospheric agents, such as air or moisture, causes *rocks* and *minerals* to break down. This process takes place at or near the Earth's surface. Weathering entails little or no movement of the material that it loosens from the rocks and minerals. See also *erosion.*

wind abrasion The process by which wind erodes rock through contact between the rock and rock particles carried by the wind.

yardang A streamlined ridge of rock, found in a *desert,* produced when unidirectional winds erode soft *sedimentary rock* layers by *abrasion,* leaving behind more resistant layers.

zone of ablation The part of a *glacier* in which there is an overall loss of snow and ice. A zone of ablation can be identified in the summer by an expanse of bare ice. See also *zone of accumulation.*

zone of accumulation The part of a *glacier* in which there is an overall gain of snow and ice. A zone of accumulation can be identified by a blanket of snow that survives summer melting. See also *zone of ablation.*

zone of aeration A region below the Earth's surface that is marked by the presence of both water and air in the pores of rocks and soil. Also called *aeration zone.*

zone of saturation A region that lies below the *zone of aeration* and is marked by the presence of water and the absence of air in the pores of rocks and soil.

Credits

Maps showing topographical relief were created using Mountain High Maps software. Mountain High Maps ® Copyright © 1993 Digital Wisdom ®, Inc.

Part 1 Opener Bern Pedit c/o Breck P. Kent

Chapter 1

Figure 1-1 Tom Bean; **Figure 1-2** Matthew Naythans/Gamma Liaison; **page 5, Figure 1** James L. Amos/Photo Researchers, Inc.; **Figure 1-3a** U.S. Geological Survey W. 100th No. 47/USGS; **Figure 1-3b** R.M. Turner 17/U.S. Geological Survey.

Chapter 2

Figure 2-1 Breck P. Kent; **Figure 2-2a** Breck P. Kent; **Figure 2-2b** Kathleen Campbell/Liaison International; **Figure 2-3** Fred Hirschmann; **Figure 2-9 (photo)** M. Claye/Jacana/Photo Researchers, Inc.; **Figure 2-11 (left)** Jeffrey Scovil; **Figure 2-11 (right)** Jeffrey Scovil; **Figure 2-12a** Ken Lucas/Biological Photo Service; **Figure 2-12b** E.R. Degginger/Bruce Coleman, Inc.; **Figure 2-13a** Stuart Cohen/Comstock; **Figure 2-13b** Breck P. Kent; **Figure 2-14** Breck P. Kent; **Figure 2-15a** Ed Degginger/Bruce Coleman, Inc.; **Figure 2-15b** Breck P. Kent; **Figure 2-16** Breck P. Kent; **Figure 2-17a** Jeffrey Scovil; **Figure 2-17b** E. R. Degginger/Bruce Coleman, Inc.; **Figure 2-18a** Breck P. Kent; **Figure 2-18b** Breck P. Kent; **Figure 2-20a (photo)** American Museum of Natural History; **Figure 2-20b (photo)** Breck P. Kent; **Figure 2-20c (photo)** Breck P. Kent; **Figure 2-20d (photo)** Breck P. Kent; **Figure 2-20e (photo)** Breck P. Kent; **Figure 2-21a** Breck P. Kent; **Figure 2-21b** M. Claye/Jacana/Photo Researchers, Inc.; **Figure 2-21c** M. Claye/Jacana/Photo Researchers, Inc.; **page 36, Figure 1** Fred Ward/Black Star; **Page 36, Figure 2** E.R. Degginger/Bruce Coleman, Inc.; **Figure 2-22a** Jeffrey Scovil; **Figure 2-22b** Manfred Kage/Peter Arnold, Inc.; **page 39 (top)** Tom McHugh/Photo Researchers, Inc.; **page 39 (bottom)** Breck P. Kent.

Chapter 3

Figure 3-1 Jay Steinke; **Figure 3-2 (left)** Fred Hirschmann; **Figure 3-2 (right)** William E. Ferguson; **Figure 3-3** E.R. Degginger/Earth Scenes; **Figure 3-4 (left)** Biological Photo Service; **Figure 3-4 (right)** Breck P. Kent/Earth Scenes; **Figure 3-5** Breck P. Kent; **Figure 3-6a** Doug Sokell/Visuals Unlimited; **Figure 3-6b** Eric Schrempp/Photo Researchers, Inc.; **Figure 3-7 (left)** Breck P. Kent; **Figure 3-7 (left middle)** Breck P. Kent; **Figure 3-7 (right middle)** Breck P. Kent; **Figure 3-7 (right)** Breck P. Kent; **Figure 3-12 (photo)** Breck P. Kent; **Figure 3-13** William E. Ferguson; **Figure 3-15 (photo)** John S. Shelton; **page 52, Figure 1a** National Archives; **Figure 3-16 (photo)** Rob Badger; **Figure 3-17** Fred Hirschmann; **Figure 3-22a** John Sanford/SPL/ Photo Researchers, Inc.; **page 59** Fred Hirschmann.

Chapter 4

Figure 4-1 Wesley Bocxe/The Image Works; **Figure 4-2** George Gerster/Comstock; **Figure 4-3a** Gary Braasch; **Figure 4-3b** Ralph Perry/Black Star; **Figure 4-4** Mark A. Johnson/The Stock Market; **Figure 4-5** U.S. Geological Survey; **Figure 4-6b** Breck P. Kent; **Figure 4-7b** D.R. Stoecklein/The Stock Market; **Figure 4-8** Fred Grassle/Woods Hole Oceanographic Institute; **Figure 4-9a** William E. Ferguson; **Figure 4-9b** Francois Gohier/Photo Researchers, Inc.; **Figure 4-10 (photo)** Alberto Garcia/SABA; **Figure 4-11 (left)** Cromoc/The Gamma Liaison Network; **Figure 4-12a** Steve Kaufman/Peter Arnold, Inc.; **Figure 4-12b** John S. Shelton; **Figure 4-13 (photo)** Marc Schechter/Photo Resource Hawaii; **Figure 4-14 (photo)** Teaching Collection/University of Washington, Department of Geological Science; **Figure 4-15 (photo)** Hydrographic Department of Japan; **page 72, Figure 1 (photo)** Ray Atkeson/The Stock Market; **Figure 4-18 (photo)** Link/Visuals Unlimited; **Figure 4-21 (inset)** NASA/ERTS; **page 78, Figure 1a** James Mason/Black Star; **page 78, Figure 1b** James Sugar/Black Star; **page 78, Figure 1c** James Sugar/Black Star; **page 79, Figure 2** Gary Braasch; **Figure 4-22** Fred M. Bullard; **Figure 4-23** NASA; **Figure 4-24** US Geological Survey/NASA/Photo Researchers, Inc.; **page 83** David Ball/Allstock/Tony Stone Images.

Chapter 5

Figure 5-1 David Muench/Tony Stone Images; **Figure 5-4a** New York Public Library, Local History and Genealogy Division; **Figure 5-4b** Runk/Schoenberger/Grant Heilman; **Figure 5-5 (photo)** Phil Degginger/Earth Scenes; **Figure 5-6** Ramesh Venkatakrishnan; **Figure 5-8** Runk/Schoenberger/Grant Heilman; **Figure 5-9 (left)** John S. Shelton; **Figure 5-9 (right)** John S. Shelton; **Figure 5-11 (photo)** Paul McKelvey/Tony Stone Images; **Figure 5-12** Fletcher and Baylis/(Photo) Researchers, Inc.; **page 94, Figure 1a** Ramesh Venkatakrishnan; **page 94, Figure 1b** U.S. Department of Agriculture; **Figure 5-15 (photo)** William E. Ferguson; **Figure 5-17** NASA; **Figure 5-18** JPL/NASA; **page 99 (photo)** Stanley Chernicoff/Patrick Spencer.

Chapter 6

Figure 6-1 Joe McDonald/Earth Scenes; **Figure 6-3a** Connie Toops; **Figure 6-3b** Breck P. Kent; **Figure 6-5a** Martin Miller; **Figure 6-7 (photo)** Ann B. Swengel/Visuals Unlimited; **Figure 6-8 (photo)** Glacier National Park; **Figure 6-11a** Kurt Hollocher/Union College Geology Department; **Figure 6-11a (inset)** Kurt Hollocher/Union College Geology Department; **Figure 6-11b** Martin Miller; **Figure 6-11b (inset)** Kurt Hollocher/Union College Geology Department; **Figure 16-11c** Martin Miller; **Figure 6-11c** (inset) Kurt Hollocher/Union College Geology Department; **Figure 6-12a** John S. Shelton; **Figure 6-12b** Breck P. Kent; **Figure 6-14** Tom Bean; **Figure 6-15** J. Fennell/Bruce Coleman, Inc.; **Figure 6-16a** E.R. Degginger/Bruce Coleman, Inc.; **Figure 6-16b** John Sohlden/Visuals Unlimited; **Figure 6-17 (photo)** Martin Miller; **page 119** Betty Crowell/Faraway Places.

Chapter 7

Figure 7-1 John Sohlden–Visuals Unlimited; **Figure 7-5** Kenneth Murray/Photo Researchers; **Figure 7-6** Breck P. Kent; **Figure 7-7b** Betty Crowell/Faraway Places; **Figure 7-9 (photo)** John D. Cunningham/Visuals Unlimited; **Figure 7-10** William Felger/Grant Heilman; **Figure 7-11** Albert J. Copley/Visuals Unlimited; **Figure 7-12** Roy Jameson; **Figure 7-13** Donna L. Whitney, University of Minnesota; **Figure 7-17 (bottom left)** William E. Ferguson; **Figure 7-17 (bottom left inset)** Bernard Evans/University of Washington, Department of Geological Science; **Figure 7-17 (bottom right)** © Francis Ors Lustwerk Dudas, 1994; **Figure 7-17 (top right)** William E. Ferguson; **Figure 7-18** Scott Frances/Esto; **Figure 7-18 (inset)** William E. Ferguson; **Figure 7-19 (left)** Raphael Gaillarde–Gamma Liaison; **Figure 7-19 (right)** Raphael Gaillarde–Gamma Liaison; **Figure 7-21 (photo)** Betty Crowell/Faraway Places; **page 135 (photo)** Ramesh Venkatakrishnan.

Chapter 8

Figure 8-1 Breck P. Kent; **Figure 8-3** William E. Ferguson; **Figure 8-4a.** G. Shanmugam; **Figure 8-5** Tom Bean; **Figure 8-6** William E. Ferguson; **Figure 8-8** Edward A. Hay, De Anza College, Cupertino, CA; **page 142, Figure 1a** Breck P. Kent; **page 142, Figure 1b** David Schwimmer/Bruce Coleman, Inc.; **page 142, Figure 1c** William E. Ferguson; **Figure 8-13 (photo)** Teaching Collection/University of Washington, Department of Geological Science; **Figure 8-14** Grant Heilman; **Figure 8-16** Joy Spurr/Bruce Coleman, Inc.

Part 2 Opener Tom Bean

Chapter 9

Figure 9-1 Michael Fogden/DRK Photo; **Figure 9-3a** Martin Miller; **Figure 9-3b** Betty Crowell/Faraway Places; **Figure 9-5a** A.J. Copley/Visuals Unlimited; **Figure 9-5b** Breck P. Kent; **Figure 9-5c** Tom Bean; **Figure 9-9 (photo)** GEOPIC©, Earth Satellite Corporation; **Figure 9-11** John S. Shelton; **Figure 9-13 (photo)** courtesy Stan Chernicoff; **Figure 9-14 (photo)** Simon Fraser/Science Photo Library/Photo Researchers, Inc.; **Figure 9-15 (photo)** Breck P. Kent; **Figure 9-18 (right)** Worldsat International Inc./Photo Researchers, Inc.; **Figure 9-19** Clyde H. Smith/Allstock/Tony Stone Images; **Figure 9-20** Jet Propulsion Lab, California Institute of Technology; **page 169** Ramesh Venkatakrishnan.

Chapter 10

Figure 10-1 Terrance White/FSP/The Gamma Liaison Network; **Figure 10-9** John S. Shelton; **Figure 10-10 (photo)** Steve McCutcheon–Alaska Pictorial Services; **Figure 10-11** JP Owen/Black Star; **page 182, Figure 1 (photo)** Bates Little-Hales–National Geographic; **Figure 10-13** Piero Pomponi/ Liaison Agency; **Figure 10-14** R.E.O. Wallace, 311/US Geological Survey; **Figure 10-16 (photo)** California Institute of Technology; **Figure 10-17** John S. Shelton.

Chapter 11

Figure 11-1 Barbara Cushman Rowell–DRK Photos; **Figure 11-10** National Oceanic and Atmospheric Administration/National Geophysical Data Center; **page 205, Figure 1** Dudley Foster/Woods Hole Oceanographic Institute; **page 205, Figure 2** Robert R. Hessler; **Figure 11-16** Betty Crowell/Faraway Places.

Part 3 Opener Joel Bennett/Earth Scenes

Chapter 12

Figure 12-1 Roger J. Wyan/AP Wide World Photo; **Figure 12-4 (photo)** Tom Bean–DRK Photos; **Figure 12-6 (photo)** Betty Crowell/Faraway Places; **Figure 12-7 (inset left)** John S. Shelton; **Figure 12-7 (inset right)** Peter L. Kresan; **page 226, Figure 1 (photo)** Breck P. Kent; **Figure 12-11** M. Western/Comstock; **Figure 12-12** Gamma Liaison; **Figure 12-13a** Lloyd S. Cluff; **Figure 12-13b** Lloyd S. Cluff; **Figure 12-15a** Linda J. Moore; **Figure 12-15b** Martin Miller; **Figure 12-17a** NASA; **Figure 12-17b** U.S. Geological Survey; **page 235** Ramesh Venkatakrishnan.

Chapter 13

Figure 13-1 Eric Hylden/Grand Forks Herald; **Figure 13-7** Norman Weiser; **Figure 13-10** Art Gingert/Comstock; **Figure 13-11 (photo)** John S. Flannery/Bruce Coleman, Inc.; **Figure 13-12a (photo)** Jeff Lepore/Photo Researchers, Inc.; **Figure 13-12b (photo)** Andrew J. Martinez/Photo Researchers, Inc.; **Figure 13-12c (photo)** Art Gingert/Comstock; **Figure 13-13 (photo)** Betty Crowell/Faraway Places; **Figure 13-15** Ramesh Venkatakrishnan; **Figure 13-17** Andrew Holbrooke/Gamma Liaison; **Figure 13-18** Peter Kresan; **Figure 13-19 (photo)** G.R. Roberts; **Figure 13-20** Galen Rowell/Peter Arnold, Inc.; **page 253, Figure 1 (photo)** World Perspectives/Gamma; **page 254, Figure 1** Andrew Holbrooke/Gamma Liaison; **Figure 13-22** NASA; **page 257** Alex S. Maclean/Peter Arnold, Inc.

Chapter 14

Figure 14-1 Gene Aherns/Bruce Coleman, Inc.; **Figure 14-8 (photo)** C.E. Siebenthal/U.S. Geological Survey; **Figure 14-9 (photo)** Jose Fuste Raga/The Stock Market; **Figure 14-10 (photo)** William E. Ferguson; **Figure 14-11** Farrell Grehan/Science Source/Photo Researchers, Inc.; **Figure 14-12 (photo)** Peter Kresan; **Figure 14-16 (photo)** Tony Marshall/DRK Photo; **Figure 14-17** US Department of Interior, USGS; **Figure 14-19** Indiana Geological Survey; **Figure 14-21** Ramesh Venkatakrishnan; **Figure 14-22a** Robert and Linda Mitchell; **Figure 14-22b** Albert Copley/Visuals Unlimited; **Figure 14-22c** Jeff Lepore/Photo Researchers, Inc.; **Figure 14-23 (photo)** Ramesh Venkatakrishnan; **Figure 14-H1-1 (photo)** Stephen Alvarez/Time Magazine; **Figure 14-24 (photo)** Timothy O'Keefe/Bruce Coleman, Inc.; **Figure 14-25** A.N. Palmer, Earth Sciences, SUNY Oneonta; **Figure 14-26 (photo)** Steve Solum/Bruce Coleman, Inc.; **Figure 14-27** Betty Crowell/Faraway Places; **page 281** USDA, Soil Conservation Service.

Chapter 15

Figure 15-1 Breck P. Kent; **Figure 15-3a** Comstock; **Figure 15-3b** Tom Bean; **Figure 15-3c** Betty Crowell/Faraway Places; **Figure 15-3d** Jim Wark/Peter Arnold, Inc.; **Figure 15-4 (photo)** Tom Bean; **Figure 15-6a** Tom Bean; **Figure 15-6d** Dan Guravich/Photo Researchers, Inc.; **Figure 15-7 (photo)** William E. Ferguson; **Figure 15-8b** Bill Kamin/Visuals Unlimited; **Figure 15-8c** Betty Crowell/Faraway Places; **Figure 15-9a** © Advanced Satellite Productions, Inc. 1993; **Figure 15-9b** U.S. Department of Interior, USGS Eros Data Center, Scene ID No. E-1272-99CT; **Figure 15-10** Peter Arnold/Peter Arnold, Inc.; **Figure 15-11 (photo)** Art Gingert/Comstock; **Figure 15-12 (photo)** Tom Bean/DRK Photo; **Figure 15-13 (photo)** John S. Shelton; **Figure 15-14 (photo)** John S. Shelton; **page 295, Figure 2a** John S. Shelton; **page 295, Figure 2b** John S. Shelton; **page 295, Figure 2c** John S. Shelton; **Figure 15-18** Betty Crowell/Faraway Places; **Figure 15-21** Peter Kresan; **Figure 15-23 (left)** NASA; **Figure 15-23 (right)** NASA; **Figure 15-24** U.S. Geological Survey, Flagstaff, Arizona; **page 305** Ramesh Venkatakrishnan.

Chapter 16

Figure 16-1 J Sohm/The Image Works; **Figure 16-2a** David Barnes/Allstock/Tony Stone Images; **Figure 16-2b** Jeff Foott/Bruce Coleman, Inc.; **Figure 16-2c** Michael Fogden–DRK Photo; **Figure 16-7** Linda Waldhofer/Liaison International; **Figure 16-9 (photo)** Betty Crowell/Faraway Places; **Figure 16-10** Scott Berner/Visuals Unlimited; **Figure 16-11 (photo)** Martin Miller; **Figure 16-11 (photo inset)** Stephanie S. Ferguson; **Figure 16-12** John Gerlach/Visuals Unlimited; **Figure 16-13 (photo)** John D. Cunningham/Visuals Unlimited; **Figure 16-14** George Gerster/Photo Researchers, Inc.; **Figure 16-14 (inset)** Peter Antal/Liaison International; **page 317, Figure 1** U.S. Farm Security Administration Collection Prints and Photographs Division, Library of Congress; **Figure 16-16** Lee Rentz/Bruce Coleman, Inc.; **Figure 16-17** Carl Purcell/Photo Researchers, Inc.; **Figure 16-19 (photo)** Peter Kresan; **Figure 16-20 (photo)** Tom Bean; **Figure 16-21 (photo)** Michael E. Long, © National Geographic Society; **Figure 16-23 (photo)** Martin Miller; **Figure 16-24** Tom Bean/DRK Photo; **Figure 16-25** Stephanie S. Ferguson; **Figure 16-26 (photo)** Peter Ward/Bruce Coleman, Inc.; **Figure 16-27** George Gerster/Comstock; **Figure 16-28** NASA; **page 327** John S. Shelton.

Chapter 17

Figure 17-1 Craig Tuttle/The Stock Market; **Figure 17-9a** William E. Ferguson; **Figure 17-9b** William E. Ferguson; **Figure 17-10 (photo)** Dick Poe/Visuals Unlimited; **Figure 17-11** Donald Carter; **Figure 17-12b** John Elk/Bruce Coleman, Inc.; **Figure 17-12c** G. R. Roberts; **page 338, Figure 1** Michael J. Chrzastowski, Illinois State Geological Survey; **Figure 17-14a** Stewart Farrell, Stockton State College, NJ; **Figure 17-14b** John S. Shelton; **Figure 17-15 (photo)** Cliff Wassmann; **Figure 17-16a** Fairchild Air Photos 0-139 and E-5780, UCLA Department of Geography Aerial Photo Archives; **Figure 17-16b** John S. Shelton; **Figure 17-17a** John S. Shelton; **Figure 17-17b** Townsend P. Dickinson/Comstock; **Figure 17-20** S.J. Krasemann/Peter Arnold, Inc.; **Figure 17-21a** David Ball/The Stock Market; **Figure 17-21b** William E. Ferguson; **Figure 17-22** Peter Kresan; **Figure 17-23** John S. Shelton; **Figure 17-24a** Jack Dermid/Photo Researchers, Inc.; **Figure 17-24b** Martin Miller.

Chapter 18

Figure 18-1 David Brown/The Stock Market; **Figure 18-2** Drake Well Museum, Titusville, Pa.; **Figure 18-4 (photo)** Silvia Dinale © 1994; **Figure 18-5 (photo)** Steve McCutcheon/Visuals Unlimited; **Figure 18-7** Ben Osborne/Tony Stone Images; **Figure 18-8b** Nicholas deVore III/Bruce Coleman, Inc.; **Figure 18-10b** Martin Bond/Science Photo Library/Photo Researchers, Inc.; **Figure 18-10c** William E. Ferguson; **Figure 18-11** Kevin Schafer/Tony Stone Images; **Figure 18-12b** Courtesy of Consolidated Edison of New York; **Figure 18-13a** Peter Kresan; **Figure 18-13b** Rona/Bruce Coleman, Inc.; **Figure 18-14a** Diego Guidice/Contrasto/SABA; **Figure 18-14b** Timothy Ross/Picture Group; **Figure 18-15** William E. Ferguson; **Figure 18-16** Bruce Dale, © National Geographic Society; **Figure 18-17** Hank Morgan/Science Source/Photo Researchers, Inc.

Chapter 19

Figure 19-1 Photo courtesy of the Smithsonian Institution, Transparency No. 86-13471A; **Figure 19-3** Samuel Bowring, MIT; **Figure 19-5a** J. William Schopf, University of California at Los Angeles; **Figure 19-5b** J. William Schopf, University of California at Los Angeles; **Figure 19-6a** Breck P. Kent; **Figure 19-6b** William E. Ferguson; **Figure 19-7 (bottom)** Samuel Bowring, MIT; **Figure 19-7 (top)** Samuel Bowring, MIT; **Figure 19-8** Andrew H. Knoll, Harvard University; **Figure 19-9 (center)** Courtesy of Smithsonian Institution; **Figure 19-9a** Neville Pledge, South Australian Museum; **Figure 19-9b** Neville Pledge, South Australian Museum; **Figure 19-9c** Neville Pledge, South Australian Museum; **Figure 19-11** Transparency No. K1253(2). Courtesy Department of Library Services, American Museum of Natural History; **Figure 19-12 (photo)** John R. Clawson/Photo Researchers, Inc.; **Figure 19-14** Richard and Dorothy Norton/Science Graphics; **Figure 19-15 (left)** Bruce Selyem/Museum of the Rockies; **Figure 19-15 (right)** James L. Amos/National Geographic Image Collection; **Figure 19-19** R. Morrison/Comstock; **Figure 19-20a** John Reade/Science Photo Library/Photo Researchers, Inc.; **Figure 19-20b** Rene Burri/Magnum.

Index